MATERIALS

FOR THE

STUDY OF VARIATION.

FOUNDATIONS OF NATURAL HISTORY

SERIES EDITORS

PAULA M. MABEE, San Diego State University
KIRK FITZHUGH, Los Angeles County Museum of Natural History

Foundations of Natural History is a series from the Johns Hopkins University Press for the republication of classic scientific writings that are of enduring importance for the study of origins, properties, and relationships in the natural world.

MATERIALS

FOR THE

STUDY OF VARIATION

TREATED WITH ESPECIAL REGARD TO

DISCONTINUITY

IN THE

ORIGIN OF SPECIES.

BY

WILLIAM BATESON, M.A.
FELLOW OF ST JOHN'S COLLEGE, CAMBRIDGE

JOHNS HOPKINS UNIVERSITY PRESS
BALTIMORE AND LONDON

© 1992 The Johns Hopkins University Press
ALL RIGHTS RESERVED
Printed in the United States of America
Original edition published in 1894 by Macmillan & Co., London

The Johns Hopkins University Press
701 West 40th Street
Baltimore, Maryland 21211-2190
The Johns Hopkins Press Ltd., London

The paper used in this book meets the minimum requirements of the American National Standard for Information Sciences—Permanence of Paper for Printed Library Materials, ANSI Z39.48-1984.

LIBRARY OF CONGRESS CATALOGING-IN-PUBLICATION DATA
Bateson, William. 1861–1926.
Materials for the study of variation treated with
especial regard to discontinuity in the origin
of species / by William Bateson.
p.   cm. — (Foundations of natural history)
Originally published: London and New York : Macmillan, 1894.
Includes bibliographical references.
ISBN 0-8018-4419-3. — ISBN 0-8018-4420-7 (pbk.)
1. Variation (Biology) 2. Evolution (Biology) I. Title. II. Series.
QH401.B32   1992   575.2—dc20   91-45926

PREFACE.

THIS book is offered as a contribution to the study of the problem of Species. The reasons that have led to its production are as follows.

Some years ago it was my fortune to be engaged in an investigation of the anatomy and development of *Balanoglossus*. At the close of that investigation it became necessary to analyze the meaning of the facts obtained, and especially to shew their bearing upon those questions of relationship and descent which modern morphology has attempted to answer. To this task I set myself as I best might, using the common methods of morphological argument and interpretation, and working all the facts into a scheme which should be as consistent as I could make it.

But the value of this and of all such schemes, by which each form is duly ushered to its place, rests wholly on the hypothesis that the methods of argument are sound. Over it all hung the suspicion that they were not sound. This suspicion seemed at that time so strong that in preface to what I had to say I felt obliged to refer to it, and to state explicitly that the analysis was undertaken in pursuance of the current methods of morphological criticism, and without prejudging the question of possible or even probable error in those methods.

Any one who has had to do such work must have felt the same thing. In these discussions we are continually stopped by such phrases as, "if such and such a variation then took place and was favourable," or, "we may easily suppose circumstances in which such and such a variation if it occurred might be beneficial," and the like. The whole argument is based on such assumptions as these—assumptions which, were they found in the arguments of Paley or of Butler, we could not too scornfully ridicule. "If," say we with much circumlocution, "the course of Nature followed the lines we have suggested, then, in short, it did." That is the sum of our argument.

Were we all agreed in our assumptions and as to the canons of interpretation, there might be some excuse, but we are not agreed. Out of the same facts of anatomy and development men of equal ability and repute have brought the most opposite conclusions. To take for instance the question of the ancestry of Chordata, the problem on which I was myself engaged, even if we neglect fanciful suggestions, there remain two wholly incompatible views as to the lines of Vertebrate descent, each well supported and upheld by many. From the same facts opposite conclusions are drawn. Facts of the same kind will take us no further. The issue turns not on the facts but on the assumptions. Surely we can do better than this. Need we waste more effort in these vain and sophistical disputes?

If facts of the old kind will not help, let us seek facts of a new kind. That the time has come for some new departure most naturalists are now I believe beginning to recognize. For the reasons set forth in the Introduction I suggest that for this new start the Study of Variation offers the best chance. If we had before us the facts of Variation there would be a body of evidence to which in these matters of doubt we could appeal. We should no longer say " *if* Variation take place in such a way," or " *if* such a variation were possible;" we should on the contrary be able to say " since Variation *does*, or at least *may* take place in such a way," " since such and such a Variation *is* possible," and we should be expected to quote a case or cases of such occurrence as an observed fact.

To collect and codify the facts of Variation is, I submit, the first duty of the naturalist. This work should be undertaken if only to rid our science of that excessive burden of contradictory assumptions by which it is now oppressed. Whatever be our views of Descent, Variation is the common basis of them all. As the first step towards the systematic study of Variation we need a compact catalogue of the known facts, a list which shall contain as far as possible all cases of Variation observed. To carry out such a project in any completeness may be impossible; but were the plan to find favour, there is I think no reason why in time a considerable approach to completeness should not be made.

Difficulty has hitherto arisen from the fact that Variation is not studied for its own sake. Each observer has some other object in view, and we are fortunate if he is good enough to mention in passing the variations he has happened to see in following his own ends. From the nature of the case these observations must at first be sporadic, and, as each standing alone seems to have little value, in the end they are unheeded and lost. If there were any central collection of facts to which such observations might from time to time be added, and thus brought into relation with cognate observations, their value would at once appear and be preserved. To make a nucleus for such a collection is the object of the present work.

The subject treated in this first instalment has been chosen for the reasons given in the text. Reference to facts that could not be included in this section of the evidence has as far as possible been avoided, but occasionally such reference was necessary, especially in the Introduction.

It was my original purpose to have published the facts without comment. This course would have been the most logical and the safest, but with hesitation it was decided to add something of the nature of analysis. I do this chiefly for two reasons. First, in starting a method one is almost compelled to shew the way in which it is to be applied. If it is hoped that others may interest themselves in the facts, it is necessary to shew how and why their interest is asked. In the old time the facts of Nature were beautiful in themselves and needed not the rouge of speculation to quicken their charm, but that was long ago, before Modern Science was born.

Besides this, to avoid the taint of theory in morphology is impossible, however much it may be wished. The whole science is riddled with theory. Not a specimen can be described without the use of a terminology coloured by theory, implying the acceptance of some one or other theory of homologies. If only to avoid misconception matters of theory must be spoken of.

It seemed at first also that the meaning of the facts was so clear that all would read it alike; but from opportunities that have occurred for the discussion of these matters I have found that it is not so, and reluctantly I have therefore made such comments as may serve to bring out the chief significance of the phenomena, pointing out what they shew and what they do not shew, having regard always to deficiencies in the evidence.

That this is a dangerous course I am aware. But in any discussion of a problem in the light of insufficient knowledge the real danger is not that a particular conclusion may be wrong, for that is a transient fault, but rather that the facts themselves may be so distorted as to be valueless to others when the conclusions that they were used to shew have been discarded. This danger I have sought indifferently to avoid by printing the facts as far as possible apart from all comment, knowing well how temporary the worth of these comments is likely to be. I have thus tried to avoid general statements and have kept the descriptions to particular cases, unless the number of similar cases is great and an inclusive description is enough.

Each separate paragraph relating a fact has been as far as possible isolated and made to stand alone; so that if any one may hereafter care to go on with the work he will be able to cut out the discarded comments and. rearrange the facts in any order preferred, inserting new facts as they come to hand. Most of these facts are numbered for reference. The numbers are distributed on no strict system, but are put in where likely to be useful.

For almost every fact stated or mentioned one reference at least is given. When this is not the case the fact is either notorious, or else the result of my own observation. In collecting evidence I have freely used the collections of former writers, especially those of Geoffroy St Hilaire, Ahlfeld, and Wenzel Gruber, but unless the contrary is stated, each passage referred to has been seen in its original place. By this system I hope I have avoided evidence corrupted by repetition. Several well known conceptions, notably that of the presence of order in abnormality, first formulated by Isidore Geoffroy St Hilaire, have been developed and exhibited in their relation to recent views.

The professed morphologist will note that many of the statements are made on authority unfamiliar to him. I have spared no pains to verify the facts wherever possible, and no case has been admitted without remark if there was reason to doubt its authenticity. So long as skilled zoologists continue to neglect all forms that are abnormal the student of Variation must turn to other sources.

This neglect of the Study of Variation may be attributed in

great measure to the unfortunate circumstance that Natural History has come to be used as a vehicle for elementary education, a purpose to which it is unsuited. From the conditions of the case when very large classes are brought together it becomes necessary that the instruction should be organized, scheduled, and reduced to diagram and system. Facts are valued in proportion as they lend themselves to such orderly treatment; on the rest small store is set. By this method the pupil learns to think our schemes of Nature sufficient, turning for inspiration to books, and supposing that by following his primer he may master it all. In a specimen he sees what he has been told to see and no more, rarely learning the habit of spontaneous observation, the one lesson that the study of Natural History is best fitted to teach.

Such a system reacts on the teacher. In time he comes to forget that the caricature of Nature shewn to his pupils is like no real thing. The perspective and atmosphere that belong to live nature confuse him no more. Two cases may be given in illustration. Few animals are dissected more often than the Crayfish and the Cockroach. Each of these frequently presents a striking departure from the normal (see Nos. 83 and 625) in external characters, but these variations have been long unheeded by pupil and by teacher; for though Desmarest and Brisout published the facts so long ago as 1848, their observations failed to get that *visa* of the text-books without which no fact can travel far.

It is especially strange that while few take much heed of the modes of Variation or of the visible facts of Descent, every one is interested in the *causes* of Variation and the nature of "Heredity," a subject of extreme and peculiar difficulty. In the absence of special knowledge these things are discussed with enthusiasm, even by the public at large.

But if we are to make way with this problem special knowledge is the first need. We must know what special evidence each group of animals and plants can give, and this specialists alone can tell us. It is therefore impossible for one person to make any adequate gathering of the facts. If it is to be done it must be done by many. At one time I thought that a number of persons might perhaps be induced thus to combine; but though I hope hereafter some such organized collection may be made, it is perhaps necessary that the first trial should be single-handed.

As I have thus been obliged to speak of many things of which I have no proper knowledge each section must inevitably seem meagre to those who have made its subject their special study, and I fear that many mistakes must have been made. To any one who may be willing to help to set these errors right, I offer thanks in advance, "humbly acknowledging a work of such concernment unto truth did well deserve the conjunction of many heads."

In the course of the work I have had help from so many that I cannot here give separate thanks to each. For valuable criticisms, given especially in connexion with the introductory pages, I am indebted to Mr F. Darwin, Dr C. S. Sherrington, Dr D. MacAlister, Mr W. Heape, Mr G. F. Stout, Dr A. A. Kanthack and particularly to Mr J. J. Lister. I have to thank the authorities of the British Museum, of the Museum of the Royal College of Surgeons, of the Musée d'Histoire Naturelle in Paris, and of the Museums of Leyden, Oxford, Rouen, Newcastle-upon-Tyne, of the École Vétérinaire at Alfort, and of the Dental Hospital for the great kindness that they have shewn me in granting facilities for the study of their collections. In particular I must thank Mr Oldfield Thomas of the British Museum for much help and advice in connexion with the subject of Teeth. I am also greatly obliged to Messrs Godman and Salvin for opportunities of examining and drawing specimens in their collections. To many others who have been good enough to lend specimens or to advise in particular cases my obligations are acknowledged in the text, but I must especially express my gratitude to Dr Kraatz of Berlin, to Dr L. von Heyden of Frankfurt, and to M. H. Gadeau de Kerville of Rouen for the large numbers of valuable insects with which they entrusted me.

My best thanks are due to Dr A. M. Norman for many useful suggestions, for the loan of specimens and for the kindly interest he has taken in my work.

My friend Mr H. H. Brindley has very kindly given me much assistance in determining and verifying several points that have arisen, and I am particularly indebted to him for permission to give an account of his very interesting and as yet unpublished observations on the variation and regeneration of the tarsus in Cockroaches.

Through the help of Dr David Sharp I have been enabled to introduce much valuable evidence relating to Insects, a subject of which without his assistance I could scarcely have spoken. It is impossible for me adequately to express my obligation to Dr Sharp for his constant kindness, for the many suggestions he has given me, and for the generosity with which he has put his time and skill at my service.

It is with especial pleasure that I take this opportunity of offering my thanks to Professor Alfred Newton for the encouragement and sympathy he has given me now for many years.

As many of the subjects treated involve matters of interpretation it should be explicitly declared that though help has been given by so many, no responsibility for opinions attaches to anyone but myself unless the contrary is stated.

The blocks for Figs. 18, 19, 25, 133, 161 and 185 (from *Proc. Zool. Soc.*) were very kindly given by the Zoological Society of London; that for Fig. 28 (from *Trans. Path. Soc.*) by the Pathological Society; and for Fig. 140 which is from the *Descent of Man* I am obliged to the kindness of Mr F. Darwin. Figs. 5 B, 5 C, and 77 were supplied by the proprietors of Newman's *British Butterflies*, and Figs. 5 A, 82 and 84 by the proprietors of the *Entomologist*. The sources of other figures are acknowledged under each. Those not thus acknowledged have been made from specimens or from my own drawings or models by Mr M. P. Parker, with the exception of a few specially drawn for me by Mr Edwin Wilson.

The work was, as I have said, begun in the earnest hope that some may be led thereby to follow the serious study of Variation, and so make sure a base for the attack on the problems of Evolution. Those who reject the particular inferences, positive and negative, here drawn from that study, must not in haste reject the method, for that is right beyond all question.

That the first result of the study is to bring confusion and vagueness into places where we had believed order established may to some be disappointing, but it is best we deceive ourselves no longer. That the problems of Natural History are not easy but very hard is a platitude in everybody's mouth. Yet in these days there are many who do not fear to speak of these things with certainty, with an ease and an assurance that in far simpler problems of chemistry or of physics would not be endured. For men of this stamp to solve difficulties may be easy, but to feel difficulties is hard. Though the problem is all unsolved and the old questions stand unanswered, there are those who have taken on themselves the responsibility of giving to the ignorant, as a gospel, in the name of Science, the rough guesses of yesterday that tomorrow should forget. Truly they have put a sword in the hand of a child.

If the Study of Variation can serve no other end it may make us remember that we are still at the beginning, that the complexity of the problem of Specific Difference is hardly less now than it was when Darwin first shewed that Natural History is a problem and no vain riddle.

On the first page I have set in all reverence the most solemn enuntiation of that problem that our language knows. The priest and the poet have tried to solve it, each in his turn, and have failed. If the naturalist is to succeed he must go very slowly, making good each step. He must be content to work with the simplest cases, getting from them such truths as he can, learning to value partial truth though he cheat no one into mistaking it for absolute or universal truth; remembering the greatness of his calling, and taking heed that after him will come Time, that "author of authors," whose inseparable property it is ever more and more to discover the truth, who will not be deprived of his due.

St John's College, Cambridge.
29 *December*, 1893.

CONTENTS.

FOREWORD, by Peter J. Bowler xvii

WILLIAM BATESON AND THE SCIENCE OF FORM,
by Gerry Webster xxix

INTRODUCTION.

SECT.		PAGE
1.	The Study of Variation	1
2.	Alternative Methods	6
3.	Continuity or Discontinuity of Variation	13
4.	Symmetry and Meristic Repetition	17
5.	Meristic Variation and Substantive Variation	22
6.	Meristic Repetition and Homology	28
7.	Meristic Repetition and Division	33
8.	Discontinuity in Substantive Variation: Size	36
9.	Discontinuity in Substantive Variation: Colour and Colour-Patterns .	42
10.	Discontinuity in Substantive Variation: Miscellaneous Examples .	54
11.	Discontinuity in Meristic Variation: Examples	60
12.	Parallel between Discontinuity of Sex and Discontinuity in Variation .	66
13.	Suggestions as to the nature of Discontinuity in Variation . . .	68
14.	Some current conceptions of Biology in view of the Facts of Variation	75
	1. Heredity. 2. Reversion. 3. Causes of Variation. 4. The Variability of "useless" Structures. 5. Adaptation. 6. Natural Selection.	

PART I.

MERISTIC VARIATION.

Linear Series	87—422	
Radial Series	423—447	
Bilateral Series	448—473	
Secondary Symmetry and Duplicity . .	474—566	

CHAPTER I. ARRANGEMENT OF EVIDENCE 83

CHAPTER II. SEGMENTS OF ARTHROPODA 91

CONTENTS.

	PAGE
CHAPTER III. VERTEBRÆ AND RIBS	102
CHAPTER IV. SPINAL NERVES	129
CHAPTER V. VARIATION IN ARTHROPODA	146
CHAPTER VI. CHÆTOPODA, HIRUDINEA AND CESTODA	156
CHAPTER VII. BRANCHIAL OPENINGS OF CHORDATA AND STRUCTURES IN CONNEXION WITH THEM	171

 1. Ascidians. 2. Cyclostomi. 3. Cervical Fistulæ and Supernumerary Auricles in Mammals

CHAPTER VIII. MAMMÆ	181
CHAPTER IX. TEETH	195

 Preliminary. Primates. Canidæ. Felidæ. Viverridæ. Mustelidæ. Pinnipedia. Marsupialia. Selachii. Radulæ of *Buccinum*.

CHAPTER X. TEETH—RECAPITULATION	265
CHAPTER XI. MISCELLANEOUS EXAMPLES	274

 Scales. Kidneys; Renal Arteries; Ureters. Tentacles and Eyes of Mollusca. Eyes of Insects. Wings of Insects. Horns of Sheep, Goats and Deer. Perforations of shell of Haliotis.

CHAPTER XII. COLOUR-MARKINGS	288

 Ocellar Markings. Simultaneity of Colour-Variation in Parts repeated in Linear Series (Larvæ of Lepidoptera: Chitonidæ)

CHAPTER XIII. MINOR SYMMETRIES: DIGITS	311

 CAT. Pp. 313—324.
 MAN AND APES. Pp. 324—360.
  Increase in number of digits, p. 324.
  Cases of Polydactylism associated with change of Symmetry. A. Digits in one Successive Series, p. 326. B. Digits in two homologous groups forming "Double-hands," p. 331. Complex cases, p. 338.
  Polydactylism not associated with change of Symmetry, p. 344. (1) A. Single extra digit external to minimus, p. 345. (1) B. Single extra digit in other positions, p. 349. (2) Duplication of single digits, p. 349. (3) Combinations of the foregoing, p. 352. (4) Irregular examples, p. 353.
  Reduction in number of phalanges, p. 355. Syndactylism, p. 356. Absence of digits and representation of two digits by one, p. 358.
 HORSE, pp. 360—373. Extra digits on separate metacarpal or metatarsal, p. 361. More than one digit borne by metacarpal III., p. 369. Intermediate cases, p. 371.
 ARTIODACTYLA, pp. 373—390. Polydactylism in Pecora, p. 373. Polydactylism in Pig, p. 381. Syndactylism in Artiodactyla, p. 383.
 BIRDS, pp. 390—395.
  Possibly Continuous numerical Variation in digits: miscellaneous examples, pp. 395—398 (*Chalcides. Cistudo. Rissa. Erinaceus. Elephas.*) Inheritance of Digital Variation. Association of Digital Variation with other forms of Abnormality.

CONTENTS.

	PAGE
CHAPTER XIV. DIGITS: RECAPITULATION	400
CHAPTER XV. MINOR SYMMETRIES: SEGMENTS IN APPENDAGES	410
CHAPTER XVI. RADIAL SERIES	423
Cœlenterata. Pedicellariæ of Echinoderms. Cell-Division.	
CHAPTER XVII. RADIAL SERIES: ECHINODERMATA	432
CHAPTER XVIII. BILATERAL SERIES	448
CHAPTER XIX. FURTHER ILLUSTRATIONS OF THE RELATIONSHIP BETWEEN RIGHT AND LEFT SIDES	463
CHAPTER XX. SUPERNUMERARY APPENDAGES IN SECONDARY SYMMETRY	474
Introductory.—The Evidence as to Insects.	
CHAPTER XXI. APPENDAGES IN SECONDARY SYMMETRY	525
The Evidence as to Crustacea.	
CHAPTER XXII. DUPLICITY OF APPENDAGES IN ARTHROPODA	539
CHAPTER XXIII. SECONDARY SYMMETRY IN VERTEBRATES. REMARKS ON THE SIGNIFICANCE OF REPETITION IN SECONDARY SYMMETRY: UNITS OF REPETITION	553
CHAPTER XXIV. DOUBLE MONSTERS	559
CHAPTER XXV. CONCLUDING REFLEXIONS	567

INDEX OF SUBJECTS, p. 576.

INDEX OF PERSONS, p. 593.

FOREWORD

PETER J. BOWLER

WILLIAM Bateson is remembered as one of the founders of modern genetics. He provided the first substantial account of Mendel's laws in English[1] and coined the term *genetics* in 1905.[2] But Bateson was an atypical geneticist, not least because for many years he refused to accept the chromosomal basis for Mendelian heredity, which had been demonstrated by the observations of T. H. Morgan's school.[3] The idea that the gene was a physical entity seemed to him an oversimplified form of materialism, inconsistent with the complex processes that governed the development of the organism. It is this concern with the problems of development which some biologists see as an enduring legacy, supplementing his achievements in the establishment of Mendelism. Bateson may not have provided the answers, but he asked questions that still seem relevant today, although they were to some extent set aside during the heyday of classical genetics. *Materials for the Study of Variation* sets out some of these problems and outlines a program of study that biology has still to complete.

From the viewpoint of the historian of science, Bateson's *Materials* is a key document in the crisis that shook biologists' thinking on both heredity and evolution in the decades around 1900. Its emphasis on discontinuous variation in evolution certainly helped to create a context within which a theory of discontinuous heredity could be articulated. But this emphasis on what Darwin's contemporaries would have called evolutionary saltations also represented a challenge to the prevailing belief that the development of life had been a slow and gradual process directed by the changing character of the physical environment. Bateson rejected both Darwinian gradualism and the view that evolution is driven by the demands of adaptation. As a result of this, he came into conflict with the biometrical school of Darwinism led by Karl Pearson and W. F. R. Weldon. The resulting antagonisms continued into the era of Mendelism, poisoning the atmosphere of British biology in a way that would significantly affect the emergence of the new science.[4] Like Bateson, many early geneticists were hostile to the Darwinian selection theory, thus delaying the advent of the "modern synthesis" of Darwinism and genetics for several decades. Looking back from the perspective of the 1940s, Julian Huxley, one of the founders of the synthesis, wrote of an "eclipse of Darwinism" at the turn of the century.[5] Bateson's *Materials* is a classic, but also somewhat anomalous, product of that eclipse.

Bateson began his study of biology at Cambridge in 1879 under the charismatic Francis Balfour. As a protégé of T. H. Huxley, Balfour modernized the teaching of biology at Cambridge, using morphology as the key to establishing the science's credentials.[6] He built up a core of enthusiastic students and followers, including Weldon (with whom Bateson was at first on good terms) and Adam Sedgwick. Their self-conceived task was to use comparative morphological, and especially embryological, evidence to assess degrees of relationship between animal groups and to reconstruct the course of life's history on earth. Although ostensibly a Darwinian, Balfour and the morphologists had little interest in the study of adaptation and biogeography through fieldwork. (The Cambridge professor of zoology, Alfred Newton, was an ornithologist who expressed grave reservations about the program of evolutionary morphology, although he did not block Balfour's appointment.) Balfour was a Darwinian in the sense that he accepted that all life had evolved through a process of gradual, continuous development; his goal was to study the pattern of development, not the processes governing evolution.

Balfour died in a climbing accident in 1882; once deprived of his leadership, the Cambridge school of evolutionary morphology disintegrated. Sedgwick stuck to morphology but increasingly despaired of using it to reconstruct evolutionary relationships. Weldon was attracted to the statistical techniques being pioneered by Francis Galton and Karl Pearson to study the variation of wild populations; he went on to do fieldwork which, he believed, demonstrated the effect of natural selection upon populations exposed to different environments. Bateson, after an unsuccessful field trip to central Asia, abandoned both morphology and Darwinism. He had no mathematical ability, and thus could not adopt the biometrical approach to the study of variation. Instead, he built upon his interest in the process of individual development in an entirely different way, rejecting the microscopical observation of embryos in favor of a direct study of how variations appeared within species.

Bateson's rejection of evolutionary morphology did not flow directly from the removal of Balfour's influence. His major project had been a study of the enteropneust *Dolichoglossus kowalevskii* (then known as *Balanoglossus*), with a view to assessing its relationship to the chordates. The origin of the chordates was a major debating ground for evolutionary morphologists.[7] Balfour had noted resemblances between *Balanoglossus* and the chordates but had insisted that the chordates must have evolved from a segmented ancestor. Bateson established a number of homologies between *Balanoglossus* and the chordates, enough to open up the possibility of an evolutionary relationship, provided one could account for the origin of the segmented chordates from an unsegmented *Balanoglossus*-like ancestor. Bateson's answer was to assume that the potential to become segmented, through the duplication or repetition of parts in the course of individual development, was a universal character of living things. He was, however, aware of the totally hypothetical nature of this conjecture. All he had done was add another suggestion to the list of possibilities that had been debated for decades by biologists addressing the question of the origin of the chordates. In the absence of any fossil evidence, he saw little likelihood of his (or anyone else's) theory ever being confirmed. Thus he began to doubt the basic assumptions underlying the use of morphology to reconstruct ancestral relationships.

These problems were articulated in his paper "The Ancestry of the Chordata," of 1886. Here he noted that the search for genealogical trees had already become "the subject of some ridicule, perhaps deserved."[8] It was also clear that his suggestion of a tendency for the repetition of parts struck at the heart of Darwinian gradualism. The duplication of an existing structure was self-evidently a discontinuous process, and he suspected that it would occur whether or not the results were useful. In effect, evolution would be driven by a process arising from within the organism, forcing the species to evolve in a certain direction whatever the environment to which it was exposed. This was not an unusual position at the time. Many paleontologists and some field naturalists accepted the theory of "orthogenesis" in which nonadaptive evolutionary trends were driven by forces arising within the process of individual development.[9] The paleontologists, however, preferred to think of orthogenetic evolution as a continuous process (e.g., the gradual development of antlers in the so-called Irish elk until they became too large and caused the species' extinction). Bateson's choice of problem led him to a different alternative, equally dismissive of utility and equally dependent on the developmental process, but based on discontinuous steps.

The alternative of saltative evolution had never been absent from the Darwinian debate, although it had been driven

underground by the success of the gradualistic approach. Many
of the first generation of Darwinists had accepted that there
was a role for the inheritance of acquired characteristics,
or Lamarckism, normally conceived as a continuous process.
Nevertheless, there had been some expressions of support for
saltations even from within the Darwinian camp. T. H. Huxley
himself had mentioned both saltations and directed variation,
and responded to Bateson's work in a letter thus: "I see you
are inclined to advocate the possibility of considerable 'saltus'
on the part of Dame Nature in her variations. I always took
the same view, much to Mr. Darwin's disgust, and we used often to debate it."[10] Francis Galton (Darwin's cousin and the
founder of biometry) had raised the possibility of saltations in
both his *Hereditary Genius*, of 1869, and his *Natural Inheritance*, of 1889.[11] He used the analogy of a solid polygon,
which may rock about on a single face (equivalent to normal
continuous variation) or tip suddenly onto a different face (a
saltation). This analogy was taken up by one of Darwin's
most vociferous scientific opponents, St. George Jackson Mivart, whose *Genesis of Species*, of 1870, offered a cornucopia
of anti-selectionist and anti-adaptationist arguments.[12]

Perhaps the strongest influence on Bateson's thought was
the American biologist W. K. Brooks, with whom he studied
in the summers of 1883 and 1884. Unlike the morphologists,
Brooks was interested in the process of variation and evolution, and his 1883 book *The Law of Heredity* cites Huxley,
Galton, and Mivart in support of saltative evolution.[13] As a
result of these influences, Bateson had already become disenchanted with morphology before he finished "The Ancestry of
the Chordata." Once the abortive trip to Asia was over, he
struck out on the study of variation that culminated in the
publication of *Materials*.

The preface and the introductory section of the book are
perhaps of most interest to the historian, since it is here that
we see most plainly Bateson's disenchantment with evolutionary morphology and his rejection of Darwinian utilitarianism
and gradualism. In the preface, Bateson rejects the search
for evolutionary pathways, claiming that it is based on assumptions about the appearance of potentially useful variations that
are as groundless as the assumptions of William Paley and
other exponents of natural theology. Since there are at least
two alternative theories of the origin of the chordates, and no
one can decide between them, this whole approach must be
flawed (pp. v–vi). There is an explicit critique of the style
of education in biology promoted by Huxley and Balfour, based
on the dissection of specimens following a textbook which shows
the student what he or she is supposed to find and deflects
attention from variations (p. ix).

Section 1 of the Introduction begins by acknowledging Darwin's role in pioneering the study of variation and evolution and makes clear Bateson's commitment to the theory of "Common Descent" (p. 4). The two chief explanations of descent with modification are those offered by Lamarck and Darwin, but the theories suffer from a common failing. They assume that change is produced continuously in response to environmental pressures, but in nature we find that environments vary continuously (temperature, humidity, etc.) while species and varieties seldom show intermediate forms. Bateson now reiterates his criticism of evolutionary morphology: the morphologists assumed that individual development recapitulates the phylogenetic history of the species, but experience has shown that the degree to which ancestral structures are repeated is so limited that no reliance can be placed on the method (pp. 8–10). The study of adaptations is equally flawed, because we can never know the extent to which an organ is or is not adapted to its use. In many cases the characters used to distinguish closely related species are precisely those to which it is most difficult to assign any utility (pp. 10–12).

Bateson then distinguishes between continuous and discontinuous variation, pointing out that most evolutionists have assumed that new characters are built up by the continuous accumulation of minute differences. He repeats the old argument (found in Mivart and other critics of Darwin) that selection could hardly favor the incipient stages of structures that would only be useful when fully developed (pp. 15–16). This leads directly to a summary of his views on meristic or repetitive variation, and here we see how the particular research problems that had engaged Bateson as a morphologist must have predisposed him to see discontinuous variation as more important. The production of multiple units of a single original structure in variant individuals will almost certainly be discontinuous, since a fraction of the original could not be incorporated into a viable organism. There is a warning (p. 29) about the implications of metameric segmentation for phylogeny: if repetitive variation is common in nature, we cannot assume that homologous patterns of segmentation indicate community of descent. Bateson goes on to argue that there are many cases in which characters which could, in principle, vary continuously (size and color, for instance) do in fact vary discontinuously. This strikes directly at the biometrical school's assumption that variation normally consists of a continuous range centered on a mean value for that particular character.

The final section of the introduction (section 14) allows Bateson to offer cautionary conclusions about the use of common terms such as *heredity*, *reversion*, and *adaptation*. He makes it clear that the study of heredity should be separated from the study of variation and that in both cases biologists are a long way short of understanding what is really involved in these processes. He returns to his attack on the claim that adaptation is the driving force of evolution, insisting that, while we cannot rule out a role for natural selection, it is fruitless to speculate about the extent to which any structure is adapted to the organism's environment and hence about whether the degree of utility is crucial for determining the survival of variant individuals.

Bateson returns to these points in the concluding chapter of the book (chapter 25). He draws a link between the origin and inheritance of discontinuous variations, pointing out that discontinuity eliminates the much-discussed problem of new variants' being swamped by interbreeding with unchanged individuals (p. 573).[14] Here he shows himself to be moving toward a view of inheritance as a discontinuous process, and he concludes that "the only way in which we may hope to get at the truth is by the organization of systematic experiments in breeding, a class of research that calls perhaps for more patience and more resources than any other form of biological inquiry" (p. 574). He goes on to lament the lack of communication between laboratory workers and field naturalists, the former delving into underlying causes without taking into account the complexity of the real world, while the latter use the inadequacy of the laboratory work as an excuse to retreat into mere collecting. In view of the subsequent gulf which opened up between the first generation of geneticists and the field naturalists, these were prophetic words. The evidence presented in *Materials* is not, of course, experimental in character. Bateson gathered his evidence for the existence of discontinuity from the study of natural varieties and occasional monstrosities. But from this he would go on to begin the kind of breeding experiments postulated in the sentence just quoted, and to a recognition of the opportunities offered by the rediscovery of Mendel's laws in 1900.

What then was the immediate outcome of the publication of *Materials*? The book was certainly not a publishing success; sales were very limited and it was soon remaindered.[15] Galton made a favorable reference to it, and we have already noted that Huxley wrote Bateson a supportive letter.[16] Perhaps the most substantial favorable review came—not surprisingly—from Mivart, but he probably did more harm than good by presenting the idea of internally directed variation as a product of the old view that evolution was the unfolding of the Creator's plan.[17] More significant may be a brief reference to "the recent great work of BATESON" added to a note in Arthur Willey's classic *Amphioxus and the Ancestry of the Vertebrates*, of

1894.[18] Willey's theory, like Bateson's own, required that segmentation evolved separately in different phyla, and he was evidently predisposed to accept studies which favored internally programed variation. Unfortunately, Bateson's own work now formed part of a general trend among biologists away from the old concentration on phylogenetic reconstruction.

The neo-Darwinians were inevitably opposed to the book. Alfred Russel Wallace, by then become the grand old man of British Darwinism, wrote a lengthy critique of Bateson's and Galton's views in the *Fortnightly Review*.[19] He derided Bateson's claim that a continuously varying environment should give rise to a continuous spectrum of organic forms, and defended the utility of specific characters. He also argued that a theory based on monstrosities could not explain natural evolution. More significant for the future was the bitterly critical review written for *Nature* by Weldon, which signaled the end of the two men's friendship.[20] Although Weldon conceded that Bateson had uncovered an important class of variations, he totally repudiated the claim that these saltations could have any role to play in evolution. The biometrical study of continuous variation was the only method that would yield data of relevance to evolution. Within a year the two men were engaged in a violent controversy which continued until Weldon's death in 1906.

The resulting ill-feeling ensured that Mendelism would be drawn into the debate, blocking any immediate chance of a synthesis with Darwinism. Bateson fitted Mendelism into his scheme of discontinuous variation and assumed that new characters were both produced and inherited as units. He continued to insist that the degree of adaptation had no influence on whether or not a character was transmitted. Weldon and Pearson saw Mendelism as just another plank in the saltationist platform and were thus led to dismiss the Mendelian effects as irrelevant to the inheritance of continuous variation in large populations. The position staked out by Bateson in *Materials* thus played a major role in determining the character of the British debate over Mendelism. Bateson himself remained something of an outsider in the British scientific community for several years. Although he got a readership in zoology at Cambridge in 1907 and a chair in genetics the following year, he still found it difficult to fund his research and soon moved to the John Innes Horticultural Institution.[21]

Although Bateson switched his attention to genetics, he retained his interest in the processes governing individual development, and this ensured that his position would continue to remain idiosyncratic.[22] We have already noted his suspicion of the chromosome theory, and unlike other geneticists he did not back away from his rejection of the Darwinian selection theory in the 1920s (he died in 1926). What is perhaps more surprising is that he remained isolated from the main source of enthusiasm for the saltationist position to emerge in the early years of the twentieth century: Hugo De Vries's "mutation theory."[23] De Vries tried to provide direct evidence for the instantaneous production of new species by discontinuous variation. We now know that the apparently new characters he observed in the evening primrose, *Oenothera lamarckiana*, were not due to genetic mutations; but in the first decade of the new century many biologists (especially in America) believed that the production of new species could now be observed directly.

Like Bateson, many supporters of the mutation theory insisted that natural selection played no role in determining which new characters would survive and reproduce. Under the influence of this theory, Thomas Hunt Morgan wrote one of the most vitriolic anti-Darwinian and anti-adaptationist books of the era, his *Evolution and Adaptation*, of 1903.[24] Later, of course, Morgan would accept Mendelism and go on to establish the modern concept of genetic mutation and the chromosomal theory of heredity. Unlike Bateson, he would gradually abandon his hostility to selectionism, accepting a form of Darwinism in which only those mutated characters conferring an adaptive advantage would spread into a population.

Despite the pioneering role *Materials* had played in making the case for saltationism and anti-selectionism, Bateson showed no enthusiasm for the mutation theory. He doubted (correctly, as it turned out) that De Vries's observations threw any light on the appearance of new genetic characters, and continued to insist that this process lay outside the scope of existing scientific competence. He offered the suggestion that the appearance of new characters was due to the loss by mutation of inhibiting factors which had originally masked them.[25] On this model, all evolution would consist of the delayed manifestation of characters present in the species since some unknown creative event in the past. Most biologists were by this time convinced that Bateson was grossly exaggerating the problems besetting Morgan's chromosomal theory of mutation and inheritance. Some even accused Bateson of adopting so negative a tone that nonscientific critics were seizing upon his words to discredit the whole idea of evolution.[26]

Bateson thus remains a paradoxical figure—a geneticist who rejected what came to be seen as some of the most exciting opportunities offered by the new science. For the historian, *Materials for the Study of Variation* offers important insights into the motivations which led Bateson toward Mendelism but also led him to oppose both the original and the modernized form of Darwinism. Biologists who retain an interest in the developmental process and its relationship to evolution may find his work illuminating for its ability to identify topics and problems which have been set aside in the construction of the genetical theory of natural selection. The few who wonder if a full recognition of the role played by development will threaten the basic principles of Darwinian adaptationism can see how far one of the pioneers of genetics was willing to carry this line of argument.

NOTES

1. William Bateson, *Mendel's Principles of Heredity: A Defence* (Cambridge: Cambridge University Press, 1902). See Robert Olby, *Origins of Mendelism*, 2nd ed. (Chicago: University of Chicago Press, 1985), and Olby, "William Bateson's Introduction of Mendelism to England: A Reappraisal," *British Journal for the History of Science* 20 (1987): 399–420. For a survey of current thinking on the origins of genetics see Peter J. Bowler, *The Mendelian Revolution: The Emergence of Hereditarian Concepts in Modern Science and Society* (Baltimore: Johns Hopkins University Press; London: Athlone, 1989).

2. Bateson coined the term *genetics* in a letter to Adam Sedgwick in 1905 (see Beatrice Bateson, *William Bateson, F.R.S., Naturalist* [Cambridge: Cambridge University Press, 1928], p. 93) and used it at an international congress the following year.

3. See William Coleman, "Bateson and Chromosomes: Conservative Thought in Science," *Centaurus* 15 (1970): 228–314, and A. G. Cock, "William Bateson's Rejection and Eventual Acceptance of Chromosome Theory," *Annals of Science* 40 (1983): 19–59.

4. See William B. Provine, *The Origins of Theoretical Population Genetics* (Chicago: University of Chicago Press, 1971), and Robert C. Olby, "Dimensions of a Scientific Controversy: The Biomentic-Mendelian Debate," *British Journal for the History of Science* 22 (1989): 299–320.

5. See Julian Huxley, *Evolution: The Modern Synthesis* (London: Allen and Unwin, 1942), pp. 22–28. For a survey of the anti-Darwinian theories see Peter J. Bowler, *The Eclipse of Darwinism: Anti-Darwinian Evolution Theories in the Decades around 1900* (Baltimore: Johns Hopkins University Press, 1983), and more generally Bowler, *The Non-Darwinian Revolution: Reinterpreting a Historical Myth* (Baltimore: Johns Hopkins University Press, 1988).

6. See Mark Ridley, "Embryology and Classical Zoology in Britain," in T. J. Horder, J. A. Witkowsky and C. C. Wylie, eds., *A History of Embryology* (Cambridge: Cambridge University Press, 1986), pp. 35–67, and Peter J. Bowler, "Development and Adaptation: Evolutionary Concepts in British Morphology, 1870–1914," *British Journal for the History of Science* 22 (1989): 283–97. On the use of morphology to create the image of a "modern" biology in the late nineteenth century see Joseph A. Caron, " 'Biology' in the Life

Sciences: A Historiographical Contribution," *History of Science* 26 (1988): 223–68.

7. On the controversy over the origin of the chordates, the best source is still E. S. Russell, *Form and Function: A Contribution to the History of Animal Morphology* (London: John Murray, 1916), chap. 15.

8. William Bateson, "The Ancestry of the Chordata," (1886), reprinted in R. C. Punnett, ed., *The Scientific Papers of William Bateson*, 2 vols. (Cambridge: Cambridge University Press, 1928), vol. 1, pp. 1–31, see p. 1.

9. On orthogenesis see Bowler, *Eclipse of Darwinism* (note 5), chaps. 6 and 7.

10. Huxley to Bateson, 20 February 1894, in L. Huxley, ed., *The Life and Letters of T. H. Huxley*, 3 vols. (London: Macmillan, 1903), vol. 3, p. 320. For an example of Huxley's support for saltations, see his review of Darwin's *Origin of Species*, reprinted in Huxley, *Collected Essays*, 9 vols. (London: Macmillan, 1893–94), vol. 2, *Darwiniana*, pp. 22–79, see p. 77.

11. Francis Galton, *Hereditary Genius: An Inquiry into its Laws and Consequences* (London: Macmillan, 1869), p. 369; Galton, *Natural Inheritance* (London: Macmillan, 1889), p. 27. Enthusiastic recollections by Bateson of reading Galton appear in a letter to the latter's great-niece, Evelyn Biggs, 7 July 1909, in Karl Pearson, ed., *The Life, Letters and Labours of Francis Galton*, 3 vols. (Cambridge: Cambridge University Press, 1914–30), vol. 3A, p. 287.

12. St. George Jackson Mivart, *On the Genesis of Species* (London: Macmillan, 1871), p. 261.

13. W. K. Brooks, *The Law of Heredity: A Study of the Cause of Variation and the Origin of Living Organisms* (Baltimore: John Murphy, 1883), pp. 301–2.

14. The classic expression of the "swamping argument" was in Fleeming Jenkin's review of the *Origin of Species*, *North British Review* 46 (1867): 277–318. Although we know that Darwin was disturbed by this review, other advocates of the selection theory, including Pearson, evaded it by emphasizing the continuity of variation.

15. See Beatrice Bateson, *William Bateson* (note 2), pp. 57–58.

16. Francis Galton, "Discontinuity in Evolution," *Mind*, n.s. 3 (1894): 362–72, see p. 369; for Huxley's letter see above, note 10.

17. [Mivart], "Bateson on Variation of Organic Life," *Edinburgh Review* 182 (1895): 78–105.

18. Arthur Willey, *Amphioxus and the Ancestry of the Vertebrates* (New York: Macmillan, 1894), p. 291.

19. A. R. Wallace, "The Method of Organic Evolution," *Fortnightly Review* n.s. 57 (1895): 211–24 and 435–45. See also Wallace's letter to E. B. Poulton of 8 September 1894, in James Marchant, ed., *Alfred Russel Wallace: Letters and Reminiscences* (New York: Harper, 1916), p. 313.

20. W. F. R. Weldon, "The Study of Animal Variation," *Nature* 50 (1894): 25–26.

21. See Robert C. Olby, "Scientists and Bureaucrats in the Establishment of the John Innes Horticultural Institution under William Bateson," *Annals of Science* 46 (1989): 497–510.

22. See for instance Bateson's *Problems of Genetics* (1913), reprinted with a historical introduction by G. Evelyn Hutchinson and Stan Rachootin (New Haven: Yale University Press, 1979).

23. Hugo De Vries's *Die Mutationstheorie* (1901–1903) was translated as *The Mutation Theory: Experiments and Observations on the Origin of Species in the Vegetable Kingdom*, 2 vols (London: Kegan Paul, Trench and Trubner, 1910). See Bowler, *Eclipse of Darwinism* (note 5) chap. 8.

24. Thomas Hunt Morgan, *Evolution and Adaptation* (New York: Macmillan, 1903). On Morgan's career see Garland E. Allen, *Thomas Hunt Morgan: The Man and His Science* (Princeton: Princeton University Press, 1978).

25. William Bateson, "President's Address," *Report of the British Association for the Advancement of Science* (1914 meeting): 3–38, see pp. 17–18.

26. See for instance Henry Fairfield Osborn, "William Bateson on Darwinism," *Science* 55 (1922): 194, a reply to Bateson's "Evolutionary Faith and Modern Doubts," *ibid.*: 55–61.

WILLIAM BATESON
AND THE SCIENCE OF FORM

GERRY WEBSTER

"If law is anywhere, it is everywhere."
E. B. TYLOR (1871)

THE most important, and the most difficult, task faced by any would-be science is that of delimiting a domain of phenomena which will form the specific object of that science, for a science is not determined by some a priori method but by the specific nature of its object. It is necessary to discover what is relevant and what is irrelevant and to find the appropriate and specific concepts in terms of which the phenomena may be classified, described, and explained. These tasks are not independent, for explanation and description are relative to each other and the development of a science involves an ongoing historical dialectic between taxonomic and explanatory concepts. The goal of any science is the identification and definition of the natural kinds of causally active things (particulars) that exist within a specific domain and the discovery and explanation of what they do and how they act.[1]

Materials for the Study of Variation is concerned with the science, more exactly the possibility of a science, of biological form. Bateson's title, perhaps unintentionally, is a nice conceit, since the "materials" discussed are of two kinds. The bulk of the book is "an imaginary catalogue of a Museum" of forms (p. 83), but the most significant parts of the book, the introductory sections, and the comments interspersed in the "catalogue" are concerned with materials of a different sort: the materials for thought, concepts for the study of form. In these sections he presents a critique of received morphological concepts. As he ruefully notes, such a critique is required even with respect to the earliest phases of morphological investigation, since, in our fallen state, we cannot avoid the "taint of theory" even in the initial description of forms (p. vii).

The enterprise of constructing a science of biological form is certainly far from completion at present. Indeed, it is arguable that it is still in the earliest stages. In this respect, *Materials* is not of historical interest only, for the fundamental conceptual issues relating to classification, description, and explanation which Bateson addresses are still not completely resolved. Even more important, perhaps, is the fact that Bateson discusses these questions in terms of a critique of Darwinian concepts of morphology, since, from the perspective of Darwinian theory, the enterprise of constructing a science of biological form is one whose very possibility, let alone completion, might be regarded as questionable at a fundamental level. To put the matter briefly and schematically. The forms we are concerned with are the morphologies of individual organisms, and the possibility of a general *science* of form presupposes that these individual organisms (considered as forms) are members of natural kinds.

Now, traditionally and conventionally in biology, the "significant" kinds, of which individual organisms are members, are taxonomic kinds, species (and higher) taxa, and these kinds are recognized empirically, in terms of morphology. It is not unreasonable, therefore, to suppose, at least initially, that taxonomic kinds are the appropriate natural kinds in relation to a science of form. Bateson is aware of the significance of kinds in this sense, as is shown by his reference in the first sentence of the Preface to the "problem of Species" (p. v) and even more, perhaps, by his choice of 1 Cor. 15:39 as an epigraph to the Introduction (p. 1). Now, these presuppositions appear to be called into question by the Darwinian theory of evolution which, on some interpretations, has been thought to entail the total rejection of the notion of individual organisms as being members of natural kinds, insofar as these kinds are identified with species (or other) taxa. On this view species taxa are historical entities; ontologically they are individuals, not kinds. The individual organisms which are "parts" of these "species individuals" are effectively "sediments" of history and hence properly explained in terms of historical narrative rather than scientific law. For some, therefore, the specificity of biology is regarded as being grounded in the historical nature of its objects. Mayr expresses a view of the nature of organisms which is consistent with this position in the course of an argument to the effect that biology "requires concepts that have no analog in the physical sciences . . . [because] organisms contain a historically evolved genetic program in which the results of 3 billion years of natural selection are incorporated."[2] The question of the ontological status of species taxa and the explanatory role of historical narrative in biology is by no means an issue of merely historical interest but is very much a live, contemporary debate.[3]

A convenient point of entry into the matters with which Bateson concerns himself is provided in the title of the book: the study of variation. In a broad sense, the problem of variation is effectively coterminous with the problem of form conceived in empirical terms, since the forms of organisms are, superficially at any rate, extraordinarily variable and diverse. However, biologists have traditionally refused to regard this diversity as either irreducible or totally contingent and have attempted, with a measure of success, to systematically identify and classify the diverse empirical forms. From this perspective, the goal of traditional, pre-Darwinian comparative anatomy and embryology is rationally reconstructed by Driesch:

> The old morphology had sought by means of anatomy and embryology to establish the laws, if any, which actually controlled morphological phenomena. It sought, in fact, to discover what morphogenesis really was. It sought, moreover, to construct what was typical in the *varieties* of forms, into a system which should be not merely historically determined, but which should be intelligible from a higher and more rational standpoint.[4]

As reconstructed by Driesch, Rational Morphology addresses two distinct but related problems. These seem to correspond approximately to the two basic tasks of any science as characterized by Bhaskar (see note 1): the explanatory and the taxonomic. According to Driesch, the first problem is an explanatory or causal problem: how to establish the link between the General and the Particular, the (empirically) typical and the individual. The "laws" Driesch refers to are those governing the process "by which the type is realised for the time being in the individual"—this is what morphogenesis really is—or, "those governing how it [the type] changes its *specificity*, if such a change, i.e., a descent, is . . . assumed."[5] The second problem is a systematic problem. Is it possible to discover or construct an "intelligible" or "rational" system of forms? By this Driesch seems to mean an attempt to discover a form of law or necessity in the system which provides its raison d'être. That is, to see to what extent, if at all, the *purely empirical* classification can be theoretically elucidated in such a way that it, or some aspects of it, have a formal structure, analogous to that of a logical system, so that the diverse forms, or better, laws, can themselves be intelligibly related in terms of some kind of law. On this basis we might be able to say either that there cannot exist more than a certain number of diverse forms or that there can be an indefinite number which follow a *definite law* with regard to their differences. In either case, we have a closed set.[6] The question of whether there are constraints on what forms can exist or whether, in the Darwinist view, "everything is possible," is one of Bateson's central concerns.

The concept of the "typical" employed here is no more than a concept of empirical regularities in morphologies, albeit rather complex regularities—sequences of forms or "life cycles." We compare individual forms in order to determine regularities and thereby subsume the diverse individuals under empirical concepts of natural kinds of things: *Fritillaria meleagris, Fritillaria, Liliaceae; Canis lupus, Canis, Canidae*, etc. Thus, this "classificatory *preparation* for the knowledge of . . . the rational in the forms of nature"[7] represents, from a realist perspective, the first, the *Humean*, stage which occurs in the development of any science, that is, the descriptive characterization of distinct kinds in terms of empirical regularities or "proto-laws."[8] It is only on this basis that we can, in the first instance, suppose that there is something to be *scientifically* explained.

Now, in nearly all sciences, a taxonomic enterprise of this typological kind, since it is merely empirical, generates an unclassifiable residue: the irregular, the atypical, the abnormal, the monstrous. If this empirically unclassifiable, apparently "unlawful," residue is small in magnitude, it tends to be ignored. However, if the irregularity is of such magnitude that it cannot be ignored, a typical solution is to multiply kinds. Since this multiplication process can, in principle, continue until there are as many kinds as there are individuals, it is conceivable that, at some point well before this logical terminus, the whole classificatory enterprise might be thought to be called into question; perhaps there are no kinds in nature, therefore no "laws of form," but merely unique individuals, each a law unto itself. Such a conclusion is likely to be drawn by those who subscribe to an empiricist philosophy of science in which causal laws are equated with empirical regularities and natural systems are regarded as closed. From a transcendental realist perspective, however, such a conclusion is unwarranted, since natural systems are conceived as open and kinds are not defined empirically but in terms of underlying generative mechanisms. Thus, while the presence of empirical regularities provides the prima facie grounds for supposing the existence of causal laws, the absence of such regularities does not necessarily provide grounds for supposing their nonexistence.[9]

Now, while it is variation of this residual or apparently individual kind with which Bateson is primarily concerned in *Materials*, he is well aware that this is only one aspect of the question of variation in the larger sense, that is, one aspect of the general problem of form. And it is in relation to this general problem that he organizes his discussion of the empirical material. The study of "freaks of nature" is merely a means towards an end. In *Materials* he attempts the analysis and classification of the empirical forms of "individual and residual variation" as a preparation for, and in relation to, a causal analysis of form in general. His approach is similar to that of the earlier rational morphologists: namely, to look for regularity in the apparently irregular, typicality in the apparently atypical, law and order in apparent chaos. In short, Bateson's aim is to demonstrate that *all* empirical variations, that is, all empirical forms, are specific or law governed and can, therefore, be regarded as the empirical manifestations of the inherent causal powers of natural kinds of things.

The particular task which Bateson undertakes in *Materials* is necessary because of the conceptualization and role of such variation in Darwinian theory. For, while the rational morphologists had conceived residual variation as contingent, as sheer individuality, and concentrated on the typical or empirically "lawful" aspects of form (for methodological and epistemological reasons valid at that stage of investigation), the Darwinists, empiricists to the core, went to the opposite extreme. For them, individuality and variation, inherently unconstrained and contingent, was all, the epistemological problems raised by this view being blithely ignored.[10] Species (and other taxa) were no longer conceptualized as natural kinds but as historical entities. Therefore any empirical order discoverable in nature must be understood as having been imposed from without rather than being the expression of the inherent and specific causal powers of individual organisms.

In order to understand the context in which Bateson develops his arguments, it is necessary to briefly summarize the morphological concepts employed in Darwinism,[11] and the best source of these is the *Origin* itself.

Darwin asserts that it is widely acknowledged that "all organic beings have been formed on two great laws—Unity of Type, and the Conditions of Existence."[12] Let us consider these in turn (but in reverse order).

The concept of conditions of existence is attributed to Cuvier.[13] Darwin summarizes it thus:

> Every detail of structure in every living creature . . . may be viewed, either as having been of special use to some ancestral form, or as being now of special use to the descendents of this form.[14]

Darwin explains that "mere chance"[15] might cause variation of the offspring from the parental type, and any "slight modification" which in any way favored these individuals by better adapting them to their conditions of life would tend to be preserved[16] and slowly accumulated by means of natural selection.[17] This utilitarian view is summarized thus:

> If variations useful to any organic being do occur, assuredly individuals thus characterised will have the best chance of being preserved in the struggle for life; and from the strong principle of inheritance they will tend to produce offspring similarly characterised. This principle of preservation, I have called . . . Natural Selection . . . small differences distinguishing varieties of the same species, will steadily tend to increase till they come to equal the greater differences between species of the same genus, or even of distinct genera.[18]

The second concept, unity of type, is interpreted by Darwin as meaning that organisms of the same class are formally related, constructed on the same morphological plan; it is on the basis of these formal relations that empirical classifications are constructed, as noted above. Now, Darwin takes these formal relations as indicative of genetic (i.e., material, historical) relations; the reasoning here is analogical and based on the empirical association of formal and genetic connection within a single species. The decisive move is to claim that the empirical, formal relations can be explained in terms of the hypothetical, genetic relations: "On my theory, unity of type is explained by unity of descent";[19] "the chief part of the organisation of every being is simply due to inheritance" from a "common progenitor."[20] Therefore the nature of classification changes; it becomes, indeed must become, genealogical:

> As all the organic beings, extinct and recent, . . . have to be classed together, and as all have been connected by the finest gradations, the best, or indeed, if our collections were nearly perfect, the only possible arrangement, would be genealogical. Descent being on my view the hidden bond of connexion which naturalists have been seeking under the term of the natural system.[21]

Thus, a particular individual form is no longer to be regarded as a member of a kind or class but as a part of a lineage.

Darwin elaborates on the problems posed for an empirical taxonomy by what I have termed residual variation and draws the typical empiricist conclusion, that there are no kinds in nature. This view is presented as consistent with the genealogical interpretation:

> No clear line of demarcation has as yet been drawn between species and sub-species . . . between sub-species and well marked varieties or between lesser varieties and in-

dividual differences. These differences blend into each other in an insensible series; and a series impresses the mind with the idea of an actual passage. . . . From these remarks it will be seen that I look at the term species, as one arbitrarily given for the sake of convenience to a set of individuals closely resembling each other, and that it does not essentially differ from the term variety. . . . The term variety, again, in comparison with mere individual differences, is also applied arbitrarily.[22]

Darwin concludes that the utilitarian concept of functional adaptation is the fundamental concept: "the law of the Conditions of Existence is the higher law; as it includes, through the inheritance of former adaptations, that of Unity of Type."[23] Thus, in the final analysis, form is to be understood in terms of one rational principle, functional adaptation, which is to be elucidated in terms of the causal mechanism of natural selection.[24]

From the perspective of Darwinism, the task of comparative anatomy and embryology can no longer be the preparatory work for a rational morphology since the latter enterprise is now inconceivable. There are no appropriate kinds in nature and therefore the empirical order of forms cannot be seen as the expression of the specific inherent powers of the individuals which are members of these kinds. Rather, this order must be conceived as having been imposed from without on organisms which are inherently plastic (*Materials*, p. 80), this imposition being a gradual process, taking long periods of time and being determined in relation to functional adaptation. From this perspective, the task of the two traditional disciplines becomes the reconstruction of pedigrees, for it is in terms of function and descent that form is to be explained. As Darwin tells us: we must learn to "regard every production of nature as one which has had a history . . . every complex structure . . . as the summing up of many contrivances, each useful to the possessor."[25]

Darwin's overall conclusion with regard to form is not modest: "On this . . . view of descent with modification, all the great facts in morphology become intelligible."[26] Clearly this is a conclusion with which Bateson disagrees, and in the course of articulating his disagreement he arrives at his own characterization of the problem of form and of the nature of the organism. Let us examine the nature of his disagreement.

I will firstly deal rather briefly with Bateson's views on functional adaptation and the explanation of forms in terms of utility. The treatment will be brief for reasons which will become apparent.

It is clear from the *Origin* that natural selection is advanced as a causal mechanism to explain the stability of an individual form as a *given* member of a population of forms; it is concerned with the *persistence* of forms in terms of their degree of functional adaptation—the "preservation of favoured races." Now what *persists* is clearly a subset of what *exists*, so the major question is, Are there any inherent constraints on what can exist, constraints on the stability of individual forms *as forms*, irrespective of function? The Darwinists apparently answer this question in the negative; variation is "random," forms arise by "mere chance"; in principle, everything is possible. Now, all that is really implied here is that in relation to functional adaptation, that is, *under a particular description*, variation is unconstrained; if it were not so, then there would be no need for a process of natural selection. But in terms of Darwin's paradigm, functional adaptation is, in the final analysis, the *only* manner in which the problem of form is conceived, and for many Darwinists (including contemporary ones) it appears to be the only manner in which it can or should be conceived; that is, forms can only be known under functional descriptions which for some are tantamount to descriptions of "life" itself.[27] In relation to the formal aspects of form—that is, form considered under a formal description—the question of constraints on what can exist is not simply left unanswered, it is not even asked.[28] That it can have been supposed to have been asked, and answered in the negative, is simply a consequence of the fact that the formal has uncritically been subsumed under and conflated with the functional.

Elsewhere,[29] Bateson characterizes descriptively an empirical individual (either a whole organism or a part of an organism insofar as it has relative autonomy) in purely formal terms as a pattern: "an individual is a group of parts differentiated in a geometrically interdependent order." The two aspects singled out here, differentiation and order, correspond to the two main classes of variation considered in *Materials*: substantive and meristic (the latter term Bateson claims as his own invention). Under this mode of description there is a whole realm of questions undreamt of in the Darwinist philosophy.[30] Purely formal, morphological questions which can, in principle, be posed quite irrespective of *any* functional, adaptive considerations; and it is these questions in which Bateson is interested, questions concerning the possibility of inherent constraints on what forms of order can exist and therefore on what forms comprise the subset which persists, assuming a process of natural selection (see p. 69).

However, it would appear either that Bateson has not entirely grasped the full nature and significance of his own position, namely, that the formal aspects of morphology can, in principle, be studied as an autonomous domain, or that he has difficulty in consistently maintaining this position. For instead of being content simply to argue that formal and functional conceptions of organisms are epistemologically distinct and should not be conflated,[31] he feels it necessary in *Materials* to attack the functionalist position and advance anti-adaptationist and anti-utilitarian arguments. But to argue whether the similarities and differences of form which are used in constructing taxonomies are or are not useful to the organisms (p. 11), or whether it is reasonable or unreasonable to see the constancy or variability of the coloration of ladybirds in terms of adaptation (p. 572), is to argue whether or not particular forms *are* knowable under functional descriptions. It is to conduct a discussion *within* the very paradigm from which Bateson is attempting to liberate himself. The point is not that such speculation is "barren and profitless" (p. 79) but that he does not need, and should not attempt, to address questions such as, Is it necessary "to suppose that every part an animal has, and everything which it does, is useful and for its good?" (p. 12). In the context of Bateson's project, in which forms are to be formally described, the question (and the answer—if there is one) is simply irrelevant. It is, in a sense, understandable that Bateson should have difficulties here, since a particular organismic ontology specific to the descriptive-explanatory scheme of Darwinism has been improperly generalized. But a critique of functionalism is not the appropriate way to deal with this issue.[32] More importantly, such discussions are likely to perpetuate the confusion and conflation of the formal and the functional and to distract attention from what is really important and novel in his argument. To this I now turn.

As I noted above, consequent upon Darwin's theory of common descent, the whole point of research in comparative embryology and comparative anatomy changed. These methods now became the tools for reconstructing genealogies and discovering common ancestors, for it is in these terms that unity of type is to be explained. Bateson criticizes this whole project, firstly on epistemological and methodological grounds, secondly in terms of the explanatory value of the theory itself, thirdly, and perhaps most important, in terms of the ontological assumptions about the nature of the organism and the nature of species which are implicit in the theory.

His methodological criticisms can be dealt with rapidly at this stage. They pertain, firstly, to the use of comparative embryology as a means of establishing particular genealogies, where he simply argues that the proposition that "the development of a form is a record of its descent" (ontogeny recapitulates phylogeny) is effectively circular (p. 8). His second point in this regard pertains to comparative anatomy and the nature of the formal relations (homologies) between organisms which are used as a basis for constructing taxa and (in Darwinism) as a means both of inferring and of characterizing genetic relations. As Bateson notes (p. vii), homology is a theoretical concept and he wishes to argue that this concept is constructed and used in an inconsistent and arbitrary way. The fact that the nature of the formal relations is not uniquely specified leads to inconsistency in application. The fact that usually only some instances of formal relations are subsumed under the concept of homology he sees as the construction of an arbitrary category by the abstraction of these instances from a more general and more "natural" category, the class of "formal repetitions." However, in discussing these issues he raises far more than epistemological and methodological problems, and they are best considered along with the concepts of inheritance, variation, and pattern, which I discuss below.

I consider now Bateson's views on the explanatory value of the theory of descent. From the new Darwinist perspective, as noted above, the point of classification changes. For example, to assert that a number of extant species can be included in a "higher" taxon is to formulate a hypothesis about the past, that is, that these species have a common ancestor.[33] Now the question is, What sort of hypothesis is this? Is it one merely about matters of (historical) fact: that there are common ancestors? If so, there is perhaps no problem since common descent is at least a plausible notion. Is it, on the other hand, a hypothesis which purports to have *explanatory* import, that is, one intended to explain the *existence* and the *distribution* of the morphological characteristics which are shared by the diverse extant forms and are the empirical grounds for including them in the same taxon? If so, then there do seem to be real problems.

It might be argued that the concept of common descent (i.e., filiation), *if supplemented* by the concept of inheritance, renders unity of type intelligible in a way that the concept of special creation does not, since it could account for the pattern of *distribution* of the characteristics in question. However, Bateson denies that there is any such phenomenon as inheritance (in Darwin's sense) in the organic world. This will be considered below. More importantly, it is difficult to see how the *existence* of characteristics shared by a group of organisms (living in the present) is in any sense *explained* by positing another group of organisms (living in the past) which also has the same characteristics. This point was made by Bateson, prior to *Materials*, in the course of a discussion of the ancestry of the chordates and the significance of a kind of repetitive morphological organization (metameric segmentation) shared by vertebrates:

> This much alone is clear, that the meaning of cases of complex repetition will not be found in the search for an

ancestral form, which, itself presenting this same character, may be twisted into the representation of its supposed descendant. Such forms there may be, but in finding them the real problem is not even resolved a single stage; for from whence was their repetition derived? The answer to this question can only come in a fuller understanding of the laws of growth and of variation which are as yet merely terms.[34]

The thrust of Bateson's argument is clear; the common ancestor is just one more (albeit hypothetical) form added to the world of forms. If it exists, it itself stands in need of explanation and therefore cannot serve as a means of explanation of other forms. The whole attempt to explain empirical morphologies in terms of historical origin and descent seems to be caught between two equally unacceptable alternatives. Either we have to accept a potentially infinite regress in which explanation is perpetually deferred, or explanation (insofar as it can be regarded as such) has to be abandoned at an arbitrary point of origin by taking some ancestral form as simply given. The latter is, in effect, Darwin's position: "The theory of descent with modification embraces all the members of the same class";[35] that is, it is primarily a concept of variation on a *given* theme (see *Materials*, p. 33).

If the past (hypothetical or real) cannot provide the means or terms to explain the present, then the present must be explained in terms of itself, that is, in terms of the "laws of growth and of variation" to which Bateson refers. And the past too, the "real past," insofar as it can be reconstructed from "traces" existing in the present and insofar as we can assume a uniformity of nature, has to be explained in the same terms. The notion of common origin has to be replaced by the notion of common nature. In other words, the study of morphology must become a synchronic study, a science.[36]

In his 1886 paper Bateson prefigures many of the issues which are central to *Materials* and already hints that the phenomena of repetition are to be understood in terms of the *intrinsic nature* of the organism—"the power to repeat" (see also *Materials*, p. 71, par. 3)—rather than in terms of history and genealogy (the reconstruction of which is rendered problematic from this perspective):

"Segmentation" . . . if resolved into its elements will be found to be by no means a peculiar feature of a few groups but rather the full expression of a tendency which is almost universally present. . . . greater or less repetition of various structures is one of the chief factors in the composition of animal forms . . . [and many instances are] examples of the recurrence of parts . . . in some more or less definite relation to the axis of symmetry of the animals.[37]

As I read it, the central issue in *Materials*, about which all else revolves, is the question of the "intrinsic nature" of the organism referred to above. This involves a consideration of what organisms are, of whether organisms can be regarded as members of natural kinds, and consequently a consideration of the ontological status of species (and perhaps other) taxa as the putative natural kinds and the relevance of this in relation to the characterization and causal explanation of empirical morphologies. This is the central issue because Darwinism, in its concepts of inheritance, homology, variation, and species, presupposes or embodies a more-or-less coherent ontology of the organism which, in Bateson's view, is not one which can sustain an adequate science of form.

The philosophical question of an ontology of natural kinds is of great complexity and in the space at my disposal I can only dogmatically assert a realist view.[38] From this perspective, a natural kind is viewed as a class of things (particulars) which share a common nature or essence, construed as an underlying generative or causal mechanism. Thus, although natural kinds may be tentatively *recognized* in terms of empirical characteristics such as morphologies and dispositional properties, they are *defined* in terms of some explanatory theory which accounts for the existence of these properties in terms of the "hidden" structure possessed by all individuals of that kind. However, although these empirical properties (or some of them) are necessary properties of the kind in question, their manifestation or actualization is a contingent matter. Thus, for example, in the light of current physico-chemical theory, water, which is a natural kind, is necessarily "polymorphic"; that is its nature. However, this necessity has to be understood in terms of potentiality, for it is entirely contingent which of the three possible "forms" is actually manifested by any particular sample of water, since it is contingent that a particular sample is subjected to particular conditions of temperature and pressure. Conversely, a particular sample of water, because it is a sample of a kind with a specific nature, will respond *specifically* to particular environmental conditions and to a change in these conditions. The distinction between potentiality (natural possibility) and actuality is central to a realist view of natural kinds.

The analogy here with a physico-chemical system is not entirely beside the point, and elsewhere[39] Bateson himself makes use of such an analogy in discussing variation. As he points out, the empirical variation, polymorphism, and irregularities of forms, which Darwinists interpret as impugning the ontological status of species taxa as natural kinds, are brute facts, unanalyzed and unexplained. He compares this state of affairs with that which pertained in the early, purely empirical, stages of the investigation of the nonliving world and notes that as these sciences developed, the varieties and diversities, regularities and irregularities, of empirical phenomena were gradually (and I would add, theoretically) distinguished and systematized in such a way that they could be interpreted as affording a key to the intrinsic nature of matter and thereby an understanding of its "essential diversity."

It is in these terms that Bateson conceives his own analytical study of variation, for a study of variant forms reveals the "inner nature" of organisms which, he believes, will provide the basis for an explanatory theory of all forms, "the physiological foundation or causation" of specific diversity.[40] *Materials* contains the purely observational and comparative beginnings of such a study, but even here there are considerations of the *type* of causal explanation which is required to relate the variant forms. Genetics, the study of "Heredity and Variation,"[41] he conceives as pursuing this same end of analysis, with a view to explanation but experimentally, by means of breeding.

Thus we can say that *if* there are relevant natural kinds (which might, but need not, be taxonomic species and other taxa), then the individual organisms which are their members would not be expected to be totally plastic but might be expected to exhibit empirical properties which are *specific* to that kind because produced by a common generative mechanism (and therefore necessary). These properties might include specific dispositional properties and the properties which we recognize as specific morphologies or aspects of these morphologies. However, we need not expect that all individual members of the kind will actually manifest *identical* empirical properties (either dispositionally or morphologically), since the actual manifestation of properties is a contingent matter.

In pursuing this issue, however, it is crucial to keep in mind the dialectic of taxonomic and explanatory knowledge which occurs during the (historical) course of a scientific investigation. Ontologies (membership of natural kinds) are relative to explanatory theories. It is on this basis that, for example, mechanics can quite legitimately ignore natural kind distinctions which are crucial to chemistry. But explanatory theories are, in the first instance, developed in relation to entities distinguished empirically under particular descriptions; and with the development of theories, the empirical criteria for identity of entities may undergo change. Bateson is concerned with organisms considered under formal (morphological) descriptions, and hence it is in relation to these descriptions and putative explanatory theories that questions of ontology and natural kinds must be posed. Now, as I have noted above, since empirical taxonomies in biology are traditionally constructed on the basis of morphology, it is legitimate, in the first instance, to consider, as Bateson does, whether conventionally determined species taxa should be regarded as the natural kinds in the context of morphology. However, a question which might arise in the course of this consideration is whether traditional morphological descriptions ("typologies") are the correct kinds of descriptions in this context or whether some other mode of description (e.g., in terms of symmetry), and hence some nontraditional classification, might be more appropriate in relation to the question of natural kinds; this is a question I will defer to the end.

It will be clear that the issues here are complex and interrelated, and the difficulties of discussion are enhanced by the impossibility of saying everything at once. In what follows I shall attempt a rational reconstruction of Bateson's central argument, which will involve a consideration of seemingly disconnected notions and then the attempt to bring them together.

A convenient starting point is provided by a consideration of the concepts of heredity and inheritance, which play such a central role in Darwinian theory. As Bateson notes elsewhere,[42] the empirical phenomenon subsumed under the concept of heredity is repetitive "truth to type," that is, the *reproduction* of a typical form. This particular empirical phenomenon is, therefore, simply one mode or instance of the general class of formal repetitions which, as noted above, is conceived by Bateson in terms of the intrinsic nature of the organism, the "power to repeat." It is clear, however, that "heredity" and "inheritance" are not used by Darwin simply as descriptive concepts but as explanatory concepts; and, as Bateson observes (p. 75), this use is based on a metaphorical extension from the "descent of property" in human society. As he further notes, the metaphor has "had a mischievous influence on the development of biological thought," not simply in the sense that it "misrepresents the essential phenomenon of reproduction" (p. 76), but in that it embodies a misleading conception of the nature of the organism; and he notes that this misleading organismic ontology pervades Darwinist discussions of morphology. It is arguable that this is still, to some extent, the case today. It is worth exploring this metaphor as a means of clarifying what is at issue.[43]

In ordinary English usage, *inheritance* refers to the acquisition of anything from a progenitor or a previous generation. In legal usage its meaning is more restricted and refers to property devolved in accordance with law or legal right as opposed to will. The term can also be used metaphorically, as when we might say: "Margaret has inherited her father's nose and authoritarian personality." In the *Origin* Darwin plays upon, or equivocates between, these different usages. In the social domain a system of inheritance (in the literal sense) presupposes (at least) three things: firstly, the existence of private property in durable forms—material goods or immaterial rights or titles; secondly, since persons cannot inherit from themselves, there must be a distinction between individuals as proprietors—the original possessor and the heir—and this presupposes a temporal (diachronic) distinction; thirdly, there must be some relation between the successive owners which determines the transmission of the property in question—in the context of this discussion, filiation is the only relevant relation.

Within the social world, the inheritance of private property is a major cause of the existence and perpetuation of inequality in the ownership of property, since it functions to restrict the distribution of property by permitting it to be preserved within a single line of descent over the generations. A corollary of this is that a lineage can maintain its *identity* over time, defined in terms of its heritable estate (the specific items of property it owns), and by the same token, its difference from other lineages. If new items of property are progressively acquired by any lineage and accumulated by inheritance, difference (inequality) will progressively increase over time.

It is important to note that the correct use of the literal concept of inheritance to describe a state of affairs involves a presupposition about the ontological status of the entities concerned. I can only refer to a gold watch as inherited from my grandfather if it is numerically the same watch as the one he owned. Thus, the items of material property which are inherited must have the ontological status of *individuals* and the relevant identity is *numerical identity*. A numerical individual (a historical entity) maintains its identity via spatiotemporal continuity and a form of spatial unity or cohesiveness which distinguishes it from other individuals. A similar ontological status must pertain to rights or titles, like Fellowships (*Materials*, p. 32).

As Bateson notes (p. 32), an ontology of individuals entails that each item of a heritable estate has its individual and proper history. It is this which accounts for the identity of the estate (and therefore the lineage) with itself over time. This ontology also suggests how we should conceive change. Since an estate of discrete individuals is simply an aggregate of spatially juxtaposed entities, it has no systemic unity and consequently is totally plastic. Change in such an aggregate is inherently unconstrained and indefinite in scope and magnitude; it will occur by the addition or loss of individuals in a piecemeal and largely contingent fashion. Change in each individual will again be largely (but not entirely) contingent, and unless this change is of a more-or-less *continuous* nature it will be difficult to keep track over time of the individual in question. In sum, a heritable estate is almost entirely a sediment of history. As such it, and its constituents, can be explained only in terms of historical narrative.

It is evident that the concept of inheritance is literally inapplicable to the domain of morphology, since it pertains to individuals (spatiotemporally localized particulars) whereas here we are concerned with entities which are generated, not transmitted, and where the relevant identity is, therefore, qualitative or formal not numerical; *petal* or *tibia* are kind names, not proper names. All Darwin's discussions which employ or presuppose the concept of inheritance are therefore metaphorical, and this accounts for their incoherence. Nevertheless, the employment of this concept is consistent with the overall organismic ontology to which Darwin subscribes and which by the same token is untenable. With regard to variation, *Natura non facit saltum* is not simply an empirical claim but the very principle of intelligibility, since this emphasis on continuity is grounded in the necessary presupposition that the morphological entities he wishes to talk about in terms of inheritance must be ontological individuals. For Darwin, the only way we can intelligibly relate or connect two distinct empirical forms is to identify them as two "moments" in the proper history of a spatiotemporally continuous individual. Thus, he discusses descent in terms of a change in the (empirical) properties of morphological elements, what Bateson refers to as substantive variation, but this change is conceived as the continuous variation of an individual. He refers to a gradual and *actual* modification (of vertebrae into skulls) or a *real* metamorphosis (of crab legs into mouthparts), and he is emphatic that this talk is *not* metaphorical even though von Baer had pointed out some thirty years earlier that such a metamorphosis "could only be conceptual."[44] (See *Materials*, pp. 31, 33.) When Darwin discusses change or variation in morphological composition or organization—what Bateson calls meristic variation—this is, as we might expect, conceived in terms of piecemeal change: "the atrophy . . . abortion . . . soldering together . . . and doubling" of parts.[45] As Bateson notes (p. 32), in all cases, "it is assumed that . . . the individuality of each member of a meristic series is respected."

We are now in a position to consider how Bateson deals with the question of homology, "the basis of all morphological study" (p. 30). As he observes (p. 33), the assumption that members of a meristic series behave, however obscurely, as individuals leads to problems concerning the morphological correspondence of elements when the number of elements differs. Such problems are impossible to solve, because, he argues, they are misconceived in the first place; variation does not necessarily respect this supposed individuality in all instances. Moreover, the problems are arbitrarily conceived, since no one supposes the existence of such individuality in a series of parts which manifest little in the way of differentiation. Yet, he wishes to insist, meristic repetition is the same phenomenon wherever it is found, and he implies that there is something arbitrary about singling out some forms of repetition as comprising a distinctive class to be explained in a distinctive way, for example, in terms of Darwinian inheritance.

He develops this argument over the following pages (section 7) in such a way that the class of morphologies conceived as "homologues" appears to lose its traditional distinctive status and becomes a subclass of the general class of formal repetitions. Thus, he argues, it is difficult in many organisms to draw any clear distinction between metameric segmentation and asexual reproduction or between budding or strobilization and colony formation. Asexual reproduction by division and the process by which ordinary meristic series are produced may be similar processes of repetition. If they are, we may expect to find an analogy between the differentiation exhibited by the members of a meristic series and the variation between parent and offspring. And if we regard bilateral symmetry as a mode of repetition, then we might expect to find variation between the two sides analogous to that exhibited by two distinct individuals. Bateson claims that the empirical material fulfills all these expectations.

In these pages, it is apparent that Bateson is moving away from a concern with questions of evolution or descent per se and towards a unified conception of apparently disparate phenomena in terms of a general notion of the orderly generation of division and repetition—in effect, a notion of morphogenesis in the most comprehensive sense. This view finds complete, if speculative, expression in the remarkable passage on the problem of order which concludes section 7 (p. 36).

It is in the terms of the generation of division or repetition that Bateson conducts his discussion of the empirical material on meristic variation. Two simple examples (from section 11) will suffice to illustrate the major points he wishes to make; variation in the number of petals in actinomorphic flowers and variation in the number of tarsal segments in the insect limb. In both cases, as Bateson notes, the variant forms are as "definite" and well formed as the typical forms; in the case of the flower, the variant form "possesses the character of division into four no less completely and perfectly than its parent possessed the character of division into three" (p. 61). He makes a similar point regarding the cockroach tarsus, but in addition is here able to argue (p. 417) that variation does not seem to respect the (supposed) *individuality* of the tarsal segments. There are no empirical grounds for believing that one of the "original" tarsal segments is missing in the variant and hence for supposing that the discrete segments behave in any sense (however obscure) as individuals. Rather, Bateson suggests, we should regard the whole tarsus as an individual, in the sense that it behaves as a unit or totality during division. Further, we should regard the four-jointed and the five-jointed forms as the two alternative states of division which are possible for the tarsus; in terms of the concept introduced by Galton,[46] they are two "Positions of Organic Stability" and "into either of these the tarsus may fall" (p. 65). It is to this complex of characteristics that Bateson is referring when he speaks of variation as being "discontinuous."

Substantive variation can also be discontinuous, and a particularly relevant set of examples is provided by what Bateson proposes to call "homoeotic variation" (p. 85), in which one member of a meristic series assumes the properties which are normally characteristic of another member. Homoeosis is particularly common in arthropods (chapter 5), and the example (p. 149) of the alternative forms (maxillipede or chela) displayed by a particular appendage of the crab is critically and self-evidently relevant to Darwin's discussion of this mode of variation, which I outlined above. As in the case of meristic variation, Bateson here invites us to think in terms of alternative positions of organic stability rather than in terms of "moments" in the history of an individual, as Darwin proposes.

Bateson's belief that the empirical variations of form are to be understood as the actualizations of alternative possibilities or potentialities which are inherent in the *specific* nature of the organism explains why he was so immediately receptive to the work of Mendel, when he read it for the first time in 1900. Given the empirical concepts of dominance and recessivity, the alternative forms—allelomorphs—fall out naturally from factorial analysis: AA or Aa versus aa. However, Bateson's notion of genetics was somewhat different from that of most of his contemporaries (let alone our contemporaries), and it was precisely his concern with the problem of form, as he construed it in terms of meristic organization—"the *essential* phenomena of heredity"[47]—which resulted in his scepticism concerning some theoretical developments in genetics. In particular he rejected the view that the "factors" of Mendelian analysis should be regarded as "particle[s] of a specific material,"[48] and subsequently (as Bowler notes in his foreword to this republication of *Materials*) he held out for a long time against Morgan's chromosomal theory of the gene, insofar as a chromosomal locus was to be identified with a Mendelian factor.[49]

It is clear that what exercised Bateson as regards the nature of the Mendelian factors was the problem of the causal explanation of form. Although his intuitions here were sound, he had difficulty in articulating the problem in a precise and explicit fashion, and I suspect that this is a consequence of his inadequate grasp of the nature of causal explanation. Assuming a generative (realist) concept of causation, any causal explanation necessarily involves an analytical distinction between causal agent or stimulus and causal mechanism (construed as inherent power or tendency). The concept of a Mendelian "factor," however, subsumes both these analytically distinct notions, and they are empirically inseparable in the simple breeding experiments of the early geneticists. In fact, subsequent empirical discoveries in genetics, such as pleiotropy, variable penetrance and expressivity, position effects, etc., rendered the causal problem even more intractable. It was not until the appearance of Waddington's *Organisers and Genes*[50] that a modicum of clarification was achieved with respect to the causal role of chromosomal genes and environmental factors in the generation of form, and the final story yet remains to be told. In this context, it is noteworthy that Waddington, like Bateson, sees genetics as addressing the same problems as experimental embryology and experimental morphology; and the concept of the epigenetic landscape, which he elaborates here, is effectively a temporalized version of the Galton-Bateson concept of positions of organic stability.

Bateson addresses the causal problem rather briefly in section 13, where he considers the nature of discontinuity in variation and its relation to the discontinuity of biological species. Since, at least in some instances, "variation may occur by steps which are integral and total," we have to look for the "factors which determine this totality and define the forms assumed in Variation" (p. 70). Bateson suggests that substantive variation may be "chemical" in nature and meristic variation "mechanical" and that the variant forms (and possibly species) "arise through discontinuous transition from one state of mechanical or chemical stability to another state of stability" (p. 74). Although the discussion of substantive variation is relatively clear, I suspect that the contemporary reader may well find the discussion of the "mechanical" nature of meristic variation somewhat obscure. The reason for this is that his entire discussion is informed by a speculative idea concerning the generative mechanism involved in pattern formation, which he chooses not to reveal to his readers, who have to make what sense they can of a cryptic reference to an analogy "between the symmetry of bodily Divisions and that of certain mechanical systems" (p. 71).

The speculative idea (rather, "the IDEA") is expounded in a series of letters to his sister Anna[51] which display none of the positivist timidity of *Materials*. Here it is revealed that the analogous "mechanical system" is that of "sand figures made by sound"—Chladni figures. Even quite primitive systems of this type can produce relatively complex, regular, and stable figures; and more sophisticated systems, in which liquids are employed, can produce extremely elaborate figures. In such systems, variation in the frequency of excitation while other factors (shape and size of plate, amplitude, etc.) are held constant results in the production of a family of patterns which differ in the number of individual elements (nodal lines) but retain the same basic form or structure.[52] The possible significance of such observations in relation to the explanation of organic form is clear; we might conceive two distinct but related variant forms as being produced by a *common generative mechanism* in which the difference is the result of different values of some parameter analogous to frequency. This is Bateson's view, and he is convinced that more than analogy is involved here. As he observes: "an eight-petalled form stands to a four-petalled form as a note does to the lower octave," and "Of course Heredity becomes quite a simple phenomenon in the light of this."

Although Bateson eventually indulged in public speculation[53] and throughout his life returned again and again to the "Undulatory Hypothesis" in a variety of contexts, including the mode of action of the embryonic "organizer," it never, in his hands, progressed beyond its original analogical formulation. Bateson's failure to develop his intuitions in a systematic fashion is explicable in terms of the antitheoretical positivism which characterized (British) biology until relatively recent times, compounded by his own lack of mathematical ability. In a letter to Hardy, in 1924, (and with a sideways swipe at his old enemies, the biometricians) he expresses the hope that he may live to see "mathematics applied to biology properly" and the belief that "the most promising place for a beginning . . . is the mechanism of pattern."[54] Some sixty years elapsed before Bateson's "idea for an idea" came to fruition with the publication, in 1952, of the now-celebrated mathematical paper by A. M. Turing,[55] but Turing makes no reference to Bateson.

In retrospect, it can be seen that what Bateson is struggling to articulate in his discussion of form and merism is a concept of the morphogenetic field. Indeed, *Materials* may be one of the earliest, if not the earliest, of systematic attempts to formulate field concepts which at least some contemporary biologists would argue are the appropriate concepts for understanding the generation of form. The conception of a spatial domain, whether a whole organism or a part of it, as being ontologically an individual, that is, behaving as a unit or totality, and the notion of well-formed variant patterns as being produced by a common generative mechanism, hence as being alternative positions of organic stability, are aspects of organismic behavior which are subsumed under the field concept. Morphogenetic fields may be natural kinds of things.

There is, at present, no generally accepted theory of morphogenetic fields. However, Goodwin[56] has pointed out that although in physics and chemistry there are many different types of fields, each characterized by different equations, all field theories have certain common properties:

> The solutions of such equations in their linearised forms are known as harmonic functions. These differ in wavelength, and any pattern of the field is initially described by some set of such harmonic functions. However, as pattern develops, the non-linear features of any particular field are expressed and distinctive wave shapes emerge. In general this results in a discrete set of stable forms, solutions of the field equations, which characterise the set of possible spatial patterns which a particular system can display.[57]

Further, Goodwin argues that genetic and environmental factors determine parametric values in the equations describing the field and therefore act to "select" and stabilize one form—as a position of organic stability—from the set of forms which are possible (generic) for that type of field. Goodwin's view of the nature of morphogenesis is, in effect, a systematization of Bateson's intuitions. For if the patterns of Chladni figures are the empirical manifestations of a sonically excited dynamical system, then organismic, morphological patterns are the empirical manifestations of a "self-exciting" dynamical system, a morphogenetic field.

In a summary of his conclusions, Bateson rejects "the crude belief that living beings are plastic conglomerates of miscellaneous attributes and that order of form or Symmetry have been impressed upon this medley by Selection alone" (p. 80). Rather, he suggests, we have to recognize that "the system of an organised being is such that the result of its disturbance may be *specific*" (p. 74, emphasis added). This remarkable conclusion implies that the response, in terms of the form produced, of a biological system to perturbation—the action of a causal agent—is determined by its intrinsic repertoire of possibilities, that is, the inherent nature of the particular causal mechanism implicated. While this conclusion may be underdetermined by the evidence which Bateson presents (but what scientific generalization is not?), subsequent observational and experimental work has only served to reinforce it. For example, the variations of form which Bateson characterizes as homoeotic can be produced by either environmental or genetic perturbations;[58] the four-jointed tarsal form which is produced by regeneration in the cockroach limb (and which Bateson notes [p. 416] is typical for

Locustodea) can be produced genetically in *Drosophila*, where the five-jointed form is typical.[59] Perhaps the most striking support for Bateson's conclusion is provided by the phenomenon of phenocopying investigated by Goldschmidt.[60] Goldschmidt claims that practically any kind of environmental perturbation of developing wild-type *Drosophila*, applied at the appropriate time, will result in a specific morphology which is normally associated with the presence of a mutant allele. Goldschmidt also discusses these phenomena in terms of the range of forms which is inherently possible for the organism and which "determines the possibility of appearance of both mutational and environmental effects."

In sum, *Materials* provides suggestive grounds for supposing that organismic forms are members of natural kinds and, therefore, that a science of form may, in principle, be possible. At times, Bateson himself seems to imply that these natural kinds correspond to species taxa, as conventionally recognized, although of course he does not put the matter in exactly this way, since he does not explicitly recognize the distinction. In *Problems of Genetics*[61] he presents a thorough and wide-ranging discussion of the whole question. He argues vigorously that we have to recognize the "reality of species" despite any difficulties we may have in empirically distinguishing them. However, as I read this discussion, *species*, in the final analysis, seems to have the general sense of natural kinds, "the universal presence of specificity . . . [which] manifests itself in respect of the most diverse characteristics which living things display." Bateson further argues that we have no good reason to suppose that the specific diversity of the organic world is of a type radically different from that present in the inorganic world; some forms of diversity may be "essential." From this perspective, it is at least a good working hypothesis to suppose that the specificity of biological kinds is to be understood in terms of the intrinsic natures of individual organisms; specificity has a "physiological foundation or causation,"[62] and it is in these terms that kinds must eventually be distinguished.

It is with this issue in mind that I turn finally to a consideration of what is, perhaps, the most interesting issue raised in *Materials*, namely symmetry, which Bateson himself regards as a matter of central importance. Indeed he suggests that "the road through the mystery of Species may . . . be found in the facts of Symmetry" (p. 36). If we construe *species* here in the general, philosophical sense of natural kinds, rather than in the conventional, taxonomic sense, then the suggestion acquires great interest and demands far more attention than it seems to have attracted.

I have noted above that Bateson characterizes organisms in purely formal terms as patterns—repeated parts arranged in specific geometrical relations—and that he uses these formal notions of repetition, geometrical disposition, and relation as a basis, "in the absence of a more natural classification" (p. 87), for the arrangement and classification of meristic variations. The arrangement thus involves three geometrical categories of repetition: linear series, bilateral series, and radial series. As Bateson notes, this arrangement "emphasises the fact that . . . the *Meristic Variations of such parts depend closely on the geometrical relation in which they stand*" (p. 89), and he believes that this relation between the type of pattern and the possible variations which it can exhibit supports his view of the essentially "mechanical" nature of meristic division which I outlined above.

The most striking (and famous) examples of the geometrically regular ordering of variation are those observed in the supernumerary appendages of arthropods and vertebrates discussed in chapters 20 to 23. Here Bateson shows that there is an almost invariable tendency for supernumerary appendages to be paired, for these pairs to be mirror images of each other, and for one of them (the nearest) to be a mirror image of the primary limb; a textbook example from *Carabus scheidleri* is illustrated on page 483. On the basis of these observations, Bateson formulated two empirical generalizations (p. 479), the second of which (pertaining to mirror symmetry) subsequently became known as Bateson's Rule.

These observations and attempted generalizations concerning the geometrical regularity of variation inspired the classic experimental work of Harrison and Swett on the determination of symmetry in amphibian limbs,[63] and they form part of the intellectual ancestry of the recent remarkable experimental and theoretical work of French, Bryant, and Bryant[64] on the behavior of limb fields in insects and amphibians, which has already achieved the status of a modern classic.

This important work notwithstanding, it is arguable that the potential role of symmetry as an *analytical concept* in the study of morphology, a role which is strongly suggested but not developed in *Materials*, has received insufficient attention as compared with that devoted to it in other disciplines concerned with form. Bateson's failure to elaborate the concept and his somewhat confusing discussion of the relation between organic symmetry and repetition, which he regards as distinct, though related, phenomena (see for example, p. 569), can be seen, retrospectively, to stem from his inadequate and restricted notion of the actual concept of symmetry. Along with the majority of his contemporaries (and many of our contemporaries), Bateson regards symmetry as a concept which is applicable only to finite figures and therefore admits of only two forms: bilateral symmetry (mirror reflection) and radial (rotational) symmetry (see pp. 19–21, 87–89). In the modern view, however, the basis of symmetry lies in the possibility of the superimposition of repeated parts, figures, or structures and the consequent idea of the generation of symmetries in plane patterns by means of the four basic rigid motions which are possible for plane figures: reflection, translation, rotation, and glide reflection.[65] From this perspective, the concept of symmetry is extended to encompass infinite (or potentially infinite) patterns, so rather than symmetry being regarded as a form of serial repetition (*Materials*, p. 569), repetition in a linear series is a form of symmetry, namely translation symmetry. Thus the three geometrical categories of repetition, referred to above, in terms of which Bateson organizes his presentation of the empirical material, correspond to three of the four basic categories of rigid motions (interestingly, the fourth category, glide reflection is apparently the most difficult to recognize empirically.[66]

Since symmetry analysis provides a means of describing and classifying patterns with respect to their structure and organization and of precisely specifying the order in repetition, it is clear that Bateson's doubts regarding the "naturalness" of his classificatory principles are unjustified. What more natural method could there be when we are concerned with patterns? Indeed it would be interesting to attempt to refine Bateson's basic classification. In principle, it should be feasible to determine whether and how variation was constrained, within a conventionally defined taxon say, by analyzing the variant forms in terms of motion classes. For example, there are seven possible one-color, one-dimensional patterns and seventeen possible one-color, two-dimensional patterns, and flow charts for their determination are available.[67] It would be interesting to know whether any of these forms was preferred or proscribed within any given taxon. Of course, as is often the case in biology, translating principle into practice might well prove difficult and would depend upon finding an organism which possessed a reasonably complex pattern and a range of variant forms.

In this context it is worthwhile considering the virtues of symmetry as the basis for a rational systematics of the kind envisaged by Driesch and referred to above. In the first place, as Washburn and Crowe point out,[68] classification of entities by means of symmetry is superior to traditional typological classification because it makes use of only one property as opposed to several different properties (number, shape, size, color). Furthermore, symmetry is a universal characteristic of patterns, and the symmetry classes can, at least in principle, be specified precisely and objectively. Recognition of a type, however, always involves (at least some) entity-specific properties and an element of subjective judgment of the degree of similarity and difference between entities, since it is generally impossible to precisely specify a set of necessary and sufficient conditions for membership of a class defined typologically. So questions of weighting become significant. Lastly, classification of patterns by symmetry generates no unclassifiable residue; there can be no "monsters." In sum, symmetry seems to have a number of distinct virtues as an analytical and descriptive concept, and an attempt at a systematic description of forms in such terms might be a valuable enterprise, bearing in mind that explanation is always relative to description.

As Driesch observes,[69] the general logical type of all rational systems can be specified a priori. Rational systematics is always possible when there exists a fundamental concept of the categorical class or some proposition based upon such a concept "which carries with it a principle of division [or] . . . which could allow an inherent sort of evolution of latent diversities." In other words, a concept or proposition in which the peculiarities of the "species" are represented potentially in the "genus." The concept of symmetry would seem to be a concept of this kind since the four basic rigid motions (and their combinations) function as operators to generate all the possible patterns. In fact, Driesch regards the classification of spatial form in terms of symmetry relations (as in crystallography) as a paradigm of a rational system, but he somewhat tersely asserts, for reasons which do not seem particularly compelling (in effect, that organisms are not like crystals) that symmetry cannot provide a basis for a biological systematics.[70] In the light of *Materials*, such a dismissive conclusion seems premature, and it may turn out that the symmetrical rather than the typical provides the empirical road to the rational.

Classification in terms of symmetry relations is, of course, only one of the many ways in which organisms can be classified. Nowadays there would probably be general agreement amongst philosophers of science regarding the dialectical relation between description and explanation, a view which is sometimes expressed in the demand that classifications be "theoretically significant," that is, related to some explanatory theory or model. While I hope I have said enough in this brief outline to indicate that symmetry as a "natural" analytical concept for the study of form is worthy of further exploration, a convincing demonstration that this is the case, and more significantly that symmetry could provide a basis for natural kinds in the domain of morphology, requires the development of explanatory (i.e., generative) models of forms explicitly described in terms of symmetry relations. Some preliminary essays have been made in this direction,[71] but much more needs to be done.

I would also urge the need for more discussion of fundamental conceptual principles in this context. Elsewhere, I have suggested that the structuralists' logico-mathematical concept of transformation (which is a central concept in Group Theory) might be one that is fundamental to the study of form.[72] Symmetries are a species of transformations (isometric transformations), but they are a different kind from the famous coordinate transformations of D'Arcy Thompson[73] (which are topological transformations). It would be extremely valuable to have a detailed discussion (preferably in terms intelligible to the nonmathematician) of the concept of transformation, in relation to biological form, of the different categories of transformations, and of their relations to each other. A preliminary essay in this direction by Woodger[74] discusses the theory of morphological correspondence, "Bauplans," and systematics in terms of transformations; but this important, though difficult, paper seems to have been largely ignored.

Some sixty years ago, Woodger[75] noted that "there are biological theories in abundance but no theoretical biology." Woodger is referring here to the absence of a systematic set of biological concepts pertaining to the nature of the organism and "a type of logical order . . . exemplified in the organic realm." In my view, *Materials for the Study of Variation* is a contribution to the development of a theoretical biology in this sense and should take its place alongside the other classic contributions of the last one hundred years: parts of D'Arcy Thompson's *On Growth and Form* and the analytical and critical sections of Driesch's *The Science and Philosophy of the Organism*. In relation to the problem of form, these texts may be regarded as contributions to the necessary work of Lockean "underlabouring," in which the ground is cleared of the inadequate concepts of Darwinism and some seeds are sown which may germinate and develop into a more adequate set of concepts and eventually result in a general theory of morphogenesis. It is arguable that such a theory, if and when it materializes, will be a more fundamental biological theory than the theory of evolution.

Those who believe that biology has already elaborated its fundamental concepts will no doubt regard *Materials* as of merely historical interest. Others, however, may well find that in reading something old, they learn something new. And there may even be a few readers who, in the words of that master of form and pattern, Vladimir Nabokov, "will jump up ruffling their hair."

NOTES

The formulation of the arguments in this essay has been greatly assisted by discussions with Brian Goodwin and John Maynard Smith. I am indebted to them both.

1. See Roy Bhaskar, *A Realist Theory of Science* (Brighton: Harvester, 1978). The philosophy of science which informs this essay is Bhaskar's transcendental realism.
2. Ernst Mayr, *Evolution and the Diversity of Life* (Cambridge: Harvard University Press, 1976), p. 425.
3. See David L. Hull, "A Matter of Individuality," *Philosophy of Science* 45 (1978): 335–60. D. B. Kitts and D. J. Kitts, "Biological Species as Natural Kinds," *Philos. Sci.* 46 (1979): 613–22. Hull, "Discussion: Kitts and Kitts and Caplan on Species," *Philos. Sci.* 48 (1981): 141–52. Hull, "Historical Entities and Historical Narratives," in *Minds, Machines and Evolution*, ed. C. Hookway (Cambridge: Cambridge University Press, 1984). Note that the debate here is not one of nominalism versus realism. That species taxa are real is not in dispute. What is disputed is whether they have the ontological status of natural kinds, the traditional view, or that of individuals, which is, according to Hull, how evolution theory requires us to conceive them. The question of what one is to understand by "scientific law" is also at issue.
4. See Hans Driesch, *The History and Theory of Vitalism* (London: Macmillan, 1914), pp. 149–50, [my emphasis]. Driesch's position here is Kantian. Whether transcendental idealism provides an adequate philosophy of science is debatable; it is at least an advance on the empiricism in terms of which the majority of discussions are conducted.
5. Driesch, *History and Theory of Vitalism*, pp. 94–95.
6. See Hans Driesch, *The Science and Philosophy of the Organism*, 2nd ed. (London: Black, 1929), pp. 243–49; 293–96. As examples of "rational systematics" or systems which approximate this ideal, Driesch cites the theories of solid geometry and of conic sections in mathematics and the theories of crystallography, of the periodic arrangement of the elements, and of the homologous series in the physical and chemical sciences. Driesch does not suppose that the biological "system," if it exists, must necessarily be entirely "rational"; the project is to discover the rational kernel.
7. Driesch, *History and Theory of Vitalism* (note 4), p. 140, my emphasis.
8. See Bhaskar, *A Realist Theory* (note 1), pp. 163–75. He distinguishes three levels in our knowledge of nature, which correspond to three phases of investigation. The first level, the *Humean*, is where an empirical regularity or pattern is recognized or experimentally produced. This is explained in terms of a second level of knowledge, the *Lockean*, which involves an account of the essential structure or constitution (i.e., the nature) of the thing which produces the effect in question. This structure is the causal or generative mechanism, which ensures the necessity of the thing's behaving in the way that it does. At the third, or *Leibnizian*, level, the possession of that structure comes to be regarded as defining the natural kind of which the thing we are concerned with is a member. Whatever has that structure (real essence) is a member of that kind. It should be noted, though (and will be discussed further below), that in the study of morphology we may

have to deal with a distinctive kind of empirical regularity, namely transformation.

9. See Bhaskar, *A Realist Theory* (note 1).

10. See, for example, "Typological versus Population Thinking," in Mayr, *Evolution* (note 2), pp. 26–29. Mayr unfavorably contrasts typological thinking with Darwinian population thinking. The populationist is credited with believing that the world consists of "unique" individuals composed of "unique features." Mayr does not seem to appreciate that the very activity of the populationist to which he refers (statistical description) is rendered impossible by this radically empiricist "belief," for the "unique" cannot be described. The populationist's activities, if they are to be possible and coherent, necessarily presuppose that individual organisms and individual features be conceptualized as kinds. Whatever else it may be, the concept of the "typical" is an indispensable epistemological concept, though one which requires more philosophical analysis than it seems to have received.

11. I take *Darwinism* to mean the core of more or less systematic explanatory theory which can be extracted from, or appears consistent with, the arguments contained in the *Origin of Species*.

12. Charles Darwin, *The Origin of Species by Means of Natural Selection* (1859; reprint, Harmondsworth, England: Penguin Books, 1968) p. 233.

13. E. S. Russell has pointed out how impoverished Darwin's interpretation of the concept of conditions of existence is as compared with Cuvier's original. See E. S. Russell, *Form and Function: A Contribution to the History of Animal Morphology* (London: John Murray, 1916).

14. Darwin, *Origin of Species* (note 12), p. 228.

15. *Ibid.*, p. 155.

16. *Ibid.*, p. 131.

17. *Ibid.*, p. 232.

18. *Ibid.*, p. 170.

19. *Ibid.*, p. 233.

20. *Ibid.*, p. 228.

21. *Ibid.*, p. 427. Note also his comment on p. 233: " 'Natura non facit saltum' . . . must by my theory be strictly true."

22. *Ibid.*, pp. 107, 108. The apparent nominalism here is misleading. See note 3.

23. *Ibid.*, p. 233.

24. Darwin of course makes much reference to other "laws": of variation, of growth, of correlation, and so on. However, it is impossible to extract any systematic core of theory from such references, which are nearly always employed as qualifications of the basic theory or fall-back positions when difficulties arise.

25. Darwin, *Origin of Species* (note 12), p. 456.

26. *Ibid.*, p. 433.

27. See Richard Dawkins, *The Blind Watchmaker* (Harlow, England: Longman, 1986).

28. The *Origin*, of course contains a discussion of "laws of varia-

tion," albeit an unsatisfactory and inconclusive one. However, there is really no place for such "laws" in Darwin's system. To talk of laws here (in any relevant sense) presupposes that we can talk about kinds of organisms and Darwin has eliminated kinds; the differences between individuals and the differences between species are differences of the same sort, that is, merely factual and contingent.

29. William Bateson, *Problems of Genetics* (New Haven: Yale University Press, 1913), p. 58.

30. Of course there are contemporary Darwinists who recognize that there are important formal problems in biology. See, for example, J. Maynard Smith, "Continuous, Quantized and Modal Variation," *Proceedings of the Royal Society*, no. 152 (1960): 397–409.

31. The argument for autonomy is, in the first instance, a purely epistemological argument; that is, in principle, forms can be known under *either* formal *or* functional descriptions. As such, it does not entail that form and function are necessarily always and completely ontologically independent.

32. Attacks on functionalism persist throughout Bateson's later writing. It seems possible that Bateson's inability to let go here may stem from his deep antipathy to utility in any shape or form, which manifests itself in all his writing, especially that concerned with social and educational matters. See the essays and addresses collected in Beatrice Bateson, *William Bateson, F. R. S., Naturalist* (Cambridge: Cambridge University Press, 1928).

33. I owe this formulation of the problem to John Maynard Smith, who attributes it to the late Helen Spurway.

34. William Bateson, "The Ancestry of the Chordata," *Quarterly Journal of Microscopical Science* 26 (1886): 535–71, see p. 548, reprinted in R. C. Punnett, ed., *The Scientific Papers of William Bateson*, 2 vols. (Cambridge: Cambridge University Press, 1928), vol. 1, pp. 1–31.

35. Darwin, *Origin of Species* (note 12), p. 454. Darwin suggests "four or five progenitors" for animals and "an equal or a lesser number" for plants. He speculates that "probably all the organic beings which have ever lived . . . have descended from some one primordial form" but recognizes that this view of total genetic connection rests upon a very general analogical argument rather than the existence of *formal* relations. It is sometimes suggested that "common descent" is an advance on "special creation," since it reduces the number of [kinds of] "progenitors"; but of course a "reduction of kinds" is implicit in the taxonomic hierarchy, and the structure of empirical taxonomies provides part of the empirical grounds for the notion of descent.

36. Driesch puts the matter more tersely: "far-seeing persons, philosophers especially, had realised that historical information could never constitute an explanation, and that in comparison to real science it was of very secondary importance." Driesch, *History and Theory of Vitalism* (note 4), p. 140. See also p. 150.

37. W. Bateson, "Ancestry of the Chordata" (note 34), pp. 538, 545.

38. See Bhaskar, *A Realist Theory* (note 1), and Hilary Putnam, "Is Semantics Possible?" in *Mind, Language and Reality: Philosophical Papers* (Cambridge: Cambridge University Press), vol. 2.
39. W. Bateson, *Problems of Genetics* (note 29), chap. 1.
40. *Ibid.*, p. 2, 11.
41. Bateson coined the term *genetics* in 1905 in the course of an (unsuccessful) attempt to achieve a secure institutional base for his work. See B. Bateson, *William Bateson* (note 32), p. 93.
42. *Ibid.*, p. 257.
43. Even subsequent to the "rediscovery" of Mendel, Bateson feels it necessary to remind his audience of the metaphorical nature of the concepts of inheritance and heredity; Bateson, *Problems of Genetics* (note 29), p. 35.
44. Darwin, *Origin of Species* (note 12), pp. 418–19. On von Baer, see E. S. Russell, *Form and Function* (note 13), p. 120.
45. Darwin, *Origin of Species* (note 12), p. 417.
46. Francis Galton, *Natural Inheritance* (London: Macmillan, 1889), chap. 3.
47. W. Bateson, *Problems of Genetics* (note 39), p. 33.
48. *Ibid.*, p. 35. See also various remarks in B. Bateson, *William Bateson* (note 32).
49. W. Bateson, review of *The Mechanism of Mendelian Heredity*, by T. H. Morgan, A. H. Sturtevant, H. J. Muller, and C. B. Bridges, *Science*, n.s., 44 (1916): 536–43.
50. C. H. Waddington, *Organisers and Genes* (Cambridge: Cambridge University Press, 1940). The intellectual affinities between Waddington and Bateson seem obvious, but it appears that their relationship has not yet been investigated by any historian of science (Peter Bowler, personal communication). I suspect that it would be an interesting and rewarding topic.
51. See B. Bateson, *William Bateson* (note 32), pp. 42–46, and W. Bateson, *Problems of Genetics* (note 29), p. 60.
52. Hans Jenny, (1967 and 1974). *Kymatik/Cymatics*, 2 vols. (Basel, Switzerland: Basilius Presse AG). See, for example, vol. 1, figs. 27–30.
53. W. Bateson, *Problems of Genetics* (note 29), chaps. 2 and 3.
54. B. Bateson, *William Bateson* (note 32), p. 45.
55. A. M. Turing, "The Chemical Basis of Morphogenesis," *Philosophical Transactions of the Royal Society*, Ser. B, 237 (1952): 37–72.
56. Brian C. Goodwin, "A Structuralist Research Programme in Developmental Biology," in *Dynamic Structures in Biology*, ed. B. Goodwin, A. Sibatani, and G. Webster (Edinburgh: Edinburgh University Press, 1989).
57. *Ibid.*, p. 52.
58. See A. E. Needham, *Regeneration and Wound Healing* (London: Macmillan, 1952), and W. J. Ouweneel, "Developmental Genetics of Homoeosis," *Advances in Genetics* 18 (1976): 179–236.
59. See Waddington, *Organisers and Genes* (note 50).
60. Richard B. Goldschmidt, *Theoretical Genetics* (Berkeley: University of California Press, 1958). Ouweneel, "Developmental Genetics" (note 58).

61. W. Bateson, *Problems of Genetics* (note 29), chap. 1.
62. *Ibid.*, p. 11.
63. R. G. Harrison, "On Relations of Symmetry in Transplanted Limbs," *Journal of Experimental Zoology* 32 (1921): 1–118; F. H. Swett, "On the Production of Double Limbs in Amphibians," *J. Exp. Zool.* 44, (1926): 419–72.
64. V. French, P. J. Bryant, and S. V. Bryant, "Pattern Regulation in Epimorphic Fields," *Science* 193 (1976): 969–81.
65. D. K. Washburn and D. W. Crowe, *Symmetries of Culture* (Seattle: University of Washington Press, 1988).
66. *Ibid.*, p. 50.
67. *Ibid.*, chaps. 4 and 5.
68. *Ibid.*, pp. 33–41.
69. Driesch, *Science and Philosophy* (note 6), pp. 244–47.
70. *Ibid.*, pp. 19, 247.
71. See French, Bryant, and Bryant, "Pattern Regulation" (note 64); B. C. Goodman and S. A. Kauffman, "Deletions and Mirror-Symmetries in *Drosophila* Segmentation Mutants Reveal Generic Properties of Epigenetic Mappings," in *Principles of Organisation in Organisms*, ed. J. Mittenthal (New York: Addison Wesley, in press).
72. Gerry Webster, "Structuralism and Darwinism: Concepts for the Study of Form," in *Dynamic Structures in Biology*, ed. B. Goodwin, A. Sibatani, and G. Webster (Edinburgh: Edinburgh University Press, 1989).
73. D'Arcy W. Thompson, *On Growth and Form*, 2nd ed. (Cambridge: Cambridge University Press, 1942).
74. J. H. Woodger, "On Biological Transformations," in *Essays on Growth and Form Presented to D'Arcy Wentworth Thompson*, ed. W. E. Le Gros Clark and P. B. Medawar (Oxford: Oxford University Press, 1945).
75. J. H. Woodger, "The 'Concept of Organism' and the Relation between Embryology and Genetics," *Quarterly Review of Biology* 5 (1930): 1–22.

CORRIGENDA.

p. 23, line 5. For " and that in " read " and in."
p. 27, line 29. For " appear " read " appears."
p. 37, line 18. For " their " read " the."
p. 54. Note 2. For " xxviii " read " xx."
p. 55. *Parra* is now known not to have affinities with the Rallidæ.
p. 141. In description of Fig. 15 insert " After Solger."
p. 151, line 2 and p. 153, Note. For " W. B." read " G. B."
p. 198. For " Pinnipediæ " read " Pinnipedia." For " *Dent.*" read " *Deut.*"
p. 212. In description of Fig. 40 delete " p^1 of the left side is in symmetry with two teeth on the right side." The figure is correct.
p. 281, 15th line from bottom. Delete " and perhaps all."
p. 382. For " W. H. Benham " read " W. B. Benham."
p. 429. For " Banyul's " read " Banyuls."
p. 473, 4th and 6th lines from bottom. For " *Tornaria* " read " *Balanoglossus.*"
p. 526. Delete the heading " (1) *Clear cases of Extra Parts in Secondary Symmetry.*"

Note to p. 461, Note 718. As to union of eyes in Bees, see further, Dittrich, *Zeit. f. Ent.*, Breslau, 1891, xvi. p. 21, and Cook, A. J., *Proc. Amer. Ass.*, 1891, p. 327.

Note to p. 468, Note 2. In connexion with Giard's observation the following fact should be given. Since this Chapter was printed I have had an opportunity of examining a sample of Flounders taken in the shallow water off Bournemouth. Of 23 specimens seen alive, all but about half a dozen were more or less blotched with shades of brown on the " blind " side. In five the brown was more extensive than the white. The eyes and dorsal fins were normal. The fishmonger who shewed them to me said that the Flounders in that place were generally thus blotched, and that those seen were a fair sample. In estimating the significance of Cunningham's experiment (p. 467) this fact should be remembered.

INTRODUCTION.

All flesh is not the same flesh: but there is one kind of flesh of men, another flesh of beasts, another of fishes, and another of birds.

SECTION I.

THE STUDY OF VARIATION.

To solve the problem of the forms of living things is the aim with which the naturalist of to-day comes to his work. How have living things become what they are, and what are the laws which govern their forms? These are the questions which the naturalist has set himself to answer.

It is more than thirty years since the *Origin of Species* was written, but for many these questions are in no sense answered yet. In owning that it is so, we shall not honour Darwin's memory the less; for whatever may be the part which shall be finally assigned to Natural Selection, it will always be remembered that it was through Darwin's work that men saw for the first time that the problem is one which man may reasonably hope to solve. If Darwin did not solve the problem himself, he first gave us the hope of a solution, perhaps a greater thing. How great a feat this was, we who have heard it all from childhood can scarcely know.

In the present work an attempt is made to find a way of attacking parts of the problem afresh, and it will be profitable first to state formally the conditions of the problem and to examine the methods by which the solution has been attempted before. This consideration shall be as brief as it can be made.

The forms of living things have many characters: to solve the problem completely account must be taken of all. Perhaps no character of form is common to all living things; on the contrary their forms are almost infinitely diverse. Now in those attempts to solve the problem which have been the best, it is this diversity of form which is taken as the chief attribute, and the attempt to solve the general problem is begun by trying to trace the modes by which the diversity has been produced. In the shape in which it has been most studied, the problem is thus the problem of Species. Obscurity has been brought into the treatment of the question through want of recognition of the fact that this is really only a part of the general problem, which would still remain if there were only one species. Nevertheless the problem of Species is so tangible a part of the whole that it is perhaps the best point of departure. For our present purpose we cannot begin better than by stating it concisely.

The forms of living things are diverse. They may nevertheless be separated into Specific Groups or Species, the members of each such group being nearly alike, while they are less like the members of any other Specific Group. [The Specific Groups may by their degrees of resemblance be arranged in Generic Groups and so on.]

The individuals of each Specific Group, though alike, are not identical in form, but exhibit differences, and in these differences they may even more or less nearly approach the form characteristic of another Specific Group. It is true, besides, that in the case of many Specific Groups which have been separated from each other, intermediate forms are found which form a continuous series of gradations, passing insensibly from the form characteristic of one Species to that characteristic of another. In such cases the distinction between the two groups for purposes of classification is not retained.

The fact that in certain cases there are forms transitional between groups which are sufficiently different to have been thought to be distinct, is a very important fact which must not be lost sight of; but though now a good many such cases are known, it remains none the less true that at a given point of time, the forms of living things may be arranged in Specific Groups, and that between the immense majority of these there are no transitional forms. There are therefore between these Specific Groups differences which are Specific.

No definition of a Specific Difference has been found, perhaps because these Differences are indefinite and hence not capable of definition. But the forms of living things, taken at a given moment, do nevertheless most certainly form a discontinuous series and not a continuous series. This is true of the world as we see it now, and there is no good reason for thinking that it has ever been otherwise. So much is being said of the mutability of species that this, which is the central fact of Natural History, is almost lost sight of, but if ever the problem is to be solved this fact must be boldly faced. There is nothing to be gained by shirking or trying to forget it.

The existence, then, of Specific Differences is one of the characteristics of the forms of living things. This is no merely subjective conception, but an objective, tangible fact. This is the first part of the problem.

In the next place, not only do Specific forms exist in Nature, but they exist in such a way as to fit the place in Nature in which they are placed; that is to say, the Specific form which an organism has, is *adapted* to the position which it fills. This again is a relative truth, for the adaptation is not absolute.

These two facts constitute the problem:

I. *The forms of living things are various and, on the whole, are Discontinuous or Specific.*

II. *The Specific forms, on the whole, fit the places they have to live in.*

How have these Discontinuous forms been brought into existence, and how is it they are thus adapted? This is the question the naturalist is to answer. To answer it completely he must find (1) *The modes* and (2) *The causes* by which these things have come to pass.

Before considering the ways in which naturalists have tried to answer these questions, it is necessary to look at some other phenomena characteristic of Life. We have said that *at a given moment*, or point of time, the specific forms of living things compose a discontinuous series. The element of time thus introduced is of consequence, and leads to important considerations. For the condition of the organized world is not a fixed condition, but changes from moment to moment, and that which can be predicated of its condition at one moment may not at any other point of time be true. This process of change is brought about partly by progressive changes in the bodies of the individuals themselves, but chiefly by the constant succession of individuals, the parents dying, their offspring succeeding them. It is then a matter of observation that the offspring born of parents belonging to any one Specific Group do as a rule conform to that Specific Group themselves, and that the form of the body, the mechanisms and the instincts of the offspring, are on the whole similar to those which their parents had. But like most general assertions about living things this is true not absolutely but relatively only. For though on the whole the offspring is like the parent or parents, its form is perhaps never identical with theirs, but generally differs from it perceptibly and sometimes materially. To this phenomenon, namely, the occurrence of differences between the structure, the instincts, or other elements which compose the mechanism of the offspring, and those which were proper to the parent, the name **Variation** has been given.

We have seen above that the two leading facts respecting the forms of living things are first that they shew specific differentiation, and secondly that they are adapted. To these we may now add a third, that in the succession from parent to offspring there is, or may be, Variation. It is upon the fact of the existence of this phenomenon of Variation that all inductive theories of Evolution have been based.

The suggestion which thus forms the common ground of these theories is this:—May not the Specific Differences between Species and Species have come about through and be compounded of the individual differences between parent and offspring? May not Specific Differentiation have resulted from Individual Variation? This suggestion has been spoken of as the Doctrine of Common Descent, for it asserts that there is between living things a community of descent.

In what follows it will be assumed that this Doctrine of Descent is true. It should be admitted from the first that the truth of the doctrine has never been proved. There is nevertheless a great balance of evidence in its favour, but it finds its support not so much in direct observation as in the difficulty of forming any alternative hypothesis. The Theory of Descent involves and asserts that all living things are genetically connected, and this principle is at least not contrary to observation; while any alternative hypothesis involves the idea of Separate Creation which by common consent is now recognized as absurd. In favour of the Doctrine of Common Descent there is a balance of evidence: it is besides accepted by most naturalists; lastly if it is not true we can get no further with the problem: but inasmuch as it is unproven, it is right that we should explicitly recognize that it is in part an assumption, and that we have adopted it as a postulate.

The Doctrine of Descent being assumed, two chief solutions of the problem have been offered, both starting alike from this common ground. Let us now briefly consider each of them.

A. *Lamarck's Solution.* So many ambiguities and pitfalls are in the way of any who may try to re-state, in a few words, the theory propounded in the *Philosophie Zoologique*, that it is with great diffidence that the following account of it is given.

Lamarck points out that living things can in some measure adapt themselves both structurally and physiologically to new circumstances, and that in certain cases the adaptability is present in a high degree. He suggests that by inheritance and perfection of such adaptations they may have become what they are, and that thus specific forms and mechanisms have been produced, as it were, by sheer force of circumstances. On this view it is assumed that to the demands made on it by the environment the organism makes an appropriate structural and physiological response; in other words, that there is in living things a certain *tension*, by which they respond to environmental pressure and fit the place they are in, somewhat as a fluid fits a vessel.

This is not, I think, a misrepresentation of Lamarck's theory. It amounts, in other words, to a proposal to regard organisms as machines which have the power of Adaptation as one of their fundamental and inherent qualities or attributes.

Without discussing this solution, we may note that it aims at being a *complete* solution of both

(1) The *existence* and *persistence* of differing forms,
(2) The fact that the differing forms are *adapted* to different conditions;

and (3) The *causes of the Variation* by which the diversity has occurred.

B. *Darwin's Solution.* Darwin, without suggesting causes of Variation, points out that since (1) Variations occur—which they are known to do—and since (2) some of the variations are in the direction of adaptation and others are not—which is a necessity—it will result from the conditions of the Struggle for Existence that those better adapted will *on the whole* persist and the less adapted will on the whole be lost. In the result, therefore, there will be a diversity of forms, *more or less* adapted to the states in which they are placed, and this is very much the observed condition of living things.

We may note that this solution does not aim at being a complete solution like Lamarck's, for as to the *causes* of Variation it makes no suggestion.

The arguments by which these several solutions are supported, and the difficulties which are in the way of each, are so familiar that it would be unprofitable to detail them. On our present knowledge the matter is talked out. Those who are satisfied with either solution are likely to remain so.

It may be remarked however that the observed cases of adaptation occurring in the way demanded on Lamarck's theory are very few, and as time goes on this deficiency of facts begins to be significant. Natural Selection on the other hand is obviously a 'true cause,' at the least.

In the way of both solutions there is one cardinal difficulty which in its most general form may be thus expressed. According to both theories, specific diversity of form is consequent upon diversity of environment, and diversity of environment is thus the ultimate measure of diversity of specific form. Here then we meet the difficulty that diverse environments often shade into each other insensibly and form a continuous series, whereas the Specific Forms of life which are subject to them on the whole form a Discontinuous Series. The immense significance of this difficulty will be made more apparent in the course of this work. The difficulty is here put generally. Particular instances have been repeatedly set forth. Temperature, altitude, depth of water, salinity, in fact most of the elements which make up the physical environment are continuous in their gradations, while, as a rule, the forms of life are discontinuous[1]. Besides this, forms which

[1] It may be objected that to any organism the other organisms coexisting with it are as serious a factor of the environment as the strictly physical components; and that inasmuch as these coexisting organisms are discontinuous species, the

are apparently identical live under conditions which are apparently very different, while species which though closely allied are constantly distinct are found under conditions which are apparently the same. If we would make these facts accord with the view that it is diversity of environment which is the measure of diversity of specific form, it is necessary to suppose either (1) that our estimate of similarity of forms, or of environment, is wholly untrustworthy, or else (2) that there is a wide area of environmental or structural divergence within which no sensible result is produced: that is to say, that the relation between environment and structure is not finely adjusted. But either of these admissions is serious; for if we grant the former we abrogate the right of judgment, and are granting that our proposed solutions are mere hypotheses which we have no power to test, while if we admit the latter, we admit that environment cannot so far be either the directing cause or the limiting cause of Specific Differences, though the first is essential to Lamarck's Theory, and the second is demanded by the doctrine of Natural Selection.

Such then, put very briefly, are the two great theories, and this is one of the chief difficulties which beset them. We must now pass to our proper work.

We have to consider whether it is not possible to get beyond the present position and to penetrate further into this mystery of Specific Forms. The main obstacle being our own ignorance, the first question to be settled is what kind of knowledge would be of the most value, and which of the many unknowns may be determined with the greatest profit. To decide this we must return once more to the ground which is common to all the inductive theories of Evolution alike. Now all these different theories start from the hypothesis that the different forms of life are related to each other, and that their diversity is due to Variation. On this hypothesis, therefore, Variation, whatever may be its cause, and however it may be limited, is the essential phenomenon of Evolution. Variation, in fact, *is* Evolution. The readiest way, then, of solving the problem of Evolution is to study the facts of Variation.

SECTION II.

ALTERNATIVE METHODS.

The Study of Variation is therefore suggested as the method which is on the whole more likely than any other to give us the kind of knowledge we are wanting. It should be tried not so much in the hope that it will give any great insight into those

element of discontinuity may thus be introduced. This is true, but it does not help in the attempt to find the cause of the original discontinuity of the coexisting organisms.

relations of cause and effect of which Evolution is the expression, but merely as an empirical means of getting at the outward and visible phenomena which constitute Evolution. On the hypothesis of Common Descent, the forms of living things are succeeding each other, passing across the stage of the earth in a constant procession. To find the laws of the succession it will be best for us to stand as it were aside and to watch the procession as it passes by. No amount of knowledge of individual forms will tell us the laws or even the manner of the succession, nor shall we be much helped by comparison of forms of whose descent we know nothing save by speculation. To study Variation it must be seen at the moment of its beginning. For comparison we require the parent and the varying offspring together. To find out the nature of the progression we require, simultaneously, at least two consecutive terms of the progression. Evidence of this kind can be obtained in no other way than by the study of actual and contemporary cases of Variation. To the solution of this question collateral methods of research will not contribute much.

Since Darwin wrote, several of these collateral methods have been tried, and though a great deal has thus been done and a vast number of facts have been established, yet the advance towards a knowledge of the steps by which Evolution proceeds has been almost nothing. It will not perhaps be wandering unduly if we consider very shortly the reason of this, for the need for the Study of Variation will thereby be made more plain.

Before the publication of the *Origin of Species* the work of naturalists was chiefly devoted to the indiscriminate accumulation of facts. By most the work was done for its own sake in the strictest sense. In the minds of some there was of course a hope that the gathering of knowledge would at last lead on to something more, but this hope was for the most part formless and vague. With the promulgation of the Doctrine of Descent the whole course of the study was changed. The enthusiasm of naturalists ran altogether into new channels; a new class of facts was sought and the value of Zoological discovery was judged by a new criterion. The change was thus a change of aim, and consequently a change of method. From a large field of possibilities the choice fell chiefly upon two methods, each having a definite relation to the main problem. The first of these is the Embryological Method, and the second may be spoken of as the Study of Adaptation. The pursuit of these two methods was the direct outcome of Darwin's work, and such great hopes have been set on them that before starting on a new line we shall do well to examine carefully their proper scope and see whither each of them may reasonably be expected to lead.

It is besides in the examination of these methods and in observing the exact point at which they have failed, that the need for the Study of Variation will become most evident.

When the Theory of Evolution first gained a hearing it became of the highest importance that it should be put to some test which should shew whether it was true or not. In comparison with this all other questions sank into insignificance.

Now, the principle which has been called the Law of von Baer, provided the means for such a test. By this principle it is affirmed that the history of the individual represents the history of the Species. If then it should be found that organisms in their development pass through stages in which they resemble other forms, this would be *prima facie* a reason for believing them to be genetically connected. The general truth of the Theory of Descent might thus be tested by the facts of development. For this reason the Study of Embryology superseded all others. It is now, of course, generally admitted that the Theory has stood this test, and that the facts of Embryology do support the Doctrine of Community of Descent.

But the claims of Embryology did not stop here. In addition to the application of the method to the general Theory of Descent, it has been sought to apply the facts of Embryology to solve particular questions of the descent of particular forms. It has been maintained that if it is true that the history of the individual repeats the history of the Species, we may in the study of Development see not only that the various forms are related, but also the exact lines of Descent of particular forms. In this way Embryology was to provide us with the history of Evolution.

The survey of the development of animals from this point of view is now complete for most forms of life, and in all essential points; we are now therefore in a position to estimate its value. It will, I think, before long be admitted that in this attempt to extend the general proposition to particular questions of Descent the embryological method has failed. The reason for this is obvious. The principle of von Baer was never more than a rough approximation to the truth and was never suited to the solution of particular problems. It is curious to notice upon how very slight a basis of evidence this widely received principle really rests. It has been established almost entirely by inference and it has been demonstrated by actual observation in scarcely a single instance.

For the stages through which a *particular* organism passes in the course of its development are admissible as evidence of its pedigree only when it shall have been proved as a *general* truth that the development of individuals does follow the lines on which the species developed. The proof, however, of this general proposition does not rest on direct observation but on the indirect evidence that particular organisms at certain stages in their development resemble other organisms, and hence it is assumed that they are descended from those forms. Thus the truth of the general proposition is established by assuming it true in special cases, while its applicability to special cases rests on its having been accepted as a general truth.

Probably however the apologists of this method would maintain that the principle of von Baer, though its truth has not been demonstrated directly, yet belongs to the class of "True Hypotheses." To establish the truth of a hypothesis in a case like the present in which the number of possible hypotheses is not limited, it should at least be shewn that its application in all known instances is so precise, so simple, and in such striking accordance with ascertained facts, that its truth is felt to be irresistible.

Nothing like this can be said of the principle of von Baer. Even if it be generally true that the development of a form is a record of its descent, it has never been suggested that the record is complete.

Allowance must constantly be made for the omission of stages, for the intercalation of stages, for degeneration, for the presence of organs specially connected with larval or embryonic life, for the interference of yolk and so forth. But what this allowance should be and in what cases it should be made has never been determined.

More than this: closely allied forms often develop on totally different plans; for example, *Balanoglossus Kowalevskii* has an opaque larva which creeps in the sand, while the other species of the family have a transparent larva which swims at the surface of the sea; the germinal layers of the Guinea-pig when compared with those of the Rabbit are completely inverted, and so on. These are not isolated cases, for examples of the same kind occur in almost every group and in the development of nearly all the systems of organs. When these things are so, who shall determine which developmental process is ancestral and which is due to secondary change? By what rules may secondary changes be recognized as such? Do transparent larvæ swimming at the surface of the sea reproduce the ancestral type or does the opaque larva creeping in the mud shew us the primitive form? Each investigator has answered these questions in the manner which seemed best to himself.

There is no rule to guide us in these things and there is no canon by which we may judge the worth of the evidence. It is perhaps not too much to say that the main features of the development of nearly every type of animal are now ascertained, and on this knowledge elaborate and various tables of phylogeny have been constructed, each differing from the rest and all plausible; but it would be difficult to name a single case in which the immediate pedigree of a species is actually known.

The Embryological Method then has failed not for want of knowledge of the visible facts of development but through ignorance of the principles of Evolution. The principle of von Baer, taken by itself, is clearly incapable of interpreting the phenomena of development. We are endeavouring by means of a mass of conflicting evidence to reconstruct the series of Descent, but of the laws which govern such a series we are ignorant. In the interpretation of Embryological evidence it is constantly necessary to make certain hypotheses as to the course of Variation in the past, but before this can be done it is surely necessary that we should have some knowledge of the modes of Variation in the present. When we shall know something of the nature of the variations which are now occurring in animals and the steps by which they are now progressing before our eyes, we shall be in a position to surmise what their past has been; for we shall then know what changes are possible to them and what are not. In the absence of such knowledge, any person is at liberty to postulate the occurrence of variations on any lines which may suggest themselves to him, a liberty which has of late been freely used. Embryology has provided us with a magnificent body of facts, but the interpretation of the facts is still to seek.

The other method which, since Darwin's work, has attracted most attention is the study of the mechanisms by which organisms are adapted to the conditions in which they live. This study of Adaptation and of the utility of structures exercises an extraordinary fascination over the minds of some and it is most important that its proper use and scope should be understood.

We have seen that the Embryological Method owed its importance to its value as a mode of testing the truth of the Theory of Evolution: in the same way the Study of Adaptation was undertaken as a test of the Theory of Natural Selection.

Amongst many classes of animals, complex structures are present which do not seem to contribute directly to the well-being of their possessors. By many it has been felt that the persistent occurrence of organs of this class is a difficulty, on the hypothesis that there is a tendency for useful structures to be retained and for useless parts to be lost. In consequence it has been anticipated that sufficient research would reveal the manner in which these parts are directly useful. The amount of evidence collected with this object is now enormous, and most astonishing ingenuity has been evoked in the interpretation of it. A discussion of the truth of the conclusions thus put forward is of course apart from our present purpose, which is to examine the logical value of this method of research as a means of attacking the problem of Evolution. With regard to the results it has attained it must suffice to notice the fact that while the functions of many problematical organs have been conjectured, in some cases perhaps rightly, there remain whole groups of common phenomena of this kind, which are still almost untouched even by speculation, and structures and instincts are found in the best known forms, as to the "utility" of which no one has made even a plausible surmise. All this is familiar to every one and every one knows the various answers that have been made.

It is not quite fair to judge such a method by the imperfection of its results, but in one respect the deficiency of results obtained by the Study of Adaptation is very striking, and though this has often been recognized it must be again and again insisted on as a thing to be kept always in view. The importance of this consideration will be seen when the evidence of Variation is examined. The Study of Adaptation ceases to help us at the exact point at which help is most needed. We are seeking for the cause of the differences between species and species, and it is precisely on the utility of Specific Differences that the students of Adaptation are silent. For, as Darwin and many others have often pointed out, the characters which visibly differentiate species are not as a rule capital facts in the constitution of vital organs, but more often they are just those features which seem to us useless and trivial, such as the patterns of scales, the details of sculpture on chitin or shells, differences in number of hairs or spines, differences between the sexual prehensile organs, and so forth. These differences are often complex and are strikingly constant, but their utility is in almost every case problematical. For example, many suggestions have been made as to the benefits which edible moths may derive from their protective coloration, and as to the reasons why unpalatable butterflies in general are brightly coloured; but as to the particular benefit which one dull moth enjoys as the result of his own particular pattern of dullness as compared with the closely similar pattern of the next species, no suggestion is made. Nevertheless these are exactly the real difficulties which beset the utilitarian view of the building up of Species. We knew all along that Species are *approximately* adapted to their circumstances; but the difficulty is that whereas the differences in adaptation seem to us to be approximate, the differences between the structures of species are frequently precise. In the early days of the Theory of Natural Selection it was hoped that with searching the direct utility of such small differences would be found, but time has been running now and the hope is unfulfilled.

Even as to the results which rank among the triumphant successes of this method of study there is need for great reserve. The adequacy of such evidence must necessarily be a matter for individual judgment, but in dealing with questions of Adaptation more than usual caution is needed. No disrespect is intended towards those who have sought to increase our acquaintance with these obscure phenomena; but since at the present time the conclusions arrived at in this field are being allowed to pass unchallenged to a place among the traditional beliefs of Science, it is well to remember that the evidence for these beliefs is far from being of the nature of proof.

The real objection however to the employment of the Study of Adaptation as a means of discovering the processes of Evolution is not that its results are meagre and its conclusions unsound. Apart from the doubtful character of these inferences, there is a difficulty of logic which in this method is inherent and insuperable. This difficulty lies in the fact that while it is generally possible to suggest some way by which in circumstances, known or hypothetical, any given structure may be of use to any animal, it cannot on the other hand ever be possible to prove that such structures are not on the whole harmful either in a way indicated or otherwise. There is a special reason why the impossibility of proving the negative applies with peculiar force to the mode of reasoning we are now considering. This is due to the fact that whereas the only possible test of the utility of a structure must be a quantitative one, such a quantitative method of assessment is entirely beyond our powers and is likely to remain so indefinitely. The students of Adaptation forget that even on the strictest application of the theory of Selection it is unnecessary to suppose that every part an animal has, and every thing which it does, is useful and for its good. We, animals, live not only by virtue of, but also in spite of what we are. It is obvious from inspection that any instinct or any organ *may* be of use: the real question we have to consider is of *how much* use it is. To know that the presence of a certain organ *may* lead to the preservation of a race is useless if we cannot tell how much preservation it can effect, how many individuals it can save that would otherwise be lost; unless we know also the degree to which its presence is harmful; unless, in fact, we know how its presence affects the profit and loss account of the organism. We have no right to consider the utility of a structure *demonstrated*, in the sense that we may use this demonstration as evidence of the causes which have led to the existence of the structure, until we have this quantitative knowledge of its utility and are able to set off against it the cost of the production of the structure and all the difficulties which its presence entails on the organism. No one who has ever tried to realize the complexity of the relations between an organism and its surroundings, the infinite variety of the consequences which every detail of structure and every shade of instinct *may* entail upon the organism, the precision of the correlation between function and the need for it, and above all the marvellous accuracy with which the presence or absence of a power or a structure is often compensated among living beings—no one can reflect upon these things and be hopeful that our quantitative estimates of utility are likely to be correct. But in the absence of such correct and final estimates of utility, we must never use the *utility* of a structure as a point of departure in considering the manner of its origin; for though we can see that it is, or may be, useful, yet a little reflexion will shew that it is, or may be, harmful, but whether on the whole it is useful or on the whole harmful, can only be guessed at. It thus happens that we can only get an indefinite knowledge of Adaptation, which for the purposes of our problem is not an advance beyond the original knowledge that organisms are all *more or less* adapted to their circumstances. No amount of evidence of the same kind will carry us beyond this point. Hence, though the Study of Adaptation will always remain a fascinating branch of Natural History, it is not and cannot be a means of directly solving the problem of the origin of Species.

SECTION III.

Continuity or Discontinuity of Variation.

What is needed, then, is evidence of a new kind, for no amount of evidence of the kinds that have been mentioned will take us much beyond our present position. We need more knowledge, not so much of the facts of anatomy or development, as of the principles of Evolution. The question to be considered is how such knowledge may be obtained. It is submitted that the Study of Variation gives us a chance, and perhaps the only one, of arriving at this knowledge.

But though, as all will admit, a knowledge of Variation lies at the root of all biological progress, no organized attempt to obtain it has been made. The reason for this is not very clear, but it apparently proceeds chiefly from the belief that the subject is too difficult and complex to be a profitable field for study. However this may be, the fact remains, that since the first brief treatment of the matter in *Animals and Plants under Domestication* no serious effort to perceive or formulate principles of Variation has been made, and there is before us nothing but the most meagre and superficial account of a few of its phenomena. Darwin's first collection of the facts of Variation has scarcely been increased. These same facts have been arranged and rearranged by each successive interpreter; the most various and contradictory propositions have been established upon them, and they have been strained to shew all that it can possibly be hoped that they will shew. Any one who cares to glance at the works of those who have followed Darwin in these fields may assure himself of this. So far, indeed, are the interpreters of Evolution from adding to this store of facts, that in their hands the original stock becomes even less until only the most striking remain. It is wearisome to watch the persistence with which these are revived for the purpose of each new theorist. How well we know the offspring of Lord Morton's mare, the bitch 'Sappho,' the Sebright Bantams, the Himalaya Rabbit with pink eyes, the white Cats with their blue eyes, and the rest! Perhaps the time has come when even these splendid observations cannot be made to shew much more. Surely their use is now rather to point the direction in which we must go for more facts.

The questions which by the Study of Variation we hope to answer may be thus expressed. In affirming our belief in the doctrine of the Community of Descent of living things, we declare that we believe all living things to stand to each other in definite genetic relationships. If then all the individuals which have lived on the earth could be simultaneously before us, we believe that it would be possible to arrange them all, so that each stood in its own ordinal position in series. We believe that all the secondary series together make up one primary series from which each severally arises. This is the fundamental conception of Evolution and is represented figuratively by the familiar image of a genealogical tree. If then all the individual ancestors of any given form were before us and were arranged in their order, we believe they would constitute a series. This view of the forms of organisms as constituting a series or *progression* is the central idea of modern biology, and must be borne continually in mind in the attempt to apply any principle to the Study of Evolution.

Each individual and each type which exists at the present moment stands, for the moment, therefore, as the last term of such a series. The problem is to find the other terms. In the case of each type the question is thus stated in a particular form, and it is a somewhat remarkable circumstance that it is in its particular forms that this problem has been most studied. The same problem is nevertheless capable of being stated in the general form also. Instead of considering what has been the actual series from which a specified type has been derived, we may consider what are the characters and attributes of such series in general. It may indeed be contended that it is scarcely reasonable to expect to discover the line of descent of a given form, for the evidence is gone; but we may hope to find the general chararacteristics of Evolution, for Evolution, as we believe, is still in progress. It is really a strange thing that so much enterprise and research should have been given to the task of reconstructing particular pedigrees—a work in which at best the facts must be eked out largely with speculation—while no one has ever seriously tried to determine the general characters of such a series. Yet if our modern conception of Descent is a right one, it is a phenomenon now at this time occurring, which by common observations, without the use of any imagination whatever, we may now see. The chief object, then, with which we shall begin the Study of Variation will be the determination of the nature of the Series by which forms are evolved.

The first questions that we shall seek to answer refer to the manner in which differentiation is introduced in these Series.

All we as yet know is the last term of the Series. By the postulate of Common Descent we take it that the first term differed widely from the last, which nevertheless is its lineal descendant: how then was the transition from the first term to the last term effected? If the whole series were before us, should we find that this transition had been brought about by very minute and insensible differences between successive terms in the Series, or should we find distinct and palpable gaps in the Series? In proportion as the transition from term to term is minimal and imperceptible we may speak of the Series as being **Continuous**, while in proportion as there appear in it lacunae, filled by no transitional form, we may describe it as **Discontinuous**. The several possibilities may be stated somewhat as follows. The Series may be wholly continuous; on the other hand it may be sometimes continuous and sometimes discontinuous; we know however by common knowledge that it is never wholly discontinuous. It may be that through long periods of the Series the differences between each member and its immediate predecessor and successor are impalpable, while at certain moments the series is interrupted by breaches of continuity which divide it into groups, of which the composing members are alike, though the successive groups are unlike. Lastly, discontinuity may occur in the evolution of particular organs or particular instincts, while the changes in other structures and systems may be effected continuously. To decide which of these agrees most nearly with the observed phenomena of Variation is the first question which we hope, by the Study of Variation, to answer. The answer to this question is of vital consequence to progress in the Study of Life.

The preliminary question, then, of the degree of continuity with which the process of Evolution occurs, has never been decided. In the absence of such a decision there has nevertheless been a common assumption, either tacit or expressed, that the process is a continuous one. The immense consequence of a knowledge of the truth as to this will appear from a consideration of the gratuitous difficulties which have been introduced by this assumption. Chief among these is the difficulty which has been raised in connexion with the building up of new organs in their initial and imperfect stages, the mode of transformation of organs, and, generally, the Selection and perpetuation of minute variations. Assuming then that variations are minute, we are met by this familiar difficulty. We know that certain devices and mechanisms are useful to their possessors; but from our knowledge of Natural History we are led to think that their usefulness is consequent on the degree of perfection in which they exist, and that if they were at all imperfect, they would not be useful. Now it is clear that in any continuous process of Evolution such stages of imperfection must occur, and the objection has been raised that Natural Selection cannot protect such imperfect mechanisms so as to lift them into perfection. Of the objections which have been brought against the Theory of Natural Selection this is by far the most serious.

The same objection may be expressed in a form which is more correct and comprehensive. We have seen that the differences between Species on the whole are Specific, and are differences of kind, forming a discontinuous Series, while the diversities of environment to which they are subject are on the whole differences of degree, and form a continuous Series; it is therefore hard to see how the environmental differences can thus be in any sense the directing cause of Specific differences, which by the Theory of Natural Selection they should be. This objection of course includes that of the utility of minimal Variations.

Now the strength of this objection lies wholly in the supposed continuity of the process of Variation. We see all organized nature arranged in a discontinuous series of groups differing from each other by differences which are Specific; on the other hand we see the divers environments to which these forms are subject passing insensibly into each other. We must admit, then, that if the steps by which the divers forms of life have varied from each other have been insensible—if in fact the forms ever made up a continuous series—these forms cannot have been broken into a discontinuous series of groups by a continuous environment, whether acting directly as Lamarck would have, or as selective agent as Darwin would have. This supposition has been generally made and admitted, but in the absence of evidence as to Variation it is nevertheless a gratuitous assumption, and as a matter of fact when the evidence as to Variation is studied, it will be found to be in great measure unfounded.

In what follows so much will be said of discontinuity in Variation that it will not be amiss to speak of the reasons which have led many to suppose that the continuity of Variation needs no proof. Of these reasons there are especially two. First there is in the minds of some persons an inherent conviction that *all* natural processes are continuous. That many of them do not appear so is admitted: it is admitted, for example, that among chemical processes Discontinuity is the rule; that changes in the states of matter are commonly effected discontinuously, and the like. Nevertheless it is believed that such outward and visible Discontinuity is but a semblance or mask which conceals a real process which is continuous and which by more searching may be found. With this class of objections we are not perhaps concerned, but they are felt by so many that their existence must not be forgotten. Secondly, Variation has been supposed to be always continuous and to proceed by minute steps because changes of this kind are so common in Variation. Hence it has been inferred that the mode of Variation thus commonly observed is universal. That this inference is a wrong one, the facts will shew.

To sum up:

The first question which the Study of Variation may be expected to answer, relates to the origin of that Discontinuity of which Species is the objective expression. Such Discontinuity is not in the environment; may it not, then, be in the living thing itself?

The Study of Variation thus offers a means whereby we may hope to see the processes of Evolution. We know much of what these processes *may* be: the deductive method has been tried, with what success we know. It is time now to try if these things cannot be seen as they are, and this is what Variation may shew us. In Variation we look to see Evolution rolling out before our eyes. In this we may fail wholly and must fail largely, but it is still the best chance left.

SECTION IV.

Symmetry and Meristic Repetition.

Having thus indicated some of the objects which we may hope to reach by the Study of Variation, we have next to consider the way in which to set about this study.

The Study of Variation is essentially a study of differences between organisms, so for each observation of Variation at least two substantive organisms are required for comparison. It is proposed to confine the present treatment of the subject to a consideration of the integral steps by which Variation may proceed; hence it is desirable that the two organisms compared should be parent and offspring, and if, as is often the case, the actual parent is unknown, it is at least necessary that the normal form of the species should be known and that there must be reasonable evidence that the varying offspring is actually descended from such a normal. For this reason, evidence from a comparison of Local Races, and other established Varieties, though a very valuable part of the Study, will for the most part not be here introduced. For the belief that such races are descended from the putative normal scarcely ever rests on proof, and still more rarely is there evidence of the number of generations in which the change has been effected.

For our purpose we require actual cases of Variations occurring as far as possible in offspring of known parentage; and if, failing this, we make use of cases occurring in the midst of normal individuals of known structure, it must in such cases be always remembered that we cannot properly assume that the varying form is the offspring of such individuals, though special reasons may make this likely in special cases.

Since the structure of the offspring is perhaps in no case

identical with that of the parent, observation of any parent and its offspring is to the point; but such a field as this is plainly too wide to be studied with profit as a whole, and it is necessary from the first, that attention should be limited to certain classes of such phenomena. With this object certain limitations are proposed, and though confessedly arbitrary, they will be found on the whole to work well.

The first limitation thus introduced concerns the *magnitude* of Variations. We have seen above that the assumption that Variation is a continuous process lands us in serious difficulties in the application of a hypothesis which, on general grounds, we nevertheless are prepared to receive. If then we can shew that Variation is to some extent discontinuous, a road will be opened by which these difficulties may perhaps be in part avoided.

Species are discontinuous; may not the Variation by which Species are produced be discontinuous too? It may be stated at once that evidence of such Discontinuous Variation does exist, and in this first consideration of the subject attention will be confined to it. The fact that Continuous Variation exists is also none the less a fact, but it is most important that the two classes of phenomena should be recognized as distinct, for there is reason to think that they are distinct essentially, and that though both may occur simultaneously and in conjunction, yet they are manifestations of distinct processes. The attempt to distinguish these two kinds of Variation from each other constitutes one of the chief parts of the study. It will not perhaps be possible to find any general expression which shall accurately differentiate between Variations which are Discontinuous and those which are Continuous, but it is possible to recognize attributes proper to each and to distinguish changes which are or may be effected in the one way from other changes which are or may be effected in the other.

For the present we shall treat only of the evidence of Discontinuous Variation.

In order to explain the second limitation which is to be introduced it is necessary to refer to some phenomena which are characteristic of the forms of organisms, and to separate from them the group with which we shall deal first.

It was stated above that perhaps no character of form is common to all living things, but nevertheless there is one feature which is found in the great majority.

In the first place, the bodies of organisms are not homogeneous but heterogeneous, consisting of organs or parts which in substance and composition differ from each other. This heterogeneity in composition is of course an objective expression of the process of Differentiation, and it is further recognized that such structural heterogeneity of material corresponds with a physiological Differentiation of function. This Differentiation or Heterogeneity is found in the bodies of all organisms, even in the simplest.

Now in a wide survey of the forms of living things there is a fact with regard to the presence of this Heterogeneity which to the purpose of our present consideration is of the highest consequence. This may perhaps be best expressed by the statement that in the bodies of living things Heterogeneity is generally orderly and formal; it is cosmic, not chaotic. Not only are the bodies of all organisms heterogeneous, but in the great majority the Heterogeneity occurs in a particular way and according to geometrical rule. This character is not peculiar to a few organisms, but is common to nearly all. We will now examine this phenomenon of geometrical order in Heterogeneity and try to see some of the elements of which it is made up.

Order of form will first be found to appear in the fact that in any living body the Heterogeneity is in some degree symmetrically distributed around one or more centres. In the great majority of instances these centres of symmetry are themselves distributed about other centres, so that in one or more planes the whole body is symmetrical.

The idea of **Symmetry** which is here introduced is so familiar that it is scarcely necessary to define it, but as all that follows depends entirely on the proper apprehension of what is meant by Symmetry it may be well to call attention to some of the phenomena which the term denotes.

In its simplest form the Symmetry of a figure depends on the fact that from some point within it at least two lines may be taken in such a way that each passes through parts which are similar and similarly disposed. The point from which the lines are taken may be called a centre of Symmetry and the lines may be called lines of Symmetrical Repetition.

Commonly the parts thus symmetrically disposed are related to each other as optical images [in a plane mirror passing through the centre of Symmetry and standing in a plane bisecting the angle which the lines of Symmetrical Repetition make with each other]. For a figure to be symmetrical, in the ordinary sense of the term, it is not necessary that the relation of optical images should strictly exist, and several figures, such as spirals, &c., are accordingly described as symmetrical. But since the relation of images exists in all cases of bilateral and radial symmetry, which are the forms most generally assumed in the symmetry of organisms, it is of importance to refer particularly to this as one of the phenomena often associated with Symmetry.

In the simplest possible case of Symmetry there is a series of parts in one direction corresponding to a series of parts in another direction. Perhaps there is no organism in which such an arrangement does not at some time and in some degree exist. For even in an unsegmented ovum or a resting *Amœba* there is little doubt that Symmetry is present, though owing to the slight degree of Differentiation, its presence may not be clearly perceived. In the manifestations, however, in which it is most familiar, Symmetry is a decided and obvious phenomenon.

Symmetry then depends essentially on the fact that structures found in one part of an organism are repeated and occur again in another part of the same organism. Symmetrical Heterogeneity may therefore be present in a spherical body having a core of different material, and it is possible that in an unsegmented ovum for example a Symmetry of this simple kind may exist. But Symmetry, as it is generally seen in organisms, differs from that of these simplest cases in the fact that the organs repeated are separated from each other by material of a nature different from that of the organs separated. Repetitions of this kind are known in almost every group of animals and plants. The parts thus separated may belong to any system of organs. There is no known limit to the number of Repetitions that may occur.

This phenomenon of Repetition of Parts, generally occurring in such a way as to form a Symmetry or Pattern, comes near to being a universal character of the bodies of living things. It will in cases which follow be often convenient to employ a single term to denote this phenomenon wherever and however occurring. For this purpose the term **Merism** will be used. The introduction of a new term is, as a practice, hardly to be justified; but in a case like the present, in which it is sought to associate divers phenomena which are commonly treated as distinct, the employment of a single word, though a new one, is the readiest way of giving emphasis to the essential unity of the phenomena comprised.

The existence of patterns in organisms is thus a central fact of morphology, and their presence is one of the most familiar characters of living things. Anyone who has ever collected fossils, or indeed animals or plants of any kind, knows how in hunting, the eye is caught by the formal regularity of an organized being, which, contrasting with the irregularity of the ground, is often the first indication of its presence. Though of course not diagnostic of living things, the presence of patterns is one of their most general characters.

On examining more closely into the constitution of Repetitions, they may be seen to occur in two ways; first, by Differentiation within the limits of a single cell, as in the *Radiolaria*, the sculpture of egg-shells, nuclear spindles, &c., to take marked cases; and secondly, by, or in conjunction with, the process of Cell-Division. The Symmetry which is found in the Serial Repetitions of Parts in unicellular organisms does not in all probability differ essentially from that which is produced by Cell-Division, for, though sufficiently distinct in outward appearance, the two are almost certainly manifestations of the same power.

Such patterns may exist in single cells or in groups of cells, in separate organs or in groups of organs, in solitary forms or in colonies and groups of forms. Patterns which are completed in the several organs or parts will be referred to as **Minor Symmetries.** These may be compounded together into one single pattern, which includes the whole body: such a symmetry will be called a **Major Symmetry.** In most organisms, whether colonial or solitary, there is such a Major Symmetry; on the other hand organisms are known in which each system of Minor Symmetry is, at least in appearance, distinct and without any visible geometrical relation to the other Minor Symmetries. Examples of this kind are not common, for, as a rule, the planes about which each Minor Symmetry is developed have definite geometrical relations to those of the other Minor Symmetries. It is possible, even, that in some if not all of these, the planes of division by which the tissues composing each system of Minor Symmetry are originally split off and differentiated, have such definite relations, though by subsequent irregularities of growth and movement these relations are afterwards obscured.

The classification of Symmetry and Pattern need not now be further pursued. The matter will be often referred to in the course of this work, when facts concerning Variations in number and patterns are being given, for it is by study of Variations in pattern and in repetition of parts that glimpses of the essential phenomena of Symmetry may be gained.

That which is important at this stage is to note the almost universal presence of Symmetry and of Repetition of Parts among living things. Both are the almost invariable companions of division and differentiation, which are fundamental characters without which Life is not known.

The essential unity of the phenomenon of Repetition of Parts and of its companion-phenomenon, Symmetry, wherever met with, has been too little recognized, and needless difficulty has thus been introduced into morphology. To obtain a grasp of the nature of animal and vegetable forms, such recognition is of the first consequence.

To anyone who is accustomed to handle animals or plants, and who asks himself habitually—as every Naturalist must—how they have come to be what they are, the question of the origin and meaning of patterns in organisms will be familiar enough. They are the outward and visible expression of that order and completeness which inseparably belongs to the phenomenon of Life.

If anyone will take into his hand some complex piece of living structure, a Passion-flower, a Peacock's feather, a Cockle-shell, or the like, and will ask himself, as I have said, how it has come to be so, the part of the answer that he will find it hardest to give, is that which relates to the perfection of its pattern.

And it is not only in these large and tangible structures that the question arises, for the same challenge is presented in the most minute and seemingly trifling details. In the skeleton of a Diatom or of a Radiolarian, the scale of a Butterfly, the sculpture on a pollen-grain or on an egg-shell, in the wreaths and stars of nuclear division, such patterns again and again recur, and again and again the question of their significance goes unanswered. There are many suggestions, some plausible enough, as to why the tail of a Peacock is gaudy, why the coat of a pollen-grain should be rough, and so forth, but the significance of patterns is untouched by these. Nevertheless, repetitions arranged in pattern exist throughout organized Nature, in creatures that move and in those that are fixed, in the great and in the small, in the seen and in the hidden, within and without, as a property or attribute of Life, scarcely less universal than the function of respiration or metabolism itself.

Such, then, is Symmetry, a character whose presence among organisms approaches to universality.

SECTION V.

Meristic Variation and Substantive Variation.

It is to the origin and nature of Symmetry that the first section of the evidence of Variation will relate. That a knowledge of the modes of Variation of so universal a character is important to the general study of Biology must at once be evident, but to the particular problem of the nature of Specific Differences this importance is immense. This special importance comes from two reasons. As it is the fact first that Repetition and Symmetry are among the commonest features of organized structure, so it will be found next that it is by differences in those features that the various forms of organisms are very commonly differentiated from each other. Their forms are classified by all sorts of characters, by shape and proportions, by size, by colour, by habits and the like; but perhaps almost as frequently as by any of these, by differences in number of parts and by differences in the geometrical relations of the parts. It is by such differences that the larger divisions, genera, families, &c. are especially distinguished. In such cases of course the differences in number and Symmetry do not as a rule stand alone, but are generally, and perhaps always, accompanied by other differences of a qualitative kind; nevertheless, the differences in number and Symmetry form an integral and very definite part of the total differences, so that in any consideration of the nature of the processes by which the differences have arisen, special regard must be had to these numerical and geometrical, or, as I propose to call them, **Meristic**, changes.

In the present Introduction I do not propose to forestall the evidence more than is absolutely necessary for the purpose of making clear the principles on which the facts are grouped, but it will do the evidence no wrong if at the present stage it is stated that Meristic Variation is frequently Discontinuous, and that in the case of certain classes of Repetitions is perhaps always so.

The nature of Merism and the manner in which Meristic Variations occur will be fully illustrated hereafter, but it is necessary to say this much at the present stage, since it is from this Discontinuity in the occurrence of Meristic Variations that the phenomena of Symmetry and Repetition derive their special importance in the Study of Variation.

The importance of the phenomena of Merism to the Study of Variation is thus, in the first instance, a direct one, for the Variations which have resulted in the production of Meristic Systems are a direct factor in Evolution. In addition to this direct relation to the Study of Variation, the phenomena of Merism have also an indirect relation, which is scarcely less important; for they are a factor in the estimation of the magnitude of the integral steps by which Variation proceeds. This will be more evident after the second group of Variations has been spoken of.

We have thus far spoken only of the processes by which the living body is divided into parts, and we have thus constituted a group which is to include Variations in number, Division, and geometrical position. From these phenomena of Division may be distinguished Variations in the actual constitution or substance of the parts themselves. To these Variations the name **Substantive** will be given. Under this head several phenomena may be temporarily grouped together, which with further knowledge will doubtless be found to have no real connexion with each other. For the present, however, it will be convenient to constitute such a temporary group in order to bring out the relative distinctness of Variations which are Meristic.

These two classes of Variation, Meristic and Substantive, may be recognized at the outset of the study. There can be no doubt that they are essentially distinct from each other, and the proof that they are thus distinct will be found in the evidence of Variation, for it will be seen that either may occur independently of the other. An appreciation of this distinction is a first step towards a comprehension of the processes by which the bodies of organisms are evolved.

A few simple illustrations may make the nature of these two classes of Variations more clear. For example, then, the flower of a *Narcissus* is commonly divided into six parts, but through Meristic Variation it may be divided into seven parts or into only four. Nevertheless there is in such a case no perceptible change in the tissues or substance of which the parts are made up. All belong to and are recognizable as belonging to the same sort of *Narcissus*. On the other hand many *Narcissi*, *N. corbularia*, for example, are known in two colours, one a dark yellow and the other a sulphur-yellow, though the number of parts and pattern of the flowers are identical. This is, therefore, an example of a Substantive Variation.

Again, the foot of a Pig may, through Meristic Variation, be divided into five or six toes instead of into four; or, on the other hand, the number may, by absence of the median division between the digits III and IV, be reduced to three, though the tissues composing the toes may not in structure differ from the normal.

Again, the tarsus of a Cockroach (*Blatta*) may, through Meristic Variation, be divided into only four joints instead of into five, the normal number, but the joints are still in substance or quality those of a Cockroach.

I am aware that Meristic and Substantive Variations often occur together, and that there is a point at which it is not possible to separate satisfactorily the changes which have come about by the one process from those which have come about by the other. Instances of this kind occur especially in the case of series of parts such as Teeth or Vertebræ, in which individual members or groups of members of the series are differentiated from the others. For example, we may see that it is through Meristic Variation that the vertebral column of a Dog may be divided into a number of Vertebræ greater or less than the normal; and though in such cases all the Vertebræ have distinctively canine characters, yet there are nearly always Substantive Variations occurring in correlation with the Meristic Variations, manifesting themselves in a re-arrangement of the points of division between the several groups of Vertebræ, and causing individual Vertebræ to assume characters which are not proper to their ordinal positions.

Further inquiry into the questions thus raised cannot at this stage be profitably undertaken, though when the evidence has been considered it will perhaps be advisable to recur to them; all that is now intended is to indicate broadly the general scope of Meristic and Substantive Variation respectively.

As has already been stated, it is proposed to begin the Study of Variation by an examination of Variations which are Meristic, leaving the consideration of Substantive Variation to be undertaken hereafter. But nevertheless in the consideration of Meristic Variation it will be necessary to refer to phenomena of Substantive Variation in so far as their occurrence or distribution in the body are affected by Meristic phenomena. For in the determination of the magnitude of the integral steps by which Variation proceeds, the existence of Merism plays a conspicuous part, and it is in consequence of this that the subject of Symmetry and Repetition of Parts has a second and indirect bearing on the Study of Variation which is scarcely less important than the direct bearing of which mention has been made above.

This indirect bearing on the manner of origin of Specific Differences arises from a circumstance which in treatises on Evolution is commonly overlooked. In comparing a species in which parts are repeated, with an allied species in which the same parts are repeated, it commonly occurs that each of the repeated parts of the one have some character by which they are distinguished from the like parts of the other. This differentiating character may be a qualitative one, or a numerical one, or both. In such cases it very frequently happens that this character occurs in each member of the series of Repetitions. For example, the tarsi of the Weevils have only four visible joints, while those of the majority of beetles have five; but the characteristic division into four joints occurs in each of the legs. Before the four-jointed character as seen in the Weevils could be produced it was necessary that not one but all of the legs should vary from the five-jointed form, and in this particular way. The leaves on a beech tree are all beech leaves, and if the tree is a fern-leaved beech, they *may*, and generally speaking do, all shew the characters of that variety; and so on with other particular species and varieties.

The limbs of a bilaterally symmetrical animal, in which the right side is the image of the left, are of course alike, and any specific character which is present in the limbs of the one side must in such an animal be normally present in those of the other side.

The same is true of many forms in which appendages are repeated in series, as for example, the fore-legs and hind-legs of the Horse, the fore- and hind-wings of the Brimstone Butterfly (*Gonepteryx rhamni*); of the patterns on several segments of many caterpillars; of the patterns of the segmental setæ of many worms, and so forth. In series whose members are differentiated from each other, it of course frequently happens that the same specific characters are not present in all the members of the series, and in nearly all such cases these characters are not presented by all in equal degree; nevertheless substantially the phenomenon remains that similar characters often are presented by the several members of a series of repeated organs.

To many this will seem little better than a truism, nevertheless I offer no apology for its introduction; for though, as a common and obvious fact, it is a truism, it is besides a truth, the far-reaching significance of which is scarcely appreciated. For, in the consideration of the magnitude of the integral steps by which Variation proceeds, we shall have this to remember: that to produce any of the forms of which we have spoken, by Variation from another form, it is not enough that the particular Variation should occur and become fixed in one member of the series, but it is necessary that the character should sooner or later be taken on by *each* member of the series which exhibits it. In such cases therefore, this question is raised. Did the Variation come in first in one member of the Series and then in another? Did it occur, for example, simultaneously on the two sides of the body? Did the right and left fore-legs of the Horse cease to develop more than the present number of digits simultaneously or separately? Was the similar form of the hind-legs assumed before, after, or simultaneously with that of the fore-legs? Were the orange markings which are present on both fore- and hind-wings of the Brimstone, or the ocellar markings of the Peacock (*V. Io*), and of the Emperor (*Saturnia carpini*), assumed by both wings at once? Were the four wings of the Plume Moths split simultaneously into the characteristic "plumes"? Did the brown spots on the three leaflets of *Medicago*, the fimbriation of the petals of Ragged Robin (*Lychnis flos-cuculi*), the series of stripes on the Zebra, the pink slashes on the segments of *Sphinx* larvae, the eyes on the scutes of Chitons, and the thousand other colour-marks, sense-organs, appendages and structural features, which throughout organized Nature occur in Series, vary to their present state of similarity by similar and simultaneous steps, or did each member of such Series take these characters by steps which were separate and occurring independently? To this question, which lies at the root of all progress in the knowledge of Evolution, the Study of Variation can alone reply. That in the facts which follow, the answer to this question will be found, cannot of course be said; but these facts, few though they are, do nevertheless answer it in part, and they suggest that more facts of the same kind would go far towards answering it completely. But beyond this, the facts are of value as an indication of the part which the phenomenon of Merism may play in determining the magnitude of Variations and the manner of their distribution among the several parts of the body. On examining the evidence it will be found that between parts related to each other in the way that has been described, there is a certain bond or kinship, by virtue of which they *may* and often do vary simultaneously and in similar ways, though the fact that they may also vary independently, and in different ways, will of course also appear.

The phenomenon of the Similar Variation of parts which are repeated Meristically in Series is a fact which will be found to have important bearings on several distinct departments of biological study.

As was shewn above, it is by recognition of the existence of such similar and simultaneous Variation that the manner of origin of the similar complexity of several organs belonging to the same system or series becomes comparatively comprehensible; for it is not then necessary to conceive a separate origin for the complexity of each member of the series. For example, it is difficult to conceive the manner of evolution of an eye of a vertebrate; nevertheless, for each vertebrate *two* eyes have been evolved. If it were necessary to suppose that each arose by separate selections of separate variations, the difficulty would be thus doubled. If, however, it is recognized that the complexity of both arose simultaneously, the phenomenon becomes the more intelligible as the number of integral variations and selections demanded is reduced.

The case chosen, of paired organs in bilateral symmetry, is a very simple one, but it will be found that similar relations hold between other parts repeated in series. For in the same way it is not necessary to suppose an independent evolution for each of the tail-feathers of the Peacock, for the legs of the Horse, and the like, since in the light of the facts of Variation it is as easy for all to take on the new characters as for one.

If the manner of development of Repeated Parts is considered, this fact will not seem surprising. For all these parts arise from the undifferentiated tissues by a process of Division, and whatever characters were potentially present in the undifferentiated tissues may appear in the parts into which it subsequently divides. A somewhat loose illustration will perhaps make this more clear. Everyone knows the rows of figures which children cut out from folded paper. There are as many figures as folds, each figure being alike if the folds coincide. If the paper is pink, all the figures are pink; if the paper is white, all the figures are white, and so on. If blotting-paper is used, and one blot is dropped on the folded edges, the blot appears symmetrically in all the figures. So also any deviation in the lines of cutting appear in all the figures; a whole row of soldiers in bearskins may be put into helmets by one stroke of the scissors. Of course it is not meant to suggest that the process of division by which parts of the body are produced bears any resemblance to that by which the figures are cut out, but merely to illustrate the fact that since it is by a process of Division of an undifferentiated mass that the Repeated Parts are produced, so the characters of these Repeated Parts depend upon the characters which were present in the original mass and upon the modes by which the parts were divided out from it.

Summary of Sections I to V.

At this point it will be well briefly to recapitulate the preceding Sections.

We are proposing to attack the problem of Species by studying the facts of Variation. Of the facts of Variation in general we have selected a particular group upon which to begin this study. The group of variations thus chosen are those which relate to Number of parts, Division, Repetition, and the other phenomena which are to be included under the term Meristic. With variations in quality and Substance it is not at present proposed to deal, except in so far as it is necessary to refer to them in their relation to the phenomena of Merism, and in illustration of the structural possibilities or necessities which in the body follow as corollaries upon the existence of Meristic Repetition.

It has also been proposed to limit the consideration to Variations which are Discontinuous. As has been already stated, Discontinuous Variations may belong to the Meristic Group or to the Substantive, but it is to the former that attention will first be directed.

SECTION VI.

Meristic Repetition and Homology.

In what has gone before, the two conceptions now introduced, namely the distinction of Variations into Meristic and Substantive, and into Continuous and Discontinuous, have been sketched in outline. The significance of the facts which follow will be made more evident if these two conceptions are now more fully developed in some of their aspects.

Under the name Merism I have proposed to include all phenomena of Repetition and Division, whenever found and in whatever forms occurring, whether in the parts of a body or in the whole. The consequences of the admission of this proposition are considerable and should be fully realized; for on recognition of the unity of these phenomena it is possible to group together a number of facts whose association will lead to simplification of some morphological conceptions, and to other results of utility.

That the phenomena of Merism form a natural group is in some respects a familiar idea, but in its fullest expression it is as yet not generally received, still less have the consequences which it entails been properly appreciated. Every one who has gone even a little way into morphological inquiry has met some of the difficulties to which we shall now refer.

It is with respect to the phenomena of Segmentation that these difficulties are most familiar, and it is in this connexion that they may be best discussed. Segmentation is a condition which reaches its highest development in Vertebrates, the Annelids, and the Arthropods, and it is in these groups that it has been most studied. In them it appears as a more or less coincident Repetition of elements belonging to most of the chief systems of organs along an axis corresponding to the long axis of the body. To segmentation of this kind the name 'Metameric' has been given, and by many morphologists the attempt has been made, either tacitly or in words, to separate such Metameric Segmentation from other phenomena of Repetition elsewhere occurring.

It has thus been attempted to distinguish the Repetitions which occur along the long axis of the body from those occurring along the long axis of appendages, such as for example the joints of antennæ or of digits, and some have even gone so far as to regard the Segmentation of the Vertebrate tail as a thing different in kind from that of the trunk itself. It would be apart from our present purpose to recur to these subjects, were it not that this suggestion of the existence of a difference in kind between Metameric Segmentation and other Repetitions has led to several notable errors in the interpretation of the facts of morphology and in the application of these facts to the solution of the problems of Descent. In order to lay a sound foundation for the study of Meristic Variation these errors must be cleared away, and to do this it is necessary to break down the artificial distinction between the phenomena of Metameric Segmentation and other cases of Repetition of Parts, so that the whole may be seen in their true relations to each other. When this is done, the mutual relations of the facts of Meristic Variation will also become more evident.

The first difficulty which has been brought into morphology by the suggestion that Metameric Segmentation is a phenomenon distinct in kind, is one which has coloured nearly all reasoning from the facts of Morphology to problems of phylogeny. For the existence of Metameric Segmentation in any given form is thus taken to be one of its chief characters, and, as such, is allowed predominant weight in considering the genetic relations of these forms. By the indiscriminate though logical extension of this principle the conclusion has been reached that Vertebrates are immediately connected with, or have arisen by Descent from Annelids, or from Crustacea and the like, for the Repetition in these forms is closely similar. Others again, being struck with the resemblance between the Repetition of Parts along the radial axes of Starfishes and those which occur along the long axes of Annelids have hazarded the conjecture that perhaps this resemblance may indicate the actual phylogenetic history of these Repetitions. Though such speculations as these are little better than travesties of legitimate theory, some of them still command interest if not belief.[1] All alike are founded on the assumption that resemblances between the manner and degree in which Repetition occurs are unlikely to have arisen save by community of Descent. A broader view of Meristic phenomena will shew that

---

[1] These modern "Instances" recall many that once were famous but are now forgotten. For example: *Item non absurda est similitudo et conformitas illa, ut homo sit tanquam planta inversa. Nam radix nervorum et facultatum animalium est caput; partes autem seminales sunt infimæ, non computatis extremitatibus tibiarum et brachiorum. At in planta, radix (quæ instar capitis est) regulariter infimo loco collocatur; semina autem supremo.* BACON, *Nov. Org.* Lib. II. 27. In *non computatis extremitatibus*, amateurs of INSTANTIÆ CONFORMES may still find matter for warning.

this assumption is unfounded; for so far are the expressions of it which are called Metamerism from standing alone, that it is almost impossible to look at any animal or vegetable form without meeting phenomena of Repetition which differ from Metamerism only in degree or in extent. Between these Repetitions and Metameric Repetitions it is impossible to draw any line, and the Meristic Variations of all will therefore be treated together.

This error in the estimate of the value of Metamerism as a guide to phylogeny is one by which the evidence of Variation is only indirectly affected. The other errors now to be mentioned are of a much more serious nature, for they concern the general conception of the nature of Homology which is the basis of all morphological study.

In introducing the method of the Study of Variation I have said that it can alone supply a solid foundation for inquiry into the manner by which one species arises from another. The facts of Variation must therefore be the test of phylogenetic possibility. Looking at organs instead of species, we shall now see that the facts of Variation must also be the test of the way in which organ arises from organ, and that thus Variation is the test of Homology. For the statement that an organ of one form is homologous with an organ of another means that there is between the two some connexion of Descent, and that the one organ has been formed by modification of the other, or both by modification of a third. The precise way in which this connexion exists is not defined, and indeed has scarcely ever been considered, though such a consideration must sooner or later be attempted. We must for the present be content with the belief that in some undefined way there is a relationship between 'homologous' parts, and that this is what we mean when we affirm that they are homologous.

We have however assumed that the transition from one form to another takes place by Variation. If therefore we can see the variations we shall see the precise mode by which the descent is effected, and this must be true of the parts or organs as it is true of the whole body. In like manner then as the Study of Variation may be hoped to shew the way by which one form passes into another, so also may it be hoped that it will shew how the organs of one form take on the shape of the homologous organs of another.

In the absence of the evidence of Variation reasoning as to Homology rests solely on conjecture, and assumptions have thus been made respecting the nature of Homology which have coloured the whole of morphological study. Of these, two demand attention now.

I. *As to Homology between the Members of one Series.* We saw above (page 29) how the resemblance between Repetitions occurring in divers forms has led to the belief that those forms had a common descent: in a somewhat similar way it has happened that the resemblance between individual members of a series of Repeated Parts has led to the belief that they must originally have been alike, and that they have been formed by differentiation of members originally similar. Many who would hesitate thus to formulate such a belief nevertheless have taken part in inquiries which can succeed only on the hypothesis that this has been the history of such parts. Of this nature are the old attempts to divide the skull into vertebræ, recognizing the several parts of each; the modern disquisitions on the segmentation of the cranial nerves; the attempts to homologize the several phalanges of the vertebrate pollex and hallux with the several phalanges of the other digits; similar attempts to trace the precise equivalence of the elements of the carpus and tarsus, and many other quests of a like nature. In all these it is assumed that there is a precise equivalence to be found with enough searching, and that all the members of such series of Repetitions were originally alike. If the series of ancestors were before us it is expected that this would be seen to have been the case. In the light of the facts of Variation this assumption will be seen to be a wrong one, and these simple views of the Repetition and Differentiation of members in Series must be given up as inadequate and misleading, even though there be no other to substitute.

II. *As to the individuality of Members of Series.* In seeking to homologize a series of parts in one form with a series of parts in another, cases often occur in which the whole series of the one is admittedly homologous with the whole series of the other. In such cases the question arises, can the principle of Homology be extended to the individual members of the two Series? If the two Series each contain the same number of members this question is a comparatively simple one, for it may be assumed that each member of the Series is the equivalent or Homologue of the member which in the other Series occupies the same ordinal position. If however the number of members differs in the two Series, how is the equivalence to be apportioned? This is a question again and again arising with regard to Meristic Series such as teeth, digits, phalanges, vertebræ, nerves, vessels, mammæ, colour-markings, the parts of the flower, and indeed in almost every system whether of animals or plants. To decide this question there are still no general principles. But though we yet know nothing as to the steps by which Meristic Variation proceeds, there is nevertheless a received view by which the interpretation of the phenomena is attempted, and though in the case of each system of organs the application of the principle is different, yet the principle applied is essentially the same.

Thus to compare the members of Series containing different members it is first assumed that the series consisted ancestrally of some maximum number, from which the formula characteristic of each descendant has been derived by successive diminution. Here, again, I do not doubt that many who employ this assumption would hesitate to make it in set terms, but nevertheless it is the logical basis of all such calculations.

Now this hypothesis involves a definite conception of the mode in which Variation works, and it is most important that this should be realized clearly. For if it is true that each member of a Series has in every form an individual and proper history, it follows that if we had before us the whole line of ancestors from which the form has sprung, we should then be able to see the history of each member in the body of each of its progenitors. In such a Series the rise of an individual member and the decline of another should then be manifest. Each would have its individual history just as a Fellowship in a College or a Canonry in a Cathedral has an individual history, being handed on from one holder to his successor, some being suppressed and others founded, but none being merged into a common fund. In other words, according to the received view of the nature of these homologies, *it is assumed that in Variation the individuality of each member of a Meristic Series is respected.*

The difficulty in applying this principle is notorious, but when the evidence of Variation is before us the cause of the difficulty will become evident. For it will be found that though Variation may sometimes respect individual homologies, yet this is by no means a universal rule; and as a matter of fact in all cases of Meristic Series, as to the Variation of which any considerable body of evidence has been collected, numerous instances of Variation occur, in which what may be called the stereotyped or traditional individuality of the members is superseded.

This error in the application of the principle of Homology to individual members of Meristic Series has arisen almost entirely through want of recognition of the unity of Meristic Repetition, wherever found. In the case of a series of parts among which there is no perceptible Differentiation, no one would propose to look for individual Homologies. For example, no one considers that the individual segments in the intestinal region of the Earthworm have any fixed relations of this kind; no one has proposed to homologize single leaves of one tree with single leaves on another; it is not expected that the separate teeth of a Roach have definite homologies with separate teeth of a Dace, for such expectations would be plainly absurd. But in series whose members are differentiated from each other the existence of such individuality is nevertheless assumed. To take only one case: a whole literature has been devoted to the attempt to determine some point in the vertebral column or in the spinal nerves from which the homologies of the segments may be reckoned. This is a problem which in its several forms has been widely studied. Some have attempted to solve it by starting from the lumbar plexus, while others have begun from the brachial. In the case of Birds this question is reduced to an absurdity. Which vertebra of a Pigeon, which has 15 cervical vertebræ, is homologous with the first dorsal of a Swan which has 26 cervicals? To decide these questions the only possible appeal is to the facts of Variation, and judged by these facts the whole inquiry comes to an end, for it is seen at once that the expectation is founded on a wrong conception of the workings of Variation. No one, as has been said above, would attempt such an inquiry if the series were undifferentiated, for this individuality would not be expected in such a Series; but to suppose that it does exist in a differentiated Series of parts, is to suppose that with Differentiation the ordinal individuality of the members has become fixed beyond revision. This supposition the Study of Variation will dispel.

Here, as in the preceding case of the theoretical doctrine of Serial Homology, the current view is far too simple and far too human. Though the methods of Nature are simple too, yet their simplicity is rarely ours. In these subjective conceptions of Homology and of Variation, we have allowed ourselves to judge too much by human criterions of difficulty, and we have let ourselves fancy that Nature has produced the forms of Life from each other in the ways which we would have used, if we had been asked to do it. If a man were asked to make a wax model of the skeleton of one animal from a wax model of the skeleton of another, he would perhaps set about it by making small additions to and subtractions from its several parts; but the natural process differs in one great essential from this. For in Nature the body of one individual has never *been* the body of its parent, and is not formed by a plastic operation from it; but the new body is made again new from the beginning, just as if the wax model had gone back into the melting-pot before the new model was begun.

SECTION VII.

Meristic Repetition and Division.

Before ending this preliminary consideration of Merism it is right that we should see other aspects of the matter. What follows is put forward in no sense as theory or doctrine, but simply as suggesting a line of thought which should be in the minds of any who may care to pursue the subject further or to study the evidence. It is perhaps only when it is seen in connexion with its possible developments that the magnitude of the subject can be fully felt.

In the treatises on Comparative Anatomy which belong especially to the beginning of this century, the idea constantly recurs that the series of segments of a metamerically segmented form do in some sort represent a series of individuals which have not detached themselves from each other. Seen in the light of the Doctrine of Descent this resemblance or analogy has been taken as a possible indication that the segmented forms may actually have had some such phylogenetic history as this. By similar reasoning the Metazoa have been spoken of as "Colonies" of Protozoa. Now though we need not allow ourselves to be drawn away into these and other barren speculations as to phylogeny, we may still note the substance of fact which underlies them. For it is now recognized that between the process by which the body of a *Nais* is metamerically segmented, and that by which it divides into a chain of future "individuals," no line can be drawn: that the process of budding, or of strobilization, by which one form gives rise to a number of detached individuals, is often indistinguishable from the process by which a near ally gives rise to a connected colony, and that the two processes may even be interchangeable in the same form; finally that the process of division of a fertilized ovum by the first cleavage plane may be in some essentials comparable with the division of a Protozoon into two new individuals. All these are now commonplaces of Natural History.

With what justice these considerations may have been applied to the problems of phylogeny we need not now inquire, but to the interpretation of the facts of Variation they have an application which ought not to be neglected.

If, then, as is admitted, there is a true analogy between the process by which new organisms may arise asexually by Division, and the process by which ordinary Meristic Series are produced, it follows that Variation, in the sense of difference between offspring and parent, should find an analogy in Differentiation between the members of a Meristic Series. Applied to the case of asexual reproduction there seems no good reason for denying this analogy. It is of course an undoubted fact that in the asexual reproduction of many forms Variation is rare, though the sexually produced offspring of the same forms are very variable. In plants this is familiar to everyone, though the extension of the same principle to animals rests chiefly on inference. Nevertheless in plants bud-variation, both Meristic and Substantive, happens often, and the division of a plant into two dissimilar branches may well be compared to the production of dissimilar offspring by one parent; indeed, if the processes of Division are admitted to be fundamentally the same, this conclusion can scarcely be escaped.

In one more aspect this subject may be considered with profit. It is, as we have seen, believed that the division of an ovum into two segmentation-spheres is not a process essentially different from the division of certain Protozoa into two "individuals." In conceiving the manner of Variation in such Protozoa we have little or no fact to guide us, but this much is obvious: that for the introduction of a variety as the offspring of a given species, it is necessary either that the two parts into which the unicellular organism divided should have varied equally, and that the division should thus be a symmetrical division (in the full sense of qualitative as well as formal symmetry); or that the division should be asymmetrical, the resulting parts being dissimilar, in which case one may conceivably belong to the type and the other be a Variety. If Variation has ever occurred in the reproduction of animals of this class it must have occurred on one or both of these plans.

Returning to the segmentation of the Metazoan ovum we have the well-known results of Roux and others, shewing that, in certain species, the first[1] cleavage-plane divides the body into the future right and left halves. In such cases then on the analogy of the Protozoon, the right and left halves of the body are in a sense comparable with the two young Protozoa, and though each half is hemi-symmetrical, it is in this way the equivalent of a separate organism. This suggestion, which is an old one, receives support from many facts of Meristic Variation, especially from the mode of formation of homologous Twins and "double Monsters" which are now shewn almost beyond doubt, to arise from the division of one ovum[2]. But besides the evidence that each half of the body may on occasion develop into a whole, evidence will be given that one half may vary in its entirety, independently of the other half. Such Variation may be one of sex, taking the form of Gynandromorphy, so well-known among Lepidoptera, in which the secondary sexual characters of one side are male, those of the other being female; or it may happen that the difference between the two sides is one of size, the limbs and organs of one side being smaller than those of the other; or lastly the Variation between the two sides may be one that has been held characteristic of type and variety or even of so-called species and species[3].

These matters have been alluded to here as things which a student of the facts of Variation will do well to bear in mind. It is difficult to see the facts thus grouped without feeling the

[1] Often it is the second cleavage-plane (if any) which corresponds with the future middle line.

[2] The well-known evidence relating to this subject will be spoken of later. The view given above, which is now very generally received, finds support in the striking observations of DRIESCH, lately published (*Zt. f. w. Zool.*, 1891, LIII. p. 160). Working with eggs of *Echinus*, Driesch found that if the first two segmentation-spheres were artificially separated, each grew into a separate *Pluteus*; if the separation was incomplete, the result was a double-monster, united by homologous surfaces. Similar experiments attended by similar results have since been made on *Amphioxus* by E. B. WILSON, *Anat. Anz.*, VII. 1892, p. 732.

[3] Evidence of such abrupt Variation between the two sides of the body belongs for the most part to the Substantive group.

possibility that the resemblance between the two sides of a bilaterally symmetrical body may be in some essentials the same as the resemblance between offspring of the same parent, or to use an inclusive expression, that the resemblance between the members of a Meristic Series may be essentially the same as the resemblance and relationship between the members of one family; that the members of a row of teeth in the jaw, of a row of peas in a pod, of a chain of Salps, or even a litter of pigs, all resulting alike from the processes of Division, may stand to each other in relationships which though different in degree may be the same in kind.

If reason shall appear hereafter for holding any such view as this, the result to the Study of Biology will be profound. For if it shall ever be possible to solve the problem of Symmetry, which may well be a mechanical one, we shall thus have laid a sure foundation from which to attack the higher problem of Variation, and the road through the mystery of Species may thus be found in the facts of Symmetry.

SECTION VIII.

Discontinuity in Substantive Variation: Size.

From the subject of Merism and the thoughts which it suggests, we now pass to another matter. The first limitation by which we proposed to group Variations was found in the characters which they affect: the second relates to the magnitude, or as I shall call it, the **Continuity** of the variations themselves. And though for many a conception has no value till it be cast in some finite mould, my aim will be rather to describe than to define the meaning of the term Continuity as applied to Variation. In dealing with a subject of this obscurity, where the outlines are doubtful, an exact mapping of the facts cannot be made and ought not to be attempted; but I trust that from the present indications, vague though they are, some larger and more definite conception of Discontinuity in Variation may shape itself hereafter by a process of natural growth. For this reason I shall as far as possible avail myself of examples rather than of general expressions, whether inclusive or exclusive.

To those who have studied the recent works of Galton, the conceptions here outlined will be familiar. In the chapter on "Organic Stability" in *Natural Inheritance*, the matter has been set forth with charming lucidity, and what follows will serve chiefly to illustrate the manner in which the facts of Natural History correspond with the suggestions there made.

In the case of most species it is a matter of common knowledge that though no two individuals are identical, there are many which in the aggregate of their characters nearly approach each other, constituting thus a normal, from which comparatively few differ widely. In such a species the magnitude of these differences is proportional to the rarity of their occurrence. Now this, which is a matter of common experience, has been shewn by Galton to be actually true of several quantities which in the case of Man are capable of arithmetical estimation. In the cases referred to it has thus been established that these quantities when marshalled in order give rise to a curve which is a normal curve of Frequency of Error. Taking for instance the case of stature, Galton's statistics shew that for a given community there is a mean stature, and the distribution of the statures of that community around the mean gives rise to a Curve of Error. In this case the individuals of that community in respect of stature form one group. Now in the case of a collection of individuals which can be separated into two species, there is some character in respect of which, when arranged by their statistical method, the individuals do not make one group but two groups, and the distribution of each group in respect of that character cannot be arranged in one Curve of Error, though it may give rise to two such curves, each having its respective mean. For example, if in a community tall individuals were common and short individuals were common, but persons of medium height were rare, the measurements of the Stature of such a community when arranged in the graphic method would not form one Curve of Error, though they might and probably would form two. There would thus be a normal for the tall breed, and a normal for the short breed. Such a community would, in respect of Stature, be what is called *dimorphic*. The other case, in which the whole community, grouped according to the degrees in which they display a given character, forms one Curve of Error, may conveniently be called *monomorphic* in respect of that character. By considering the possible ways in which such a condition of dimorphism may arise in a monomorphic community, one of the uses of the term Discontinuity as applied to Variation will be made clear.

Considering therefore some one character alone, in a species which is monomorphic in respect to that character, individuals possessing it in its mean form are common while the extremes are rare; while if the species is dimorphic the extremes are common and the mean is rare. Now the change from the monomorphic condition to the dimorphic may have been effected with various degrees of rapidity: for the frequency of the occurrence of the mean form may have gradually diminished, while that of the extremes gradually increased, through the agency of Natural Selection or otherwise, in a long series of generations; or on the other hand the diminution in the relative numbers of the mean individuals may have been rapid and have been brought about in a few generations by a few large and decisive changes, whether of environment or of organism.

Referring to the curve of Distribution formed in the graphic method of displaying the statistics, during the monomorphic period the curve has one apex corresponding with the greatest frequency of one normal form, but in the dimorphic period the curve has two apices, corresponding with the comparative frequency of the two extremes, and the comparative rarity of the mean form. The terms Continuous or Discontinuous are applicable to the process of transition from the monomorphic to the dimorphic state according as the steps by which this change was effected are small or large.

The further meanings of Discontinuous Variation will be explained by the help of examples. The first cases refer to Substantive Variation[1], and we may conveniently begin by examining a case of Variation in a character which is easily measured arithmetically.

Among beetles belonging to the Lamellicorn family there are numerous genera in which the males may have long horns arising from various parts of the head and thorax[2]. These horns may be

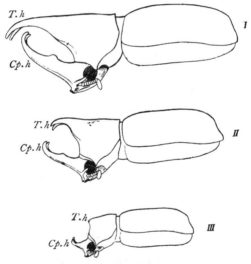

Fig. 1. Side-views of the Lamellicorn beetle, *Xylotrupes gideon*. Legs not represented. I, High male, II, Medium male, III, Low male.

[1] In referring thus to evidence as to Substantive Variation, I find myself in the difficulty mentioned in the Preface. For it is necessary to allude to matters which cannot be properly treated in this first instalment of facts. In order, however, that the one introductory account may serve for all the evidence together, such allusion is inevitable and I can only trust that full evidence as to Substantive Variation may be produced before long.

[2] For particulars of this subject with illustrations, see *Descent of Man*, 1st ed., vol. I. pp. 369—372. A detailed account of this and the succeeding example in the case of the Earwig was given by Mr Brindley and myself in *P. Z. S.*, 1893.

of very great size, as in the well-known Hercules beetle (*Dynastes hercules*) and others. The females of these forms are usually without horns. In such genera it is commonly found that the males are not all alike, but some are of about the size of the females and have little or no development of horns, while others are more than twice the size of the females and have enormous horns. These two forms of male are called "low" and "high" males respectively.

In many places in the Tropics such beetles abound, both "high" and "low" males occurring in the same locality. An admirable example of this phenomenon is seen in *Xylotrupes gideon*, of which a "high," "low," and medium male are shewn in profile in Fig. 1. Of this insect a very large number were kindly given to me by Baron Anatole von Hügel, who collected them at one time, in one locality, in Java. In this species there is one cephalic and one thoracic horn, placed in the positions shewn in the figure. Fig. 1, I shews a "high" male, II is a medium, and III a "low" male. In the gathering received there were 342 males. My friend, Mr H. H. Brindley, has made careful measurements of the lengths of the horns of these specimens and has constructed the diagram, Fig. 2. In this each dot represents an individual, and the abscissæ shew the measurements of the length of the cephalic horn. For clearness these measurements are represented as of twice the natural size. So far as the numbers go the result shews that the most frequent forms are

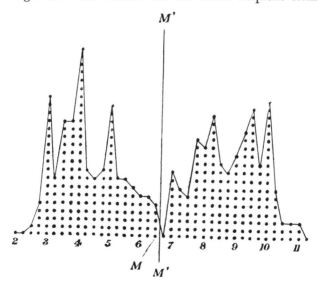

FIG. 2. Diagram representing the frequency of the lengths of cephalic horn in male *Xylotrupes gideon*. M, the mean case; M' the mean value. The abscissæ give lengths of cephalic horn in lines.

the moderately low and the moderately high, the forms of mean measurement being comparatively scarce. It is true that the numbers are few, but so little heed is paid to phenomena of this kind that material is difficult to obtain and the present opportunity was indeed wholly exceptional[1]. But taking the evidence for what it is worth, the comparative scarcity of "medium" males in that particular sample is clear, and so far the form is dimorphic, and has two male normals.

Now such a condition may have arisen in several ways. First, in the past history of the species there may have been a time when the males were horned and were monomorphic, the "medium" form being the most frequent, and the present dimorphic condition may have been derived from this, either continuously or discontinuously as described above for the case of Stature. Secondly, the dimorphism may date from the first acquisition of the horns, and this character may perhaps have always been distributed in the dimorphic way. In this case the term Discontinuous would be applicable to the Variation by which the groups of "high" and "low" males have been severally produced. I am not acquainted with evidence as to the course of inheritance in these cases, and I do not know therefore whether both "high" and "low" males may be produced by one mother. If this should be shewn to be the case, it would suggest that the separation of the males into two groups was a case of characters which do not readily blend, and are thus exempt from what Galton has called the Law of Regression[2].

In the case of a somewhat similar structure found in the Common Earwig (*Forficula auricularia*) the dimorphism is still more definite. In the autumn of 1892 on a visit to the Farne Islands, a basaltic group off the coast of Northumberland, it was found that these islands teem with vast quantities of earwigs. The abundance of earwigs was extraordinary. They lay in almost continuous sheets under every stone and tussock, both among the sea-birds' nests and by the light-keepers' cottages. Among them were males of the two kinds shewn in Fig. 3; the one or high male having forceps of unusual length, the other or low male, being the common form. It appears that the high male is known from many places in England and elsewhere and that it was made into a distinct species, *F. forcipata*, by

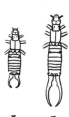

FIG. 3. I, High male, II, Low male of Common Earwig (*Forficula auricularia*) from the Farne Islands.

---

[1] In the Lucanidæ, of which the Stagbeetle (*L. cervus*) is an example, a similar phenomenon occurs, the "high" and "low" males being distinguished by the degree of development of the mandibles. No sufficient number of male Stagbeetles has yet been received to warrant any statement as to the frequency of the various types of males.

[2] *Natural Inheritance*, pp. 88—110.

STEVENS[1] though by later authorities[2] the species has not been retained. A large sample of Earwigs collected in a Cambridge garden contained 163 males of which 5 would come into the high class, but the great abundance of high males at the Farnes seems to be quite exceptional.

With a view to a statistical determination of the frequency of the high and low forms 1000 of these Earwigs were collected by Miss A. Bateson, the whole being taken at random on one day from three very small islands joined to each other at low tide. Of the 1000 specimens 583 proved to be mature males with elytra fully developed, no specimen with imperfect elytra being included in this number[3]. On measuring the length of the forceps to the nearest half mm. and grouping the results in the graphic method the curve shewn in Fig. 4 was produced. The figures on the

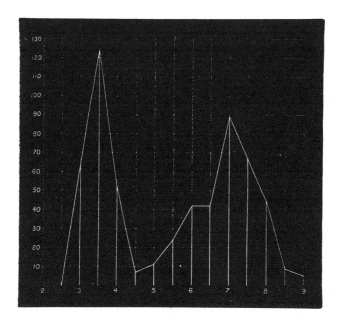

FIG. 4. Curve shewing frequency of various lengths of forceps of male Earwigs (*F. auricularia*) from the Farne Islands. Ordinates, numbers of individuals: abscissæ, lengths of forceps in mm.

ordinates here shew the numbers of individuals, those on the abscissæ giving the length of the forceps in millimetres. As there

[1] STEVENS, *Brit. Ent.* 1835, VI. p. 6, Pl. XXVIII. fig. 4.
[2] FISCHER, *Orthop. Europ.*, 1853, p. 74; BRUNNER VON WATTENWYL, *Prodr. d. europ. Orthop.*, 1882, p. 12.
[3] For particulars in evidence of the maturity of these specimens see *P. Z. S.*, 1893.

shewn the smallest length of forceps was 2·5 mm., and the greatest 9 mm., the greatest frequency being grouped about 3·5 mm. and 7 mm. respectively. The mean form having forceps of moderate length is comparatively rare. The size of the forceps of the females scarcely varies at all, probably less than 1 mm. in the whole sample.

The number of cases is enough to fairly justify the acceptance of these statistics and it is not likely that a greater number of cases would much alter the shape of the curve. Here, therefore, is a group of individuals living in close communion with each other, high and low, under the same stones. No external circumstance can be seen to divide them, yet they are found to consist of two well-marked groups.

Before leaving these examples special attention should be directed to the fact that the existence of a complete series of individuals, having every shade of development between the "lowest" and the "highest" male, does not in any way touch the fact that the Variation may be Discontinuous; for we are concerned not with the question whether or no all intermediate gradations are possible or have ever existed, but with the wholly different question whether or no the normal form has passed through each of these intermediate conditions. To employ the metaphor which Galton has used so well—and which may prove hereafter to be more than a metaphor—we are concerned with the question of the positions of Organic Stability; and in so far as the intermediate forms are not or have not been positions of Organic Stability, in so far is the Variation discontinuous. Supposing, then, that the "high" and "low" males should become segregated into two species—a highly improbable contingency—these two species would have arisen by Variation which is continuous or discontinuous according to the answer which this question may receive.

SECTION IX.

Discontinuity in Substantive Variation: Colour and Colour-Patterns.

From the consideration of Discontinuity in the Variation of a character, size, which may be readily measured arithmetically, we pass to the more complex subject of Discontinuous Variation in qualities which are not at once capable of quantitative estimation. In this connexion the case of colour-variation may be profitably considered. Nature abounds with examples of colour-polymorphism, and in numerous instances such Variation is discontinuous. Of such discontinuous Variation in colour I shall speak under two heads, considering first variations in colours themselves and secondly variations in colour-patterns. As it is not proposed to give the evidence as to Substantive Variation in this volume, a few examples must suffice to shew the use of the term Discontinuity as applied to these Colour-variations.

I. *Colours.* The case of the eye-colour of Man may well be mentioned first, as it has been studied statistically by Galton. In this case the facts clearly shewed that certain types of eye-colour are relatively common and that intermediates between these types are comparatively rare. The statistics further shewed that in this respect inheritance was alternative, and that the different types of eye-colour do not often blend in the offspring. "If one parent has a light eye-colour and the other a dark eye-colour, some of the children will, as a rule, be light and the rest dark; they will seldom be medium eye-coloured, like the children of medium eye-coloured parents.[1]"

Colour dimorphism of this kind is very common among animals and plants. It is well known, for example, among beetles. Several metallic blue beetles have bronze varieties of both sexes, living together in the same locality. A familiar instance of this dimorphism occurs in the common *Phratora vitellinæ*. Again in the Elaterid beetle, *Corymbites cupreus*, there is a similar dimorphism in both sexes, the one variety having elytra in larger part yellow-brown, while the elytra of the other are metallic blue. This blue variety was formerly reckoned a distinct species, *C. æruginosus*. In the latter case I am informed by Dr Sharp, who has had a large experience of this species, that no intermediate between these two varieties has been recorded, and in the case of the *Phratora* the occurrence of intermediates is very doubtful. Another common example of colour dimorphism is seen in *Telephorus lividus*, the "sailor" of "soldiers and sailors." This beetle may be found in large numbers, about half being slaty in colour (var. *dispar*), while the remainder have the yellowish colour which coleopterists call "testaceous." Such instances may be multiplied indefinitely. When the whole evidence is examined it will be found that different colours are liable to different discontinuous variations; as instances may be mentioned black and tan in dogs; olive-brown or green and yellow in birds, &c.[2]; grey and cream-

[1] *Natural Inheritance*, p. 139.

[2] A specimen of the green Ring Parakeet (*Palæornis torquatus*) at the Zoological Society's Gardens was almost entirely canary-yellow in 1890. Since that date it has become more and more "ticked" with green feathers. A Green Woodpecker (*Picus viridis*) is described, having the feathers of the rump edged with red instead of yellow, the normally green feathers of the three lower rows of wing-covers and the back were pointed with yellow. J. H. GURNEY, *Zoologist*, XI. p. 3800. I am indebted to Mr Gurney for the loan of a coloured drawing of this specimen. Another example is described as being entirely canary-yellow, with the exception of a few feathers on the cap, which were purple-red. DE BETTA, *Mater. per una fauna Veronese*, p. 174. For this reference I am indebted to Prof. Newton. Specimen of Common Bunting whitish yellow. EDWARD, *Zool.*, 6492; Sedge Warbler canary-yellow. BIRD, *Zool.*, 3632. The Canary itself is a similar case. An Eel gamboge-yellow. GURNEY, *Zool.*, 3599.

colour in mice and cygnets[1]; red and blue in the eggs of many Copepoda[2], the tibiæ of Locusts[3], the hind wings of the Crimson Underwing (*Catocala nupta*)[4], &c. Another case of blue as a variety of scarlet is the familiar one of the flower of the Pimpernel (*Anagallis arvensis*). Discontinuous colour-variation of this kind is one of the commonest phenomena in nature, but to advance the subject materially it is necessary for a large mass of evidence to be produced. This cannot now be attempted, but in order to bring out the close relation between these facts and the problem of Species I propose to dwell rather longer on one special section of the evidence which must serve to exemplify the rest. The case which I propose to take is that of certain yellow, orange, and red pigments. For brevity I shall present the chief facts in the first instance without comment.

1. *Colias edusa* (Clouded Yellow) is usually orange-yellow, having a definite pale yellow female variety, *helice*, which is not recognized as occurring in the male form. A specimen is figured having the right side *helice* and the left *edusa*. FITCH, E. A., *Entomologist*, 1878, XLI. p. 52, Pl. fig. 11. This was an authentic specimen, for Mr Fitch tells me that it was taken by his son and seen alive by himself.

A specimen having one wing white and the rest orange is recorded by MORRIS, *Brit. But.*, p. 13.

Intermediates between *edusa* and *helice* must be exceedingly rare. OBERTHÜR records two such specimens and says that STAUDINGER took a similar one at Cadiz. For this intermediate he proposes a new name, *helicina*. *Bull. Soc. Ent. Fr.* (5), x. p. cxlv.

[1] In this case I can affirm the alternative character of the inheritance. For several years a pair of swans kept by St John's College, Cambridge, have produced cygnets, some of which have been of the normal grey, while others have been fawn-colour, a condition which Prof. Newton tells me has been thought characteristic of the "Polish" swan, a putative species. None of these cygnets are intermediate in colour, and all acquire the full white adult plumage, but the feet of the fawn-coloured cygnets remain pale in colour. Now the father of these has pale feet and was doubtless himself a fawn-coloured cygnet; the hen is normal. The cock formerly belonged to Dr Gifford, who kindly told me that the cygnets of this bird by a different hen were also thus diverse. A pair of these were given to Sir John Gibbons, who informs me that "from these there has been a brood every year, and always I think *one* of the cygnets has been white or nearly so, the others being of the usual colour." One of Dr Gifford's birds was also given to the late Mrs Gosselin of Blakesware, to whom I am indebted for descriptions of and feathers from several fawn-coloured cygnets which were its offspring. A similar case on the Lake of Geneva is recorded by FAUVEL, *Rev. Zool.*, 1869, p. 334, and another in the Zool. Gardens at Amsterdam, by NEWTON, *Zool. Rec.*, 1869, p. 99.

[2] This is well known to collectors of fresh-water fauna, and I have repeatedly seen the same phenomenon in species of *Diaptomus*, especially *D. asiaticus*, in the lakes of W. Siberia. Among thousands of individuals with red-brown egg-sacs, will often occur a few specimens having the egg-sacs of a brilliant turquoise-blue. In this connexion compare the case of the Crayfish (*Astacus fluviatilis*), which turns scarlet on being boiled, and which, like the Lobster, not uncommonly appears in a full blue variety.

[3] *Caloptenus spretus* with hind tibiæ blue instead of red, DODGE, *Can. Ent.*, 1878, x. p. 105; *Melanoplus packardii*, having hind tibiæ red instead of bluish, BRUNER, *Can. Ent.*, 1885, XVII. p. 18. For reference to these observations I am indebted to COCKERELL, *Ent.*, 1889, XXII. p. 127.

[4] WHITE, *Ent.*, 1889, XXII. p. 51. Compare the fact that in another species of *Catocala* (*C. fraxini*), the Clifden Nonpareil, the hind wings are normally bluish.

A curious specimen, apparently a *male*, having the colour of *helice* was kindly shewn me by Mr F. H. WATERHOUSE. The light marks which in the female are present on the dark borders of the fore-wing are only represented by one minute light mark on each fore-wing.

In most if not all of the *edusa* group of *Colias*, there is a pale aberration of the female, corresponding to the *helice* variety of *edusa*. ELWES, *Tr. Ent. Soc.*, 1880, p. 134. In the same paper is a full account of the geographical distribution of the several species and colour-varieties of *Colias*.

Colias hyale (Pale Clouded Yellow) is normally sulphur-coloured. Nearly white varieties and a variety with the field rich sulphur colour, and the apical marginal patches red, are recorded in several works.

2. *Gonepteryx rhamni* (The Brimstone) is sulphur-yellow in the male, and greenish-white in the female. There is a spot in each wing, and the scales covering this on the upper side are bright orange.

Gonepteryx cleopatra, a S. European species, is like the above in the hind-wings, while the field of the fore-wings is flushed with orange of exactly the tint of that on the spots of *G. rhamni*.

There are several records in entomological literature alleging the capture of "*G. cleopatra*" in Britain, e.g. *Proc. Ent. Soc.*, 1887, p. xliii.

In addition to these there are records of specimens of *G. rhamni* more or less flushed with orange; e.g., a specimen at Aldershot with orange spots on fore-wings as in *cleopatra*, *Proc. Ent. Soc.*, 1885, p. xxiv. Mr Jenner Weir said he had seen a specimen in Ingall's collection, intermediate between *rhamni* and *cleopatra*. *ibid.*

A male of *G. rhamni* taken at Beckenham had the costal margin of each fore-wing broadly but unequally suffused with bright rose-colour or scarlet, and the right posterior wing was marked in like manner. The insect was thus marked when captured. BICKNELL, *Proc. Ent. Soc.*, 1871, p. xviii.

3. *Anthocharis (Euchloe) cardamines* (The Orange Tip), in the male has the fore-wings tipped with orange on both sides, while in the female these orange tips are absent. The field in both is white. In entomological literature are many records of variations in the extent and depth of the orange markings on upper or under side, or both (cp. *Zoologist*, xiii. 4562; *Proc. Ent. Soc.*, 1870, p. ii.; MOSLEY, *Illustrations of British Lepidoptera;* HAWORTH; BOISDUVAL and many others), but with these we are not immediately concerned.

A specimen is figured in which the orange spots were completely represented by yellow. MOSLEY, *Illustrated Brit. Lep.*

The white of the field is replaced by primrose or lemon yellow in several Continental forms. These have been described as species under the names *eupheno, belia, euphenoides, gruneri,* &c.

A local variety of *A. eupheno* is described from Mogador, where it was found common at a little distance from the town. The female was much larger than the type, resembling the male in markings and in shape of the fore-wings. The orange blotch, instead of being confined to tip of the fore-wing as normally, extends to the discoidal spot and is usually bounded by a black band, sometimes suffusing the whole tip of the wing. The colour of the field varies from pure white to pale lemon: the hind-wings are always yellower than in the type, in some specimens being nearly as yellow as those of the male. Mr M. C. Oberthür supplied a specimen from Central Algeria which was intermediate between the type and this variety. LEECH, J. H., *P. Z. S.*, 1886, p. 122.

4. Amongst Lepidoptera the change from red to yellow is very common. A case of *Vanessa atalanta*, having the red partially replaced by yellow, is figured in *Entom.*, 1878, XI. p. 170, *Plate*. Varieties of *Arctia caja*, *Callimorpha dominula*, *C. hebe*, *C. hera*, *C. jacoboeæ*, *Zygæna filipendulæ*, *Z. minos*, &c., with yellow instead of red, are to be seen in many collections. See especially OCHSENHEIMER, *Schm. v. Europa*, 1808, II. p. x, also p. 25, and many other authors. A chalk-pit at Madingley, Cambridge, has long been known to collectors as a locality for the yellow *Z. filipendulæ* (Six-spot-Burnet); see *Ent. Mo. Mag.* XXV. p. 289. In some of these the yellow is tinged with red, but it is commonly a very distinct variety. A variety of the Red Underwing (*Catocala nupta*) with brownish-yellow in the place of the red, is figured by ENGRAMELLE, *Papill. d'Eur.*, Pl. CCCXXII. The evidence relating to this subject is very extensive, and concerns many genera and species besides those named above.

5. *Pericrocotus flammeus* (an Indian Fly-catcher) is grey and yellow in the female, and black and orange-red in the male. The young male is grey and yellow like the female. An adult male is described in which the grey had been fully replaced by black, but the yellow remained, not having been replaced by red. R. G. WARDLAW RAMSAY, *P. Z. S.*, 1879, p. 765. See also LEGGE, *Birds of Ceylon*, I. p. 363, for description of male in transitional plumage.

Curiously enough the change from red to yellow and from light yellow to dark is no less common among plants, though it can scarcely be supposed that the substances concerned are similar.

1. *Narcissus corbularia* and other species are known in sulphur-yellow and in full yellow[1].

2. The Iceland Poppy (*P. nudicaule*) is very common in gardens under three forms, white, yellow and orange. Intermediate and flaked varieties occur, but are less common than the three chief forms. Respecting this species Miss Jekyll of Munstead, who first brought out the varieties, kindly gives me the following information. She writes:—"I began with one plant of the yellow colour that I take to be the type-colour. It was then new as a garden plant, so I saved the seed. The first sowing gave me various shades of orange, as well as the type, in different shades. In the 3rd and 4th years I got buffs, whites, and very pale lemon colourings. As there was only one plant to begin with there was no question of cross-fertilization. A white appeared in the 3rd year of sowing and I kept on selecting for 2 or 3 years......and gave it to a friend in Ireland, who returned it to me 2 years later still more improved. This strong white seems now to be fixed and quite unwilling to revert to the yellow colourings, and is a rather stouter and

---

[1] Mr P. Barr, who has collected these forms in Portugal, tells me that he believes the pale ("*citrina*") varieties of *N. ajax* and *N. corbularia* to be confined to calcareous soils.

handsomer plant altogether." In seedlings from the orange or yellow form grown in separate beds the proportion of seedlings true to their parent colour would not be nearer than about 60 or 70 per cent., but in the case of the white form Miss Jekyll considers that 95 per cent. may be expected to come true.

The yellow Horned Poppy (*Glaucium luteum*) is normally of a lemon yellow very like that of *P. nudicaule*. Of this species also there is an orange cultivated variety. The varieties of the tomato offer a similar series of colour-variations.

3. Fruits of many kinds are known in red and yellow forms. For instance the yellow berried Yew is well known. It is described under the name *Taxus baccata fructu-luteo*, LOUD. "It appears to have been discovered about 1817 by Mr Whitlaw of Dublin, growing in the demesne of the Bishop of Kildare, near Glasnevin; but it appears to have been neglected till 1833 when Miss Blackwood discovered a tree of it in Clontarf churchyard near Dublin. Mr Mackay on looking for this tree in 1837 found no tree in the churchyard, but several in the grounds of Clontarf Castle, and one, a large one, with its branches overhanging the churchyard, from which he sent us specimens. The tree does not differ, either in its shape or foliage, from the common yew, but when covered with its berries it forms a very beautiful object, especially when contrasted with yew trees covered with berries of the usual coral colour." LOUDON, *Arb. et Frut. Brit.*, IV. 1838, p. 2068.

4. The Raspberry (*Rubus idæus*) is another fruit which is known wild in both the red and yellow forms, though the latter is less common. According to BABINGTON, it has pale prickles, and leaflets rather obovate. *Brit. Rubi*, p. 43. (See RIVERS, *Gard. Chron.*, 1867, p. 516.)

Any person who has opportunities of handling animals and plants in numbers can add many similar cases. These few are taken more or less at random, as illustrations of the frequency with which red, orange, and yellow may vary to each other. It is of course not necessary to say that in numerous instances both among animals and plants, the same parts which in one species are yellow, in an allied species or in a geographically distinct race are represented by orange or by red. To an appreciation of the rapidity with which such changes may have come about, facts like the foregoing contribute.

The frequency of such variations suggest that many of these yellow and red pigments are either closely allied bodies or different forms of the same body. Until the chemistry of these substances has been properly investigated nothing can be definitely stated as to this, but the fact that vegetable yellows are very sensitive to reagents is familiar. The lemon variety of the Iceland Poppy treated with ammonia turns to a colour almost identical with that of the orange variety, while the white variety so treated goes primrose yellow. The lemon variety when boiled, or treated with alcohol yields an orange solution, which is of the same tint. This returns to lemon-colour if treated with ammonia or acids. The wings of *G. rhamni* when boiled yield a soluble yellow, which according to HOPKINS (*Proc. Chem. Soc.*, reported *Nature*, Dec. 31, 1891) is a derivative of mycomelic acid, allied to uric acid. This substance turns orange with reagents. The wings of *G. rhamni* turn orange-red when exposed to wet potassium cyanide (*Proc. Ent. Soc.*, 1871, p. xviii) as may be easily seen.

When these facts, meagre though they are, are considered together with the evidence of variability, the suggestion is very strong that the discontinuity between these several characteristic colours is of a chemical nature, and that the transitions from one shade of yellow to another, or from yellow to orange or red is a phenomenon comparable with the changes of litmus and some other vegetable blues from blue to red or of turmeric from yellow to brown. If such a view of these phenomena were to be accepted, it would, I think, be simpler to regard the constancy of the tints of the several species and the rarity of the intermediate varieties as a direct manifestation of the chemical stability or instability of the colouring matters, rather than as the consequences of environmental Selection for some special fitness as to whose nature we can make no guess. For we do know the phenomenon of chemical discontinuity, whatever may be its ultimate causes, but of these hypothetical fitnesses we know nothing, not even whether they exist or no.

II. *Colour-patterns.* Thus far I have spoken only of discontinuous variations in colours themselves, but there are no less remarkable instances of discontinuous variations in the distribution of colours in particoloured forms. By a combination of these modes, variations of great magnitude may occur.

One of the most obvious cases of this phenomenon is that of the Cat. In European towns cats are of many colours, but they nevertheless fall very readily into certain classes. The chief of these are black, tabby, silver-grey and silver-brindled, sandy, tortoiseshell, black and white, and white. Of course no two cats have identical colouring, but the individual variations group very easily round these centres, and intermediate forms which cannot at once be referred to any of these groups are immediately recognized as something out of the common and strange. Yet it is almost certain that cats of all shades breed freely together, and there is no reason to suppose that the discontinuity between the colour-groups is in any way determined by Natural Selection.

Another example may be seen in the Dog-whelk (*Purpura lapillus*). This animal occurs on nearly the whole British coast, wherever there are rocks or even clay hard enough to form definite crevices. Like most littoral animals, the Dog-whelks of each locality differ more or less from those of other localities, and these differences may be differences of size, texture of shell, degree of calcification, amount of "frilling," &c. The peculiarities may be so striking that each individual can at once be recognized as belonging to a given locality, or they may be trifling, and appreciable only when a large number of individuals are gathered. But apart from these differences of form and texture there are a great number of colour-varieties of which the following are the three chief whole-coloured forms, viz. white, dark purple-brown, and yellow. In addition to these there are banded forms, and the bands may be coloured with any two of the three colours mentioned above. Among the banded forms there are two distinct sorts of banding, in the one there are very many fine bands and in the other there are a few broad bands. In most localities these colour-varieties may all be found; though in some places, especially where the water is foul, as at Plymouth, the shells are greatly corroded and the colours, if originally present, are obscured. Speaking however of localities in which colour-varieties are to be seen at all, several may generally be found together. If any one will take the trouble to gather a few hundreds of these shells and will set himself to sort them into groups according to their colours, he will find that the majority fall naturally into groups of this kind; and that those which cannot be at once assigned to groups but fall intermediately between the groups are comparatively few. I have seen this at many places on the English coast; in Yorkshire, Norfolk, Suffolk, Kent, Sussex, Dorsetshire, Devonshire, Cornwall, &c. In several localities I have found pairs belonging to different colour-varieties breeding together, and there is therefore no reasonable doubt that these colour-variations do not freely blend, but are discontinuous.

The statements here made with regard to *P. lapillus* hold in almost the same way for *Littorina rudis*, but in this case the number of colour-types is larger. In *L. rudis* I have occasionally seen specimens of which the upper part belonged to one colour-type, and the lower to another, the transition occurring sharply at one of the varices. In these cases the shell appears to have been injured and is possibly renewed.

One of the commonest British Lady-birds (*Coccinella decempunctata*) is an extremely variable form. A great number of its varieties may be found together, ranging from forms with small black spots on a red field to forms in which the field is black with a few red spots. But in spite of the great diversity there are certain types which are again and again approached, while the intermediates are comparatively scarce.

The following case, well known to entomologists, may be mentioned here. The Painted Lady (*Pyrameis cardui*) is found in the typical form over the entire extent of every continent, with the exception of the Arctic regions and possibly S. America. A special form of it (var. *kershawi*) is found in Australia and New Zealand, but the other large islands south of Asia possess the normal type. The latter is also found in the Azores, Canaries, Madeira and St Helena. This butterfly has been taken on the snow-level in the Alps; and in N. America, though it may be regarded as one of the commonest butterflies in the elevated central district, it is most abundant at a level of 7000—8000 feet. It has been taken on Arapahoe Peak, between 11,000 and 12,000 feet (from SCUDDER, *Butterflies of N. America*, I. pp. 477—480). Of this insect, which is a very constant one, a certain striking aberration has been found, always as a great rarity, in many lands. In this aberration the markings are almost entirely rearranged. It is said to have been first described by RAMBUR under the name var. *Elymi*, but this description I have never found. (The reference quoted is *Annales des Sci. d'observation*, Paris, 1829, Vol. II. Pl. v.) As often happens with Variation, without coloured figures description is almost useless, but the figures referred to are very accessible. In a British specimen of this aberration the white bars are absent from the anterior costæ and a series of white fusiform blotches are present along the marginal border; two abnormal white spots are also present near the anal angle, thus continuing the series down the wing (*fig.* 5, A.). The hind-wings are equally aberrant. The two large dark spots which are usually on the disk between the median nervure and the inner margin are altogether wanting. Between each of the nervures of the hind-wing is a white spot, whereas in the normal form there is no white spot at all on the hind-wings. These white spots on the hind-wings form a row parallel to the border of the wing and, as it

FIG. 5. A. Clark's specimen of *P. cardui*, var. *elymi* from *Ent.* 1880.
B. Newman's specimen. *Brit. But.*, p. 64.
C. *P. cardui*, normal, also from NEWMAN. *Brit. But.*, p. 64.

were, continue the series of white spots borne by the anterior wings. [Underside not described.] This specimen was reared from a larva found near the river Lea, Clapton Park. CLARK, J. A., *Entomologist*, 1880, XIII. p. 73, *fig*. A coloured figure of the same specimen, MOSLEY, S. L., Pl. 8, *fig*. 3.

A form very closely similar to the above is figured in black and white by Newman from a specimen in Ingall's collection (*fig*. 5, B). [This is apparently the specimen given in *Zoologist*, p. 3304.] NEWMAN, *British Butterflies*, p. 64, *fig*. A British specimen which nearly approaches this aberration in the absence of the white bars on the costæ and in the absence of the black transverse bar is recorded. In it each of the sub-marginal rows of black spots on the posterior wings is drawn, containing a white spot. In this specimen the brown-red of the type was represented by rose-colour. NEWMAN, *Entomologist*, 1873, p. 345, *fig*.

Another specimen closely resembling this aberrant form is described from New South Wales. OLLIFF, A. S., *Proc. Linn. Soc.*, N. S. W., S. 2, III. p. 1250.

Another specimen closely resembling the above was taken at Graham's Town, S. Africa, and is mentioned by JENNER WEIR, *Entomologist*, 1889, XXII. p. 73.

Another specimen is figured in which the hind-wings are marked as in the above, but the anterior wings, though strongly resembling this aberration in the general disposition of the colours, yet differ in details, the chief points of difference being that the white costal bar is only partially obliterated and the white spots on the anal angles of the fore-wings are not developed.

[This specimen was in Kaden's collection and was presumably European.] HERRICH-SCHÄFFER, Bd. I. p. 41, *Pl*. 35, figs. 157 and 158.

A description is given of an aberrant form taken at King William's Town, S. Africa, which "closely resembled that figured by Herrich-Schäffer." TRIMEN, R., *South-African Butterflies*, I. p. 201.

A specimen (British) resembling the above, but lacking the white spots on the anal angles of the fore-wings and having the marginal row on the hind-wing light-coloured, but not quite white, is figured by MOSLEY, Pt. III. Pl. 3, *fig*. 3.

Two specimens were taken in New Jersey, U.S.A., which are stated to have conformed to this aberration. STRECKER, *Cat. N. Amer. Macrolepidop.*, p. 137.

Another British specimen generally resembling Herrich-Schäffer's figure is represented by MOSLEY, Pl. 8, fig. 4.

In all the above specimens the resemblance, as far at least as the upper surface is concerned, is considerable. With the exception of Herrich-Schäffer's example, the undersides are not figured, but from the descriptions it may be gathered that they also resembled each other though probably not so closely as the upper surfaces. The resemblance between the underside of the Australian specimen and that figured by Herrich-Schäffer must have been very close.

"Intermediate between these extreme sports and the normal form are three examples taken at Cape Town in 1866, 1873 and 1874—the first by myself—in which the fore-wing markings are scarcely affected, and the hind-wing spots are minutely ocellate and externally prolonged, so as to be confluent with the succeeding row of lunules." TRIMEN, *ibid.* pp. 201, 202.

Another aberration, a Belgian specimen, resembles "*Elymi*" in kind but differs from it in degree. In it also the white bars are absent from the costæ, and the brown and black markings of the anterior wings are rearranged in almost exactly the same manner. The posterior wings are modified to a much less extent and the normal row of black spots between the nervures remains, while only the first and second of the series of white spots is present, the former being very slight. In this individual the markings of the underside also resemble the aberration generally, but it retains the four ocelli of the type. DE DONCEEL, H. DONCKIER, *Ann. Soc. d'Ent. Belge*, 1878, XXI. p. 10, *Plate*.

A specimen, also Belgian, is described in which the two anterior wings resemble Herrich-Schäffer's figure in lacking the white bars on the costæ and in the arrangement of the black and ground colour. In neither of them are the white spots of the anal angles (found in the British and Australian specimens) present. The white markings at the apex of the anterior wings differ on the two sides, being in both of them unlike the type and an approach to the aberrations in question, but the degree to which they are developed differs markedly, being greatest on the right side. The *left* posterior wing resembles the aberration in having the six abnormal white spots, but less emphasized than in the figures quoted above; in general colour this wing is darker than the type. The *right* posterior wing, however, has none of the white spots of the aberration, and differs from the type only in being more suffused with

4—2

black. To recapitulate, the two anterior and the left posterior wing resemble *generally*, though not entirely, the aberration, while the right posterior wing is nearly normal.

A specimen is described from Ekaterinoslav, S. Russia, which resembles this aberration in wanting the black transverse band and in the disposition of the apical white spots. A trace of the white costal bar remains on the costal border. On the underside of this specimen the ocelli were placed in a pale rose-coloured band. (Name proposed, aberration, *inornata*). BRAMSON, K. L., *Ann. Soc. Ent. France*, S. 6, VI. 1886, p. 284.

Besides the rare aberration "var. *Elymi*," there is a variety sometimes found in Europe, which in Australia is so constant and definite that it has been regarded as a species. The following may be quoted respecting its occurrence in Australia, where it is common :

"There is in abundance about Melbourne and in many other parts of Australia a *Cynthia* with the general appearance and habit of *C. cardui*, so closely represented that every entomologist I know refers it to that species. The Australian species differs from the European one constantly, however, in having the centres of the three lower round spots on the posterior wings bright blue, and having two other blue spots on the posterior angles of the same wings, the corresponding parts of the European form being black." For this form the name *C. kershawi* is proposed. M'COY, F., *Ann. and Mag. of Nat. Hist.*, Ser. 4, I. 1868, p. 76. See also OLLIFF, A. S., *Proc. Linn. Soc., N. S. W.*, Ser. 2, III. p. 1251. The notices of its occurrence in Europe are as follows. In 1884 Mr Jenner Weir exhibited a specimen of *P. cardui*, taken in the New Forest. Three of the five black spots in the disk of the upper side of the hind-wings had blue pupils ; he pointed out that the specimen thus approached the Australian form, *P. kershawi*. *Proc. Ent. Soc.*, 1884, p. xxvii.

OLLIFF, *loc. cit.*, states that he has taken a specimen having these blue markings at Katwijk, in Holland.

In the case given, the evidence certainly suggests that these various forms of aberration are grouped round a normal form of aberration, just as the individuals of the type are grouped round its normal.

One example of a similar discontinuity in a melanic variation may profitably be given. I have taken this opportunity of referring to such a case, as the general evidence of melanic variations goes on the whole to shew that they are not commonly discontinuous, and further evidence on this point would be most valuable. To appreciate the evidence BUTLER'S coloured plate should be referred to.

Terias. A well-marked group of butterflies of this genus allied to *T. hecabe*, is found in Japan. It contains forms of great diversity in amount of black border which occurs on the outer margins of the fore- and hind-wings. The remainder of the wings is lemon-yellow. The black border may be confined to the tip of the fore-wings, or may there occupy a considerable area and be extended along the whole outer margin of both wings. The form with the least black is called *T. mandarina*, that with the most, is called *T. mariesii*, and the intermediate form is called *T. anemone*. Upwards of 150 specimens, all from Nikko, were examined; these ranged between the two extremes, and were found to form a continuous series. Butler states that "the absence of six of them, referable only to two gradations, would at once leave the three species as sharply defined as any in the genus."

[In the case of these butterflies, there are thus three groups of varieties, two extreme groups and one mean group ; intermediates between these are comparatively rare. Butler suggests that these intermediate forms should be regarded as hybrids, even in the absence of experimental evidence. This view is of course dependent on the truth of the belief that such a discontinuous occurrence of variations is anomalous.]

Twenty specimens of the species *T. betheseba* and thirty-nine of *T. jaegeri* (both from Japan), were also examined. The former presented no variations whatever, and the latter only vary in the yellower or redder tint on the under surface of the secondaries. BUTLER, A. G., *Trans. Ent. Soc.*, 1880, p. 197, Pl. VI.

Compare the following:

Terias constantia. Twenty-five pupæ, all found together on the same twigs at Teapa, Tabasco, Mexico, by Mr H. H. Smith. The butterflies from these are in Messrs Godman and Salvin's collection, who kindly allowed me to examine them. The amount of black border on both wings varies much, nearly though not quite so much as in the cases figured by BUTLER. In the lightest the apex of the fore-wing alone is black, and there is no black on the hind-wing in 9 specimens; of the remaining 16 some have a well-defined black border to the hind-wing, while in the rest (about 6) this border is slight. This case is a particularly interesting one, as the specimens were associated and presumably belonged to one brood.

For another beautiful case of discontinuous Variation in pattern I am indebted to Dr D. Sharp. The Cambridge University Museum lately received a series of 38 specimens of *Kallima inachys*, the well-known butterfly whose folded wings resemble a dead leaf with its mid-rib and veinings. The underside of this butterfly is sometimes marked with large blotches and flecks of irregular shape, which, as has often been noted, resemble the patches of discoloration caused by fungi in decaying leaves. Dr Sharp pointed out to me that the specimens examined fell naturally into four groups according to the coloration of the underside. In the first group the field is nearly plain, though the tint varies in individuals. The "mid-rib" is strongly marked in this and all the groups, but the "veinings" are absent or very slightly marked in the first group: 18 specimens. In the second group the ground is almost plain, but it bears numerous strongly marked black-speckled spots, of forms which though irregular in outline are closely alike, and occupy the same positions in all the six specimens, being scarcely if at all represented in any of the others. In the third group the dark bars representing "veins" are strong, but the field is nearly uniform: 10 specimens. In the fourth group, of four specimens, the ground-colour is darkened in such a way as to leave large and definite blotches of light colour in particular places. Of these specimens three have the veinings very strongly marked, while the fourth is without them.

Into these four groups the specimens could be unhesitatingly separated, though in each group many individual differences occurred. No marked variation in the upper-sides was to be seen. These specimens were all from the Khasia hills, Assam, but there was of course no evidence that all were flying together.

One of the most interesting examples of discontinuous Variation in colour-patterns is the case of ocellar markings or eye-spots. Upon this subject nothing need here be said as the evidence will be given in detail in the course of this volume (see Chap. XIII.).

SECTION X.

DISCONTINUITY IN SUBSTANTIVE VARIATION.—MISCELLANEOUS EXAMPLES.

Of the discontinuous occurrence of Substantive Variation, the manifestations are many and diverse. We have seen that in such features as size, colour, and colour-patterns, Variations may be discontinuous, and a form may thus result, differing markedly from the type which begot it. Variation in the proportions or the constitution of essential parts may no less suddenly occur. The range of these phenomena is a large one, but for the purposes of this Introduction a few examples must suffice in general illustration of their scope.

A discontinuous variation which is familiar to all is that of "reversed" varieties, especially of Molluscs and Flat-fishes. Such varieties are formed as optical images of the body of the type. In both of the groups named, some species are normally right-handed, others being normally left-handed, while as individual variations reversed examples are found. In Molluscs this is not peculiar to Gasteropods with spiral shells, but may occur also both in Limacidæ (slugs)[1] and in Lamellibranchs[2]. Such variation is commonly discontinuous, and the two conditions are alternative. The fact that the reversed condition may become a character of an established race is familiar in the case of *Fusus antiquus*. This shell is found in abundance as a fossil of the Norwich Crag, such specimens being normally left-handed, though the same species at the present day is a right-handed one. Of the left-handed form a colony was discovered by MACANDREW on the rocks in Vigo Bay[3]. It was there associated with certain other shells proper to the Norwich Crag. This discovery seemed to Edward Forbes to be so remarkable that he looked on it as corroborative evidence of a special connexion between the fauna of Vigo Bay and the Crag fossils[3]. Jeffreys had the same variety from Sicily[4].

[1] For example, a sinistral *Arion*, BAUDON, *Jour. de Conch.*, XXXII. 1884, p. 320, and many others.
[2] Sinistral *Tellina*, FISCHER, P., *Jour. de Conch.*, XXVIII. 1880, p. 234. The same is recorded in several other genera.
[3] Seven specimens, *Ann. N. H.*, 1849, p. 507.
[4] *Brit. Conch.*, I. p. 326.

That they may the better serve to bring out the significance of Discontinuity in Variation to the general theory of Descent, it may be well to choose some examples with reference to characters which when seen in domestic animals are looked on as especially the result of Selection.

In exoskeletal structures several of this kind are known. From time to time there have been records of captures of the "hairy variety" of the Moorhen (*Gallinula chloropus*), in which the feathers were destitute of barbules and consequently had a hairy texture, greatly changing the general appearance of the bird.

Of the "hairy" variety twelve specimens were recorded, five from Norfolk, and the rest from Cambridgeshire, Hampshire, Sussex (2), Suffolk, Nottinghamshire and Athlone in Ireland. The tips of the barbs and shafts of the feathers have been broken off and the barbules are entirely wanting, giving a hairy appearance. This appearance was found in the whole of the plumage. Owing to the absence of barbules, the general coloration is tawny. A few feathers of this kind have been found in Hawks and Gulls, and in the case of a *Parra* (a bird which bears considerable resemblance to a Moorhen), lent to Mr Gurney by Professor Newton, a great portion of the body feathers were in this condition. The feathers of the *Apteryx* and Cassowary are also partially destitute of barbules. Mr Gurney was informed of a single case of a Grey Brahma hen which shewed the same peculiarity which appears otherwise to be without parallel. The case of the Silky Fowl is similar in the absence of most of the barbules, but in it the point of the shaft is produced to a delicate point, and the barbs are fine and sometimes bifid or trifid at the apex. From J. H. GURNEY, *Trans. Norwich Nat. Soc.*, III. p. 581, *Plate.* [Bibliography given.] [If another "hairy" Moorhen is found, note of the colour of the skin and bones should be made, for, as is well known, in the Silky Fowl they are purplish blue.]

The following may be compared : "Cochins are now and then met with in which the webs of the feathers having no adhesion, the whole plumage assumes a silky or flossy character like that of the Silky Fowl. It usually occurs quite accidentally, and in every case we have met with, the variety has been Buff. By careful breeding the character can be transmitted, but we have only known *one* case in which there had been this hereditary character, the others having been of accidental occurrence. Such birds are sometimes called 'Emu' fowls." LEWIS WRIGHT, *Illust. Book of Poultry*, 1886, p. 230.

Of many domestic animals, for example, the goat, cat and rabbit, varieties with long, silky hair are familiar under the name of "Angoras." Very similar breeds of guinea-pigs are kept, to which the name "Peruvian" is given. In this connexion the capture of a mouse (*Mus musculus*) with long, black, silk-like hair is interesting[1], as shewing that such a total variation may occur as a definite phenomenon without Selection.

[1] COCKS, W. P., *Trans. Cornwall Polytech. Soc.*, 1852. Like other animals, mice have of course often been found black. For instance, a number of black mice were found in Hampstead-down Wood. HEWETT, W., *Zool. Jour.* IV. p. 348.

As to the partial nakedness of the skin of many animals (Man, &c.), several suggestions have been made. It has been variously supposed that the covering of hair has been gradually lost by Man, in correlation with the use of clothes; with the heat of the sun; for ornamental purposes under sexual selection[1]; or perhaps as a protection from parasites[2]. Various suggestions have also been made to explain the persistence of hair at the junction of the limbs and on the head and face. To a consideration of the origin of nakedness, the evidence of Variation in some measure contributes, and though the bearing is not very direct, it may illustrate the futility of inquiries of this kind made without regard to the facts of Variation.

MOUSE (*Mus musculus*): male and pregnant female found in a straw-rick at Taplow; both were entirely naked, being without hairs at all, excepting only a few dark-coloured whiskers. The skin was thrown up into numerous prominent folds, transversely traversing the body in an undulating manner. This condition of the skin obtained for them the name of "Rhinoceros mice." The ears were dark or blackish, the tail ash-coloured, and the eyes black, indicating that they were not albinos. The exfoliations from the skin were examined microscopically but no trace of hair-follicles was found, nor any suggestion of disease. The animals were active and healthy.

The young ones, when born, were similar to the parents. The teeth were normal.

In the Museum of the College of Surgeons is a precisely similar specimen which was found in a house in London. GASKOIN, *Proc. Zool. Soc.*, 1856, p. 38, *Plate*.

Three specimens of the common Mouse (*Mus musculus*) were caught in the town of Elgin. The whole bodies of these three creatures "were completely naked—as destitute of hair and as fair and smooth as a child's cheek. There was nothing peculiar about the snout, whiskers, ears, lower half of the legs and tail, all of which had hair of the usual length and colour. They had eyes as bright and dark as in the common variety......At least two others were killed in the same house where these were found." GORDON, G., *Zoologist*, 1850, VIII. p. 2763.

SHREW. (*Sorex* sp.) "whole of upper surface of head and body destitute of hair, and skin corrugated like that of Naked Mice figured in *P. Z. S.*, 1856;" sent to Brit. Mus. by Mr P. Garner. GRAY, J. E., *Ann. and Mag. of N. H.*, 1869, S. 4, IV. p. 360.

In connexion with these cases, the following fact is interesting:

Heterocephalus is a genus of burrowing rodent from S. Africa. It contains two species, of which one is about the size of a mouse and the other is rather larger. They are characterized by possessing an apparently hairless skin which is on the head and body of a wrinkled and warty nature. On closer inspection the skin is seen to be furnished with fine scattered hairs, but there is no general appearance of a hairy covering. There is no external ear in these animals. OLDFIELD THOMAS, *P. Z. S.*, 1885, p. 845, *Plate LIV*.

Naked horses have often been exhibited. Such a horse caught in a

[1] C. DARWIN, *Descent of Man*, I. p. 142.
[2] BELT, *Naturalist in Nicaragua;* see also HUDSON, *Naturalist in La Plata*, 1892.

semi-feral herd in Queensland was described by TEGETMEIER, *Field*, XLVIII. 1876, p. 281. The skin was black and like india-rubber. Careful examination shewed no trace of hair, or any opening of a hair-follicle. In Turkestan, in the year 1886, I heard of one thus travelling, but failed to see it. 'Hairless' dogs in S. America remain distinct (BELT, *l. c.*).

Of discontinuous Substantive Variation in bodily proportions a single example must suffice. Among domestic animals of many kinds, races are known in which the bones of the face do not grow to their full size, while the bones of the jaw are, or may be, of normal proportions. Familiar examples of this are the bull-dog, the pug, the Japanese pug, the Niata cattle of La Plata[1], some short-faced breeds of pigs, and others. In the case of these domestic animals the part which Selection has taken in their production is unknown, and the magnitude of the original variations cannot be ascertained. It is nevertheless of interest to notice that parallel variations have occurred in distinct forms, and I think that this is to some extent evidence that the variations were from the first definite and striking. As regards the dogs even, there is a presumption that the short face of at least the Japanese pug arose independently from that of the common, or Dutch pug (as it used to be called), but as to this the evidence is insufficient. Among the dogs' skulls found in ancient Inca interments, a skull was found having the form of the bull-dog. NEHRING, *Kosmos*, 1884, XV. As these remains belong to a period before the European invasion, it is most probable that this bull-dog breed arose independently of ours.

Apart however from domestic animals there is evidence as to the origin of short-faced breeds. This evidence, which is not so well-known as it deserves to be, is provided by the occurrence of a similar variation in fishes. Darwin in speaking of the evidence as to Niata cattle makes allusion to the case of fishes in a note[2], quoting WYMAN as to the cod, which occurs in a form known to fishermen as the "bull-dog" cod. The interest of this observation is increased by the fact that it does not stand alone, but similar variations have been seen in the carp, chub, minnow, pike, mullet, salmon and trout. In the last-named there is even evidence of the establishment of a local race having this singular character.

CARP (*Cyprinus carpio*). "Bull-dog"-headed Carp have often been described. The face ends more or less abruptly in front of the eyes, while the lower jaw has almost its normal length. The front part of the head is bulging and prominent, giving the fish an appearance which several authors compare to that of a monumental dolphin. A good figure of such a specimen is given by G. ST HILAIRE, *Hist. des Anom.*, ed. 1837, I. p. 96, where a full account of the older literature of the

[1] C. DARWIN, *Animals and Plants under Domestication*, 2nd edition, I. p. 92.
[2] *Ibid.*, p. 93, *note*.

subject may be found. Inasmuch as carp are largely bred in ponds on the continent, there is in this case some suggestion that unnatural conditions may be concerned, but this suggestion does not apply to other cases of the same Variation. OTTO, *Lehrb. path. Anat.*, I. § 129, states that in the ponds of Silesia such fish are not rare. See also VOIGT, *Mag. f. d. Naturk.*, III. p. 515.

Cyprinus hungaricus: specimen from the Danube similarly formed. The forehead was protuberant and bulged in front of the eyes so that its anterior border was almost vertical. The attachments of the mandible are carried forward in such a manner that the mandible itself was directed upwards almost at right angles to the body. [Good figure.] STEINDACHNER, *Verh. zool.-bot. Ges. Wien*, 1863, XIII. p. 485, Plate.

[Several other types of Variation in the heads of Cyprinoids occur, but cannot be described here.]

CHUB (*Leuciscus dobula = cephalus*): specimen having anterior part of head rounded "like a monumental dolphin." The body was normal, measuring 33 cm. in length. LANDOIS, *Zool. Garten*, 1883, XXIV. p. 298.

MINNOW (*Phoxinus lævis*) specimen having a snout like a pug ("*museau du mopse*") [no description]. LUNEL, *Poiss. du lac Léman*, p. 96.

MULLET (*Mugil capito*): specimen having both jaws directed upwards, and the upper and anterior parts of the skull greatly elevated and protuberant: the appearance of the head was like that of a pug dog. Full measurements given. CANESTRINI, R., *Atti della soc. Ven.—Trent. di. sci. nat. in Padova*, 1884, IX. p. 117 [Bibliography given].

PIKE (*Esox lucius*) described as like a pug, *ibid.*, p. 124; see also VROLIK's *Atlas*, 1849, Tab LXI. *fig.* 6.

SALMON (*Salmo salar*): specimen having front part of face little developed, the supra-maxillaries being asymmetrical. Lower jaw projects far in front of upper jaw. Animal of fair size, and not meagre. VAN LIDTH DE JEUDE, *Notes from Leyden Mus.*, VII. p. 259, Plate. [Curious malformation of *S. trutta* ibid.], see also *Jahrb. Ver. vaterl. Nat. Württ.* XLII. p. 345.

TROUT (*S. fario*): several specimens having bull-dog heads were taken in Lochdow, near Pitmain, Inverness-shire. Heads short and round; upper jaw truncated like a bull-dog. This variety does not

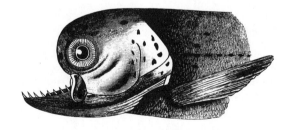

FIG. 6. Bull-dog-headed TROUT after CARLET.

occur in neighbouring lochs. None weighed more than ½ lb. YARRELL, *Brit. Fishes*, I. p. 286, figure given.

Another specimen (Fig. 6), agreeing closely with Yarrell's figure, was taken in a lake at an altitude of over 6000 ft. in the valley of Sept-Laux (Isère). Saving the head it was in all respects normal. This specimen is described and figured by CARLET, M. G., *Journ. de l'Anat. et Phys.*, 1879, xv. p. 154. [It is declared that the fishermen who took it, having previously met with similar specimens, supposed that they had found a new species, but it is not expressly stated that these other specimens were from the same locality.]

Before ending this preliminary glance at Discontinuity in Substantive Variation, allusion must be made to a case which is at once more famous and more instructive than any other. I refer to the celebrated phenomenon of the production of nectarines by peaches, or conversely. Upon the subject of almond, peach and nectarine, Darwin produced a body of facts which, whether as an example of a method or for the value of the facts themselves, form perhaps the most perfect and the most striking of all that he gave.

The evidence which is there collected is known to all, and though similar observations have been made since by many, there is I believe nothing of importance to add to Darwin's statement. The bearing of these phenomena on the nature of Discontinuity in Variation is so close that Darwin's summary may with profit be given at length.

"To sum up the foregoing facts; we have excellent evidence of peach-stones producing nectarine-trees, and of nectarine-stones producing peach-trees—of the same tree bearing peaches and nectarines—of peach-trees suddenly producing by bud-variation nectarines (such nectarines reproducing nectarines by seed), as well as fruit in part nectarine and in part peach,—and, lastly, of one nectarine-tree first bearing half-and-half fruit and subsequently true peaches"[1]. After disposing of alternative hypotheses he concludes that "we may confidently accept the common view that the nectarine is a variety of the peach, which may be produced either by bud-variation or from seed."

In this case the evidence is complete. The variation from peach to nectarine or from nectarine to peach may be *total*. If less than total, the fruit may be divided into either halves or quarters[2], so that for each segment the Variation is total still. Of intermediate forms other than these divided ones, we have in this case

---

[1] *Animals and Plants under Domestication*, ed. 2, I. p. 362.
[2] *Ibid.*, p. 362, quoting from *Loudon's Gard. Mag.* 1828, p. 53. The case of a Royal George peach which produced a fruit, "three parts of it being peach and one part nectarine, quite distinct in appearance as well as in flavour." The lines of division were longitudinal.

no evidence: it is therefore a fair presumption that they are either rare or non-existent; and that the peach-state and the nectarine-state are thus positions of "Organic Stability," between which the intermediate states, if they are chemical and physical possibilities, are positions of instability.

These examples of Discontinuity in Substantive Variation must suffice to illustrate the nature of the phenomena. It will be seen that the matters touched on cover a wide range, and the evidence relating to them must be considered separately and at length. Such a consideration I hope in a future volume to attempt.

SECTION XI.

DISCONTINUITY IN MERISTIC VARIATION: EXAMPLES.

Inasmuch as the facts of Meristic Variation form the substance of this volume, it is unnecessary in this place to do more than refer to the manner in which they exhibit the phenomenon of Discontinuity. One or two instances must suffice to give some suggestion of this subject, detailed consideration being reserved.

Parts repeated meristically form commonly a series, which is either radial or linear, or disposed in some other figure derived from or compounded of these. For the purpose of this preliminary treatment an instance of Discontinuous Variation in each of these classes may be taken.

1. *Radial Series.*

Variations in the number of petals of actinomorphic flowers exhibit the Discontinuity of Meristic Variation in perhaps its simplest form.

Phenomena of precisely similar nature will hereafter be described in animals, but such variations in flowers are so common and so accessible that reference to them may with profit be made. In Fig. 7 such an example is shewn.

It represents a Tulip having the parts of the flower formed in multiples of four, instead of in multiples of three as normally. Variation of this kind may be seen in any field or hedgerow[1].

Meristic Variation is here presented in its greatest simplicity. Such a case may well serve to illustrate some of the phenomena of Discontinuity.

---

[1] For full literature and lists of cases see especially MASTERS, *Vegetable Teratology*, s. v. *Polyphylly*. It is perhaps unnecessary to refer to the fact that the numerical changes here spoken of are quite distinct from those which result from an assumption by the members of one series or whorl of the form and characters proper to other whorls.

A form with four segments occurs as the offspring of a form with three segments. Such a Variation, then, is discontinuous

FIG. 7. Diagram of the flower of a Tulip having all the parts in -4.

because a new character, that of division into four, has appeared in the offspring though it was not present in the parent. This new character is a definite one, not less definite indeed than that of division into three. It has come into the strain at one step of Descent. Instances in which there is actual evidence of such descent are rare, but there can be no question that these changes do commonly occur in a single generation, and, indeed, in many plants, as for example *Lysimachia* (especially *L. nemorum*), flowers having all the parts in -4 or in -6 may be frequently seen on plants which bear likewise normal flowers with the parts in -5.

Now such a variation as this of the Tulip illustrates a phenomenon which in the Study of Variation will often be met.

We have said that the variation is discontinuous, meaning thereby that the change is a large and decided one, but it is more than this; it is not only large, it is *complete*.

The resulting form possesses the character of division into four no less completely and perfectly than its parent possessed the character of division into three. The change from three to four is thus perfected: from the form with perfect division into three is sprung a form with perfect division into four. This is a case of a *total* or *perfect* Variation.

This conception of the totality or perfection of Variation is one which in the course of the study will assume great importance, and it may be best considered in the simple case of numerical and Meristic Variation before approaching the more complex question of the nature of totality or perfection in Substantive Variation.

The fact that a variation is perfect at once leads to the question as to what it might be if imperfect. Between the form in -3 and the form in -4 are intermediates possible? and if possible, do they exist? Now by choosing suitable species of regular flowers, individual flowers may no doubt be found in which there are three large segments and one small one; or two normal segments and a third divided into two, making four in all. Such flowers are firstly rare, while cases of perfect transformation are common. But besides their rarity there is, further, a grave doubt whether they are in any true sense *intermediate* between the perfect form in -3 and the perfect form in -4. After this again it must be asked whether or no they do as a matter of fact occur as intercalated steps in the descent of the form in -4 from the form in -3? To the last question a general negative may at once be given; for though there is abundant evidence that Meristic Variations of many kinds and in several degrees of completeness may be seen in the offspring of the same parent, yet any one member of such a family group may shew a particular Variation in its perfection, and the occurrence of any intermediate in the line of Descent is by no means necessary for the production of the perfect Variation.

To answer the former question, whether or no forms imperfectly divided into four parts are in reality intermediate between those in -3 and those in -4, a knowledge of the mechanics of the process of Division is required. Such knowledge is as yet entirely wanting, and discussion of this matter must therefore be premature. With much hesitation I have decided to make certain reflexions on the subject, which will be found in an Appendix to this work. These may perhaps have a value as suggestions to others, though from their theoretical nature they can find no place here.

There is however another class of cases which are intermediate in a different way. In the Tulip described above the quality of division into 4 was present in all the floral organs. This is not always the case, for a Meristic Variation may be present in one series of organs, though it is absent in some or all of the others, and this is a phenomenon frequently recurring. Nevertheless, though only partially distributed, a Variation may still be displayed in its totality in the parts wherein it is present. The parts of a single whorl, the calyx for example, may undergo a complete Variation, while the corolla and other parts are unchanged. In the same way single members of a radial series, as a petal for example, may undergo a complete Variation while the other members of the series are unchanged. The same will be shewn hereafter to be true of animals also.

For instance, the normal number of the parts in the disc of *Aurelia* is four, but the whole body may be divided instead into six or some other number of parts. Examples are also found in which the parts of one-half or of one quadrant are arranged in the new number, while the remainder is normal; and, as in flowers, this new number may prevail in some or in all of those systems of organs which are disposed around the common centre.

2. *Linear Series.*

Before speaking further of the totality or perfection of Variation it will be well to give an illustration of Discontinuous Meristic Variation as it occurs in the case of a linear series of parts. As such an illustration the case of the variation in the number of joints in the tarsus of the Cockroach (*Blatta*) may be taken. This variation has been the subject of very full investigation by Mr H. H. Brindley. The tarsus of the Cockroach is normally divided into five joints, but in about 25 per cent. of *B. americana* (and in a smaller proportion of several other species) the tarsus of one or more legs is divided into only four joints, though the total length may be the same as that of the corresponding leg of the other side, Fig. 8. Between the five-jointed form of tarsus and the four-jointed form no single case in any way intermediate was seen. The whole

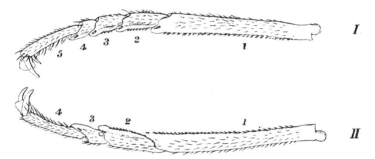

Fig. 8. Tarsi of the third pair of legs in a specimen of *Blatta americana*, I. the left tarsus, having the normal, or 5-jointed form; II. the right tarsus, having the 4-jointed form.

evidence will be given in full in the proper place and raises many questions of great interest; but that which is important to our present consideration is the fact that the Variation is here undoubtedly discontinuous, arising suddenly as a total or perfect Variation, from the five-jointed form to the four-jointed. Here the variation, though total as regards the limb in which it is present, is not total as regards all the legs taken together. For commonly only a single leg had a four-jointed tarsus, and only one specimen was met with in which all six legs thus varied, and one specimen only shewed the variation in five legs.

In speaking of such a Variation as a *perfect* Variation several things are meant.

First, it is meant that the tarsus of the new pattern is as distinctly divided into four joints as the normal is into five. In addition to this the statement that the varying limb is perfect conveys a number of ideas that cannot be readily formulated; for example, that the joints are to all appearance properly proportioned and serviceable, shewing no sign of unfitness: they have in fact much the same appearance as they have in those of the Orthoptera in which the tarsus is normally four-jointed. But besides these attributes, which though useful enough for ordinary description are still in their nature formless and of no precise application, there is another which in the case of these varying legs we are entitled to make. We have said that these four-jointed tarsi are to all appearance normal, save for the number of the joints. Now the measurements which, at my suggestion, Mr Brindley has been kind enough to make, entitle us to go beyond this, and to assert that the four-jointed tarsus has another character by reason of which it is actually in a sense a "normal" form. A brief consideration of this will clearly illustrate the meaning of the term "perfection" applied to Variation.

We saw above that in a monomorphic form, the frequency with which, in respect of any given character, it departs from its mean condition follows a curve of Frequency of Error. This is, indeed, what is meant by the statement that the mean condition is a normal.

Taking the five-jointed tarsus, measurements shewed that the ratio of the length of any given joint to the length of the whole tarsus varied in this way about a mean value. Measurement of the joints of the four-jointed form shewed that the ratios which they bear to the total length of their respective tarsi vary in a similar way about their mean values, and that there is thus a "normal" four-jointed condition just as there is a "normal" five-jointed condition. In the same way, then, that the ratio of the length of each of the five joints to that of the whole tarsus is not always identical but exhibits small variations, so the ratios of the several joints of the four-jointed tarsus to the length of the whole tarsus also vary, but in each case the ratio has a mean value which is approached with a frequency conforming to a curve of Error.

The measurements established also another fact which is of consequence to an appreciation of the nature of totality in Variation. It not only appeared that the departures from the mean value of these ratios in the four-jointed variety were distributed about the mean in the same way as those of the five-jointed form, but it was also shewn that the absolute variations from the mean values of these ratios were not on the whole greater in the four-jointed tarsi than in the five-jointed tarsi. In other words, the four-jointed tarsus occurring thus sporadically, as a variety, is not less definitely constituted than the five-jointed type, and the proportions of its several joints are not less constant. It is scarcely necessary to point out that these facts give no support to the view that the exactness or perfection with which the proportions of the normal form are approached is a consequence of Selection. It appears rather, that there are two possible conditions, the one with five joints and the other with four, either being a position of Organic Stability. Into either of these the tarsus may fall; and though it is still conceivable that the final choice between these two may have been made by Selection, yet it cannot be supposed that the accuracy and completeness with which either condition is assumed is the work of Selection, for the "sport" is as definite as the normal.

This interesting case of Meristic Variation in the tarsus of the Cockroach illustrates in a striking way the principle which is perhaps the chief of those to which the Study of Variation at the outset introduces us. We are presented with the phenomenon of an organ existing in two very different states, between which no intermediate has been seen. Each of these states is definite and in a sense perfect and complete; for the oscillations of the four-jointed form around its mean condition are not more erratic than those of the normal form. Now when it is remembered that just such a four-jointed condition of the tarsus is known as a normal character of many insects and especially of some Orthoptera, it is, I think, difficult to avoid the conclusion that if the four-jointed groups are descended from the five-jointed, the Variation by which this condition arose in them was of the same nature as that seen as an individual Variation in *Blatta*; that as the modern phenomenon of the individual Variation which we see, so that past phenomenon of the birth of a four-jointed race, was definite and complete, and that the change whose history is gone, like the change to be seen to-day, was no gradual process, but a Discontinuous and total Variation[1].

[1] Since this Section was written it has seemed possible that the account given above may be found to need an important modification. It is well known that *Blatta*, in common with many other Orthoptera, has the power of reproducing the antennæ and legs after amputation or injury, and we have made some observations shewing that the tarsi of these regenerated legs sometimes, if not always, contain *four* joints. The question therefore arises whether the 4-jointed tarsus is a truly congenital variation, and not rather a variation introduced in the process of regeneration, somewhat after the manner of a bud-variation. To determine this point a considerable number of immature specimens were examined, and it was found that the percentage of individuals with 4-jointed tarsi is considerably less in the young than in the adult. These facts lend support to the view that the 4-jointed condition is not congenital. A quantity of individuals were also hatched from the egg-cocoons, and among them there has thus far been found no case of 4-jointed tarsus. On the other hand the total number thus hatched is not yet sufficient to create any strong probability that none are ever hatched in the 4-jointed state. We have also seen the 4-jointed tarsus in three very young individuals, which, to judge from their total length, must have been newly hatched. The statistics shew besides that the abnormality is distinctly commoner in females than in males, and that it is commoner in the legs of the 2nd pair than in the 1st, and much more common in the 3rd pair of legs than in the 2nd. These facts somewhat favour the view that the variation may be congenital. It seems also exceedingly improbable that in the specimen with all the tarsi 4-jointed, the six legs could each have been lost and renewed. There seems on the whole to be a pre-

SECTION XII.

Parallel between Discontinuity of Sex and Discontinuity in Variation.

The application of the term Discontinuity to Variation must not be misunderstood. It is not intended to affirm that in discontinuous Variation there can be between the variety and the type no intermediate form, or that none has been known to occur, and it is not even necessary for the establishment of Discontinuity that the intermediate forms should be rare relatively to the perfect form of the variety, though in cases of discontinuous Variation this is generally the case; but it is rather meant that the perfect form of the variety *may* appear at one integral step in Descent, either without the occurrence of intermediate gradations, or at least without the intercalation of such graduated forms in the pedigree.

In the case of the tarsus of *Blatta* we have seen an example of a total and complete Variation affecting single members of a series of repeated parts, not collectively, but one or more at a time[1]. Such an instance of a Meristic Variation occurring in a state which is total as regards members of a series but not total as regards the whole series finds many parallels among Substantive Variations, as, for example, that of the Crab (*Cancer pagurus*) bearing the right third maxillipede fashioned as a chela, while the left third maxillipede was normal. Variations of this nature in plants are of course well known to all.

At a previous place (Section VII.) allusion was made to the familiar but very curious analogy between members of a series of Meristic parts and separate organisms. The facts of Variation bring out this analogy in many singular ways, and in speaking of the totality of Variation it is necessary to bear these facts in mind. Not only are there abundant instances of independent division or multiplication of single members of Meristic series, but as has been said, single members of such series may thus independently and singly undergo qualitative or Substantive Variation, being treated in the physical system of the body as though they were separate units. In Variation, therefore, though it will be

sumption that the variation may at least sometimes be congenital. Supposing however that this shall be found hereafter not to be the case, I do not think that the deductions drawn from the facts will be less valid. The conclusions as to the definiteness of the two types, and the relationships of the several parts of each to the several parts of the other, would still hold good. There are besides in other forms, instances of similar numerical Variation, as for example, in the number of joints in the antennæ of Prionidæ, where the hypothesis of change on renewal is impossible, from which a similar argument might be drawn; but on the whole I have preferred to leave the account as it stands, taking the case of *Blatta* as an example, because it is easily accessible and because, from the fewness of the joints concerned, the issues are singularly clear.

[1] See Note at the end of Section XI.

found that members of Meristic series *may* vary simultaneously and collectively—and this is one of the most important generalizations which result from the Study of Variation—yet it is also true that in Variation single members of such series *may* vary independently and behave as though they possessed an "individuality" of their own. If ever it shall be possible to form a conception of the physical processes at work in the division and reproduction of organisms, account must be taken of both of these phenomena.

I know no way in which the nature of Discontinuity in Variation and the position of intermediate forms may be so well illustrated as by the closely parallel phenomenon of Sex. In the case of Sex in the higher animals we are familiar with the existence of a race whose members are at least dimorphic, being formed either upon one plan or upon the other, the two plans being in ordinary experience alternative and mutually exclusive. Between these two types, male and female, there are nevertheless found intermediate forms, "hermaphrodites," occurring in the higher animals at least, as great rarities. Now though these intermediate forms perhaps exist in gradations sufficiently fine to supply all the steps between male and female, it cannot be supposed that the one sex has been derived from the other, and still less that the various stages of hermaphroditism have been passed through in such Descent. Besides this, even though there is an accurate correspondence or homology between the several organs which are modified upon the one plan in the male and upon another in the female, and though this homology is such as to suggest, were we comparing two species, that the one had been formed from the other, part by part, yet by the nature of the case such a view is here inadmissible: for firstly it is impossible to suppose that either sex has at any time had the organs of the other in their completeness, and secondly it is clear that any hypothetical common form, by modification of which both may have arisen, must have been indefinitely remote and could certainly not have possessed secondary sexual organs bearing any resemblance to those now seen in the higher forms. All this has often been put, but the application of it to Variation is of considerable value. For in the case of Sex there is an instance of the existence of two normals and of many forms intermediate between them, occurring in a way which precludes the supposition that the intermediates represent stages that have ever occurred in the history of the two forms.

In yet another way Sex supplies a parallel to Variation. As we know, the sexes are discontinuous and occur commonly in their total or perfect forms. Now just as members of a Meristic series may present total variations independently of each other, so may single members of such a series present opposite secondary sexual characters, which may nevertheless be in each case complete.

The best known instance of this is that of gynandromorphic insects, in which the characters of the whole or part of one side of the body, wings and antennæ, are male, while those of the other side are female. Remarkable instances of a similar phenomenon have been recorded among bees and will be described later. As is well known, the organs and especially the legs of the sexless females or workers are formed differently from those of the drones, but there are cases of individuals having some of the parts and appendages formed on the one plan and some on the other. Thus in these individuals, which are in a sense intermediate between workers and drones, the characters of the two sexes may still be not completely blended, the male type prevailing in some parts, and the female in others. In the Discontinuity of Substantive Variation will be found examples of imperfect blending of variety and type closely comparable with this case of the imperfect blending of Sex.

SECTION XIII.

SUGGESTIONS AS TO THE NATURE OF DISCONTINUITY IN VARIATION.

The observations at the end of Section XI, regarding the Discontinuity of Meristic Variation lead naturally to certain reflexions as to the nature of Discontinuous Variation in general. In the case of the Cockroach tarsus, there given, it appeared that just as the structure of the typical form varies about its mean condition, so the structure of the variety varies about another mean condition. This fact, which in the given instance of Meristic Variation is so clear, at once suggests an inquiry whether this is not the usual course of Discontinuous Variation, and, indeed, whether Discontinuity in Variation does not mean just this, that in varying the organism passes from a form which is the normal for the type to another form which is a normal for the variety. Such transitions plainly occur in many cases of Meristic Variation, and in a considerable number of Substantive Variations there will be found to be indications that the phenomenon is similar. It is true that at the present stage of the inquiry the evidence has the value rather of suggestion than of proof, but the suggestion is still very decided and it is scarcely possible to exaggerate the importance of even this slender clue.

In stating the problem of Species at the beginning of this inquiry it was said that the forms of living things, as we know them, constitute a discontinuous series, and it is with the origin of the Discontinuity of the series that the solution of the main problem is largely concerned. Now the evidence of Discontinuous Variation suggests that organisms may vary abruptly from the definite form of the type to a form of variety which has also in some measure the character of definiteness. Is it not then possible that the Discontinuity of Species may be a consequence and expression of the Discontinuity of Variation? To declare at the present time that this is so would be wholly premature, but the suggestion that it is so is strong, and as a possible light on the whole subject should certainly be considered.

In view of such a possible solution of one of the chief parts of the problem of Species it will be well to point out a line of inquiry which must in that event be pursued. If it can be shewn that the Discontinuity of Species depends on the Discontinuity of Variation, we shall then have to consider the causes of the Discontinuity of Variation.

Upon the received hypothesis it is supposed that Variation is continuous and that the Discontinuity of Species results from the operation of Selection. For reasons given above (pp. 15 and 16) there is an almost fatal objection in the way of this belief, and it cannot be supposed both that all Variation is continuous and also that the Discontinuity of Species is the result of Selection. With evidence of the Discontinuity of Variation this difficulty would be removed.

It will be noted also that it is manifestly impossible to suppose that the perfection of a variety, discontinuously and suddenly occurring, is the result of Selection. No doubt it is conceivable that a race of Tulips having their floral parts in multiples of four might be raised by Selection from a specimen having this character, but it is not possible that the perfection of the nascent variety can have been gradually built up by Selection, for it is, in its very beginning, perfect and symmetrical. And if it may be seen thus clearly that the perfection and Symmetry of a variety is not the work of Selection, this fact raises a serious doubt that perhaps the similar perfection and Symmetry of the type did not owe its origin to Selection either. This consideration of course touches only the part that Selection may have played in the first building up of the type and does not affect the view that the perpetuation of the type once constituted, may have been achieved by Selection.

But if the perfection and definiteness of the type is not due to Selection but to the physical limitations under which Variation proceeds, we shall hope hereafter to gain some insight into the nature of these limitations, though in the present state of zoological study the prospect of such progress is small. In the observations which follow I am conscious that the bounds of profitable speculation are perhaps exceeded, and I am aware that to many this may seem matter for blame; but there is, in my judgment, a plausibility in the views put forward, sufficient at least to entitle them to examination. They are put forward in no sense as a formulated theory, but simply as a suggestion for work. It is, besides, only in foreseeing some of the extraordinary possibilities that lie ahead in the Study of Variation, that the great value of this method can be understood.

It has been seen that variations may be either Meristic or Substantive, and that in each group discontinuous and definite variations may occur by steps which may be integral or total. We are now seeking the factors which determine this totality and define the forms assumed in Variation. In this attempt we may, by arbitrarily confining our first notice to very simple cases, recognize at least two distinct factors which may possibly be concerned in this determination. Of these the first relates to Meristic Variation and the second to Substantive Variation.

1. *Possible nature of the Discontinuity of Meristic Variation.*

Looking at simple cases of Meristic Variation, such as that of the Tulip or of *Aurelia*, or of the Cockroach tarsus, there is, I think, a fair suggestion that the definiteness of these variations is determined *mechanically*, and that the patterns into which the tissues of animals are divided represent positions in which the forces that effect the division are in equilibrium. On this view, the lines or planes of division would be regarded as lines or planes at right angles to the directions of the dividing forces; and in the lines of Meristic Division we are perhaps actually presented with a map of the lines of those forces of attraction and repulsion which determine the number and positions of the repeated parts, and from which Symmetry results. If the Symmetry of a living body were thus recognized as of the same nature as that of any symmetrical system of mechanical forces, the definiteness of the symmetry in Meristic Variation would call for no special remark, and the perfection of the symmetry of a Tulip with its parts divided into four, though occurring suddenly as a "sport," would be recognized as in nowise more singular than the symmetry of the type. Both alike would then be seen to owe their perfection to mechanical conditions and not to Selection or to any other gradual process. If reason for adopting such a view of the physics of Division should appear, the frequency with which in any given form a particular pattern of Division or of Symmetry recurs, would be found to be determined by and to be a measure of the stability of the forces of Division when disposed in that particular pattern. It will of course be understood that in these remarks no suggestion is offered as to the causes which determine whether a tissue shall divide into four or into three, but merely as to the conditions of perfection of the division in either case. It will also be clear that though the symmetry of a flower or of any other tissue depends also on symmetrical growth, it is primarily dependent on the symmetry of its primary divisions, upon which symmetrical growth and secondary symmetrical divisions follow.

It would be interesting and I believe profitable to examine somewhat further the curiously close analogy between the symmetry of bodily Division and that of certain mechanical systems by which close imitations both of linear and of radial segmentation can be produced; and though to some this might seem overdaring, the possibility that the mechanics of bodily Division are in their visible form of an unsuspected simplicity is so far-reaching that it would be well to use any means which may lead others to explore it.

And even if at last this suggestion shall be found to have in it no other element of truth, it would still be of use as a forcible presentation of the fact, which when realized can hardly be doubted, that among the factors which combine to form a living body, the forces of Division may be distinguished as in their manifestations separable from the rest and forming a definite group. For, already (Section v.) it has been pointed out that the patterns of Division or Merism may be changed, while the Substance of the tissues presents to our senses no difference. The recognition of this essential distinctness of the Meristic forces will, I believe, be found to supply the base from which the mechanics of growth will hereafter be attacked.

The problems of Morphology will thus determine themselves into problems in the physiology of Division, which must be recognized together with Nutrition, Respiration and Metabolism, as a fundamental property of living protoplasm.

To sum up: there is a possibility that Meristic Division may be a strictly *mechanical* phenomenon, and that the perfection and Symmetry of the process, whether in type or in variety, may be an expression of the fact that the forms of the type or of the variety represent positions in which the forces of Division are in a condition of Mechanical Stability.

2. *Possible nature of the Discontinuity of Substantive Variation.*

Passing from the phenomena of Division and arrangement to those of constitution or substance we are, as has been said, again presented with the phenomenon of discontinuous or total Variation, and we must seek for causes which may perhaps govern and limit this totality, and in obedience to which the Variation is thus definite. Now as in the case of Meristic Variation, by arbitrarily limiting the examination to those cases which seem the simplest it appears that there is at least an analogy between them and certain mechanical phenomena, so by similarly restricting ourselves to very simple cases there will be seen to be a similar analogy between the discontinuity of some Substantive Variations and that of *chemical* discontinuity. It is on the whole not unreasonable to expect that the definiteness of at least some Substantive Variations depends ultimately on the discontinuity of chemical affinities. To take but one instance, that of colour, we are familiar with the fact that the colours of many organic substances undergo definite changes when chemically acted on by reagents, and it is not suggested that the definiteness and discontinuity of the various colours assumed is dependent on anything but the definiteness of the chemical changes undergone. The changes of litmus and many vegetable blues to red on treatment with acids, of many vegetable yellows to brown on treatment with alkalies, the colours of the series of bodies produced by the progressive oxidation of biliverdin are familiar examples of such definite colour-variations.

With facts of this kind in view, the conclusion is almost forced on us that the definiteness of colour-variation is a consequence of the definiteness of the chemical changes undergone. No one doubts that the orange colouring matter of the variety of the Iceland Poppy (*P. nudicaule*) is a chemical derivative from the yellow colouring matter of the type. It is not questioned that in such cases a definite alteration in the chemical conditions in which the pigment is produced determines whether the flower shall be orange or yellow; and I think it is reasonable to expect that the frequency with which the flowers are either yellow or orange as compared with the rarity of the intermediate shades is an expression of the fact that the yellow and orange forms of the colouring matter have a greater chemical stability than the intermediate forms of the pigment, or than a mixture of the two pigments. If then it should happen, as we may fairly suppose it might, that the orange form were to be selected and established as a race, it would owe the definiteness of its orange colour and the precision of its tint, not to the precision with which Selection had chosen this particular tint, but to the chemical discontinuity of which the originally discontinuous Variation was the expression.

To pass from the case of a sport to that of Species, it is well known that of the many S. African butterflies of the genus *Euchloe* (= *Anthocharis*, Orange-tips), some have the apices or tips of the fore-wings orange-red (for example, *E. danae*), while in others they are purple (for example, *E. ione*). Upon the view that the transition from orange to purple, or *vice versa*, had been continuously effected by the successive Selection of minute variations, we are met by all the difficulties we know so well. Why is purple a good colour for this creature? If purple is a good colour and red is a good colour, how did it happen that at some time or other all the intermediate shades were also good enough to have been selected? and so on. These and all the cognate difficulties are opened up at once, and though they have been met in the fashion we know, they have scarcely been overcome. But at the outset this view assumes that every intermediate may exist and has existed, an assumption which is gratuitous and hardly in accordance with the known fact that chemical processes are frequently discontinuous. When besides this it is known that Variation *may* be discontinuous, I submit that it is easier to suppose that the change from red to purple was from the first complete, and that the choice offered to Selection was between red and purple; and that the tints of the purple and of the red were determined by the chemical properties of the body to which the colour is due. This case is a particularly interesting one in the light of the fact that, as Mr F. G. Hopkins has lately shewn me, this purple colour, dissolved in hot water, leaves on evaporation a substance which gives the murexide reaction and cannot as yet be distinguished from the substance similarly derived from the orange or yellow colouring matters of Pieridæ in general. As was stated above, Mr Hopkins has shewn that these yellows are acids, allied to mycomelic acid, a derivative of uric acid, and therefore of the nature of excretory products. Whether the purple body is related to the yellow or to the orange as a salt is to an acid, or otherwise, cannot yet be affirmed; but if the difference between them is a chemical difference, which can hardly be doubted, there is at least a presumption that the discontinuity of these colours in the several species, is an expression of the discontinuity of the chemical properties of this body. The possibility that from such bodies a series of substances might perhaps by suitable means be prepared in such a way as to represent many or even all intermediate shades, does not greatly affect the suggestion made; for even in such series it is almost certain that points of comparative stability would occur, and Discontinuity would be thus introduced.

The case of Colour has been taken in illustration because it is the simplest and most intelligible example of the possibility that the Discontinuity of some Substantive Variations is determined by the Discontinuity of the chemical processes by which the structures are produced. It is true that perhaps no species has been rightly differentiated by colour alone, but colour is still one of the many characters which go to the distinguishing of a species, and it is precisely one of the characters whose significance and delimitation by Natural Selection is most obscure. Moreover by the fact that in the case of these yellow and red Pieridæ the colours are of an excretory nature, we are reminded that Variation in colour may be an index of serious changes in the chemical economy of the body, and that when an animal is said to be selected because it is red or because it is purple, the real source of its superiority may be not its red colour or its purple colour, but other bodily conditions of which these colours are merely symptoms. By those who have attempted to reconcile the phenomena of Colour with the hypothesis of Natural Selection this fact is too often overlooked.

But though it may reasonably be supposed that much of the Discontinuity of Variation and some of the Discontinuity of

Species arise through discontinuous transition from one state of mechanical or chemical stability to another state of stability, there nevertheless remain large classes of discontinuous variations, and of Specific Differences still more, whose Discontinuity bears no close analogy with these. To these phenomena inorganic Nature offers no parallel. We may see that they are discontinuous and that their course is in some way controlled, but as to the nature of this control we can make no guess.

Though the resemblance may be misleading, it is nevertheless true that in *living* Nature there are other phenomena, those of disease, which present a Discontinuity closely comparable with that of many variations. In problems of disease we meet again the same problem which we meet in Variation, namely, changes which may be complete or specific, though occurring so suddenly as to exclude the hypothesis that Selection has been the limiting cause. All this is familiar to everyone who has considered the problem of Species.

For though, like discontinuous variations, the manifestations of specific disease are not always identical, but differ in intensity and degree, varying about a normal form, still these manifestations may be specific in the sense in which the term is used with reference to the characters of Species. If we exclude those diseases whose specific characters are now known to be the result of the invasion of specific organisms, there still remain very many which are known and recognized by definite and specific symptoms produced in the body, though there is as yet no evidence that they are due to specific organisms. [Of course if it were shewn that these diseases also result from the action of specific organisms, they then only present to us again the original problem of Species; for if the definiteness, or Species, of a disease is due to the definiteness, or Species, of the micro-organism which causes it, the cause of that definiteness of the micro-organism remains to be sought, and we are simply left with a particular case of the general problem of Species.] But in the meantime we can see that the manifestations are specific; and while we do not know that they result from causes themselves specific, the nature of the control in obedience to which they are specific is unknown.

The parallel between disease and Variation may be misleading, but this much at least may fairly be learned from it: that the system of an organized being is such that the result of its disturbance may be specific. And in the end it may well be that the problem of Species will be solved by the study of pathology; for the likeness between Variation and disease goes far to support the view which Virchow has forcibly expressed, that "every deviation from the type of the parent animal must have its foundation on a pathological accident[1]."

[1] R. VIRCHOW, *Journal of Pathology*, I. 1892, p. 12.

SECTION XIV.

Some current conceptions of Biology in view of the facts of Variation.

Enough has now been said to explain the aim of the Study of Variation, and to shew the propriety of the choice of the facts of Meristic Variation as a point of departure for that study. Before leaving this preliminary consideration, reference to some cognate subjects must be made.

It has been shewn that in view of the facts of Variation, some conceptions of modern Morphology must be modified, while others must be abandoned. With the recognition of the significance of the phenomena of Variation, other conceptions of biology will undergo like modifications. As to some of these a few words are now required, if only to explain methods adopted in this work.

1. *Heredity.*

It has been the custom of those who have treated the subject of Evolution to speak of "Heredity" and "Variation" as two antagonistic principles; sometimes even they are spoken of as opposing "forces."

With the Study of Variation, such a description of the processes of Descent will be given up, even as a manner of speaking. In what has gone before I have as far as possible avoided any use of the terms Heredity and Inheritance. These terms which have taken so firm a hold on science and on the popular fancy, have had a mischievous influence on the development of biological thought. They are of course metaphors from the descent of property, and were applied to organic Descent in a time when the nature of the process of reproduction was wholly misunderstood. This metaphor from the descent of property is inadequate chiefly for two reasons.

First, by emphasizing the fact that the organization of the offspring depends on material transmitted to it by its parents, the metaphor of Heredity, through an almost inevitable confusion of thought, suggests the idea that the actual body and constitution of the parent are thus in some way handed on. No one perhaps would now state the facts in this way, but something very like this material view of Descent was indeed actually developed into Darwin's Theory of Pangenesis. From this suggestion that the body of the parent is in some sort remodelled into that of the offspring, a whole series of errors are derived. Chief among these is the assumption that Variation must necessarily be a continuous process; for with the body of the parent to start from, it is hard to conceive the occurrence of discontinuous change. Of the deadlock which has resulted from the attempt to interpret Homology on this view of Heredity, I have already spoken in Section VI.

Secondly, the metaphor of Heredity misrepresents the essential phenomenon of reproduction. In the light of modern investigations, and especially those of Weismann on the continuity of the germ-cells, it is likely that the relation of parent to offspring, if it has any analogy with the succession of property, is rather that of trustee than of testator.

Hereafter, perhaps, it may be found possible to replace this false metaphor by some more correct expression, but for our present purpose this is not yet necessary. In the first examination of the facts of Variation, I believe it is best to attempt no particular consideration of the working of Heredity. The phenomena of Variation and the *origin* of a variety must necessarily be studied first, while the question of the perpetuation of the variety properly forms a distinct subject. Whenever in the cases given, observations respecting inheritance are forthcoming they will be of course mentioned. But speaking of discontinuous Variation in general, the recurrence of a variation in offspring, either in the original form or in some modification of it, has been seen in so many cases, that we shall not go far wrong in at least assuming the possibility that it *may* reappear in the offspring. At the present moment, indeed, to this statement there is little to add. So long as systematic experiments in breeding are wanting, and so long as the attention of naturalists is limited to the study of normal forms, in this part of biology which is perhaps of greater theoretical and even practical importance than any other, there can be no progress.

2. *Reversion.*

Around the term Reversion a singular set of false ideas have gathered themselves. On the hypothesis that all perfection and completeness of form or of correlation of parts is the work of Selection it is difficult to explain the discontinuous occurrence of new forms possessing such perfection and completeness. To account for these, the hypothesis of Reversion to an ancestral form is proposed, and with some has found favour. That this suggestion is inadmissible is shewn at once by the frequent occurrence by discontinuous Variation, of forms which though equally perfect, cannot all be ancestral. In the case of *Veronica* and *Linaria*, for example, a host of symmetrical forms of the floral organs may be seen occurring suddenly as sports, and of these though any *one* may conceivably have been ancestral, the same cannot be supposed of all, for their forms are mutually exclusive. On *Veronica buxbaumii*, for instance, are many symmetrical flowers, having *two* posterior petals, like those of other Scrophularineæ: these may reasonably be supposed to be ancestral, but if this supposition is made, it cannot be made again for the equally perfect forms with three petals, and the rest[1].

The hypothesis of Reversion to account for the Symmetry and perfection of modern or discontinuous Variation is made through a total misconception of the nature of Symmetry.

There is a famous passage in the *Descent of Man*, in which Darwin argues that the phenomenon of double uterus, from its perfection, must necessarily be a Reversion.

> ...." In other and rarer cases, two distinct uterine cavities are formed, each having its proper orifice and passage. No such stage is passed through during the ordinary development of the embryo, and it is difficult to believe, though perhaps not impossible, that the two simple, minute, primitive tubes could know how (if such an expression may be used) to grow into two distinct uteri, each with a well-constructed orifice and passage, and each furnished with numerous muscles, nerves, glands and vessels, if they had not formerly passed through a similar course of development, as in the case of existing marsupials. No one will pretend that so perfect a structure as the abnormal double uterus in woman could be the result of mere chance. But the principle of reversion, by which long-lost dormant structures are called back into existence, might serve as the guide for the full development of the organ, even after the lapse of an enormous interval of time[2]." *Descent of Man*, vol. I. pp. 123 and 124.

This kind of reasoning has been used by others again and again. It is of course quite inadmissible; for by identical reasoning from the perfect symmetry of double monsters, of the single eye of the Cyclopian monster, and so on, it might be shewn that Man is descended from a primitive double vertebrate, from a one-eyed Cyclops and the like. For other reasons it is likely enough that double uterus was a primitive form; but the perfection and symmetry of the modern variation to this form is neither proof nor indication of such an origin. Such a belief arises from want of knowledge of the facts of Meristic Variation, and is founded on a wrong conception of the nature of symmetry and of the mechanics of Division. The study of Variation shews that it is a common occurrence for a part which stands in the middle line of a bilaterally symmetrical animal, to divide into two parts, each being an optical image of the other: and that conversely, parts which normally are double, standing as optical images of each other on either side of such a middle line may

---

[1] For a full account of such facts, see a paper by Miss A. BATESON and myself On Variations in Floral Symmetry. *Journ. Linn. Soc.*, XXVIII. p. 386.

[2] This extraordinary passage is scarcely worthy of Darwin's penetration. If read in the original connexion it will seem strange that it should have been allowed to stand. For in a note to these reflexions on Reversion (*Descent*, I. p. 125) Darwin refers to and withdraws his previously expressed view that supernumerary digits and mammæ were to be regarded as reversions. This view had been based on the perfection and symmetry with which these variations reproduce the structure of putative ancestors. It was withdrawn because Gegenbaur had shewn that polydactyle limbs often bear no resemblance to those of possible ancestors, and because extra mammæ may not only occur symmetrically and in places where they are normal in other forms, but also in several quite anomalous situations. In the light of this knowledge it is strange that Darwin should have continued to regard the perfection and symmetry of a variation as evidence that it is a Reversion.

be compounded together in the middle line forming a single, symmetrical organ.

It would probably help the science of Biology if the word 'Reversion' and the ideas which it denotes, were wholly dropped, at all events until Variation has been studied much more fully than it has yet been.

In the light of what we now know of the process of reproduction the phrase is almost meaningless. We suppose that a certain stock gives off a number of individuals which vary about a normal; and that after having given them off, it begins to give off individuals varying about another normal. We want to say that among these it now and then gives off one which approaches the first normal, that shooting at the new mark it now and then hits the old one. But all that we know is that now and then it shoots wide and hits *another* mark, and we assume from this that it would not have hit it if it had not aimed at it in a bygone age. To apply this to any other matter would be absurd. We might as well say that a bubble would not be round if the air in it had not learned the trick of roundness by having been in a bubble before: that if in a bag after pulling out a lot of white balls I find a totally red one, this proves that the bag must have once been full of red balls, or that the white ones must all have been red in the past.

Besides the logical absurdity on which this use of the theory of Reversion rests, the application of it to the facts of Variation breaks down again and again. I have already mentioned some cases of this, but there are many others of a different class. For instance, it will be shewn that the percentage of extra molars in the Anthropoid Apes is almost the highest reached among mammals. On the usual interpretation, such teeth are due to Reversion to an ancestral condition with 4 molars, and on less evidence it has been argued that a form frequently shewing such "Reversion" is older than those which do not. From this reasoning it should follow that the Anthropoids are the most primitive form, at least of monkeys. It is surely time that these brilliant and facile deductions were no more made in the name of science.

3. *Causes of Variation.*

Inquiry into the causes of Variation is as yet, in my judgment, premature.

4. *The Variability of "useless" Structures.*

The often-repeated statement that "useless" parts are especially variable, finds little support in the facts of Variation, except in as far as it is a misrepresentation of another principle. The examples taken to support this statement are commonly organs standing at the end of a Meristic Series of parts, in which there is a progression or increase of size and degree of development, starting from a small terminal member. In such cases, as that of the last rib in Man, and several other animals, the wisdom-teeth of Man, etc., it is quite true that in the terminal member Variation is more noticeable than it is in the other members. This is, I believe, a consequence of the mechanics of Division, and has no connexion with the fact that the functions of such terminal parts are often trifling. Upon this subject something will be said later on, but perhaps a rough illustration may make the meaning more clear at this stage. If a spindle-shaped loaf of bread, such as a "twist," be divided with three cuts taken at equal distances, in such a way that the two end pieces are much shorter than the middle ones, to a child who gets one of the two large middle pieces the contour-curves of the loaf will not matter so much; but to a child who gets one of the small end bits, a very slight alteration in the curves of the loaf will make the difference between a fair-sized bit and almost nothing, a difference which the child will perceive much more readily than the complementary difference in the large pieces will be seen by the others. An error in some measure comparable with this is probably at the bottom of the statement that useless parts are variable, but of course there are many examples, as the pinna of the human ear, which are of a different nature. It is unnecessary to say that for any such case in which a part, apparently useless, is variable, another can be produced in which some capital organ is also variable; and conversely, that for any case of a capital organ which is little subject to Variation can be produced a case of an organ, which though trifling and seemingly "useless," is equally constant. With a knowledge of the facts of Variation, all these trite generalities will be forgotten.

5. *Adaptation.*

In examining cases of Variation, I have not thought it necessary to speculate on the usefulness or harmfulness of the variations described. For reasons given in Section II, such speculation, whether applied to normal structures, or to Variation, is barren and profitless. If any one is curious on these questions of Adaptation, he may easily thus exercise his imagination. In any case of Variation there are a hundred ways in which it may be beneficial, or detrimental. For instance, if the "hairy" variety of the moorhen became established on an island, as many strange varieties have been, I do not doubt that ingenious persons would invite us to see how the hairiness fitted the bird in some special way for life in that island in particular. Their contention would be hard to deny, for on this class of speculation the only limitations are those of the ingenuity of the author. While the only test of utility is the success of the organism, even this does not indicate the utility of one part of the economy, but rather the nett fitness of the whole.

6. *Natural Selection.*

In the view of the phenomena of Variation here outlined, there is nothing which is in any way opposed to the theory of the origin of Species "by means of Natural Selection, or the preservation of favoured races in the struggle for life." But by a full and unwavering belief in the doctrine as originally expressed, we shall in no way be committed to representations of that doctrine made by those who have come after. A very brief study of the facts will suffice to gainsay such statements as, for example, that of Claus, that "it is only *natural selection which accumulates those alterations, so that they become appreciable to us* and constitute a variation which is evident to our senses[1]." For the crude belief that living beings are plastic conglomerates of miscellaneous attributes, and that order of form or Symmetry have been impressed upon this medley by Selection alone; and that by Variation any of these attributes may be subtracted or any other attribute added in indefinite proportion, is a fancy which the Study of Variation does not support.

Here this Introduction must end. As a sketch of a part of the phenomena of Variation, it has no value except in so far as it may lead some to study those phenomena. That the study of Variation is the proper field for the development of biology there can be no doubt. It is scarcely too much to say that the study of Variation bears to the science of Evolution a relation somewhat comparable with that which the study of affinities and reactions bears to the science of chemistry: for we might almost as well seek for the origin of chemical bodies by the comparative study of crystallography, as for the origin of living bodies by a comparative study of normal forms.

---

[1] *Text-book of Zoology*, Sedgwick and Heathcote's English translation, vol. I. p. 148. In the original the passage runs: "*erst die natürliche Zuchtwahl häuft und verstärkt jene Abweichungen in dem Masse dass sie für uns wahrnehmbar werden und eine in die Augen fallende Variation bewirken.*" C. Claus, *Lehrb. d. Zool.*, Ed. 2, 1883, p. 127, and *Grundzüge der Zoologie*, 1880, Bd. I. p. 90. The italics are in the original.

PART I.

MERISTIC VARIATION.

CHAPTER I.

ARRANGEMENT OF EVIDENCE.

THE cases of Meristic Variation, here given, illustrate only a small part of the subject. The principles upon which these have been chosen may be briefly explained. It was originally intended to give samples of the evidence relating to as many different parts of the subject as possible, so that the ground to be eventually covered might be mapped out, leaving the separate sections of evidence to be amplified as observations accumulate. This plan would be the most logical and perhaps in the end the most useful, but for several reasons it has been abandoned. I have chosen a different course, first, because during the progress of the work opportunities occurred for developing special parts of the evidence; secondly, since isolated observations have no interest for most persons, it is more likely that the importance of the subject will be appreciated in a fuller treatment of special sections, than in a general view of the whole; and lastly, because as yet the attempt to make an orderly or logical classification of the phenomena of Merism, however attractive, must be so imperfect as to be almost worthless. For these reasons I have decided to treat more fully a few sections of the facts, hoping that in the course of time similar treatment may be applied to other sections also. The sections have been chosen either because there is a fairly large body of evidence relating to them, or on account of the importance or novelty of the principles illustrated.

As far as possible I have described each case separately, in terms applicable specially to it, deductions or criticism being kept apart. The descriptions are written as if for an imaginary catalogue of a Museum in which the objects might be displayed[1]. This system, though it entails repetition, has, I believe, advantages which cannot be attained when the descriptions are given in a comprehensive and continuous form. In speaking of subjects, such as supernumerary mammæ, or cervical fistulæ, where the evidence has been exhaustively treated by others, and upon which I can add nothing, it has not seemed necessary to follow this system, and in such cases connected abstracts are given.

[1] Cases of special importance are marked by an asterisk.

As the evidence here presented consists, as yet, only of specimen chapters in the Natural History of Meristic Variation, and does not offer any comprehensive view of the whole subject, no strict classification of the facts is attempted. The evidence of Meristic Variation relates essentially to the manner in which changes occur in the number of members in Meristic series. Such numerical changes may come about in two ways, which are in some respects distinct from each other. For instance, the number of legs and body-segments in *Peripatus edwardsii* varies from 29 to 34[1]: here the variation in number must be a manifestation of an original difference in the manner of division or segmentation in the progress of development. The change is strictly Meristic or divisional. On the other hand, change in number may arise by the Substantive Variation of members of a Meristic series already constituted. For example, the evidence will shew that the number of oviducal openings in *Astacus* may be increased from one pair to two or even three pairs. Here the numerical variation has come about through the assumption by the penultimate and last thoracic appendages, of a character typically proper to the appendages of the antepenultimate segment of the thorax alone. Now there is here no change in the number of segments composing the Meristic series, but by Substantive Variation the number of openings has been increased.

The case of the modification of the antenna of an insect into a foot, of the eye of a Crustacean into an antenna, of a petal into a stamen, and the like, are examples of the same kind.

It is desirable and indeed necessary that such Variations, which consist in the assumption by one member of a Meristic series, of the form or characters proper to other members of the series, should be recognized as constituting a distinct group of phenomena. In the case of plants such Variation is very common and is one of the most familiar forms of abnormality. MASTERS, in his treatise on Vegetable Teratology[2], recognizes this phenomenon and gives to it the name "Metamorphy," adopting the word from Goethe. As Masters says, so long as it is only proposed to use the word in Teratology, no great confusion need arise from the fact that the same term and its derivatives are used in a different sense in several branches of Natural History. But if, as I hope, the time has come when the facts of what has been called "Teratology" will be admitted to their proper place in the Study of Variation, this confusion is inevitable. In this study, besides, this particular kind of variation will be found to be especially important and I believe that in the future its significance and the mode of its occurrence will become an object of high interest. For this reason it is desirable that the term which denotes it should not lead to misunderstanding, and I think a new term is demanded.

---

[1] SEDGWICK, A., *Quart. Jour. Micr. Sci.*, 1888, XXVIII. p. 467.
[2] MASTERS, M. T., *Vegetable Teratology*, p. 239.

For the word 'Metamorphy' I therefore propose to substitute the term **Homœosis,** which is also more correct; for the essential phenomenon is not that there has merely been a change, but that something has been changed into the likeness of something else.

In the cases given above, the distinction between Homœotic Variation and strictly Meristic Variation is sufficiently obvious, but many numerical changes occur which cannot be referred with certainty to the one class rather than to the other. Such cases are for the most part seen in Vertebrates: for in them what may be called the *fundamental* numbers of the segments are not constituted with the definiteness found in Arthropods or in the Annelids, and several Meristic series of organs are disposed in numbers and positions independent of, or at least having no obvious relation to those of the other Meristic series. The number and positions of mammæ, or stripes, for instance, need not bear any visible relation to the segmentation of the vertebræ &c. The repetition of members of such a series may thus not coincide with, or occur in multiples of the segmentation of other parts in the same region. When such is the case, when the segmentation of one series of organs bears no simple or constant geometrical relation to the segmentation of other systems, it is not always possible to declare whether a numerical change in one of the systems of organs belongs properly to the first or the second of the classes described above. It is likely enough that in such a case as that of mammæ, there may sometimes be an actual Meristic division and subsequent separation of the tissues already destined to form the mammæ, occurring in such a way that each comes to take up its final position, and indeed the numerous cases in which such division has been imperfectly effected go far to prove that this is the case. But, on the other hand, it is not possible to know that the division did not occur before any tissue was specially differentiated off to form mammæ, and that the separation may be as old even as the division of the mammæ of the right side from those of the left, a process which almost beyond question occurs in the segmentation of the ovum. The distinction between these two alternatives is thus one rather of degree than of kind, and it is only in such forms as the Arthropods, the floral organs of some Phanerogams and the like, where the members of the several Meristic series have definite numbers, or coincide with each other, that this distinction is easily recognized. For this reason I do not think it well to attempt to carry out any classification of the evidence based on this distinction.

In the foregoing remarks I am aware that a very large question, which lies at the root of all accurate study of Meristic Variation, has been passed over somewhat superficially, but I scarcely think a fuller treatment possible in the present state of knowledge of the physics of Division, and in the absence of thorough observation of the developmental history of those tissues which ultimately become differentiated to form members of such non-coincident or independent Meristic series.

Some years ago[1], in the course of an argument that *Balanoglossus* should be considered as representing some of the ancestral characters of Chordata, I had occasion to refer to some of these difficulties, and especially to the different characters of the two kinds of segmentation; that of the Annelids, in which the repetitions of the organs belonging to the several systems are coincident, and, on the other hand, that of the Chordata, for example, in which this coincidence may be irregular or partial. At that time I was of opinion that these two sorts of segmentation may, in certain cases, have had a different phylogenetic history, and have resulted from processes essentially distinct. It appeared to me that we should recognize that, in the Annelids on the one hand, segmentation of the various systems of organs had been coincident from the beginning, while in the Chordata the segmentation had been progressive and had arisen by segmentation or repetition of the organs of the several systems independently. The reasons for this view were derived chiefly from the fact that it is possible to arrange the lower Chordata in order of progressive segmentation of the several systems. In particular such treatment was shewn to be applicable to the central nervous system, the vertebral column and the mesoblastic somites, and in these cases it was maintained that the evidence of the lower forms of Chordata goes to shew that segmentation had occurred in these systems one after another, and that their segmentation was not derived from a form having a complete repetition of each part in each segment: that these forms, in fact, shewed us the history of this progress from a less segmented form to one more fully segmented.

The views then set forth have met with little acceptance. Those who are occupied with the search for the pedigree of Vertebrates still direct their inquiries on the hypothesis, expressed or implied, that in the ancestral form there was a series of complete segments, each containing a representative of each system of those organs which in the present descendants appear in series. It is thus supposed that each segment of the primitive form must have been a kind of least common denominator of the segments of its posterity. The possibility that the segmentation of Vertebrates may have arisen progressively is, indeed, scarcely considered at all.

Though in the light of the study of Variation, it now seems to me that the discussion of these questions must be indefinitely postponed, and that there are radical objections to any attempt to interpret the facts of anatomy and development in our present ignorance of Variation, I have seen no reason to depart from the view expressed in the paper referred to: that interpreted by the current methods of morphological criticism, the facts go to shew that the segmentation of the Chordata differs essentially from that of the Annelids &c., and that it has arisen by progressive segmentation of the several systems of an originally unsegmented form. To those who hold as Dohrn, Gaskell, Marshall and others have done, that the evolution of Vertebrates has

[1] *Quart. Jour. Micr. Sci.*, 1886.

been a progress from a more fully segmented form to forms less segmented, I would again point out that this view is in direct opposition to the indications afforded by the lower Chordata, which are *less and not more segmented* than the higher forms.

The hypothesis of an ancestor made up of complete segments is resorted to because it is felt to be difficult to conceive the progressive building up of a segmented form, but on appeal to the facts of Variation the evidence will clearly shew that Repetition of parts previously existing is a quite common phenomenon; that such repetition may occur in almost any system of organs; and lastly that such new repetitions may be coincident in the several systems. To argue moreover that these repetitions, for instance that of oviducal apertures in *Astacus*, of mammæ or cervical ribs in mammals are "reversions," leads to absurdity, for on the same reasoning, the occurrence, in the Crab, of a third maxillipede formed as a chela, would shew that these appendages had been originally chelæ, that the occurrence of petaloid sepals shews that the sepals had originally been petals, and so forth.

These considerations will suffice to illustrate the great difference of degree, if not of kind, which probably exists between these two kinds of segmentation, that which arises by the repetition of bud-like segments, each containing parts of many systems on the one hand, and the progressive and separate segmentation of the several systems on the other. For reasons already given, however, I shall not attempt in this first collection of evidence to separate the facts on these lines. Though some cases can at once be seen to be strictly Meristic while others are plainly Homœotic, many cannot be affirmed to belong to the one group rather than to the other. There is, besides, a serious doubt whether perhaps after all, Homœotic Variation even in its most marked forms, may not ultimately rest on and be an expression of a change in the processes of Division, and be thus, at bottom, strictly Meristic also. In our present ignorance of the physics of Division, this doubt cannot be satisfied, and therefore it will be best to make no definite separation between the two classes of variations, though whenever the nature of a given variation is such that it may at once be recognised as Homœotic, it will be well to specify this.

In the absence of a more natural classification, the material has been roughly arranged with reference to the geometrical disposition and relations of the structures concerned. In the Introduction, Section IV. p. 21, reference was made to the fact that the Symmetry of an organism may be such as to include all the parts into one system of Symmetry, and for such a system the term **Major Symmetry** was proposed. Systems of this kind are seen in the Vertebrates and Echinoderms, for example. On the other hand systems of Symmetry occur in limbs and other separate parts of organisms, in such a way that each such system is either altogether or partially geometrically complete and symmetrical in itself. For example, the toe of a Horse, the arm of a Starfish, the eye-spots of some Satyrid butterflies, &c., are each in themselves nearly symmetrical. To these separate systems of Symmetry the term **Minor Symmetry** will be applied. Minor Symmetries may or may not be compounded into a Major Symmetry. Between these there is of course no hard and fast line.

In each class of Symmetry, Meristic Repetition may occur, and the repeated parts then stand in either

I. Linear or Successive Series.
II. Bilateral or Paired Series.
III. Radial Series.

Parts meristically repeated may thus stand in one or more geometrical relations to each other, and the first part of the evidence of Meristic Variation will be arranged in groups according as it is in one or other of these relations that the parts are affected. In each group cases affecting Major Symmetry will be given first, and those affecting Minor Symmetries will be taken after.

As it is proposed to arrange the facts of Meristic Variation in groups corresponding with these three forms of Meristic Repetition, it will be useful to consider briefly the nature of the relation in which the members of such series stand to each other, and the characters distinguishing the several kinds of series. Reduced to the simplest terms, the distinction may be thus expressed.

In the Linear or Successive series *the adjacent parts of any two consecutive members of the series are not homologous, but the severally homologous parts of each member or segment form a successive series, alternating with each other.* For example, the anterior and posterior surfaces of such a series of segments may be represented by the series

$$A......, AP, AP, AP, ...P.$$

The relation of any pair of organs in Bilateral Symmetry differs from this, for in that case *each member of the pair presents to its fellow of the opposite side parts homologous with those which its fellow presents to it, each being, in structure and position, an optical image of the other.* The external and internal surfaces of such a pair may therefore be represented thus:

$$E......I, I......E.$$

If the manner of origin of these two kinds of Repetition be considered, it will be seen that though both result from a process of Division, yet the manner of Division in the two cases is very different. For in the case of division to form a paired structure, the process occurs in such a way as to form a pair of images, of which similar and homologous parts lie on each side of the plane of division; while, in the formation of a chain of successive segments, each plane of division passes between parts which are dissimilar, and whose homology is alternate. The distinction between these two kinds of Division is of course an expression of the fact that the attractions and repulsions from which Division results are differently disposed in the two cases. It is further to be observed that the distinction, though striking, is nevertheless one of degree, for the two kinds of Division pass gradually into each other. By one or other of these two modes, or by a combination of both, all Meristic Series of Repetitions are formed.

In Radial series, the *Major Symmetry* is built up by radial divisions of the first kind, producing segments whose adjacent parts are homologous, and related to each other as images. Each of these segments is therefore bilaterally symmetrical about a radial plane. There is no succession between the segments, and in a perfectly symmetrical series, Successive or Linear repetitions can only occur in Minor Systems of Symmetry.

The considerations here set forth, though well known, have an importance in the interpretation of the evidence, for the connexion between the geometrical relations of organs and their Meristic Variations is intimate.

An arrangement of the facts with reference to these geometrical relations cannot, of course, be absolute, for it is clear that a Bilateral Symmetry, containing Linear Repetitions may be derived from a Radial Symmetry, and that these figures · cannot be precisely delimited from each other; nevertheless this plan of arrangement has still several advantages. Chief among these is this: that it brings out and emphasizes the fact that the possible, or at least the probable *Meristic Variations of such parts depend closely on the geometrical relation in which they stand.* This is, perhaps, in a word, the first great deduction from the facts of Meristic Variation. The capacity for, and manner of Meristic Variation appear to depend not on the physiological nature of the part, on the system to which it belongs, on the habits of the organism, on the needs or exigencies of its life, but on this fact of the geometrical position of the parts concerned. Linear series are liable to certain sorts of Variation, Bilateral Series are liable to other sorts of Variation, and Radial Series to others again. As I have ventured to hint before, the importance of all this lies in the glimpse which is thus afforded us of the essential nature of Meristic Division and Repetition. Such interdependence between the geometrical relations, or pattern, in which a part stands, and the kinds of Variation of which it is capable, is, I think, a strong indication that in Meristic Division we are dealing with a phenomenon which in its essential nature is mechanical. Since this is a thing of the highest importance, it will be useful to employ a system which shall give it full expression.

Evidence as to Meristic Variation in cell-division and in the segmentation of ova will be spoken of in connexion with the Variation of Radial and Bilateral series.

The second section of evidence is less immediately relevant to the problem of Species; nevertheless it bears so closely on the nature of Merism and on the mechanics of Physiological Division, that in any study of this subject reference to it cannot be omitted. The evidence in question relates first to abnormal repetition of limbs or other peripheral structures, (which in the normal form are grouped into and form part of a system of Symmetry,) such abnormal repetitions *occurring in such a way as to lie outside this normal system of Symmetry and unbalanced by any parts within it.* This phenomenon occurs in many forms, especially in bilateral animals, and may be exceptionally well studied in the case of supernumerary limbs in Insects and in supernumerary chelæ in Crabs and Lobsters. It will be shewn that such extra parts generally, if not always, make up a **Secondary system of Symmetry** in themselves; and the way in which such a Secondary system is related to the normal or **Primary system of Symmetry** of the body from which they spring, constitutes an instructive chapter in the study of Meristic Variation.

More extensive repetitions of this class, when affecting the axial parts of the body, give rise to the well-known Double and Triple Monsters, which, as has often been said, reproduce in the higher animals phenomena which, under the name of fission, are commonly seen in the lower forms. The general evidence as to these abnormalities is so accessible and familiar that it need not be detailed here, and it will therefore be enough to give an outline of its chief features and to point out the bearing of this class of evidence on the subject of Meristic Variation in general.

CHAPTER II.

MERISTIC VARIATION OF PARTS REPEATED IN LINEAR OR SUCCESSIVE SERIES.

SEGMENTS OF ARTHROPODA.

INDIVIDUAL Variation in the fundamental number of members constituting a Linear Series of segments can only be recognized in those forms which at some definite stage in their existence cease to add to the number of the series. Hence in a large proportion of the more fully segmented invertebrates this phenomenon cannot be studied, for in many of these, as for instance in Chilognatha, and in most of the Chætopoda the formation of new segments is not known to cease at any period of life, but seems to continue indefinitely. On the other hand, while in Insecta, and in Crustacea excepting the Phyllopods, the fundamental numbers are definite, no case of individual Variation in them has been observed.

Between these two extremes, there are animals in certain classes, for example, *Peripatus*, some of the Chilopoda among Myriapods, Aphroditidæ among Annelids, and some of the Branchiopoda among Crustacea, in which the number of segments does not increase indefinitely during life, but is nevertheless not so immutable as in the Insects and the majority of Crustacea. In the forms mentioned, certain numbers of segments, though not the same for the whole family, are characteristic of certain genera, as in the case of the Chilopoda (excepting Geophilidæ), or of certain species, as in some of the *Peripati*. But besides this, in some of the forms named, *e.g.*, the *Geophili* and *Peripatus edwardsii*, individual Variation has been recorded among members of the same species. It is unfortunate that for many of the forms in which Variation of this kind possibly takes place, no sufficient observation on the point has been made, but as examples of a phenomenon which, on any hypothesis, must have played a chief part in the evolution of these animals, the few available instances are of interest.

*1. **Peripatus.** The number of segments which have claw-bearing ambulatory legs differs in different species of this genus. While, moreover, in some of the species the number appears to be very constant for the species, in the case of others, great individual variation is seen to occur. SEDGWICK'S observations in the case of *P. edwardii* shew conclusively that these variations cannot be ascribed to difference in age. There is besides no ground for supposing that increase in the number of legs occurs in any species after birth, and it is in fact practically certain that this is not the case. In *Peripatus capensis*, which was exhaustively studied by Sedgwick, the appendages arise in the embryo successively from before backwards, the most posterior being the last to appear, and the full number is reached when the embryo arrives at Sedgwick's Stage G. The following is taken from the list constructed by Sedgwick from all sources, including his own observations. As the bibliography given by him is complete and easily accessible it is not repeated here, and the reader is referred to Sedgwick's monograph for reference to the original authorities.

SEDGWICK, A., *Quart. Jour. Micr. Sci.* XXVIII., 1888, pp. 431—493. Plates.

SOUTH AFRICAN SPECIES.

P. capensis: 17 pairs of claw-bearing ambulatory legs (Table Mountain, S. Africa).

P. balfouri: 18 pairs of legs, of which the last pair is rudimentary (Table Mountain, S. Africa).

Sedgwick has examined more than 1000 specimens from the Cape, and has only seen one specimen with more than 18 pairs of legs. This individual had 20 pairs, the last pair being rudimentary. It closely resembled *P. balfouri*, but differed in the number of legs and in certain other details (*q. v.*); Sedgwick regarded this form provisionally as a variety of *P. balfouri*.

P. mosleyi: 21 and 22 pairs of legs: near Williamstown, S. Africa. The specimens with 22 legs were two in number and were both females. They differed in certain other particulars from the form with 21 legs, but on the whole Sedgwick regards them as a variety of the same species.

P. brevis (DE BLAINVILLE): 14 pairs of legs. (This species not seen by Sedgwick.)

Other species from S. Africa which have been less fully studied are stated to have 19, 21 and 22 pairs of legs respectively.

In all South African forms, irrespective of the number of legs, the generative opening is subterminal and is placed behind the last pair of fully developed legs (between the 18th or rudimentary pair in *P. balfouri*). SEDGWICK, pp. 440 and 451.

AUSTRALASIAN SPECIES.

P. novæ-zealandiæ. 15 pairs of legs. New Zealand.
P. leuckartii. 15 pair of legs. Queensland.

In both of these species the generative opening is between the last pair of legs. (SEDGWICK, p. 486.)

NEOTROPICAL SPECIES.

In all the Neotropical Species which have been at all fully examined, the number of legs varies among individuals of the same species.

P. edwardsii: number of pairs of legs variable, the smallest number being 29 pairs, and the greatest number being 34. Males with 29 and 30 pairs of legs. The females are larger, and have a greater number of legs than the males.

The new-born young differ in the same way. From 4 females each having 29 legs, seven embryos were taken which were practically fully developed. Of these, 4 had 29 legs, 2 had 34, 1 had 32. An embryo with 29 and one with 30 were found in the same mother. An embryo, quite immature, but possessing the full number of legs, was found with a larger number of legs than one which occupied the part of the uterus next to the external opening. (Caracas.)

Peripatus demeraranus: 7 adult specimens had 30 pairs of legs; 6 had 31 pairs; 1 had 27 pairs. Out of 13 embryos examined, 7 have 30 pairs and 6 had 31. (Demerara.)

Peripatus trinidadensis: 28 to 31 pairs of ambulatory legs. (Trinidad.)

Peripatus torquatus: 41 to 42 pairs. (Trinidad.)

Specimens of other less fully known species are recorded as having respectively, 19, 28, 30, 32, 36 pairs of legs, &c.

In the Neotropical Species, irrespective of the number of legs, the generative opening is placed between the legs of the *penultimate pair*. (SEDGWICK, p. 487.)

Peripatus (*juliformis?*) from St Vincent: six specimens examined. Of these, 1 specimen had 34 pairs of legs, 2 had 32 pairs, 1 had 30 pairs, and 1 had 29 pairs. POCOCK, R. I., *Nature*, 1892, XLVI. p. 100.

In connexion with the case of *Peripatus*, the following evidence may be given, though very imperfect and incomplete.

2. **Myriapoda.** CHILOGNATHA. Variation in the number of segments composing the body in this division of Myriapods cannot be observed with certainty; for it is not possible to eliminate changes in number due to age, nevertheless the manner in which this increase occurs has a bearing on the subject.

In *Julus terrestris* the number of segments is increased at each moult by growth of new segments between the lately formed antepenultimate segment and the permanent penultimate segment. At each of the earlier moults *six* new segments are here added: in *Blaniulus* the number thus added is *four*, and in *Polydesmus?* two fresh segments are formed at each of the earlier moults. In each of these forms the number added is the same at each of the earlier moults. NEWPORT, G., *Phil. Trans.*, 1841, pp. 129 and 130.

CHILOPODA. The number of leg-bearing segments differs in the several genera of Chilopoda, but except in the Geophilidæ the number proper to each genus is a constant character. For instance in *Lithobius* this number is 15; in *Scolopendra* it is 21; in *Scolopendrops*, 23; in *Cryptops* 21, &c.

In Geophilidæ, however, the total number of moveable segments is much larger, ranging from about 35 to more than 200. Though not characteristic of genera, the number seems within limits to mark each particular *species*. It was found that male *Geophili* have fewer segments than the female. The males of *Arthronomalus longicornis* have 51 or 52 leg-bearing segments, while females usually have 53 or 54. Full-grown females of *Geophilus terrestris* have 83 or 84 pairs of legs and segments, and the males of the same species have 81 or 82. In a large Neapolitan species, *Geophilus lævigatus* BRUHL.? the variation is rather greater. In eight males the number varied between 96 and 99; in eleven females, between 103 and 107. Of two female *Geophilus sulcatus* one individual had 136 and the other 140. NEWPORT, G., *Trans. Linn. Soc.*, XIX. 1845, p. 427, &c.

[In some of the Chilopoda[1] an increase in the number of segments takes place after the larva hatches, but the variations mentioned above are recorded as occurring in fully formed specimens independently of changes due to age.]

In the foregoing cases, a fact which is often met in the Study of Variation is well seen. It often happens that in particular genera or in particular species, a considerable range of Meristic Variation is found, while in closely allied forms there is little or none. Examples of this are seen in the variability of the Geophilidæ as compared with the other Chilopoda, and in the neo-tropical species of *Peripatus* which vary in the number of legs, while *P. balfouri*, for instance, is very constant. It will be noticed that in both these cases, the absolute numbers of parts repeated are considerably higher in the variable than in the constant forms. But though such cases have given rise to general statements that series of organs containing a small number of members are, as such, less variable than series containing more members, these statements require considerable modification; for it is not difficult to give instances both in plants and in animals, where series made up of a small number of members, shew great meristic variability.

The bearings of these cases on the nature of Meristic Repetition and the conception of Homology will be considered hereafter. Here, however, it may be well to call attention to the fact that we have now before us cases in which various but characteristic numbers of legs or segments differentiate allied species or genera; that in assuming the truth of the Doctrine of Descent, we have expressed our belief that in each case the species with diverse numbers are descended from some common ancestor. In the evolution of these forms, therefore, the number has varied: this on the one hand. On the other hand, in *Geophilus* and in *Peripatus*, we see

[1] According to NEWPORT (*Trans. Linn. Soc.* XIX. 1845, p. 268), *all* Myriapoda acquire a periodical addition of segments and legs, but according to later observers this is not true of all the Chilopoda.

contemporary instances of the way in which such a change at its origin may be brought about. Though there are several things to be gained by study of these instances, one feature of them calls for attention now, namely, the *definiteness* of the variations recorded. The change from a form with one number to a form with another number here shews itself not as an infinitesimal addition or subtraction, but as a definite, discontinuous and *integral* change, producing it may be, as in *Peripatus edwardsii*, a variation amounting to several pairs of legs, properly formed, at one step of Descent. This will not be seen always to be the case, but it is none the less to be noted that it is so here.

Among Insects I know no case of such individual variation in the fundamental number of segments composing the body. Among Crustacea two somewhat remarkable examples must be mentioned, though it will be seen that both of them belong to categories very different from that with which we are now concerned. But inasmuch as they relate to the general subject of Meristic Variation they should not be omitted.

3. **Carcinus mænas.** The abdomen of these crabs consists normally of seven segments, including the last or telson. In the female the divisions between all these seven are very distinct. The abdomen of the normal male is much narrower than that of the female, and in it the divisions between the 3rd, 4th and 5th segments are obliterated. Males, however, which are inhabited by the Rhizocephalous parasite *Sacculina* do not acquire these sexual characters, and in them there are distinct divisions between the 3rd, 4th and 5th segments. (Fig. 9 c.)

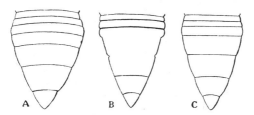

FIG. 9. A. Abdomen of *Carcinus mænas*, female, normal.
B. Abdomen of male, normal.
C. Abdomen of male infested by *Sacculina*. After GIARD and BONNIER.

In male *Carcinus mænas* inhabited by the Entoniscian parasite, *Portunion*, a similar deformity may occur, but is often very much less in extent, sometimes being only apparent in a slight alteration in the contour of the sixth abdominal somite. In specimens of *Portunus, Platyonychus, Pilumnus* and *Xantho* inhabited by Entoniscians, no change was observed. GIARD and BONNIER comment on the remarkable fact that the change in the sexual characters effected by *Sacculina* is greater than that resulting from the presence of Entoniscians; for since the latter are more internal parasites, preventing the growth of and actually replacing generative organs entirely or in part, it might have been expected that the consequences of their presence would be more profound. GIARD, A., and BONNIER, J., Contrib. à l'étude des Bopyriens, *Travaux de l'inst. zool. de Lille et du laboratoire zool. de Wimereux*, 1887, tom. V. p. 184.

4. **Branchipus and Artemia.** As it has been alleged that variation may be produced in the segmentation of the abdomen of these animals by changes in the waters in which they live, it is necessary here to give the facts on which this statement rests. The further question of the relation of *Artemia salina* to *A. milhausenii* is so closely connected with this subject, that though not strictly cognate, some account of the evidence on this point also must be given.

Some years ago SCHMANKEWITSCH[1] published certain papers on variations of *Artemia salina* induced by changes in the salinity of the water in which the animals lived. The statements there made excited a great deal of interest and have often been repeated both by scientific and popular writers. The facts have thus at times been somewhat misrepresented, and so much exaggeration has crept in, that before giving any further evidence it will be well to give Schmankewitsch's own account. It is frequently asserted that Schmankewitsch observed the conversion of *Branchipus* into *Artemia* and of *Artemia salina* into *A. milhausenii* following upon the progressive concentration of the waters of a salt lake. Strictly speaking however this is not what was stated by Schmankewitsch. His story is briefly this: That the salt lagoon, Kuyalnik, was divided by a dam into an upper and a lower part; the waters in the latter being saturated with salt, while the waters of the upper part were less salt. By a spring flood in the year 1871 the waters of the upper part of the lake swept over the dam and reduced the density of the lower waters to 8° Beaumé (=about sp. g. 1·051), and in this water great numbers of *A. salina* then appeared, presumably having been washed in from the upper part of the lake, or from the neighbouring salt pools. After this the dam was made good, and the waters of the lower lake by evaporation became more and more concentrated, being in the summer of 1872 14° B (about sp. g. 1·103); in 1873, 18° B (about sp. g. 1·135); in August 1874, 23·5° B (about sp. g. 1·177) and later in that year the salt began to crystallize out. In 1871 the *Artemiæ* had caudal fins of good size, bearing 8 to 12, rarely 15, bristles, but with the progressive concentration of the water the generations of *Artemia* progressively degenerated, until at the end of the summer of 1874 a large part of them had no caudal fins, thus presenting the character of *A. milhausenii* Fischer and Milne Edw. The successive stages of the diminution of the tail-fins and of the numbers of the bristles are shewn in the figures, with which all are now familiar.

A similar series was produced experimentally by gradual concentration of water, leading to the extreme form resembling *A. milhausenii*. It was found also that if the animals without caudal fins were kept in water which was gradually diluted, after some weeks a pair of conical prominences, each bearing a single bristle, appeared at the end of the abdomen.

It is further stated that the branchial plates[2] of the animals living in the more highly concentrated water were materially larger than those of animals living in water of a less concentration.

Schmankewitsch next goes on to say that by artificially breeding *Artemia salina* in more and more diluted salt water he obtained a form having the characters of SCHÄFFER's genus *Branchipus*, and that he considers this form as a new species of *Branchipus*. He explains this statement thus: In the normal *Artemia*, the last segment of the post-abdomen is about twice as long as each of the other segments, while the corresponding part in *Branchipus* is divided into two segments. He states that in his opinion the condition of the last segment of the post-abdomen constitutes the essential difference between *Artemia* and *Branchipus*, and that such division of the last segment occurred in the third generation of the form produced by him from *Artemia* by progressive dilution of the water. A second distinction between the genera is found in the fact that *Artemia* is reproduced parthenogenetically, while *Branchipus* is not known to be so reproduced. As to the condition of his new form in this respect, Schmankewitsch had no evidence.

In a subsequent paper, *Z. f. w. Z.*, 1877, further particulars are given, respecting especially the natural varieties of *A. salina*. Of these he distinguishes two, var. *a* and var. *b*. The first of these is distinguished by its greater size (8 lines instead of 6 lines, the average for the type) and by the greater length of the post-abdomen. In the type the bristles on each caudal fin are generally 8—12, and in

---

[1] *Z. f. w. Z.*, xxv., 1875, 2, p. 103 and xxix., 1877, p. 429; also in several Russian publications, to which references will be found *l. c.*

[2] Upon this point a good deal of interesting evidence is given in Schmankewitsch's papers, but as it does not bear immediately on the question of the specific differences, it has not been introduced here.

var. *a*, 8—15, rarely more. Amongst specimens of var. *a*, as also among those of the type, specimens may be found having three, two, or even only one bristle on the caudal fin. The second antennæ of the male are less wide in var. *a* than in the type, and the knobs on the inner border are rather larger than in the type.

The variety *b* was found in pools of a concentration of 4° Beaumé. It differs from the type in having the post-abdomen shorter in proportion, though the whole length is about the same. The number of bristles on the caudal fins is greater in the variety. The second antennæ of the male are narrower in the variety than in the type, and bear a tooth and a thickening of the skin internal to the rough knob-like projections. But the most important difference characterizing var. *b* is the appearance of transverse segmentation in the last (8th) post-abdominal segment. This, according to Schmankewitsch, does not amount to an actual segmentation, but is really a transverse annulation, which may be more or less conspicuous, and suggests an appearance of segmentation. Schmankewitsch looks on this second variety as a transitional form between *Artemia* and *Branchipus*.

Before going further it may be remarked that Schmankewitsch gives no figures of these varieties, except in so far as they are represented in the well-known series of sketches of the caudal forks with varying numbers of bristles. No analysis of the waters is given.

It will be seen that two principal and distinct statements are made:

(1) That *A. milhausenii* may be reared from *A. salina* by gradually raising the concentration of the water.

(2) That by diluting the water a division is produced in the last (8th) segment of *A. salina:* that this is a character, or, as Schmankewitsch says, the chief character, of the genus *Branchipus*.

First as to the relation of *A. salina* to *A. milhausenii*. The species *milhausenii* was made by G. FISCHER DE WALDHEIM[1] on spirit specimens sent to him, and the absence of caudal fins and bristles was taken as the diagnostic character. Fischer's figures are very poor, and indeed are scarcely recognizable: they are also incorrect in several points, giving for instance 12 pairs of swimming feet instead of 11. The description is also very imperfect. In the course of this he speaks of the male, saying that its second antennæ are larger than those of the female, in which he declares the second antennæ may be sometimes absent. From Fischer's account it is quite clear that his material was badly preserved, and indeed, as Schmankewitsch says, specimens of these animals preserved with spirit only are of little use.

In 1837 RATHKE[2] gave a better figure of *A. milhausenii* ♀ from the original locality of Fischer's specimens. The tail, ending in two plain lobes, is shown. The male is not mentioned. The following analysis of the water is given:

Potassium Sulphate	0·7453
Sodium Sulphate	2·4439
Magnesium Chloride	7·5500
Calcium Chloride	0·2760
Sodium Chloride	16·1200
	27·1352

in 100 of the water.

Other authors mention *A. milhausenii*, but there is, so far as I am aware, no special account of the male, or any material addition to the above.

I will now give an abstract of such further evidence on this subject as I have been able to collect.

In the course of a journey in Western Central Asia and Western Siberia I collected samples of Branchiopods from a great variety of localities. Of these two consist of *Branchipus ferox* (Milne Edwards), one of *Branchipus spinosus* (Milne Edwards), three of a species of *Branchipus* not clearly corresponding with any species of which a description is known to me, and the remainder of *Artemia*. All the species of *Branchipus* collected are quite clearly defined both in the male and the female, and have certainly nothing to do with the *Artemia*. Of the latter some preliminary account may now be given, as the facts bear on Schmankewitsch's problem. Omitting those which were badly preserved and those which do not contain adults, there remain twenty-eight samples, satisfactorily preserved with corrosive sublimate, from as many localities. Of these, eight contain males, all of them having the

---

[1] *Bull. Imp. Soc. Nat. Moscou*, 1834, VII. p. 452.
[2] *Mém. Ac. Sci. Pét.*, 1837, III. p. 395.

distinctive characters of *A. salina*. It is difficult to speak with confidence as to the species of an *Artemia* from the female alone, but by careful comparison I can find no point of structure which differentiates any of the remainder from the females found with males, and I therefore regard them as all of the same species, *A. salina*. The waters were of many kinds, some being large salt lakes, while others were small salt ponds or even pools. The specific gravities of these waters varied from 1·030 to 1·215, and judging from the results of the analysis of six samples, the composition of the waters is also very different. The specific gravities were measured in the field with a hydrometer reading to ·005, and on comparing these readings with the determinations of the Sp. G. of the samples brought home it appears that they were approximately correct, and I think therefore that these rough readings are fairly trustworthy. As to the composition of the waters not analyzed, nothing can be said with much confidence. As the analyses shew, some of these lakes contain chiefly chlorides, others chiefly sulphates, and so on. In a few (*e.g.* xxix) there is a great quantity of sodium carbonate, so much that the water was strongly alkaline and felt soapy to the hands. This can generally be recognized on the spot in various ways.

The first point raised by Schmankewitsch's work is that of the caudal fins. Among my samples I have every stage between the large fins with some twenty bristles, down to the condition with no distinct fin or bristles. The following table gives the results as regards the number of bristles on the caudal fins, and this

No. in Catalogue	Sp. G.	Bristles on single caudal fin. Eggbearing ♀♀ only	Remarks
XXIX.	1·030	10 to 24	Analyzed. Strongly alkaline. ♂♂ present.
LI.	1·050	11—13	
XXXIV.	1·056	9—17	♂♂ present.
XXV.	1·065	2— 7	♂♂ present.
XLII.	? 1·070		
XXXVII.	1·075	8—13	
XXXIX.	1·075	5— 7	
XLI.	1·085	13—15	
IV.	1·095	20—28	♂♂ present. This and III. both pools in one dry stream-bed.
XIV.	1·100	8—14	Analyzed.
XLV.	1·100	8—12	
XXVII.	1·100	4—10	
XXXI.	1·105	5— 9	♂♂ present.
XXXV.	1·105	4— 8	
XLIII.	1·115	1— 6	Analyzed.
XIX.	1·115	5— 9	
XL.	about 1·130	12—16	Pool in a stream-bed. ♂♂ present
LII.	1·140	3— 7	
XXXVI.	? 1·150	4—10	
XLIV.	1·150	7— 8	Analyzed.
XVI.	1·150	0— 1	
III.	1·160	16—19	♂♂ present. This and IV. both pools in one dry stream-bed.
XII.	1·165	1— 3	
XXII.	1·165	1— 5	
XVIII.	1·170	6— 8	
XXIII.	1·175	1— 5	
XXVI.	1·179	4— 9	Analyzed.
XXIV.	1·204	2— 5	Analyzed.
XXXII.	1·215	2— 4	
XXXIII.	1·215	2— 7	

number is a fair guide to the size of the fins, large fins for the most part having many bristles and small fins having few. In the third column the range of this number in several individuals is shewn, and for this purpose only *adult females bearing eggs in the ovisac* are reckoned, as with sex and age there are changes in respect of the number of bristles.

ANALYSIS OF WATER FROM SIX LOCALITIES CONTAINING *ARTEMIA SALINA*.

Catalogue Number	XXIX.	XIV.	XLIII.	XXVI.	XLIV.	XXIV.
Chlorine Cl_2	2·6950	24·8646	54·7793	70·8130	57·6653	61·0830
Sulphuric anhydride SO_3	5·9105	13·3585	30·3797	53·8150	71·8775	74·4463
Carbonic anhydride CO_2	7·0125	·3185	·3926	·2398	·3231	·2451
Lime CaO	·0311	·2256	·0678	·2266	·1466	·5175
Magnesia MgO	·0384	3·3561	6·0367	4·7514	4·5115	9·8394
Soda and Potash Na_2O, K_2O	16·7471	27·4589	63·6088	96·7906	100·0803	97·2084
Total	32·4346	69·5822	155·2649	226·6364	234·6043	243·3397
Oxygen equivalent to the Chlorine	·6082	5·6112	12·3620	15·9804	13·0133	13·7846
Total solids in 1000 grams	31·8264	63·9710	142·9029	210·6560	221·5910	229·5551
Sp. G. compared with Water at 20°	1·03074	1·05196	1·11787	1·17999	1·19586	1·20441

These analyses were undertaken for me by Mr H. ROBINSON, of the Cambridge University Chemical Laboratory, and my best thanks are due to him for the care with which he has conducted them.

The table shews the great variability in the development of the tails and bristles. In specimens from the same locality there is generally great difference, and even the numbers on the two fins of the same individual are rarely the same. It will be seen that on the whole the forms with few bristles came from waters of high specific gravity, thus generally agreeing with Schmankewitsch's statement. This relation to the salinity is not however very close, but Schmankewitsch never asserted that it was. He frequently refers to the existence of individuals with tails in several conditions of degeneration in the same water, and especially (*Z. f. w. Z.*, 1877, p. 482) he expressly states that in the original locality of *A. milhausenii* he found this form and with it several others intermediate between it and *A. salina*.

It will also be seen in the Table, that the three samples, IV, XL and III stand out as having far more bristles than other samples from waters of equal specific gravity. Each of these localities was exceptional, and all belong to one class. III and IV were pools in the dry bed of a stream in the Ḳara Ḳum, near the Irghiz river. They were close together, and must be joined in each spring. XL. was a pool in a somewhat similar dried stream-bed, coming down to the lake Tulu Bai in the district of Pavlodar. The conditions in these pools must be very different from those of the large, shallow, permanent salt lakes from which the other samples mostly came, and it is only fair to Schmankewitsch's case to remember that the water in such pools must be almost fresh during the early part of each summer.

On the whole, then, it seems satisfactorily shewn that the tailless form is connected by intermediate stages with the fully-tailed *A. salina*, and that this transition is at all events partly connected with the degrees of salinity of the water in which it lives. Almost each locality has its own pattern of *Artemia*, which differs from those of other localities in shades of colour, in average size, or in robustness, and in the average number of spines on the swimming feet, but none of these differences seem to be especially connected with the degree of salinity.

Passing now to the question of the distinctness of *A. milhausenii*, it seems clear that, as Rathke said, it should never have been considered a distinct species. The character of the finless tail, which is now seen to be one of degree, does not differentiate it satisfactorily, and, as Schmankewitsch found, it is to be seen swimming with fin-bearing individuals. It has never been shewn that there is a male *A. milhausenii*, with distinctive sexual characters, and among the Branchiopoda the various sexual characters of the second antennæ in the male are most strikingly distinctive of the several forms. While being in no sense desirous of disparaging the value of Schmankewitsch's very interesting observation, I think it is misleading to describe the change effected as a transformation of one species into another. Schmankewitsch himself expressly said that he did not so consider it, and it is unfortunate that such a description has been applied to this case.

The question of the division of the 8th post-abdominal segment of *Artemia*, stated to occur on dilution of the water, directly concerns the subject of Meristic Variation. As to the facts, there is no doubt that the tail of *Branchipus* appears to be made up of seven segments besides the two which bear the external generative organs, in all, nine, while in the commonest forms of *A. salina* there are only eight such segments; and that the difference lies in the fact that in the long terminal segment of *A. salina* there is generally no appearance of division. But as Claus[1] has shewn, the last apparent division in *Branchipus* is of a different character from that of the other abdominal segments. This is indeed easily seen in *B. ferox*, *B. stagnalis*, *B. spinosus*, &c., in which the appearance of the last division is very different from that of the other divisions. It appears, in fact, to be rather an annulation than a segmentation. In longitudinal sections the distinction is quite clear. Such a division, according to Schmankewitsch, appears in the third generation of *A. salina* bred in diluted salt water.

Among my own specimens an appearance of division in the last segment occurs in a considerable number, and these are not by any means from the most dilute waters alone, some of them being from waters of great concentration. For instance, the specimens in XXIX, LI, XXXVII, XXXIX and XIV, all have no trace of such division. On the other hand, it was found in several specimens from XXVI (Sp. G. 1·179) and XLIII (Sp. G. 1·115), while others from these localities did not shew it. These facts relate to adult females bearing eggs. I do not think, therefore, that the relation of this appearance of division to the salinity of the water is a constant one.

Lastly, as regards the relation of *Artemia* to *Branchipus*, Schmankewitsch has maintained that the division of the last abdominal segment is the only structural character really differentiating *Branchipus*. Claus (*l. c.*) pointed out that there are many other points of difference, and that the supposed division is not a structural character of great moment. But above all these, it should be remembered that by the sexual characters of the males, *Branchipus* is absolutely separated from *Artemia*. No *Branchipus* has any structure at all resembling the great leaf-like second antennæ of the male *A. salina* or *A. gracilis*[2] Verrill. Schmankewitsch remarks (*Z. f. w. Z.*, 1877, p. 492) that there are species of *Branchipus* (e.g. *B. ferox*) without the appendages characterizing the second antennæ of *B. stagnalis* ♂, &c., and that the males of *Artemia* bear on the second antennæ a knob, which is possibly the representative of the appendages of *Branchipus*, but nevertheless there is no resemblance whatever between the males of *B. ferox* or of any other *Branchipus* and those of *Artemia*, and there is no reason to suppose that these sexual characters are modified by the degree of concentration of the water. The statement that the descendants of an *Artemia* can be made to assume the characters of *Branchipus* Schäffer, depends entirely on the acceptance of Schmankewitsch's criterion of that genus, which is set up in practical disregard of the far more distinctive sexual characters. It is, besides, as has already been stated, only an irregular and possibly misleading relation which subsists between this appearance of segmentation and the salinity of the water[3].

---

[1] *Anz. Ak. Wiss. Wien*, 1886, p. 43; see also *idem*, *Abhandl. Göttingen*, 1873, Taf. III. Fig. 10, Taf. v. Fig. 16.

[2] For two samples of this American form I am indebted to Dr A. M. Norman, who received them from Professor Packard.

[3] I cannot leave this subject without expressing astonishment at the comparatively slight and evasive differences in the structure of *Artemiæ* and other Crustacea inhabiting waters of different salinity and composition. It is not a little

surprising that the animals living in No. XIV, for example, are scarcely distinguishable from those in No. XXIX, though the water in the latter was so strongly alkaline as to feel soapy. The conditions of animal life in these two waters must surely be very different, and yet no visible effect is produced. It is of course certain that there are great differences in the physiology of these forms, for, as I have often seen, animals (Copepoda, Cladocera, &c.) transferred from one water to another of materially different composition, die in a few minutes, though the second water may be inhabited by the same species; but in visible structure, the differences are for the most part trifling and equivocal.

CHAPTER III.

Linear Series—*continued*.

Vertebræ and Ribs.

The Meristic Variations of the vertebral column constitute a subject of some complexity. In considering them it must be remembered that numerical change may be brought about in the series of vertebræ by two different processes: first, by Variation in the total number of segments composing the whole column, in which case the variation is truly Meristic; and secondly by Variation in the number or ordinal position of the vertebræ comprised in one or more regions of the column, not necessarily involving change in the total number of segments forming the whole series, and in this case the variation is Homœotic. Though Homœotic Variation is often associated with change in the total number of segments, from the nature of the case it is rarely possible in any given instance to distinguish clearly whether such change has occurred or not. This arises largely from the fact that while to find the total number of vertebræ it is necessary to know the exact number of caudal vertebræ, in many specimens these are incomplete, and even if present their number cannot often be given with confidence. For these reasons the chief interest of this section of the facts arises in connexion with Homœotic Variation, and the modes in which it occurs; but it must be constantly borne in mind that in almost any given case there may be Meristic Variation also, though the evidence of this may be obscured.

True Meristic Variation in Vertebræ and Ribs.

I. *Vertebræ.*

True Meristic Variation, that is to say, change in the total number of segments composing the whole column, may nevertheless be plainly recognized in certain animals. Among some of the lower vertebrates, Fishes and Snakes, for example, the range of such Variation may be very great. Among Mammals the following may be given as an example of considerable Variation in the number of præsacral vertebræ in a wild animal, and such evidence may be multiplied indefinitely.

5. **Erinaceus europæus** (the Hedgehog).

	C	D	L	S	C	Total
No. 1	7	14	6	4	11	42
2	7	15	6	3	10 +	
3	7	16	6	3	9 +	
4	7	15	6	4	12	44
5	7	15	6	4	11	43
6	7	14	6	3	9 +	
7	7	15	6	3	11 or 12	
8	7	15	6	3	13	44
9	7	15	6	3	12 or 13	

Nos. 1—5 in Mus. Coll. Surg., see *Catalogue*, 1884, pp. 645 and 646; No. 6 in Cambridge Univ. Mus.; Nos. 7—9 in British Museum.

6. **Man.** The simplest form of true Meristic Variation in the total number of vertebræ may occur in Man by the formation of an extra coccygeal vertebra, making five coccygeals in addition to five sacrals, *i.e.* ten pelvic vertebræ in all. Instances of this are rare (STRUTHERS), though in many tailed forms such Variation is common. Two cases, in both of which the sixth piece (1st coccygeal) was partially ankylosed to the sacrum, are fully described by STRUTHERS, J., *Journ. Anat. Phys.*, 1875, pp. 93—96.

In the presence of cases like that last given, there is a strong suggestion that the number of vertebræ has been increased by simple addition of a new segment behind, after the fashion of a growing worm: the variation of vertebræ thus seems a simple thing. But there is evidence of other kinds which plainly shews this view of the matter to be quite inadequate. Some of these facts may now be offered, and in them we meet a class of fact which will again and again recur in other parts of the study of Repeated Parts.

IMPERFECT DIVISION OF VERTEBRÆ.

*7. **Python tigris**[1]. This is a case of great importance as illustrating several phenomena of Meristic Division. In a skeleton of *Python* in the *Mus. Coll. Surg.*, No. 602, the following peculiarities of structure are to be seen. Up to the 147th inclusive the vertebræ are normal, each having a pair of transverse processes and a

[1] This and the following cases of *Pelamis* and *Cimoliasaurus* are discussed by BAUR, G., *Jour. of Morph.*, IV. 1891, p. 333.

104   MERISTIC VARIATION.   [PART I.

pair of ribs. The appearance of the next vertebra is shewn in the figure (Fig. 10, I.). Anteriorly, and as far as the level of the posterior surface of the transverse processes, it is normal, save that its neural spine is rather small from before backwards. The transverse processes bear a pair of normal ribs. But behind this pair of transverse processes the parts, so to speak, begin again, rising again into a neural spine, and growing outwards into a second pair of transverse processes, with a second pair of normal ribs. Posteriorly again the parts are normal. This specimen is described in the *Catalogue* of 1853, as "148th and 149th vertebræ ankylosed," but upon a little reflexion it will be seen that this account misses the essential point. For the bone is not two vertebræ simply joined together as bones may be after inflammation or the like, but it is two vertebræ whose *adjacent parts are not formed*,

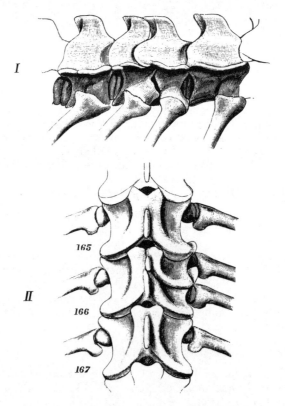

Fig. 10. Two examples of imperfect division of vertebræ in one specimen of *Python tigris*. I. The vertebræ 147—150 seen from the right side, shewing the imperfect division between the 148th and 149th. The condition on the left side is the same. II. View of dorsal surface of vertebræ 165—167, shewing duplicity of 166th vertebra on the right side. On this side it bears two ribs. The left side is normal. (From a skeleton, in Coll. Surg. Mus., No. 602.)

and between which the process of Division has been imperfect. With more reason it may be spoken of as one vertebra partly divided into two, but this description also scarcely recognizes the real nature of the phenomenon.

Further on, in the same specimen, at the 166th vertebra, there is an even more interesting variation. This vertebra is represented in Fig. 10, II. As there seen, it is normal on the left side, bearing one transverse process and one rib, while on the right side there are *two* complete transverse processes and *two* ribs. The 185th vertebra is also in exactly the same condition, being double on the right side and single on the left.

8. **Python sebæ :** a precisely similar case (Brussels Museum, No. 87, I. G.), in which the 195th vertebra is single on the right side and double with two ribs on the left, is described by ALBRECHT, P., *Bull. Mus. Nat. Hist. Belg.*, 1883, II. p. 21, Plate II.

9. **Python** sp.: a precisely similar case of duplicity in the 168th vertebra, on the left side, in a mounted skeleton in the *Camb. Univ. Mus.*

It is to be especially noticed that in each of these four cases of lateral duplicity, the degree to which the process of reduplication has gone on is the same.

10. **Pelamis bicolor** [= *Hydrophis*]. The 212th vertebra simple on the left side, and double on the right. It bears one rib on the left side and two ribs on the right side. *Yale Univ. Mus.*, No. 763. BAUR, G., *Jour. of Morph.*, IV. 1891, p. 333.

11. **Cimoliasaurus plicatus** (a Plesiosaur). " Centrum of a small and malformed cervical vertebra from the Oxford Clay near Oxford. This specimen is immature, and on one side is divided into two portions, each with its distinct costal facet." LYDEKKER, R., *Cat. Fossil Rept. and Amph. in Brit. Mus.*, Pt. II. 1889, p. 238, No. 48,001.

A case somewhat similar to the above is recorded in the Rabbit by BLAND SUTTON, *Trans. Path. Soc.*, XLI., 1890, p. 341. See also certain cases of a somewhat comparable variation in Man, considered in connexion with the variations of Bilateral Series.

II. *Ribs.*

12. **Man.** Partial division of ribs is more common than that of vertebræ. Five cases are given by STRUTHERS. 1. Fourth rib becoming broad, and bifurcated in front. Male, aged 93. From about middle of shaft these ribs gradually increase in length from 7 lines to $1\frac{1}{2}$ inch on the left side, $1\frac{1}{4}$ on right. They then fork, the left $1\frac{1}{4}$ inch, the right $\frac{1}{4}$ inch from where they join their cartilages. Cartilage of right forks close to rib, enclosing a space which admits little finger; cartilage of left lost, but the diverging bony divisions, each of good breadth for a rib (6 to 7 lines) enclose an intercostal space $1\frac{1}{2}$ inch long, attaining a breadth of $\frac{3}{4}$ inch, which was probably continued forwards by the division of the

cartilage or by two cartilages. The cartilage of the left 7th rib is also double for 1½ inch, all the others are normal. 2. Left fourth rib becoming very broad and bifurcating in front; two large spaces, one in the bone, one at the bifurcation. 3. Left fourth rib becoming broad towards sternal end, where it joins bifurcated cartilage. In these three cases the division affected the 4th rib. Three others are given in which the rib affected was probably the 4th or 5th. STRUTHERS, J., *Jour. Anat. Phys.*, Ser. 2, VIII. 1875, p. 51. Such cases are often recorded and preparations illustrating them may be seen in most museums.

Besides these cases of obviously Meristic Variation, there are many which are combined with Homœosis so as to produce far greater anatomical divergence. Though in some of these examples there may be change in the total number of vertebræ shewing that true Meristic change has occurred, they cannot well be treated apart from the more distinctly Homœotic cases.

HOMŒOTIC VARIATION IN VERTEBRÆ AND RIBS.

Homœosis in vertebræ may be best studied in Mammals, and the following account in the first instance relates chiefly to them. Before considering the details of such variations in vertebræ, it may be useful to describe briefly the ordinary system of nomenclature which is here followed. In treating this subject it is impossible to employ a terminology which does not seem to imply acceptance of the view that there is a true homology between the individual vertebræ of two spines containing different total numbers, for all the nomenclature of Comparative Anatomy is devised on this hypothesis. This difficulty is especially felt in regard to vertebræ, and at this point it should be expressly stated that in using the ordinary terms no such assent is intended. This matter has already been referred to in Section VI. of the Introduction, and will be discussed in relation to the facts to be given.

The vertebral column[1] is divided into five regions:—cervical, dorsal, lumbar, sacral and caudal. None of these regions can be absolutely defined, but the following features are generally used to differentiate them.

Cervical vertebræ are those of the anterior portion of the column, which either have no moveable ribs, or else have ribs which do not reach the sternum. *Dorsal* vertebræ are those which lie posterior to the cervicals and have moveable ribs. *Lumbar* vertebræ are those which succeed to the dorsals and have no moveable ribs. *Sacral* vertebræ cannot be defined in terms applicable even to the whole class of mammals, but, for the purpose of this consideration, it will be enough to use the term in the sense ordinarily given to it in human anatomy, to mean those vertebræ which are ankylosed together to form a sacrum. *Caudal* vertebræ are vertebræ posterior to the sacrum.

The characters thus defined are distributed among the several vertebræ according to their ordinal positions. Among mammals the number of vertebræ which develop the characters of each re-

[1] Abridged from FLOWER, W. H., *Mammals, Living and Extinct*, 1891, p. 41.

gion, though differing widely in different classificatory divisions, are as a rule maintained with some constancy within the limits of those divisions, which may be species, genera or larger groups, so that vertebral formulæ are often of diagnostic importance. Changes in the numbers of vertebræ composing the several regions must therefore have been an important factor in the evolution of the different forms.

Homœotic Variation in the spinal column consists in the assumption by one or more vertebræ of a structure which in the type is proper to vertebræ in a different ordinal position in the series. Examples of this are seen in the case of the development of ribs on a vertebra which by its ordinal position should be lumbar; or in the occurrence of a vertebra, normally lumbar, in the likeness of a sacral vertebra, having its transverse processes modified to support the pelvic girdle, &c. Variations of this kind have one character in common, which though at first sight obvious, will help us in interpreting certain other cases of Homœosis. In all cases of development of a vertebra normally belonging to one region, in the likeness of a vertebra of another region, this change always takes place in vertebræ adjacent to the region whose form is assumed. For example, if one vertebra, normally cervical, bears ribs, it is always the last cervical; if two cervicals bear ribs, they are the last two, and so on. No gaps are left.

Homœotic Variation in the spinal column may occur by the assumption of

(1) *dorsal* characters by a vertebra in the ordinal position of a *cervical*,

(2) *lumbar* characters by a vertebra in the ordinal position of a *dorsal*,

(3) *sacral* characters by a vertebra in the ordinal position of a *lumbar*,

(4) *coccygeal* characters by a vertebra in the ordinal position of a *sacral*,

or by the reverse of any of these. Since almost any of these changes may occur either alone or in conjunction with any of the others, it is not possible to group cases of such Homœosis under these heads, but the consideration of the more complex cases will be made easier if simple examples of each class are first described as seen in Man.

I. *Simple cases.*—**Man.**

(1) *Homœosis between cervical and dorsal vertebræ.*

(*a*) From cervical towards dorsal type.

The chief character distinguishing dorsal vertebræ is the possession of moveable ribs. This character may to a greater or less extent be assumed by cervicals.

*13. Cases of the development of ribs on the 6th cervical seem to be extremely rare. One is given by STRUTHERS in a young spine, æt. 4. The ribs were present as rudiments only, being the same on both sides in the 6th vertebra, and on the left side in the 7th. Each of these rib-elements was $\frac{5}{12}$ inch long. In the 6th the ribs rested on the body of the vertebra, but in the 7th the rib did not reach so far. Full details, q. v., STRUTHERS, *J. Anat. Phys.*, 1875, p. 32.

Cervical ribs on the 7th vertebra are comparatively common, being sometimes moveable and sometimes fixed. The literature of this subject up to 1868 is fully analyzed by WENZEL GRUBER, *Mém. Ac. Sci. Pét.*, Ser. VII. T. XIII., 1869, No. 2, who refers to 76 cases of such ribs, occurring in 45 bodies, being all that were known to him in literature or seen by himself. In addition to these 12 cases are described (10 in detail) by STRUTHERS (*l. c.*). Some of the results of an analysis of these cases are important to the study of Variation.

Of 57 cases, the ribs were present on both sides in 42 cases and on one side only in 15.

According to the degree of completeness with which the cervical ribs are developed, GRUBER divided them into four classes[1].

1. *Lowest development.* Cervical rib not reaching beyond the transverse process; corresponding to the vertebral end of a true rib with *capitulum* and *tuberculum*, and articulating by both of them. *Rare form.*

2. *Higher development.* Cervical rib reaching beyond the transverse process for a greater or less extent, either ending freely or joining with the first true rib. *Commonest form.*

3. *Still higher development.* Cervical rib reaching still further, and joining the cartilage of the first true rib either by its cartilaginous end or by a ligament continued from this. *Rarest form.*

4. *Complete development.* Cervical rib resembling a true rib, having a cartilage (generally for a greater or less part of its length united with the cartilage of the first true rib) connecting it with the sternum. *Less rare form.*

Gruber states, as the result of an analysis of 47 cases, that the third of these states is very rare, that the second condition is the common one, and that the fourth or complete condition is commoner than the first or least state of development, which is also rare. Of Struthers' cases the majority seem to belong to Gruber's second class, while that on the left side in Struthers' Case 4 must have approached Class 1, and that on the left side in Case 10 belonged to Class 3.

Two features in this evidence are of especial consequence: first

---

[1] Gruber considered that cervical ribs in Man are probably of two kinds, the one arising by development of an "epiphysis" on the superior transverse process, and the other by development of the "rib-rudiment" contained in the inferior transverse process. It is of cases of the latter kind that he is here speaking.

that the variation is more common on both sides than on one side; secondly, that it is not in its lowest development that it is most frequent, but rather in a condition of moderate completeness, having the proper parts of a true rib.

(b) From dorsal towards cervical type.

14. Reduction of ribs in the first dorsal is described by Struthers in a specimen in the Path. Mus. of Vienna. "The whole of the cervical vertebræ being present[1] there is no doubt as to the case being one of imperfect first rib. On left side rib goes about $\frac{2}{5}$ round, and articulates with a process of the second rib. On right side it joins second rib at from $\frac{1}{2}$ to 1 inch beyond tubercle, but again projects as a curved process where the subclavian artery has passed over it. The manubrium sterni first receives a broad cartilage, as if from one rib only, and secondly a cartilage at the junction of the manubrium and body which is the cartilage of the *third* thoracic rib." STRUTHERS, *J. Anat. Phys.*, 1875, p. 47, *Note*. (See also Nos. 24 and 25.)

(2) *Homœosis between dorsal and lumbar vertebræ.*

15. (a) From dorsal towards lumbar type. The characters chiefly distinguishing dorsal vertebræ from lumbars are the presence of ribs attached to the former, and of long, flat transverse processes in the latter. Secondly, the articular processes of lumbar vertebræ generally differ from those of most of the dorsal series, each pair of articular surfaces facing inwards and outwards respectively instead of upwards and downwards as they do in the dorsal region. The transition from the one type of process to the other, in passing down the column, is generally an abrupt and not a gradual one. In Man it occurs between the 12th dorsal and 1st lumbar, but in most Mammals it takes place more or less in front of the last dorsal, leaving several dorsal vertebræ with articular processes of the lumbar type. (STRUTHERS, *l. c.*, p. 59.)

Cases of rudimentary 12th rib in Man are not rare. When the last dorsal in this respect approaches to the lumbar type, the change of the articular process from dorsal to lumbar may take place higher than normally, as in STRUTHERS' Cases 1 and 2 (*l. c.* p. 54 and p. 57). In both of these the change was symmetrical, and in the first case it was abrupt and completed between the 11th and 12th dorsals, but in the second it was less complete. Though the place at which the change of articular processes takes place here varies in correlation with the diminution of the last ribs, both being higher than usual, such correlation is not always found, change in respect of either of those characters sometimes occurring alone.

---

[1] Struthers points out that unless the cervical vertebræ above the rudimentary ribs are counted there can be no certainty that in any given case these ribs are not extra cervical ribs.

(b) *From lumbar towards dorsal type.*

16. The formation of moveable ribs upon vertebræ normally belonging to the lumbar groups is in Man rarer than reduction of the 12th ribs. In these cases the ribs may or may not coexist with transverse processes of considerable size. In a case of 13th rib in Man, given by Struthers (*l. c.*, p. 60), the change of articular processes occurred a space lower than usual, being thus correlated with the appearance of ribs at a lower point.

(3) AND (4). *Homœosis between lumbar, sacral and coccygeal vertebræ.*

17. The differences between the vertebræ of these regions are far more matters of degree than those between the members of other vertebral regions. By detachment of the 1st sacral (25th vertebra) the lumbars may become 6, and in this case the 2nd sacral wholly or partially takes the characters proper to the 1st sacral, but this change is not necessarily accompanied by union between the last sacral and the 1st coccygeal (see, for example, STRUTHERS, *l.c.*, p. 68). On the other hand, the last lumbar may unite with the 1st sacral, and such union may be either symmetrical or unilateral only. The amount to which the ilium articulates with these vertebræ and the degree to which their processes are developed to support it also present many shades of variation. Similarly the last sacral may be free, or the 1st coccygeal may be united to the sacrum.

Since all these changes are manifestly questions of degree it would be interesting to know whether any particular positions in the series of changes are found more frequently than others, but I know no body of statistics from which this might be determined. In the absence of such determination there is no reason to suppose the existence of Discontinuity in these variations.

HOMŒOTIC VARIATION, VERTEBRÆ AND RIBS.

II. *More Complex Cases.*—**Man.**

From examples of the occurrence of Homœosis between members of the several regions we have now to pass to the more interesting question of the degree to which Homœosis in one part of the column may be correlated with similar Homœotic variation in the other parts. For, though each of the particular changes in the various regions may occur without correlated change in other regions, such correlation nevertheless often occurs, and in any consideration of magnitude of Variation it is a factor of importance. In several of the examples to be given it will be seen that the redistribution of regions is also associated with Meristic change in the total number of segments in the column. It is obvious that in the present place only the most summary notice of the various cases can be given.

Amongst them can be recognized two groups, the first in which the Homœosis is *from before backwards*, the second in which it is *from behind forwards*.

A few words in explanation of the use of these terms are perhaps needed.

In describing cases of such transformation in the series, it is usual to speak of structures, the pelvis for example, as "travelling forwards" or "travelling backwards." These modes of expression are to be avoided as introducing a false and confusing metaphor into the subject, for there is of course no movement of parts in either direction, and the natural process takes place by a development of certain segments in the likeness of structures which in the type occupy a different ordinal position in the series. In using the expression, Homœosis, we may in part avoid this confusion, and we may speak of the variation as occurring from before backwards or from behind forwards, according as the segment to whose form an approach is made stands in the normal series behind or in front of the segment whose variation is being considered. The formation of a cervical rib on the 7th vertebra is thus a backward Homœosis, for the 7th vertebra thus makes an approach to the characters of the 8th. On the other hand development of ribs on the 20th vertebra (1st lumbar), is a forward Homœosis, for the 20th vertebra then forms itself after the pattern of the normal 19th[1].

A. *Backward Homœosis.*

If each segment in the series of vertebræ were to be developed in the likeness of that which in the normal stands in the position next posterior to its own, we should expect the whole series to be one less than the normal. The following case makes an approach to this condition.

*18. Skeleton of old woman. C 7, D 11, L 5, S 5, C 4 (5th and 6th cervicals partially ankylosed). The 7th cervical bore a pair of cervical ribs [of Gruber's class 2, see p. 108], that on the left being ankylosed to the 7th cervical. There were only 11 pairs of thoracic ribs. The 1st lumbar was a true lumbar. GRUBER, WENZEL, *Mém. Ac. Sci. Pét.*, 1869, *Sér.* VII., XIII., No. 2, p. 23. Here the 7th vertebra resembles a dorsal in having ribs, the 19th, which in the type is the last dorsal, resembles a lumbar in all respects, the 24th is the 1st sacral, and there is no 33rd vertebra.

---

[1] The same terminology may conveniently be adopted in the case of the parts of flowers. Development of petals in the form of sepals being an *outward* Homœosis, while the formation of sepaloid petals would be thus called an *inward* Homœosis, and so forth.

*19. Male, in Cambridge Univ. Mus., No. 78. Preparation shews C 7, D 11, and the 19th vertebra formed as the 1st lumbar: remainder not preserved, but Professor A. Macalister kindly informs me that there were 5 lumbars and 5 sacrals, giving C 7, D 11, L 5, S 5. The 7th vertebra has cervical ribs, the left being large and articulating with a tuberosity on 1st thoracic rib, the right being considerably smaller, but now broken at the end. Only 11 pairs of thoracic ribs. Change of articular process from dorsal to lumbar begins partially on the left side between 17th and 18th vertebra (instead of between 19th and 20th) and is complete on both sides between 18th and 19th. The 19th bears no rib. [Backward Homœosis, greater on left side than on right, as seen in the greater size of the left rib on the 7th vertebra, and in the change of processes beginning at a higher level on this side. As the coccyx is not preserved it cannot be seen whether there is one segment less in the whole column, which would be the case were the backward Homœosis complete.]

20. Female, æt. 40. C 7, D 12, L 5, S 6, C 3. The 7th vertebra bore cerv. ribs, free on left, ankylosed to vertebra on right. Change of artic. processes partially on left side between 18th and 19th (instead of between 19th and 20th). Twelfth thoracic ribs short, being $1\frac{1}{8}$ in. long on left, $1\frac{3}{4}$ in. on right. STRUTHERS, J., *J. Anat. Phys.*, 1875, pp. 53 and 35. [There is therefore backward Homœosis, greater on the left side than on the right.]

21. Vertebræ C 7, D 11 or 12, L 5 or 4, S 6, C lost. Eleven pairs of ribs. The 19th vertebra having a transverse process on the left side resembling that of the vertebra next below it, as regards place of origin and its upward slope, but is longer than it by $\frac{1}{8}$ in. and is nearly a third broader and also thicker. On *right* side corresponding part is in two pieces. Change of articular processes complete between 18th and 19th (instead of between 19th and 20th). The 24th vertebra is united to sacrum, but is of unusual shape, differing greatly from a normal 1st sacral (25th vertebra). The 29th vertebra is nevertheless not detached from sacrum. STRUTHERS, *l. c.*, pp. 70 and 57.

22. Adolescent subject. 7th cervical, 12 dorsals and ribs, and 3 lumbars preserved. 11th ribs reduced, 4 in. long, $4\frac{1}{2}$ in. with cartilage. 12th ribs rudimentary, left 1 in., right $\frac{3}{4}$ in. long, breadth of each about $\frac{1}{8}$ in. Artic. processes change chiefly between 18th and 19th vertebræ. STRUTHERS, *l. c.*, p. 55.

23. Male, æt. 47. C 7, D 12, L 5, S 5, C 4. Twelfth ribs very unequal; right scarcely 2 in., left $3\frac{1}{2}$ in. The 5th lumbar ankylosed to sacrum by its right transverse process. STRUTHERS, *l. c.*, p. 57. [Backward Homœosis on right side in respect of reduction of 12th rib and union of 24th vertebra to sacrum on that side.]

B. *Forward Homœosis.*

As was remarked in the case of backward Homœosis, if each vertebra were to be developed in the likeness of the one which in the normal stands next behind it in ordinal sequence, we should expect such backward Homœosis to be accompanied by *reduction* in the total number of vertebræ; so, conversely we should expect forward Homœosis to be accompanied by an *increase* in total number. This will be found to be sometimes the case (*e.g.* No. 26).

*24. Male. C 7, D 13, L 5, S 5 [C not recorded]. 13 ribs on each side. The right side differed considerably from the left.
Right side. 1st rib resembled the usual supernumerary cervical, being moveable and extending $\frac{3}{4}$ in. from its tubercle. Greater part of ixth nerve crossed the neck of the rib; just before doing so *it was joined by large branch of* xth. The 2nd rib, borne by ninth vertebra, in all respects resembled a normal 1st rib. The 3rd rib articulated with sternum like a normal 2nd rib. In all, 8 ribs articulated with sternum on right side, as usual. The 13th rib (on 20th vertebra) was $4\frac{1}{2}$ in. long.
Left side. The 1st rib articulated with body and transverse process of 8th vertebra, connecting with sternum in normal position, but differing much from a normal 1st rib, being nearly straight with very slight horizontal curve. 2nd rib normal in form and direction; articulates with sternum $\frac{1}{3}$ in. higher than right 3rd rib, owing to the lower margin of manubrium being directed slightly obliquely upwards and to the left. In all, 8 ribs articulated with sternum, all below the first being at a level slightly higher than that of the right ribs. The 13th rib (on 20th vertebra) was $4\frac{3}{4}$ in. long. LANE, W. ARBUTHNOT, *J. Anat. Phys.*, 1885, p. 267 [full description and discussion].

In this remarkable case, by the reduction of the 1st rib on the right side, the 8th vertebra shews a forward Homœosis so far as that side is concerned. The 20th vertebra, bearing a pair of 13th ribs, also shews a forward Homœosis, but this seems to have been a little greater on the left than on the right (cp. No. 20), the right rib being a $\frac{1}{4}$ in. less in length. The fact that a large branch of the xth nerve on the right side joined the brachial plexus instead of the usually minute fibre is specially noteworthy, as shewing a forward Homœosis in the brachial plexus on the right side in correlation to the similar Homœosis appearing in the reduction of the 1st rib on the same side. (Compare Nos. 14 and 25.)

25. Skeleton C 7, D 12, L 6 [S and C not recorded]. First pair of ribs rudimentary, about $1\frac{1}{2}$ in. long, exactly alike, as small horns attached to 8th vertebra. Scalene muscles were inserted into 2nd rib. The 25th vertebra was free, but the first lumbar (20th vertebra) had no trace of a rib. BELLAMY, E., *J. Anat. Phys.*, 1885, p. 185.

[In this case there is forward Homœosis in the reduction of the first ribs and in the formation of the 25th vertebra as a lumbar, but there were no ribs on the 1st lumbar, which would

114   MERISTIC VARIATION.   [PART I.

have been expected had there been an even Homœosis throughout the dorso-lumbars.]

*26. Male, æt. 50. C 7, D 12, L 6, S 5, C "3 or 4, probably 4." Thirteen pairs of ribs, 13th ribs on 20th vertebra, nearly symmetrical, right 2 in. long; left 1⅞, and in breadth a little less than the right. The 6th lumbar, 25th vertebra, had the characters of a normal last lumbar (sc. 24th vertebra), including normal transverse processes. Coccyx in 3 moveable pieces, the 3rd apparently composed of two. There is therefore probably one more than the normal number in the whole series. STRUTHERS, *J. Anat. Phys.*, 1875, p. 62.

27. Male, æt. 56. C 7, D 12, L 6, S 5, C 3. Dorsal vertebræ and ribs normal. 20th vertebra normal, except that it has no trace of transverse processes; ribs have perhaps been present on it. 25th vertebra quite free from sacrum, but articulating with ilium by small facet on each side. The 1st coccygeal joined to sacrum. STRUTHERS, *l. c.*, p. 66 and p. 91. [Homœosis in absence of trans. processes in 20th vertebra, in separation of 25th from sacrum, and in union of 30th with sacrum.]

28. Skeleton C 7, D 12, L 6, S 5, C lost. The 25th vertebra is separate from the ilium and the sacrum, but the 30th is united to the latter. STRUTHERS, *l.c.*, p. 69.

29. Male, æt. 29. C 7, D 12, L 6 (1st bearing ribs—6th partially joined to sacrum), S 5 (exclusive of 5th lumbar), C 4. 20th vertebra bearing ribs; 25th partially free from sacrum but partly supporting the ilium, and one extra vertebra in the series. STRUTHERS, *l. c.*, p. 64 and p. 92.

30. Skeleton D 12, L 6, S 4, C 4. The 25th vertebra by right transverse process articulates with sacrum and on the same side with the ilium; the 30th, however, though moveable on the sacrum, has characters transitional between those of a 5th sacral and a 1st coccygeal. STRUTHERS, *l.c.*, p. 68 and p. 91.

31. Male. C 7, D 13, L 5, S and C ankylosed together of uncertain number. Articular processes change between 20th and 21st, i.e. a space lower than usual, but the processes between 19th and 20th are smaller than those higher up and are not quite symmetrical. The 20th vertebra bore rib on left side and rib has apparently been present on right, but probably not so much developed. STRUTHERS, *l. c.*, p. 64, *note.* [Forward Homœosis in development of ribs on 20th and in detachment of 25th.]

But though the variations of the vertebræ may thus in great measure be reduced to system, there remain other cases, rare in Man but not very uncommon in lower forms, which cannot be brought into any system yet devised. Such cases shew that the limits imposed by a system of individual homologies, between which we conceive the occurrence of Variation, are not natural limits, and that they may be set aside in nature. In the following case it may be especially noted that Variation in the segmentation of the spinal nerves does not necessarily coincide with that of the vertebræ. This fact will be more fully illustrated in the section of evidence respecting the spinal nerves.

*32. Female, æt. 40. As it stands, the grouping is C 6, D 12, L 6, S 5, C 3; in all 32, viz. one less than usual. The vertebral artery did not enter till 5th cervical (instead of 6th) on left side. The 7th vertebra bore a pair of ribs, left small, ceasing at middle of shaft; right has been sawn off, but has all the appearance of a rib that would have reached the sternum. The 19th vertebra bore no ribs, and has transverse processes like those of a normal 1st lumbar. 23rd has transverse processes triangular and sloping upwards, like those of normal last lumbar but one (sc. 23rd), though in a less degree: pedicle thicker than usual for this vertebra.

The articular processes change in the normal space, between 19th and 20th vertebræ. Sacrum 5; Coccyx represented by 3 pieces ankylosed together.

Two entire lumbar nerves went down from the lumbar region to the sacral plexus. [Bones described in detail, *q. v.*] STRUTHERS, *J. Anat. Phys.* 1875, p. 72 and p 29.

Here then the 7th vertebra shews backward Homœosis, imperfect on left side, but more complete on right. 19th having no ribs, shews the same, and this also appears in the absence of a 4th coccygeal. The fact that two entire lumbar nerves join the sacral plexus is also a variation of the same kind. But if the backward Homœosis were complete, the 24th vertebra should be the 1st sacral, and the 29th should be joined to the coccygeal. The change of articular processes moreover is in the normal place.

An example like this brings out the difficulty that besets the attempt to find an individual homology for each segment. If the characters proper to each segment in the type may be thus redistributed piecemeal amongst a different total number of segments, the question, which in this body corresponds to any given vertebra, say the 25th, in a normal body, cannot be answered. The matter is thus clearly summed up by STRUTHERS (*l. c.* p. 75):

" The variation in this case presents some complexity. To which region is the suppression of the vertebra to be referred? The lumbo-sacral nerves would seem to indicate that the lowest lumbar vertebra is the usual 1st sacral set free, thus accounting for the seemingly deficient pelvic vertebra, and leaving 23 instead of 24 vertebræ above. The appearance of suppression of a vertebra in the neck, is met by the consideration that the 7th vertebra carries ribs, imperfectly developed on one side, like cervical ribs.

"Then, although only 11 ribs remain, the next vertebra below, though rib-less, has the normal articular processes of a 12th dorsal (19th vertebra). If it is to be regarded as such, and not as the 1st lumbar, then the suppressed vertebra would be really a lumbar, although there are six free vertebræ between the thorax and the pelvis. Whichever view be taken, this case is an interesting one, as exhibiting variation in every region of the spine, and as shewing the importance of examining the entire spine before deciding as to a variation of any one part of it."

To the question, which vertebra is missing, there is no answer; or rather the answer is that there is *no* segment in this body strictly corresponding to the normal 7th, 20th, 25th, &c.; that the characters of these several segments are distributed afresh and upon no strict, consistent plan among the segments of this body, and that, therefore, there is no one segment missing from the body. Surely further efforts to answer questions like these can lead to no useful result.

Attempts to interpret Variation by the light of simple arithmetic serve only to obscure the real nature of Repetition and segmental differentiation; for by constantly admitting to the mind the fancy that this simple, subjective representation of these processes is the right guide, and that the tangible complexity in which they present themselves is a wrong one, we only become used to an idea which is not true to the facts and the real difficulty is shirked.

ANTHROPOID APES.

Though adding little that is new in kind to the foregoing specimen-cases occurring in Man, the following instances of Variation in the vertebræ of the Anthropoid Apes are of some interest if only as illustrations of the fact that the frequency of such Variation has no necessary relation to the conditions of civilization or domestication. (On the subject of Variation in the vertebræ of Anthropoids, see especially ROSENBERG's list, *Morph. Jahrb.* I. p. 160.)

Troglodytes niger (the Chimpanzee).

[In considering cases of variation in the Chimpanzee it should be borne in mind that there are several races and perhaps species included under this name, which have not been clearly distinguished. It is possible, therefore, that some of the variations recorded may be characteristic of these races and not actually individual variations.]

C 7, D 13, L 4, S 5.

This is the formula in the great majority of Skeletons (*v. auctt.*).

33.   An adult female having C 7, D 12, L 4, S 5, C 5, viz. one vertebra and one pair of ribs less than usual. This is a specimen of DU CHAILLU's *T. calvus*. It was received united by the natural ligaments and no vertebra therefore is lost. *Cat. Coll. Surg.*, 1884, II. No. 4.

34.   Specimen having rudimentary ribs unequally developed on the 21st vertebra. The 25th vertebra was transitional or lumbo-sacral in character. The 26th—30th formed the sacrum and there were 6 caudals, while other specimens had from 2 to 4. For the lumbo-sacral plexus of this specimen, see No. 71. ROSENBERG, *Morph.*

Jahrb., I. p. 160. *Tables, Note* 19. This case therefore shews forward Homœosis in the presence of ribs on the 21st, also in the transitional character of the 25th, together with increase in total number. This increase is however not always found when the 25th is lumbo-sacral, for, on the contrary one such case quoted by Rosenberg had only 4 caudals (*q.v.*).

In this form the number of vertebræ articulating with the ilium varies, and the number uniting with the sacrum is also liable to alterations probably connected with age. ROSENBERG, *l. c.*; *Cat. Coll. Surg.*, 1884, II. p. 3.

Gorilla savagii. C 7, D 13, occur in all skeletons of which I have found descriptions, making therefore one pair of ribs more than in Man[1].

The number of vertebræ articulating with the ilium and the number joining with the sacrum vary, perhaps with age. Cf. ROSENBERG, *l.c.*; *Cat. Coll. Surg.*; STRUTHERS, *J. Anat. Phys.*, 1875, p. 79 *note*, &c.

*35.   Adult female. C 7, D 12, L 4, S 5, C 3. This is a remarkable case. There is one rib-bearing vertebra less than usual, while the number of lumbo-sacrals is nine, as in the normal cases collected by Rosenberg. In a normal skeleton in the Camb. Mus. the articular processes change from the dorsal to the lumbar type between the 20th and 21st, but in this abnormal specimen the change is completed on the right side between the 19th and 20th as in Man, and on the left side, though the change has there also taken place, there is a curious irregularity in the fact that the posterior zygapophysis of the 19th is divided to form two processes which fit into two similar processes of the left anterior zygapophysis of the 20th vertebra. The rest is normal. *Cambridge Univ. Mus.*, 1161, F. [There is here, therefore, a backward Homœosis of all vertebræ from the 19th onwards; perhaps also an absolute diminution in the total number of segments. The simultaneous variation of both the number of ribs and the position of the

---

[1] Since this account was written, STRUTHERS has published a valuable paper (*Journ. Anat. Phys.*, 1892, XXVII. p. 131), giving particulars of twenty Gorilla skeletons. Of these the following are especially remarkable.

Female, C 8, D 13, L 3. The seventh cervical is formed like a sixth, and the eighth is formed as a seventh, bearing no rib. The vertebræ 9 to 21 bear ribs, those of the 21st being well formed and coming close to iliac crest. The change of articular processes from dorsal to lumbar type occurred between 21st and 22nd, namely, one vertebra lower than usual. There is thus a forward Homœosis in absence of ribs on 8th, in presence of ribs on 21st, and in the variation of position of the articular change.

Out of 20 skeletons 3 have 14 pairs of ribs (on 8th to 21st) instead of 13 pairs. In one of these the articular change also occurred one vertebra lower than usual. On p. 136 a case is described in which there was a remarkable asymmetry in the structure of the articular processes, which as Dr Struthers has pointed out to me, is in some respects like that here described as No. 35 in the text.

Struthers points out that it would be better in all cases to speak of the change of processes as from lumbar to dorsal instead of from dorsal to lumbar. I regret that this suggestion comes too late for me to adopt.

change of articular processes to the human numbers is especially worthy of notice.]

* **Simia satyrus** (Orang-utan). Out of eight skeletons in the Mus. Coll. Surg., C 7, D 12, L 4 occurs in seven. In young specimens the distinction between the last lumbar and the first sacral is clearly shewn by presence of pleurapophysial ossifications in the transverse processes of the latter. Thus though *Simia* resembles Man in the number of ribs, it differs in the total number of praesacral vertebræ. *Cat. Mus. Coll. Surg.*, 1884, II. p. 10.

The arrangement C 7, D 12, L 4, S 5 occurs in a great number of specimens (for cases quoted, see ROSENBERG, *Morph. Jahrb.*, I. p. 160, Tabellen; *Cat. Mus. Coll. Surg.* &c.)

36. Adult male, Sumatra. C 7, D 11, L 5, S 5, C 2. *Mus. Coll. Surg.*, No. 37.
37. Fœtal skeleton. C 7, D 11, L 5, S 5, C 2. TRINCHESE, S., *Ann. Mus. civ. Storia nat. Genova*, 1870, p. 4.
38. Adult. C 7, D 11, L 4, S + C, ankylosed together, containing 8 ? pieces. *Camb. Univ. Mus.*, 1160, A.
39. Adult. C 7, D 12, L 4, S 4, C 3. The last lumbar shared in supporting iliac bones. DE BLAINVILLE, *Ostéogr.*, Primates, *Fsc.* I. p. 29.
40. A young specimen, well preserved: there were certainly L 4, S 3, C 4, but in the adult mentioned above, one of the coccygeal was joined to the sacrum. DE BLAINVILLE, *ibid.*
41. Young specimen in spirit, C 7, D 12, L 4, S 5, C 2. ROSENBERG, E., *Morph. Jahrb.* I. p. 160.
42. Specimen in spirit, not full grown, C 7, D 12, L 4, S 5, C 1. There was no doubt that only one coccygeal was present. ROSENBERG, *ibid.*
43. [**Hylobates.** Considerable differences in the number of vertebræ and ribs found in this genus are recorded in the Catalogue of the Museum of the College of Surgeons, &c.; since however the specific divisions of the genus are very doubtful (see *Catalogue*, II. p. 15), it is not possible to consider these as necessarily individual variations. See also ROSENBERG, *l.c.*, *Tables.*]

BRADYPODIDÆ.

To the study of Variation of the vertebral regions the phenomena seen in the Sloths are of exceptional importance, and in attempts to trace the homologies of the segments special attention has always been paid to them. The following table contains brief particulars of 11 specimens of *Bradypus* and 11 of *Cholœpus* seen by myself in English museums, and of a few others of which descriptions have been published. To these is added a summary of 40 specimens of *Bradypus* and 9 of *Cholœpus* in German museums[1] examined by WELCKER. His account is unfortunately not given in detail, but I have tabulated his results so far as is

[1] viz. Göttingen, Tübingen, Marburg, Leipzig, Frankfurt, Berlin, Giessen, Jena and Halle.

possible. Welcker's list does not, I believe, include any of the specimens separately given in No. 44.

The determination of the *species* is quite uncertain. Welcker in his analysis does not divide the species of *Bradypus*. In the other cases I have simply taken the name given on the labels. As regards *Cholœpus* the confusion of species is much to be regretted, for according to the received account[1] the more northern species, *C. hoffmanni*, has only 6 cervicals, while *C. didactylus* has 7. In the table it will be seen that four specimens in different places have C 6, though generally marked *C. didactylus*. Possibly some or all of these are *C. hoffmanni*, and I have therefore entered them as *Cholœpus* sp. In the case of *Bradypus* it has not been alleged that the number of cervicals characterizes particular species, so the fact that the species are confused is of less consequence.

*44. **Bradypus.**

	C	D	L	S	C	
B. tridactylus	9	15	4	6	5+	C^8 minute c. r. rt. C^9 large c. r. both sides (one lost). D^{15} moveable r. rt., fixed on l. Camb. Mus.
,,	9	15	4	6	8+	C^9 no rib. Coll. Surg. 3427.
,,	9	15	4	6	9?	Brit. Mus. 919 a.
,,	9	15	4	5	12	Brit. Mus. 52. 9. 20. 5.
,,	9	15	4[5]	5	10	C^9 c. r. $\frac{1}{2}$ in. long. Univ. Coll. Lond.
,,	9	14	4	5	11	C^9 c. r. $\begin{cases} \text{l. } \frac{3}{4} \text{ in.} \\ \text{rt. } \frac{1}{2} \text{ in.} \end{cases}$ Coll. Surg. 3428.
sp.?	9	14	4	5	9	Oxford Mus.
,,	9	16	3	6	11	Coll. Surg. 3422.
,,	8	15	3	7	9?	7th sacral only ankylosed in part. Brit. Mus. 46. 10. 16. 14.
,,	9	15	4	5	11	C^9 small rib-like horn on l. Mus. Med.-Chir. Acad. Pétersb. Gruber[2].
sp.?	9					Gruber's *private collection*[2].
,,	9		Struthers[3]			C^8 may have borne rib on rt. C^9 $\begin{cases} \text{l. free c. r. 1st thoracic complete.} \\ \text{rt. c. r. ankylosed. 1st thor. } \frac{1}{2} \text{ in.} \\ \text{long, like a c. r.; ankylosed.} \end{cases}$
B. cuculliger	9	15	4	6	9	C^9 has pair short c. r. Brit. Mus. 921 b.
B. torquatus	9	14	4			Gruber[2].
ditto				5	6	10 Brit. Mus. 47. 4. 6. 5.
Bradypus sp.	8	15				⎫ 3 specimens from Brazil said to have
,, sp.	8	15				⎬ 8 cervicals. No detailed account
,, sp.	8					⎭ given. de Blainville[4].

[1] Flower, W. H., *Mammals, Living and Extinct*, 1891, p. 183.
[2] Gruber, *Mém. Imp. Ac. Sci. Pét.* Ser. vii., xiii. 1869, no. 2, p. 31.
[3] Struthers, *Jour. Anat. Phys.*, 1875, p. 48 note.
[4] de Blainville, *Ostéogr.*, Fsc. v., pp. 27, 28 and 64. In the place cited, de Blainville gives C 9, D 16, L 3, S 6, C 9—11 as the normal, but he does not say in how many specimens this formula was seen. I have therefore been unable to tabulate this observation. It will be seen that D 16 is quite exceptional, but as it occurred in the Coll. Surg. specimen no. 3422 it was described by Owen as the normal, and this statement has been copied by many authors, perhaps by de Blainville.
[5] Fourth lumbar ankylosed to sacrum by tr. proc.

120   MERISTIC VARIATION.   [PART I.

SUMMARY OF 40 CASES: WELCKER[1].

Bradypus	C	D	L				
	10	14	4	C^{10} no c. r.	2 cases[2].		
	10	14 or 15	4 or 3	C^{10} with c. r. of fair size.	3 cases.	On C^9 c. r. very small or absent. 29th is 1st sacral.	
	9	15 or 16	4 or 3		9 cases.		
	9	15	3	15 cases.	C^9 usually with c. r. 21 cases.	28th is 1st sacral.	
	9	14	4	6 cases.			
	9 or 8	14 or 15	3	C^9 has either large c. r. or complete r. 5 cases. (This normal in *B. torquatus*: once in *B. cuculliger*.)		27th is 1st sacral.	

(c. r., cervical rib. C^6, C^7, &c., sixth, seventh cervical vertebra, &c.)

*45. **Cholœpus.**

	C	D	L	S	Cd	
C. didactylus	7	23	3	8	4	Coll. Surg. 3435.
,,	7	24	3	7		Oxford.
,,	7	23	4	5		Coll. Surg. 3427 (Catalogue).
,,	7	23	3	7	6	Coll. Surg. 3424.
sp.	6	24	3	6	5	Cambridge.
sp.	6	23	3	9	3 or 4	Brit. Mus. 65. 3. 4. 5.
sp.	6	22	4	8	5	Univ. Coll. Lond.
sp.	6	21	3	8	5?	Brit. Mus. 1510 b.
C. hoffmanni	6	22	5	8	5?	Brit. Mus. 1510 c.
,,	6	21	4	7	5	Coll. Surg. 3439.
C. hoffmanni?	6^3	23	2	7	4?	Brit. Mus. 80. 5. 6. 84.

SUMMARY OF 9 CASES: WELCKER[4].

	C	D+L		
C. didactylus	7	27	1st sacral is the 35th.	2 cases.
,,	7	26	1st sacral is the 34th.	2 cases.
C. hoffmanni	6	27	1st sacral is the 34th.	1 case.
,,	6	26	1st sacral is the 33rd.	1 case.
,,	6^5	25	1st sacral is the 32nd.	3 cases.

[1] WELCKER, *Zool. Anz.* 1878, I. p. 294.
[2] This includes the celebrated specimen (in natural ligaments) described by RAPP, *Anat. Unters. d. Edent.*, Tübingen, ed. 1843, p. 18.
[3] This specimen is labelled *C. didactylus*, but coming from Ecuador and having this formula is probably *C. hoffmanni*. (Compare THOMAS, O., *P. Z. S.*, 1880, p. 492.) In it C^6 bears cervical rib articulating with shaft of the first thoracic rib.
[4] *Zool. Anz.* 1878, I. p. 295.
[5] In a specimen in Leipzig Museum, no. 459, the 6th cervical bears large ribs, of which the right nearly reaches the sternum, so that Welcker says that there are only 5 true cervical vertebræ. In another of these specimens there is a cervical rib on C^6 measuring 19 mm.

On this evidence several comments suggest themselves. First it should be noted that the Bradypodidæ strikingly exemplify the principle which Darwin has expressed, that forms which have an exceptional structure often shew an exceptional frequency of Variation. Among Mammals the Sloths are peculiar in having a number of cervicals other than 7, and from the tables given it will be seen that both the range and the frequency of numerical Variation is in them very great, not only as regards the cervicals, but as regards the vertebræ generally.

As concerning the correlation between Variation in the several regions, WELCKER points out that his results go to shew that there is such a relation, and that when the sacrum is far back, the ribs also begin further back, or at least are less developed on the cervicals. As he puts it. with a long trunk there is a long neck. This is a very remarkable conclusion, and it must be admitted that it is, to some extent, borne out by the additional cases given above. The connexion, however, is very irregular. For instance, the Cambridge specimen of *Bradypus*, though the 29th is the 1st sacral, has had cervical ribs of good size on the 9th vertebra, and even has a small one on the 8th. But taking the whole list together, Welcker's generalization agrees with the great majority of cases. Expressed in the terms defined above, we may therefore say that backward Homœosis of the lumbar segments is generally, though not quite always, correlated with backward Homœosis of the cervicals, and *vice versa*.

It will be seen further that this Variation concerns every region of the spine, and that even in the total number of præ-sacral vertebræ there is a wide range of variation, viz. from 27 to 29 in *Bradypus* (52 specimens) and from 30 to 34 in *Cholœpus* (20 specimens). Perhaps no domestic mammal shews a frequency of variation in the fundamental number of segments comparable with this. In this connexion it may be observed that the absolute number of dorso-lumbars in *Cholœpus* (25—27) is exceptionally large amongst mammals; but this is not the case in *Bradypus*.

If the case of *Bradypus* stood alone, some would of course recognize the occurrence of cervical ribs on the 9th and 8th vertebræ as an example of atavism, or return to the normal mammalian form with 7 cervicals. The occurrence of normal ribs on the 7th in *Cholœpus* and the occasional presence of cervical ribs on the 6th vertebra in this form, even reaching nearly to the sternum as in Welcker's Leipzig case, obviate the discussion of this hypothesis.

We have, then, in the Bradypodidæ an example of mammals in which the vertebræ undergo great Variation as regards both their total number and their regional distribution. As the tables shew this is no trifling thing, concerning merely the number of the caudal vertebræ, the detachment of epiphyses which may then be called ribs, or some other equivocal character, but on the contrary it effects besides changes in the number of præ-sacral segments, that is to say, of large portions of the body, each with their proper supply of nerves, vessels and the like, producing material change in the mechanics and economy of the whole body: this moreover in wild animals, struggling for their own lives, depending for their existence on the perfection and fitness of their bodily organization.

CARNIVORA.

The following cases, though few, have an interest as exemplifying vertebral Variation in another Order.

*46.   **Felis domestica.**   In all the skeletons of FELIDÆ that I have examined the formula is C 7, D 13, L 7, S 3. A specimen of the domestic Cat having C 7, D 14, L 7 is described by Struthers. The change of articular processes from dorsal to lumbar was completed between the 18th and 19th vertebræ but the posterior zygapophyses of the 17th, though of the dorsal type, have to some extent the characters of a transition-joint. As is stated below, the change in the domestic Cat normally occurs between the 17th and 18th. In this case therefore with increase in numbers of ribs the position of the articular change has varied. This case is described by STRUTHERS, *J. Anat. Phys.*, 1875, p. 64, *Note*, but the description there given differs in some respects from that stated above, which is taken from a letter kindly written by Professor Struthers in answer to my inquiries.

There is here forward Homœosis in the development of ribs on the 21st vertebra, in the alteration in position of the articular change, and in the fact that the 28th is not united to the sacrum.

As seen in some other cases, therefore, with forward Homœosis the number of præ-sacral vertebræ is increased; but as usual owing to the equivocal nature of caudal vertebræ it is not possible to state that the total number of vertebræ is greater.

Canis vulpes.   Normally, C 7, D 13, L 7; articular change from dorsal to lumbar between 17th and 18th.

47.   Specimen having C 7, D 14, L 6, in which further the articular change occurs partly between the 17th and 18th, and partly between the 18th and 19th. In *Mus. Coll. Surg. Edin.* Information as to this specimen was kindly sent me by Professor STRUTHERS.

48.   **Jackal.**   Specimen having C 7, D 13, L 8 instead of 7. Articular change as usual between 17th and 18th. STRUTHERS *in litt.*

49.   **Canis familiaris.**   Case of cervical rib on left side borne by 7th cervical. This rib was 1½ in. long and articulated with a tubercular elevation on the 1st thoracic rib of the same side. The remaining ribs and vertebræ were normal. [fully described] GRUBER, W., *Arch. f. Anat. Phys., u. wiss. Med.*, 1867, p. 42, Plate.

[In connexion with the foregoing observations it may be mentioned that the articular change does not take place in the same place in all Felidæ. In 4 specimens of *F. leo*, 2 of *F. tigris* and 2 of *F. pardus*, in Edinburgh, and in one Lion and one Tiger in Cambridge the lumbar type begins between the 18th and 19th as in Struther's abnormal Cat above described; but in 4 *F. domestica*, and 2 *F. catus* in Edinburgh, 1 *F. domestica*, 1 *F. catus*, 1 *F. concolor* and 1 *Cynœlurus jubatus* in Cambridge the change is between the 17th and 18th. For information as to the Edinburgh specimens, I am indebted to Professor STRUTHERS.]

50. **Galictis vittata.** Specimen from Paraná had 16 pairs of ribs, 11 true and 5 false; 5 lumbar, 2 sacral and 21 caudal vertebræ.

A specimen from Brazil had only 15 pairs of ribs and the same number of lumbar and sacral vertebræ. BURMEISTER, *Reise durch d. La Plata-Staaten*, Halle, 1861, II. p. 409.

[This is therefore another case of forward Homœosis, (as manifested in the presence of an additional pair of ribs) associated with an *increase* in the number of præsacral vertebræ.]

51. **Halichœrus grypus.** Phocidæ generally have C 7, D 15, L 5. Specimen of *H. grypus* having C 7, D 15, L 6 at Berlin. The anterior of the six lumbars bears a rudimentary rib about 5 cm. in length on the left side. The 28th vertebra is here detached from the sacrum giving S 3, but generally it is united to it, giving S 4. NEHRING, A., *Sitzb. naturf. Fr. Berlin*, 1883, pp. 121 and 122. There is here therefore a forward Homœosis in the development of a rib on the 23rd, and also in the detachment of the 28th from the sacrum.

REPTILIA.

52. Mr Boulenger kindly informs me that though the number of ventral shields (which is the same as that of the vertebræ) is as a rule very variable in the several species of Snakes as a whole, there is nevertheless great difference in the degree of variability. A case of maximum variation is that of **Polyodontophis subpunctatus,** in which the number of ventral shields has been observed to vary from 151 to 240 (BOULENGER, *Fauna of Brit. India; Reptilia* &c. 1890, p. 303).

53. On the other hand the range of variation in **Tropidonotus natrix** is unusually small. Among 141 specimens examined the number of ventral shields varied from 162 to 190 (STRAUCH, *Mém. Ac. Sci. Pét.*, 1873, XXI., No. 4, pp. 142 and 144).

*54. **Gavialis gangeticus.** In this animal there are normally present 24 præsacral vertebræ and 2 sacrals, the first caudal being the 27th. This vertebra has a peculiar form, being biconvex. Specimen described having 25 præsacrals, 2 sacrals, the 28th being the first caudal. BAUR, G., *J. of Morph.*, IV., 1891, p. 334. In this case Baur argues that since the first caudal is clearly recognizable by its peculiar shape, this vertebra must be "homologous" in the two specimens and he considers that a vertebra must have been "intercalated" at some point anterior to the first caudal by a process similar to that seen in *Python* (see No. 7). In his judgment this has occurred between the 9th and 10th vertebræ, but no reason for this view is given. On the system here adopted, this would be spoken of as a case of forward Homœosis.

55. **Heloderma.** The first caudal in the normal form may be distinguished by having a perforation in the small rib connected with

it. In this it is peculiar. Four specimens shewed the following arrangements:—

H. horridum No. 1. First caudal is the 36th vertebra (Troschel).
ditto No. 2. ....37th...... .....(Baur).
H. suspectum No. 1. ....38th............(Shufeldt).
ditto No. 2. ....39th............(Baur).

BAUR, G., *J. of Morph.* IV. 1891, p. 335.

. BATRACHIA.[1]

*56. **Rana temporaria.** In the normal frog there are nine separate vertebræ in addition to the urostyle. A specimen is described by BOURNE having 10 free vertebræ (Fig. 11, III.). The axis and third vertebra bore tubercles upon the transverse processes, perhaps representing a partial bifurcation of the kind described in No. 58. The ninth vertebra was abnormal in having zygapophyses, and in that its centrum presented two concavities

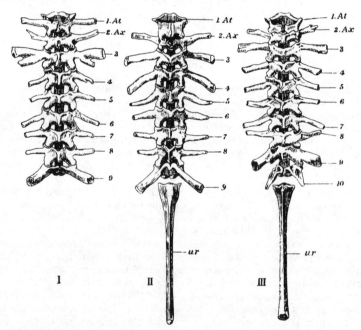

FIGURE 11. Vertebral columns of Frog (*Rana temporaria*), after BOURNE.
 I. Specimen having transverse processes borne by the atlas, together with other abnormalities described in text No. 58.
 II. Normal Vertebral Column.
 III. Specimen having ten free vertebræ, described in text, No. 56.

[1] I regret that the paper bearing on this subject lately published by ADOLPHI, *Morph. Jahrb.*, 1892, XIX. p. 313, appeared too late to permit me to incorporate the valuable facts it contains.

for articulation with a tenth vertebra. The right zygapophysis was well formed and articulated with the tenth, but the left was rudimentary. The tenth vertebra itself had an imperfect centrum and the neural arch though complete was markedly asymmetrical. Posteriorly its centrum presented two convexities for articulation with the urostyle. [For details see original figures.] BOURNE, A. G., *Quart. J. Micr. Sci.*, XXIV. 1884, p. 87.

This is a case of some importance as exhibiting Meristic Variation in a simple form. Of course, as Bourne says, we may say that in this specimen the end of the urostyle has been segmented off and that it is composed of "potential" vertebræ, and as he also remarks, it is interesting in this connexion to notice that some Anura, *e.g. Discoglossus*, present one or two pairs of transverse processes placed one behind the other at the proximal end of the urostyle. But this description is still some way from expressing all that has happened in this case; for beyond the separation of a tenth segment from the general mass of the urostyle there is Substantive Variation in the ninth vertebra in correlation with this Meristic Variation. For the ninth has developed a zygapophysis and has two *concavities* behind, like the vertebræ which in the normal frog are anterior to the ninth. There is therefore a forward Homœosis, associated with an increase in number of segments, just as there is in such a case as that of Man (No. 26) or in that of *Galictis vittata* (No. 50).

It is also interesting in this case to see that the actually last free vertebra here, though it is the 10th, has two *convex* articular surfaces behind like the 9th, which is the last in the normal frog, thus shewing a similar forward Homœosis. Now applying the ordinary conception of Homology to this case, we may, as Bourne says, prove that the 9th in it is homologous with the 9th in a normal frog for its transverse processes are enlarged in the characteristic manner to carry the pelvic girdle. But similarly we may prove also that the *tenth* in this case is homologous with the ninth of the normal, for its centrum has the peculiar convexities characterizing the last free vertebra. Baur's proof that the first caudal was homologous in the two specimens of *Gavialis* (see No. 54) rested on the same class of evidence, and for the moment is satisfying, but as here seen this method though so long established leads to a dead-lock. Upon this case it may be well to lay some stress, for the issues raised are here so easily seen. Besides this the imperfect condition of the extra vertebra enables us to see the phenomenon of increase in a transitional state, a condition rarely found. In the instances recorded in *Gavialis* (No. 54), owing to the perfection and completeness of the variation, the characters of the 1st caudal are definitely present in the 28th though normally proper to the 27th, and therefore it may be argued that the 28th here *is* the 27th of the type. The frog here described shews that in this conclusion other possibilities are not met. On the analogy of several cases already given, it is not impossible that if the variation seen in this frog had gone further, the 10th vertebra might alone support the ilium (cp. Nos. 57 and 60) and thus present the characters of the normal 9th in their completeness. If this change had taken place, we should have a case like that of *Gavialis*, and there would be nothing to shew that the new 10th vertebra was not the 9th of the normal. The truth then seems to be that owing to the correlation between Meristic Variation producing change in number, and simultaneous Substantive Variation producing a change of form or rather a redistribution of characters, the attempt to trace individual homologies must necessarily fail; for while such determination must be based either on ordinal position or on structural differentiation, neither of these criterions are really sound. As I have tried to shew, the belief that they are so depends rather on preconception than on the facts of Variation.

*57. A male specimen of *R. temporaria* ♂ with *ten* free vertebræ is described by Howes. In this case the 9th had a posterior zygapophysis on the left side only. Upon the left side the transverse process of the 9th was not larger than that of the 8th and did not support the ilium, which on the left side was entirely borne by the large transverse process of the 10th. On the right side the transverse processes of both 9th and 10th were developed to support the ilium, neither being in itself so large as that of the 10th on the left side. The 9th was concave in front instead of convex as usual, and thus the 8th which is normally biconcave is convex behind. The posterior faces of both 9th and 10th bore two convexities such as are normal to the 9th. The urostyle was normal, having well-developed apertures for exit of the last pair of spinal nerves. Howes, G. B., *Anat. Anz.*, I. 1886, p. 277, *figures.*

In this case the departure from the normal, exemplified by No. 56, has gone still further, and the new 10th vertebra bears the ilium wholly on the left side and in part on the right. The condition is thus again intermediate between the normal and a complete transformation of the 9th into a trunk vertebra and the introduction of a 10th to bear the ilium (as in No. 60). As regards the homologies of the vertebræ, the same issues are again raised which were indicated in regard to No. 56.

58. **Rana temporaria:** Case in which transverse processes were present in the atlas vertebra and the transverse processes of several of the vertebræ were abnormal (Fig. 11, I.). The atlas possessed well-developed transverse processes.

In the axis the transverse processes are directed forwards instead of backwards, and that of the *left* side presents an indication of bifurcation at its extremity.

The third vertebra possessed two pairs of transverse processes which are joined together for two-thirds of their length. The fourth vertebra presents a transverse process on the right side which is bifurcated at its extremity.

The remaining vertebræ, though slightly asymmetrical, present no special peculiarity, except that the neural arch of the ninth vertebra is feebly developed. BOURNE, A. G., *Quart. Journ. Micr. Sci.*, 1884, XXIV., p. 86, *Plate.*

There is here backward Homœosis of the atlas, the only case of the kind I have met with[1]. The reduplication of the transverse processes of the third vertebra should be studied in connexion with the cases of double vertebræ in *Python* (No. 7) and the cases of bifid rib (in Man, No. 12), for they present a variation perhaps intermediate between these two phenomena.

Bombinator igneus. In this form there is a considerable range of variation in the development of the transverse processes for the attachment of the pelvic girdle.

59. GÖTTE figures a specimen in which the flat expanded transverse processes have a similar extent on the two sides, but while that on the right side is made up of the processes of the 9th and 10th vertebræ (in about the proportions of two to one), that on the left side is entirely formed by the transverse process of the 10th vertebra. GÖTTE, *Entw. d. Unke, Atlas*, Pl. XIX., fig. 346.

*60. Sardinian specimen figured in which the processes for the attachment of the pelvic girdle seem to be composed entirely by those of the 10th vertebra while those of the 9th are not developed. GENÉ, J., *Mem. Reale Ac. di Torino*, S. 2, I., Pl. v., fig. 4.

61. Specimen figured in which both transverse processes of 9th and of the 10th are almost equally developed to carry the pelvic girdle. CAMERANO, L., *Atti R. Ac. Sci. Torino*, 1880, XV., *fig.* 3.

62. Specimen in which the *left* transverse process of the 9th bears the pelvic girdle on the left side, and the *right* transverse process of the 10th bears it on the right side, while the corresponding processes of the opposite sides were not developed. Similar case recorded in **Alytes obstetricans** by LATASTE, *Rev. int. des Sci.*, III., p. 49, 1879 [not seen, W.B.]; *ibid. fig.* 4.

63. Specimen in which the transverse processes of the 9th alone were developed to carry pelvic girdle, but the proximal end of the urostyle was laterally expanded more than usual, *ibid.* p. 7, *fig.* 3.

[Case of hypertrophy of coccyx, *ibid. fig.* 6; *ad hoc v.* BEDRIAGA, *Zool. Anz.*, 1879, II., p. 664; CAMERANO, *Atti R. Ac. Sci. Torino*, XV., p. 8.]

Recapitulation of important features of Variation as seen in the vertebral column.

I. *As regards fact.*
1. The magnitude of the variations.
2. The rarity of imperfect vertebræ.
3. The phenomenon of imperfect Division of vertebræ and ribs.

ADOLPHI, *l. c.*, p. 352, Pl. XII. fig. 3 gives an account of a specimen of **Bufo variabilis** in which the atlas bore a transverse process on the left side only. In this specimen the first two vertebræ were united and their total length was reduced.

4. The frequency of substantial if imperfect bilateral symmetry in the variations, but the occasional occurrence of asymmetry also.
5. The special variability of some types, e.g. *Simia satyrus*; the Bradypodidæ; *Bombinator igneus*.
6. The evidence that this variability may occur without the influence of civilization or domestication.

II. *As regards principle.*
1. The occasional, though not universal, association of forward Homœosis with increase in number and of backward Homœosis with reduction in number.
2. The frequent correlation between Variation in several regions, such correlated Variation being sometimes unilateral.
3. The impossibility of applying a scheme of Homology between individual segments.

CHAPTER IV.

LINEAR SERIES—*continued.*

SPINAL NERVES.

THE spinal nerves compose a Meristic Series in many respects similar to that of the vertebræ. As between the vertebræ, so between the spinal nerves, there is differentiation according to the ordinal succession of the members, certain distributions and functions being proper to nerves in certain ordinal positions. The study of the way in which Variation occurs in this series is one of great interest, but unfortunately it is extremely complicated. For while as regards vertebræ the distribution of structural differentiation can be recognized on inspection, in the spinal nerves to obtain a true knowledge of the arrangement in any one case physiological investigation or at least elaborate and special methods of dissection are needed. Though it is therefore impossible to introduce any account which should at all adequately represent the great diversity of possible arrangements, it is nevertheless necessary to refer briefly to the chief results attained by these methods and to the principles which have been detected in the Variation of the nerves. It must of course be foreign to our purposes to examine the many diversities of pattern produced by the divisions and anastomoses of nerve-cords in the formation of plexuses, &c., and we must confine our consideration to cases of Variation in the distribution of differentiation among the spinal nerves, that is to say, in the segmentation of the nervous system in so far as it may be judged from the arrangement of spinal nerves.

Some conception of the magnitude and range of Variation found in single species of Birds may be gained by reference to the beautiful researches of FÜRBRINGER[1]. A table is given by Fürbringer, shewing the number and serial position of the spinal nerves which take part in the formation of the brachial plexus in 67 species of

[1] Fürbringer's memoirs are of such magnitude and completeness that I have felt it to be somewhat of an impertinence to attempt to make selection from them; and it must be remembered that from the isolated and typical cases here given, only a distorted view of the evidence can be gained. As regards this subject, therefore, reference to the original work is especially needed.

130 MERISTIC VARIATION. [PART I.

64. In the following Table the Roman numbers are the ordinal numbers of the spinal nerves forming the brachial plexus.

The * marks the division between cervical and dorsal regions. In cases where the last cervical nerve is the last of the plexus the fact is not specially indicated.

	X.	XI.	XII.	XIII.	XIV.	XV.	XVI.	XVII.	XVIII.	XIX.	XX.	XXI.	XXII.	XXIII.	XXIV.	XXV.	XXVI.
Struthio camelus																	
Rhea americana								XVII.	XVIII.	XIX.	XX.	XXI.					
Apteryx australis					XIV.	XV.	XVI.	XVIII.	XVIII.								
Colymbus arcticus				XIII.	XIV.	XV.	XVI.	........*									
Chroicocephalus ridibundus			XII.	XIII.	XIV.	XV.	XVI.										
Puffinus obscurus				XIII.	XIV.	XV.	XVI.										
,, ,,				XIII.	XIV.	XV.	XVI.										
Pelecanus rufescens						XV.	XVI.	XVIII.	XVIII.	XIX.							
Anser cinereus						XV.	XVI.	XVII.	XVIII.	XIX.	XX.						
,, ,,							XVI.	XVII.	XVIII.	XIX.*	XX.						
,, ,,							XVI.	XVII.	XVIII.	XIX.	XX.						
Cygnus atratus										XIX.	XX.	XXI.	XXII.	XXIII.	XXIV.	XXV.	XXVI.
Phœnicopterus ruber				XIII.	XIV.	XV.	XVI.	XVII.	XVIII.	XIX.	XX.*	XXI.					
Grus canadensis			XII.	XIII.	XIV.	XV.	XVI.*	XVII.*									
Charadrius pluvialis						XV.	XVI.	XVII.									
Gallus domesticus						XV.	XVI.	XVII.	XVIII.	XIX.	........*						
,, ,,							........*										
Opisthocomus cristatus	XI.																
Goura coronata		XII.	XIII.	XIV.	XV.												

SPINAL NERVES: BIRDS.

	X.	XI.	XII.	XIII.	XIV.	XV.	XVI.	XVII.	XVIII.	XIX.	XX.	XXI.	XXII.	XXIII.	XXIV.	XXV.	XXVI.
Columba livia	X.	XI.	XII.	XIII.	XIV.	XV.											
,, ,, ...			XII.	XIII.	XIV.	XV.											
,, ,, ...			XII.	XIII.	XIV.	XV.											
,, ,, ...		XI.	XII.	XIII.	XIV.	XV.											
Uraetos audax			XII.	XIII.	XIV.	XV.*	*XVI.										
Buteo vulgaris			XII.	XIII.	XIV.	XV.	*XVI.										
Caprimulgus europaeus		XI.	XII.	XIII.	XIV.	*XV.	*XVI.										
Podargus humeralis				XIII.	XIV.	XV.											
,, ,, ...		XI.	XII.	XIII.	XIV.*												
Bucorvus abyssinicus		XI.	XII.	XIII.	XIV.	*XV.											
Cypselus apus	X.	XI.	XII.	XIII.	XIV.	XV.											
Picus medius		XI.	XII.	XIII.	XIV.	XV.											
Geeinus viridis			XII.	XIII.	XIV.	XV.											
,, ,, ...		XI.	XII.	XIII.	XIV.	XV.											
Garrulus glandarius			XII.	XIII.	XIV.	XV.											
,, ,, ...		XI.	XII.	XIII.	XIV.	XV.											
Corvus corone			XII.	XIII.	XIV.	XV.											
Turdus pilaris			XII.	XIII.	XIV.	XV.											

Selected from the table given by FÜRBRINGER, M., *Unters. zur Morph. u. Syst. der Vögel*, 1888, pp. 240 and 241.

(For drawings of the arrangements see the original work.)

Birds investigated by himself. He also gives particulars of the individual variations which were found in certain cases. From this table the following statement is compiled, shewing the most important diversities met with and the instances of individual Variation. In the majority of cases the most posterior spinal nerve of the cervical region was the most posterior nerve of the brachial plexus, but in a certain number of cases it does not join the plexus at all; in some other cases the anterior spinal nerve of the dorsal region also takes part in forming the plexus. As the table shews, each of these plans has been likewise met with as an individual variation.

Fürbringer's table shews 3 as the minimum number of spinal nerves found taking part in the formation of the plexus of any bird (*Bucorvus abyssinicus*): the same number has been found as a minimum by other observers in other birds (v. FÜRBRINGER, p. 242, *note*). The maximum number was 6, found in *Charadrius* and some specimens of *Columba*. The plexus is generally formed by 4 or 5 spinal nerves.

In cases where several individuals were examined, individual variation was generally found, as in *Anser, Podargus, Picus, Gecinus* and *Garrulus*; in these cases the number of spinal nerves which took part in forming the brachial plexus varied between 4 and 5, while in *Columba,* the number even varied between 4 and 6.

Variations also occurred in this respect between the two sides of the body. For example, in a specimen of *Anser cinereus* the plexus was formed on the right side by the nerves XVI, XVII, XVIII and XIX, while on the left side it received a strand from the XXth nerve in addition to these.

As has been stated, the last cervical nerve is generally the last nerve supplying the brachial plexus but deviations from this plan occur in both directions. These deviations may occur as individual variations and they may even be unilateral, owing to the transition between the cervical and dorsal vertebræ being effected at different points on the two sides of the body.

Particulars are given respecting the average proportions of the several roots in the different arrangements, but the arrangement or size of the roots relatively to each other was not found to bear any constant relation either to the systematic position of the bird, or to its size, or to its capacity for flight. It was however generally found that there was a certain relation between the relative size of the roots and the length of the neck in birds with a plexus composed of four roots. In this case the greatest thickness was generally either in or anterior to the middle roots of the plexus in short-necked birds, but posterior to the middle of the plexus in long-necked birds, but even this rule was not at all closely observed and many exceptions occurred. FÜRBRINGER, *l. c.* p. 243.

In Variation in the ordinal positions of the spinal nerves composing the plexus, the pattern of the plexus as newly constituted commonly bore a resemblance to the original pattern of the plexus, a phenomenon which FÜRBRINGER has called "imitatory Homodynamy" or "Parhomology" of the plexus[1] (*l. c.* p. 245).

Correlation between the constitution of the brachial plexus and the position and number of moveable cervical ribs.

65. **Anser cinereus, var. domestica.** Upon this point Fürbringer has made a series of important observations, especially in the Goose, which enabled him to state that there is, within limits, a certain correlation between the composition of the brachial plexus and the development of the ribs of this region. Speaking generally, those individuals in which the plexus was formed in a more anterior position usually shewed a fairly developed cervical rib on the 18th vertebra (*Anser*), and even as in Fig. 12, I, a very short but moveable rib on the 17th vertebra; and in such cases the 19th vertebra generally bore the first true sternal rib. On the other hand, examples with a more posterior development of the brachial plexus shewed not only an entire absence of moveable ribs on the 17th, but even a considerable reduction in the size of the ribs of the 18th and 19th vertebræ, so that these became "transitional" in character, leaving the 20th vertebra as the first vertebra bearing

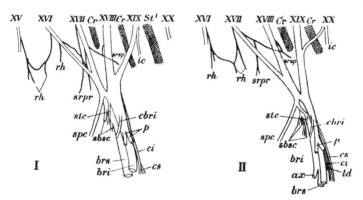

Fig. 12. Diagrams of brachial plexus and cervical ribs in two Geese (*Anser cinereus,* var. *domestica*) after Fürbringer (being his specimens D, *left*, and G, *right*).
I. Case in which the 17th and 18th vertebræ bear cervical ribs and the 19th bears the first sternal rib. II. Case in which the 17th and 18th vertebræ bear cervical ribs, and the 20th bears the first sternal rib.

ax axillaris, *bri* brachialis longus inferior, *brs* brachialis longus superior, *cbri* coraco-brachialis internus, *ci* cutaneus brachii inferior, *cs* cutaneus brachii superior, *ic* intercostals, *ld* latissimus dorsi, *p* pectoralis, *rh* rhomboideus, *sbsc* subscapulares, *srpr* nerves to levator scapulæ and serratus profundus, *srsp* nerves to serratus superficialis, *stc* sterno-coracoideus.

[1] The principle denoted by these expressions is nearly the same as that here expressed in the term Homœosis, which is perhaps more convenient as being a more inclusive expression.

true sternal ribs (Fig. 12, II.). The measurements are given by FÜRBRINGER for 7 specimens, of which those relating to two extreme cases (here figured) are appended.

		Ribs of 17th vert., length in mm.	Ribs of 18th vert., length in mm.	Ribs of 19th vert., length in mm.	Ribs of 20th vert.
I. 23 cm. long	rt.	2·5	20	23·5 (sternal)	(sternal)
	l.	2·75	21	23·75 (sternal)	(sternal)
II. 51 cm. long	rt.	—	7	51 + 13·5 ligt. and cartilage	59 (sternal)
	l.	—	12·5	51 + 15·5 ligt. and cartilage	60 (sternal)

FÜRBRINGER, M., *Morph. Jahrb.*, 1879, v. pp. 386 and 387.

66. By comparison of specimens of the Pigeon, **Columba livia**, var. **domestica,** a similar correlation was found to occur, as shewn in Fig. 13, I. and II. (Fürbringer's specimens A and E).

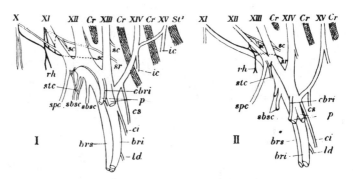

FIG. 13. Diagrams of brachial plexus and cervical ribs in two Pigeons (*C. livia*, var. *domestica*) after Fürbringer.
I. Case in which the 12th, 13th and 14th vertebræ bore cervical ribs. II. Case in which the 13th, 14th and 15th bore cervical ribs. Letters as in Fig. 12.

The measurements of the ribs of these individuals were as follows:

		Ribs of 12th vert., length in mm.	Ribs of 13th vert., length in mm.	Ribs of 14th vert., length in mm.	Ribs of 15th vert.	Ribs of 16th vert.
I.	rt.	—	18	25	1st sternal	2nd sternal
	l.	3	20	26	1st sternal	2nd sternal
II.	rt.	—	3	18	(damaged)	1st sternal
	l.	—	—	18	23	1st sternal

67. The same correlation was established in the case of the Jay **Garrulus glandarius,** but an actual variation in the number of moveable cervical ribs is not recorded in this species (see Fig. 14, I. and II., Fürbringer's specimens A and D). FÜRBRINGER, M., *Morph. Jahrb.*, 1879, v. p. 375.

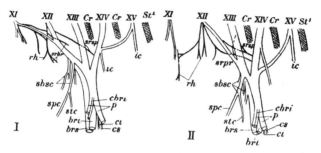

FIG. 14. Diagrams of the cervical ribs and brachial plexus in two Jays (*Garrulus glandarius*) after Fürbringer.
I. Case in which the brachial plexus began from the xith nerve, the cervical ribs of 13th and 14th vertebræ being longer than in II, a case in which the xiith is the first nerve contributing to the brachial plexus. Letters as in Fig. 12.

The measurements of the two specimens here figured were as follows:

		Ribs of 13th vert., length in mm.	Ribs of 14th vert., length in mm.	Ribs of 15th vert. (with sterno-costal parts), length in mm.
I.	rt.	5	18·5	24·5
	l.	7	20·5	26
II.	rt.	3·5	17·25	22
	l.	3·5	16·5	23

FÜRBRINGER, M., *Morph. Jahrb.*, 1879, v. p. 363.

But though this correlation between the nerves and the ribs is on the whole decided and unequivocal, it should be explicitly stated that it only occurs within certain limits and is not universal, and this statement of correlation is far from covering the whole ground. FÜRBRINGER, *l. c.* p. 387.

BRACHIAL PLEXUS.

*68. **Man and other Mammals.** By minute dissection of the brachial plexus in fifty-five subjects (32 fœtal and 23 adult) HERRINGHAM obtained important evidence as to the parts supplied by the fibres of the several spinal roots forming the plexus, and as to the considerable variation which occurs in respect of this supply. Of the facts thus arrived at, two examples may be quoted in illustration, concerning the composition of the median and ulnar nerves respectively.

The *median* is formed by two heads from the plexus; into the outer head the VIth and VIIth spinals enter, while the inner is formed by branches of the VIIIth and IXth, sometimes with the addition of some bundles of the VIIth. The presence of fibres from the VIIth depends on whether the anterior branch of the VIIth bifurcates, or goes wholly to the anterior (outer) cord of the plexus. In order to see whether both VIIIth and IXth contribute to the median, twenty-eight dissections were made, fourteen in infants, fourteen in adults. In one fœtus and in one adult no branch from the IXth was found, these being the only exceptions to the rule that both VIIIth and IXth send fibres to the median nerve. The median is then made of the VIth, VIIth, VIIIth and IXth, but these roots do not send to it a constant proportion. The bundle from the VIth varies little, that from the VIIth varies considerably, that from the VIIIth is sometimes equal to, sometimes smaller, and sometimes larger than the bundle from the IXth.

The origin of the *ulnar* nerve was traced in thirty-two cases, fourteen being adults. It was found to arise in four different ways. Most commonly it arose from the VIIIth and IXth: this occurred in twenty-three cases. With the VIIIth and IXth is sometimes combined a strand from the VIIth, as shewn in five cases (four fœtal, one adult). In three fœtal cases it arose from the VIIIth only, and in one fœtal and one adult case from the VIIth and VIIIth. The VIIth is only added to the ulnar in some of those cases in which it gives a branch to the posterior (inner) cord of the plexus. In several cases the branch from the VIIIth was much larger than that from the IXth, but the reverse was never met with.

Evidence similar to the above is given respecting other nerves from the brachial plexus.

From the results of the investigation generally, it appeared that the range of Variation though considerable was not extravagant, and that when parts, usually supplied by some given nerve root, are supplied by some other root, this other root is then either the one anterior or the one posterior to the root from which the supply normally comes. Some muscles seemed to bear definite relations to each other and their nerve supply seemed also "to vary solidly," their nerve supplies remaining the same relatively to each other, though derived from a different root. "The best example of this is in the three muscles which are attached along the inner side of the bicipital groove, the subscapularis, teres major, and latissimus dorsi. The first is usually supplied by the Vth and VIth, the second by the VIth, and the last by the VIIth, and however much they may vary above and below their typical place, they do not change their relations to each other. A similar relation exists between the two supinators and the two radial extensors. These last are sometimes supplied by the VIth, sometimes by the VIIth, but they are never in any case placed above the supinators. These are always supplied by the VIth alone. The flexor group in the forearm show a similar fixed relation." Herringham concludes that "the nerve roots are not always composed of the same fibres, but that what is in one case the lower bundle of the Vth may be in another the upper bundle of the VIth, and what is now the upper bundle of the VIIIth will at another time be the lower of the VIIth root." Hence the following principle is enuntiated: "*Any given fibre may alter its position relative to the vertebral column, but will maintain its position relative to other fibres.*"

HERRINGHAM, W. P., *Proc. Roy. Soc.*, XLI., 1886 pp. 423, 427, 430, 435.

By physiological methods, SHERRINGTON working chiefly on *Macacus*, but on other animals also, found that this principle substantially holds good for the outflow of fibres throughout considerable *regions* of the cord, but that it is not always applicable to great lengths of the cord, for the brachial plexus may be constituted in a region which is near the head end in comparison with the place of origin in other individuals, while in the same individual the sciatic plexus may be constituted in a region which is for it comparatively far back. No exception to the principle was found in the sense that a given efferent fibre which in one individual is anterior to some other particular fibre is ever in any individual of the same species posterior to it. SHERRINGTON, C. S., *Proc. Roy. Soc.*, LI. 1892, p. 76. This principle of Herringham's is analogous to that which in the much simpler case of Variation in vertebræ was pointed out on p. 107. It was stated that in such Homœotic variation no gaps are left. If a vertebra assumes a cervical character, it is the 1st dorsal, and so on.

*69. The following noteworthy case is described by HERRINGHAM in an infant. It should be borne in mind that to a normal brachial plexus the IVth nerve gives a small communication, the Vth, VIth, VIIth, VIIIth and IXth give large cords, while the Xth (or IInd dorsal) gives a minute fibre only. In this abnormal specimen, on the *left* side the part from the Xth was as large as that from the IXth, and this was as large as the VIIIth, whereas the natural proportion of VIIIth to IXth is about 2 to 1. The musculo-cutaneous received from the VIIth, instead of from the Vth and VIth only as more commonly found; the median received no VIth (*v. supra*); the teres major was supplied by the VIIth alone, instead of by the VIth; the circumflex received from the VIIth, instead of Vth and VIth alone as seen in 43 cases without any other exception; the musculo-spiral was formed by the VIIth, VIIIth and IXth, instead of by the VIth, VIIth and VIIIth (and sometimes even Vth); the deep branch in the hand received from both VIIIth and IXth (instead of VIIIth alone, as seen in five cases out of six). But though in all these respects the nerve-supply of the plexus was in ordinal position posterior to the normal, nevertheless the IVth sent a communication to the Vth (as it does normally) and the suprascapular and subscapular were given off normally. Here, then, the supply to the plexus began at the normal place, though it extended further back than it normally does. On the *right* side the branch from the Xth was slightly bigger than usual, but otherwise the only abnormality noted was that the IXth sent a branch to the musculo-spiral. HERRINGHAM, W. P., *Proc. Roy. Soc.*, 1886, XLI. p. 435. In view of FÜRBRINGER'S evidence (see Nos. 65 and 67), it might be expected that the first rib would be reduced in correlation with the irregular forward Homœosis of the nerves. In reply however to a question on the subject, Dr Herringham has kindly informed me that no abnormality in the ribs was seen, but that this point was not specially considered.

Compare also LANE'S case, No. 24, in which similarly a large branch from the Xth joined the plexus on the right side and the first rib was rudimentary, both structures thus shewing a correlated forward Homœosis.

LUMBO-SACRAL PLEXUS.

*70. By physiological methods SHERRINGTON found that the supply to the lumbo-sacral plexus varied considerably with regard to its origin from the spinal nerves. This was seen in *Macacus*, in the Cat and in the Frog. In none of these animals was any one arrangement found sufficiently often to justify its selection as a "normal" type. In each case it was found convenient to divide the different forms of arrangement into two classes, the one in which the supply to the plexus was in ordinal position more anterior ("pre-axial," Sherrington), the other being more posterior ("post-axial," Sherrington). Particulars respecting the distribution of the several nerves and the movements resulting from their stimulation in the two classes, are given in detail (q. v.). In *Macacus*, 31 individuals belonged to the more anterior class, and 21 to the more posterior. In the Cat the number of individuals in the two classes was 22 and 39 respectively. It is stated generally that

"The distribution of the peripheral nerve-trunks is not obviously different, whether, by its root-formation the plexus belong to the pre-axial class, or to the post-axial. The peripheral nerve-trunks are, as regards their muscles, relatively stable in comparison with the spinal roots. When the innervation of the limb-muscles is of the pre-axial class, so also is that of the anus, vagina and bladder; and conversely." SHERRINGTON, C. S., *Proc. Roy. Soc.*, 1892, LI. pp. 70—76.

71. **Primates.** Since in examining the facts of Variation we are seeking for evidence as to the modes in which specific differences

originate, allusion may therefore be made to some facts of normal structure in differing forms in illustration of the nature of such differences, and for comparison with the differences which are seen to occur by Variation. The arrangement of the lumbo-sacral plexus in the Primates well exemplifies some of these points. In Man, Chimpanzee and Gorilla the 1st sacral vertebra is the 25th; in the Orang it is the 26th; in the Baboons, e.g. *Macacus inuus* (= *Inuus pithecus* Is. Geoff., the Barbary Ape) it is the 27th. Now, as Rosenberg says, seeing that in Man the sacral plexus receives one whole præ-sacral root, the XXVth, and part of the XXIVth, it might be supposed that this plexus in the Orang would receive *two* whole præ-sacral roots and part of a third, or that in *Macacus* it would receive *three* præ-sacral roots and part of a fourth. But, as a matter of fact, in each of these forms, Chimpanzee, Orang and *Macacus*, according to Rosenberg, only one whole præ-sacral root and part of the next above it enter the sacral plexus, just as in Man, though the ordinal positions of the nerve-roots are different.

The Chimpanzee, however, which Rosenberg examined, was the specimen described (No. 34), having the 25th as a transitional lumbo-sacral vertebra, and rudimentary ribs on the 21st. In this specimen the præ-sacral nerves received by the sacral plexus were the XXVIth and part of the XXVth, thus bearing the same ordinal relations to the sacrum that the nerves of the lumbo-sacral cord do in the other forms and in Man, though each is ordinally one lower in the whole series than it is in Man. The same was true of the spinal roots composing the obturator and crural. ROSENBERG, E., *Morph. Jahrb.*, I. 1876, pp. 148, 149 and *Tables*, note 19.

This case is interesting as an example of forward Homœosis in the vertebræ associated with forward Homœosis in the sacral plexus. When compared with the following case of a Chimpanzee[1] having normal lumbo-sacral vertebræ, several discrepancies will be seen beyond those which can be accounted for by the single change of one in the ordinal position of the roots. No doubt for the larger nerves Rosenberg's account is correct, but as he states that the specimen was so badly preserved that the nerves could not be satisfactorily traced, it is possible that some of the branches may have been missed. However this may be, the specimen dissected by Champneys had important features of difference, notably that the sacral plexus received from the XXIInd spinal, while the highest recorded as entering it in Rosenberg's case was the XXVth, a greater difference than can be accounted for on the simple hypothesis of a change of one place throughout. Though, speaking generally, Rosenberg is right in saying that the evidence of the normal condition in *Macacus* and Orang as compared with each other and with Man

[1] CHAMPNEYS, F., *Journ. Anat. Phys.*, Ser. 2, v. 1872, p. 176.

suggests that the variation of the vertebral regions goes hand in hand with that of the plexus, and though a comparison between Rosenberg's abnormal Chimpanzee with that dissected by Champneys largely bears out this suggestion, yet it is also clear that this correlation is not a precise one, as indeed has already appeared in several instances.

In giving the compositions of the several nerves of the lumbo-sacral plexus in Man and Chimpanzee, I have given the numbers of the nerves in the *whole* series for simplicity of comparison. It will be remembered that a Chimpanzee has one pair of ribs more than Man, the XXIst nerve is the 1st lumbar in Man, but is the 13th dorsal in Chimpanzee, the XXVIth nerve being the 1st sacral in both forms. The table given shews, as Champneys says, that the *general* arrangement of the nerves of the lower limb and lumbar and sacral plexuses was in Chimpanzee very similar to that in Man, but that the nerves are very differently composed.

	MAN.	CHIMPANZEE.
Ilio-hypogastric } Ilio-inguinal }	XXI.	XXI.
Genito-crural	XXI.—XXII.	XXI.
External cutaneous	XXII. XXIII.	XXI., XXII.
Obturator	XXIII. XXIV.	XXI.—XXIII.
Anterior crural	XXII.—XXIV.	XXI.—XXIV.
Superior gluteal	XXIV.—XXVI.	XXIV.—XXVI.
Sacral plexus	XXIV.—XXIX.	XXII.—XXVII.
Small sciatic	XXIV.—XXIX.	XXIV.—XXVI.

(From CHAMPNEYS, *l.c.* p. 210.)

The origin of the nerves is therefore in several cases lower in Man than in the Chimpanzee, although in the absence of ribs on the 20th vertebra Man shews a character which, as compared with the presence of ribs in this position in the Chimpanzee represents a backward Homœosis.

Man. With the foregoing, compare the case mentioned above (No. 32) in which two entire lumbar nerves joined the sacral plexus in a human subject having no ribs on the 19th vertebra, &c. STRUTHERS, *J. Anat. Phys.*, 1875, p. 72 and p. 29.

72. For information as to the variations of the lumbo-sacral plexus in the Primates see also ROSENBERG, *Morph. Jahrb.*, I. 1876, p. 147 *et seqq.*; and as to cases in Primates and in other vertebrates compare VON JHERING, *Das peripherische Nervensystem der Wirbelthiere, &c.*, Leipzig, 1878. Of these, two cases of partial backward Homœosis in the lumbo-sacral plexus of the Dog are perhaps noteworthy, as being represented and described in greater detail than many of von Jhering's cases. In *one* of these the rib of the 13th dorsal (20th vertebra) was not developed, this vertebra being formed as a lumbar and thus itself shewing a backward Homœosis in correlation to that of the nerves

(VON JHERING, *l. c.* p. 182, pl. IV. fig. 2). Descriptions and diagrams of similar cases are given throughout the work, but as some of them represent specimens described by others (e.g. STRUTHERS and ROSENBERG) originally without diagrams, it is difficult to know how far the accounts given are schematic. For this reason reference to the original work must be made.

*73.   **Bradypodidæ.** *Brachial plexus.* As examples of normal differences the Sloths are especially interesting, but unfortunately an extended investigation of the nerves in several individuals has not been made. The results found by SOLGER relate to one specimen of *B. tridactylus* and one of *C. didactylus*. The latter was a perfect specimen, but the former had been partially dissected and the details of the nerves were largely imperfect. The *Cholœpus* was a specimen with seven cervicals, and the *Bradypus* had nine, the last bearing rudi-

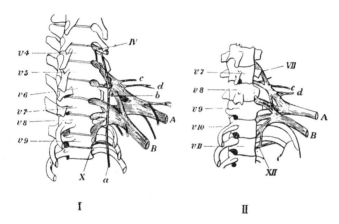

FIG. 15. Diagrams shewing the composition of the brachial plexus in I. a *Cholœpus*, II. a *Bradypus*. v^1—v^{11}, the vertebræ. IV, VII, X, XII, fourth, seventh, tenth and twelfth cervical nerves. *A*, dorsal cord. *B*, ventral cord. *a*, phrenic. *b*, dorsalis scapulæ. *c*, suprascapular. *d*, subscapular.

mentary ribs. As the figure shews (Fig. 15), there was a close but not a perfect resemblance between the composition of the plexus in the two cases, that of *Bradypus* being in nearly each case two roots lower than that in *Cholœpus*. In the latter the IVth nerve gave a branch to the Vth, but whether in *Bradypus* the VIth gave a branch to the VIIth was not determined with certainty owing to the condition of the specimen. [For details see original paper] SOLGER, B., *Morph. Jahrb.*, 1875, I. p. 199, Pl. VI.

One more case may be given in illustration of the kind of difference which normal forms may present.

74.   **Pipa** (the Surinam Toad). In the majority of the Batrachia, the most anterior pair of spinal nerves leaves the vertebral column between the first and second vertebræ, no sub-occipital being present. The

second pair leaves between the second and third vertebræ, and the third pair leaves between the third and fourth vertebræ. The brachial plexus is formed by the whole of the second pair together with parts of the first and third pairs. (The details of the arrangement are complicated and vary greatly in different forms.) In *Pipa* a different arrangement exists. The most anterior pair of nerves leaves the spinal column by *perforating* the first vertebra, and the pair which leaves between the first and second vertebræ is therefore ordinally the *second* pair of spinal nerves in this form; the pair which leaves between the second and third vertebræ is the *third*, and so on. The brachial plexus is made up of the whole of the second nerve, nearly the whole of the third nerve and of a branch of the first.

If then it were to be supposed that the pair of nerves which leaves the column between the first and second vertebræ in *Pipa* is homologous with the pair of nerves which leaves in the same place in *Rana*, &c., it is clear that between the skull and the 2nd vertebra of *Pipa*, there is an extra pair of nerves not found in *Rana*. The number of free vertebræ in *Pipa* is however less than in *Rana*. For in the former there are only seven of these, making with the united sacral vertebra and urostyle eight pieces in all; but in *Rana* there are eight præsacrals, one sacral, and counting the urostyle, ten pieces in all. In *Rana* only one spinal nerve, the 10th, leaves the urostyle, while in *Pipa* two pairs, the 9th and 10th, pass out through the terminal piece of the vertebral column, suggesting that the diminution in the number of vertebræ is due to the absence of separation between the 9th vertebra and the urostyle. The whole number of spinal nerves is therefore the same in both *Rana* and *Pipa*, but in the latter the 1st pair perforate the 1st vertebra in addition to the 2nd pair which pass out between the 1st and 2nd vertebræ. FÜRBRINGER[1], M., *Jen. Zt.*, 1874, VIII. p. 181 and *Note*, Pl. VII. fig. 37; also *Jen. Zt.*, 1873, VII. Pl. XIV. figs. 5 and 6.

It was suggested by STANNIUS (*Lehrb. d. vergl. Anat.*, p. 130, *Note*) that perhaps the 1st vertebra of *Pipa* represents two coalesced vertebræ, but in an anatomical examination of two specimens of *Pipa*, Fürbringer (*l. c.* 1874, p. 180), found no confirmation of this suggestion, and developmental evidence also went to shew that no such fusion occurs in the ontogeny at least[2]. KÖLLIKER, A., *Verh. phys.-med. Ges. Würzburg*, 1860, x. p. 236.

As Fürbringer says there is no satisfactory way of bringing this case of *Pipa* into accord with the condition seen in *Rana*. In the Urodela there is of course a suboccipital nerve between the skull and the 1st vertebra which is not present in *Rana*, and some resemblance to *Pipa* is thus suggested; but in the Urodela the 1st spinal does not actually

---

[1] Compare VON JHERING, H., *Morph. Jahrb.*, 1880, VI. p. 297. The statement made by von Jhering that the nerves of *Pipa and Rana* correspond nerve for nerve, though in different positions relative to the vertebræ, if established would be important; but from the want of detailed description it is not clear whether this conclusion was arrived at by actual dissection.

[2] This is questioned by ADOLPHI, *Morph. Jahrb.*, XIX. 1892, p. 315, *note*. The same paper contains much important matter bearing on the variation of the nerves of Amphibia. I regret that this paper did not appear in time to enable me to incorporate the facts it contains.

anastomose with the plexus, though it gives off the superior thoracic which in both *Rana* and *Pipa* comes off at a point peripheral to the formation of the plexus (Fürbringer).

If the two spinal nerves which come out of the urostyle in *Pipa* may be taken to shew that this bone contains $n + 2$ vertebræ while the single pair in *Rana* shews the urostyle to consist of $n + 1$, there is in *Pipa* (as compared with *Rana*), a diminution of one in the total number of vertebræ, together with a backward Homœosis, which is seen in the fact that the 8th vertebra bears the pelvic girdle. Turning now to the nervous system, the fact that the last spinal nerves to join the brachial plexus in *Pipa* are the IIIrd, while in *Rana* they are the IVth, is again an evidence of backward Homœosis. But if this process were completely carried out, the pair of nerves which in *Pipa* pass out through the 1st vertebra should pass out between this vertebra and the skull, i.e. in the position of the suboccipital of the Urodela. Beyond this analysis cannot be carried, and this case is a good illustration of the fact that the hypothesis of an individual homology between the segments does not satisfy all the conditions of the problem.

Relation between the ordinal position of spinal nerves and their distribution to the limbs.

This subject is introduced partly because it further illustrates the nature of the relations which the spinal nerves maintain towards each other, and thus bears indirectly on the phenomena of their Variation; but chiefly because it presents a view of some of the complexities which arise in the apportionment of organs centrally disposed in Meristic Series, to the parts of peripheral appendages having no clear or coincident relation to the primary or fundamental segmentation of the body. The facts have thus a value as furnishing a kind of commentary on the nature of Meristic Repetitions in vertebrates. In any attempt to interpret or comprehend Meristic Repetition as a whole, they must be taken into account.

The principles of the distribution of the spinal nerves *to the muscles of the fore-limb* have been thus enuntiated by HERRINGHAM.

1. "Of two muscles, or of two parts of a muscle, that which is nearer the head-end of the body tends to be supplied by the higher, that which is nearer the tail-end by the lower nerve.

2. "Of two muscles, that which is nearer the long axis of the body tends to be supplied by the higher, that which is nearer the periphery by the lower nerve.

3. "Of two muscles, that which is nearer the surface tends to be supplied by the higher, that which is further from it by the lower nerve." HERRINGHAM, W. P., *Proc. Roy. Soc.*, XLI. 1886, p. 437.

Details are given shewing the manner in which the innervation of the muscles in Man bears out these principles.

FORGUE and LANNEGRACE[1], who worked with dogs and monkeys by physiological methods, arrived at conclusions identical with those which Herringham came to by human dissection.

[1] *Distrib. des racines motrices*, &c., Montpellier, 1883, p. 45 [quoted from Herringham: not seen, W. B.]; also *Comptes Rendus*, 1884, CXVIII. p. 687.

As regards *the sensory nerves in the fore-limb*, the following principles were similarly established by dissection in Man.

1. "Of two spots on the skin, that which is nearer the pre-axial border tends to be supplied by the higher nerve.

2. "Of two spots in the pre-axial area the lower tends to be supplied by the lower nerve, and of two spots in the post-axial area the lower tends to be supplied by the higher nerve."

"Thus, if the limb be seen from the front, the two highest nerves on the outer and inner sides respectively are the IVth and Xth. Lower than these the Vth and VIth take the outer, the IXth and Xth the inner side. Below the elbow the VIth alone takes the outer, and the IXth alone the inner. In the hand, while the VIth and IXth continue their positions, the VIIth and VIIIth for the first time join in the supply." Particulars from which this general statement is made are given. HERRINGHAM, *l.c.* p. 439.

According to subsequent investigations of SHERRINGTON's on the *hind-limb*, the innervation of the muscles of the posterior aspect of the thigh and leg do *not* follow the third of Herringham's principles, for in their case the deep layer of muscles is innervated by roots anterior to those which innervate the superficial muscles. The same experiments also, though clearly shewing that the nerve-supply of the *skin* of the hallux is anterior to that of the 5th digit, gave only equivocal evidence that the same was true of the musculatures of these two digits; and in the thigh the gracilis is not supplied before the vastus externus, whose relation is rather that of ventral to dorsal than of anterior to posterior. SHERRINGTON, C. S., *Proc. Roy. Soc.*, 1892, LI. p. 77.

RECAPITULATION.

Some features in the Meristic Variation of the spinal nerves, as illustrated by the foregoing evidence, may be briefly summarized.

In the first place, as might be anticipated from the compound nature of a spinal nerve, when Homœotic Variation takes place, it does not commonly occur by the transformation of entire nerves, but rather by change in the distribution and functions of parts of nerves. In this respect, therefore, there is a difference between Homœosis in spinal nerves and that in vertebræ, for in the latter, Homœosis is often complete.

A rough illustration may make this more clear.

Just as in making up the chapters of a book into volumes, whole chapters may be put into one volume or into the next, and the following chapters renumbered, so it may be with the Variation of vertebræ, for these may belong wholly to one region of the spine or to another. But the nerves are like chapters made up of sections; particular sections or groups of sections may come in an earlier chapter or in a subsequent one, and the places of those that have been moved on may be filled up consecutively, but it seldom happens that whole chapters are renumbered. Nevertheless it is clear from such a case as that of *Bradypus* and *Cholœpus*, on the hypothesis that both forms are descended from a common ancestor, that such changes and renumbering of whole nerves must have happened, though there is evidence to shew that this may happen piecemeal, as in cases given.

Of course in speaking of such changes among the vertebræ it will not be forgotten that partial changes occur too, but there is still greater Discontinuity in their case than in that of the nerves.

But that there is Discontinuity in the case of nerves also is clear; for a given fibre, supplying a given muscle, must leave the spinal cord either by one foramen and one spinal nerve, or by another. Conversely the nth motor nerve must supply either one muscle or another, and the transition between the two, however finely it may be subdivided, must ultimately be discontinuous in the case of individual fibres. It would be interesting to know to what extent fibres vary in bundles, but this can hardly be determined.

There is, however, some evidence that the group of fibres supplying a limb does to some extent vary up and down the series as a group, though much rearrangement may occur also within the limits of the group itself.

Lastly, there is important evidence that Variation in other parts *may be* correlated with change in the ordinal positions at which nerves with given distributions emerge from the spinal cord. With Variation in the ordinal positions at which the nerves come out, change in other parts, notably in the ribs, *may* happen too; so that we may say that in a sense there may be, at least within the limits of single species (see cases Nos. 24, 65 and 71), a correlation between the apportionment of their functions among the nerves and the contour of the body, both changing together, the ribs rising and falling with the rise and fall of the brachial plexus. The nerves do not merely come out through the foramina like stitches through the welt of a shoe, the shape of the shoe remaining the same wherever the threads pass out. The arrangement is, rather, like that of the strings of such an instrument as a harp or piano, in which there is a correlation between the curves of the frame and the positions of the several notes: so long as the frame is the same, the strings cannot be moved up or down, the instrument still retaining the same compass and the same number of notes.

CHAPTER V.

LINEAR SERIES—*continued.*

HOMŒOTIC VARIATION IN ARTHROPODA.

THE occurrence of Homœosis among the appendages of Arthropoda is illustrated by a small but compact body of evidence. To this evidence special value may be attached, not because it is likely that in the evolution of the Arthropods variations have really taken place, in magnitude comparable with those now to be described, but rather because these cases give a forcible illustration of possibilities that underlie the common and familiar phenomena of Meristic Repetition. Of these possibilities they are indeed "Instances Prerogative," salient and memorable examples, enuntiating conditions of the problem of Variation in a form that cannot be forgotten. Facts of this kind, so common in flowering plants, but in their higher manifestations so rare in animals, hold a place in the study of Variation comparable perhaps with that which the phenomena of the prism held in the study of the nature of Light[1]. They furnish a test, an *elenchus*, which any hypothesis professing to deal with the nature of organic Repetition and Meristic Division must needs endure.

INSECTA.

*75. **Cimbex axillaris** (a Saw-fly), having the peripheral parts of the left antenna developed as a foot. The right antenna is normal, ending in a club-shaped terminal joint. In the left antenna the terminal joint is entirely replaced by a well-formed foot, having a pair of normal claws and the *plantula* between them (Fig. 16). This foot is rather smaller than a normal foot, but is perfectly formed. The rest of the antenna, so far as the point at which the club should begin is normal in form, but is a little smaller and thinner than the same parts in the right antenna. KRAATZ, G., *Deut. ent. Ztschr.*, 1876, xx., p. 377, *Pl.*

---

[1] See the well-known passage in *Nov. Org.*, II. xxii.

This specimen was most kindly lent to me for examination by Dr Kraatz, but to this description I am unable to add anything[1].

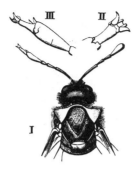

Fig. 16. *Cimbex axillaris:* right antenna normal; left antenna bearing a foot. II. the left antenna seen from in front. III. the same from above. After Kraatz.

It should be noted that the plantar surface of the foot was turned rather forwards as shewn in the figure, and not downwards like the normal feet.

*76.   **Bombus variabilis** ♂ (a Humble-bee). A specimen taken beside the hedge of a park in Munich, having the left antenna partially developed as a foot. The first two joints were normal. They were followed by two joints which were rather compressed and increased in thickness and breadth. Of these the first was oblong and somewhat narrowed towards its apex by two shallow constrictions, giving it an appearance as of three joints united into one; below it presented a projecting and tooth-like point. This joint was only slightly shiny. The next joint to it was almost triangular, and was reddish-brown, shiny, and having hairs on its lower surface. Posteriorly it was prolonged inwards, covering the previous joint so that both seemed to form one joint: the posterior edge was somewhat thickly covered with hairs. The upper part of the first of these two joints and the prolongation of the second were together covered by a hairy, scale-like third joint, which seemed to be only attached at its base. From the apex of the second joint arose a shortened claw-joint, like the claw-joint of a normal foot. This joint was reddish-brown and shiny, bearing a pair of regularly formed claws, like the claws of the foot. Kriechbaumer, *Entom. Nachr.*, 1889, xv. No. 18, p. 281.

[1] Some to whom I have spoken of this specimen, being unfamiliar with entomological literature, and thus unaware of the high reputation of Dr Kraatz among entomologists, have expressed doubt as to its genuineness. I may add therefore that the specimen, when in Cambridge, was illuminated as an opaque object and submitted to most careful microscopical examination both by Dr D. Sharp, F.R.S., and myself, and not the slightest reason was found for supposing that it was other than perfectly natural and genuine. The specimen was also carefully relaxed and washed with warm water, but no part of it was detached by this treatment.

148   MERISTIC VARIATION.   [PART I.

The two following cases must be given here, inasmuch as they relate to Homœosis of the appendages in Insects; but in the case of the first the evidence is unsatisfactory, and in the case of the second there is considerable doubt whether the variation is really of the nature of Homœosis.

77. **Prionus coriarius** ♂ : having elytra represented by legs.

The following is a translation of an announcement in the *Stettiner Ent. Ztg.*, 1840, vol. I. p. 48, which is copied from the original communication to the *Preussische Provinzial-Blätter*, Bd. XX. [The latter journal not seen, W. B.]:—",One of my pupils brought me to-day a male *Prionus coriarius*, Fbr., the thorax of which is remarkably constructed. The horny covering of the mesothorax is absent, and in place of the elytra is a pair of fully developed legs which are directed upwards and backwards. These legs are inserted at the points of articulation of the elytra. The metathorax supports the wings as usual and the abdomen is not hardened more than it usually is. In trying to fly, the creature moved these upwardly directed legs simultaneously with its wings. The scutellum is absent and the prothorax has only two spines; other parts normally developed." Dr SAAGE, Braunsberg, 1839:—Hagen, in quoting this case, mentions that the specimen was afterwards seen by von Siebold, but gives no reference to any writing of von Siebold on the subject.

[If this specimen still exists, it is to be hoped that a description of it may be published. In the absence of further information there seems to be no good reason for accepting the case as genuine.]

*78. **Zygæna filipendulæ** ♂. Specimen possessing a supernumerary wing arising in such a position as to suggest that it replaced a leg. This specimen was originally described by RICHARDSON, N. M.,

FIG. 17. *Zygœna filipendulæ*, ♂ , having a supernumerary wing on the left side. The upper figure shews the neuration of the supernumerary wing. From drawings by Mr N. M. RICHARDSON.

Proc. Dorset Field Club, 1891, and was exhibited at a meeting of the Entomological Society of London, 1891, *Proc.* p. x. The extra wing was in general form and appearance like a somewhat folded hind wing but its colour was rather yellower, though it was more red than yellow. I have to thank Mr Richardson for allowing me to examine this specimen in company with Dr Sharp. In compliance with Mr Richardson's wish we did not strip the wing or remove the thick hairs which surrounded its base, and it is therefore not possible to speak with certainty as to its precise point of origin. The following description of it was drawn up for me by Dr Sharp: " The supernumerary wing projects on the under side of the body, and at its base there intervenes a space between it and the dorsal region of the body about equal to the length of the metathoracic side-piece. The exact attachment of the base of the supernumerary wing cannot be seen owing to the hairiness of the body, but so far as can be seen it is to be inferred that the wing is attached along the length of the posterior coxa, the outer edge of the point of attachment may be inferred to extend as far as the suture between the coxa and thoracic side-piece; if this view be correct the abnormality may be described as the absence of the hind femur and parts attached to it, and the addition of a reduced wing to the hind-margin of the coxa. It is, however, just possible that if the parts could be clearly distinguished it might be found that the real point of attachment of the abnormal wing is the suture between the metathoracic side-piece and the hind coxa."

It should be distinctly stated that there is no empty socket or other suggestion that the rest of the leg had been lost, and it was in fact practically certain that it had never been present. There is thus a strong *prima facie* case for the view that the leg has been developed as a wing, however strong may be the theoretical objections to this conclusion. On the other hand, as will be shewn in a later chapter, supernumerary wings are known in specimens having a full complement of legs, and it is conceivable that one of these supernumerary wings may have arisen in such a way as to prevent the proper development of the leg from the imaginal disc. If the specimen were carefully stripped of hairs some light might perhaps be thrown on this question. The figure (Fig. 17) is from a drawing kindly lent me by Mr Richardson.

CRUSTACEA.

*79. **Cancer pagurus.** Specimen having the right third maxillipede developed as a chela. This animal was brought by a fisherman to the Laboratory of the Marine Biological Association at Plymouth. It is a male, measuring five inches from one side of the carapace to the other. All the parts appear to be normal with the exception of the third maxillipede of the right side. This structure, however, has the form shewn in Fig. 18, A, differing entirely from the ordinary condition of the appendage. Fig. 18, B, is taken from the third maxillipede of the left side and shews the ordinary structure of the same parts. On comparing the two figures it will be seen that the protopodite does not differ in the limbs of the two sides; that the exopodite of the right side is

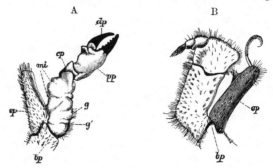

Fig. 18. *Cancer pagurus* ♂; the right and left third maxillipedes, that of the right side having the endopodite in the likeness of the endopodite of a chela. *bp.* basipodite, *cp.* carpopodite, *dp.* dactylopodite, *ep.* epipodite, *g.* groove between parts representing ischiopodite and meropodite, *g'.* groove representing the suture at which a normal chela is thrown off if injured. From *P. Z. S.*, 1890.

essentially like that of the left, but that it lacks the inner process and the flagellum which are borne by the normal part. There was some indication that this branch of the limb had been injured, and perhaps the flagellum may have been torn away, but the appearances were not such as to warrant a conclusion on this point. The branchial epipodites (not shewn in the figures) were normal in both cases. The endopodite of the right side was entirely peculiar, and was, in fact, literally transmuted into the likeness of one of the great chelæ. It consists of a single joint (*mi*), articulating with the basipodite centrally and bearing the carpopodite. This single joint represents, as it were, the ischiopodite and meropodite of an ordinary chela, but these two parts are ankylosed together and the articulation between them is only represented by a groove (*g*). Another groove (*g'*) represents the groove upon the ischiopodite of the chela, at which the limb is commonly thrown off by the animal if it is injured. The carpopodite, propodite and dactylopodite are freely moveable on each other and hardly differ, save in absolute size, from those of the normal chelæ. The shape, proportions and texture are all those of the chela. BATESON, W., *Proc. Zool. Soc.*, 1890, p. 580, fig. 1.

80. A similar case[1] of *Cancer pagurus* ♀. 4 inches across carapace, mature, right pedipalp [*i.e.* 3rd maxillipede] normal, left pedipalp modified into a chela having all the joints clearly defined, CORNISH, T., *Zoologist*, S. 3, VIII. p. 349.

*81. **Palinurus penicillatus**. The left eye bearing an antenna-like flagellum, growing up from the surface of the eye as shewn in the figure (Fig. 19). The eye-stalk and cornea, as represented, appear to have been of the normal shape but reduced in size.

[1] Similar cases since published by RICHARD, *Ann. Sci. Nat., Zool.*, 1893.

MILNE-EDWARDS, A., *Comptes Rendus*, LIX. 1864, p. 710; described and figured by HOWES, W. B., *Proc. Zool. Soc.*, 1887, p. 469.

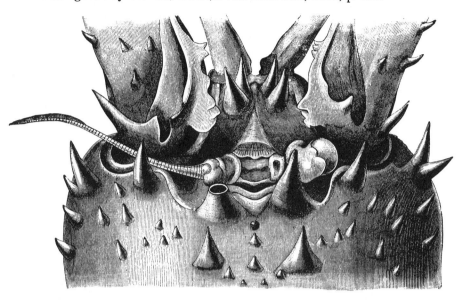

FIG. 19. *Palinurus penicillatus*, the left eye bearing an antenna-like flagellum. After HOWES.

82. **Hippolyte fabricii** differs from other species of the genus in being usually without epipodites at the bases of all the cephalothoracic legs except the first pair, while in the other species these appendages are usually present upon the bases of the first and second, or upon the first, second and third pairs, and on this character it was placed by KRÖYER in a separate section of the genus.

Of 52 individuals (18 males varying in length from 27 mm. to 39 mm. and 34 females varying from 16·5 mm. to 50 mm.), from various localities on the New England coast, 47 had the normal number of epipodites, while 5 had epipodites on one or both of the second pairs of legs. Of the latter 3 were from the Bay of Fundy; one ♂, 35 mm. long, has well-developed epipodites on each side of the 2nd pair of legs; another ♂, 36 mm. long, has a short epipodite on the left side and none on the right; the other, ♀, 47 mm. long, has a well-developed epipodite on the left side and none on the right. The two others were from Casco Bay; a ♀, 36 mm. long, with a short epipodite on the left side, and a ♂, 28 mm. long, with a rudimentary one on the right side. As the measurements shew, the presence of these epipodites is not characteristic of the young. SMITH, S. J., *Trans. Connecticut Acad.*, v. 1879, p. 64.

Variation in the number of generative openings in Crayfishes.

*83.   **Astacus fluviatilis.** A female having the normal pair of oviducal openings on the bases of the *antepenultimate* pair of walking legs, and in addition to them another pair of similar openings placed upon the corresponding joints of the *penultimate* pair of walking legs. On dissection it was found that the ovary was normal, and that from each side of it a normal oviduct was given off; but each of these oviducts divided a little lower down to form two smaller oviducts, one of which went to each of the four oviducal openings. DESMAREST[1], E., *Ann. Soc. Ent. France*, 1848, Ser. 2, VI. p. 479, *Pl.*

*84.   **Astacus fluviatilis** ♀, having a supernumerary pair of oviducal openings placed *on the last pair* of thoracic legs. The normal oviducal openings were in the usual position and of the usual shape and size, but in addition to them there was an extra pair placed on the last thoracic legs. It should be remarked that though these are the appendages upon which the openings of the male organs are placed, the oviducal openings were not in this case situated at the posterior surface of the joint as the male openings are, but were placed relatively to the leg in the same situation as the female openings on the antepenultimate legs. The penultimate legs and the abdominal appendages were normal. On dissection it was found that each oviduct after passing for the greater part of its course as a single tube, divided into two parts, one of which went to each oviducal opening. The ovary itself was normal. BENHAM, W. B., *Ann. Mag. N. H.*, 1891, Ser. 6, VII. p. 256, *Pl.* III. [I am greatly obliged to Mr Benham for an opportunity of examining this specimen. Attention is called to the fact that in this specimen Homœosis occurs in an unusual way, leaving a gap in the series; for the openings are on the *antepenultimate* and *last* thoracic legs respectively.]

Desmarest's observation stood apparently alone until lately, when the specimen just described and several others presenting the same or similar variations were observed by BENHAM. Mr Benham was kind enough to send me the following specimens for examination: one female having a *single extra* oviducal opening on the left side upon the penultimate thoracic leg (Fig. 20 *C*), and two females having a similar extra opening in the same place on the right (Fig. 20, *B*); in both of these the normal oviducal openings were unchanged. Together with these Mr Benham also sent a female having *only one* oviducal opening on the right side and another having *only the left* oviducal opening (Fig. 20, *A*), the corresponding leg of the other side having no trace of an opening.

---

[1] DESMAREST had this specimen from ROUSSEAU (*l. c.*, p. 481 *note*): Faxon quoting the case (*Harv. Bull.*, VIII.) accidentally represents it as two cases, but the note to Desmarest's paper shews that the description referred to a single specimen only.

*85. After receiving these specimens I made an attempt to ascertain the degree of frequency with which such variations occur in

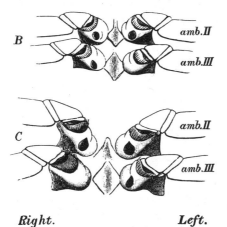

Right.   Left.

Fig. 20. Females of *Astacus fluviatilis* having an abnormal number of oviducal openings. N.B. The form with *three* pairs of openings is not figured.
A. Right oviducal opening absent.   *B.* Extra opening in right penultimate leg.   *C.* Extra opening in left penultimate leg.

the Crayfish, and though the total number examined is too small to give a percentage of much value it may be well to record the result.

In all, 586 female *A. fluviatilis* have been examined: of these 563 were normal in respect of the number of oviducal openings, and 23 were abnormal, as follows:

1. Extra oviducal opening on left penult. leg 7
2. ditto.........................right........................ 10
3. ditto.....................on both penult. legs 1
4. ditto.....................on both penult. & last legs 1
5. Single oviducal opening on left side only 3
6. ditto.............................right[1].................... 1

Total abnormal specimens 23

[1] Mr R. Assheton sends me word of a similar specimen found among 80 of both sexes; Prof. W. B. Howes of another among 144 of both sexes.

In all cases of supernumerary oviducal openings the normal openings were also present.

These cases are in addition to those received from Mr Benham. So far, therefore, the cases of extra opening amount to over 3 *per cent.* of females examined.

Of 714 *males* examined, only one was abnormal, having no trace of a generative opening on the right side, the vas deferens ending blindly and hanging free in the thoracic cavity. There was no female opening in this specimen, and the abdominal appendages had the form characteristic of the male on both sides. The base of the last thoracic leg on the right side bore no enlargement for the genital opening, but was plain and like that of the penultimate leg[1].

In cases of females which lacked one of the openings, the basal joint for the leg which should have been dilated and perforated for the opening, was undilated and resembled the basal joint of a penultimate leg. The oviduct upon the imperforate side was more or less aborted and hung loosely in the thoracic cavity.

In the abnormal females with extra oviducal openings, the oviduct divides generally into two just before it enters the legs, the fork being placed at the level between them. In some few cases no branch of the oviduct could be traced to the extra opening. In one specimen the extra opening led into a short tube which ended blindly, not communicating with the oviduct. The specimen (4) with extra openings on the penultimate and last legs had thus in all six oviducal openings. Those in the normal position on the antepenultimate legs were of normal size, those on the next pair were smaller but still of fair size, while those on the last pair of thoracic legs were very small, that on the left side being the smallest and admitting only a fair-sized bristle. In this specimen the single oviduct of each side forked in its peripheral third, giving a duct to each of the first two pairs of openings, but I failed to find any connexion between it and the openings on the last thoracic legs, which were very short blind sacs.

In all cases of extra oviducal opening the basal joint of the leg is expanded like those of the normal antepenultimate legs, the degree of expansion being proportional to the size of the opening. The normal openings are always the largest, but the extra ones are sometimes almost as large and would easily allow the passage of ova, but occasionally they are too small to let an egg through.

As regards principles of Homœotic Variation illustrated by these cases, three points should be especially remarked:

[1] Compare the following: **Astacus fluviatilis.** Amongst 1500 specimens 3 were found in which the tubercle through which the green gland opens was entirely absent. The opening itself was not formed and the green gland of the same side was absent. In another specimen the opening was deformed, probably owing to some mutilation. In this and the previous cases the green gland of the other side was considerably enlarged. STRAHL, C., *Müller's Archiv für Anat. u. Phys.*, 1859, p. 333, *fig.*

1. That this Variation may be bilaterally symmetrical, but that the evidence goes to shew *that it is more often unilateral*.

2. That there is a clear *succession* between the several oviducal openings, those of the antepenultimate legs being the largest, the penultimate the next, and those of the last legs the smallest.

3. That Homœosis may occur between segments which are *not* adjacent, as in the case of extra oviducal openings on the last thoracic legs, none being formed on the penultimate (No. 84).

4. That the Variation may be *perfect*.

With the foregoing, the following evidence may be compared, though it is very doubtful whether it properly belongs here[1].

86.   **Cheraps preissii** [an Australian freshwater Crayfish, nearly allied to *Astacus*]. Of seven specimens received one was a normal male and three were normal females. The other three had on the basal joint of the third [antepenultimate] pair of legs a round opening, having the size and shape and situation of the normal female openings. These apertures were closed with soft substance. The fifth legs bore the usual male openings, from which the ends of the *ductus ejaculatorius* protruded. The coiled spermatic ducts were normal; but no ovary was found and no internal structure was connected with these female openings. VON MARTENS, E., *Sitzb. Ges. naturf. Fr. Berlin*, 1870, p. 1.

87.   **Astacus pilimanus** ♂, a single specimen, and **A. braziliensis** ♂, a specimen collected by HENSEL in Southern Brazil, a similar opening was found on the third pair of legs; but in other specimens of these forms there was only a slight though sharply defined depression in the chitinous covering at this point. VON MARTENS, E., *l.c.*

---

[1] See also NICHOLLS, R., *Phil. Trans.*, 1730, xxxvi. p. 290, figs. 3 and 4 describing a Lobster (*Homarus vulgaris*) having male organs on the left side and female organs on the right.

CHAPTER VI.

LINEAR SERIES—*continued*.

CHÆTOPODA, HIRUDINEA AND CESTODA.

Imperfect Segmentation[1].

Though from the circumstance mentioned at the beginning of Chapter II, that the total number of segments in the Annelids is generally indefinite, true Meristic Variation cannot be easily recognized in this group, there is nevertheless a remarkable group of cases of *imperfect* segmentation, in which by reason of the incompleteness of the process of Division, the occurrence of Variation is at once perceived. The following cases were all originally described by CORI, who speaks of them as instances of "intercalation" of segments. For reasons sufficiently explained in the Chapter on Vertebræ, there are objections to the use of this term, if only as a mode of expression, and the evidence concerning these cases has therefore been re-cast.

*88.   **Lumbricus terrestris**: the 46th segment having the form shewn in Fig. 21, I. being normal on the right side, but double on the left. Internally a septum divided the two parts a and a' from each other. Each of them contained a nephridium, setæ, &c. CORI, C. J., *Z. f. w. Z.*, LIV. 1892, p. 571, fig. 1.

*89.   Specimen having, in the region close behind the clitellum, three consecutive segments, each resembling that just described. Of these the first was double on the right side, the second on the left, and the third on the right again. Fig. 21, II. shews the internal structure, the nephridia and other parts having doubled in each of the doubled half-segments. CORI, *l. c.*, p. 572, fig. 2.

90.   **Lumbriconereis**: case similar to the first case in *Lumbricus*, Fig. 21, III. CORI, C. J., *l. c.*, p. 572, fig. 4.

91.   **Halla parthenopeia.** A specimen 50 cm. long presented numerous abnormalities of which two are represented in Fig. 21, IV. At the point marked a' the lines of division between the segments

---

[1] Numerous facts illustrating this subject are given in a recent paper by BUCHANAN, F., *Q. J. M. S.*, 1893.

enclose a small spindle-shaped island of tissue. Three segments lower a wedge-shaped half-segment is similarly formed. At

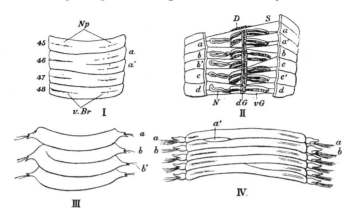

Fig. 21. Examples of imperfect segmentation in Annelids (after Cori).
I. *Lumbricus terrestris* (No. 88). II. *L. terrestris* (No. 89), as seen when laid open on the dorsal side. III. *Lumbriconereis* (No. 90). IV. *Halla parthenopeia* (No. 91).
N, nephridium; Np, nephridial pores; D, alimentary canal; dG, dorsal vessel; vG, circular vessel.
The letters a, b, c, &c. indicate the parts belonging to the respective segments.

another point in the same animal (not shewn in Fig. 21) one of the segments was partly divided into two in the right dorso-lateral region. Cori, p. 572, figs. 8 and 9.

Spiral Segmentation[1].

92. **Lumbricus terrestris.** Fig. 22, I. A shews a part of an Earthworm seen from the dorsal side, the ventral side being normal in appearance. By following the groove indicating the plane of the septum between b and c on the right side to the ventral surface, it could be traced to the left side between b and c, so across the dorsal surface, between c and d on the right side, across the ventral surface and between c and d on the left, reaching nearly to the middle dorsal line again. This is shewn diagrammatically in Fig. 22, I. B.

93. A simpler case affecting one segment only is shewn in Fig. 22, II.

94. Another specimen exhibited a similar arrangement near the tail-end (Fig. 22, III.). The lettering of the figure sufficiently explains the course of the spiral septal plane. [Cori does not state that the septa internally formed a spiral division, but it can scarcely be doubted that they did so, following the external groove,

[1] Further observations on this subject have been lately published by Morgan, T. H., *Journ. of Morph.*, 1892, p. 245, and by Buchanan, F., *Q. J. M. S.*, 1893.

like the spiral valve of an Elasmobranch's intestine.] CORI, Z. f. w. Z., LIV. 1892, p. 573, figs. 5, 6 and 7.

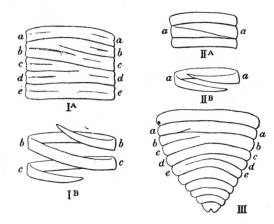

FIG. 22. Spiral segmentation in *Lumbricus terrestris*.
I, A, the case No. 92; I, B, diagrammatic representation.
II, A, the case No. 93; II, B, diagrammatic representation.
III, the case No. 94.   (After CORI.)

Two other cases described by CORI may be mentioned here, though there is a presumption that they are not really examples of Variation in the segmentation along the axis of a Primary Symmetry, but rather belong to the class of Secondary Symmetries. They are alluded to here as it is convenient to illustrate this distinction by taking them in connexion with the examples just given.

*95.   **Hermodice carunculata.** (Fig. 23, III.) Between two normal segments is what seems at first to be a segment double on the left side with two complete sets of parapodia, but imperfectly divided on the right (left of figure), the septal groove stopping short before it reaches the parapodial region. The lower half on this side is represented with a normal ventral ramus of the parapodium, but the ventral ramus in the upper was itself *partially doubled*, having in particular two cirri Cv. I. and Cv. II. and two branches of setæ. The condition of the dorsal ramus is not described. Of course without seeing this specimen it is impossible to say more than this, but the figure strongly suggests that the division between the two halves of this parapodium was a *division into images* and not into successive segments. The figure represents the lower cirrus Cv II. as standing in the normal position for the cirrus, on the posterior limb of the parapodium, but the anterior cirrus is distinctly shewn as placed on the *anterior* limb of the elevation and anterior to the bristles. If this were actually the case, this double parapodium must be looked on as a kind of bud, with a distinct Secondary Symmetry of its own. Described afresh from CORI, C. J., Z. f. w. Z., LIV. 1892, p. 574, fig. 3.

96. **Diopatra neapolitana.** In the middle of a specimen 35 cm. long was an arrangement somewhat similar to the above. The part marked

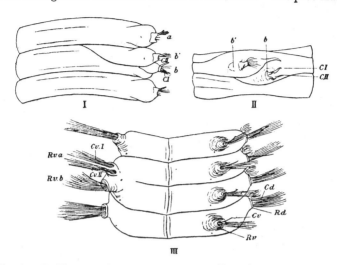

Fig. 23. I. The case of *Diopatra neapolitana* (No. 96) from the side. II, the same looking upon the parapodia. *C* I. *C* II, the two supernumerary cirri.
III. The case of *Hermodice carunculata* No. 95. *Cv*, cirrus of ventral branch of parapodium ; *Cd.* dorsal cirrus; *Cv.* I, *Cv.* II, the two cirri borne on the supernumerary parapodium. (After CORI.)

b' was cut off as shewn in Fig. 23, I., it bore a normal cirrus, and the other part of the segment, marked b, bore *two* cirri and two bunches of bristles. The figure does not indicate that there was any relation of images between these two parts, but this would scarcely appear in this case unless specially looked for. Described afresh from CORI, C. J., *l.c.*, p. 573, figs. 10 and 11.

In considering the evidence as to Secondary Symmetries reference to these cases will again be made.

GENERATIVE ORGANS OF EARTHWORMS[1].

The number and ordinal positions of the primary and accessory generative organs and of their ducts differ in the several classificatory groups of Earthworms. In the evolution of these forms it may therefore be supposed that Variation in these respects has occurred. To this subject the following evidence relates. The difficulty which was mentioned in the case of Variation in vertebræ, that there is no clear distinction between Homœotic and strictly Meristic Variation, will here also be met, inasmuch as the total number of segments in these forms is indeterminate; but

[1] For information and references on this subject I am indebted to Mr F. E. Beddard and Mr W. B. Benham.

probably we shall be right in regarding the majority of these variations as Homœotic.

LUMBRICUS. Throughout this genus there is normally a single pair of ovaries, placed in the 13th segment, on the posterior surface of the septum between the 12th and 13th segments. The following cases of supernumerary ovaries are recorded:

97. **Lumbricus turgidus**: specimen having an extra pair of ovaries in the 14th[1] segment.
98. Specimen having an extra ovary on the *right* side in the 14th segment.
99. **L. purpureus**: specimen having an extra ovary on the *left* side in the 14th segment.

In all these cases the extra ovaries were in size, form and position like the normal ovaries. There was no extra oviduct or receptaculum ovorum, but the normal ovaries and oviducts were present as usual. BERGH, R. S., *Zeit. f. wiss. Zool.*, XLIV. 1886, p. 308, *note*.

*100. **Allolobophora** sp. [partly = *Lumbricus*, the common Earthworm]: specimen having, in all, seven pairs of ovaries; viz. a pair in the 12th, 13th, 14th, 15th, 16th, 17th and 18th segments. Of these all except the pair of the 13th segment are supernumerary. Each of these ovaries was placed on the posterior face of a septum in the usual position. The three anterior pairs in shape, structure and position closely resembled the normal structures. Of these the most anterior were slightly the largest. The four posterior pairs were smaller and resembled the ovaries of a very young or immature worm, but on examination all were found to contain ova. The normal pair of oviducts were present and no extra oviducts could be found, though carefully sought for. WOODWARD, M. F., *P. Z. S.*, 1892, p. 184, *Plate* XIII.

*101. **Lumbricus herculeus,** Savigny (= *L. agricola*, Hoffmeister), having an asymmetrical arrangement of the generative organs, &c. On the *left* side the arrangement was normal; the ovary being in the 13th segment, the oviducal opening in the 14th, and the opening of the vas deferens in the 15th segment (Fig. 24).

On the *right* side each of these structures was placed in the segment anterior to that in which it is normally found: the right ovary was in the 12th, the external opening of the right oviduct was in the 13th, and the external opening of the right vas deferens was in the 14th segment. The spermathecæ were normal on the left side, being placed in the 9th and 10th segments, but on the right side one spermatheca only was present, that of the 9th segment. The vesiculæ seminales were present as usual in the 9th and 11th segments, but there was no vesicula in the 12th

---

[1] In BERGH's enumeration the ordinal number of these segments is one less than in that commonly used: the latter system is adopted above.

segment on the right side, while that of the left side was fully

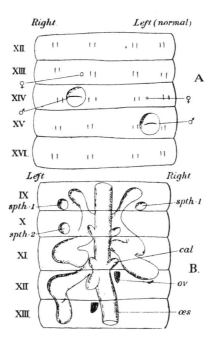

Fig. 24. *Lumbricus herculeus*, having the generative organs of the *right* side one segment higher than usual. A, external view from below. B, view of the organs from above. *spth*, spermathecæ; *ov*, ovary; *œs*, œsophagus; *cal*, calciferous glands. After Benham.

formed. It is remarkable that in this case, the calciferous gland of the 12th segment was absent on the right side. [I am indebted to Mr Benham for an opportunity of examining this specimen.] Benham, W. B., *Ann. & Mag. N. H.*, 1891, Ser. 6, VII. p. 257, Pl. III.

102. Another specimen presented the same variations as the foregoing, both as regards the asymmetrical arrangement of the genital pores and the absence of the calciferous gland: but in it there were vesiculæ seminales on the right side in segments 10 and 11, but none in segment 9; and there was a spermatheca on the right side in segments 8 and 9. [In the normal form the spermathecæ are in segments 9 and 10, so that, in this individual in the matter of the spermathecæ as well as of the genital pores, structures were formed in particular segments which are normally found one segment lower down.] Benham, W. B., *in litt.*, March, 1891.

103.   *Table shewing position of ovaries in forms having two or more pairs of ovaries, and in the Variations found* (slightly altered from M. F. WOODWARD, *P. Z. S.*, 1892):

Segments	9	10	11	12	13	14	15	16	17	18
Acanthodrilus					×	×?				
Eclipidrilus	×	×	×							
Eudrilus					×	×				
Lumbricus terrestris (normal)					×					
L. herculeus (? = *terrestris*) Benham's l. 2 specimens.............................. rt.				×	×					
L. turgidus Bergh's spec. (abnorm.)........						× ×	× ×			
do. do. do. (abnorm.)... l. rt.					×	× ×	×			
L. purpureus do. do. (abnorm.)... l. rt.					×	× ×	×			
Allolobophora sp. (Common Earthworm, abnorm.)..					×	×	×	×	×	×	×
Perionyx (two pairs, varying from 9—16)..	←————————————————→									
Phreodrilus				×	×					
Phreoryctes[1]					×	×				
Urochæta					×	×				

104.   **Allolobophora** sp. [♂ pores normally in 15th and ♀ pores in 14th, as in common Earthworm]: specimen having on the *right* side ♂ pore in 20th and ♀ pore in 19th; on the left side, ♂ pore in 17th and ♀ pore in 16th. MICHAELSEN, W., *Jahrb. Hamburg. wiss. Anstalt,* 1890, VII. p. 8. In each case the ♂ pore is in the segment behind the ♀ pore, as normally. The position of ovaries not given.

105.   **Lumbricus agricola** Hoffm. (= *terrestris* L.) : amongst 230 specimens in which the position of the male pores are determined, 6 specimens were found in which these openings were not normally placed (viz. one on each side in the 15th segment). In two of these specimens, both pores were in the 14th segment; in one case the left pore was in the 14th segment and the right was in the 15th : these three worms were German. One specimen was found in Savigny's collection in Paris which had two pores on the left side [and none on the right (?)]. In one English specimen the " vulva " [*sc*. the two male pores] was in the 14th segment and in another it was in the 16th. [The author speaks sometimes of both pores as the " vulva," and at other times he uses this term for one pore only, but the meaning is plainly that given above.] HOFFMEISTER, W., *Uebersicht aller bis jetzt bekannten Arten a. d. Familie Regenwürmer,* Braunschw., 1845, p. 7.

---

[1] *Phreoryctes*, a N. Zealand Oligochæt, has 2 pairs of testes and 4 vasa deferentia opening separately; 2 pairs of ovaries and 4 oviducts. BEDDARD, F. E., *Ann. and Mag.,* 1888, I. p. 339, *Pl.*

*106.   **Perionyx excavatus.** In this earthworm a very remarkable series of variations has been observed by Beddard. The accompanying table shews the varieties in number of spermathecæ and position of the generative openings which were found. The spermathecæ are generally 4, and are placed in the 7th and 8th segment, but in several specimens there were 8 and their position varied from the 6th to the 11th segment. In all the varieties, however, they were in segments adjacent to each other. In four specimens the spermathecæ were in the 8th and 9th segment on the right side and in the 9th and 10th on the left. In normal specimens the male pores are 2, but individuals with 4 (and perhaps 6) were found. There are *generally* 2 pairs of ovaries and oviducts. In Var. No. 11 an additional ovary was found on the right-hand side in the 11th segment and in Var. No. 10 there were three pairs of ovaries.

Table of Variations seen in P. excavatus (from BEDDARD).

	Spermathecæ	♀ pores	♂ pores	Clitellum
Normal (412 specs.)	8, 9	14	18	14—17
Var. 1 (1 spec.)	7, 8	11	16	12—15
,, 2 ,,	......	13, 14	18	
,, 3 ,,	8, 9	13, 14	17	13—17
,, 4 ,,	......	15, 16	20	
,, 5 (2 specs.)	8, 9	14, 14	18	13—17
,, 6 (1 spec.)	6, 7	10	14, 15	
,, 7 ,,	7, 8, 9, 10	15, 16	18	
,, 8 ,,	......	14, 15	18	
,, 9 ,,	7, 8, 9	14	17	
,, 10 ,,	8, 9, 10, 11	15, 16	19	15—18
,, 11 ,,	6, 7, 8	13, 14	16	
,, 12 (2 specs.)	8, 9, rt.; 9, 10, l.	14	18	
,, 13 ,,	8, 9, rt.; 9, 10, l.	14, 15	18	
,, 14 (1 spec.)	8, 9	15, 17	21	
,, 15 ,,	......	15, 16	18	

Though the position of both varied greatly, the male pores were always posterior to the female ones.

In some specimens certain of the segments were only divided from each other on one side of the body, being confluent on the other. For example in Var. No. 14, segments 11 and 12 and also segments 18 and 19 were only divided from each other on the left side (cp. Nos. 88—91).

Out of 430 individuals 15 variations in these structures were seen; of 12 of these variations single specimens only were found, but two specimens occurred with each of the other three forms of variation. In a single case a nephridium was found nearer to the dorsal line in one segment than in the adjacent segment. Many of the conditions here occurring as variations are found *normally* in other genera and species. BEDDARD, F. E., *Proc. Zool. Soc.*, 1886, p. 308, figs.

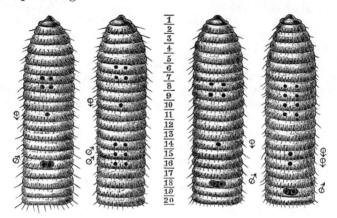

FIG. 25. *Perionyx excavatus*. Diagrams shewing some of the variations in respect of the number and positions of the openings of the spermathecæ and generative pores. From BEDDARD, *P. Z. S.*, 1886.

Perionyx grünewaldi, Michaelsen. Normally a pair of male genital pores on the 18th segment, and a single oviducal opening for the two oviducts in the middle line of the 14th segment.

107.   In two specimens a different arrangement was found. One of these had the oviducal opening in the 15th segment [position of male openings not specified and presumably normal].

108.   The other had two oviducal openings, one in the 13th and one in the 14th segment [not stated whether these openings were median or lateral, nor whether each of them was a double structure as of course the normal female opening is]. In this specimen the male openings also were placed anteriorly to their normal position, being in the 17th segment. MICHAELSEN, *Jahrb. d. Hamburg. wiss. Anstalt*, 1891, VIII., p. 34.

ALLURUS. In Terricolæ generally, the ♂ pores are on the 15th, and the ♀ pores on the 14th, as in the common Earthworm.

109.   **Allurus teträedrus,** a widely distributed form, has ♂ pores on the 13th and ♀ pores on the 14th, the ♂ pores being thus *in front* of the ♀ pores as a specific character. Under the name *Allurus dubius* Michaelsen described two specimens having the male pores on the 14th instead of on the 13th, and the ♀ pores on the 15th instead of on the 14th, each being thus one segment in advance of its normal place [backward Homœosis]. MICHAELSEN, W., *Jahrb. Hamb. wiss. Anst.*, 1890, VII., p. 7; see also *Arch. f. Naturg.*, 1892, LVIII., p. 251. Compare No. 111.

110.   Besides these is a batch of 8 specimens of *A. teträedrus*, loc. unknown, 6 specimens had both ♂ and ♀ pores in the 14th. Clitellum began in 23rd, tuberc. pubert. in 24th. These specimens are thus intermediate between *A. hercynius*, which has the pores as in *Lum-*

bricus, and *A. teträedrus*. MICHAELSEN, W., *Arch. f. Naturg.*, 1892, LVIII., p. 251.

111. **Allurus putris**: specimen having ♂ pores on 13th (instead of 15th) as an abnormality; in it the other external generative organs (and doubtless the internal also) were 2 segments higher than usual, the ♀ pore being on the 12th instead of 14th. Tuberc. pubert 26—28. MICHAELSEN, *Jahrb. Hamburg. wiss. Anst.*, 1891, VIII., p. 8. Compare No. 109.

112. **Allurus** sp.: specimen having l. side normal; *right* side, ♂ pore in 12th, ♀ in 11th, clitellus and tuberc. pubert. one segment higher than usual. *Ibid.*

*113. ENCHYTRŒIDÆ. ♂ pore generally in the 12th segment. In *Buchholzia appendiculata* (Buch.) it is on the 8th, as a specific character. In **Pachydrilus sphagnetorum** (Vejd.) it is either on the 8th or on the 9th, according to individual variation, the other parts being then disposed as follows:

	♂ pore on 8th	♂ pore on 9th
Testes on dissep.............	7/8	6/7
Ovaries on dissep.	8/9	7/8
Vas def. in front of dissep.	8/9	7/8
♂ pore	9	8
Oviduct on dissep.	9/10	8/9
♀ pore	10	9
Clitellum	9 and ½ 10	8 and ½ 9

MICHAELSEN, W., *Arch. f. mikr. Anat.*, 1888, XXXI. p. 493; see also *Jahrb. Hamb. wiss. Anst.*, VII. p. 8.

114. **Perichæta hilgendorfi**, n. sp. Mich. 7 specimens. Variation in number of spermathecal openings, as follows. 5 specimens had 2 pairs in the groove between segments 6/7 and 7/8; 1 specimen had 3 pairs, between 5/6, 6/7 and 7/8; 1 specimen had only one, on the left side between 6/7, which corresponded internally to a single spermatheca [other variations also observed in these specimens, q. v.]. MICHAELSEN, W., *Arch. f. Naturg.*, 1892, LVIII. p. 236.

115. **Perichæta forbesi** (an Earthworm from New Guinea). In this animal a pair of spermathecæ is placed in the 8th segment and another pair in the 9th. Two specimens only have been examined and in both of these an additional spermatheca was found on the left side, internal to the other. In one individual the 5th spermatheca was in the 8th segment, and in the other it occurred in the 9th. BEDDARD, F. E., *Proc. Zool. Soc.*, 1890, p. 65, Plate.

116. **Allolobophora lissäensis.** Similar variation in spermathecæ, MICHAELSEN, W., *Jahrb. Hamb. wiss. Anst.*, VIII., 1891, p. 19.

HIRUDINEA.

117. **Hirudo medicinalis.** The number of pairs of testes is variable, but 9 pairs most often found. Of 31 specimens of this species, 21 had 9 pairs, 6 had 10 pairs, and 4 had 9 on one side and 10 on the other. CHWOROSTANSKY, C., *Zool. Anz.*, 1886, p. 446.

118. **Hirudo officinalis:** of 7 specimens, 5 had 9 pairs of testes, 1 had 10 pairs, and though in the 7th specimen there were 2 pairs, the vas deferens of the last pair of testes ended blindly. *Ibid.*

119. **Hirudo medicinalis.** Fairly often the vas deferens is prolonged beyond the 9th testis, and having passed through five annuli, ends in a glandular mass of irregular form. Case given in which the 7 last testes of right side were absent or only represented by amorphous material, the testes of the left side being abnormally large. ÉBRARD, *Nouv. monogr. des Sangsues méd.*, Paris, 1857, p. 99.

*120. **Hirudo officinalis:** an individual having a supernumerary penis, and vesicula seminalis of the right side, in the 5th somite.

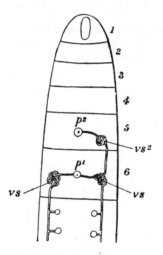

FIG. 26. Case of *Hirudo officinalis*, No. 120.
p^1, penis in normal position; p^2, supernumerary penis; *vs*, the usual vesiculæ seminales; vs^2, supernumerary vesicula seminalis. (From a diagram sent to me by Mr Gibson.)

The normal penis in the sixth segment was fully formed and into it opened on either side a vas deferens, provided with a vesicula seminalis as usual. But the vesicula of the right side gave off in addition a vas deferens, which passed forwards into the fourth segment and there enlarged into another vesicula seminalis. This additional vesicula was connected by a duct with a supernumerary penis placed and opening in the middle of the fifth segment. The parts of the left side as well as the female organs were normal. [I have to thank Mr Gibson for furnishing me with a diagram (Fig. 26) supplementing the published account.] GIBSON, R. J. HARVEY, *Nature*, 1887, XXXV., p. 392.

Aulastoma gulo (Horse Leech). In this form as is usual among the *Gnathobdellidæ* there are from 9 to 12 pairs of testicular sacs which communicate with a tortuous vas deferens on each side which together enter a single penis. The paired ovaries are placed behind this and the oviducts unite to form a common vagina.

*121. In a specimen found amongst a large series investigated, each vas deferens opened by a separate penis, of which the most anterior opened in the 20th annulus and the posterior in the 25th. The female apparatus was similarly divided. One ovary was placed near the penis in the 25th annulus and from it a vagina passed down to open with the penis. The other ovary, with a similar vagina, lay in the 30th annulus. ASPER, G., *Zool. Anz.*, 1878, I., p. 297.

Recapitulation of evidence as to OLIGOCHÆTA *and* HIRUDINEA.

Variation in these two groups appears in such similar modes that points of special consequence in both may conveniently be spoken of together.

1. As elsewhere seen, so here, there are forms, *e.g.*, *Perionyx excavatus* or *Pachydrilus sphagnetorum*, shewing great variability, while others, the common Earthworm for instance, rarely vary.

2. Both forward and backward Homœosis may occur; a form normally having the ♂ pores, for instance, on the 15th segment, may as an individual variation have them on the 16th (No. 105), while an individual of another genus, starting from the same normal, may have them on the 13th (No. 111).

3. As in other cases of Homœosis, when a member of a Meristic Series, in this case a segment, develops an organ proper to another segment, this organ is formed in a place serially homologous with its normal place. (To this principle certain limitations must hereafter be introduced.)

4. Variation may, or may not, be simultaneous and correlated in the several systems. The position of the ♀ openings, for example, may or may not vary similarly and simultaneously with that of the ♂ openings, though on the whole the evidence suggests that such correlation is not uncommon. The facts seen in the genus *Allurus*, in which one species (*A. tetraedrus*) has the ♂ pore normally in front of the ♀ pore, sufficiently indicate that the variation in the position of these two openings is not always so correlated. It may be further mentioned that variation in number of ovaries seems to occur generally without correlated variation in the number of oviducts.

5. Such Variation may or may not be simultaneous on the two sides of the body. When not thus bilaterally symmetrical, there may nevertheless be a full correlation between the parts of the same side.

6. The evidence does not indicate any limit to the number of segments which may take on a certain character, or approxi-

mate to a given pattern. The highest number of ovaries, for instance, recorded, is 7 pairs; but there is nothing to shew that more segments might not undergo similar Homœosis. (The progressive diminution in size of these ovaries from before backwards in this case is worth noticing.)

7. The principle so often manifested in the evidence of Variation, that the magnitude, completeness, and symmetry of a variation bears no necessary proportion to the frequency of occurrence of that variation, is here strikingly exemplified.

8. The evidence as to the existence of two varieties of *Pachydrilus sphagnetorum*, the one with all the organs a segment higher than their place in the other variety may be well compared with SHERRINGTON'S observation, that in the Frog and in several Mammals (see No. 70) the individuals could be roughly divided into two classes according as the lumbo-sacral plexus was formed more anteriorly ("preaxial class") or more posteriorly ("postaxial class").

9. In the evidence as to *Perionyx*, it was seen that many of the arrangements found occurred in single specimens only, suggesting the inference that the systems do not fall into one of these conditions more easily than into others; nevertheless of each of three abnormal arrangements two examples were found, a circumstance hardly to be expected on the hypothesis of fortuitous Variation.

10. It is perhaps unnecessary to point out that the examples of Variation given are in their several degrees Discontinuous, and that by the nature of the case the Variation by which the several specific forms have attained their particular numbers and characteristic disposition of organs, must almost of necessity have been thus Discontinuous.

CESTODA.

The following facts respecting Variation in Cestoda are chiefly taken from LEUCKART, *Parasiten des Menschen*[1].

Besides the variations here enumerated, abnormalities of several other kinds (variation in number of suckers, prismatic segments, bifurcation, &c.) are known in this group, but as these do not directly illustrate the Variation of Linear Series, consideration of them must be deferred.

The degree to which the parts bearing sexual organs are separated from each other differs greatly in the various groups of Cestodes. In some (*Triænophorus*) the segmentation amounts to an inconsiderable constriction, while in *Ligula* the generative organs are repeated several times in a common body. L., p. 347.

---

[1] In what follows the letter L. is used in reference to this work.

122. Even in the groups whose segmentation is commonly perfect, variations in the degree of separation between the proglottides are not rare. It frequently happens that specimens of *Tænia* are found in which the external segmentation is partial, being only found on half of the contour. This abnormality, which does not affect the internal organs, occurs several times in the same chain. MONIEZ, R., *Bull. Sci. du Nord*, x., p. 200.

123. **Tænia saginata.** Cases of the "intercalation" of a triangular, wedge-like segment between two proglottides are recorded. In such cases the generative opening is on the same side as in an adjacent segment, not taking part in the alternation. L., p. 572. Compare with similar phenomena in Chætopoda (p. 156).

The evidence of abnormal repetition of parts occurring in single proglottides bears on the question of the relationship of the perfectly segmented forms to the less fully segmented.

124. **Tænia saginata:** a specimen 128 mm. long, wanting the head, without any division into segments. The longitudinal vessels were seen, but no transverse vessels were discovered. On the margins were numerous genital openings, of which 41 were counted, each leading from a genital organ. There was no regular lateral alternation between the genital papillæ, but they were disposed without uniformity of pattern, and several were closely approximated to each other. In no part was there any trace of division into proglottides. From the characters of the genital openings and from the number and size of the calcareous bodies together with other histological details, the specimen was determined without much doubt as *Tænia saginata*. GROBBEN, C., *Verh. zool.-bot. Ges. Wien*, 1887, Bd. XXXVII., p. 679, fig.

Such repetition of the generative openings in single segments is very common, especially in *Tænia saginata*, and indeed examples of it may be seen in most chains of segments. Usually such repetition is confined to one segment and is not striking. *Five* generative papillæ have been seen by L. in one segment, and COLIN [ref. not found, W. B.] described 25—30 genital pores in an unsegmented piece measuring 15 cm. L, p. 571.

125. Repetitions are not confined to the generative openings, but the generative organs themselves are also thus abnormally repeated. In cases in which several sets of generative organs occur in the same segment it is found that those near the middle of the segment are the least developed. In these cases, though the different organs frequently cross each other, Leuckart found no anastomoses between them, but the number of distinct sets of generative organs was the same as the number of pores.

It was not found that the length of the segments increased in the same ratio as the number of the pores they contain. For example, a segment with two pores measured 18 mm. in length (instead of about 20 mm.), and one with five pores measured 28 mm. (instead of 50 mm.). L., p. 571.

126. **Tænia.** Case quoted by Leuckart from HELLER of a *Tænia* having generative openings placed on the *surface* of the segments. Leuckart himself has never seen an example of this variation. [Original reference not found] L., p. 570, *Note*.

127. **Tænia solium** and **T. saginata.** Specimens are known having *two generative pores opposite each other at the same level*. In such cases each leads to a male and a female duct with cirrus-sac and receptaculum seminis; but the organs for preparing the ova are normal in construction, as the two vaginæ lead to a common uterus and shell-gland. Two cases only have been seen by LEUCKART and he cites another from WERNER. L., pp. 529 and 571.

128. **Tænia solium** in which the pores are normally alternating, may be found with symmetrically developed pores; and on the contrary, **T. elliptica** in which they are normally symmetrical, may occur with an asymmetrical arrangement. L., pp. 353 and 529.

129. **T. saginata:** in a chain of about 6·5 metres in length, and containing some 650 joints, there was found a single, heart-shaped, supernumerary joint like those described; a single joint was found with two genital pores, one being on each lateral border at about the same level.

The largest number of consecutive joints having the genital pores on the same side was six. TUCKERMAN, F., *Zool. Anz.*, XI., 1888, p. 94.

130. **Tænia cœnurus.** Specimen observed by Leuckart in which the last 8 or 10 segments shewed a *transposition* of the generative organs, those which usually lie at the distal end being placed at the proximal. This change of position was especially seen in the case of organs engaged in the preparation of the ova. The proximal proglottides of this individual were normal. The transition segment between these two regions contained two simple vesiculæ seminales and two marginal papillæ which were on opposite sides; but in spite of the resemblance of these structures to genital pores, neither opening, nor cirrus, nor vasa deferentia could be distinguished. L., p. 504.

131. Amongst chains of normal proglottides it is not rare to find a segment containing male organs only. L., p. 504.

Speaking generally, slight abnormalities are far more common than great ones. Nearly every specimen of Tapeworm has individual peculiarities, and these generally repeat themselves in the same chain of proglottides. This repetition of the same abnormality in different parts of the chain is also the rule for the greater abnormalities also. L., pp. 529, 572 and 573.

CHAPTER VII.

LINEAR SERIES—*continued.*

BRANCHIAL OPENINGS OF CHORDATA AND STRUCTURES IN CONNEXION WITH THEM.

UNDER the general heading of Variation of branchial openings facts will be given relating to the following subjects.
 I. Variation in the patterns formed by the bars, vessels and stigmata of the branchial sac in Ascidians.
 II. Variation in the number of gill-sacs in Cyclostomi.
 III. Abnormal openings in the cervical region of Mammals, known as "cervical fistulæ," and external appendages called "cervical auricles," or "supernumerary ears," present sometimes in connexion with such openings.

With reference to the two first subjects the evidence is only fragmentary, but the instances recorded seem to be of sufficient consequence to warrant their introduction in illustration especially of the magnitude and definiteness of Variation.

Variations affecting the opercular opening in Amphibia are mentioned in connexion with Bilateral Series.

I. ASCIDIANS.

Transverse vessels of Branchial Sac.

132. **Ascidia scabra.** Branchial sac in one specimen shewing abnormal and irregular structure owing to branching of transverse vessels. The resulting appearance is entirely peculiar. HERDMAN, W. A., *J. Linn. Soc.* (Zool.), 1881, xv., p. 284, Pl. XVII., fig. 3; also p. 330.

133. **Ascidia virginea** (O. F. Müller): a case of great irregularity exactly similar to the above. *Ibid.*, p. 330.

134. **Ctenicella lanceplani.** Branchial sac may present characters due to variations in disposition of transverse vessels &c., which assume three distinct patterns or marked varieties. LACAZE-DUTHIERS, *Arch. Zool. Exp.*, S. 1, Vol. VI., p. 619, Vol. XXXIII., figs. 9—11.

*135.   **Ascidia plebeia** (Alder): branchial sac has very characteristic appearance and is very constant in the size of meshes, papillæ &c. One point is liable to variation: as a rule the transverse vessels are of the same calibre, but in several specimens *every fourth* vessel is much wider than the intervening three. HERDMAN, p. 331.

Stigmata and Meshes.

136.   **Ciona intestinalis:** meshes vary but according to no apparent method: 5 stigmata in a mesh normal; 4 and 6 met with frequently; 10 the utmost seen. HERDMAN, p. 332.

137.   **Ascidia aspersa.** In typical specimens, transverse vessels all same size, the meshes being square and undivided, but individuals occur in which many (not all) of these square meshes are divided by delicate transverse vessels into pairs of oblong areas. HERDMAN, p. 332.

138.   **Styela grossularia.** The genus *Styela* is characterized by the presence of branchial folds, normally four on each side, but in this species the folds are almost obsolete, being entirely wanting on the left side and reduced to a single slight inward bulging on the right side, bearing internal longitudinal bars. This fold is separated from the dorsal lamina by a broad space without internal longitudinal bars. A similar wide space is present on the left side of the dorsal lamina, and two others on the vertebral edge of the sac, one on each side of the endostyle. These spaces vary in size in individuals. They commonly contain 16 stigmata, but numbers down to 12 were frequent and in one case 10 only were present: only once more than 16 observed, and in that case *there were 23*. Number of internal longitudinal bars on fold varies from 6 to 9, generally 8 or 9. HERDMAN, p. 330.

In considering the significance of these cases with reference to the origin of Species it is to be remembered that the characters of the branchial sac, the sizes of the transverse vessels, shape of meshes and the number of stigmata they contain are held to be of the first importance for the classification of Ascidians; but HERDMAN finds that while they are highly characteristic in some species they are not so in others[1].

II.   CYCLOSTOMI.

*139.   **Myxine glutinosa.** In this genus there are normally six pairs of branchial pouches. I am indebted to Professor Weldon for an account of a specimen dissected by him in which there were *seven* pairs of these pouches. On the left side all the seven pouches were distinct and separate, each having a separate open-

[1] The olfactory tubercle in Ascidians may have a different form and position in different individuals of the same species, but the range of variation changes according to the species. *Molgula* was found to be the most constant, *Ascidia virginea* and *A. plebeia* the most variable forms. HERDMAN, *Proc. R. Phys. Soc. Edin.*, VI., p. 267, figs.; also *id., Proc. Lit. Phil. Soc. Liverpool*, XXXVIII. p. 313, Pls. I. and II. Variation respecting the atrial pore will be considered in connexion with Bilateral Repetition.

ing from the œsophagus and a separate aortic arch supplying it. On the right side the sixth and seventh pouches were practically

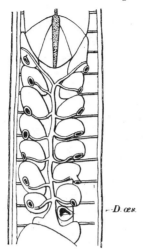

Fig. 27. *Myxine glutinosa*; specimen having seven pairs of branchial sacs. Diagram shewing branchial sacs, heart and aortic arches from the dorsal surface. On the right side the sixth and seventh branchial sacs were partially confluent.
D. œs., *ductus œsophageus.*
(From a drawing kindly lent by Prof. Weldon.)

confluent though each had a separate œsophageal opening and a separate arch from the aorta. In the drawing, for which I am also indebted to Professor Weldon, the œsophageal openings are not shewn.

*140. **Bdellostoma.** In this genus the number of branchial sacs is variable, different numbers being found in different species and individual variation also occurring.

The generic name *Heptatrema* was originally given by DUMÉRIL from the presence of seven gill-sacs. In 1834 JOH. MÜLLER, finding that this character is not constant proposed the name *Bdellostoma*. Of three Cape specimens examined by him one had seven gill-sacs on each side, one had six on each side, and one had six on the right side and seven on the left. To these he gave the names *B. heptatrema*, *B. hexatrema* and *B. heterotrema* respectively (*Abh. k. Ak. Wiss. Berlin*, 1834, pp. 66, 67 and 79, Taf. VII.). Further observation has shewn that the number of gill-sacs in the Cape *Bdellostoma* is liable to individual variation, some specimens having six while others have seven. The name *B. cirrhatum* (GÜNTHER, *Cat. Brit. Mus.*, VIII. 1870, p. 511) includes these and the New Zealand specimens. As to the relative frequency of specimens with six or seven pairs or with an asymmetrical arrangement I have no information. A collection lately brought from the Cape by Sedgwick includes one specimen with six pairs and several with seven pairs.

141.　**B. polytrema:** single specimen from Chili, badly preserved but apparently having fourteen pairs of gill-openings. GÜNTHER *l. c.*, p. 512.

Specimen having 14 gill-openings on left side and 13 on right. SCHNEIDER, A., *Arch. f. Naturg.*, XLVI. 1880, p. 115 (cp. PUTNAM, *Proc. Bost. N. H. S.*, XVI. 1873, p. 160).

B. bischoffii: single specimen, 10 gill-openings on each side. *ibid.*

Ammocœtes: having eight branchial openings on each side instead of seven, the normal number. The shape of the mouth of this specimen was also abnormal, being described as somewhat square. [No satisfactory description.] EDWARD, THOMAS, *Zoologist*, XVI., p. 6097.

142.　In connexion with individual Variation in the number of gill-sacs in Myxinoids it should be borne in mind that in *Petromyzon* there are *normally seven* pairs of gill-sacs. The case of the Notidanidæ may also be mentioned in this connexion. Among Selachians the Notidanidæ are peculiar in having a number of gill-slits other than five, and of them *Hexanchus* has six pairs, while *Heptanchus* has seven[1].

III.　CERVICAL FISTULÆ AND SUPERNUMERARY AURICLES IN MAMMALS.

Though the evidence of this subject is well known and has often been collected, it may be convenient to give here some abstract of the facts in so far as the phenomena of Variation are illustrated by them. Since cervical fistulæ have been believed to result from the persistence of the embryonic branchial clefts, they may properly be considered in relation to the general question of Variation in the number of gill-slits, while the development of external appendages, perhaps serially homologous with the external ears, directly concerns the subject of Meristic Variation.

Man. The subject has been studied by many observers, especially by ASCHERSON[2], and by HEUSINGER[3], who brought together and abstracted 46 cases, being all that had been described in Man up to 1864. G. FISCHER[4] gives a full list of the literature of the subject up to 1870, with an analysis of 65 cases. A further paper by HEUSINGER[5] contains a general account of these structures as they are found in Man and in the domestic animals. Additional cases, together with a general discussion of the subject, especially in relation to fistulæ on the external ears, were given by Sir JAMES

---

[1] **Balanoglossus.** In five species with which I am acquainted, the number of gill-bars and slits varies in proportion to the size of the body, and as it is not unlikely that these animals continue to grow throughout life, it is probable that the number of branchiæ is always increasing by formation of new gill-slits at the posterior end of the branchial region. The same is probably true of *Amphioxus*.
[2] ASCHERSON, *De fistulis colli congenitis*, Berlin, 1832.
[3] HEUSINGER, *Arch. f. Path. Anat. u. Phys.*, 1864, XXIX.
[4] FISCHER, G., *Deut. Ztsch. f. Chirurg.*, 1873.
[5] HEUSINGER, *Deut. Ztsch. f. Thierm.*, 1878.

PAGET[1] in 1878. Lastly, the whole evidence as to cervical fistulæ and the structures associated with them has been fully collected up to 1889 and tabulated by KOSTANECKI and MIELECKI[2], who also discuss in detail the relations of these abnormalities to the facts of development. The following account is taken from these sources. For figures the reader is referred to the original memoirs.

*143. Cervical fistulæ are generally known as orifices placed in the region of the neck, leading into a sinus of greater or less extent, varying in size from a mere pit to a duct some inches in length. In the greater number of cases the sinus ends blindly, but in about a third of recorded cases (K. and M.) it passes inwards to open into the pharynx, forming thus a communication between the pharyngeal cavity and the exterior. Such passages are spoken of as *complete* cervical fistulæ, those which have an external but no internal opening being *external incomplete* fistulæ. Besides these there are cases of diverticula from the pharynx or œsophagus which do not reach the exterior, and these are known as *internal incomplete* fistulæ.

Cervical fistulæ are more commonly present on one side only, but in a good many cases they have occurred on both sides. According to Fischer they are more common on the right side than on the left. The following statistics are given by him. 65 persons had 79 fistulæ: 51 unilateral, 14 bilateral: 20 complete, 53 without an opening to the pharynx: of the unilateral cases 33 were on the right and 13 on the left: 34 in males, 30 in females. There was evidence of heredity in 21 cases.

The external opening is very small and may either be on the surface of the skin or elevated on a minute papilla. Sometimes it is covered by a small flap of skin as with a valve, in other cases it is placed as a fissure between two lips. The positions in which the external openings of cervical fistulæ are found are very variable, but in the great majority of cases the opening is close to the middle line in the neighbourhood of the sterno-clavicular articulation, generally from a few lines to an inch above it, on either the inner or the outer border of the sterno-cleido-mastoid muscle. In rarer cases the external opening is placed at the level of the middle of the cricoid cartilage, and is sometimes just behind the angle of the jaw. These positions are not however at all precisely maintained, but vary a good deal in different cases. When the external opening is in the higher situation and the fistula is complete, a sound may then be passed into the pharynx, but when the external opening is low, the duct when present passes upwards covered by skin only, in a straight line so far as the upper limit of the larynx, at which point it turns at a sharp angle upwards and inwards. For this reason it is not possible in such cases to follow the course to the pharynx by means of a sound, but in some of them the presence of an internal opening has been proved by the injection of fluids having colour or taste. The position of the internal openings is also variable, and from the nature of the case has been accurately

---

[1] PAGET, Sir J., *Trans. Med. Chir. Soc.*, LXI., 1878.
[2] VON KOSTANECKI und VON MIELECKI, *Arch. f. path. Anat. u. Phys.*, cxx. and cxxi.

determined in comparatively few instances. In a case dissected by
NEUHÖFER[1] there was a fistula on each side, the external opening
of the right was ½ in. from the middle line and 7 lines above the
clavicle, that of the left was 3—4 lines higher and further from the
middle line. The right internal opening was on the posterior border of
the pharyngo-palatine muscle, behind the cornu of the hyoid near the
tonsil, the left internal opening being rather higher than the right.
Internal openings of such fistulæ have also been seen on the edge of
the arcus pharyngo-palatinus, also in the neighbourhood of the root of
the tongue. SEIDEL[2] gives a case in which there were two fistulæ, the
one on the right side in the upper position, and the other in the middle
line at about the same level, but whether either of these communicated
with the pharynx could not be made out. The twin-brother of the
same infant had a single minute fistula.

The ducts of cervical fistulæ are usually of greater calibre than the
external openings but they are rarely wider than a fine quill. The
walls are tough and the lining epithelium is sometimes flat and sometimes ciliated. The degree to which the walls are sensitive differs in
different cases. The external opening is described in several instances
as having a reddish colour. In three cases of the presence of branchial
fistulæ in female patients, it is recorded that the external openings
became inflamed during the menstrual periods.

From the point of view of the naturalist the chief interest of
cervical fistulæ arises in connexion with the question of their morphology. Since the time of Ascherson the view has been commonly
accepted that these structures arise by persistence of embryonic gill-
clefts, and some of the recent writers[3] on the subject have gone so far
as to apportion the various forms of cervical fistulæ among the several
gill-clefts from the first to the fourth, according to the situations of
the external openings, giving diagrams shewing the regions occupied by
each. As KOSTANECKI and MIELECKI point out, this apportionment is
quite arbitrary; for in the development of the neck the external in-
vaginations for all the clefts behind the hyoid arch become included in
the sinus cervicalis of Rabl (sinus præcervicalis of His), which is
eventually closed by the growth of the opercular process from the
hyoid arch. The external opening of a cervical fistula may thus
represent a part of the sinus cervicalis still left open, but it cannot on
the ground of its position be referred to any gill-cleft in particular.
Such reference could only be properly made on the ground of the
position of the internal opening and the course of the duct in relation
to structures whose relation to the visceral clefts is known. Moreover owing to the way in which the 3rd and 4th clefts are shifted
inwards by the formation of the sinus cervicalis, Kostanecki and
Mielecki consider that they are practically excluded. The same
authors after an analysis of the cases in which the position of the
internal opening has been properly ascertained, come to the conclusion
that in all these it falls within the region of the 2nd visceral sac

---

[1] NEUHÖFER, M., *Ueb. d. angeb. Halsfistel*, Inaug. Diss., Munich, 1847.
[2] SEIDEL, J., *De fist. colli congen.*, Inaug. Diss., Breslau, 1863.
[3] SUTTON, J. BLAND, *Lancet*, 1888, p. 308; CUSSET, *Étude sur l'appareil branchial*, &c., Paris, 1887.

(hyo-branchial). Besides they point out that the evidence in the few cases in which the course of the duct has been traced, shewed that it passed between the external and internal carotids. In their judgment, therefore, cervical fistulæ are all to be referred to the second (hyo-branchial) cleft.

Next it is to be remembered that according to many observers (especially His) there is at no period a complete connexion between the outer gill-clefts and the evagination from the pharynx or branchial sacs, but the membrane separating these chambers is stated by them never to be broken down. If this account is accepted, it is, as Kostanecki and Mielecki have said, necessary to suppose that in the case of any complete cervical fistula a communication between the exterior and the pharynx has arisen by some abnormal occurrence. This is illustrated by reference to the normal condition of the first or hyo-mandibular cleft. Here the auditory meatus represents an external incomplete fistula, and the Eustachian tube an internal incomplete fistula, the two being separated by the tympanic membrane. In a single case given by Virchow[1] a complete passage existed congenitally in this position, together with great abnormality in position and form of the external ear.

From the evidence it may thus on the whole be concluded that incomplete external fistulæ result from imperfect closure of the sinus cervicalis, and that incomplete internal fistulæ may arise by persistence of one of the branchial sacs, but it is doubtful whether many cases of the latter properly belong to the category of branchial fistulæ at all.

Supernumerary Auricles.

144. Abnormal appendages attached to the neck have been described by several observers, and by those who have discussed the subject of cervical fistulæ some account of these appendages is generally given. In the neighbourhood of the external ears, especially near the antitragus, such structures having the form of small warts or flaps of skin are not very uncommon. Their presence is generally associated with deformity of the external ear, and often with what are known as "aural fistulæ[2]." In the region of the neck, supernumerary auricles

[1] Virchow, *Arch. path. Anat. u. Phys.*, 1865, xxxii.
[2] *Aural* fistulæ are spoken of by many writers as being of the same nature as cervical or branchial fistulæ. They are blind ducts or pits, opening on some part of the external ear and are nearly always associated with other abnormalities either in the form of the ear or defective hearing, &c. (Schmitz, *De fist. colli congen.*, Inaug. Diss., Halle, 1873 [not seen, W. B.]; Urbantschitsch, *Monatsch. f. Ohrenh.*, 1877, transl. *Edin. Med. Jour.*, xxiii. 1878, p. 690.) They may be either unilateral or on both sides of the body. Sir James Paget (*Trans. Med. Chir. Soc.*, lxi., p. 41) described the occurrence of such fistulæ in the ears of several members of a family, many of whom were affected with deafness. The supposed connexion of these fistulæ with cervical fistulæ was in this case suggested by the fact that several cases of actual cervical fistulæ occurred in the same family, several of its members having both cervical and aural fistulæ. From the evidence of the not infrequent association of the two kinds of malformation most writers (Paget, Urbantschitsch, &c.) consider that the aural fistulæ must be branchial in origin and may be taken to represent the first (hyo-mandibular) cleft.

Kostanecki and Mielecki (*l. c.*), following His, point out that since in no case has an aural fistula ever been known to communicate with the auditory meatus or tympanic cavity, this belief is unsupported; and in addition, that from the mode of development of the external ear from a number of tubercles, it is

are much rarer, but in several instances they have attained a considerable development. Of this class of variation the following well-known case is one of the most remarkable.

*145. A healthy female infant was brought to Guy's Hospital in 1851 on account of two projecting growths about the middle of the lateral cervical regions. The growths were not removed until February 1858, when they were found to have increased slightly. They were situated over about the centre of the sterno-cleido-mastoid muscles. To the touch they resembled the tissue of the lobe of the auricle, and they contained within them a firm resisting nucleus like the cartilage of the same organ. They were also covered with peculiarly delicate, soft, downy hairs, like the lobe of the ear. They were excised without difficulty. Each was supplied with a small artery. They appeared to be intimately associated with the fibres of the platysma myoides, not dipping deeper than this structure, and to be entirely cutaneous appendages. (Fig. 28.)

Fig. 28. Child having a well-developed supernumerary auricle on each side of the neck (from BIRKETT).

A vertical section was made in the long axis of each growth; and the tissues of the lobe and of the fibro-cartilage of the auricle were clearly distinguished. The shape of the fibro-cartilage resembled more or less closely in parts, the outline of the proper auricle, and its tissues were the same. BIRKETT, J., *Trans. Path. Soc. Lond.*, IX., 1858, p. 448, fig.[1].

sufficient to suppose that aural fistulæ arise by the imperfect union of these tubercles. The fact, however, that these various defects in development of the branchial apparatus and its derivatives are frequently associated together is well established. As indicating the frequency of association with disease of the ear, Urbantschitsch mentions that in 2000 aural cases, 12 instances of aural fistulæ were seen. The same author gives a remarkable case of the occurrence of aural fistula on the *right side only* in many members of the same family with other important particulars (*l. c.*).

[1] In *Lancet*, 1858, II. p. 399 (quoting HARVEY), and in a paper by VIRCHOW (quoting WILDE), *Arch. path. Anat. Phys.*, 1864, XXX. p. 225, reference is made to a case of CASSEBOHM, *Tract. sextus, de aure monstri hum.*, Norimb., 1684, pp. 36 *et seqq.*, describing a child with "four ears." On referring to the original however it appears that this was merely a double monster, having two incomplete heads, and thus bears no analogy with the present examples.

Several cases analogous to the above, though differing in the extent of the development, are on record [1]. KOSTANECKI and MIELECKI (*l. c.*), who give references to the literature of the subject, consider together with VIRCHOW and others, that there is no doubt that these supernumerary auricles may properly be regarded as "heterotopic" partial repetitions of the external ears. According to a view which has been held by the majority of writers on the subject, and which is in part alternative to that given above, it is suggested that the cartilages contained in these appendages are in reality parts of one or other of the usually undeveloped branchial arches behind the hyoid. As against this suggestion it is to be remembered that in the subsequent development of the neck these arches are pushed in far from the surface, whereas the cartilages in question are always superficial. The usual histology of these bodies is in favour of the view that they are repetitions of the ear-cartilages, but on the other hand a specimen of cervical auricle in Mus. Coll. Surg. (No. 373, *c*) contains not only cartilage but also a small bone of complex form. But whether or not any part of such cervical auricles truly represents any part of the gill-bars, it is clear that these external projections having the structure of the ear, considered from the point of view of Variation must be regarded as partial repetitions of the ears, and there is a considerable probability that they stand to the sinus cervicalis in a relation similar to that which the normal external ear bears to the hyo-mandibular cleft, being according to the terminology here proposed, examples of repetition by forward Homœosis.

In this connexion the question of correlation between supernumerary auricles of the neck and cervical fistulæ is especially important. If it is true that such auricles are repetitions of the ears, it might, on the analogy of other cases of repetition, be expected that they would usually be found bounding the external openings of fistulæ. As a matter of fact they have several times been found in such a position, but the connexion between these two variations is by no means a close one, for cervical fistulæ are not as a rule accompanied by cervical auricles, nor are cervical auricles generally associated with cervical fistulæ, such collocation being on the whole exceptional. It should also be mentioned that in a few cases small cartilaginous or bony structures have been found imbedded in the neighbourhood of cervical fistulæ, but that similar structures have also occurred independently of any fistula [2].

In many domestic animals both cervical fistulæ and auricles are well known and have been described by Heusinger [3] from whom the following account is chiefly taken.

146.   **Pig.**   Cervical auricles are not uncommon and have been referred

---

[1] A figure is given by SUTTON, J. B., in *Ill. Med. News*, 1889 (repeated in "*Evolution and Disease*," by the same author 1890, p. 83), representing a large supernumerary auricle on the right side of the neck of a girl. The structure is represented as helicoid in form, closely resembling the normal ear. It is unfortunate that no description of this specimen is given: in the absence of such description this quite unprecedented case cannot be accepted without reserve.

[2] Dermoids of many kinds occurring in the cervical region of Man and other animals are by many writers considered to arise by modification of tissues occluded from the walls of the branchial clefts.

[3] HEUSINGER, *Deut. Arch. f. Thiermed.*, 1876, II.

to as distinguishing particular local breeds. They are generally paired structures. The following case is exceptional in the fact that the auricle was present on the left only, and that it was associated with an opening possibly of a cervical fistula. A pig having a single appendage about 7 cm. long attached under the lower jaw on the left, is described by EUDES-DESLONGCHAMPS [1]. It contained a stalk of cartilage stated to have resembled the cartilage of the ear. To this on either side was attached a small muscle. Unfortunately the appendage had been cut off close to the skin. A small opening (*pertuis*) was present on the skin near the appendage, and from this opening a small brush or tuft of bristles protruded.

Fistulæ in the neck of swine are well known as giving rise to a disease called *weisse Borste* in Germany (Fr. *la Soie* or *poil piqué*) from the fact that certain white bristles are found at the opening of the duct. In the popular fancy it is supposed that the bristles themselves bore the perforation, but according to ZÜNDEL[2] they are congenital and often bilateral. Heusinger agrees with Zündel in regarding such openings as branchial fistulæ.

147. **Sheep** and **Goats.** Cervical fistulæ unknown, but appendages on the neck common. The sheep of the Wilster marshes are described[3] as having the neck bare of wool, and an appearance as of a fur-collar. Above the collar and below the pharynx they have a pair of appendages about the size of an acorn. Such appendages are said to be not uncommon in Merinos[4]. Among the Kalmuck and Kirghiz sheep and goats such auricles are said by Pallas[5] to be common. In many foreign races of goats these auricles seem to be a constant character. In position they may vary from the angle of the jaw to the middle of the neck. The length is usually about 3 in. but they are recorded as reaching 15 cm. Figures of goats having such auricles are given by SUTTON[6]. The anatomy of one of these bodies is described by GOUBAUX[7], and it is mentioned that a plate of cartilage was found in the interior. A similar cartilage was found by STEWART[8] together with striped muscular fibre. Goubaux gives a case of two she-goats on a farm, one having cervical appendages, the other having none. Each gave birth to a pair of kids at the same time. Each pair was a male and female, and in the one the male only had the appendages and in the other the female only. The characters of the father of these kids were not known.

Ox. Neither cervical fistulæ nor auricles known.

148. **Horse.** Cervical auricles unknown. Fistulæ (in the position considered by Heusinger to indicate the first branchial cleft) are common and are recognized by their action in soiling the hair near the external opening.

RECAPITULATION. The evidence as a whole goes to shew that structures, sometimes of large size, having several essential features of the external ear, that may in fact be fairly spoken of as repetitions of the ear, may by Homœotic Variation appear on the neck of Man and other animals: further, that these repetitions have been known to occur at the openings of cervical fistulæ, suggesting a comparison with the relation of the external ear to the hyo-mandibular cleft, but that such a relation to cervical fistulæ is exceptional.

[1] *Mém. Soc. Linn. de Norm.*, 1842, VII. p. 41, Pl. IV. fig. 3.
[2] *Deut. Arch. f. Thierm.*, I. 1875, p. 175.
[3] VIBORG, *Samtl. Vet.-Afhandl.*, I. p. 148 [Heusinger].
[4] SCHMALZ, *Thierveredlungskunde*, p. 223 [Heusinger].
[5] *Spicileg. Zool.*, XI. p. 172 (two figures).
[6] SUTTON, J. B., *Evolution and Disease*, pp. 84 and 85.
[7] GOUBAUX, *Rec. de Méd. Vétér.*, Ser. 3, IX. p. 335.
[8] Figured by SUTTON, *l. c.*, p. 87.

CHAPTER VIII.

LINEAR SERIES—*continued*. MAMMÆ.

Some of the phenomena of Meristic Variation are well seen in the case of mammæ[1], and especially in the modes by which increase in the number of these organs takes place.

The facts regarding these variations in Man have so often been collected that it is scarcely necessary to detail them again. For our present purposes it will be sufficient to give a recapitulation of the chief observations in so far as they illustrate the phenomena of Variation.

The most important collections of the evidence on this subject are those of Puech[2], Leichtenstern[3], and Williams[4], from whose papers references to all cases recorded up to 1890 may be obtained. Besides these, Bruce[5] has given a valuable account of a considerable number of new cases together with measurements and statistical particulars. These accounts contain almost all that is known on the subject but additional reference will be made to original authorities in a few special cases.

In Man supernumerary mammæ or nipples nearly always occur on the front of the trunk, being usually placed at points on two imaginary lines drawn from the normal nipples, converging in the direction of the pubes. These lines may thus be spoken of as the "*Mammary lines.*" It is with reference to supernumerary mammæ occurring on these lines that the subject of mammary variations is chiefly important to the study of Meristic Variation. In addition to these, however, there are a few well authenticated examples of mammæ placed in parts of the body other than the mammary lines and of these some mention must be made hereafter.

---

[1] It will be understood that facts as to variations consisting in absence of mammæ or nipples and other such changes do not come within the scope of this volume, but belong rather to the province of Substantive Variation.

[2] Puech, *Les Mamelles et leurs anomalies*, Paris, 1876.

[3] Leichtenstern, *Virch. Arch. f. path. Anat. u. Phys.*, 1878, LXXIII. p. 222. This collection was apparently made independently from that of Puech.

[4] Williams, W. Roger, *Jour. Anat. Phys.*, 1891, xxv. p. 225.

[5] Bruce, J. Mitchell, *Jour. Anat. Phys.*, 1879, XIII. p. 425.

In the great majority of cases (over 90 per cent., LEICHTEN-STERN [1]) of mammæ placed on the mammary lines, the supernumerary structures are *below* the normal ones, being then as a rule *internal* to them, while those found *above* the normal mammæ are less common and are external to the normal mammæ. The distance separating the normal from the supernumerary mammæ differs greatly in different cases, and most conditions have been seen intermediate between a stage in which the nipple is bifid, and that in which completely separate supernumerary mammæ are presented. It is of consequence to observe that there appears to be no case in which a supernumerary mamma is so large as the normal mamma of the same individual.

The degree to which supernumerary structures of this nature are developed is very various. They may be fully formed mammæ with nipples, in the female capable of function; while in other cases, on the contrary, they may either consist of nipples only, having no distinguished glandular tissue of mammary character in connexion with them, or they may be tumours of mammary character without nipples or even definite ducts. Between these several conditions there is no sharp distinction. It appears therefore that there are *two* rudimentary or imperfect conditions possible: either supernumerary nipples without recognizable mammary glands, shading off into small warty elevations of uncertain character, and on the other hand redundant portions of mammary gland without nipples. The latter may be partially connected with the normal mammæ or quite separate from them. All these states of imperfection are much more common than the complete supernumerary mammæ.

Fully formed supernumerary mammæ have been found above the normal mammæ and also below them, the latter being the more frequent position. For those found on the mammary lines the axilla is the highest position and the upper part of the abdominal wall the lowest. Of the rudimentary forms, the mammary tumours without nipples occur usually if not always above and external to the normal mammæ, being generally in or near the axilla. The supernumerary nipples however are in the great majority of cases below and internal to the normal ones.

Small supernumerary nipples are quite common in Man, but the statistics of different observers give various results. Bruce found in 2311 females 14 cases (·605 per cent.), and in 1645 males 47 cases (2·857 per cent.). These persons were patients at the Brompton Hospital for Consumption and were not specially examined with a view to this inquiry. Among 315 such persons examined for the purposes of these statistics, 24 cases were seen (7·6 per cent.), 19 being male and 5 female. In 8 cases two extra nipples were present, and one doubtful case of three extra nipples

---

[1] Not including mammary tumours without nipples in the axillæ.

was seen. Bruce regards 7·6 as for various reasons rather too high a proportion. In a recent paper BARDELEBEN however states that among 2736 recruits examined with regard to supernumerary nipples, 637 cases (23·3 per cent.) were seen, 219 being on right side, 248 on left side, and 170 on both sides. The discrepancy between these statistics no doubt arises through want of agreement as to the inclusion of cases in which the extra nipples are very rudimentary.

It seems to be clearly shewn that the abnormality is commoner in men than in women, and there is some evidence that it is more frequent on the left side than on the right (BRUCE, LEICHTENSTERN and BARDELEBEN). It is also well established that supernumerary nipples are much more commonly present as single than as paired structures, and that when paired they are by no means always at the same level on the two sides. Cases of the presence of supernumerary mammæ as paired structures symmetrically placed are nevertheless sufficiently numerous. Organs of this nature may also occur simultaneously on the same side of the body at different levels. For example in one of LEICHTENSTERN'S cases, a small secreting supernumerary mamma with a nipple was present in the left axilla, while there was also another supernumerary nipple on the lower border of the left breast. The greatest number of supernumerary nipples occurred in a case described by NEUGEBAUER[1], represented in Fig. 29. In this patient there were on each side three supernumerary nipples above the normal ones, and

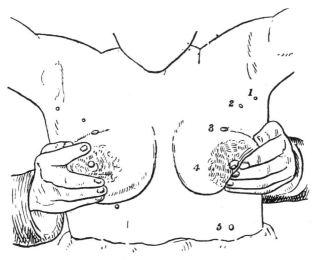

Fig. 29. Diagram of a case of four pairs of supernumerary nipples in human female. The normal breasts raised to shew the lowest pair. (After NEUGEBAUER.)

[1] NEUGEBAUER, F. L., *Centralb. f. Gynäk.*, 1886, p. 729.

one on each side below them. The latter were concealed by the pendent breasts. When the child was being suckled milk oozed from each of the uppermost or axillary nipples, but from the remaining six supernumerary nipples milk could only be extracted by pressure. The flowing of milk from supernumerary nipples when the child is at the normal breasts, has often been observed.

A few references to cases exhibiting the several features above mentioned may be of use.

149. Bifid nipple, the same on each breast [plane of division not specified]. DUVAL, *Du Mamelon et de son auréole*, Paris 1861, p. 90.

150. Two nipples on the same areola, bilaterally symmetrical. The two nipples stood in the mammary line defined above. TIEDEMANN, *Ztsch. f. Physiol.*, v., 1833, p. 110, Taf. I. fig. 3.

151. Cases are given by CHARCOT and LE GENDRE, *Gaz. méd. de Paris*, 1859, p. 773, in which an extra nipple was placed *external* to the normal one on the same breast. In one of these the extra nipple had no areola. Leichtenstern (p. 253) in quoting these cases, speaks of them as instances of supernumerary nipples on the same level as the normal ones, but this is not expressly stated in the original account, which does not, as I think, exclude the possibility that the supernumerary nipples were *above* and external to the normal ones. Two functional nipples with separate areolæ on the left breast, which nevertheless was not larger than that of the right side, *ibid.* The same authors mention another case in which such a second nipple had no areola; the mother of patient stated to have been the same. See also SINÉTY, *Gaz. méd. de Paris*, 1887, p. 317 (full description and measurements). In this case the supernumerary nipple was placed below the normal one.

152. A case in which three nipples were placed on each breast is given by PAULLINUS, *Miscell. Curios.*, &c., 1687, Decur. ii. Ann. v. Append. p. 40. The case is given on the authority of PRACKEL and the three nipples are said to have been arranged in an equilateral triangle, the normal being above at the apex, and the two others at the same level below. The description and the figure accompanying it do not however justify complete confidence in this observation, and indeed the contributions of Paullinus to the *Miscellanea Curiosa* contain so much of the marvellous that they should not be accepted without hesitation. The same may be said of the case of *five* nipples each having an areola quoted by PERCY and LAURENT, *Dict. Sci. méd.*, XXXIV. p. 517, *s. v.* "Multimamme." The authority for this case is a letter of Hannæus to Borrichius, dated 1675. I have not found any observation of this class of abnormality later than the seventeenth century, but it is of course quite possible that cases may occur in which the nipples are distributed on the breast otherwise than along the mammary lines.

153. Supernumerary mamma with nipple in axilla, LEICHTENSTERN, p. 245, and others.
154. Supernumerary mamma above and external to the normal ones. Numerous cases; see especially case of two bilaterally symmetrical mammæ in this position, SHANNON, *Dubl. Med. Jour.*, 1848, v. p. 266, *fig.* [figure repeated by AHLFELD, WILLIAMS &c.]; also similar case, QUINQUAUD, *Rev. photogr. des hôp.*, 1870, p. 19.
155. Supernumerary mammæ below and internal to normal ones: numerous cases, see LEICHTENSTERN, &c. In nearly all these the

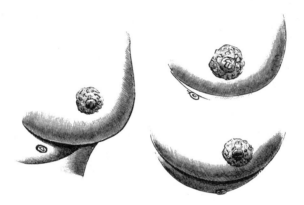

FIG. 30. Supernumerary nipples and mammæ of different sizes in human female. (After BRUCE.)

supernumerary organs are close to the normal mammæ. A few examples of such structures on the upper part of the abdominal wall are known, *e.g.*, TARNIER in his edition of CAZEAUX, *Traité de l'art des Accouchements*, 1870, ed. 8, p. 86. In the male several such cases are recorded, *e.g.*, BRUCE, *J. Anat. Phys.*, XIII. 1879, p. 446, *Pl.* Examples of this kind in the female are shewn in Fig. 30 (after Bruce) and in the male in Fig. 31 (after Leichtenstern).
156. Mammary tumours in the axilla are described by CHAMPNEYS, *Med. Chir. Trans.*, 1886, LXIX. p. 419, as of common occurrence in lying-in women. These structures are of various sizes and without any nipple, pore, or duct. The secretion was obtained by squeezing the lump and oozed through the skin at the situations of the sebaceous follicles. In this manner both colostrum and milk were obtained, following each other as in the normal mammæ. Similar observations in single cases have been made by many writers.
157. Redundant mammary tissue of this kind connected with, and thus forming an axillary extension of the normal mammæ, CAMERON, *Jour. Anat. Phys.*, 1879, XIII. p. 149; also NOTTA, *Arch. de Tocologie*, 1882, p. 108.

*158. *Two pairs* of supernumerary mammæ below the normal ones, DE MORTILLET, *Bull. Soc. d'Anthrop.*, 1883, Sér. 3, VI. p. 458. An

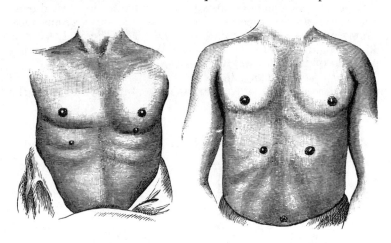

FIG. 31. Supernumerary mammæ in the male, symmetrical and asymmetrical. (After LEICHTENSTERN.)

important case of a man having two pairs of supernumerary mammæ on the mammary lines. There was a gradual diminution in size from the highest to the lowest, the latter being a little above the level of the umbilicus. Each pair was at the same level.

*159. *Four pairs* of supernumerary nipples (ten in all) are recorded only in NEUGEBAUER's patient, already mentioned. Three of the supernumerary pairs were above the normal ones, and the other pair below them. As seen in Fig. 29 the nipples of each pair did not stand in the case of each pair at precisely the same levels, and between those of the lowest pair there was a considerable difference of level, that on the left side being at some distance below the normal breast, while that on the right side was on its lower border.

In a few cases the supernumerary nipple is described as having been perpendicularly below the normal one, and it is likely that such cases must be looked on as exceptions to the general rule that the mammary lines converge posteriorly; but it is not impossible that even in some of them the supernumerary nipple might have been found to be rather nearer the middle line if this point had been specially inquired into.

The foregoing examples are given as selected illustrations of the several facts, and for full lists of cases the reader is referred to the works already mentioned.

160. Of supernumerary mammæ placed in parts of the body other than the mammary line some mention must be made, though those of them

that are authentic have no close bearing on the subject of Meristic Variation. There are firstly two often quoted cases[1] in the *Miscellanea Curiosa* in which mammæ are said to have been present on the back, but as has already been remarked, many of the stories told in this collection are clearly fabulous, and this is especially true of the contributions of Paullinus. Both these records are given at second hand and the first case (Paullinus) is said to have been seen in 1564, more than a hundred years before the date of the account. Helbig's accounts of things seen by himself are generally trustworthy, but in this case he is only repeating what was told to him by a Polish noble about a woman seen in Celebes. There are no modern cases on record. There is however indisputable evidence of the presence of a mammary gland on the thigh (especially ROBERT's case; for references to several accounts of this see Leichtenstern, p. 255); on the cheek, BARTH, *Arch. f. path. Anat. u. Phys.*, 188, p. 569; on the acromion, KLOB, *Ztsch. f. K. K. Ges. d. Aerzte in Wien*, 1858, p. 815; in the labium majus, HARTUNG, *Ueb. einen Fall von Mamma Accessoria*, Inaug. Diss., Erlangen, 1875. In the two last cases the mammary nature of the gland was proved by microscopic examination. In Barth's case of a mamma on the cheek the microscopical investigation did not give a certain result (*q. v.*).

As Leichtenstern shewed, the case of *inguinal* mamma, mentioned by Darwin and others, really related to Robert's case of a femoral mamma. In 1885, however, BLANCHARD (*Bull. Soc. d'Anthrop.*, 1885, p. 230) stated that TESTUT had lately seen such a case and was about to publish an account of it, but this has not yet appeared (1892).

Most writers on the subject have accepted cases of supernumerary mamma placed *anteriorly in the middle line*. These are given by PERCY and LAURENT, *Dict. Sci. méd.*, XXXIV., 1819, on the authority of several different persons. One case was seen by themselves (p. 526), and in it the third mamma stood below and between the other, forming a triangle with them. In another case given on the authority of GORRÉ there are said to have been a pair of extra mammæ below the normal ones, and a fifth between the supernumeraries. In view of the fact that many paired organs may by Variation occur compounded in the middle line, there is nothing incredible in these accounts, nevertheless there is, so far as I know, no recent observation of such an occurrence in the case of mammæ, and with the one exception (which is very briefly described), the accounts given are at second hand[2]. It is moreover not clear that the words used "*au-dessous et au milieu des deux autres*" do not mean simply below and *between* the other two. The case contributed by Gorré is nevertheless given in great detail and cannot lightly be set aside.

Before speaking of the bearing of these facts on morphological conceptions it is necessary to refer to some of the phenomena of

---

[1] PAULLINUS, *Miscell. Curios.*, &c., Dec. ii., Ann. IV. 1686, p. 203, *Appendix*, giving a case said to have been seen in 1564; also OTTO HELBIG, *ibid.*, Dec. i., Ann. IX. and x., pubd. 1693, p. 456.

[2] Williams (p. 235) quotes BARTELS, *Arch. f. Anat.*, 1872, p. 306, as alluding to such a case, but I do not think that the passage is meant to convey this meaning.

mammary Variation in other mammals. In connexion with the case of Man it may be mentioned that supernumerary mamma below and internal to the normal ones has been seen in *Macacus* and in *Cercopithecus patas*, SUTTON. J. B., *Intern. Jour. of Med. Sci.*, 1889, XCVII. pp. 252 and 253; in the Orang-utan, OWEN, *Comp. Anat.*, iii. p. 780. In many mammals the number of the mammæ is very inconstant even within the limits of species and from the facts seen in such cases deductions may be drawn which are at once instructive as to the nature of mammary Variation and have an application to the morphology of Meristic Series in general. Of these I shall give examples taken from three species.

*161. The first is that of the cow's udder. Normally the cow has four teats of about equal size. Not unfrequently there are six teats, of which four are large and may be said in the usual parlance to be the "normal" ones, and two are small and placed posteriorly to the others. A case of this kind is shewn in Fig. 32, II. Commonly these extra teats give no milk, but in many cases they have been known to do so. Their size and position vary greatly; sometimes they are placed near the other teats as shewn in the figure, but I have seen them very high up, almost in the fold between the udder and the thighs.

Very frequently, however, there is only one extra teat making five in all, such an extra teat being so far as I know, always on

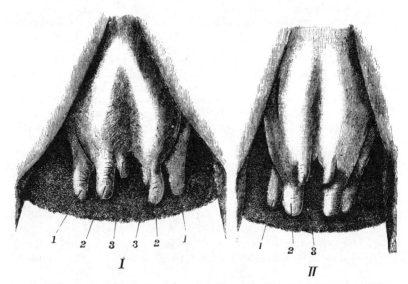

FIG. 32. Supernumerary teats in two heifers. I. The third teat is completely separate on the left side, but on the right side is united with the second. (The cleft between the two is incorrectly represented as a sharp line; there was no such sharp line of demarcation; the skin being very slightly depressed in this place.) II. Teats of the third pair both completely separate.

*162. one side of the udder. The sketch given in Fig. 32, I. was taken from a heifer having an arrangement intermediate between the condition with four teats and that with six. As the figure shews, on the left side there were three complete teats but on the right side the third teat was incompletely separated from the second. This third teat was joined to the second for its whole length but had a separate pore. The animal which belonged to the St John's College Dairy Farm was unfortunately sold before the first calf was born, so I had no opportunity of seeing whether milk was given by both these teats. The significance of such a case will afterwards appear.

In many mammals, such as the pig, rabbits, cats and dogs, the mammæ are distributed in two mammary lines along the ventral surface. The number of the mammæ in such cases is notoriously variable, and in some respects this variation is interesting and has a bearing on questions of the nature of Meristic Repetition. If a number of such animals be examined it will be found that as a rule there are the same number of glands on the two sides, and that they are arranged in pairs, those of each pair standing at the same level or nearly so. Nevertheless departures from this arrangement are very frequent. Individuals are in the first place commonly found with a different number of mammæ on the two sides, and in such cases it is interesting to observe that together with the difference in the number of mammæ on the two sides

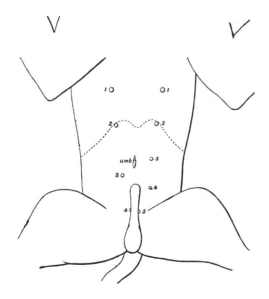

Fig. 33. Diagram of nipples in a male Bull-dog. On right side, four; on left, five; the two anterior and two posterior being almost at the same levels. *umb.* umbilicus. The dotted line shews the outline of the thorax.

there is generally if not always a disturbance in the paired arrangement. A simple case of this kind occurring in the dog is represented in Fig. 33. The animal is a male bulldog lately in my possession. On the right side there are four nipples, while on the left there are five. The most anterior on each side stand almost at the same level on the thorax. The second on each side are almost at the same distance below them, that on the left side being $\frac{1}{4}$ in. higher. Similarly the most posterior nipples stand on each side at almost exactly the same level on the sheath of the penis, the total length from the first to the last nipple on each side being practically the same. On the left however there are *two* nipples placed between the second and the last, but on the right there is only one. This one nipple stands at a level not far from the middle between the 3rd and 4th of the other side, making as it were a complement to or balance with them.

*164. Thirty-five young pigs examined with regard to these questions gave the following results. They belonged to five litters (30 pure-bred Tamworths; 5 cross-bred, out of Berkshire sow, sire unknown). These pigs were all quite young, about a fortnight old, and consequently there was no displacement due to functional development of the glands.

MAMMÆ.

	Right.	Left.			Pigs.
A.	6 —	6	regularly arranged in pairs		3
B.	7 —	7	ditto		10
C.	7 —	7	ditto	5th rudimentary	1
D.	8 —	8	ditto	4th rudimentary	3
E.	7 —	8	all paired exc. l. 4th rudimentary		2
F.	7 —	8	l. 4th rudimentary, l. 3rd and 5th displaced		1
G.	7 —	8	rt. 2nd balances l. 2nd and 3rd		1
H.	8 —	7	l. 2nd balances rt. 2nd and 3rd		1
I.	8 —	7	l. 3rd balances rt. 3rd and 4th		1
K.	6 —	7	rt. 1st balances l. 1st and 2nd		1
L.	6 —	7	l. 2nd rudimentary l. 1st and 3rd displaced		1
M.	6 —	7	all paired exc. l. 4th rudimentary		1
N.	7 —	7	altogether irregular		4
O.	6 —	6	ditto		2
P.	7 —	6	ditto		1
Q.	7 —	8	ditto		1
R.	8 —	7	ditto		1
				Total	35

The animals in groups D and E, except one of the latter, belonged to the same litter. In them a small rudimentary nipple stood between the 3rd and 5th, but the latter were not spaced out for it, being no further apart than any of the others. The measurements of the distances between the nipples on one side in one of these cases were, in inches, $1\frac{1}{16}$, 1, $\frac{7}{16}$, $\frac{7}{16}$, $\frac{14}{16}$, $\frac{13}{16}$, $\frac{15}{16}$, the rudimentary nipple standing $\frac{7}{16}$ in. from either of its neighbours. In the D

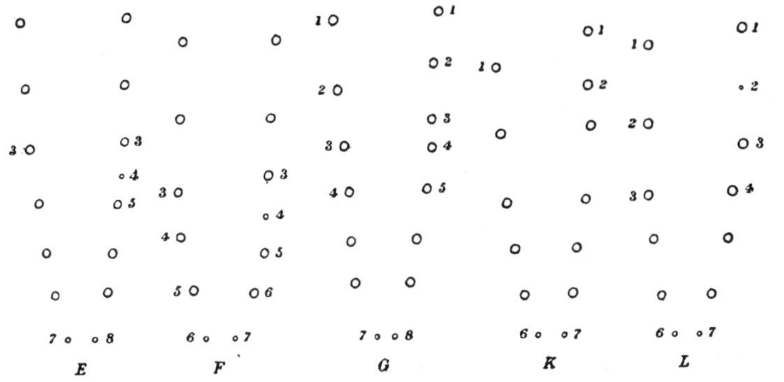

Fig. 34. Diagrams of nipples in very young pigs. Letters refer to groups in No. 164.

group this was found on both sides, but in the E group on one side only, as in the figure (Fig. 34).

Comment on foregoing evidence.

On looking at a series of cases like those roughly illustrated in the diagrams, one is tempted to inquire as to the factors which determine the positions of these mammæ and nipples. Though such an inquiry must lead to small definite results it may not be unprofitable to point out some deductions which may be made from the facts. I take this opportunity as a good one for illustrating the position here adopted with respect to the theory of Reversion, and for discussing certain features of the phenomena of Division.

The mammary glands form an example of a class of Meristic organs which are distributed in series along a body already segmented, but whose positions have no obvious coincidence with the fundamental segmentation. In the case of the pig, for instance, it would doubtless be found that the mammæ bear more or less definite relations to particular vertebræ, but they are not limited to such positions as the ribs or spinal nerves must be. The segmentation of the mammæ is thus a segmentation, or serial arrangement, superadded upon that of the vertebræ. The question to be considered is, what determines the points at which mammæ are to be formed?

In the paper to which reference has been made, WILLIAMS has contended for the view that each somite bore originally a pair of mammæ; and we may remark that if this were so the problem of the segmentation of the mammæ would be the same as that of the general segmentation of the trunk. The same author then argues that the appearance of supernumerary nipples or mammæ along the mammary lines is a reversion to an ancestral condition, and a figure is given, shewing the places at which mammæ are on this view believed to have been placed, definite ordinal numbers being assigned to each. Against this suggestion may be urged those objections to appeals to the hypothesis of reversion which were mentioned in the Introduction (Section XII.), but in addition to these there are a number of objections applying specially in the case of mammary Variation. The view that supernumerary mammæ are reversions rests on the frequency and definiteness with which they occupy certain positions. But though they do occur more often in some positions than in others they are in no sense limited to these positions, for they may stand anywhere, at least upon the mammary lines. To justify the view that the positions of supernumerary mammæ are definite it is necessary to exclude the cases of bifid nipple, of multiple nipples on the same breast, and of axillary extensions of the mammæ, all which phenomena would then be looked on as belonging to a class different from that of actual supernumerary mammæ. In the argument referred to, this course is actually adopted. The acceptance of such a view leads to great difficulty. For example, in Neugebauer's case (see Fig. 29), Williams considers that the posterior nipples of the two sides belong to different pairs, and have consequently different homologies, because they stand at different levels.

Such distinctions are, I believe, unreal. It is surely impossible to suppose that the Repetition seen in the udders of the two cows in Fig. 32 is a phenomenon different in the two cases. In the one there are two extra teats in symmetrical positions, equally spaced out from the second teats; in the other there is a third teat on one side and a double second or posterior teat on the other. Surely it is clear that the double condition of this teat represents an imperfect phase of a process perfected on the other side. If further proof were needed it may be found in the fact already mentioned, that the mammæ of the pig and other such animals, may be the same in number even on the two sides, but nevertheless stand quite irregularly and without any visible arrangement into pairs.

The existence of these cases in which no order of form or regularity can be traced may seem at first sight to be an insuperable objection to any attempt at the detection of principles in the arrangement of the mammæ. There is however the fact that many, and indeed in most forms the majority of individuals do shew an orderly and paired arrangement, and the further fact that of those cases which depart from this, a certain number present appearances which suggest that this departure has come about in a regular way. Though the irregular cases remain, something would be gained if we could comprehend any of the elements on which the regularity depends. The case of regularity and symmetry, in a sense, includes the cases of irregularity. The difficulty is to understand the causes of regularity and of symmetry; but if we could be sure of these it would not be hard to conceive disturbances resulting in irregularity.

In the pigs are found, first, cases of six on both sides in pairs, and also of seven on both sides in pairs; besides these there were cases of 6—7 and of 7—8. Of these there were some in which two on one side stood in positions which geometrically balanced that of one of the other side, the others being arranged in pairs. In such cases the appearances suggest that there has been a division of one mamma to form two, and that the two have then separated or travelled apart. The division of organs into two is of course a common occurrence, and may naturally be supposed to be a phenomenon of the same nature as the division of single cells. The case of mammæ is perhaps instructive inasmuch as it bears witness to the fact that such division must take place at a remotely early period in development. For while in cases to be given hereafter of division, for example, between teeth, it may be supposed that the travelling apart of the two resulting teeth is mechanical, in the sense that the two growing teeth may simply push apart from each other just as two cartilage-cells, &c., may separate by the concentric deposition of material, the separation cannot be supposed to occur in the mammæ by these *late* changes, but the process of mechanical separation, though the same in kind as that in the case of teeth, must be conceived as beginning early in the history of segmentation.

At this point a circumstance, very often to be seen in other cases, should be mentioned. When an organ, single on one side, corresponds geometrically with two organs on the other side, each of the latter is frequently of the same size and developed to a like extent as the single one of the other side. This of course would be expected on the hypothesis that the division of organs is a phenomenon similar to the division of cells, that is to say, not merely a *division*, but a *reproduction*.

But the supposition of division of single members of the series is not sufficient to account for all the facts of Variation seen. We have to consider not only the case in which one organ of one side balances two of the other. We have to deal also with the cases of six on each side and seven on each side all corresponding in pairs. In these there is no indication that there has been a division of a single member on each side. The spacing is regular in each case and there is no obvious crowding at any part of the series. Even if therefore in the former case there is a suggestion that the germs of single mammæ have divided into two at a period of development after the series of mammæ was constituted as a series, there is no such suggestion in the present case. We must, I think, in the latter suppose that the existences of all the mammæ, whether six or seven, are *determined together*. How or at what stage such determination is made, there is no direct evidence to shew.

The various arrangements seen suggest then that the relative positions occupied by the mammæ depend partly on the number that are present, and that the position of each mamma is to some extent dependent on the position of other mammæ, especially of its neighbours. In this connexion the cases F and L are interesting ones (Fig. 34). In L for example, the 1st on the left is at a higher level than the 1st on the right. It is succeeded by a rudimentary 2nd having none on the same level on the other side. The left 3rd is behind the right 2nd, but posterior to this point the nipples are approximately paired. These appearances suggest that the displacement of the 1st and 3rd on the left are in some way connected with the presence of the rudimentary left 2nd. Similarly in F the left 3rd and 5th are spaced out for the rudimentary 4th. From its position and small size it might fairly be supposed that this is a "supernumerary" organ, for at all events it is visibly different from the others: but in the case of seven on each side in pairs, no one mamma rather than another can be pointed out as obviously supernumerary when compared with a similar series of six. It seems therefore that of the factors determining the relative positions of the mammæ along the mammary lines, the number of the mammæ is one, and that the positions of the mammæ are in some way and to a limited extent correlated with each other. That there are other factors at work, also, is sufficiently shown by the existence of cases of apparently utter irregularity.

In seeking to go beyond this and inquire as to the way in which this correlation is brought about there is, in the present state of knowledge of the mechanics of Division, not much to be gained. Reference may be made to recent observations published in abstract by O. SCHULTZE[1]. According to him there is in young embryos of several mammals (Pig 1·5 cm. long; Rabbit 13—14 days, &c.) a ridge running along the *dorso-lateral* aspect on each side and at points upon this the mammæ and nipples are eventually formed. (The formation of the true nipples is preceded by the raising of the epidermis into small elevations, "primitive teats," which afterwards disappear.) The two mammary lines are by subsequent changes and growth of the body brought into the ventro-lateral position. The question of the position of the mammæ therefore resolves itself into this: what determines the positions at which mammary centres, to borrow the word used in the case of bone, are to be formed on the mammary lines? In a subsequent place I shall contend that the facts given are only intelligible on the view that the forces determining the points of growth of mammæ are compounded into one system of forces. But to the question what are these forces there is no answer.

---

[1] O. SCHULTZE, *Anat. Anz.*, 1892, VII. p. 265, since published in full (*Verh. d. phys.-med. Ges. zu Würzburg*, XXVI. 1893, p. 171, *Pls.*).

CHAPTER IX.

LINEAR SERIES—*continued*.

TEETH.

FROM the consideration of numerical Variation in mammæ we may proceed to an examination of like phenomena in the case of the teeth of vertebrates. The modes of Variation in these organs are, as might be expected, in many ways similar, but several circumstances combine to make the Variations of teeth more complicated than those of mammæ.

Teeth arise developmentally by special differentiation at points along the jaws, much as the mammæ arise by differentiation at points along the mammary lines; and as in the case of mammæ, so in the case of teeth, we are concerned first with changes in the number of points at which such differentiation takes place, and next with the general changes or accommodations which occur in the series in association with numerical changes. As in mammæ, so also in teeth, numerical Variation may occur sometimes by the division of a single member of the series into two, and sometimes by a reconstitution of at least a considerable part of the series.

Between the case of mammæ and that of teeth, there is however an important point of distinction. The series of mammæ is practically an undifferentiated series. There is between mammæ standing in one mammary line no obvious qualitative differentiation. Though not all identical in structure, the differences between them are of size and of quantity, not of form or quality. If such qualitative difference is present it must be trifling. In considering Variation in mammæ we have thus to deal only with changes in number, and with the geometrical and perhaps mechanical question of the relative positions of the mammæ. The teeth of most Vertebrates, however, are differentiated to form a series of organs of differing forms and functions, and the study of Variation in teeth may thus be complicated by the occurrence of qualitative changes in addition to simply numerical ones. In teeth, in fact, there are not only Meristic variations, but Substantive variations also; and thus, as in the case of vertebræ, for instance, in any given example of a numerical change qualitative changes must be looked for too.

As a preliminary to the consideration of evidence relating to the Variation of teeth it may be useful to call attention to certain peculiarities of teeth considered as a Meristic Series. In the Introduction, Section V, it was pointed out that in order to get any conception of the Evolution of parts repeated in an animal, the fact of this Repetition must be recognized, and it must be always remembered that we are seeking for the mode in which not one part but a series of similar parts has been produced. The simplest case to which this principle applies is that of organs paired about the middle line, and in the steps by which such parts have taken on a given form it is clear that similar variations must have occurred on the two sides. In the absence of evidence it might be supposed either that such variations had occurred little by little on the two sides independently, or on the other hand, that Variation had come in symmetrically and simultaneously on the two sides. Upon the answer given to this question the success of all attempts to form a just estimate of the magnitude of the integral steps of Variation depends. In many examples already given it has now been shewn that though in the case of paired organs Variation may be asymmetrical, yet it is not rarely symmetrical, and in part the question has thus been answered.

In the evidence that remains many more cases of such symmetrical variations will be described, and it may be taken as established that when the organs stand in bilateral symmetry, that is to say, as images on either side of a middle line, their Variation *may* be similar and symmetrical.

The teeth present this problem of the Variation of parts standing as images, in an unusual and peculiar way. For in the case of teeth we have to consider not only the steps by which the right and left sides of each jaw have maintained their similarity and symmetry, but in addition the further question as to the relation of the teeth in the upper jaw to those in the lower jaw. There are many animals in which there is very great difference between the upper and lower rows of teeth, and it must of course be remembered that perhaps in no animal are the teeth in the upper jaw an exact copy of those in the lower, but nevertheless there is often a substantial similarity between them, and in such cases we have to consider the bond or kinship between the upper and lower teeth whereby they have become similar or remained so. For it may be stated at once that there is some evidence that the teeth in the upper and lower jaws may vary similarly and simultaneously, though such cases are decidedly rare, especially in numerical Variation, and are much less common than symmetrical Variation on the two sides of the same jaw.

In speaking of the relation of the series of the upper jaw to that of the lower jaw as one of images, it must be remembered that the expression is only very loosely applicable. In particular it should be noticed that though in so far as the lower teeth are a copy of the upper ones the resemblance is one of images, yet the teeth which resemble each other do not usually stand opposite to each other in the bite, but members of the upper series alternate with those of the lower. The incisors, as a rule, however, and the back teeth of a certain number of forms do bite opposite each other, and in them the relation of images is fairly close.

The importance of the recognition of the relation of images as subsisting between the teeth of the upper and lower jaws will be seen when this case is compared with that of the two sides of the body. For ordinary bilateral symmetry is, as has already been suggested, an expression of the original equality and similarity of the two halves into which the ovum was divided by the first cleavage-plane, or by one of the cleavages shortly succeeding upon this. The fact that the two halves of the body are images of each other is thus both an evidence and a consequence of the fact that the forces dividing the ovum into two similar halves are equal and opposite to each other. The bilateral symmetry of Variation is thus only a special case of this principle.

In view of the fact that the teeth in the upper and lower jaws may vary simultaneously and similarly, just as the two halves of the body may do, it seems likely that the division of the tissues to form the mouth-slit must be a process in this respect comparable with a cleavage along the future middle line of the body. It is difficult, however, to realize the actual occurrence of such a process of division in the case of the slit forming the original stomodœum, and this difficulty is increased by the recent observations of SEDGWICK[1] to the effect that in the Elasmobranchs examined by him the mouth-slit first appears as a *longitudinal* row of pores. If this is so the relation of images must exist in the case of the mouth, not only in respect of the two sides of the slit, but also in respect of the anterior and posterior extensions of the slit. But whatever may be the processes by which the tissues bounding the mouth of a vertebrate come apart from each other, the result is clearly in many cases to produce an anterior series of organs in the upper jaw, related to a posterior series of organs in the lower jaw, much in the same way that the right side of a jaw is related to the left of the same jaw. This relation may appear as has been stated, not only in the normal resemblances between the upper and lower teeth, but also in the fact that similar and simultaneous Variation is possible to them.

In another respect the Repetition of teeth may differ from that of other Linear Series already considered. In many animals, the

---

[1] SEDGWICK, A., *Quart. Jour. Micr. Sci.*, 1892, p. 570.

Pike, the Alligator, or the Toothed Whales, for example, the teeth stand in a regular and usually continuous series, differing from each other chiefly in size, ranging from small teeth in front, through large teeth, and often down to small teeth again at the back of the jaw. Such a 'homodont' series as a rule passes through only one maximum. Most mammals, however, are 'heterodont,' that is to say, the teeth can be distinguished into at least two groups, the incisors and canines on the one hand, and the premolars and molars on the other; and in a large number of animals having this arrangement the anterior members of the series of premolars and molars are small, increasing regularly in size from before backwards, reaching a maximum usually in some tooth anterior to the last. Though instances will be given of Variation, and especially of reduplication, occurring in most of the teeth, even in those which stand well in the middle of the series of back-teeth, such as the upper carnassials of the Cat, or the fourth premolars of the Seal, yet on the whole Variation in heterodont forms is more common at the anterior and posterior ends of the series of back-teeth. In view of this fact it is of some importance to recognize that the small members at the beginning of the premolar series are as regards their relatively small size, in the condition of terminal members of series, and exhibit the variability of terminal members almost as much as the last molars.

With these remarks by way of preface, evidence as to the numerical Variation of teeth in certain groups will be given in full. This account will for the most part be confined to a brief description of the conditions presented by the specimens. In the next chapter the principles which may be perceived to underlie these facts and the general conclusions to which they appear to lead will be separately discussed.

The evidence here given relates to certain selected groups[1] of Mammals, and chiefly to the Primates (excepting Lemuroidea), Carnivora (Canidæ, Felidæ, Viverridæ, Mustelidæ and Pinnipediæ), and Marsupialia (Phalangeridæ, Dasyuridæ, Didelphyidæ, part of Macropodidæ, &c.).

The facts to be given relate chiefly to increase in number of teeth. In the case of terminal members of series, such as the most anterior premolar or the last molar, some reliable facts as to cases of absence were found, but for the most part the evidence as to the absence of teeth is ambiguous and each case requires separate treatment.

The evidence is in this chapter arranged according to the

---

[1] Evidence as to the dental variations of Man is not here introduced. Considerable collections of such facts have been made by MAGITOT (*Anom. du syst. dent.*), BUSCH (*Dent. Monats. f. Zahnh.* 1886, IV.), and others, and illustrative specimens are to be found in most museums. I do not know that among these human variations are included phenomena different in kind from those seen in other groups, except perhaps certain cases of teeth united together, a condition rarely if ever recorded in other animals.

zoological position of the groups concerned. In several cases variations of similar nature were seen in different groups; cases of this kind will be brought into association in the next chapter.

As regards nomenclature I have in the main followed the common English system, numbering both the premolars and molars from in front backwards. In one respect I have departed from the practice now much followed. It has seemed on the whole better that the premolar which in any given jaw stands first, should be called p^1, even though in certain cases there may be reasons for doubting whether it is the true homologue of the p^1 of other cases[1]. Theoretical views of this kind can only at best be used as a substitute for the obvious nomenclature in a few restricted cases, such as that of the Cat, in which by the application of the methods of reasoning ordinarily adopted in Comparative Anatomy the first upper premolar would be looked on as the equivalent of $\underline{p^2}$ in the Dog. There are, however, few who would feel confident in extending this reasoning to many other cases, that of Man, for instance, and I believe it is on the whole simpler to number the teeth according to their visible and actual relations. As I have already attempted to shew in another place[2], in the light of the facts of Variation, it is to be doubted whether in their variations teeth do follow those strict rules of individual homology by which naturalists have sought to relate the arrangements in different types with each other.

The material examined has consisted chiefly of specimens in the British Museum and the Museums of the College of Surgeons, Leyden, Oxford and Cambridge, the Paris Museum of Natural History, and some smaller collections. I have to thank the authorities of these several museums for the great kindness I have received from them; and in particular I must express my indebtedness to Mr Oldfield Thomas, of the British Museum, for the constant help and advice which he has given me, both as regards the subject of teeth generally and especially in examining the specimens in the British Museum[3].

PRIMATES.

SIMIIDÆ. The Anthropoid Apes (Orang, Chimpanzee, and Gorilla).

*165. The teeth of the three large Anthropoids are perhaps more variable, both in number and position, than those of any other

---

[1] In cases where confusion might arise any change from common nomenclature is notified in the text.
[2] *Proc. Zool. Soc.*, 1892, p. 102.
[3] In the following descriptions B.M. stands for British Museum; C.S.M. for Museum of the Royal College of Surgeons; C.M., O.M., U.C.M., Leyd. M., P.M., for the Cambridge, Oxford, University College London, Leyden and Paris Museums respectively.

group of mammals of which I have been able to examine a considerable number. In different collections 142 normal adult skulls were seen and 12 cases of extra teeth. Of these one was a case of extra incisor (Gorilla, No. 186), one of anomalous teeth (Gorilla, No. 187), and the remainder molars. Thus far therefore there are nearly 8 per cent. cases of extra teeth. This figure is remarkable in comparison with the rarity of such cases in *Hylobates* (51 skulls seen, all normal), and the like rarity in other Old World monkeys (423 normals and 2 cases of extra teeth).

Simia satyrus (Orang-utan).

Normal adult skulls seen, 52.

Supernumerary molars.

*166. Adult male having additional posterior molar (m^4) behind and in series with the normal teeth, on *both* sides in upper jaws and on left side in lower jaw. In each case the m^4 is rather smaller than m^3, but all are well formed, having each four cusps and the normal complement of fangs, viz., one in front and one behind in the lower jaw, and two on outer and one on inner side in upper jaw. On right side of lower jaw there is no trace of additional molar, though there is almost as much room for it as on the left side. C. M., 1160, *D*, described by HUMPHRY, G. M., *Jour. Anat. Phys.*, 1874, p. 140, *Plate*.

167. Female (Borneo) having six cheek-teeth in each upper jaw and in right lower jaw [doubtless a case like the foregoing] mentioned by PETERS, W., *Sitzungsb. naturf. Fr. Berlin*, 1872, p. 76.

168. Specimen with large alveolus on each side for $\underline{m^4}$. L. M., 24.

169. Specimen (Borneo) having $\overline{m^4}$ in right lower jaw, behind and in series with the normal teeth. The tooth is of rather small size, but is regular in position and form. B. M., 3, *m*.

170. Specimen having a right $\overline{m^4}$ more than half the size of $\overline{m^3}$. U. C. M., *E*, 253.

171. Specimen having supernumerary molar on each side in lower jaw. MAYER, *Arch. f. Naturg.*, 1849, 1. xv. p. 356.

172. Similar case. FITZINGER, *Sitzungsb. math.—nat. Cl. Ak. Wien*, 1853, I. p. 436.

Similar case. BRÜHL, *Zur Kenntniss des Orangkopfes*, Wien, 1856. [? refers to the case described by FITZINGER.]

Molar absent.

173. Specimen "remarkable for absence of the upper right third molar and for absence of nasal bones, which are greatly reduced in some other specimens." C. S. M., 44. See *Catalogue Mus. Coll. Surg.* 1884. The other teeth are all normal and fully formed.

Variations in position of teeth. Though not directly pertaining to the subject here considered, the following examples of considerable departure from the normal arrangement may be perhaps usefully introduced in illustration of the peculiar variability of the dentition of the group.

*174. A skull from Borneo in the Oxford University Museum (numbered 2043 a) has the following extraordinary arrangement. All the teeth are normal and in place except the second premolar of each side in the upper jaw. On both sides there is a large diastema between $\underline{p}^1$ and $\underline{m}^1$. The diastema on the left side is of about the same size as the normal second premolar, but that on the right side is considerably too small for a normal tooth. The singularity of this specimen lies in the fact that the missing tooth of the right side is present in the skull, but instead of being in its proper place it stands up from the roof of the mouth within the arcade immediately *in front of the right canine and almost exactly on the level of the second incisor*, being *in the premaxilla*, at some distance in front of the maxillary suture.

That this tooth is actually the second premolar which has by some means been shifted into this position there can be no doubt whatever. It has the exact form of the normal second premolar, and is of full size. It stands nearly vertically but is a little inclined towards the outside. The canine is by the growth of this tooth slightly separated from the second incisor, and the first premolar is consequently pushed also somewhat further back. Hence it happens that the diastema for the second premolar on the right side is not of full size. This should be understood, as it might otherwise be imagined that the contraction was due to a complementary increase in the size of the other teeth, of which there is no evidence.

On the *left* side of the palate there was a very slight elevation at a point homologous and symmetrical with that at which the second premolar of the right side was placed. As it seemed possible that the missing tooth of the left side might be concealed beneath this elevation, a small piece of bone was here cut away, with the result that a tooth of about the same size and formation as $\underline{p}^2$ was found imbedded in the bone. In this case therefore the second premolar of the right side and of the left side have travelled away from their proper positions and taken up new and symmetrical positions in the palate, anterior to the canines. The facts of this case go to shew that the germ of a tooth contains within itself all the elements necessary to its development into its own true form, provided of course that nutrition is unrestricted. This might no doubt be reasonably expected; but since the forms of organs and of teeth in particular are by some attributed to the mechanical effects of growth under mutual pressure, it may be well to call special attention to this case, which goes far to disprove such a view.

175. Specimen having the teeth of the two sides in the lower jaw in extraordinarily asymmetrical disposition. The bone of the jaw does

not seem to have been broken, but there appears to have been disease of the articulations of the mandibles. B. M., 86, 12, 20, 10.

176. Specimen in which "position of the left upper canine is abnormal. It is displaced backwards and lies to the outer side of the first premolar, which it has pushed towards the middle line." C. S. M., 41 (see *Catalogue*).

177. Case in which upper right canine occupies a position within and on a level with the first premolar, which is pushed outwards. C. S. M., 40, *A*.

Troglodytes niger, calvus, &c. (Chimpanzee).

Normal adult skulls seen, 35.

Supernumerary molars.

*178. Specimen having on right side in upper jaw a very small square tooth behind m^3, in the arcade (Fig. 35); and in the left upper

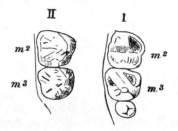

FIG. 35. Posterior right upper molars of Chimpanzee.
I. The case No. 178 (*Coll. Surg. Mus.*, No. 1).
II. A normal Chimpanzee of approximately the same size.

jaw an empty alveolus in the similar place, shewing clearly that a similar tooth has been present: lower jaw normal. C. S. M., 1.

179. Specimen in which teeth all gone, but alveoli exist behind those of the normal teeth on both sides' in upper jaw, and there is little doubt that there was here a fourth molar on each side. C. S. M., 9.

180. Specimen in which teeth all gone, but alveoli shew clearly that there was a fourth upper molar on right side; evidence on left side inconclusive: lower jaw gone. C. S. M., 12.

181. Specimen of *T. calvus* having an extra $\overline{m^4}$ in lower jaw on right side. This tooth is about one quarter of the size of $\overline{m^3}$, resembling that in case No. 178. This specimen is in the private collection of Prof. MILNE EDWARDS, who was so kind as to shew it to me.

Gorilla savagei (Gorilla).

Normal adult skulls seen, 55.

Supernumerary molars.

*182. Specimen having m^4 behind and in series with the others on *both* sides in lower jaw and on *right* side in upper jaw. On left side both teeth are square and somewhat worn, but the right $\overline{m^4}$ is a curious conical tooth. Gallery of P. M., A, 505, described by GERVAIS, P., *Journ. de Zool.*, III. p. 164. *Pl.*

183. Two cases of four molars in each upper jaw. MAGITOT, *Anom. du syst. dent.*, p. 100, Pl. v. fig. 8. [Of these one is in collection of Dr Auzoux; the other is No. 121 in P. M., but as I did not see it when examining the collection it is not reckoned in the statistics given above.]

Similar case, HENSEL, *Morph. Jahrb.*, v. p. 543.

184. Specimen having supernumerary molar which had not quite pierced bone [no statement as to position]. WYMAN, JEFFRIES, *Proc. Boston N. H. S.*, v. p. 160.

185. Specimen having extra molar in crypt on each side in upper jaw behind $\overline{m^3}$. L. M., 3.

Supernumerary incisor.

*186. Fully adult male from Congo having an extra incisor in lower jaw. There are thus five incisors in lower jaw (Fig. 36), of which

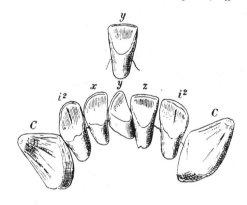

FIG. 36. Lower incisors and canines of Gorilla No. 186. x, y and z are three central incisors. The upper figure shews the tooth y as seen from the side. (Specimen in *Coll. Surg. Mus.*, 21, *A*.)

one, presumably the supernumerary, stands almost exactly in the middle line. This tooth is turned half round, so that the plane of its chisel stands obliquely. The teeth are all well formed and none belong to the milk-dentition, for the milk-teeth are much smaller and of different form. I did not succeed in satisfying myself that the central tooth is certainly the supernumerary. The second incisors are in place on each side and are quite distinct, and the right first incisor is similarly normal.

But whether the oblique tooth, or the tooth between it and the right i^1, should be rather considered supernumerary cannot be declared with certainty. Probably this is one of the cases, of which more will be said hereafter, in which *both* teeth replace the normally single i^1. C. S. M. 21, *A*.

187. *Anomalous extra teeth.* A lower jaw in the Museum of the Odontological Society "having two supernumerary teeth embedded in the bone beneath the coronoid process and sigmoid notch. Originally only a small nodule of enamel was visible on the inner surface of the right ascending ramus, just external to the upper extremity of the inferior dental canal. On cutting away the bone this nodule was found to be a portion of a supernumerary tooth having a conical crown and a single tapering root. Lying above it, another supernumerary tooth was discovered, of which there had previously been no sign whatever. This was likewise exposed by removing the superjacent bone, and found to be a larger tooth with a conical crown and three long narrow roots. The teeth were lying parallel to each other, with their crowns pointing upwards and backwards, so that they could hardly under any circumstances have been erupted in the alveolar arch." *Trans. Odont. Soc.*, 1887, xix. p. 266, fig.

Specimen having fragment of a tooth imbedded in bone between left lower canine and p^1; perhaps a fragment of a milk-tooth P.M., *A*, 506.

[Two specimens in the stores of the P.M. shew great irregularities in the arrangement of the teeth; but in both cases so many teeth had been lost during life that a satisfactory description of the abnormalities cannot now be given.]

Hylobates (Gibbons).

Normal specimens seen, 51. No abnormal case known to me.

OLD WORLD MONKEYS other than Anthropoid Apes.

188. Of the genera *Semnopithecus, Colobus, Nasalis, Cercopithecus, Cercocebus, Macacus* and *Cynocephalus;* 419 normal specimens examined. Only two had definite supernumerary teeth, but in one other case it was possible that extra molars had been present.

Supernumerary molars.

189. **Cynocephalus porcarius,** having large extra molar behind and in series in each upper jaw. The two teeth are of the same pattern precisely. In lower jaw there is on each side a large space behind m^3, but there is no tooth in it. O. M., 2011, *b*.

190. **Macacus rhesus,** old male, having a fourth molar in place in right *lower* jaw. The tooth does not stand up fully from the bone. On the same side in the *upper* jaw there is also a fourth molar, but was entirely enclosed in bone and was only found by cutting away the side of the maxilla by way of exploration. B.M., 30, *c*.

191. **Macacus radiatus,** having small and fairly definite depression behind m^3 in each jaw. These depressions seem to be perhaps the alveoli of teeth but it cannot be positively stated that extra molars have been present. C.S.M., 145.

192. *Abnormal arrangement.* Only one case of considerable irregularity of arrangement seen, viz., **Cercopithecus lalandii** (C. S. M. 113), case in which lower canines are recurved and pass *behind* the upper ones. See *Cat. Mus. Coll. Surg.*

New World Monkeys.

*193. In the species of Cebidæ and especially in *Ateles* supernumerary teeth are rather common, eight cases being found in 284 skulls, or nearly 3 per cent. (in addition to cases recorded by others). Of American monkeys belonging to other genera 92 skulls were seen, all being normal. Some cases of absence of the third molar were seen in *Ateles*, which are interesting in connexion with the fact that there are normally only two molars in Hapalidæ.

CEBIDÆ: normal formula $i\frac{2}{2}, c\frac{1}{1}, p\frac{3}{3}, m\frac{3}{3}$.

Chrysothrix, normal adults, 5.

Cebus, normal adults belonging to about ten species, 66.

Supernumerary molars.

194. **Cebus robustus:** supernumerary molar in each upper jaw giving $p\frac{3}{3}, m\frac{4}{3}$; DE BLAINVILLE, Laurent's *Annal. d'Anat. et Phys.*, 1837, I. p. 300, Pl. VIII. fig. 6.

195. **C. variegatus:** small tubercular molar in right lower jaw behind $\overline{m^3}$. The extra tooth is cylindrical and peg-like, having about ⅙th the diameter $\overline{m^3}$. *Leyd. Mus.* 8, *Cat.* 11.

Ateles: normal adult skulls, belonging to several species, 60.

Supernumerary molars.

*196. **A. pentadactylus:** extra molar in series behind m^3 *in both upper and lower jaws on right side*, in each case a small round tooth. P. M., *A*, 1505. This specimen described by DE BLAINVILLE Laurent's *Ann. d'Anat. et Phys.*, 1837, I. p. 300, Pl. VIII. fig. 5; mentioned also by GEOFFROY ST HILAIRE, *Anom. d'Organ.*, I. p. 660.

197. **A. vellerosus:** extra molar on left side in lower jaw behind $\overline{m^3}$, as a fully-formed and well-shaped tooth, but not so large as $\overline{m^3}$. B. M., 89. 12. 7. 1.

198. **Ateles sp.:** extra molar on left side in lower jaw. MAGITOT, *Anom. du syst. dent.*, p. 101, No. 6.

Supernumerary premolars.

*199. **Brachyteles hemidactylus** [a genus doubtfully distinct from *Ateles*]: specimen from S. America having l. upper series and all lower series normal. In place of right upper p^1 are *two* teeth (Fig. 37). These two teeth are similar to each other and for want of space they bulge a little out of the arcade. Each is in size and shape very like normal p^1, having a sharp cusp and a flat internal part to the crown. Both teeth are slightly rotated in opposite directions, so that the cusp of the anterior is anterolateral instead of lateral, while the cusp of the posterior is postero-

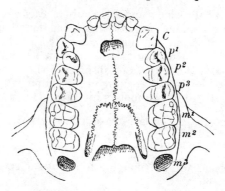

Fig. 37. Surface view of upper jaw of *Brachyteles hemidactylus*, described in No. 199. From skull in *Brit. Mus.*, 42, *a*.

lateral. These two teeth stand thus in somewhat complementary positions. B. M., 42, *a*.

*200. **Ateles marginatus**: wild specimen from river Cupai, has

$$p\frac{4-4}{3-3}, \quad m\frac{3-3}{3-3},$$

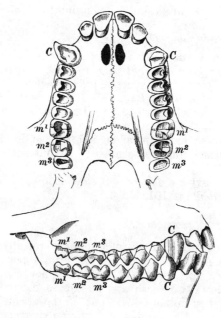

Fig. 38. Surface view of upper teeth of *Ateles marginatus*, specimen described in No. 200, and side view of both jaws together. The specimen is in *Brit. Mus.*, 1214, *b*.

that is to say, an extra premolar on each side in the upper jaw, the lower jaw being normal. The four upper premolars are perfectly formed, large teeth, in regular series on both sides. As a consequence the lower canines bite on and *partly behind* the upper canines. There was nothing to suggest that any one of these teeth was supernumerary, rather than another (Fig. 38). B. M., 1214, *b*.

Supernumerary incisor.

201. **Ateles ater:** specimen from Peruvian Amazon: in right upper jaw there is a large alveolus for i^2, which is gone, while a *third* incisor stands between this and the canine. This third incisor bites on lower canine, and lower p^1 of the same side bites *in front* of the upper canine. B. M., 1108, d.

202. **Ateles paniscus:** extra incisor in upper jaw. RUDOLPHI, *Anat.-phys. Abh.*, 1802, p. 145.

Absence of molars (cp. No. 209). Inasmuch as $p\frac{3}{3}$, $m\frac{2}{2}$ is the normal formula for the Hapalidæ, the following cases of absence of m^3 in *Ateles* are interesting. There was in no case any doubt that the skulls were fully adult, and there was no suggestion that the absent tooth had been lost.

*203. **Ateles marginatus:** specimen from the Zoological Society's menagerie, bones rough and unhealthy-looking, but skull well formed and certainly not very young, has no m^3 in either jaw, giving the formula $p\frac{3-3}{3-3}$, $m\frac{2-2}{2-2}$, as in Hapalidæ. There is no space in the jaw behind m^2, and in the upper jaw the bone ends there almost abruptly.

204. **A. melanochir:** Caraccas specimen, having no posterior m^3 on either side in upper jaw. The lower series normal, but the jaws are somewhat asymmetrical, so that the lower posterior right $\overline{m^3}$ is behind the level of its fellow of the other side. B. M., 48. 10. 26. 3.

205. **A. variegatus:** wild specimen, having lower $\overline{m^3}$ absent on both sides. Left i^2 is also absent, but has been almost certainly present. C. M., 1098, *B*.

Mycetes: of various species, adult normals, 81.

Supernumerary molar.

206. **M. niger:** supernumerary molar in the right upper jaw. The arrangement is peculiar. So far as m^2 the teeth are normal. Behind and in series with m^2 there is a large tooth, a good deal larger than the normal m^3, and having rather the form of m^2 than of m^3. Its form is, however, not precisely that of m^2, for the middle or fifth cusp is rather *anterior* to the centre of the tooth,

instead of being posterior to it as usual. *Outside* this tooth is another, standing out of the arcade, having the size and almost the form of normal m^3. B. M., 749, *c*. (Fig. 39).

This case may be an example of one of two principles which will be in the next chapter pointed out as operating in the case

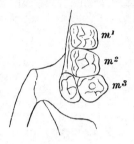

FIG. 39. *Mycetes niger*, No. 206; right upper molars. *Brit. Mus.*, 749, *c*.

of dental Variation. *Either* m^2 may have divided into two, both standing in series, and the normal m^3 may have been pushed out of the arcade in connexion with this reduplication; *or* the tooth standing outside may represent an addition to the normal series, and in that case the tooth standing as m^3 in the series may be a representation of m^3, raised to the normally higher condition of m^2 in correlation with the presence of an extra tooth in the series, in the way shewn to occur in other cases (see Chapter X., Section 7). Between these alternative possibilities I cannot decide.

Supernumerary premolar.

207. **Mycetes niger**: between and internal to p^1 and p^2 on left side there is a premolar. This is probably a supernumerary one, but the jaw is so much diseased that the relations are not distinct. B. M., 749, *d*.

208. **Callithrix**, normal adults, 22. (In B.M., 51, *b* on both sides m^3 is separated by a narrow diastema from m^2. The appearances suggest that possibly a small rudimentary tooth may have stood between them, but this is quite uncertain).

Nyctipithecus: 11 normals.

209. **Pithecia**: 11 normals.

Specimen having no right m^3, and apparently this tooth was not about to be formed, for the dentition is otherwise complete. C. M., 1094, *a*. (Cp. No. 202.)

Lagothrix, 6, **Chiropotes,** 1, **Ouakaria,** 3 normals respectively.

HAPALIDÆ. In this group m^3 is normally absent; and no specimen having this tooth or any other dental abnormality was seen. Of adult normal skulls 33 were seen, belonging to various species.

CARNIVORA.

CANIDÆ.

The evidence of the Variation of teeth in Canidæ is divided into three groups according as it concerns (1) incisors, (2) premolars, (3) molars. No case specially relating to the canines is known. In each of these groups the cases relating to (A) *wild* Canidæ are taken first, and those relating to (B) domestic Dogs afterwards.

Of wild specimens of the genus *Canis* (including the Fox) 289 skulls were seen, and amongst them were 11 cases of supernumerary teeth, about 3·5 per cent. (besides many recorded cases). Of 216 domestic Dogs (including Pariahs, Esquimaux, &c.) 16 had supernumerary teeth, or 7·4 per cent. (besides many recorded cases). I have not included skulls of edentulous breeds, in which the original condition of the teeth cannot be told with certainty.

Statistics of the occurrence of supernumerary teeth are given by HENSEL, *Morph. Jahrb.*, 1879. Among 345 domestic Dogs in his collection there are 28 cases of one or more extra molars, 12 cases of extra premolar, and 5 cases of extra incisor. [If therefore no two of these cases refer to the same skull, there were in all 45 cases of extra teeth in 345 skulls, or 13 per cent. It is not stated that the collection was not strictly promiscuous, but it may be anticipated that this figure is rather high.] An analysis of Hensel's cases will be given in the sections relating to the particular teeth.

The usual dentition of the genus *Canis* is $i\frac{3}{3}$, $c\frac{1}{1}$, $p\frac{4}{4}$, $m\frac{2}{3}$. The Wild Dog of Sumatra, Java and India, *C. javanicus* and *C. primævus* (by some considered as one species) have $m\frac{2}{2}$ and have been set apart as a genus under the name *Cuon* (HODGSON, *Calcutta Jour. N. H.*, 1842, ii. p. 205). The genus *Icticyon* differs in having normally $m\frac{1}{2}$. The genus *Otocyon* on the contrary has usually $m\frac{3}{4}$.

Of the variations to be described in *Canis* the most notable are (1) cases of $i\frac{4-4}{}$; (2) cases of extra premolar, common in upper, very rare in lower jaws; (3) cases of $\underline{m^3}$ or $\overline{m^4}$, and one case of $m\frac{3}{4}$ giving the formula characteristic of *Otocyon*. In several instances a considerable increase in the size of $\underline{m^2}$ or $\overline{m^3}$ is found associated with the presence of $\underline{m^3}$ or $\overline{m^4}$ respectively. An interesting group of cases of extra molars was found in *C. cancrivorus*, in which this abnormality seems to be common.

The frequent absence of p^1 in the Esquimaux dogs is worth notice. Absence of $\overline{m^3}$ is common in Dogs, but absence of $\underline{m^2}$ is rare.

In *Otocyon* one case of $m\frac{4}{4}$ is recorded, and in *Icticyon* one example has $m\frac{2}{2}$ instead of $m\frac{1}{2}$.

I. *Variation in Incisors and Canines.*

A. WILD CANIDÆ.

No case of *extra* incisor known to me.
Two cases of *absent* incisor, viz.

210. **[Canis] Vulpes pennsylvanica,** Brit. Columbia, having $i\frac{3-3}{2-2}$; apparently $\overline{i^1}$ has not been present on either side. B. M., 1402, b.

211. **Canis vulpes:** only 5 incisors in lower jaw, with no trace of alveolus for the sixth. SCHÄFF, E., *Zool. Gart.*, 1887, XXVIII. p. 270.

B. DOGS.

212. **Dog** (resembling Bloodhound): *four* incisors on each side in upper jaw. The externals, i^3, normal, but no evidence as to which of the other teeth supernumerary. *Leyden Mus.*

213. **Thibetan Mastiff,** Nepal: sockets for four teeth on each side in pmx. Teeth all gone. Alveoli of two sides nearly symmetrical. In absence of the teeth it cannot be positively stated that this is not a case of persistent milk-teeth, but this seemed unlikely. B. M., 166, *g*.

214. **Mastiff:** four teeth on each side in front of canines; from form of teeth probably case of persistent milk-canines. Lower jaw gone. O. M., 1749.

215. **Dog:** on right, sockets for three teeth in addition to i^3 which is in place. *These three sockets all smaller than the normal ones*, and socket for upper right canine also slightly reduced in size. *Odont. Soc. Mus.*

216. **Dog:** small skull in my possession, has in place of right i^3 *two alveoli, both at the same level*, divided by a thin bony septum, the one internal to the other: left $\underline{i^3}$ is in place and normal: lower jaw gone.

217. Among 345 Dogs' skulls four had extra upper incisor on *one* side, and one skull had perfectly formed fourth upper incisor on *both* sides. This tooth smaller than third incisor. HENSEL, *l. c.*, p. 534. Several cases of 7 or 8 incisors in upper jaw, teeth being usually asymmetrical. NEHRING, *Sitzb. nat. Fr. Berl.*, 1882, p. 67.

218 In *lower jaw* such cases much rarer. Supernum. lower incisor on *one* side, one case [? in 650 skulls], NEHRING, *ibid.;* also a Dog (*chien chinois-japonais*), 4 incisors in each lower jaw. MAGITOT, *An. syst. dent.*, p. 81.

Case of divided incisor.

219. **Bulldog:** right i^2 with very wide crown; main cusp partially bifid, as if intermediate between single and double condition, *Morph. Lab. Cambridge.*

Similar case kindly sent to me by Prof. G. B. HOWES.

220. *Absence of incisor* is very rare in Dog. One case of $i\frac{3-3}{2-2}$ given by HENSEL, *l. c.* p. 534. (Hensel observes that this gives the formula for incisors of *Enhydris* [*Latax*]; he also calls attention to fig. of **Enhydris** with *three lower incisors* in OWEN, *Odontogr.*, Pl. 128, fig. 12, but as this is not mentioned as an anomaly in text, it is very doubtful.)

221. **Dog** having the upper canine imperfectly divided into two on each side as shewn in Fig. 39. The plane of division was at right

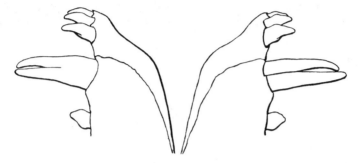

FIG. 39. Right and left profiles of Dog having the canines partially divided.

angles to the line of the alveolus so that the two parts of each canine stood in the plane of the series of teeth. The division was more complete on the right side than on the left. The lower canines were normal. This specimen was kindly sent to me by Mr J. Harrison.

II. *Variation in Premolars.*

Several distinct variations were found in the premolars of Canidæ. A number of cases shew five upper premolars instead of four, and the question then arises whether the extra tooth is due to the division of a single tooth, or to reconstitution of the series[1]. The occurrence of a fifth premolar in the *lower* jaw is much rarer, only three or four cases (Wolf (2) and Greyhound (? 2)) being known to me. The following other forms of Variation occurred. In *C. mesomelas*, No. 228, an extra tooth stood internal to p^3, and was perhaps a duplicate of this tooth. One case of bifid p^1 was seen, and two cases in which p^2 had apparently divided to form two single-rooted teeth (*C. viverrinus*, No. 227 and a Sledge-dog, No. 237). A few examples of absence of p^1 deserve notice. Lastly, though really an example of Substantive Variation, I have included a curious case of possibly Homœotic variation of p^3 into the partial likeness of the carnassial (No. 245).

[1] On this point see Chapter x. Sections 3 and 5.

Increase in number of Premolars.

A. Wild Canidæ.

222. **C. dingo:** specimen having two closely similar teeth between p^2 and the canine in each upper jaw[1]. Both the teeth had the form and size of a premolar. This not a case of persistent milk-tooth, NEHRING, A., *Sitzb. naturf. Fr. Berlin*, 1882, p. 66.

223. **C. dingo:** on right side p^1 is in place, and there is an alveolus for second tooth of about same size. On left side p^1 is rather *small*. L. M.

*224. **C. lateralis,** Gaboon. On l. side p^1 is single, but on rt. side there are two almost identical teeth between p^2 and the canine: of these the most anterior is level with, but slightly smaller than, left p^1. (Fig. 40) B. M., 1689, *a*. (See MIVART, *P. Z. S.*, 1890, p. 377.)

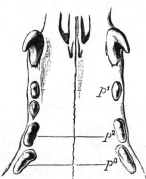

FIG. 40. *Canis lateralis*, No. 224. View of canines and front premolars of the upper jaw. p^1 of the left side is in symmetry with two teeth on the right side.

225. **C. vulpes:** in 142 skulls, one case of two teeth between p^2 and canine (sc. five premolars) in left upper jaw. HENSEL, *l. c.*, p. 548.

In *C. vulpes* the root of p^1 is not rarely partly divided into two by a groove of variable depth. The division is sometimes nearly complete, as in C. S. M., 651.

226. **C. mesomelas:** two teeth between p^2 and canine in left lower jaw, anterior the larger. DÖNITZ, *Sitzb. naturf. Fr. Berl.*, 1869, p. 41.

Division of p^2

227. **C. viverrinus:** left p^2 represented by two teeth, each having one root. Of these the anterior is tubercular, while the posterior is rather long from before backwards. Anterior premolars normal. L. M. (Compare **Sledge-dog,** No. 237.)

Reduplication of p^3.

*228. **C. mesomelas:** *inside* right upper p^3 is a supernumerary tooth which nearly resembles p^3, but is a little smaller; lower jaw normal. C. S. M., 643. (See Nos. 226 and 247.)

229. **C. lupus:** in addition to irregularities in position of teeth, there is a doubtful appearance as of an alveolus inside left p^2 which is displaced outwards. C. S. M., 624.

[1] MIVART, *l.c.*, by mistake quotes this case as one of extra teeth above and below.

CHAP. IX.]   TEETH : CANIDÆ.   213

Partially bifid premolar.

230.   **C. vulpes**: right p^1 has *three* roots and a partially double crown with two cusps (Fig. 41). The whole crown is pyramidal, the labial face being parallel to the arcade and the three roots stand each at one angle of the base: left p^1 normal; lower jaw missing. B. M., 175, o.

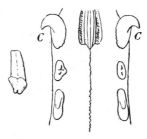

Fig. 41. Teeth of Fox (*C. vulpes*) described in No. 230. The separate view shews the right first premolar removed, seen from the labial side.

Extra premolar in lower *jaw.*

231.   **C. lupus**: two teeth between p^2 and canine in lower jaw on *right* side, one case: and the same on *left* side also, one case. These two occurred in 27 Wolf skulls seen by HENSEL, *Morph. Jahrb.*, 1879, v. p. 548.

B. DOMESTIC DOGS.

*232.   **Dog**: between p^2 and canine on rt. side there are two teeth, each shaped like a normal p^1, the anterior being somewhat the larger. This seen in two cases, figured in Fig. 42, II. and III.

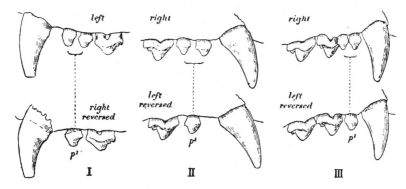

Fig. 42. Profiles of canines and anterior premolars in three dogs having two teeth on one side in symmetry with one tooth on the other.
I. C. S. M., 570.   II. and III. Skulls in Cambridge Univ. Morph. Lab.

Lower jaws absent. The property of the Zool. Lab., Cambridge (cp. *C. lateralis*, No. 224).

233.   **Spaniel**: similar case, left side, Fig. 42, I.   C. S. M., 570.

214   MERISTIC VARIATION.   [PART I.

234.   **Dog**: large skull, having $p\dfrac{5-5}{5-5}$, all the normal teeth being in place, of proper form and size, standing evenly without crowding. O. M., 1780.

235.   **Dogs.** In 345 skulls were 11 cases of supernumerary premolar in the upper jaw, viz.
> on both sides, 1 case,
> on right side, 7 cases,
> on left side, 3 cases.

These were all cases described by HENSEL as instances of the presence of "p^5" of his notation, *i.e.* a tooth between p^1 and canine. HENSEL, *Morph. Jahrb.*, 1879, v. p. 546. Out of 650 skulls, *including* Hensel's 345, 18 had two anterior premolars as described, on both sides in upper jaw. NEHRING, *Sitzb. naturf. Fr. Berl.*, 1882, p. 66.

> **English Spaniel**: *outside* and anterior to right $\underline{p^1}$ is a worn stump, probably of an extra tooth (?). B. M., 166, *j*.

236.   **Deerhound**: *two* alveoli where $\bar{p}^1$ should be; probably two distinct teeth stood here, but it is possible that the two alveoli were for distinct roots of a single tooth. C. M., 991, *B*.

Division of p^2.

237.   **Sledge-dog,** Greenland: all teeth normal except left upper p^2. This tooth normally of course has two roots. Here it is represented by two distinct teeth, each having one root. The anterior has a fairly sharp cusp, but the posterior has a rounded crown. The teeth are in perfectly good condition and do not look worn. They are separated from each other by a considerable diastema. It appears clear that instead of the normal p^2, two distinct teeth have been formed.   O. M., 1787 (compare **C. viverrinus,** No. 227).

Absence of Premolars.

A. WILD CANIDÆ.

238.   **C. corsac:** $\bar{p}^1$ absent on both sides without trace.   GIEBEL, *Bronn's Kl. u. Ord.*, Mamm. p. 196, *Note*.

239.   **C. occidentalis:** $\bar{p}^1$ absent on both sides.   C. S. M., 629.

240.   **C. vulpes:** in 142 skulls:

> $\underline{p^1}$ absent from both sides 1 case,
> do.   ,,   ,,   left   ,,   1   ,,
> do.   ,,   ,,   right   ,,   1   ,,
> $\bar{p}^1$,,   ,,   both   ,,   1   ,,
> do.   ,,   ,,   left   ,,   2   ,,
> do.   ,,   ,,   right   ,,   2   ,,

HENSEL, *Morph. Jahrb.*, 1879, p. 548. A doubtful case of absence of left $\bar{p}^1$. B. M., 175, *c*.

241.   C. (Nyctereutes) procyonoides: $\underline{p^1}$ absent on both sides without trace in B. M., 186, *e*; and absent on right side in B. M., 186, *d*. On the contrary B. M. 186 *a* and *b* and C. S. M., 672, are normal.

The following cases of absent premolars were doubtful: **C. dingo:** right $\underline{p^1}$ and left $\overline{p^1}$. C. S. M. **C. antarcticus:** p^1 above and below on left side. C. S. M., 635.

B. Domestic Dogs.

242.   From the nature of the case it is not often possible to say with confidence that p^1 has *not* been present in a given skull, but from the material examined this variation appears to be rather rare. In 216 skulls, excepting those of Esquimaux dogs, I only saw two clear cases in which the bones were smooth, without trace of alveolus, viz. **"Danish" Dog:** $\overline{p^1}$ absent on both sides, O. M., 1786. **Terrier:** $\underline{p^1}$ absent on both sides. C. S. M., 579. Many others doubtful.

According to HENSEL, however, absence of p^1 is common, and he states that in 345 skulls the following occurred:

$\underline{p^1}$ absent on both sides		5 cases,
do. „ „ one		4 „
$\overline{p^1}$ „ „ both		frequently,
do. „ „ one		9 cases,
$\overline{p^1}$ absent on both sides and $\underline{p^1}$ on one side, 1 case.		

Morph. Jahrb., 1879, p. 546. [This is of course a far higher frequency than was found by me, but perhaps discrepancy arises from difference in reckoning the evidence of absence.]

Two doubtful cases of absence of $\overline{p^2}$ were seen in Dogs.

*243.   **Esquimaux Dogs:** absence of p^1 quite common, the following skulls being all of the breed that I have seen.

Normals, with p^4_4, only two specimens. Specimens with no p^1, above or below, the canines in such cases standing close to p^2, three cases, viz. B. M., 58. 5. 4. 96; B. M., 166, *a*; C. S. M., 542. $\underline{p^1}$ absent on left side and $\overline{p^1}$ on both sides, C. M., 1000, *c*. $\overline{p^1}$ absent both sides and $\overline{p^1}$ absent on left side, L. M. $\underline{p^1}$ and $\overline{p^1}$ both absent from right side; left normal, O. M., 1789. $\overline{p^1}$ absent on left side, B. M., 166, *r*, 3. $\overline{p^1}$ absent on right side, B. M., 166, *t*, 2.

The partial establishment of a character of this kind in a breed, which, if selected at all, has been selected for very different qualities, is rather interesting. It need scarcely be remarked that the partial loss of this tooth cannot in the Esquimaux dog have occurred in connexion with an enfeebled habit of life, as might perhaps be supposed by some in the case of the edentulous lap-dogs.

As will be shewn in the next section, absence of the front premolars is a common character in the dogs of the ancient Incas, but in them the posterior molars are also frequently absent. There is no special reason for supposing that the Esquimaux dogs came originally from America, but it may be worth recalling as a suggestion, that according to anthropologists the relations of the Esquimaux are rather with American tribes than with Europeans. If this were established, it would be not unlikely that the Esquimaux dogs might be descended from dogs domesticated in America before the coming of Europeans, and so far belong rather with the Inca dogs than with ours[1].

*244.   **Inca Dogs.** The domestic dogs from the Inca interments, belonging to a period before the coming of the Spaniards, have been investigated by NEHRING. Of nine skulls not one had the full number of teeth and there was no case of supernumerary teeth. Sometimes the anterior premolar was absent, sometimes a posterior molar, and in some cases both. The formulæ were as follows:

$$p \frac{4-4}{4-3},\ m \frac{2-2}{2-3}\ \text{1 case.}$$

$$p \frac{4-4}{3-3},\ m \frac{2-2}{3-2}\ \text{1 case.}$$

$$p \frac{4-4}{3-3},\ m \frac{2-2}{3-3}\ \text{3 cases.}$$

$$p \frac{4-4}{3-3},\ m \frac{2-1}{2-2}\ \text{1 case.}$$

$$p \frac{4-3}{4-4},\ m \frac{2-2}{2-2}\ \text{1 case.}$$

$$p \frac{3-3}{3-3},\ m \frac{2-2}{3-3}\ \text{2 cases.}$$

The dogs were all of moderate size, and none shewed any defects in the form of teeth, which were all strong and sound. NEHRING, A., *Kosmos*, 1884, xv. p. 94.

Variation (? Homœotic) in form of third Premolar.

*245.   **Dog**: large breed. In the upper jaw *on both sides* the third premolar, instead of having only two roots, has a third internal root, thus somewhat resembling the carnassial. The crown of the tooth very slightly changed. This is not a case of persistent milk-tooth, which though a three-rooted tooth, is very different. C. S. M., 558.

III.   *Variation in Molars.*

Supernumerary molars are not rare in Canidæ. In all cases seen by me these teeth are single-rooted, round-crowned, rather tubercular teeth, placed behind m^2 or $\overline{m}^3$ as the case may be. HENSEL[1] has observed that if $\overline{m}^4$ occurs, then $\overline{m}^3$ which is normally single-rooted, not infrequently has a double root, though the same variation may occur when there is no $\overline{m}^4$ present. Conversely, when $\overline{m}^3$ is absent, not a rare variation, then $\overline{m}^2$ is often of a

---

[1] BARTLETT, arguing chiefly from habits, considers the Esquimaux dogs to be domesticated wolves, and says that they often breed with the wolf. P. Z. S., 1890, p. 47.
[2] HENSEL, *Morph. Jahrb.*, 1879, v. p. 539.

size below the normal, having a single root and a crown slightly developed, like that of $\overline{m^3}$. This reduced condition of $\overline{m^2}$ may also occur in cases in which $\overline{m^3}$ is not absent. These observations of Hensel's, which are of great consequence to an appreciation of the nature of Repetition, I can fully attest, and similar cases of Variation in adjacent teeth associated with the presence of a supernumerary were seen in other animals also.

A. Wild Canidæ.

Supernumerary Molars.

246.   **C. lupus:** 26 normals seen. Specimen from Courland having supernumerary $\underline{m^3}$ on left. In this specimen $\underline{m^2}$ is rather abnormally large on *both* sides, and the lower third molar, on the *left* side, viz. that on which the upper jaw has an extra tooth, is larger than right $\overline{m^3}$, but it is not larger than usual. C. M., 976, *M*.

Hensel, *l. c.*, p. 548, saw 27 skulls, none having extra molar, but one specimen known to him had a right $\underline{m^3}$.

247.   **C. mesomelas** ♀ (a Jackal): small, bitubercular left $\underline{m^3}$. Dönitz, *Sitzb. naturf. Fr. Berlin*, 1872, p. 54. (See Nos. 226 and 228.)

The S. American *Canidæ* (*Lycalopex* group) are remarkable for the frequency with which they possess extra molars, as the following cases (*C. azaræ, vetulus, magellanicus* and *cancrivorus*) testify. Flower and Lydekker[1] speak of the occasional presence of $\underline{m^3}$ in *C. cancrivorus*, but the evidence taken together seems rather to shew that there is a general variability at the end of the molar series in both jaws in these species; for not only is $\underline{m^3}$ found, but in some cases $\overline{m^4}$ also, while in one instance there was an 'odontome,' or rather a complex of 4 small teeth attached to $\overline{m^3}$.

248.   **C. vetulus**, Brazil: specimen having an extra molar in right lower jaw (Fig. 44, I.). The posterior part of $\overline{m^3}$ is slightly pushed outwards and a very small extra tooth stands behind and partly internal to it. Right $\overline{m^3}$ is slightly larger than left $\overline{m^3}$ and differs from it also a little in pattern. The extra tooth has one large and about three smaller blunt cusps on its crown, and might be described as a small representation of the larger $\overline{m^4}$ seen in other cases. B. M., 84. 2. 21. 1 (mentioned by Mivart[2], *Monogr. Canidæ*).

*249.   **Canis azaræ:** Brazilian specimen having a large supernumerary molar ($\underline{m^3}$) in *each* upper jaw placed in series with the others. In this specimen the great enlargement of $\underline{m^2}$ is very

---

[1] *Mammals, Living and Extinct*, 1891, p. 546.
[2] In the same place Mivart mentions a case of $\underline{m^3}$ in "*C. cancrivorus*," but I have not seen it. Perhaps this reference is to van der Hoeven's case (No. 249) which was by Burmeister named *C. cancrivorus* (see Huxley, *P. Z. S.*, 1880, p. 268).

noticeable on both sides, and this tooth is present as a large tooth with apparently *three* roots. In the lower jaw there is no extra tooth, but the molars are considerably larger than those of a

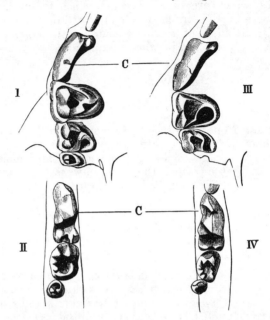

FIG. 43. *Canis azaræ*. I. and II. Right upper and lower jaws of the specimen described in No. 248, shewing the extra upper molar and the correlated enlargement of $\overline{m^2}$ and $\overline{m^3}$. III. and IV. are taken from a normal specimen of slightly larger size. *C*, carnassial teeth.
This figure was kindly drawn for me by Mr J. J. Lister.

normal specimen (Fig. 43). In the figure, side by side with the teeth of the abnormal form, are shewn the teeth of a normal skull which was slightly larger than the abnormal one, for comparison. *Leyden Mus.* [1]

250.   **C. magellanicus:** specimen having $\overline{m^4}$ on both sides. B. M., 46. 11. 3. 9 (mentioned by HUXLEY, *l. c.*).

*251.   **C. cancrivorus.** The only skulls of this species seen by me are those in B. M. Of these one skull with lower jaw, one skull without lower jaw, and one lower jaw without skull, have numerically the normal dentition of *Canis*, but of these, one has right $\overline{m^3}$ much larger than corresponding left tooth. The following were abnormal: small tubercular $\overline{m^4}$ on both sides, upper series normal, B. M., 1033, *b*, and also B. M., 1033, *c*, (Fig. 44, II.) mentioned by HUXLEY, *l. c.*

252.   Specimen having upper series and left lower series normal. On inner side of right $\overline{m^3}$ and as it were growing out from this tooth is a

---

[1] This is no doubt the skull described by van der HOEVEN, *Verh. k. Ak. Wet.*, Amst., iii. 1856, *Pl.* See HUXLEY, *P. Z. S.*, 1880, p. 268.

large 'odontome' composed of four small tubercular teeth. Each of these has a distinct crown and neck, but apparently the necks join with

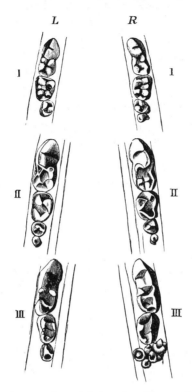

FIG. 44. Posterior lower molars of S. American Foxes. I. *C. vetulus* No. 249. II. *C. cancrivorus* No. 251. III. *C. cancrivorus* No. 252. In each case the right and left sides are shewn. R, right. L, left.

each other and with the neck of $\overline{m^3}$, which is displaced (Fig. 44, III.). B. M., 1033, *a*. (mentioned by HUXLEY, *P. Z. S.*, 1880, p. 268; figured by MIVART, *P. Z. S.*, 1890, p. 377).

In answer to an inquiry, Prof. NEHRING informs me that he has three skulls of *C. cancrivorus* Desm. (= *C. braziliensis* Lund.) from the province of S. Paolo, Brazil, which are normal, except that in one $\overline{p^2}$ has never replaced $\overline{d^2}$, which is in place; and that another Venezuelan skull of this species is also normal. [Whether the B. M. specimens are really of the same species as these I do not know.]

*253.   The rarity of supernumerary molars in *C. vulpes*, the common Fox, is remarkable in contrast with the foregoing evidence. In 142 cases (to which I can add 37), HENSEL, *Morph. Jahrb.*, 1879, found no single case.

Absence of Molars.

254. $\underline{m^2}$ is very rarely absent in Canidæ, and among the wild forms no case seen in 289 skulls (except a doubtful case in *C. occidentalis*, right

side, C. S. M., 628). $\overline{m^3}$ was observed to be absent in the following : **C. lagopus,** from Kamtschatka, absent on both sides in two cases received in same consignment with 4 normal skulls. *B. M.*, 88. 2. 20. 9 and 10; another case from Norway. *Leyd. Mus.* **C. zerda:** on left side. C. S. M., 671. **C. vulpes:** ditto, 2 cases. B. M., 177, *a* and 175, *b*. **C. viverrinus:** on right side. *Leyd. Mus.* **C. procyonoides:** ditto. *Leyd. Mus.* HENSEL, *l. c.*, gives the following: **C. vulpes:** 142 skulls; $\overline{m^3}$ absent on both sides, 5 cases; on left side, 3 cases. **C. lupus:** $\overline{m^3}$ absent on left side, 2 cases; on right side, 1 case.

ICTICYON AND OTOCYON.

* It is remarkable that in each of the two genera *Icticyon* and *Otocyon*, which are especially distinguished from *Canis* by the possession of unusual dental formulæ, numerical Variation in the teeth has been recorded, though the number of skulls of these forms in Museums is very small. The two forms, besides, differ from *Canis* in opposite ways, the one having a tooth less in each jaw while the other has in each jaw a tooth more, so that the presence of extra teeth in the two species is all the more important.

255. **Icticyon venaticus:** according to the authorities has $p\frac{4}{4}$, $m\frac{1}{2}$, viz. a molar less than the Dog in each jaw. The following skulls are all that I have seen. The carnassials did not vary appreciably in the three skulls. Each skull differs from the others, as follows.

$p\frac{4}{4}$, $m\frac{1}{2}$, B. M., 185, *a*.
$p\frac{4}{4}$, $m, \frac{1}{1}$, B. M., 185, *b*.
$p\frac{4}{4}$, $m\frac{2}{2}$, C. S. M., 533. (See FLOWER, *P. Z. S.*, 1880, p. 71.)

256. **Otocyon megalotis** [= *lalandii* and *caffer*]: the usual formula is $p\frac{4}{4}$, $m\frac{3}{4}$, that is, one molar more than the Dog in each jaw. It occurs in 4 skulls at B. M. and in 2 at C. S. M. One specimen has in addition an extra molar of good size in each upper jaw, giving $m\frac{4-4}{4-4}$. In this case m^3 is enlarged also on both sides. C. S. M., 675 (see *Cat. Mus. Coll. Surg.*, &c.). Three specimens having $m\frac{3}{4}$ mentioned by DÖNITZ, *Sitzb. naturf. Fr. Berlin*, 1872, p. 54.

B. DOMESTIC DOGS.

Supernumerary Molars.

257. **Dogs.** In 345 skulls the following 28 cases occurred, chiefly in large breeds:

m^3 on both sides and $\overline{m^4}$ on one side,	1 case.
m^3 on both sides	2 cases.
m^3 on one side	9 cases.
m^3 and $\overline{m^4}$ on one side only	2 cases.
$\overline{m^4}$ on both sides	6 cases.
$\overline{m^4}$ on one side only	8 cases.

HENSEL, *Morph. Jahrb.*, 1879, v. p. 538.

In addition to these,

m^3 and $\overline{m^4}$ absent on both sides, 1 case.

This was the only case in 860 skulls of *Canis*, of which about 650 were Dogs. The formula in it is thus that of *Otocyon* or the fossil *Amphicyon*. NEHRING, *Sitzb. naturf. Fr. Berlin*, 1882, p. 66.

In 216 skulls seen by me there were 8 cases of extra molars, viz.:—

258. **Sheep dog**: left m^3. C. S. M., 587; **Bulldog**: left m^3. B. M., 166 *s;* **Dog** from New Zealand, having left m^3, left m^2 being larger than right m^2. C. M., 1000; Bhotea **Mastiff**: $\overline{m^4}$ on right B. M., 166, f.; **Pointer**: left $\overline{m^4}$. C. M., 1000, *A*; **Dog**: right $\overline{m^4}$ *Camb. Morph. Lab.*; **Pariah**: $\overline{m^4}$ has been present on both sides, also a small stump below p^1 and p^2, possibly part of a milk-tooth. B. M., 166, *d*.

259. **Mastiff**: supernumerary $\overline{m^4}$ on right. The right $\overline{m^3}$ materially larger than left $\overline{m^3}$ (Fig. 45). C. S. M., 555.

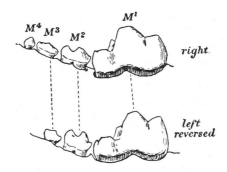

FIG. 45. Posterior molars of lower jaw of Mastiff No. 259, having an extra $\overline{m^4}$ on the right side. Right $\overline{m^3}$ is materially larger than left $\overline{m^3}$.

260. **Dog**, large size, supernumerary $\overline{m^4}$ on right side. On both sides $\overline{m^3}$ is two-rooted[1] and of large size. *Leyd. Mus.*, 258.

261. WINDLE and HUMPHREYS, *P. Z. S.*, 1890, p. 27, give an account of extra molars in the Dog, speaking of *upper jaws only*, and some of the foregoing are mentioned by them. As they do not specify the collection in which each is found the identity of the cases is not easy to tell. The following cases given by them are, I believe, all in addition to those already specified:—**Bulldog, Lurcher, Pointer** and **Terrier**, m^3 on both sides. **Bulldog** m^3 on left side; **Esquimaux, Pug, Spaniel, West Indian Dog**, m^3 on right side.

Coach-dog: $\overline{m^4}$ on both sides, MAGITOT, *Anom. Syst. dent.*, p. 103.

Absence of Molars.

262. **Dog**: in 345 skulls the following seen: m^2 and $\overline{m^3}$ absent on both sides, 2 cases; $\overline{m^2}$ and $\overline{m^3}$ absent on both sides, 1 case; $\overline{m^3}$ absent on both sides, 25 cases; $\overline{m^3}$ absent on one side, 9 cases. HENSEL, *l. c.*

In 216 seen by me the following occurred: $\overline{m^3}$ absent on both sides, 7 cases; C. M., 993 and 978; C. S. M. (*Store*), 65 and 67; two skulls marked "Skye Terrier,"

[1] It generally has a simple, conical root, but not rarely it has an imperfectly divided root, *e.g.* Newfoundland dog, O. M., 1778.

probably both of the same strain, C. M. 991, *F* and *G*; and Fox Terrier, C. M., 991, *R*; $\overline{m^3}$ absent on left side, 2 cases. Irish Wolf-dog, B. M. 82. 11. 11. 1; Fox Terrier, C. S. M., 580, *A* ; $\overline{m^3}$ absent on right side, 1 case, Bloodhound, B. M., 166, *t*. besides a few doubtful cases.

Inca dogs : for evidence as to absence of molars, see No. 244.

FELIDÆ.

The following evidence relates to the genera *Felis* and *Cynælurus*. The usual formula is $i\frac{3}{3}$, $c\frac{1}{1}$, $p\frac{3}{3}$, $m\frac{1}{1}$. Of *wild* species, 278 adult skulls having no extra teeth were seen, and 8 cases of extra teeth (nearly 3 per cent.): of domestic Cats, 35 adults without, and 3 cases with extra teeth (so far, about 9 per cent.). As in Canidæ so in Felidæ, there is a remarkable group of cases of variation in the anterior premolars. In the normal a small anterior premolar stands in the upper jaw, and commonly it is one-rooted, sometimes two-rooted (cases given); but there is no small anterior premolar in the lower jaw.

Cases of variation consisting in the presence of *two* small premolars above are common[1], just as there are often two small anterior premolars in the Dog. There are besides a few cases of the presence of a small anterior premolar in the *lower jaw*, but they are rather rare, and curiously enough there seems to be no case of the coincidence of these two variations in the same skull.

As already stated, in describing cases, the small anterior premolar in the upper jaw will be here spoken of as $\underline{p^1}$, though no suggestion that it is the homologue of the Dog's $\underline{p^1}$ is meant.

In a few species $\underline{p^1}$ is most commonly absent (cases given). There are some curious cases of *duplicates* of large premolars (Cat) and one of *duplicate canine* (Tiger), also a few of supernumerary molar. Though so small, and biting on no tooth of the lower jaw, $\underline{m^1}$ is nearly always in place even in old skulls (HENSEL).

Variation in Incisors.

No quite satisfactory case of numerical variation in incisors of Felidæ known to me. The following should however be mentioned.

263.   **F. lynx :** two extra teeth in premaxillæ. Right incisors normal; sockets for left incisors normal. Outside left $\underline{i^3}$ and close to canine is an extra tooth of good size, and in same place on right is a socket for a similar tooth. Since they are in premaxillæ these teeth are *probably not* persistent milk-canines. Lower canines bite *in front* of the extra teeth. B. M., 1156, *a*.

Incisors absent.

264.   **F. pardalis :** $\underline{i^1}$ and $\underline{\bar{\imath}^1}$ absent on left side. As regards the lower jaw the tooth *may* have been present, and been lost, but left $\underline{i^1}$ has probably never been present. It is especially notable that left $\underline{i^2}$ is larger than right $\underline{i^2}$, but there is no indication that $\underline{i^1}$ is compounded with it. B. M., 1068, *a*.

F. chate [?=*pardalis*]: doubtful if $\underline{i^1}$ has been present on either side. B. M., 55. 12. 26. 178.

265.   **Cynælurus jubatus :** no trace of right $\underline{i^3}$; same skull has no $\underline{p^1}$; lower jaw normal. B. M., 135, *f*.

---

[1] For discussion of such cases see Chapter x. Section 5.

Anterior Premolars (supernumerary).

Upper Jaw.

*266. **F. pardus:** right p^1 single and normal; on l. side two such teeth, both standing at level anterior to right p^1. The anterior is of same size as right p^1, the posterior is rather smaller. B. M., 87. 4. 25. 1.

267. **F. eyra:** two small anterior premolars in left upper jaw, BAIRD, *U. S. and Mex. Bound. Surv.*, Pt. 2, Pl. XIII. figs. 2, *a* and 2, *c* [anomaly not mentioned in text].

*268. **F. catus,** Athens. Two small anterior premolars in upper jaw both sides (Fig. 46, I.), small and standing close together. On rt. anterior the larger, on l. posterior the larger. B. M., 47. 7. 22. 2.

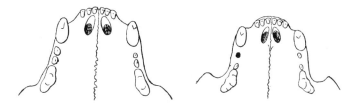

Fig. 46. Left-hand figure: upper jaw of *F. catus*, No. 268. Right-hand figure: upper jaw of *F. inconspicua*, No. 269.

*269. **F. inconspicua** (= *torquata*). Rajpootana. Two small anterior premolars in upper jaw both sides; both small, diastema between them. Posterior is nearly in contact with "p^3", while anterior is only a little behind canine (Fig. 46, II.). B. M., 85. 8. 1. 26. (Another specimen has p^1 as large single-fanged tooth.)

270. **F. domestica** (out of 38 skulls): *internal* to and rather behind left p^1 is an almost identical copy of it, though rather smaller. Not a milk-tooth. C. S. M., 414.

271. Out of 252 skulls two anterior premolars on *both* sides, 4 cases; on right, 2 cases; on left, 1 case [none specially described]. HENSEL, *Morph. Jahrb.*, 1879, v. p. 553.

272. **F. caligata,** Socotra: *outside* right p^1, a small extra tooth. In this specimen p^1 on each side has two roots. B. M., 857, *b*.

Doubtful cases of extra upper anterior premolar, *F. pardus*, C. S. M. 365; *F. leo*, C. S. M. 308.

Lower Jaw.

273. **F. concolor:** a supernumerary anterior premolar on both sides present, *Berl. Anat. Mus.*, 3678. HENSEL, *ibid.* **F. catus** or **maniculata:** ditto. *Frankf. Mus.,* HENSEL, *ibid.* **F. catus:** ditto. on left side, closely resembling p^1. Two cases, B. M., 1143 and 1143, *a*.

274. **F. domestica:** (in 252 skulls) a supernumerary premolar on both sides, just in front of and nearly same size as the usual "p^2," one case; on left, as a very small tooth midway between canine and "$\overline{p^2}$," one case; on right, rather larger than in foregoing and nearer to "$\overline{p^2}$," one case. HENSEL, *ibid.*

275. **F. tetraodon:** alveolus for small anterior premolar in right lower jaw; but as this fossil form very rare, uncertain whether normally

present in the species, DE BLAINVILLE, *Ostéogr.*, Atlas, Pl. XVI. *Feles fossiles.*

Variations in size of p^1.

276. **F. pardus**: p^1 sometimes two-rooted, as C. S. M., 360 (African); more often one-rooted, as C. S. M., 364, &c.: many gradations between these. In B. M., 115, *q* right p^1 extraordinarily large, left normal. Minute alveolus external and posterior to each of these; on left side a small worn stump [? of milk-tooth] in this alveolus.

277. **F. domestica**: p^1 two-rooted C. S. M. 409 and B. M., 127, *q* ; on right side two-rooted B. M. 127, *s.* **F. catus** C. S. M. 401 and **F. minuta** (Borneo) B. M., 122, *f*, p^1 partially two-rooted. **F. caligata**, see above, No. 272. **F. chaus**: left p^1 very small, right p^1 fair size. B. M. 131, *e.* **F. jaguarondi**, ditto, B. M.

Absence of p^1.

In the following cases it appeared that p^1 had not been present.

278. **F. catus**, both sides, a cave-skull, HENSEL, *l. c.*; left side only, Caucasus, B. M. 1143, *m*; **F. tigris**, HENSEL, **F. onca**, both sides, B. M.,117, *c*; **F. manul**, ditto, B. M. 1863, *a*; **F. nebulosa**, ditto, two cases [? normal for species] B. M.; **F. rubiginosa**, Malacca, ditto, B. M., 1856, *a*; **F. chaus**: both sides in domesticated specimen from India, B. M.; and in B. M. 58. 5. 4. 69, similar specimen, this tooth is small on left, absent on right; **F. brachyurus**, absent both sides, B. M.; **F. chinensis**, right absent, B. M., 70. 2. 18. 25; **F. javanensis**, left absent, B. M. 1641, *a* (but in B. M. 1309, *b*, p^1 is particularly large). **F. domestica**: in 252 skulls p^1 absent both sides 6 times, and one side, once (in 2 cases anterior deciduous tooth remained on both sides in upper jaws of adults) HENSEL, *l. c.* p. 552; in 38 skulls seen by me, p^1 absent both sides, 2 cases; right side in one case (Manx, C. S. M., 428, A).

In the following species the absence of p^1 was so frequent as to call for special notice.

279. **Cynælurus jubatus**: of 8 skulls 3 (2 African) were like Cat, having p^1 both sides; p^1 absent both sides, 3 cases, B. M., 135, *f.* and C. S. M.—; left p^1 absent, right very small, C. S. M., 441; right p^1 absent [?]. B. M., 135, *c*.

280. **F. caracal**: out of 8 skulls only 4 had any indication that p^1 might have been present, and in these it was doubtful.

281. **Lynx**: of Lynxes of possibly different species, 17 skulls have no p^1, a skull marked "*Lynx borealis*," B. M., 1230, *a* has a small, worn stump as p^1 on each side.

282. **F. pajeros** (= *pampana*), Chili: 2 skulls only known to HENSEL, *l. c.*, both without p^1. This tooth absent in B. M. 126 and 126, *c*; but in one specimen seen, right p^1 absent but left p^1 of good size.

Partial division (?) of lower premolar.

Two cases relate to this subject. The first lower premolar of Felidæ is a two-rooted tooth of well-known form. In the first of the following cases it bore an extra talon and root; in the second there was a small extra root on the internal face. (Cp. **C. vulpes**, No. 230.)

283. **F. tigris**: anterior right lower premolar has a thin supernumerary root on internal side of the tooth at the level between the two normal roots. This tooth in form resembled a milk-carnassial to some extent, but it was certainly not one of the normal milk-teeth. C. S. M., 333.

*284.   **F. fontanieri** (see No. 290): anterior premolar of right lower jaw has additional talon on internal and anterior surface (Fig. 47). This

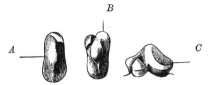

Fig. 47. *Felis fontanieri*, No. 284.
A. The normal anterior premolar of the left lower jaw. B. The corresponding tooth of the right side from above. C. The same from the lingual side.

portion has a separate root, and stands somewhat apart from rest of crown, looking like a partially separated tooth. B. M., 90. 7. 8. 1.

Duplicate Teeth.

285.   **F. tigris:** on right side, two canines in the same socket, both of large size, the anterior being the smaller; neither is a milk-tooth. *Mus. Odont. Soc.*

286.   **F. domestica:** having a large supernumerary tooth in each upper jaw. The extra tooth was in each case a small but accurate copy of the carnassial tooth (Fig. 48) of its own side. In each

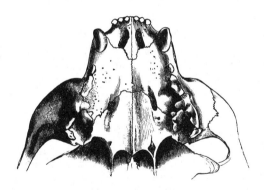

Fig. 48. The teeth in upper jaw of Cat, No. 286.

case the extra tooth stood internally to the carnassial tooth, extending from the level of the middle of the carnassial tooth to the level of the middle of the molar. B. M., 83. 3. 10. 1.

287.   Specimen having a tooth in the upper jaw closely resembling the second premolar ("p^3" *auctt.*) internal to and between it and the carnassial. The internal tooth is slightly smaller than the second premolar[1] (Fig. 49). C. S. M., 414.

[1] In this case, it is not possible to say strictly that either of the two teeth "is" the normal second premolar, rather than the other.

226   MERISTIC VARIATION.   [PART I.

288.   Specimen having a small tooth internal to the middle of the lower [?side] carnassial ($\overline{m^1}$): the extra tooth was here divided into two cusps so that it was a copy of the carnassial. HENSEL, *l. c.*

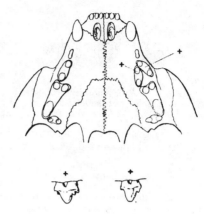

FIG. 49. Plan of teeth in upper jaw of Cat, No. 287. The two teeth marked with crosses are separately shewn, that on the right being the external.

289.   Specimen having a tooth like the last, but not so distinctly divided into two cusps, internal to posterior end of lower carnassial [?side]. *ibid.*

Supernumerary Molars.

Cases like the last cannot be clearly separated from cases of true extra molars in series, such as the following.

It is remarkable that no case of supernumerary *upper* molar in series seems to be known in Felidæ. In the Tiger and other species the upper molar is sometimes single- and sometimes double-rooted.

*290.   **F. fontanieri**: a species nearly allied to the Leopard (*F. pardus*), inhabiting the Kiu-Kiang, a geographically isolated region of N. China. Only two skulls are known, and each of them presents an abnormality in dentition (see No. 284). Skull having supernumerary tubercular tooth in series ($\overline{m^2}$) behind the left lower molar ($\overline{m^1}$). B. M., 1490, *a*.

291.   **F. pardalis**: $\overline{m^2}$ on left side. HENSEL, *Morph. Jahrb.*, 1879, v, p. 541. **F. tigrina**: tubercular $\overline{m^2}$ on left side. SCHLEGEL, *P. Z. S.*, 1866, p. 419. **F. lynx**: ditto [? side]. MAGITOT, *Anom. syst. dent.*, p. 103. **F. domestica**: "supernumerary permanent molar in lower jaw" [no particulars]. WYMAN, J., *Proc. Boston N. H. S.*, v, p. 160. **F. pardus**: doubtful indication that a left $\overline{m^2}$ has been present. C. M., 933, F.

Absent Molar.

F. leo: $\underline{m^1}$ absent on both sides, and there is no space for it behind the upper carnassials. B. M., 3043. The only case seen in all Felidæ examined. **F. domestica**: $\underline{m^1}$ absent [? both sides]. HENSEL, *l. c.*, p. 541.

VIVERRIDÆ.

Of the Viverridæ, *Herpestes* and *Crossarchus* are the only genera represented in collections in quantity sufficient to repay study of their dental variations. In the teeth of these two genera, however, variation is considerable and appears in some interesting forms.

In *Herpestes* there is first some evidence of variability in the number of the incisors, including one case of extra incisor. Next the facts respecting the presence or absence of the antérior premolar are of some consequence, both as illustrating the general variability and modes of Variation of this tooth, and also because the normal presence or normal absence of the anterior premolar is one of the characteristics of different species, which shew a progression in this respect. There is one case which should probably be looked on as an example of duplicate anterior premolar.

There are besides two cases of duplicates of large premolars, but of true supernumerary molars in series only one case was seen. Another specimen shewed what is perhaps partial division of a molar. Of 130 skulls, five had supernumerary teeth, not including cases of unusual presence of anterior premolar.

Incisors.

The following cases shewed departure from the normal $i\frac{3}{3}$.

292. **Herpestes gracilis:** an extra incisor in lower jaw. $\bar{i}^2$ and $\bar{i}^3$ in place and clearly recognizable on both sides, but between the two second incisors are *three* small teeth, all of about the same size and shape. Neither of these is a milk-tooth, for the milk-teeth are distinctly different both in size and form. There was no evidence to shew which tooth was the supernumerary one. B. M., 826, *a*.

*293. **H. nipalensis** ♂ : only *four* incisors in lower jaws. This is a remarkably clean and sound skull. The four incisors stand close together, filling up the whole space between the two lower canines. There is no reasonable doubt that only four lower incisors have been present. It is difficult to see that any of the four incisors exactly corresponds with any of the normal teeth; for while the two lateral teeth are of about the same size as normal $\bar{i}^2$, they have a different position, arising from the outer sides of the jaw, slightly in front of the roots of the canines, whereas normal $\bar{i}^2$ arises *internal* to the other incisors. To what extent the alteration in position is correlated with the change in number cannot be affirmed. B. M. 146, *m*.

*294. **H. persicus:** only *four* incisors in lower jaw. Judging from general appearances it seemed that $\bar{i}^1$ was missing from both sides. The teeth stand in a close series between the canines, which are nearer together than in normal specimens. The consequence of this to the arrangement of the bite is curious. The left lower canine bites in its normal place, between the upper canine and i^3; but the right lower canine bites *in front of* the upper i^3, which is displaced backwards towards the right upper canine. The whole anterior part of the lower jaw is thus twisted a little towards the left side.

Besides these two definite cases of absence of incisors, in the following instances there was a presumption that the absence was due to variation, but a definite statement cannot be made.

H. smithii: only four incisors in lower jaw. B. M., 1435, *a*. **H. gracilis**: doubtful case of absence of $\bar{i}^1$ on both sides. B. M., 789, *b*. **H. nyula**: doubtful if right i^2 has been present. B. M.

Anterior Premolars.

In the great majority of both Asiatic and African species of *Herpestes* the anterior premolar (p^1) is normally present in both jaws, and in these species 6 cases of absent $\bar{p}^1$ were seen. When present it is a tooth of small but still considerable size. It appeared from the specimens that $\bar{p}^1$ in the species *H. gracilis* (Africa generally), and both p^1 and $\bar{p}^1$ in *H. galera* (E. Africa) are commonly absent. As in other cases of absence of teeth the question arises whether the absence is due to age or accident, or on the other hand to original deficiency. This question cannot be definitely answered, but some considerations touching it should be mentioned.

First, as has been said, the tooth when present is of moderate size: though small, it is quite large enough to be functional, and is in no sense rudimentary. In his synopsis of the genera, THOMAS[1] says of *Herpestes*, "Premolars $\frac{4}{4}$ (if only 3 in either jaw, a diastema always present)." There is however no reason for supposing that the presence or absence of p^1 is determined by chance. From the fact that a tooth is small, it by no means follows that it is often lost. To any one handling large numbers of skulls, instances of the contrary must be familiar. A case in the Otters well illustrates this point. In *Lutra vulgaris* upper p^1 is a small tooth, and from its singular position internal to the canine, it might be supposed that the development of the canine might easily push it out; yet in 41 skulls of *Lutra vulgaris*, only 1 case of absence of p^1 was seen. Of *L. cinerea* on the contrary six skulls are without p^1; but as in two young skulls it is present on both sides, there is thus a strong presumption that in this species the tooth is lost with maturity. The frequent absence in the one species and the constant presence in the others points to a difference in organization between them. When p^1 is missing in a skull, though we are not entitled to infer that it has not been present, still the fact of its presence in one case and of its absence in another is on the face of it an indication that between the two there is a difference or Variation, but whether the Variation lay in the number of teeth originally formed or in the mode in which they were affected by subsequent growth is uncertain. In the specimens to be described the absence of p^1 in certain individuals or species is no less definite than its presence in the others, and that which is a variation in one species will be seen to be the rule in others.

* As regards the presence of p^1 the specimens thus make a progressive series. Most species having $p\frac{4}{4}$, but $p\frac{4}{3}$ as a variation; *H. gracilis* (and *pulverulentus*) having $p\frac{4}{3}$ normally, but $p\frac{4}{4}$ as a variation and $p\frac{3-4}{3-3}$

[1] THOMAS, O., on the African Mungooses, *P. Z. S.*, 1882, p. 62.

also as a variation; and lastly *H. galera* having $p\frac{3}{3}$ normally but shewing a case of $p\frac{3-4}{3-4}$ and another of $p\frac{4-4}{4-3}$. Lastly, all specimens of *Crossarchus* seen had $p\frac{3}{3}$.

Of species commonly having $p\frac{4}{4}$, 91 such skulls and the following cases of absence of p^1 were seen:

*295. **H. ichneumon**, 9 normals: $\overline{p}^1$ absent both sides. B. M.; on left, C. M., 965, *D*. **H. griseus**, 21 normals: $\overline{p}^1$ absent on right, two cases. B. M., 145, *k* and *m*. **H. smithii**, 6 normals: $\overline{p}^1$ absent both sides. B. M., 979, *b*; on left side, B. M. 84. 6. 3. 13.

296. **H. gracilis** on the contrary shewed $p\frac{4}{4}$ in 8 specimens, $\overline{p}^1$ present both sides once, B. M., 789, *a*; left $\overline{p}^1$ absent once. B. M., 789, *b*.

H. pulverulentus: $p\frac{4}{4}$ in 2 specimens.

297. **H. galera**: $p\frac{3}{3}$ in 7 skulls, one being quite young: p^1 is present in all four places, in one young skull making $p\frac{4}{4}$, B. M., 148, *d*; $\overline{p}^1$ and $\underline{p}^1$ both present and well developed on right side in old skull. On the left there is ample room for them. B. M., 79, *a*, $\underline{p}^1$ present on both sides and alveolus for $\overline{p}^1$ on right. B. M., 148, *l*.

Crossarchus: 13 skulls assigned to 4 species, all had $p\frac{3}{3}$.

Case of two *Anterior Premolars*.

298. **H. microcephalus**: on right side *two* teeth like $\underline{p}^1$, crowded together, others normal. Leyd. M. Compare *Rhinogale melleri* (an African Mungoose) of which only known skull (in B. M.) has $p\frac{5-5}{4-4}$. The appearance here is that a tooth unlike and rather larger than $\underline{p}^1$ stands in front of it on each side (see Thomas, *l.c.*, pp. 62 and 84).

Supernumerary Large *Premolars*.

Taken together the two following cases are important as illustrating the difficulty of drawing any sharp distinction between cases of duplicates of particular teeth and cases of extra teeth in series. They should be read in connexion with the cases of *F. domestica* (No. 286), *Helictis orientalis* (No. 312), *Vison horsfieldii* (No. 311), *Ommatophoca rossii* (No. 320), *Phoca grœnlandica* (No. 324), &c.

*299. **Herpestes gracilis**: supernumerary tooth in right lower jaw (Fig. 50). On comparing the teeth of this specimen with those of

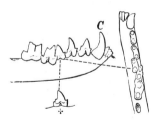

Fig. 50. Right lower jaw of *Herpestes gracilis*, No. 299. View from labial side; ground-plan of the jaw; separate view of the tooth +. *C*, the canine.

other *Herpestes* in which $\overline{p}^1$ is present it is quite certain that no tooth in the abnormal jaw corresponds with $\overline{p}^1$. The foremost of its premolars on both sides clearly has the form of $\overline{p}^2$. The next teeth have the correct form of $\overline{p}^3$. In the left lower jaw the next tooth is $\overline{p}^4$; but on the right side immediately in succession to $\overline{p^3}$ but slightly within the arcade is another tooth (marked + in the figure), which is very nearly a copy of $\overline{p^3}$, though a little smaller. On the *outside* of the jaw and behind this tooth is a normal $\overline{p^4}$. From its singular position outside the series, this tooth might easily be taken for a supernumerary one though its form clearly shews it to be a natural $\overline{p^4}$ displaced, while two teeth having the form of $\overline{p^3}$ stand in succession. B. M., 63. 7. 7. 18. (mentioned by THOMAS, *P. Z. S.*, 1882, p. 62).

300. **H. ichneumon** (Andalusia): in one of the upper jaws between and internal to $\underline{p^2}$ and $\underline{p^3}$ is a 3-rooted tooth (not a milk-tooth) which in size and shape is about intermediate between $\underline{p^2}$ and $\underline{p^3}$. *Leyd. M.*

Molars.

The only cases of noticeable variation in molars were both in the same species, *Crossarchus zebra*. Of this species six skulls were seen, four normal, and also the two following, the first being a case of extra molar on each side, the next a case of increase in size and complex variation in $\underline{m^2}$, on the left side suggesting a partial division of this tooth.

*301. **Crossarchus zebra:** small but well-formed additional molar in upper jaw on each side, making $p\,\tfrac{3}{3}$, $m\,\tfrac{3}{2}$. (Fig. 51, III.) Teeth unfortunately all much worn, so that it is not possible to determine whether any of the molars differ from their normal forms in correlation with the existence of these extra teeth; but as far as size is concerned, there was no sign of such change, $\underline{m^1}$ and $\underline{m^2}$ being of the usual size. B. M., 73. 2. 24. 18 mentioned by THOMAS, *P. Z. S.*, 1882, pp. 61 and 89.

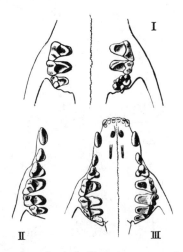

FIG. 51. *Crossarchus zebra.* I. Posterior upper molars of No. 302.
II. A normal specimen, right upper jaw.
III. Upper jaw of No. 301.

302.   **C. zebra:** all teeth normal except second molars in the upper jaws on each side, which depart from the normal in the following manner. *Right* $\underline{m^2}$ has a small extra cusp (Fig. 51, I.) on its outer side, making four instead of three as usual (cp. figure of normal, Fig. 51, II.). The *left* $\underline{m^2}$ is very extraordinary. It is rather less than twice the size of its fellow of the other side (Fig. 51, I.). The crown is of an irregularly elliptical form, the long axis being oblique. The posterior and anterior faces are marked by a shallow groove, giving an appearance of imperfect division into two teeth. The total number of cusps is greater than twice that borne by the other, but from the irregularity of the surface it is not possible to speak more precisely. For fear of injury the tooth was not extracted, so that the number of roots cannot be specified. B. M., 82. 5. 26. 1.

303.   **H. ichneumon** (Egypt), having no right $\overline{m^2}$. *Leyd. Mus.*

MUSTELIDÆ[1].

The evidence of dental Variation in this family is at present too small in amount to be of much value. It is chiefly interesting in so far as it relates to cases of the occurrence in one genus or sub-family, of a formula characteristic of another. Variations of this class, consisting in the presence of or absence of the anterior premolar or last molar, are in some of the forms very common. As will be suggested in the next chapter, some of these, for example, the variations in p^1 in the Badger, have a certain importance as giving some measure of the magnitude which a tooth may have when the species is, as it were, oscillating between the possession and loss of the tooth in question.

Amongst Mustelidæ there were two cases of supernumerary large premolars, probably reduplicatory.

Anterior Premolars.

Mustela (Martens), normally $p\frac{4}{4}$, $m\frac{1}{2}$. Seen in adult skulls of various species (*M. pennanti, martes, foina, zibellina, flavigula, americana*), 62: also the following:

304.   **M. foina** ♂ : $\overline{p^1}$ absent both sides. B. M., 1229, *k*. **M. zibellina:** p^1 absent both sides from both jaws [perhaps lost], B. M., 58. 5. 8. 189. **M. flavigula**, Madras. p^1 clearly absent from both jaws, B. M., 79. 11. 21. 621. **M. martes**? the same. C. S. M., 681. **M. melanopus:** p^1 absent, probably lost, B. M., 42. 1. 19. 100.

Putorius (Weasels, Stoats, Ferrets and Polecats), normally $p\frac{3}{3}$, $m\frac{1}{2}$. Seen in adult skulls of various species (*P. vulgaris, erminea, brasiliensis = xanthogenys, fœtidus = eversmanni = sarmaticus, lutreola, nudipes*, &c.), 105: also the following:

305.   **P. erminea:** l. $\overline{p^1}$ absent, B. M., 43. 5. 27. 11. On the other hand, **P. fœtidus**, B. M., 192 *s*, has rt. $\overline{p^1}$ as a two-rooted tooth, standing in a plane at right angles to the arcade.

Gulo: $p\frac{4}{4}$, $m\frac{1}{2}$. 5 specimens.

---

[1] Totals of normal skulls refer to *Brit. Mus.* and *Cambridge Mus.* only.

Galictis: $p\frac{3}{3}$, $m\frac{1}{2}$. Normal adults (*G. barbara* 8, *vittata* 4, *allamandi* 2), 14 specimens.

*306. **G. barbara,** having minute extra anterior premolar (making 4) in each lower jaw. B. M., 839, *f*.

In 28 skulls HENSEL found the following variations in premolars, the molars being always $m\frac{1}{2}$.

$$p\,\frac{3-3}{3-3},\text{ viz. the normal, 12 cases}$$

$$p\,\frac{3-3}{2-2} \ldots\ldots\ldots\ldots\ldots 6\ \text{,,}$$

$$p\,\frac{3-3}{3-2} \ldots\ldots\ldots\ldots\ldots 3\ \text{,,}$$

$$p\,\frac{3-3}{2-3} \ldots\ldots\ldots\ldots\ldots 2\ \text{,,}$$

also $p\,\dfrac{2-3}{2-2}$, $p\,\dfrac{3-2}{3-3}$, $p\,\dfrac{4-4}{2-2}$, $p\,\dfrac{2-2}{2-3}$, $p\,\dfrac{3-4}{3-3}$ each in one case. Taken together therefore there were 12 normals with $p\frac{3}{3}$, 16 cases of greater or less reduction, and 2 cases of increase. HENSEL[1], *Säugethiere Süd-Brasiliens*, p. 83.

307. **G. vittata**: p^1 may be absent, especially from upper jaw. BURMEISTER, *Reise durch d. La Plaata-Staten*, Halle, 1861, II. p. 409 [this variation not seen by HENSEL].

Pœcilogale: $p\frac{2}{3}$, $m\frac{1}{1}$. 3 specimens.

Mephitis: $p\frac{3}{3}$, $m\frac{1}{2}$. 9 specimens.

308. **Conepatus**: $p\frac{2}{3}$, $m\frac{1}{2}$. 12 specimens. *Conepatus* is the S. American representative of *Mephitis*, and normally differs from it in having one premolar less in upper jaw. This tooth is sometimes present as a minute tooth making $p\frac{3}{3}$. Sometimes on the contrary there is a premolar less in the lower jaw, giving $p\frac{2}{2}$. COUES, *Fur-bearing Animals of N. Amer.*, p. 192 and *Note*.

In addition to the 12 normals mentioned two cases of $p\frac{3}{3}$ were seen, viz. **C. mapurito**, B. M., 88. 11. 25. 8, and **C. chilensis**, B. M. 829, *a*. In the former the anterior premolar is of good size, but in the latter it is very rudimentary. Another case mentioned by BAIRD, *Mamm. of N. Amer.*, p. 192.

Mydaus: $p\frac{3}{3}$, $m\frac{1}{2}$. 4 specimens.

*309. **Meles**: commonly $p\frac{4}{4}$, $m\frac{1}{2}$. In **M. taxus,** the common Badger, p^1 is frequently absent from one or more places. Of 36 skulls only 16 had p^1 in all jaws, 7 have it in each lower jaw and 2 had no such tooth in either jaw. In remaining cases it was sometimes absent on right, sometimes on left, sometimes from above and sometimes from below. Some of these cases may be due to senile changes but this was certainly not so in all. Absence from lower jaw seems the most common. HENSEL, *Morph. Jahrb.*, 1879, v. p. 550.

Of genus *Meles* the following were seen by myself. + means presence, − absence of p^1.

---

[1] The numbers given by Hensel are the totals of $p+m$, but he states that the variation always concerned the small anterior premolars next the canines.

TEETH: MUSTELIDÆ.

	Upper jaw		Lower jaw		Cases
	right	left	right	left	
Meles taxus	+	+	+	+	3
,, ,,	−	−	+	+	16
,, ,,	−	−	?	?	1
,, ,,	−	−	−	−	1
,, ,,	+	−	+	+	3
,, ,,	+	+	−	+	1
					25
M. anakim } Japan	−	−	−	−	2
M. chinensis	−	−	−	−	3

Taxidea: $p\frac{3}{3}$, $m\frac{1}{2}$. 7 specimens.
Mellivora: $p\frac{3}{3}$, $m\frac{1}{1}$. 7 specimens.
Helictis: $p\frac{4}{4}$, $m\frac{1}{2}$. 6 specimens.
Ictonyx (= Zorilla): $p\frac{3}{3}$, $m\frac{1}{2}$. 14 specimens.

310.   **Lutra.** The Otters for the most part have $p\frac{4}{3}$, $m\frac{1}{2}$. The anterior premolar of the upper jaw is a small tooth standing internal to the canine, but in the common Otter its presence is most constant. In the Oriental *L. cinerea*, and the Neo-tropical *L. felina* on the contrary this tooth appears to be more frequently absent than present. The following table gives the results of examination of a series of skulls.

+ signifies presence, − absence of p^1.

	right	left	Cases
Lutra vulgaris	+	+	40
,, ,, ...	−	+	1
,, *macrodus*	+	+	11
,, ,, ...	−	−	2 (1 old; 1 young)
,, *cinerea*	+	+	2 (young)
,, ,, ...	−	−	6
,, *sumatrana*	+	+	4
,, *capensis*	+	+	1
,, ,, ...	−	−	1
,, *maculicollis*	+	+	1
,, *felina*	+	+	3
,, ,, ...	−	−	3
,, ,, ...	+	−	2
,, sp. (S. America)	+	+	14
,, ,,	−	−	1
,, ,,	+	−	1

In *L. cinerea* (= *leptonyx*) the absence of p^1 is associated with a more forward position of p^2, of which the anterior border is then level with the posterior border of the canines[1].

[1] See FLOWER and LYDEKKER, *Mammals, Living and Extinct*, p. 568, Fig. 261.

Large Premolars.

311. **Putorius** (labelled "*Vison Horsfieldii*"): at the place in which the right lower posterior premolar ("$\overline{p^4}$") should stand there are two such teeth at the same level. They are almost identical, but the inner

Fig. 52. *Putorius*, No. 311, right lower jaw, ground-plan of teeth and profile views of two teeth at the same level. Upper figure is the internal tooth.

(upper in figure) is slightly the smaller (Fig. 52). B. M., 823, *a*.

312. **Helictis orientalis,** Java: having supernumerary two-rooted tooth internal to and between p^2 and p^3. This extra tooth is almost a copy of $\underline{p}^2$ (Fig. 53). B. M., 824, *a*.

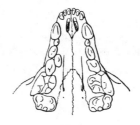

Fig. 53. *Helictis orientalis*, No. 312. Surface view of upper jaw and a representation of the right upper teeth as seen from inside.

Molars.

313. **Putorius**: Hensel, *Morph. Jahrb.*, v. 1879, p. 540, states that he has several skulls of *Fœtorius putorius* with an extra upper molar on one side in a rudimentary condition. Giebel, Bronn's *Kl. u. Ord.*, p. 186, Taf. xv. figs. 1, 2 and 3, figures a specimen of "*Putorius typus*" having a fairly well developed extra upper molar on each side making $m\frac{2}{2}$ instead of $m\frac{1}{2}$. Probably both these accounts refer to **P. fœtidus.**

314. **Lutra platensis**: supernumerary molar on one side of upper jaw. Such a tooth normally present on both sides in *L. valetoni*, a fossil form. Von Heuglin, *Nov. Act. Leop. Car. Cæs.*, xxix. p. 20. **Lutra**

sp., S. America, B. M., 85. 11. 23. 1, has small alveolus behind $\overline{m^2}$ on each side.

315. **Mellivora** (= *Ratelus*): similar case. VON HEUGLIN, *ibid.*
316. **Meles taxus** has normally $m\frac{1}{2}$. Skull from Quarternary diluvium of Westeregeln has small alveolus behind right $\underline{m^1}$ and left $\overline{m^2}$. Another fossil skull has $m\frac{2}{3}$. NEHRING, *Arch. f. Anthrop.* x. p. 20. [? Small alveolus behind left $\overline{m^2}$ in B. M., 211, *h.*]
317. **Lutra**: case of absence of $\overline{m^2}$; **Mustela**: $\overline{m^2}$ may be absent. HENSEL, *l. c.*

PINNIPEDIA.

With reference to dental Variation in Otariidæ and Phocidæ there is a considerable quantity of evidence. In some of the species the frequency of abnormalities is remarkably great. Among the most interesting examples are two cases of reduction in the number of incisors, both occurring in *Phoca barbata*. These cases are especially important in connexion with the fact that the Seals are exceptional among Carnivores in having a number of incisors other than $\frac{3}{3}$, and that among the different sub-families of Seals there is diversity in this respect.

Taken together, the cases of Variation in the premolars and molars of Seals illustrate nearly all the principles observed in the numerical Variation of teeth. In both premolars and molars there are examples of the replacement of one tooth by two, and in some of these the resulting teeth stand in series while in others they do not. Besides these there are numerous instances of extra premolars and molars belonging to various categories.

As regards the frequency of extra teeth in Seals it may be mentioned that of Phocidæ 139 normals were seen, and 11 cases of supernumerary teeth; of Otariidæ 121 normals and 5 cases of supernumerary teeth.

From the simplicity of the normal dentition and from the diversity of the variations presented, the evidence as to the teeth of Seals may conveniently be studied by those who are interested in the phenomena of Variation without special knowledge of the subject of mammalian dentition.

Incisors.

It will be remembered that of Phocidæ the sub-family Phocinæ (like Otariidæ) has normally $i\frac{3}{2}$, while the Monachinæ have $i\frac{2}{2}$ and the Cystophorinæ $i\frac{1}{2}$. Of Phocinæ of various genera and species 105 skulls having $i\frac{3}{2}$ were seen, and in addition the two following.

*318. **Phoca barbata.** Greenland: skull having $i\frac{2}{2}$ on both sides (Fig. 54). This skull is a particularly good one and is neither very old nor very young. The teeth stand regularly together and there is no lacuna between them. There is no reasonable doubt that an incisor is absent from each side of each jaw. The shape of the premaxillæ is different from that seen in other specimens of *Phoca*, and, doubtless in correlation with the absence of the two upper

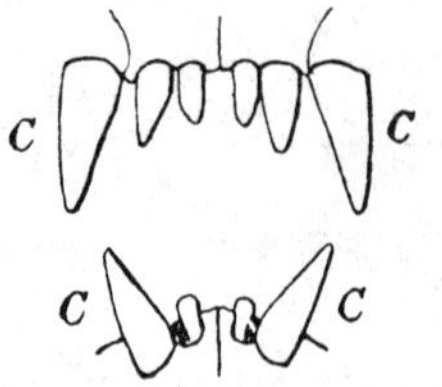

FIG. 54. Incisors and canines of *Phoca barbata*, No. 318.

incisors, the width of the premaxillæ is considerably less than in specimens having the normal dentition. B. M., 90. 8. 1. 6.

319. **P. barbata:** in left upper jaw are three normal incisors; but on the right side the incisors have been lost. The alveoli, however, shew plainly that only two incisors had been present. Of these the outer one in size agrees with i^3, being a large alveolus equal to that of i^3 of the other side, but the second alveolus, occupying the place of i^1 and i^2, *is also a large alveolus*, scarcely smaller than that for i^3. It appears therefore that in this specimen a single large tooth stood in place of i^1 and i^2. A lower jaw placed with this skull was normal, but it was not certain that it belonged to the skull. O. M., 1724.

Premolars and Molars.

Normal arrangement. In Phocidæ there are normally five teeth behind the canines in each jaw, and according to the received accounts, of these teeth 4 are premolars and one is a molar, giving $p\frac{4}{4}$, $m\frac{1}{1}$. The Otariidæ on the other hand have generally $p\frac{4}{4}$, $m\frac{2}{1}$, but *both* the two upper molars stand at a level behind that of the lower molar, so that the posterior molar, m^2 is placed so far back that it meets no tooth in the lower jaw. Some of the Otariidæ, however, as *O. californiana*, do not possess such a posterior tooth, and have only $m\frac{1}{1}$. *O. stelleri* is peculiar in the fact that it also has only one upper molar, but this tooth is separated by a large diastema from p^4, and stands in the position characteristic m^2 of the other Otariidæ. Hence it may be supposed that m^1 is really absent while m^2 is present.

Amongst the cases will be found some of the presence in Phocidæ, especially *Halichœrus*, of an extra molar placed in the usual position of m^2 in the Otariidæ. But lest any one should think it manifest that this is an example of Reversion to the Otarian condition, attention is called to cases of such an extra molar in the Otariidæ also. Similarly there are instances of absent molar in those Otariidæ which have $m\frac{2}{1}$, leaving $m\frac{1}{1}$; and of these cases one occurs in such a way as to leave the peculiar diastema between

p^4 and the molar, referred to above as characteristic of *O. stelleri* (see No. 342).

The cases are grouped in an arbitrary collocation, according as it seemed desirable that particular variations should be studied together. In the sections dealing with premolars, Phocidæ are not separated from Otariidæ.

First Premolar.

*320. **Ommatophoca rossii,** an Antarctic Seal. Of this form only two skulls are known, both in the British Museum. One of these (B. M., 324, *b.*) has the arrangement usually found in Phocidæ, namely, five teeth behind the canines in each jaw, giving the formula

$$i\frac{2-2}{2-2},\ c\frac{1-1}{1-1},\ p+m\frac{5-5}{5-5}$$

(on the analogy of other Seals $p\frac{4}{4}$, $m\frac{1}{1}$). The other specimen is exceedingly remarkable (Fig. 55). In it the incisors and canines

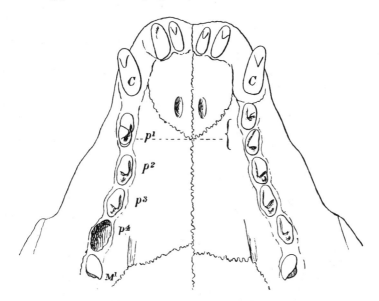

Fig. 55. *Ommatophoca rossii*, No. 320, teeth of the upper jaw.

are as in the first specimen, but the first tooth behind the canines *on both sides* in the lower jaw and on the *right* side in the upper jaw, has a very peculiar form, having a deep groove passing over the whole length of the tooth, on its outer and inner sides. These grooves extend from the tip of the root along both sides of the crown, and thus imperfectly divide each tooth into an anterior and a posterior half. The cusp of each tooth is also divided by the grooves so as to form two small cusps. Each of these teeth is therefore an imperfectly double structure, and may be described as being just half-way between a single tooth and two teeth. These teeth are shewn in Fig. 56.

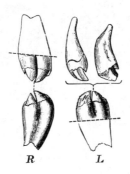

FIG. 56. *Ommatophoca rossii*, No. 320. The anterior premolars of upper and lower jaws from the side. (The left lower and right upper teeth were not extracted.)

On the left side in the upper jaw, as the *vis-à-vis* to one of these double teeth, there are actually two complete teeth, of very similar but not identical form, as shewn in Fig. 56. Each stands in a distinct alveolus, the two being separated by a bridge of bone. The dental formula of this skull, taken as it stands, is therefore $p\frac{5-4}{4-4}, m\frac{1-1}{1-1}$, for since the bigeminous teeth are not completely divided, they must be reckoned as single teeth.

321. **Cystophora cristata:** internal to and slightly in front of p^1 on each side in the upper jaws is an extra tooth. These extra teeth are alike in form but are rather smaller than p^1. C. M., 895.

322. **Cystophora cristata** (label, *Phoca cristata*): internal to *right upper* p^1 is an alveolus for a small one-rooted tooth. In the corresponding situation in the *left lower* jaw there is such an extra tooth in place. Leyd. M.

323. **Zalophus lobatus** (= *Otaria lobata*): left p^1 smaller than right p^1, and between the canine and the left p^1 there is a supernumerary tooth, smaller than left p^1. (The same skull has another extra tooth outside and between p^3 and p^4, see below No. 333.) Leyd. M.

[**P. vitulina:** alveolus for left p^1 much larger than that for rt. p^1; the latter tooth is in place, but left p^1 is missing. C. M., 902.]

Large Premolars.

324. **P. grœnlandica:** in the position in which left upper p^4 should stand there are two whole and complete teeth, each as large as normal p^4. Fig. 57). The two stand perfectly in series, and owing to the wide

gaps normally existing between the teeth in this species there is no crowding. Between these two teeth there are slight differences of

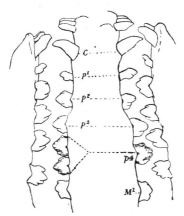

Fig. 57. *Phoca grœnlandica*, No. 324. Left and right profiles. This figure was kindly drawn for me by Mr J. J. Lister.

form, and the posterior is rather the larger. On both sides $\overline{m^1}$ is in place and at the same level. Both the two teeth in place of p^4 bite between $\overline{p^4}$ and $\overline{m^1}$ of the lower jaw. On the *right* side $\underline{p^3}$ is normal and $\overline{m^1}$ is also normal but $\underline{p^4}$ is a very large and thick tooth, and its main cusp is cloven, giving it the appearance of imperfect division into two. In this case therefore $\underline{p^4}$ on the one side may be supposed to have divided into two perfect and nearly similar teeth, while on the right side this division is begun but not completed. *Leyd. M.*

325. **Otaria ursina** ♂: supernumerary premolar in left upper jaw. This is a curious case. The right upper and both lower jaws are normal. On comparing the left upper series of 7 teeth with the right series which has 6 normal teeth, it is seen firstly that the two molars of each side are alike in form and stand at their proper levels (Fig. 58).

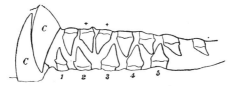

Fig. 58. *Otaria ursina*, No. 325, seen from left side. 1, 2, 3, 4, first to fourth lower premolars; 5, lower molar.

Next, the two posterior premolars of each side ($\underline{p^3}$ and $\underline{p^4}$) agree so nearly that there is no reasonable doubt that they are not concerned in the variation. Anterior to this there is difficulty, for whereas p^1 and p^2 are normal and in place on the right side, there are *three* teeth on the left side to balance them. These three teeth moreover are so nearly alike that it is impossible to say that either of them is definitely the extra tooth. The first premolars of each side are almost exactly alike, and the second and third of the left side are each very like the second on the right side (p^2), so that it might be said that p^2 was represented by two teeth on the left side; and as seen in Fig. 58 the second and third on the left side bite between $\bar{p}^2$ and $\bar{p}^3$ of the lower jaw, as the normal $\underline{p}^2$ would do. This is however accomplished by the backward displacement of $\bar{p}^3$. Probably therefore this should be looked on as a case of division of $\underline{p}^2$, but there is no proof that the *three* first premolars of the left side are not collectively equivalent to the first two of the right side. C. M., 911, f.

326. **P. grœnlandica:** the second upper right premolar is represented by two teeth, each of which has two roots; the two teeth stand at the same level in the arcade, the inner one being rather smaller. On the left side the second upper premolar is *incompletely* double, the crown being partially divided by an oblique constriction into an anterior and internal portion and a larger posterior and external part. The former has one root and the latter two. P. M., *A*, 2897.

327. **Otaria jubata:** left upper p^3 a bigeminous tooth something like the anterior premolars of *Ommatophoca* (No. 320). In this animal all the premolars and molars are one-rooted and have simple conical crowns. The abnormal tooth is formed as it were of two such simple teeth imperfectly divided from each other through their whole length, the plane of division being transverse to the jaw. The teeth of the two sides are not alike and in particular the posterior lower $\overline{m^1}$ is much smaller than the right. The skull has been much mended and the position of some of the teeth is not very certain, but the above-mentioned facts are correct. C. S. M., 975.

*328. **Otaria cinerea:** supernumerary tooth in upper jaw on both sides. The extra tooth in each case stands within the arcade, internal to the

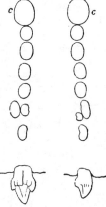

FIG. 59. *Otaria cinerea*, No. 328. A diagram of the positions of the upper teeth, and profiles of the teeth standing internal to each $\underline{m}^1$.

5th tooth behind the canine (sc. $\underline{m}^1$), which is pushed outwards by it. The extra tooth of the left side (Fig. 59) is a little larger and at a level rather anterior to that of the left extra tooth. C. M., 911**.

*329. **P. vitulina:** having a supernumerary tooth in each jaw on the right side. This is a somewhat remarkable case. In both jaws the extra tooth does not stand in series with the others but is placed within the arcade (Fig. 59, + +). That of the upper jaw is a curved tooth with one large median cusp and a small cusp anterior to and posterior to it, having somewhat the form of $\bar{p^2}$ of the lower jaw. This tooth stands within the arcade *at a level between that of* $\underline{p^2}$ *and* $\underline{p^3}$ which are pushed outward by it. The extra tooth of the lower jaw in shape closely resembles that of the upper jaw, but is slightly larger, having very much the size and shape of the lower right $\bar{p^2}$. In position this extra tooth does *not* stand between p^2 and p^3 like the upper supernumerary, but *is placed within the arcade and* $\bar{p^3}$ *and* $\bar{p^4}$ which are somewhat separated by it. C. M., 903. [Judged by the ordinary rules of dental homology, the two extra teeth are not homologous, for the upper one is between p^2 and p^3, while the lower one is between $\bar{p^3}$ and $\bar{p^4}$. But when the jaws are put together it appears that the two extra teeth are opposite to each other almost exactly, the large cusp of the lower one being in the bite scarcely at all posterior to the large cusp of the upper. The tooth of the lower jaw is thus almost exactly the image or reflexion of the tooth in the upper jaw.]

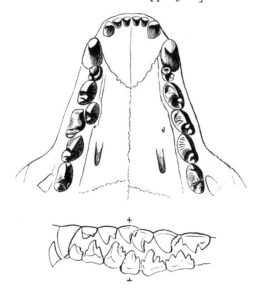

Fig. 60. *Phoca vitulina*, No. 329; view of upper teeth from the surface, and an imaginary profile of the upper and lower teeth of the right side seen from within.

330. **Otaria ursina:** this skull in bad condition. The *Catalogue* (1884) states that between $\underline{p^2}$ and $\underline{p^3}$ on both sides and between $\underline{p^4}$ and $\underline{m^1}$ on both sides there was a small supernumerary tooth, in all, four extra teeth in the upper jaw. The anterior supernumeraries are in place and one rather smaller than $\underline{p^2}$. The posterior supernumeraries are lost, but from the alveoli they must have been of fair size, though not so large

as $\underline{p}^4$. In each case the extra tooth is placed a little within the arcade though the adjacent teeth are also spaced out for it. This skull has been a good deal mended. C. S. M., 990.

331.   **Phoca grœnlandica:** in right upper jaw $\underline{p}^4$ is smaller than the corresponding tooth of the left side, though it is two-rooted as usual. Between it and $\underline{p}^3$ there is a small, peg-like, supernumerary tooth. Both $\underline{p}^4$ and the extra tooth bite between $\overline{p}^4$ and $\overline{m}^1$ of the lower jaw. *Leyd. M.*

332.   **P. grœnlandica:** supernumerary tooth with two roots placed internally to and between left $\underline{p}^4$ and $\underline{m}^1$. The last molars stand at the same level on the two sides. B. M., 328, *i.*

333.   **Zalophus lobatus:** in *right* upper jaw a supernumerary tooth placed on the *outside* of the arcade on a level with the interspace between p^3 and p^4. This tooth resembles p^4 or m^1. On the *left* side $\underline{p}^1$ is smaller than on right side and a supernumerary tooth which is still smaller stands between $\underline{p}^1$ and the canine. *Leyd. M.* [given above, No. 323].

334.   **P. vitulina:** in right lower jaw a supernumerary tooth *inside* the arcade, between $\overline{p}^3$ and $\overline{p}^4$. In size and form it agrees very nearly with the first premolar of the right lower jaw: other teeth normal. C. M., 903, *F.*

335.   **P. vitulina:** in front of $\overline{p}^3$ on left side the teeth are all lost but there has been some irregularity, probably a supernumerary tooth level with $\overline{p}^2$: also behind right $\underline{m}^1$ there is a small tubercular nodule of bone which may perhaps cover a supernumerary molar. *C. S. M.*, 1064.

Molars.

336.   **P. vitulina :** on left side there is a small supernumerary molar placed behind $\underline{m}^1$. This tooth stands in the line of the arcade (Fig. 61)

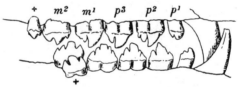

Fig. 61. *Phoca vitulina* No. 336, a profile of the left teeth in the bite as seen from within.

but is turned so that its greatest width is set transversely to the jaw. In the *lower jaw* of the same side there is a supernumerary tooth placed internally to $\overline{m}^1$. This tooth has two roots and three cusps, and is therefore not a copy of $\overline{m}^1$, which has 4—5 cusps. C. S. M., 1067.

*337.   **Halichœrus grypus :** of 47 skulls seen, 12 have one or more supernumerary molars. One case of $p\frac{4}{4}$, $m\frac{2}{2}$. NEHRING, *Sitzb. naturf. Fr. Berlin*, 1883, p. 110.

Of 34 skulls in Greifswald Museum there were 3 cases of $m\,\dfrac{2-2}{1-1}$, and five cases of $m\frac{2}{1}$ on one side only. *Ibid.*, 1882, p. 123.

Of 11 skulls seen by myself two individuals (C. M.) have an extra molar on left side. In these cases the extra teeth are placed at a considerable distance behind m^1 as they are in *Otaria.* [In addition to these GRAY figures a skull with $m\frac{2}{1}$ but without allusion to this fact in the text. *Hand-list of Seals in B. M.*, 1874, Pl. VII.]

A skull having left m^1 two-rooted, right m^1 being much less so. C. S. M., 1059.

338. **P. grœnlandica:** minute supernumerary molar on each side in upper jaw making $m\frac{2}{1}$. P. M., *A.* 2898.

339. **Zalophus californianus,** an Eared Seal not far removed from *Otaria*, but having $p + m\frac{5}{5}$ instead of $\frac{6}{5}$. The five back teeth are arranged as a rule in a continuous series, but sometimes there is a small space between the last molar and the penultimate [cp. *O. stelleri*], and occasionally they are all slightly and evenly spaced.

One case of $p + m\frac{6}{5}$ on both sides and two cases of $p + m\frac{6}{5}$ on one side only. In these the extra teeth were behind the (normally) last molar and smaller than it, being without the accessory cusps seen in that tooth. ALLEN, J. A., *N. Amer. Pinnipeds*, 1880, pp. 209, 224 and 226.

340. **Z. lobatus:** one specimen having $p + m\frac{6}{5}$ on right and $\frac{5}{5}$ on left, *Leyd. M.* [in addition to 3 specimens with the normal $\frac{5}{5}$].

341. **Callorhinus ursinus:** normally $p + m\frac{6}{5}$; one case having $\frac{7-7}{5-5}$ and one case with $\frac{7-6}{5-5}$. ALLEN, *l. c.*, p. 224 (cp. No. 343).

Reduction in numbers of molars.

342. **Arctocephalus australis,** normally $p + m\frac{6}{5}$: one case of $\frac{6-5}{5-5}$.

General statement made that in cases of absence of a tooth it is the antepenultimate molar which is missing [not described in a specific case]. ALLEN, *l. c.*, p. 224.

343. **Callorhinus ursinus,** normally $\frac{6}{5}$; 2 cases of $\frac{5}{5}$. ALLEN, *l. c.* (cp. No. 341).

344. **Otaria jubata,** normally $\frac{6}{5}$: one specimen having $\frac{6}{5}$ on both sides, *Leyd. M.*; one specimen having right $\frac{5}{5}$ left $\frac{6}{5}$. *Leyd. M.*

Cystophora cristata: only one molar, viz. left m^1 present; from the state of the bones it seemed possible that the others had not been formed, but this is quite uncertain. C. S. M., 1101. **Macrorhinus leoninus:** doubtful if the molars had been present. C. S. M. 1109.

UNGULATA.

As to the occurrence of Variation in the dentition of Ungulates I have no statistics, but a certain number of miscellaneous cases have been collected from different sources. Most of the cases relate to domestic animals and are given on the authority of MOROT and GOUBAUX.

Perhaps the most interesting evidence is that regarding the change of form in the "canines" of the Sheep. These teeth of course have normally the shape of *incisors*, but in the cases described by Morot they had more or less of the character of canines. This evidence, though belonging properly to the Substantive class, is introduced here on account of its close relation to some general aspects of variation in teeth.

* It is noticeable that there is so far no case of an incisor appearing in the upper jaw of Ruminants.

The evidence is divided into two groups, the first relating to incisors and canines, the second to premolars and molars.

Incisors and Canines.

345. **Elephas africanus** ♂ : the left tusk imperfectly doubled. The root of this tooth was double[1], one root being outer and the other inner. The half of the tusk arising from the outer root twisted round and over the other half so that at the other end it lay above and internal to it. The structure of the tusk was essentially double, but the two parts were more or less blended together in the middle third. The external ends were separate, but broken and somewhat deformed. FRIEDLOWSKY, A., *Sitzungsb. d. K. Ak. Wien*, 1868, LIX. I. p. 333. Plate.

346. **Horse.** Supernumerary incisors common. MAGITOT, *Anom. Syst. dent.*, p. 104, *Plates*. Numerous specimens in Museum of Veterinary School at Alfort.

347. Specimen having 12 upper incisors and 12 lower incisors belonging to the permanent dentition. GOUBAUX, *Rec. méd. vét.*, 1854, Ser. 4, I. p. 71. Similar observation, LAFOSSE, *Cours Hippiatrique*, 1772, p. 32.

348. Extra teeth of more or less irregular form placed behind upper incisors very common : many specimens in museum at Alfort. Specimen having left i^2 as a double structure, the two halves not being separated. (*Alfort Mus.*) MAGITOT, *l. c.*, Pl. XIX. fig. 25.

Absence of incisor in Horse is rare. GOUBAUX, who has largely studied the subject, knew no case of absence of any tooth in Horse, *l. c.*

349. Skeleton of Cart-mare in C. M. has only two incisors on the left side in the upper jaw. The teeth stand evenly and without break or trace of any other incisor having been present. There is no sufficient indication to shew which of the incisors is missing, but the two incisors present agree most nearly with i^2 and i^3. This specimen was first pointed out to me by Mr S. F. Harmer. (See also case given by RUDOLPHI, *Anat.-phys. Abh.*, 1802, p. 145.)

*350. Mare of common breed, foaled March, 1876, having in the upper jaw no i^3 in either milk or permanent dentition, and in the lower jaw no permanent i^3. In the upper jaw there were only 4 milk incisors, which were subsequently replaced by 4 permanent incisors. Animal seen by MOROT in Apr. 1880; it then had 4 permanent incisors in the upper jaw, but no i^3. In the lower jaw permanent i^1 and i^2 were in place, together with i^3 of the milk series on each side. As Morot remarks it is still possible that the other incisors might appear. Dam normal; half-sister abnormal, given in next case. MOROT, *Bull. Soc. méd. vét.*, 1885, Ser. 7, II. p. 125.

*351. Mare out of same mother as last case, by another sire, foaled Apr. 1877, had only 4 milk-incisors in upper jaw. Seen by Morot at 3 years old, had then the teeth of lower jaw normal, viz. permanent i^1, and milk i^2 and i^3 all in place. In upper jaw were permanent i^1 and milk i^2 on each side. The right milk i^2 on the external side had a light groove parallel to the long axis of the tooth, suggesting that it might be a double structure, but the groove was very slight and the crown was single. At five years old this animal had the normal 6 lower incisors, but in the upper jaw left i^3 was absent. On the other hand a well-formed supernumerary tooth stood behind right i^3, right i^2 being partly rotated. *Ibid.*, p. 127.

[1] See also a curious case of "nine tusks" imperfectly described by CHAPMAN, J., *Travels in Interior of S. Africa*, II. p. 98.

352.   **Ass:** (♀ some 20 yrs. old) on right side in upper jaw were two canines, one in front of the other in the same alveolus. MOROT, *Rec. méd. vét.*, 1889, Ser. 6, VII. p. 480.   Another somewhat similar case, *ibid.*
353.   **Cow:** in place of right i^3, two third incisors placed side by side. MOROT, *Bull. et mém. Soc. méd. vét.*, 1886, p. 321.

Goat, 4—5 weeks old; supernumerary lower incisor placed between the two median incisors which rose above it. This tooth stood transversely so that its edge lay exactly in the long axis of the head. MOROT, *l. c.*

354.   **Sheep:** extra incisor on left side. (*Alfort Mus.*). GOUBAUX, *Rec. méd. vét.*, 1854, Ser. 4, I. [Several other cases.]

Abnormal form of Canines in Sheep.

*355.   In the lower jaw of the Sheep there are on each side 4 incisiform teeth, arranged in close series without any diastema. Of these the outermost, known in veterinary works as "corner teeth," are considered by zoologists as representing canines.

The corner teeth or canines have been found in a considerable number of cases actually shaped like canines instead of like the incisors as usual. These teeth have been found presenting this modification in several degrees, but in order to gain a fair view of the matter it is necessary to read the evidence in its entirety.

The facts given were founded on 18 animals, 15 ewes and 3 males [whether rams or wethers not stated]. In these 18 cases there were 28 individual teeth of abnormal form. Of these 14 were conical with a point either sharp or rounded; 7 were conical with a bifid point; 5 were cuneiform; 1 was cylindrical with a surface shaped like an ass' hoof; 1 was pyramidal.

In 8 specimens the abnormality was unilateral and in 10 it was bilateral, but in the latter the corner teeth of the two sides were frequently of differing forms [details given]. MOROT, *Bull. Soc. méd. vét.*, 1887, p. 166.

Pig.   No case of Variation in incisors met with.

[This is perhaps singular in connexion with the fact that the Peccaries (*Dicotyles*) have $i\,\frac{2}{3}$.]

*356.   **Dicotyles torquatus** (normally $i\frac{2}{3}$): two specimens having $i\,\dfrac{3-2}{3-3}$; in one of them i^x of the side having the extra tooth is deformed. Another young skull of *Dicotyles* also had 3 incisors on left side. HENSEL, *Säugethiere Süd-Brasiliens*, p. 94.

Molars.

357.   **Horse:** supernumerary molars exceedingly rare; case of such a tooth in left upper jaw, behind and in series with the others. GOUBAUX, *Rec. méd. vét.*, 1854, Ser. 4, I. p. 71, same case, figured by MAGITOT, *l. c.*, Pl. v. fig. 9.

*358.   **Ass**: thoroughbred Spanish she-ass, in the Museum of the Royal College of Surgeons, has a large supernumerary molar on each side in series in the upper jaw, and a similar tooth in the left lower jaw. The same skull has the first premolar also present on each side in the upper jaw, as is not unfrequently the case in Equidæ. All four canines are present as minute teeth. The dental formula for this skull is therefore

$$i\,\frac{3-3}{3-3}\ c\,\frac{1-1}{1-1}\ p\,\frac{4-4}{3-3}\ m\,\frac{4-4}{4-3} = 45.$$

359.   **Auchenia lama**: specimen having a supernumerary (fourth) molar in the lower jaw [? on both sides]. This tooth was fully formed and resembled the normal last molar. In the upper jaw was a small alveolus behind m^3, for another tooth which was not present in the specimen. RÜTIMEYER, L., *Vers. einer natürl. Geschichte des Rindes*, Zurich, I. p. 55, *Note*.

360.   **Cervus axis** ♀ : specimen having a supernumerary grinder placed on the inside of the normal series on the left side of the upper jaw. In the lower jaw of the same specimen the following supernumerary teeth: (1) a small, compressed accessory tooth on both sides placed internally to $\overline{m^2}$; and (2) behind the large three-fold sixth molar was a smaller two-fold tooth which had caused a displacement of the 6th molar. DÖNITZ, *Sitzungsb. d. naturf. Fr.*, Berlin, 1872, p. 54.

361.   **Cervus rufus**: having supernumerary (4th) premolar on one side in lower jaw. HENSEL, *Morph. Jahrb.*, v. p. 555.

362.   **Ox**: supernumerary upper molar on left side. MAGITOT, *l. c.*, p. 106.

Sheep: extra molar in left lower jaw, *ibid.*, p. 105, Pl. v. fig. 10. [? some error; the figure represents a normal jaw.]

MARSUPIALIA.

The facts given in illustration of Variation in the dentition of Marsupials relate only to a part of the subject and to selected forms. Some of the cases to be given are however of exceptional importance. Evidence is offered in reference to the following subjects:

(1) Incisors.
(2) Premolars, and the "Intermediate" teeth (in the lower jaw), of Phalangeridæ.
(3) Premolars and Molars of Dasyuridæ and Didelphyidæ.
(4) Molars of certain Macropodidæ.

(1) *Incisors.*

The following cases of Variation in incisors are all that were met with in the Marsupials examined.

DIDELPHYIDÆ: incisors normally $\frac{5}{4}$, thus differing from the Dasyuridæ ($i\,\frac{4}{3}$) with which they have much in common[1]. Of various

---

[1] THOMAS states that the family Didelphyidæ "is, on the whole, very closely allied to the Dasyuridæ, from which, were it not for its isolated geographical position, it would be very doubtfully separable." *Cat. Marsup. Brit. Mus.*, 1888, p. 315.

species 90 adult skulls seen having this number of incisors and three cases of abnormal number of incisors. Of these the first two must not be reckoned in estimating the percentage of abnormalities in a promiscuous sample, for Mr Thomas, who kindly shewed me these specimens, informs me that they were preserved and brought to the Museum expressly as abnormalities. The existence of these variations is nevertheless particularly interesting in connexion with the exceptional number of incisors normal in *Didelphys*.

363. **Didelphys marsupialis**: in right upper jaw *six* incisors; left upper jaw and the whole lower jaw missing. B. M., 92. 11. 3. 28.

364. Another specimen has on the right side $\frac{5}{4}$ as usual, but on the left $i\frac{5}{5}$. It appears that $\bar{i}^1$ and $\bar{i}^2$ of the two sides correspond, but on the left side *three* very similar teeth stand in series behind $\bar{i}^2$. B. M., 92. 11. 3. 29.

365. **D. turneri** (= *crassicaudata*), Demerara. A single specimen of this species in collection. It has $i\frac{4-4}{4-4}$, but there is no evidence to shew which of the upper incisors were missing. B. M.

DASYURIDÆ: incisors normally $\frac{4}{3}$; of genera other than *Myrmecobius*, 63 normal skulls seen.

Dasyurus sp., having only two incisors in left lower jaw; right lower jaw normal, upper jaws missing [doubtful case]. B. M., 250.

* **Myrmecobius fasciatus**: with incisors normal 4 whole skulls,

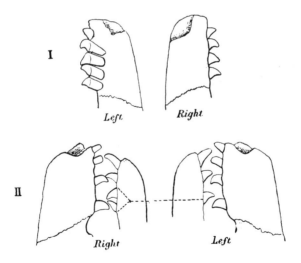

FIG. 62. *Myrmecobius fasciatus.*
I. Right and left profiles of upper jaw of No. 366.
II. Right and left profiles of the two jaws of No. 367.
(Premaxillary teeth alone shewn.)

5 skulls without lower jaws, and 1 lower jaw without skull; abnormals 2, as follows:

*366. A young skull having in the upper jaw on the left side (Fig. 62, I.) *two* teeth, both apparently in place of left $\underline{i^3}$, making $i\dfrac{5—4}{3—3}$. B. M., 314, g.

*367. A specimen having *four* incisors in the right lower jaw, the left being normal. Perhaps the two hindmost of the four represent the third lower incisor of the left side in the way suggested by the dotted lines in the figure (Fig. 62, II.). B. M., 314, b.

PHALANGERIDÆ: incisors (neglecting "intermediate" teeth of lower jaw) normally $\tfrac{3}{1}$; this seen in 209 skulls of various genera and species.

368. **Phalanger orientalis,** Solomon Islands: left $\underline{i^3}$ as an imperfectly double tooth, having two sub-cylindrical crowns and only one root (Fig. 63). The two crowns

FIG. 63. *Phalanger orientalis.* No. 368.
Upper incisors and canines. The separate figure shews the left i^3 extracted.

stand in the same transverse plane, the one being internal to the other and rather smaller than it. Lower jaw missing. B. M., 1936, c. [Two other skulls from same locality normal.]

369. **P. maculatus,** Port Moresby: only *two* incisors on each side in the upper jaw. The centrals, $\underline{i^1}$, of each side, are in place; externally to them there is on each side an alveolus for a tooth, which, judging from the size of the alveolus, was probably i^2. Immediately behind these alveoli the canines follow on each side. In this case it may be said that the missing teeth are $\underline{i^3}$ in all probability. Lower jaw normal. B. M., 79. 3. 5. 8.

370. Specimen having "in each upper jaw two incisors instead of three," [also has no left $\underline{p^1}$, see No. 377]. *Leyd. Mus.*, 55. JENTINK, F. A., *Notes Leyd. Mus.*, 1885, VII. p. 90. Two specimens, *Leyd. Mus.*, 56 and 61 are without $\underline{i^3}$ of right upper jaw, *ibid.*, p. 91. Specimen in which "five of the upper incisors are wanting [only one "intermediate" tooth in left lower jaw, see No. 377]. *Leyd. Mus.*, 63, *ibid.*, p. 91.

371. **Pseudochirus forbesi:** of this species only a single skull known; it has no upper $\underline{i^3}$ [and no upper first premolar, see No. 379]. B. M., 1943. THOMAS, O., *Cat. Marsup. Brit. Mus.*, 1888, p. 183.

(2) *Premolars, and the "Intermediate" teeth (in the lower jaw) of*
PHALANGERIDÆ.

The evidence here offered relates to the following genera:—*Phalanger, Trichosurus, Pseudochirus, Petauroides, Dactylopsila* and *Petaurus*. Before speaking of the variations seen, a few words are needed in explanation of the nomenclature adopted.

In these forms there is only one tooth having a milk-predecessor, and in all the genera here referred to this is a distinct and recognizable tooth, with a chisel-shaped crown. Following Thomas' system I shall call this tooth p^4 throughout. This name is used as being well understood and convenient, but without any intention of subscribing to the principles of homology upon which the system of nomenclature is based.

In front of p^4 there is great diversity.

In Thomas' paper[1] a careful and well-considered attempt was made to bring these anterior teeth into a formal scheme of homologies, and though the application of this method to the teeth of the lower jaw was avowedly tentative, yet at first sight the results in the case of the upper teeth were fairly satisfactory. Nevertheless it appears to me that in view of the facts of Variation about to be related, the system elaborated by Thomas breaks down; not because there is any other system which can claim to supersede it, but because the phenomena are not capable of this kind of treatment. To anyone who will carefully study the examples given in the following pages, especially those relating to the genus *Phalanger*, it will, I think, become evident that it is not possible to apply any scheme based on the conception that each tooth has an individual Homology which is consistently respected in Variation.

The evidence concerns first the premolars of the upper jaw, and secondly the lower "intermediate" teeth. Inasmuch as in several of the cases there was Variation in both these groups of teeth, the evidence relating to them cannot well be separated. As regards the upper teeth, all the cases of importance occurred in *Phalanger* and *Trichosurus*, and owing to the similarity between the dentitions of these two genera it is not difficult to employ terms which shall be distinctive, though the question of the homologies of the teeth go unanswered. In all the forms concerned there are three upper incisors, and the tooth immediately succeeding them will be called the canine, though its position and form differ greatly in the various genera; for while in *Phalanger* and *Trichosurus* it is a large caniniform tooth placed on the suture between premaxilla and maxilla, in *Pseudochirus*, for instance, it is proportionally smaller and stands in the maxilla at some distance behind the suture.

Upper jaw. As already stated, the large premolar having a milk-predecessor will be called p^4.

In *Trichosurus* between the canine and p^4 there is usually one large tooth, in shape and size much like the canine: this tooth will be called $\underline{p}^1$ as Thomas proposed. Though when present it is large, it is not rarely absent altogether (*v. infra*). In *Phalanger* there is a similar $\underline{p}^1$, though of somewhat smaller size; but besides $\underline{p}^1$ there is usually another premolar, a small tooth placed between $\underline{p}^1$ and $\underline{p}^4$. On Thomas' system this is p^3 and for purposes of description the name will be used here. In the *left* upper jaw of the skull shewn in Fig. 65 C, $\underline{p}^1$, $\underline{p}^3$ and $\underline{p}^4$ are

---

[1] *Phil. Trans.*, 1887, clxxviii. and *Cat. Marsup. Brit. Mus.*

shewn in the ordinary state. Lastly, in *Pseudochirus* behind the canine there is a very small tooth, presumably p^1, and between it and p^4 a tooth of good size, presumably p^3.

Lower jaw. In the front of the lower jaw there is on each side one long incisor. Between it and the tooth corresponding to p^4 of the upper jaw there are several small or "intermediate" teeth, whose number varies greatly throughout the group. Thomas has made a provisional attempt to find homologies for these small teeth, but in view of the facts of their Variation it seems impossible to attribute individuality to them and they will therefore be here merely numbered from before backwards.

Phalanger orientalis. In this species evidence will be offered to prove the following facts:—

(1) That between $\underline{p^4}$ and $\underline{p^1}$ there may be *two* small teeth, one or both of which may perhaps represent $\underline{p^3}$ (Fig. 65).

(2) That between $\underline{p^4}$ and the small $\underline{p^3}$ there may be a large tooth (Fig. 64, C), like the p^3 of *Pseudochirus*.

(3) That $\underline{p^3}$ may be absent.

(4) That in case of absence of p^3, $\underline{p^1}$ may be near to $\underline{p^4}$ (Fig. 64, A).

(5) That between the canine and $\underline{p^4}$ there may be on one side the usual large $\underline{p^1}$, but on the other *two* teeth, evenly spaced, each of about the proportions of $\underline{p^1}$ (Fig. 64, B).

(6) That in the lower jaw the number of intermediate teeth may vary from none to five, three being the most usual number.

*372. Specimen having left side normal, one small premolar standing between $\underline{p^1}$ and $\underline{p^4}$. In the right upper jaw $\underline{p^1}$ is normal and stands at the same level as left $\underline{p^1}$; $\underline{p^4}$ is also normal in size, form and position (Fig. 64 C). In front of $\underline{p^4}$ however there is a two-rooted tooth (marked y in the figure) having somewhat the same shape as $\underline{p^4}$, but about $\frac{2}{3}$rds the size. This tooth has not the form of the milk-predecessor of $\underline{p^4}$. A small peg-like tooth (x in the figure) matching the small premolar ("$\underline{p^3}$") is also present, but is crowded out of the arcade and stands internal to the tooth y. The lower jaw has three intermediate teeth on each side. B. M., 1780, f. The form and position of the tooth y suggest a comparison with the arrangement in *Pseudochirus*, in which "$\underline{p^3}$" is in a very similar condition. In Fig. 64, D, a profile of *Pseudochirus* is shewn, the dotted lines indicating the comparison suggested. It will thus be seen that if the tooth y corresponds to p^3 of *Pseudochirus*, the tooth x then has no correspondent.

*373. Specimen (var. **breviceps,** Solomon Islands) having in right upper jaw p^1 and p^4 but no "p^3": in left upper jaw p^1 stands at a level anterior to that of right p^1, and a small peg-shaped tooth, "p^3," is present close to and almost touching p^4. (Fig. 64, A) Lower jaw, right side, two intermediate teeth, of which the posterior stands *internal* to $\overline{p^4}$; right side three intermediate teeth. B. M., 1936, f.

*374. Specimen (var. **breviceps** ♀, Duke of York I.) having in rt. upper jaw p^1 and p^4, but no "p^3": in left upper jaw there are *two* teeth of the size and shape of p^1 (Fig. 64, B), one of them

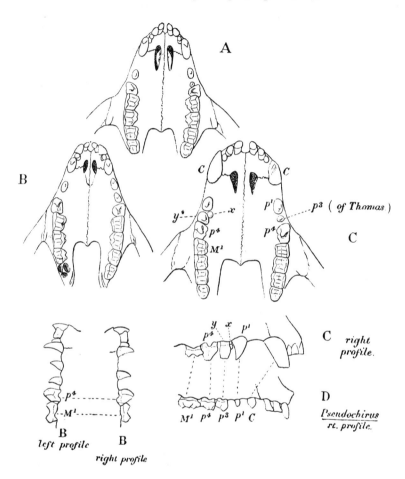

Fig. 64. Dentition of *Phalanger orientalis*.
  A. *P. orientalis*, No. 373, having no right "p^3": left p^1 in front of right p^1.
  B. *P. orientalis*, No. 374: no right "p^3"; on left, two teeth both like p^1, in symmetry approximately balancing right p^1. Below are the right and left profiles of the upper jaws of this skull.
  C. *P. orientalis*, No. 372. The left side normal, lettered on Thomas's system. Right side described in text. Below is a profile of right side.
  D. *Pseudochirus*, profile of normal upper teeth from right side enlarged to compare with C. Teeth lettered on Thomas's system.

being at a level anterior to right p^1 and the other posterior to it (see figures). On neither side is there any tooth having the size and form of "p^3." In lower jaw, right side 3 interm. teeth; left side *no* interm. tooth. B. M., 1936, *j*.

*375. Specimen having *two* small premolars on each side between p^1 and p^4. The two teeth are very small and sharply pointed (Fig. 65). In

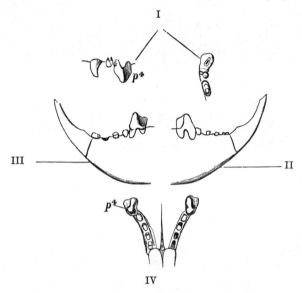

FIG. 65. Teeth of *Phalanger orientalis*. No. 375.
I. Premolars of left upper jaw, surface and side views.
II. and III. Right and left lower jaws as far as p^4.
IV. The same in surface view.
(From a drawing kindly sent by Dr JENTINK.)

the lower jaw there are on the right side five intermediate teeth between the incisor and $\overline{p^4}$, and on the left side *four* such teeth (instead of three as usual). *Leyd. Mus.*,104, JENTINK, F. A., *Notes Leyd. Mus.*, 1885, VII. p. 90.

*Statistics of the occurrence of small Premolars and lower
"Intermediate" teeth in* Phalanger orientalis *and*
Phalanger maculatus.

*376. **Phalanger orientalis.**

Statistics as to the absence of the small "p^3," and as to the number of the "intermediate" teeth, may conveniently be given together in tabular form. The species has a wide distribution and is by THOMAS divided into a larger var. *typicus*, and a smaller eastern var. *breviceps*. In the latter the small p^3 is usually absent. The Leyden specimens are not thus divided by JENTINK, and in order to include the statistics given by him (*l. c.*) the distinction into two races is not followed in the table.

When present, "p^3" generally stands at an even distance from p^1 and p^4, as in the left side of Fig. 64, C, and *not* as in the left side of Fig. 64, A. The positions of the intermediate teeth are most various, sometimes they are evenly spaced out between p^1 and p^4, but sometimes they are crowded together. The teeth in corresponding ordinal positions do not always stand at the same levels on the two sides.

Small upper premolar ("p^3")		No. of intermediate teeth in l. j.		Cases.			
right	left	right	left	Leyden (Jentink)	Other Museums	Total	
–	–	1	1		3	2	
–	–	3	0		1	1	(No. 374)
–	–	1	3		1	1	
–	–	2	3	1	1	2	
–	–	3	3	3	2	5	
+	–	3	3	2		2	
–	+	4	3	1		1	
+	+	1	2	1		1	
+	+	2	2	2	2	4	
+	+	2	3	1		1	
+	+	3	2	2		2	
+	+	3	3	44	6	50	(One of these is No. 372)
+	+	3	4	1		1	
+	+	4	3	1		1	
2	2	5	4	1		1	(No. 375)
						76	

*377. **Phalanger maculatus:** in this species the small premolar ("p^3") between upper p^1 and p^4 is generally absent, and in the lower jaw there are usually only two "intermediate" teeth. The following table shews the variations seen in 58 skulls and 7 lower jaws wanting skulls (including 43[1] Leyden skulls described by Jentink, *l. c.*).

Small upper premolar		Intermediate teeth in l. j.		Cases.			
right	left	right	left	Leyden (Jentink)	Other Museums	Total	
–	–	1	1	2	1	3	+1 lr. jaw
–	–	1	2	2		2	
–	–	2	1	2		3	+2 lr. jaws
–	–	2	2	27[1]	8	35	+4 lr. jaws
–	–	3	2		2	2	
–	–	3	3	2		2	[1] In one of these l. p^1 absent (see No. 370)
+	–	2	3	1		1	
–	+	3	3		1	1	
+	+	2	1	1		1	
+	+	2	3	1		1	
+	+	3	2	2		2	
+	+	3	3	2	1	4	
+	+	4	3	1		1	
						58	

[1] Not including the case, *Leyd. Mus.*, 153 (Jentink, *l. c.*, p. 91), in which the "small" upper premolar is stated to be absent as an abnormality. As p^3 is usually absent in the species, probably this refers to p^1.

254   MERISTIC VARIATION.   [PART I.

The above includes six skulls from Waigiu, the individual peculiarities of which are given below:

−	−	3	3		1	1	
−	+	3	3	1		1	B. M., 61. 12. 11. 18.
+	+	2	3	1		1	
+	+	3	2	1		1	
+	+	3	3		1	1	B. M., 61. 12. 11. 17.
+	+	4	3	1		1	

The great variability of these skulls from the island of Waigiu is very remarkable. The 4 Leyden specimens were described by JENTINK[1]. In one of these there was besides no left upper 2nd molar, which was entirely absent without trace, leaving a diastema between m^1 and m^3. In connexion with the variations of the dentition of *P. maculatus* in Waigiu the following singular circumstance should be mentioned. In all other localities the male *P. maculatus* alone is spotted with white, the female being without spots, but in Waigiu the females are spotted like the males[2]. This curious fact was first noticed by JENTINK (*l. c.*, p. 111).

In the other species of *Phalanger* no case of special importance met with; but since in *P. ursinus* p^1 is normally (4 skulls seen) two-rooted, it may be of interest to note that such a two-rooted condition of p^1 was seen on both sides as a variation in **P. ornatus**, B. M., 1317, *b* (2 other specimens having single-rooted p^1).

*378.   **Trichosurus vulpecula** (= *Phalangista vulpina*). The typical form of this species is Australian, while the large variety, **fuliginosa,** is peculiar to Tasmania. In the typical form no instance of absence of p^1 seen in 17 specimens examined. All possessed this tooth on each side, and though varying a good deal in size, it was in every case well-formed and functional, never being in a condition which could be called rudimentary.

Of the Tasmanian variety *fuliginosa*, 18 specimens (8 in B. M., 10 in C. M.) were examined.

In 6 p^1 was present on both sides.
1   right side only.
1   left.
2 p^1 was absent altogether. C. M., 14 *k* and *l*.

Nevertheless in every case in which this tooth is present it is a large tooth of about the size of the canines. In one case p^1 is two-rooted on each side, as (THOMAS, *Cat. Marsup.*, p. 186) in the Celebesian *Phalanger ursinus*. C. M., 14 *a*, Hobart Town, Tasmania.

Of the "intermediate" teeth in lower jaw one only is usually present, being

---

[1] The small premolar was accidentally described in the paper referred to as being between the *canine* and p^4, instead of between the anterior premolar and p^4. JENTINK, *in litt.*

[2] Compare the converse case of **Hepialus humuli** (the Ghost Moth), of which, in all other localities, the males are clear white and the females are light yellow-brown with spots; but in the Shetland Islands the males are like the females, though in varying degrees. See JENNER WEIR, *Entomologist*, 1880, p. 251, Pl.

close to the large incisor. In two cases (C. M., 15 *g* and *h*, prob. both Australian) there are two intermediate teeth, one near the incisor, the other near $\overline{p^4}$.

379. **Pseudochirus.** Of various species 29 skulls shew no numerical variation in upper series. The number of "intermediate" teeth in lower jaw is very variable, 2 on each side being the most frequent, but 1 and 3 being also common. **P. peregrinus**, Upper Hunter R., B. M., 41, 1182, has 2 intermediate teeth in left lower jaw, but on the right side one *partially double* intermediate tooth. (See also No. 371.)

Petaurus: 25 skulls shew no numerical variation in upper series. In this genus the number of small teeth in the lower jaw is remarkably constant. In addition to $\overline{p^4}$ there were 3 small teeth on each side in
380. all cases seen except two, viz.—**P. breviceps** var. **papuanus** (8 normals): right side normal; left lower jaw has 4 teeth besides $\overline{p^4}$ (Fig. 66). B. M., 77. 7. 18. 19.

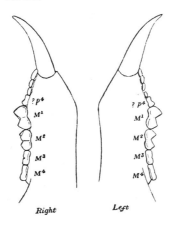

Fig. 66. *Petaurus breviceps*, No. 380. Lower jaws in profile: on right side three intermediate teeth, on left side four.

381. Another specimen has, in addition to $\overline{p^4}$, *four* small teeth in each lower jaw. There is a small diastema between the 3rd and 4th. B. M., 42. 5. 26. 1. [no skull].

382. **Dactylopsila trivirgata:** 3 specimens have upper series normal. In addition one has an extra tooth in left upper jaw between $\underline{p^1}$ and canine. This tooth somewhat resembles but is rather smaller than the canine, near and slightly internal to which it stands [?reduplicated canine]. B. M., 1197, *d*.

(3) *Premolars and molars of* Dasyuridæ *and* Didelphyidæ.

Thylacinus, 19 normals; **Sarcophilus,** 9 normals, no abnormal known to me.

Dasyurus, 37 normals (4 species).

383. **D. geoffroyi:** specimen in which p^4 in right lower jaw has its crown partially divided into two, the plane of division being at right angles to the jaw. C. M., 39, *a*.

384.   **D. viverrinus :** right upper m^4 slightly larger than the left, which is normal. C. M., 38, g.

*385.   **D. maculatus,** Tasmania, having a supernumerary molar in left upper jaw, and on both sides in the lower jaw. The fourth molar in the upper jaws is increased in size in a remarkable manner (Fig. 67, B and C).

This case requires detailed description. In Fig. 67, A, a normal right upper jaw is shewn. It belongs to a specimen considerably larger than the abnormal one, but the latter, Mr Thomas tells me, is a good deal smaller than the normal size of the species. In the normal there are two small premolars (p^1 and p^3 of Thomas), and behind these, four molars. The molars increase in size from the first to the third, which is by far the largest. Behind the third is the fourth molar, which is much smaller than the others, having the peculiar flattened form shewn in the Figure 67, A.

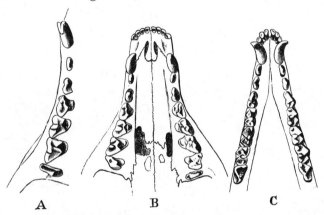

Fig. 67. A. Right upper jaw of normal *Dasyurus maculatus* (shewn as far as the canine) for comparison with the variety. (N.B. The latter is considerably smaller.)
  B. Upper jaw of *D. maculatus*, No. 385.
  C. Lower jaw of the same specimen.

On comparing the abnormal skull with a normal one it is seen that the two premolars and first three molars on each side are unchanged. Behind the third molar on the *right* side there is a single tooth; but this, instead of being a thin tooth like normal m^4, is considerably larger and the longitudinal measurement in the line of the jaw is not very much less than the transverse measurement. In the right upper jaw therefore the *number* of the teeth is unchanged.

On the *left* side, behind the third molar, there is a square tooth (m^4) of good size, about equal in bulk to half m^3, while behind this again there is another tooth, m^5, which is a thin and small tooth having nearly the form and size of normal $\underline{m^4}$. The lower series is alike on both sides, each having an extra molar behind $\overline{m^4}$ (Fig. 67, C). The two extra teeth are well formed, being as long but not quite so thick as $\overline{m^4}$. B. M., 41, 12, 2, 3.

In *Cat. Marsup. Brit. Mus.*, 1888, p. 265, *note*, THOMAS refers to this skull, and describes it as an instance of an additional molar inserted between m^3 and m^4 on the left side above and on both sides below. This view is of course based on the resemblance that the extra $\underline{m^5}$ of the left side bears to a normal $\underline{m^4}$ and on the fact that the left $\underline{m^4}$ is like no tooth normally present. In the light however of what has been seen in other cases of supernumerary molars a simpler view is possible. For in cases in which a supernumerary molar is developed behind a molar which is normally a small tooth, the latter is frequently larger than its normal size. In the present case it appears that on the right side m^4 has been thus raised from a small tooth to be a tooth of fair size, while on the left side the change has gone further, and not only is $\underline{m^4}$ promoted still more, but a supernumerary $\underline{m^5}$ is developed as well. It is interesting to note that this $\overline{m^5}$ is a small tooth, very like normal $\underline{m^4}$, and it thus may be said to be beginning at the stage which $\underline{m^4}$ generally reaches. In the lower jaw $\overline{m^5}$ is added without marked change in $\overline{m^4}$; for $\overline{m^4}$ is normally a large tooth and has, as it were, no arrears to be made up. Mr Thomas, to whom I am indebted for having first called my attention to this remarkable case, allows me to say that he is prepared to accept the view here suggested.

Phascologale. In the upper jaw normally 3 premolars, by Thomas reckoned as p^1, p^3 and p^4. Between the first and second ("p^3") there is sometimes, but not always, a small space, and in the following case a supernumerary tooth was present in this position.

386.   **Phascologale dorsalis,** (Fig. 68) having an extra premolar between the first and second in the left upper jaw: rest normal. B. M., 1868, *b.* THOMAS, O., *Phil. Trans.*, 1887, p. 447, Pl. 27, figs. 7 and 8.

In the lower jaw $\overline{p^4}$ is often small and may be absent. As Thomas has observed, the size of p^4 in the upper and lower jaws maintains a

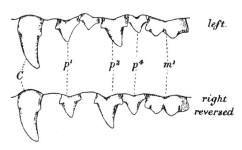

FIG. 68. *Phascologale dorsalis*, No. 386. Teeth of left upper jaw from canine to first molar; below, the teeth of the right side reversed (after THOMAS).

fairly regular correspondence. Within the limits of one species $\overline{p^4}$ may
387. shew great variation; for instance, of **Phascologale flavipes** 7 specimens were seen; in 1 $\overline{p^1}$ was absent, in 2 it was small, in 2 moderate, and in 2 it was large.

388. **Didelphys:** 79 specimens normal. One specimen alone, **D. lanigera**, Colombia, B. M., 1733, b, was abnormal, having no m^4 in either upper or lower jaws.

D. opossum (one specimen, B. M.) had right $\underline{m^4}$ larger than the left.

(4) *Molars of certain* Macropodidæ.

The following evidence relates to the genera *Bettongia*, *Potorous* and *Lagorchestes*. In these forms the molars are normally four in each jaw. As Thomas observes (*Cat. Marsup. Brit. Mus.*, p. 105, *note*), in *Bettongia* cases of fifth molar occur, but on the other hand cases of non-eruption of m^4 occur also. The variations seen in the three genera were as follows.

Bettongia penicillata: 8 specimens have $m\dfrac{4-4}{4-4}$; in 7 of them m^4 is small (in B. M., 279, j, $\underline{m^4}$ is very minute; but in B. M., 278, m, the lower $\overline{m^4}$ is large).

*389.   1 specimen ♀ has m^4 in left lower jaw only, this tooth being small. B. M., 279, a.

*390.   1 specimen has $m\dfrac{1.2.3.0.5-1.2.3.0.0}{1.2.3.4.5-1.2.3.4.5}$. In both upper jaws there is a small empty crypt behind m^3, and on right side behind this again there is a minute tubercular tooth not represented on the other side. B. M., 279, b.

B. cuniculus: 2 specimens have $m\dfrac{4-4}{4-4}$.

391.   1 specimen has no left $\overline{m^4}$. B. M., 982, c.

*392.   1 specimen has $m\dfrac{5-5}{5-5}$; in upper jaws m^5 very small in crypts, but in lower jaws they are of good size. B. M. 51. 4. 24. 7.

B. lesueri: 13 specimens have $m\dfrac{4-4}{4-4}$ (in one of them m^4 very small. B. M., 277, g).

393.   1 specimen has $m\dfrac{5-5}{4-4}$, m^5 being minute and lying in crypts. B. M., 41, 1157.

394.   **Potorous (Hypsiprymnus):** $m\dfrac{4-4}{4-4}$ in 5 specimens of *P. tridactylus* and in 2 of *P. platyops*. A single specimen of **P. gilberti** has no right upper m^4. B. M., 282, b.

395.   **Lagorchestes.** In this genus m^4 is present and is a large tooth, not materially smaller than m^3. Nevertheless it commonly falls short of the other teeth and remains partly within the jaw. This was the case in 10 skulls of *L. leporoides* and *L. conspicillatus*. In one skull of *L. leporoides*, m^4 stood at the same height as the other teeth. I see no reason to suppose that all the other skulls were young, and it seems more likely that this imperfect eruption of m^4 is characteristic.

SELACHII.

Some features characteristic of Meristic Variation are well seen in the case of the teeth of Sharks and Rays. Of these fishes there are many having little differentiation between the separate rows of teeth. In these a distinct identity cannot be attributed to the several rows, and numerical Variation is quite common. But besides these there are a few forms whose teeth are differentiated sufficiently to permit a recognition of particular rows of teeth in different specimens, and to justify the application of the term "homologous" to such rows. Nevertheless with such differentiation Meristic Variation does not cease.

In the following examples it will be seen further that in such Variation there may be not merely a simple division of single teeth but rather a recasting of the whole series, or at least of that part of it which presents the Variation, for the lines of *division* in the type may correspond with the *centres* of teeth in the variety.

These cases also exemplify the fact that variations of some kinds are often only to be detected when in some degree imperfect; for if the divisions in No. 396 for instance had taken place similarly on both sides, it would have been difficult to recognize that this was a case of Variation.

*396. **Rhinoptera jussieui** (= *javanica*): specimen in which the number and arrangement of the rows of teeth is different on the two sides, as shewn in Fig. 69, upper diagram. The disposition on the right side of the figure is normal, that on the left being unlike that of any known form. Specimen in B. M. described by SMITH WOODWARD, *Ann. and Mag. N. H.*, Ser. 6, vol. i. 1888, p. 281, fig. As Woodward points out, the rows of plates on the left side may be conceived as having arisen by division partly of the plates of the central row and partly from the lateral row, marked I. But if this be accepted as a representation of the relation of the normal to the abnormal, in the way indicated by the lettering, the plates of the row marked $0\,b$, for instance, must be supposed each to belong half to one rank and half to a lower rank. The same applies to the plates in the row $I\,b$. By whatever cause therefore the points of development of the teeth are determined, it is clear that the centres from which each of the teeth in the rows $I\,b$ and $0\,b$ was developed were not merely divided out from centres in the normal places but have undergone a rearrangement also. With change of number there is also change of pattern.

The tessellation on the abnormal side is so regular and definite that had it existed in the same form on both sides the specimen might readily have become the type of a new species.

There is indeed in the British Museum a unique pair of jaws in both of which (upper and lower) a very similar tessellation

260   MERISTIC VARIATION.   [PART I.

occurs in a nearly symmetrical way. This specimen is described as *Rhinoptera polyodon*, but it is by no means unlikely that it

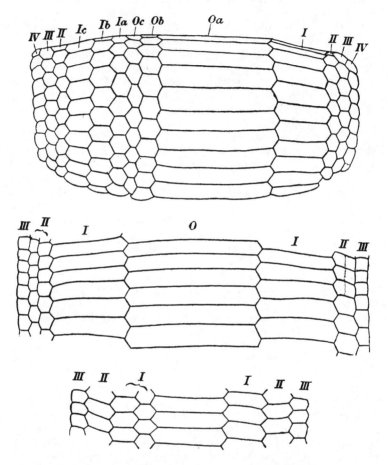

FIG. 69. Upper figure; *Rhinoptera jussieui*, No. 396, after SMITH WOODWARD, from whom the lettering is copied.
Middle figure, *Rhinoptera*, sp., No. 397.
Lower figure *Rhinoptera javanica*, No. 398, after OWEN.

is actually a Variation derived from the usual formula of *Rhinoptera*. It is figured by GÜNTHER, *Study of Fishes*, 1880, p. 346, Fig. 133.

*397.   **Rhinoptera** sp. incert.: teeth as in middle diagram, Fig. 69. On the left side *three* rows of small lateral teeth, while on the right side two of these rows are represented by one row, which in one part of the series shews an indication of division. C. S. M. (*Hunterian Specimen*).

398. **Rhinoptera javanica:** the row of teeth marked I is one side single, but on the other side is represented by two rows. Fig. 69, lower diagram. OWEN, *Odontography*, Pl. 25, Fig. 2. C. S. M. (*Hunterian specimen*).

399. **Cestracion philippi:** an upper jaw having the teeth disposed as in the figure (Fig. 70). C. S. M.

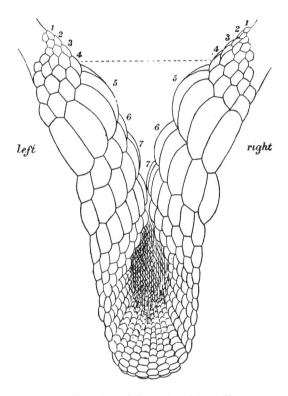

FIG. 70. Upper jaw of *Cestracion philippi*, No. 399.

On comparing the teeth of the two sides it will be seen first that the rows do not correspond individually, and secondly that they do not at all readily correspond collectively. Assuming that the rows marked 4 on each side are in correspondence (which is not by any means certain) several difficulties remain: for right 5th is larger than left 5th, but left 6th and 7th together are larger than right 6th; right 7th is about the same size as left 8th, but right 8th is larger than left 9th. The proportions in the figure were carefully copied from the specimen.

400. "**Cestracion** *sp.*" [so labelled, but probably not this genus]: lower jaw as in Fig. 71. On the right side the second row of large plates is represented by two rows, properly fitting into each other, but on the left side the plates of the inner side are completely

divided, but the division is gradually lost towards the middle of the jaw and the external plates are without trace of division.   C. S. M.

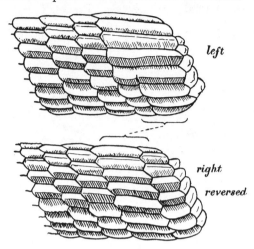

Fig. 71. The lower jaw of a Selachian, No. 400. The proximal ends shewn (enlarged). The right is reversed for comparison with the left.

RADULÆ OF A GASTEROPOD.

The following example of Meristic Variation in the teeth of a Molluscan odontophore may be taken in connexion with the subject of teeth, though the structures are of course wholly different in nature. For information on this subject I am indebted to the Rev. A. H. COOKE.

Generally speaking the number and shapes of the radular teeth are very characteristic of the different classificatory divisions. There are however certain forms in which a wide range of Variation is met with; of these the case of *Buccinum undatum* is the most conspicuous.

*401. **Buccinum undatum.** In most specimens the number of denticles on the central plate is 5—7 and on the laterals 3—4.

In 27 specimens from Hammerfest and Vardö the teeth were as follows :—

Central plate.	Lateral plates.	Cases.
5	4	8
6	4	12
7	4	2
6—8	4	1
9	4	1
6	3 & 4	1
7	3 & 4	1
8	4 & 5	1

from FRIELE, *Jahrb. deut. mal. Ges.*, VI. 1879, p. 257.

*402. The range of Variation may be still greater than this, the number of centrals being sometimes as low as 3. Fig. 72 shews the different conditions found. In it eight varieties are shewn,

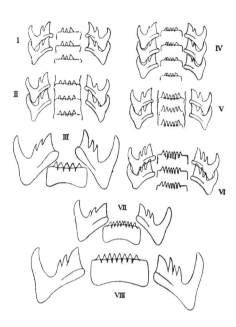

FIG. 72. Variations in odontophore of *Buccinum undatum*.
I. Three centrals (Labrador). II. Four centrals. III. Five centrals, approximately symmetrical bilaterally. IV. Five centrals, not symmetrical; the two external centrals on one side almost separate, correspond with a bifid denticle on the other side (Labrador). V. Six complete centrals (Labrador). VI. Seven centrals (Lynn). VII. Nine almost distinct centrals. VIII. Eight centrals; laterals asymmetrical (4 and 5).
I. II. IV.—VI. from photographs made and kindly lent by Mr A. H. COOKE. III. VII. VIII. after FRIELE.

I. II. IV.—VI. being taken from Mr Cooke's specimens, III. VII. and VIII. from Friele's figures.

As thus seen, in these variations considerable symmetry may be maintained. This symmetry and definiteness of the varieties in the cases with 3 and 4 centrals is especially noteworthy, inasmuch as these are abnormal forms and have presumably arisen discontinuously. As also seen in the figure, *e.g.* IV. and VI. this symmetry is not universal, and may be imperfect. The specimen shewn in VIII. is remarkable for the asymmetry of the lateral plates, which have 4 and 5 denticles respectively.

In connexion with the subject of symmetrical division interest attaches to cases like that shewn in Fig. 72, IV. in which on the outside of the central plate a pair of almost wholly separate denticles on one side correspond with a large, imperfectly divided denticle of the other side. A very similar specimen is figured by FRIELE, *Norske Nordhavs-Exp.*, VIII. Pl. v. fig. 16.

The number found in one part of the radula is usually maintained throughout the whole series, but this is not always so. A case in which the number of centrals at the anterior end of the radula was 6, and at the posterior end 8, is given by FRIELE, *Norske Nordhavs-Exp.*, 1882, VIII. p. 27, Taf. v. fig. 17.

CHAPTER X.

LINEAR SERIES—*continued.*

TEETH—RECAPITULATION.

IN this chapter I propose to speak of those matters which seem to have most consequence in the foregoing evidence as to the Variation of Teeth. Each of the following sections treats of some one such subject, specifying the cases which chiefly illustrate it. It will be understood that the sections do not stand in any logical collocation but are simply arranged consecutively. The treatment given is of course only provisional and suggestive, being intended to emphasize those points which may repay investigation.

The subjects which especially call for remark are as follows:

(1) The comparative frequency of dental Variation in different animals.
(2) Symmetry in Meristic Variation of Teeth.
(3) Division of Teeth.
(4) Duplicate Teeth.
(5) Presence and absence of Teeth standing at the ends of series (first premolars, last molars).
(6) The least size of particular Teeth.
(7) Homœotic Variation in terminal Teeth when a new member is added behind them.
(8) Reconstitution of parts of the Series.

(1) *The comparative frequency of dental Variation in different animals.*

The total number of skulls examined for the purpose of this inquiry was about 3000. From so small a number it is clearly impossible to make any definite statement as to the relative frequency of Variation in the different orders, but some indications of a general character may be legitimately drawn.

First, the statistics very clearly shew that while dental Variation is rare in some forms, it is comparatively frequent in others, but there is no indication that this frequency depends on any condition or quality common to these forms. Setting aside examples of the coming and going of certain small and variable teeth, the animals shewing the greatest frequency of extra teeth were the domestic Dogs, the Anthropoid Apes and the Phocidæ.

Attention is especially called to the fact that the variability of domestic animals is not markedly in excess of that seen in wild forms. From the hypothesis that Variation is uncontrolled save by Selection, there has sprung an expectation, now fast growing into an axiom, that wild animals are, as such, less variable than domesticated animals. This expectation is hardly borne out by the facts. It is true that, so far as the statistics go, supernumerary teeth were more common in domestic Dogs than in wild Canidæ, and though the number of Cats seen was small, the same is true in their case also as compared with wild Felidæ. But though it is true that the domestic Dog is more variable in its dentition than wild Dogs, it is not true that it is much more variable than some other wild animals, as for instance, the Anthropoid Apes or the genus *Phoca*. The doctrine that domestication induces or causes Variation is one which will not, I think, be maintained in the light of fuller evidence as to the Variation of wild animals. It has arisen as the outcome of certain theoretical views and has received support from the circumstance that so many of our domesticated animals are variable forms, and that so little heed has been paid to Variation in wild forms. To obtain any just view of the matter the case of variable domestic species should be compared with that of a species which is variable though wild. The great variability of the teeth of the large Anthropoids, appearing not merely in strictly Meristic and numerical Variation, but also in frequent abnormalities of position and arrangement, is striking both when it is compared with the rarity of variations in the teeth of other Old World Monkeys and the *comparative* rarity of great variations even in Man. If the Seals or Anthropoids had been domesticated animals it is possible that some persons would have seen in their variability a consequence of domestication.

When the evidence is looked at as a whole it appears that no generalization of this kind can be made. It suggests rather that the variability of a form is, so far as can be seen, as much a part of its specific characters as any other feature of its organization. Of such frequent Variation in single genera or species some curious instances are to be found among the facts given.

Of *Canis cancrivorus*, a S. American Fox, the majority shewed some abnormality. Of *Felis fontanieri*, an aberrant Leopard, two skulls only are known, both showing dental abnormalities. In Seals only four cases of reduplication of the first premolar were seen, and of these two were in *Cystophora cristata*. The number of cases of abnormality in the genus *Ateles* is very large. Of six specimens of *Crossarchus zebra*, two shew abnormalities. Of the very few skulls of *Myrmecobius* seen, two shew an abnormal number of incisors. Three cases of Variation were given in *Canis mesomelas*, not a very common skull in museums. On the other hand the rarity of Variation in the dentition of the Common Fox (*C. vulpes*) is noteworthy, especially when compared with the extraordinary frequency of Variation in the molars of S. American Foxes. The constant presence of the small anterior premolar in the upper jaw of Otters (*Lutra*) of most species, as compared with the great variability of the similar tooth in the Badgers (*Meles*) and in other species of Otters, may also be mentioned.

The evidence given in the last chapter should not, I think, be taken as indicating the frequency of dental Variation in Mammals generally. The orders chosen for examination were selected as being those most likely to supply examples of the different forms of dental Variation, and it is unlikely that the frequency met with in them is maintained in many other orders.

(2) *Symmetry in Meristic Variation of Teeth.*

With respect to bilateral Symmetry an examination of the evidence shews that dental Variation may be symmetrical on the two sides, but that much more frequently it is not so. The instances both of bilaterally symmetrical Variation, and of Variation confined to one side are so many that examples can be easily found in any part of the evidence.

Besides these there are a few cases in which there is a variation which is complete on one side, while on the other side the parts are in a condition which may be regarded as a less complete representation of the same variation. Such cases are *Ommatophoca rossii* No. 320, *Phoca grœnlandica* No. 324, *Dasyurus maculatus* No. 385, *Canis lupus* No. 246, *C. vetulus* No. 248, &c.

In the remarks preliminary to the evidence of dental Variation, reference was made to a peculiarity characteristic of the teeth considered as a Meristic Series of parts. As there indicated, the teeth are commonly repeated, so as to form a symmetry of images existing not only between the two halves of one jaw, but also to a greater or less extent between the upper and lower jaws. It was then mentioned that cases occur in which there is a similar Variation occurring simultaneously in the upper and lower jaws of the same individual. Such similar Variation may consist either in the presence of supernumerary teeth, or in the division of teeth, or in the absence of teeth. It should, however, be noticed that examples of Variation thus complete and perfect in both jaws are comparatively rare. Speaking generally, it certainly appears from the evidence that similar Variation, (1) on one side of both jaws, or (2) on both sides of one jaw and on one side of the other, or (3) on both sides of both jaws are all rare. Of these three the following examples may be given:—

Of (1), *Macacus rhesus* No. 190, *Ateles pentadactylus* No. 196, Esquimaux dog No. 243, *Phoca vitulina* No. 329.

Of (2), *Simia satyrus* No. 166, *Dasyurus maculatus* No. 385, *E. asinus* No. 352.

Of (3), Dog No. 257, *Bettongia cuniculus*, No. 392, *Ateles marginatus* No. 203, *Phoca barbata* No. 318, *Ommatophoca rossii* No. 320.

Of these, further examples may be seen in the evidence given regarding the anterior premolars of *Galictis barbara, Meles,* and *Herpestes*.

(3) *Division of Teeth.*

Among the cases of increase in number of teeth are many in which by the appearances presented it may be judged that two teeth in the varying skull represent one tooth in the normal, and have arisen by the division of a single tooth-germ.

Of such division in an incomplete form several examples have been given. The plane of division in these cases is usually at right angles to the line of the jaw, so that if the division were complete, the two resulting teeth would stand in the line of the arcade. Incomplete division of this kind is seen in the first premolar of *Ommatophoca rossii* No. 320, in the fourth premolar of *Phoca grœnlandica* No. 324, in the incisors of Dogs No. 219, in the canine of Dog No. 221, in the lower fourth premolar of *Dasyurus geoffroyi* No. 383. The plane of division is not however always at right angles to the jaw, but may be oblique or perhaps even parallel to it, though of the latter there is no certain case. Cases of division in a plane other than that at right angles to the jaw are seen in *C. vulpes* No. 230, *Phalanger orientalis* No. 368, *Phoca grœnlandica* No. 326 and doubtfully in a few more cases. The existence of the possibility of division in these other planes is of some consequence in considering the phenomenon of duplicate teeth standing together at the same level in relation to that of the presence of duplicate teeth in series. Beyond this also it may be anticipated that if ever it shall become possible to distinguish the forces which bring about the division of the tooth-germ, the relation of the planes of division to the axis of the Series of Repetitions will be found to be a chief element.

(4) *Duplicate Teeth.*

Teeth standing at or almost at the same level as other teeth which they nearly resemble may conveniently be spoken of as duplicate teeth, though it is unlikely that there is a real distinction of kind between such teeth and those extra teeth which stand in series. Duplicate teeth were seen in *Felis domestica* Nos. 286 and 287, *Canis mesomelas* No. 228, *Herpestes ichneumon* No. 300, [*Putorius*] *Vison horsfieldii* No. 311, *Helictis orientalis* No. 312, *Cystophora cristata* No. 322, and perhaps in some other cases. That these cases are not separable on the one hand from examples of extra teeth in series may be seen from *Herpestes gracilis* No. 299, *Cystophora cristata* No. 321 [compare with No. 322], *Brachyteles hemidactylus* No. 199 [compare with *Ateles marginatus* No. 200], *Phoca vitulina* No. 336; and that on the other hand they merge into cases of supernumerary teeth standing outside or inside the series, and whose forms do not correspond closely to those of any tooth in the series, may be seen by comparison with *Otaria ursina* No. 325, *Phoca vitulina* No. 329, *Phalanger orientalis* No. 372. Though in some cases the shapes of duplicate teeth make a near approach to the shapes of normal teeth, yet they are never exactly the same in both, and teeth whose forms approach so nearly to those of other teeth in the series as to suggest that they are duplicates of them and that they may have arisen by multiplication of the same germ, cannot be accurately distinguished from extra teeth whose forms agree with none in the normal series.

(5) *Presence and Absence of Teeth standing at the ends of Series (first premolars, last molars): the least size of particular Teeth.*

Of the cases of numerical Variation in teeth the larger number concern the presence or absence of teeth standing at the ends of Series. As was mentioned in introducing the subject of dental Variation, in many heterodont forms the teeth at the anterior end of the series of premolars and molars are small teeth, standing to the teeth behind them as the first terms of a series more or less regularly progressing in size. Not only in teeth but in the case of members standing in such a position in other series of organs, *e.g.* digits, considerable frequency of Variation is usual.

Variability at the ends of Series is manifested not only in the frequency of cases of absence of terminal members, but also in the frequency of cases of presence of an extra member in their neighbourhood. An additional tooth in this region may appear in several forms. It may be a clear duplicate, standing at the same level as the first premolar (*e.g.* Cat, No. 270). On the other hand, as seen in the Dogs (Nos. 232 and 233) there may be two teeth standing between the canine and (in the Dog) the second premolar. The various possibilities as to the homologies of the teeth may then be thus expressed. The posterior of the two small teeth may correspond with the normal first premolar, and the anterior may be an extra tooth representing the first premolar of some possible ancestor having five premolars; or, the first of the two premolars may be the normal, and the second be intercalated (see No. 224); or, both the two teeth may be the equivalent of the normal first premolar; lastly, neither of the two may be the precise equivalent of any tooth in the form with four premolars. Of these possibilities the first is that commonly supposed (HENSEL and others) to most nearly represent the truth. But the condition seen in cases where there is an extra tooth on one side only, as in the Dogs figured (Fig. 42), strongly suggests that neither of the two teeth strictly corresponds with the one of the other side. Seeing that in such cases the single tooth of the one side stands often at the level of the diastema on the other, it seems more likely that the one tooth balances or corresponds to the two of the other side, which may be supposed to have arisen by division of a single germ. On the other hand since the two anterior premolars found in such cases are not always identical in form and size, either the anterior or the posterior being commonly larger than the other, there is no strict criterion of duplicity, and it is clearly impossible to draw any sharp distinction between cases of duplicity of the first premolar and cases in which the two small premolars are related to each other as first and second. These two conditions must surely pass insensibly into each other. If the case of the teeth is compared with that of any other Linear series in which the number of members is indefinite, as for example that of buds on a stem, the impossibility of such a distinction will appear. A good illustration of this fact may often be seen in the arrangement of the thorns on the stems of briars. For large periods of the stem both the angular and linear succession of the thorns of several sizes may be exceedingly regular; but it also frequently happens that a thorn occurs with two points, and on searching, every condition may sometimes be found between such a double thorn and two thorns occurring in series, having between them the normal distinctions of form or size. Very similar phenomena may be seen in the case of the strong dermal spines of such an animal as the Spiny Shark (*Echinorhinus spinosus*). These structures are of course from an anatomical standpoint closely comparable with teeth. In them, spines obviously double, triple or quadruple, are generally to be seen scattered among the normal single spines, but between the double condition and the single condition, it is impossible to make a real distinction.

The remarks made as to the first premolars apply almost equally to the last molar. See *Phoca vitulina* No. 336, *Mycetes niger* No. 206, Man, MAGITOT, *Anom. syst. dent.*, Pl. v. figs. 4, 5 and 6, *Canis cancrivorus* Nos. 251 and 252, *Crossarchus zebra* No. 302.

(6) *The least size of particular Teeth.*

What is the least size in which a given tooth can be present in a species which sometimes has it and sometimes is without it? In other words, what is the least possible condition, the lower limit of the existence of a given tooth? This is a question which must suggest itself in an attempt to measure the magnitude or Discontinuity of numerical Variation in teeth.

The evidence collected does not actually answer this question completely for any tooth, but it shews some of the elements upon which the answer depends.

In the first place it is seen at once that the least size of a tooth is different for different teeth and for different animals.

Considered in the absence of evidence it might be supposed that any tooth could be reduced to the smallest limits which are histologically conceivable; that a few cells might take on the characters of dental tissue, and that the number of cells thus constituting a tooth might be indefinitely diminished. Indeed on the hypothesis that Variation is continuous this would be expected. Now of course there is no categorical proof that this is not true, and that teeth may not thus occur in the least conceivable size, but there is a good deal of evidence against such a view. The facts on the whole go to shew that teeth arising by Variation in particular places, at all events when standing in series in the arcade, have a more or less constant size on thus appearing. Within limits it seems also to be true that the size in which such a tooth appears has in many cases a relation to the size of the adjacent teeth and to the general curves of the series. For example in the Orang, the series of molars does not diminish in size from before backwards, and extra molars when present are, so far as I know, commonly of good size, not wholly disproportionate to the last normal molar. The same is I believe true in the case of the Ungulates. In the Dogs however the series of lower molars diminishes rapidly at the back, and the extra molars added at the posterior end of the series are of a correspondingly reduced size. As presenting some exception to this rule may be mentioned two cases in the Chimpanzee, Nos. 178 and 181 and the case of *Cebus robustus* No. 194, in each of which the extra molar is disproportionately small.

The principle here indicated is of loose application, but speaking generally it is usual for an extra tooth arising at the ends of series to be of such a size as to continue the curves of the series in a fairly regular way. It would at all events be quite unparalleled for an extra tooth arising at the end of a successively diminishing series, as the Dog's lower molars, to be *larger* than the tooth next to it, and with the exception of cases of duplicate anterior premolars (see Dogs Nos. 232 and Cat No. 268) I know no such case. In these besides, the anterior tooth is very slightly larger than its neighbour, and it should be remembered that the first premolar, though the terminal member of the series of premolars, is not actually a terminal tooth.

Examples have been given of animals which seem to be oscillating between the possession and loss of particular teeth, the first premolar of the Badgers, p^1 of some species of Otter, &c. In these cases we are not yet entitled to assume because in a given skull the tooth is absent, that it has never been formed in it, though this is by no means unlikely, but as already pointed out (p. 228), the fact of its presence or absence may still indicate a definite variation. Attention should be called to the case of *Trichosurus vulpecula*, var. *fuliginosa* No. 378, in which the first premolar is generally of good size if present, and there can be no doubt that it has never been present in those skulls from which it is absent.

Variation of unusual amplitude may be seen also in the molars of *Bettongia* Nos. 389, &c., for while on the one hand the last or fourth molar may be absent, it may on the contrary be large and may even be succeeded by a fifth molar as an extra tooth. All these conditions were seen in looking over quite a small number of specimens.

(7) *Homœotic Variation in terminal Teeth when a new member is added behind them.*

Upon the remarks made in the last Section the fact here noticed naturally follows. We have seen that there is a fairly constant relation between the size of extra teeth and that of the teeth next to which they stand, so that the new teeth are as it were, from the first, of a size and development suitable to their position. We have now to notice also that the teeth next to which they stand may also undergo a variation in correlation with the presence of a new tooth behind them.

It may be stated generally that if the tooth which is the last of a normal series is relatively a small tooth, as for example $\overline{m^3}$ or $\underline{m^2}$ in the Dog, then in cases of an addition to the series, by which this terminal tooth becomes the penultimate, it will often (though not always) be found that this penultimate tooth is larger and better developed than the corresponding ultimate tooth of a normal animal of the same size.

Of this phenomenon two striking examples (*q. v.*) have been given, *Canis azaræ* No. 249 and *Dasyurus maculatus* No. 385. Besides these are several others of a less extreme kind *e.g. Otocyon megalotis* No. 256, Mastiff No. 259, Dog No. 260. The same was also seen in the molars of *Bettongia*.

This phenomenon, of the enlargement of the terminal member of a series when it becomes the penultimate, is not by any means confined to teeth; for the same is true in the case of ribs, digits, &c., and it is perhaps a regular property of the Variation of Meristic Series so graduated that the terminal member is comparatively small. This fact will be found of great importance in any attempt to realize the physical process of the formation of Meristic Series, and it may be remarked that such a fact brings out the truth that the members of the Series are bound together into one common whole, that the addition of a member to the series may be correlated with a change in the other members so that the general configuration of the whole series may be preserved. In this case the new member of the series seems, as it were, to have been reckoned for in the original constitution of the series.

(8) *Reconstitution of parts of the Series.*

Lastly there are a few cases, rare no doubt in higher forms but not very uncommon for example in the Sharks and Rays (see pp. 259, &c.), in which the members of the series seem to have been so far remodelled that the supposed individuality of the members is superseded. In the Selachians several such cases were given, but in Mammals the most manifest examples were seen in the Phalangers and *Ateles marginatus* No. 200 (*q.v.*). In the latter specimen there were four premolars on each side in the upper jaw, and there was nothing to indicate that any one of them was supernumerary rather than any other. In such a case I submit that the four premolars must be regarded as collectively equivalent to the three premolars of the normal. The epithelium which normally gives rise to three tooth-germs has here given rise to four, and I believe it is as impossible to analyze the four teeth and to apportion them out among the three teeth as it would be to homologize the sides of a triangle with the sides of a square of the same peripheral measurement.

Such a case at once suggests this question: if the four premolars of this varying *Ateles* cannot be analyzed into correspondence with the three premolars of the typical *Ateles*, can the three premolars of this type be made to correspond individually with the two premolars of Old World Primates?

In the case of *Rhinoptera* No. 396, for the reason given in describing the specimen, there is plainly no correspondence between the rows of plates of the variety and those of the type, and the rows are, in fact, not individual, but divisible.

Though cases so remarkable as that of *Ateles marginatus* are rare, there are many examples of supernumerary teeth, in the region of the anterior premolars of the Dog or Cat for instance, which cannot be clearly removed from this category. As indicated in the fourth section of this Chapter, it is impossible to distinguish cases of division of particular teeth from cases of the formation of a new number of teeth in the series. Finally, on the analogy of what may be seen in the case of Meristic Series having a wholly indefinite number of members, it is·likely that the attempt thus to attribute individuality to members of series having normally a definite number of members should not be made.

CHAPTER XI.

LINEAR SERIES—*continued*.

MISCELLANEOUS EXAMPLES.

IN this chapter are given some miscellaneous examples. Most of them illustrate the Meristic Variation of parts standing in bilateral symmetry on either side of a median line.

Here also are included certain cases of Variation concerning the series of apertures in the shell of *Haliotis*, though probably they are of a wholly different nature.

SCALES.

Among animals possessing an exoskeleton composed of scales, the number of the scales or of the rows of scales found in particular regions is usually more or less definite. So constant are these numbers in their range of Variation that in both Reptiles and Fishes either actual numbers or certain ranges of numbers are made use of for purposes of classification.

Considerable Variation in these numbers is nevertheless well known, and many instances are given in works dealing with Reptiles or Fishes. The following cases are given as illustrations of some of the larger changes which may occur.

403*. **Clupea pilchardus** (the common Pilchard). Among the Pilchards brought to the curing factories at Mevagissey, Cornwall, specimens have from time to time been found by Mr Mathias Dunn, the director, having the scales of one side very many more in number than those of the other side. Two specimens[1] shewing this abnormality were given to me by Mr Dunn in 1889. Owing to the fact that the fresh Pilchards are shovelled wholesale into the brine-vats, it is not until the fish are picked over for packing after the salting process that any individual peculiarities are

---

[1] These specimens are now in the Museum of the Royal College of Surgeons. An account of them was published in *P. Z. S.*, 1890, p. 586. Figures of the same variation were given by DAY, F., *P. Z. S.*, 1887, p. 129, Pl. xv.

noticed. This was the case with the present specimens, which were given to me as they came salted from the presses. Nevertheless when received they were in fairly good condition.

The first specimen measured 8 in. to the base of the caudal fin. The head and opercula were normal on both sides. *The number of scales along the lateral line or the left side is* 32 *and the number on the right side is* 56 *or* 57. On the left side the scales have the size usually seen in Pilchards of this length, and on the right side for a distance of about an inch behind the operculum the scales are not much smaller than those of a normal Pilchard, but behind this point each scale is of about half the normal size.

The second specimen has a very similar length. It differs from the first in having the reduplication on the left side instead of on the right. Furthermore the scales are normal in size as far as the level of the anterior end of the dorsal fin, behind which place they are of about half the normal size. The transition in this specimen is quite abrupt. The scales had been somewhat rubbed, and the counting could not be very accurately made, but the total number along the left lateral line was approximately 48.

As these abnormal individuals were taken with the shoal there can be little doubt that they were swimming with it.

In *P. Z. S.*, 1887, p. 129, Pl. xv. Day described a specimen, also obtained from Mr Dunn, exhibiting characters similar to those above described. The number of scales along the lateral line is given as 32 on the right side and 51 on the left. In the figure no transition from normal to abnormal scales is shewn, but there is a general appearance of uniformity.

Mr Day regarded this specimen as a hybrid between the Herring (*C. harengus*) and the Pilchard, and before adopting the view that the case is one of Variation this suggestion must be discussed. This view was chiefly based on the presence of the small scales on one side, but it is added that the ridges on the operculum, which are characteristic of the Pilchard as compared with the Herring, were better marked on the right side than on the left, though they are stated to have been very distinct on the left side also. In the specimen described, the gill-rakers were 61 in the "lower branch of the outer branchial arch" (viz. the bar consisting of the first hypobranchial and ceratobranchial), and it is mentioned that this number is intermediate between that found in a Pilchard (71) and in a Herring (48); but whether this intermediate number was found on the side shewing the "Herring" characters, or on the other, or on both, is not stated. These gill-rakers are also said to have been intermediate in length between those of a Pilchard and those of a Herring. From these points of structure Mr Day concludes that the specimen was a hybrid between the Herring and the Pilchard.

As against the theory that these specimens are hybrids it may be remarked that no direct evidence is adduced which points to hybrid parentage. The suggestion is derived from (1) the condition of the scales, (2) the number of the gill-rakers, (3) the alleged difference in the opercula of the two sides. In view of the first point, viz. that the number of the scales on one side is intermediate between that of the Pilchard and that of the Herring, it seemed desirable to know whether the resemblance extended to the minute structure of the scales or was restricted to their number only. On comparing microscopically the scales of the Pilchard and the Herring, I find that those of the Herring bear concentric lines which are almost always smooth and without serrations, while those of the Pilchard are marked with lines which are waved into very characteristic crenelated serrations. On comparing the scales which are repeated, it was found that they also shew these characteristic serrations and that in pattern they differ in nowise from the scales of the Pilchard. This evidence appears to tell very strongly against the theory that the small scales are derived from a Herring parent.

The evidence from the gill-rakers seems to be also unreliable. In a normal Pilchard Mr Day found 71 on the hypo- and cerato-branchials of the first gill-bar, and in a specimen examined by me 72 were present and in normal Herrings 48. But in my two specimens shewing the repeated scales there were present, on the normal sides 79 and 67 respectively, and on the abnormal sides 78 in the one fish and 67 in the other. In size and shape the gill-rakers were like those of the Pilchard, being smooth, and unlike those of the Herring, which bear well-marked teeth.

As it is stated that the serrations characteristic of the operculum of the Pilchard were very distinct on the abnormal side, it is impossible to lay much stress on the circumstance that they were less distinct than those of the other side.

In addition to the considerations given above, there are several à *priori* objections to the hypothesis of the hybrid origin of these forms; as, for example, that unilateral division of parental characters is certainly not a common phenomenon in hybrids, if it occurs at all, and so on. But since the evidence advanced for the theory of hybrid parentage is already open to criticism, it is perhaps unnecessary to discuss these further difficulties.

On the whole, therefore, it seems simpler to look on these abnormalities as instances of the phenomenon of Meristic Variation[1].

In Ophidia the number of scales occurring in different parts of the body is constant in some genera and species, and variable in others. Variation in the number of *rows* of scales on the body may be specially referred to as an instance of a change in number occurring at right angles to that just described. The number of such rows in *Tropidonotus*, for example, is generally 19, but Mr

404. BOULENGER informs me that the Swiss **Tropidonotus viperinus** has either 21 or else 23 rows.

405. **Tropidonotus natrix** is remarkably constant in the possession of 19 rows of body scales. A specimen taken in Switzerland

[1] Compare with an interesting series of cases in **Gasterosteus** (Stickleback). BOULENGER, G. A., *Ann. and Mag. N. H.*, 1893, S. 6, XI. p. 228, see also *Zool.*, 1864, p. 9145; SAUVAGE, *Nouv. Arch. du Mus.*, 1874; DAY, *Journ. Linn. Soc.*, XIII. 1878, p. 110; &c.

is described by STUDER, *Mitth. natur. Ges. Bern*, 1869, p. 24, as having 20 rows. This specimen was unusually dark in colour. [The presence of an even number of rows is in itself remarkable, but it is not stated whether this total was reached by duplicity in the median dorsal row or by inequality on the two sides.]

406. A specimen of Snake from Morocco closely resembled *Macroprotodon mauritanicus* Guichenot (= *Lycognathus cucullatus* Dum. Bibr.), but differed from it in having 23 rows of body-scales instead of 19, being 4 rows in excess of the normal number. PETERS, W., *Sitzb. Ges. naturf. Fr. Berlin*, 1882, p. 27.

For particulars as to the range of variation in these numbers in different species, see numerous examples given by BOULENGER, G. A., *Fauna of Brit. India : Reptilia and Batrachia*, 1890.

KIDNEYS; RENAL ARTERIES; URETERS.

Meristic Variation in these organs is well known and the principal forms found are described in most text-books of anatomy. Some information as to these is given below. The examples are all from the human subject.

407. *Kidneys.* Male having three kidneys. The left kidney was normal in shape, position and consistency but was abnormally large. The right kidney was placed opposite to it and weighed only half as much as the left. From it a ureter with a small lumen arose and passed in a normal course so far as the division of the aorta. At this point its course lay along the surface of the third kidney. This third kidney lay over the whole right iliac artery, a portion of the right crural artery for the space of 9 lines, the right crural vein and the psoas major muscle. It was larger than the upper right kidney and had the form of an oval with its ends cut off. The anterior and posterior surfaces were convex. The anterior surface was grooved for the passage of the ureter mentioned above, which received the ureter of the second kidney and passed normally into the bladder. The man was a sailor and died of enteritis at the age of 39. THIELMANN, C. H., *Müller's Arch. f. Anat. u. Phys.*, 1835, p. 511.

408. *Renal Arteries.* The number of the renal arteries in Man is liable to great variation. In specimens in which the kidneys are normal in position the arteries may be (*a*) diminished or (*b*) increased in number. The latter is much more common.

Multiple renal arteries may be threefold. (*a*) Most commonly the additional branches spring from the aorta, (*b*) they may come from other sources; or (*c*) there may be a co-existence of additional vessels from both sources.

Of the first class, there have been described cases of

$$\left.\begin{array}{l}\text{one,}\\\text{two,}\\\text{or}\\\text{three}\end{array}\right\} \text{right aortic renals associated with} \left\{\begin{array}{l}\text{one,}\\\text{two,}\\\text{three}\\\text{or}\\\text{four}\end{array}\right\} \text{left aortic renals.}$$

In the commonest form, next to the normal condition of one on each side, there are two on the right side and one on the left. In the second commonest condition there are two on the left and one on the right; but among the forms with larger numbers, the greatest number is more frequently seen on the left than on the right side. In all these cases one vessel arises in the position of the normal renal; a second commonly springs from the aorta much lower down, generally on the level of, or below the inferior mesenteric; the third when present, is at a very short distance above the normal renal, very close to the supra-renal and on the level of the superior mesenteric. Cases of five on the right are described by OTTO and MECKEL, and other multiple forms are recorded by the older anatomists. MACALISTER, A., *Proc. Roy. Irish Ac.*, 1883, p. 624.

409.   Three renal arteries on each side, symmetrically placed (Fig. 73). In this case the posterior ends of the kidneys were united

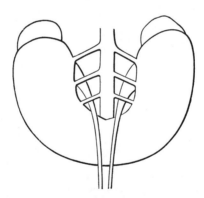

FIG. 73. Case of three renal arteries on each side combined with "horse-shoe kidney" (Man). (From *Guy's Hosp. Rep.*).

across the middle line in the condition known as "horse-shoe kidney" [see evidence as to Bilateral Series]. *Guy's Hosp. Rep.*, 1883, p. 48, fig.

410.   *Ureters.* Male. *Four* ureters emerging from the hilum of each kidney. After proceeding about four inches they became united, forming a pelvis from which sprang the proper ureter. The hilum of the kidney was found to be occupied by a quantity of

fat and connective tissue, imbedded in which the ureters could be traced to the infundibula, communicating with the calices and pyramids: thus there was no pelvis within the hilum, but the calices united to form infundibula of which these ureters seemed to be the continuation, and they became united in a pelvis some distance removed from the kidney. There were other signs of abnormal urino-genital development and the author believes that it is almost certain that the abnormality described was congenital and not a sequel of disease. RICHMOND, W. S., *Jour. Anat. Phys.*, XIX. p. 120.

411. *Two* ureters from one kidney are frequent. For an example, see *Guy's Hosp. Rep.*, 1883, p. 48.

TENTACLES AND EYES OF MOLLUSCA.

412. **Subemarginula:** specimen having a supernumerary eye on *each* eye-stalk (Fig. 74, II.). Author remarks that supernumerary eyes are common in forms having eyes borne on tentacles, but are rare in forms in which the tentacle is reduced as it is in *Subemarginula*. FISCHER, P., *Jour. de Conch.*, S. 2, I. p. 330, Pl. XI. fig. 4.

413. **Patella vulgata:** tentacle and eye repeated on left side (Fig. 74, I.). Right side normal. Supernumerary eye and tentacle of normal size. *Ibid.*, S. 3, IV. p. 89, Pl. VIII. fig. 8.

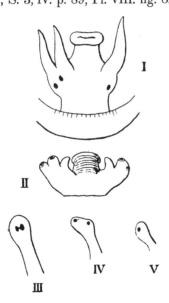

FIG. 74. Repetitions of eyes and tentacles in Molluscs. (After Fischer and Moquin-Tandon.)
I. *Patella vulgata*, No. 413. II. *Subemarginula*, No. 412. III. *Helix kermovani*, No. 416. IV. *Clausilia bidens*, abnormal, No. 417; V. normal of the same.

414. **Triopa clavigera** (a Nudibranch): adult of the usual size, having the lamellar rhinophore of the right side formed of three branches, of which the two anterior were lamellar, borne on a common peduncle, and the posterior was simple, of regular shape and probably representing the normal rhinophore of the right side. The rhinophore of the left side was normal. *Ibid.*, S. 3, XXVIII. p. 131.

415. **Physa acuta:** right tentacle bifid, left normal. MOQUIN-TANDON, *Hist. nat. des Moll. terr. et fluv. de France*, I. p. 322, Pl. XXXII. fig. 15.

416. **Helix kermorvani:** a second eye present, close to, but separate from the normal eye (Fig. 74, III.) on the left tentacle. *Ibid.*, Pl. XI. fig. 10.

417. **Clausilia bidens:** supernumerary eye on the right tentacle as shewn in Fig. 74, IV. *Ibid.*, Pl. XXIII. fig. 24.

418. **Littorina:** supernumerary eye on one tentacle. PELSENEER, *Ann. Soc. belge de microscopie*, XVI., 1891.

In examining large numbers of *Pecten* of several species, Mr BRINDLEY occasionally found one of the eyes imperfectly divided into two, the division being at right angles to the mantle-edge.

EYES OF INSECTS.[1]

The following are examples of supernumerary eyes in Insects. They are mentioned as examples of the development of tissues of the same nature as those of the normal eye in abnormal situations. All the cases known to me occur in Coleoptera.

419. **Toxotus** (= **Pachyta**) **4 – maculatus**: a normal female. On the vertex of the margin of the right eye and abutting against it is a small third eye. This third eye is round-oblong in shape. It is separated from the large eye only by the outermost margin of the eye, and though it is more convex than the latter there is nevertheless a considerable depression between the upper surfaces of the two eyes. This supernumerary eye is of a brighter colour than the normal eye, being brownish-yellow, while the latter is of a pitchy black. It is facetted in the same way as the normal eye is. LETZNER, K., *Jahresb. d. Schles. Gesell. für vaterl. Cultur.*, 1881, p. 355.

420. **Calathus fuscus:** having a third eye. On the left side of the vertex was placed a supernumerary eye. This structure was smaller and less projecting than the normal eye and was separated from it by the usual groove. It did not appear to be a part of the normal eye which had separated from it, for the normal eyes of the left and right sides were exactly alike. The integument of the head was slightly wrinkled around the supernumerary eye. DE LA BRULERIE, P., *Ann. de la Soc. Ent. de France*, S. 5, V., 1875, p. 426, *note*.

421. **Vesperus luridus** ♀ : head abnormal and bearing a third

---

[1] For cases of eyes compounded in the middle line (Bees), see evidence as to Bilateral Series.

facetted eye. The consistency of the chitinous covering of the head, its sculpture and hairs, colour, &c. are all normal and of the usual structure. The left side of the head however is rather less developed than the right, and the left eye seems to be smaller and somewhat less convex, but there is no special deformity or alteration in the facetting.

At the left side of the head arises an irregular chitinous *loop* of unequal thickness and having a diameter of about 2·5 mm. This loop is attached to the substance of the head before and behind and these two attachments are distant from each other about 1 mm. The height of this loop from the surface of the head is about 1 mm. in the highest part. Upon the upper surface of the loop is a small, irregularly rounded *eye*. The diameter of this eye is about 2·5 mm. and its convexity is considerable. It is facetted, but its facetting is not quite regular and is finer and slighter than that of the normal eyes. VON KIESENWETTER, *Berl. Ent. Ztschr.*, 1873, XVII. p. 435, *Plate*.

[A case is recorded by REITTER (*Wiener Ent. Ztg.*, IV., 1885, p. 276) of a *Rhyttirhinus deformis*, having a "complete and fully formed facetted eye placed on the left side of the thorax." Upon the request of Dr Sharp, this specimen was most kindly forwarded by Dr Reitter for our examination, when it was found that upon the application of a drop of water, the supposed abnormal eye came off. The eye appeared to be that of a fly, and had no doubt become accidentally attached to the beetle either in the collecting-box or before its capture.]

WINGS OF INSECTS.

Supernumerary parts having the structure of wings have been occasionally recorded in Lepidoptera, but their occurrence is exceedingly rare. In a subsequent chapter detailed evidence will be given respecting supernumerary legs and other of the jointed appendages of Insects and it will be shewn that in very many and perhaps all of these cases the supernumerary parts constitute a Secondary Symmetry within themselves (see p. 90). Extra wings however are of a different nature altogether, and there is so far as I am aware no indication that any of their parts are disposed as a Secondary Symmetry. In other words, an extra wing if on the left side is a left wing, and if on the right side a right wing.

In some cases the extra wing is a close copy of a normal structure, in others it seems to be more or less deformed. No genuine case of an extra wing present on both sides of the body is known to me.

From the fact that no specimen of supernumerary wing has ever been properly dissected, it is not possible to make any confident statement as to the attachments or morphology of such parts. (See also No. 78.)

The cases of *S. carpini*, No. 422, and of *Bombyx quercus*, No. 429, nevertheless suggest that Variation in number of wings is of the same nature as that seen in teeth, digits, or other parts standing in a Meristic Series. In the specimen of *S. carpini* it is especially noticeable that on the side having three wings, both the wings formed as secondaries were smaller than the secondary of the normal side; but in other cases, *G. rhamni* (No. 427) for instance, this was not the case, and the wing standing next to the extra wing was normal. Both these conditions are frequently found in cases of the occurrence of supernumerary parts in series: for two members of a varying series may clearly correspond jointly with a single member of the normal series, or on the contrary a new member may stand adjacent to members in all respects normal as in *G. rhamni* (No. 427.)

*422.   **Saturnia carpini** ♀, having a supernumerary hind wing. The specimen is rather a small female. The right wings and the left anterior wing are normal, but in the place of the left posterior wing, there are two rather small but otherwise nearly normal posterior wings. Of these the anterior is rather the larger and to some extent overlaps the posterior. The costal border of the posterior wing is folded over a little so that its width cannot be exactly measured.

	Greatest length.	Greatest width.
Right hind-wing normal	22·5 mm.	19 mm.
First left hind-wing	20·5 „	14 „
Second left hind-wing	15·5 „	11 „ about.

From the fact that the bases of these two wings are greatly overgrown with hair, it is difficult to distinguish their exact points of origin from the body, but so far as may be seen, the second arises immediately behind and on a level with the first. The neuration of each of the two small wings is identical with that of a normal hind-wing. The scaling is perfect on both surfaces of both wings, but is perhaps a little more sparse on the anterior of the two abnormal ones. In colour the anterior abnormal wing is rather light, but the posterior one is identical with that of the other side. The markings on each of the wings are normal, but are on a reduced scale in proportion to the size of the wings. This is especially remarkable in the case of the ocelli, which are both of a size greatly less than that of the ocellus of the normal hind wing of the right side.

The two wings were in every respect true left hind-wings and were in no way complementary to each other. [Specimen in collection of and kindly lent by Dr MASON.]

423.   **Bombyx rubi** ♀: 5th wing on left side. The additional wing was placed behind the left posterior wing. It was of normal structure as regards scaling and coloration. Its length was that of the hind-wing but in breadth it did not exceed 6 mm. The

insertion of this wing into the body was immediately above that of the normal hind-wing. The extra wing bore 4 nervures, of which 3 reached to the margin but one was shorter. The proper hind-wing of the same side was rather narrower than that of the other side and was not so thickly covered with scales, but its neuration was complete and normal. SPEYER, A., *Stettiner Ent. Ztg.*, 1888, XLIX. p. 206.

424.   **Samia cecropia** ♂, having a fifth aborted wing. Bred in captivity: ordinary size, expanding about $5\frac{1}{2}$ inches: a smoky variety in which red portion of transverse bands on wings is much narrowed. Right primary and both secondaries normal in shape and marking. Left primary in length from base to apex exactly the same as the right, but in width from inner angle across to the costa is $\frac{3}{16}$ of an inch less; the markings are the same, but condensed into the narrower space. Neuration normal in all wings. Left primary also somewhat narrower at base, where it joins the body. The inner margin is in exact line with its fellow; hence the costal line of the left primary is somewhat posterior to that of the right primary. The supernumerary wing emerges from the side of the collar and runs parallel to the normal left primary. It consists mainly of the costal and subcostal nervures, a small part of the median nervure and a strip of wing about $\frac{1}{4}$ inch wide which was much curled in drying. The supernumerary wing is in no way connected with the normal one.

[The author regards this supernumerary wing as a repetition of the anterior part of the left primary wing.] STRECKER, H., *Proc. Ac. Sci. Philad.*, 1885, p. 26.

425.   **Limenitis populi,** having four normal wings and a fifth wing behind the left posterior one. This supernumerary wing was 20 mm. long and 9 mm. wide. It slightly overlapped the left secondary and was attached to it for a length of 12 mm., but its outer end was free. It is described as exactly resembling the part of the secondary which bears the three anterior nervures, and it is stated that both surfaces were normal as regards scales and colouration. RÖBER, J., *Correspondenzbl. d. ent. Ver. "Isis" z. Dresden*, 1884, I. p. 31.

426.   **Vanessa urticæ,** having an additional hind-wing on the right side. This structure is inserted into the thorax *dorsal to and between* the two normal wings. It is shorter and of about $\frac{1}{3}$ the width of the normal hind-wing. In colouring it is a close copy of the anterior third of the hind-wing. WESTWOOD, *Trans. Ent. Soc.*, 1879, pp. 220 and 221, *Plate*. [Now in Brit. Mus.]

427.   **Gonepteryx rhamni** with additional imperfectly developed hind-wing on the right side. In this case the normal right hind-wing is only about two-thirds of its normal size. It overlies the additional hind-wing. The latter is coloured like the normal wing and bears an orange spot. From the neuration of the two wings Westwood considered that the supplemental wing contained missing parts of the normal wing.

Only two legs existed on the side of the abnormal wing, but for fear of injury the specimen was not sufficiently examined to shew whether the missing leg had been broken off or whether the extra wing was in its place. WESTWOOD, *ibid.*, p. 220.

<small>A specimen of *G. rhamni* having five wings was caught at Brandon, Norfolk, in Aug. 1873 by Mr J. Woodgate, and exhibited to the Ent. Soc. by Prof. Meldola, *Proc. Ent. Soc.*, 1877, p. xxvi. A similar specimen of this species was bought at Stevens's auction-rooms and exhibited to Linn. Soc. by Prof. C. Stewart, in April, 1891. This specimen is now in Mus. Coll. Surg. Whether it is the same as that taken by Mr Woodgate, or that described by Westwood, or not, I cannot say, but possibly the references are all to one individual.</small>

428. **Lycæna icarus** ♂. A coloured figure is given of a specimen of this form with 5 wings from Taurus, Asia Minor. [No further description is given. The figure is not very clear. It shews however that all the wings are normal except the right anterior. This wing is represented by two wings, which together are about a third wider than the normal wing. The costal portion of the foremost of these wings appears to be nearly normal in neuration, and the posterior part of the hindmost seems to be also normal. The two taken together shew several supernumerary nervures as compared with the normal wing, but the details are not shewn with sufficient clearness to justify a more precise statement.] HONRATH, E. G., *Berl. Ent. Ztschr.*, XXXII. 1888, p. 498, *Taf.* VII. fig. 9.

429. **Bombyx quercus** ♀ : specimen having 5 wings figured in colour by HONRATH, with statement that the left anterior wing shews a double structure. [No further description given. The figure shews the left anterior wing represented by two wings. Of these the *posterior* appears to represent a nearly complete anterior wing on a reduced scale. It bears the white ocellar mark of the anterior wing. The pale-yellow submarginal band is curved inwards over the ocellus upon the costal border as in a normal wing and thus shews that the foremost wing is not merely the separated costal part of this wing. The foremost wing is anomalous. Its central half is rather darker in colour than that of the normal wing and its peripheral half is pale in colour, deepening towards the margin. It bears no ocellus. The neurations cannot be made out from the figure with precision but the two wings together contain many more nervures than the normal anterior wing. The legs are not described.] HONRATH, E. G., *ibid.*, fig. 10.

430. <small>**Zygæna minos**, having a fifth wing on the left side, inserted above and between the normal wings. The neuration of this wing is peculiar. The colouring of the supernumerary wing was that of the anterior wing. [Dr Rogenhofer kindly informs me that the legs were normal.] ROGENHOFER, A., *Sitz.-Ber. d. zool.-bot. Ges. Wien*, 1883, XXXII. p. 34, fig.

In the same place the following instances of five-winged *Lepidoptera* are given :</small>

431. <small>**Orthosia lævis** with an additional posterior wing on the left side, in the Museum of Pesth. TREITSCHKE, Bd. VI. Abth. II. p. 407.</small>

432. <small>**Pygæra anastomosis** with a wing-like appendage to the left anterior wing in the collection of OCHSENHEIMER in Pesth.</small>

433. <small>**Nænia typica** with an additional posterior wing in the collection of NEUSTADT at Breslau.</small>

434.   **Crateronyx dumi** with five wings in the collection of WISKOTT in Breslau.

435.   **Penthina salicella**: left fore-wing about ¼ wider than the normal right fore-wing. The apical border was markedly emarginated, giving it a bilobed appearance. The nervures were as in the normal wing, except that the cells between the branches of the subcostal nervure were enlarged.   ROGENHOFER, *ibid.* [I am indebted to Dr Rogenhofer for a sketch of this specimen.]

[**Palloptera ustulata** (Diptera): specimen having a large upright scale on the thorax. This abnormal structure is like a third wing in appearance, and is fixed on the thorax, passing from the head, backwards between the wings. Its upper border is circular, and in all respects it resembles the upper wing-scale of one of the Calypterous *Muscidæ*. GERCKE, G., *Wiener Ent. Ztg.*, 1886, v. p. 168.]

HORNS OF SHEEP, GOATS AND DEER.

436.   **Sheep.** Repetition of the horns in sheep is well known. The best account is that of H. VON NATHUSIUS[1] of which the following is chiefly an abstract.

Commonly there is a pair of extra horns placed *externally* to the usual pair, but there may be three pairs in all, and even higher numbers are recorded, though Nathusius had seen no such case. The numbers on the two sides may be different, two on one side and one on the other, and three on one side and two on the other being sometimes met with.

It is noticeable that in all cases the horns stand in a *transverse* series, and not in a longitudinal series as they do in the Four-horned Antelope (*Tetraceros quadricornis*). The bases of the horn-cores are generally in contact, standing one outside the other at the same transverse level on the skull. Nathusius observed that in development the outgrowth for the horns of one side is at first single, but afterwards divides into two or more points, but he surmises that the division may appear earlier in other cases.

The external horns are generally smaller than the internal ones, but this is not universal. In some cases of two pairs of horns a small fifth horn is placed between the external and internal horns of one side.

In another form of double horn the *horn-core* of one side or other may be a double structure, both cores being enclosed in a single horn, which on being separated has a double-barrelled appearance.

Several examples of permanently four-horned breeds occur in various localities, being described as common in Cyprus and notably in Iceland and other northern islands. YOUATT (p. 169) stated that there were two breeds of sheep in Iceland, the one small and the other large, and that the greater part of both breeds

---

[1] H. VON NATHUSIUS, *Vortr. üb. Viehzucht u. Rassenkenntniss*, Th. II., *Die Schafzucht*, 1880, p. 177, fig. 47.

had more than two horns, some having eight. I am informed however by Mr E. H. Acton, who has spent some time in the country, that many-horned sheep are by no means common in Iceland at the present day. In Kishtwar (district of S.E. Kashmir) a breed of 4-horned sheep is carefully preserved, in which the horns are as a rule very symmetrical, somewhat resembling No. 438[1].

Nathusius states that a four-horned ram does not always beget four-horned offspring even when the ewe has the same character, and the variation between father and son in respect of horns is frequently considerable.

<small>The best figures of many-horned sheep are those given by BUFFON, *Hist. nat.*, Vol. XI. Pls. 31 and 32 (3-horned and 4-horned); YOUATT, *The Sheep*, pp. 141 and 171, copied from BUFFON. Numerous other figures are referred to by Nathusius, but few of them are satisfactory.</small>

437. **Goat.** A family of goats on an isolated farm near Bozen had 4 horns, which had been inherited for many generations. In most cases the two ordinary horns were typical in shape and direction; and in addition to these there were two lateral ones, which were laterally curved, being sickle-shaped and bent into a semicircle. GREDLER, V., *Korrespondenzbl. d. zool. min. Ver. Regensburg*, 1869, XXIII. p. 35.

*438. **Rupicapra tragus** (Chamois): skull bearing two well-formed and symmetrical extra horns. The cores of these horns were a little outside and posterior to the normal pair. ALSTON, E. R., *P. Z. S.*, 1879, p. 802.

439. **Capreolus caprea** (Roebuck): specimens having a supernumerary beam are probably not very rare, and a number of such antlers were shewn among the hunting-trophies exhibited by H. H. the Duke of Saxe-Coburg-Gotha, and H. S. H. the Prince of Waldeck-Pyrmont at the German Exhibition held in London in 1891. The normal antler of the roebuck has a single beam rising vertically, then bifurcating, the posterior branch again dividing. In the abnormal specimens from the single burr of one side arose a supernumerary beam in addition to the normal one. In one specimen, in which the supernumerary beam was nearly as long as the normal one, the latter bifurcated as usual but was rather more slender than that of the other side (Fig. 75 I.). In another case (Fig. 75 II.), from the left burr, which was much enlarged, arose (1) an innermost beam, in thickness and texture resembling that of the normal right horn, though it was much shorter and bore no tine; (2) an external beam at once dividing into two almost equivalent branches having about the same length as the innermost beam. In such a case I know no criterion by which one of the three beams can be certified to be the normal to the exclusion of the others. As in the sheep and goats, the several horns resulting from subdivision seem to be generally in or nearly in the same transverse plane.

---

[1] GODWIN-AUSTEN, H. H., *P. Z. S.*, 1879, p. 802.

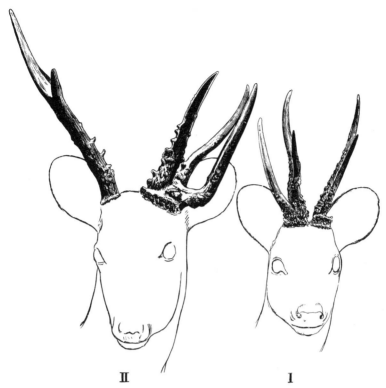

Fig. 75. Abnormal horns of Roebuck (*Capreolus caprea*), No. 438. (When seen by me the horns were fixed upon heads modelled in plaster.)

Perforations of shell of *Haliotis*.

440. **Haliotis gigantea** (Japan) having two rows of perforations in the shell. In addition to the ordinary row of perforations, of which 12 were present in this specimen, there was a series of 8 additional perforations which began within an inch of the apex. Of the normal series the last four remained open, but all the perforations in the abnormal row were closed with nacre. Specimen in Brit. Mus. SMITH, E. A., *Ann. and Mag. of N. H.*, 1888 (1), p. 419.

441. **Haliotis**: two specimens, of different species, in which the perforations were entirely absent, their place being taken by a continued convex, spiral rib, like the second rib of *Padollus*. "Probably in this individual the mantle was without any slit, and hence the malformation, the water being admitted to the gills by the slight notch in front of the ribs, as in some *Emarginulæ*, or *Scuta*." GRAY, J. E., *Proc. Zool. Soc.*, 1856, p. 149.

442. **H. albicans**: several specimens in which the perforations were united to form a continuous slit. The appearances were so uniform that Gray was disposed to think that these specimens might represent a new genus, but on comparison with types they seemed to belong to the species named. In some fossil genera (*Scissurella*) the perforations are replaced by a more or less continuous slit over the mantle. The specimens in question were greatly eroded and had a diseased appearance, *ibid*. Plate.

CHAPTER XII.

LINEAR SERIES—*continued*. COLOUR-MARKINGS.

OCELLAR MARKINGS[1], ESPECIALLY THOSE OF LEPIDOPTERA.

UPON the bodies of animals belonging to many classes are markings which consist of a central patch of colour surrounded by a variable number of concentric rings of different colours. Such markings are known as ocelli or eye-spots from their resemblance to the pupil and iris of vertebrates. Eye-spots are perhaps best known in Lepidoptera, but similar markings are not unfrequent in other groups and especially on the feathers of Birds and in Fishes.

In one of the best known chapters in the *Descent of Man*[2] the nature and mode of evolution of these markings is the subject of a full discussion, the case of eye-spots on feathers being chiefly taken in illustration. As is well known, Darwin by the comparative method, comparing the eye-spots found in different species, on the different feathers of the same bird, or on different parts of the same feather, found that it was possible to construct a complete progression from a plain spot to a fully-formed ocellus. Though no one examining such a series can possibly doubt that the simple spot and the fully-formed ocellus are really of the same nature and that the one represents a modification of the other, there remains nevertheless the difficulty that members of a series of parts cannot be assumed to represent conditions through which the other members of the same series have passed, and it is of course clear that the conditions found in some forms do not necessarily correspond with phylogenetic phases of other forms. In the present instance however Darwin is not specially urging this view, but brings forward the comparative evidence chiefly in illustration of the possibility that such structures may exist in an imperfect state and so may be conceived of as having had a gradual origin.

[1] The evidence concerning eyespots of Lepidoptera is taken here because eyespots when repeated in series, though borne on appendicular parts, are nevertheless arranged chiefly with reference to the chief axis of symmetry of the body. In some few forms, *e.g. Taygetis*, there is a conspicuous Minor Symmetry within the limits of a single wing (the posterior), but this is not often the case.

[2] *Descent of Man*, 1871, II. pp. 132—153.

Though doubtless the eye-spots of Birds are in their nature not different from those of Lepidoptera yet their manifestations in the latter are usually in some respects simpler than they are in Birds. From the abundance of material also the Variation of eye-spots is most easily studied in Lepidoptera and it is to them that the present evidence chiefly relates.

In preface to the evidence a few remarks are needed to direct attention to certain features in the mode of normal occurrence of eye-spots and in the manner of their Variation.

On a survey of the facts it is at once seen that eye-spots are extraordinarily variable both in number and size, some of the best formed being occasionally absent, and large and perfect ocelli being sometimes added in situations having normally no trace of such marks. With this fact Darwin was well acquainted and he refers to observations in illustration of it. In speaking of *Cyllo leda* he concludes that from the great variability of the eye-spots "in cases like these, the development of a perfect ocellus does not require a long course of variation and selection;" and again, that bearing in mind "the extraordinary variability of the ocelli in many Lepidoptera, the formation of these beautiful ornaments can hardly be a highly complex process, and probably depends on some slight and graduated change in the nature of the tissues." The facts to be given and the circumstances attendant on the variation of ocelli tend to support this conclusion.

Considered from the point of view of Meristic Variation the chief feature in the manner of occurrence of eye-spots in Lepidoptera is the frequency with which they are repeated. A single spot may be repeated in homologous places in both pairs of wings; in other cases there is a series along the margins of one or both wings. Besides the repetitions thus occurring it is especially worthy of notice that ocelli are very commonly repeated on *both* surfaces of the wing (Satyridæ, &c.), the centres of the upper and lower ocelli coinciding. It need scarcely be remarked that this effect is not produced by transparency of the wing-membranes and scales, but is an actual repetition, the scales of both surfaces being so coloured as to form an eye-spot on each side, the two having their centres coincident. In some cases, *e.g. Saturnia carpini* (the Emperor Moth), the rings and centres of the upper and lower ocelli have nearly the same colouring, but in the majority *e.g. Pararge megœra* (The Wall), *Erebia blandina*, &c., the upper and lower spots, though coincident, have quite different colours. In considering the Variation of the spots these facts as to the repetition of the spots should be remembered, for, as has been often insisted on in other cases of repetitions, we are concerned with the evolution of the *series* and not of one member only. Here therefore regard must be had to the degree of correspondence between the variations of the eye-spots in the fore and hind wings, on the upper and lower surfaces of the same wing, in the several eye-spots along the margin of the same wing, or in all of these, as the case may be. The evidence will shew that there is sometimes a close correspondence between the variations of eye-spots in these several positions.

But though these are the matters with which we have now the more direct concern it will be convenient to speak at the same time more generally of eye-spots. It should be remembered first that there are eye-spots of various complexity. In the simplest all the bands are circular, having one centre; the ocellus is then as a rule complete in one cell of the wing, though sometimes the outer zones of colour overspread parts of the adjacent cells. In some cases the spot is double, having two centres, the bands being disposed round them in an hour-glass shape. As to the visible structure of eye-spots it can be seen with the microscope that the colour of the eye-spot lies in the colours of the scales. The scales are arranged in parallel rows running (with little crossing or anastomosing) as nearly as possible at right angles to the nearest nervures, being disposed in regard to them much as the circular threads of a cobweb are in regard to the radial threads. Across these rows of scales run the colour-zones, in no way limited or guided by them. On the other hand it can be seen that the patterns are almost wholly made up by the colours of single scales, each having its own colour, particoloured scales being exceptional. The effect thus seen is very like that of a mosaic picture made of similar pieces, or of a design worked in cross-stitch on canvas, all the stitches being in rows and each stitch having its own colour.

As regards the position of eye-spots it should be noticed that the simpler sort, *e.g.* those of *Morpho* or of Satyridæ, *are usually placed in such a position that each of their centres is on the line of one of the creases or fold-marks of the wing*, and it sometimes happens that these creases seem to begin from the centre of an ocellus. From the fact that the creases for the most part run evenly between two nervures, bisecting a cell, it commonly results that the centre of the eye-spot is exactly halfway between two nervures. The large spots on the hind wings of some Pieridæ, *e.g. Parnassius apollo*, are an exception to this rule.

In that cell of the hind wing which lies between the submedian and first median nervures in many ocellated forms (Satyridæ, *Morpho*, &c.) there are *two* creases, and it is especially interesting to notice that in this cell there are commonly two ocelli, one on each crease; but if there is only one ocellus its centre does *not* correspond with the middle of the cell but is nearer to the first median nervure, being placed exactly on the anterior of the two creases. In spite of the excessive variability of ocelli, in for instance Satyridæ, it appears that they are not formed in situations other than these, being so far as I have seen always on one of the creases[1].

[1] These remarks refer to simple ocelli with one or more definite centres.

On looking at such a series of repeated ocelli as those on the hind wing of *Pararge megæra*, from this fact that the ocelli are on these creases or folds the question naturally arises whether the wing may not have been, in its development, folded along these creases so as to bring the ocelli into contact with each other like the fold-edges of a fan. If this were the case it might be supposed that the repetition of the ocelli was due to the action of some one cause on all the folded edges together. As a matter of fact, however, so far at least as can be judged from the condition of the wings in the pupal state before scales or pigments are excreted, there is no such folding, but each wing is laid smoothly out, and the increase in extent of the wings of the imago is attained, not by a process of unfolding, but by a stretching of the elastic wing-membranes on inflation from the tracheæ. On the whole it does not seem likely that the repetition of similar eye-spots on the Lepidopteran wing arises in any way more immediately mechanical than that by which other repeated patterns are elsewhere formed on animals.

The Variation of eye-spots as already stated may be very great, and examples are to be given both of the total absence of large eye-spots present in the normal, and of the presence of perfect eye-spots in abnormal places. Besides these extreme cases there is immense Variation in the degree to which eye-spots are developed, and such variability is nearly always to be seen in any species possessing simple ringed ocelli. In the manner of Variation of ocelli the following things are noteworthy.

(1) The *whole* of an eye-spot, centre and various concentric bands together, may be wanting; conversely a *whole* new eye-spot having the centre and all the bands pertaining to the normal eye-spot of the species may suddenly appear upon a crease normally bearing no eye-spot. Eye-spots therefore *may* come or disappear in their entirety.

(2) If a number of specimens of some much ocellated species are taken and compared, examples will be found in which some of the normal ocelli are absent altogether. But besides these there may generally be found specimens having an ocellus in a reduced and imperfect condition. Speaking generally such reduction commonly occurs by diminution of the diameter of the whole spot; but if any of its component parts are wanting the *centre* is the first to disappear, then the next innermost band, and so on. In Fig. 76 is shewn a series of specimens illustrating this fact in the case of *Hipparchia tithonus*. The eye-spot in its least form is represented by a plain black patch. In the more complete condition a white centre appears. A similar case in *Morpho* is shewn in Fig. 81. Here on the right side a certain eye-spot is absent altogether, while on the left side it is present in a reduced state; the white centre and the innermost broad black band are absent, and the actual centre is of the yellow-red colour which in the normal eye-spot of the species is the third colour from the centre. The spots

on the upper surface of the hind wings of the Wall (*P. megœra*) are an excellent illustration of these principles of Variation.

The principle here stated, though generally followed, is not absolutely universal, and in other instances it occasionally happens that even when of very minute size an eye-spot still retains all its bands; but the statement that the order of disappearance is from the centre outwards and not the reverse is substantially true. Some have expressed a belief that ocelli arise by the breaking up of bands of colour, but this view finds no support in the facts of Variation so far as the simple ocelli of such forms as *Morpho* and the Satyridæ are concerned; for in its rudimentary condition a circular eye-spot is in them a circular eye-spot still.

The fact just stated, that in the reduction of a circular ocellus its central parts are the first to disappear, recalls phenomena seen in many cases of disturbance propagated from a centre through a homogeneous medium. A whole eye-spot may come, or it may go (as seen in cases of *Morpho*), leaving the field of the cell plain and without a speck. The suggestion is strong that the whole series of rings may have been formed by some one central disturbance, somewhat as a series of concentric waves may be formed by the splash of a stone thrown into a pool. It is especially interesting to remember that the formation even of a number of concentric rings of different colours from an animal pigment by the even diffusion of one reagent from a centre occurs actually in Gmelin's test for bile-pigments. Bile is spread on a white plate and a drop of nitric acid yellow with nitrous acid is dropped on it. As the acid diffuses itself distinct rings of yellow, red, violet, blue and green are formed concentrically round it by the progressive oxidation of the bile-pigment.

If the experiment is made by letting a drop of the acid fall on a piece of blotting-paper wetted with bile, a fairly permanent imitation of an ocellar mark can be made. It will be noticed that as in the natural eye-spot, so here, the outermost zone appears first and the central colour last. As also is usually the case in the ocellus, when all the zones are formed, the centre may greatly increase in diameter without any increase in the breadths of the circular zones, which merely get larger in diameter, remaining of the same breadth.

There is of course no reason whatever for supposing that ocelli are actually formed by the oxidation or other simple chemical change of the pigments of the field, but this example is merely given as an illustration of the possibility that a series of discontinuous chemical effects may be produced in concentric zones by a single central disturbance. Indeed, that the formation of an ocellus cannot be in reality of such simplicity is shewn by the fact that the scales of the centres of ocelli generally exhibit interference-colours (usually white or blue) and are then wholly or partially without pigment, while in not a few cases the centres of ocelli are deficient in, or destitute of, scales. It must also be remembered that occasionally the colour of one of the outer zones is repeated in an inner zone, which would scarcely be expected on the analogy of the oxidation of bile-pigments.

(3) As in the case of Teeth at the ends of series, disappearance of a member of a close series of eye-spots commonly occurs by the

loss of the spot standing at one of the *ends* of the series. This is easily seen in *P. megœra*, &c. Likewise as was found in Teeth, disappearance of such a terminal eye-spot is associated with reduction in the size of the other members of the series, and especially of those nearest to the place of the absent member. If as in *Satyrus hyperanthus* and many others, the series is broken into groups, then as in the case of heterodont dentitions containing gaps, a new member may be added on to the end of either group.

(4) The condition of the ocelli may vary similarly and simultaneously in both anterior and posterior wings. In a series of *Saturnia carpini* for example I notice that the size of the ocelli varies greatly, those of a particular female specimen in the Cambridge University Museum being nearly a quarter larger than those of the specimen having the smallest ocelli; but the size of the ocelli in the hind wings of each individual varies with that of the ocelli in the fore wings not less closely than the size of the right ocelli does with that of the left.

(5) This correlation between the wings of the two pairs is seen also in the presence or absence of ocelli as exhibited for instance in *H. tithonus* (Fig. 76). It is of course often very irregular, but for our purpose it is even of consequence that such correlation may occur sometimes.

(6) As mentioned, ocelli are often coincident on the upper and lower surfaces. When this is so, the degree of development of the spots on the one surface is generally an accurate measure of the degree to which they are developed on the other surface. But in species having spots developed thus coincidently on the two surfaces it can be found that, in varying, an ocellus always first appears in its least condition *either* on one surface *or* on the other, and not indefinitely sometimes on one and sometimes on the other. In *P. megœra*, for example, ocelli of both pairs of wings can be seen on the under surface when not formed on the upper and conversely. Nevertheless there is always a close correlation between the degrees of development on the two surfaces.

(7) Lastly, attention is called to the circumstance that in two cases of great variation in ocellar markings there was a variation in the neuration. In the first case, *P. megœra*, No. 458, the second median nervure was absent from both fore and hind wings. In the fore wing upon the line where it should be there was an eye-spot: in the hind wing the eye-spots of the two cells which should be separated by the second median were partially coalescent. In the other case, *S. carpini*, No. 459, the large ocellus was *absent* from each wing, and it is stated that a nervure was also absent, but of this case no proper description has appeared, and it is uncertain which nervure was absent. When however these facts are considered in connexion with the circumstance that ocelli stand on the creases of the wings it seems likely that in some way unknown the positions and perhaps even the existence of the eye-spots may

294   MERISTIC VARIATION.   [PART I.

be determined by the manner of stretching of the wing-membranes. It must still be remembered that in the great majority of cases of ocellar variation there is no change in the neuration.

As to the *function* of ocellar markings nothing is known, and I am not aware that any suggestion has been made which calls for serious notice.

EVIDENCE AS TO VARIATION OF OCELLI IN LEPIDOPTERA.

General variability of ocelli.

The following are chosen to illustrate the general variability of ocelli in Satyridæ. Any of the common forms, such as *C. davus*, *P. megæra*, &c. shew similar variations. Generally speaking the condition is bilaterally symmetrical, but somewhat asymmetrical examples are not rare.

*443.   **Hipparchia tithonus:** from some 80 specimens taken in one

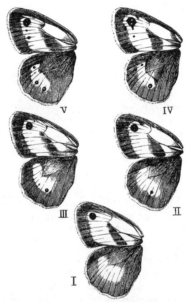

FIG. 76. *Hipparchia tithonus* ♂, cases illustrating Variation in number of ocelli. I. In f. w. the upper half of the large ocellus has a pupil, the lower has none: in h. w. no ocellus. II. Both halves of large ocellus of f. w. have pupils, and the h. w. has one ocellus. III. Pupils of large ocellus of f. w. are larger: h. w. has two ocelli. IV. F. w. has a new ocellus and the large double ocellus is half-joined to a second new ocellus. H. w. has two ocelli, one being placed otherwise than in III. V. F. w. has two ocelli without pupils as well as the large double one. H. w. has three ocelli. The wings of the other side corresponded nearly though not accurately. II. is the most frequent form.

(This figure was drawn with especial care from the specimens by Mr Edwin Wilson.)

ditch in the Cambridgeshire Fens on the same day the individuals shewn in Fig. 76 were selected. These cases especially illustrate the statements numbered (2) and (5), viz. the order of appearance of the colours and the similar Variation of the two pairs of wings.

*444. **Satyrus hyperanthus:** four specimens (Fig. 77) shewing

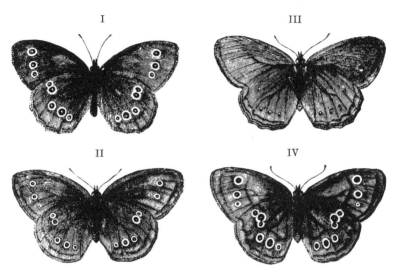

Fig. 77. *Satyrus hyperanthus.* Various conditions of ocelli. II. is the most frequent form.
(From NEWMAN's *British Butterflies.*)

different conditions of ocelli in this species from NEWMAN's *British Butterflies.* A form without ocelli is mentioned by PORRITT, *Ent.*, XVI., 1883, p. 188.

On one day I have myself taken all the forms shewn in Fig. 77 (except III.) and others in Monk's Wood, so that here no question of seasonal or local difference is necessarily involved.

445. **Chionobas.** The North-American species of this genus [in general appearance somewhat resembling the British *Hipparchia semele*, the Grayling] are of a brown colour having eye-spots on some or all of the wings. According to STRECKER the number of eye-spots varies extremely, and the following instances are given. The species *norma* may have two spots on fore wings and none on hind wings; two on f. w. and one on h. w.; one on f. w. and one on h. w.; one on f. w. and none on h. w.; three on f. w. and two on h. w. Of the species *uhleri* one of the types has three on f. w. and four on h. w., the other has four on f. w. and five on h. w., the subapical being very small; other examples have only one on f. w. and two or three on h. w. The species *chryxus* may have one on f. w. and none on h. w.; or two on f. w. and one on h. w. STRECKER, *Cat. Macrolepid.*, p. 155.

446. **Arge pherusa:** a butterfly resembling the British *Arge galathea*, the Marbled White, has a variety *plesaura*, in which the eye-spots of hind wing are wanting.
Specimen figured in which the left hind wing is a third smaller than the right and lacks the eye-spots. FAILLA-TEDALDI, *Nat. Sicil.*, I. p. 208, Pl. XI. fig. 8.

MORPHO.

A number of species of this genus, for example, *M. achilles, menelaus, octavia, montezuma,* &c. are marked upon the under surface of both pairs of wings with large ocelli having four principal zones in addition to the white central spot. Of the zones the outermost is silvery, the next dark brown, the next either red or some shade of yellow. Within this is a band of very variable width having a deep chocolate colour. When very broad, as in *M. montezuma* or *M. achilles*, the inner parts of this band are irregularly sprinkled with red scales. The centre is white or bluish-white, some of the scales in its periphery being nearly always distinctly blue. The centre is commonly not circular but is produced (especially in larger ocelli) in a direction at right angles to the crease on which it stands. Fig. 78, I, taken from a normal specimen of *M. achilles*, shews the usual positions of the eye-spots in all the species whose variations are described below. The ocelli on the fore wing are 3, on the hind wing 4. In speaking of them the letters a, b, c, d, e, f, g are used as shewn in the figure. Between a and b there is a cell normally bearing no ocellus, and between d and e there are two such cells. The spot g as described on p. 90, stands at the anterior side of its cell and not in the middle of it, and a second spot g_1 may appear behind it in the same cell.

The following examples are taken from the series in the collec-

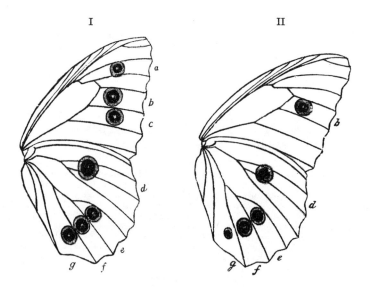

Fig. 78. *Morpho achilles*. Undersides of left wings. I. Normal. II. Specimen wanting the spots a and c on both sides.
(From specimens in the collection of Messrs Godman and Salvin.)

tion of Mr F. D. Godman and Mr O. Salvin, to whom I am much indebted for permission to examine the specimens[1].

447.   **Morpho achilles** ♂. Specimen having the spots a and c entirely absent (Fig. 78, II) and the spot g very small. This specimen occurred together with two normals from Pará. Ten other normal males seen, and also a specimen in Camb. Univ. Mus. having no c, the spot a being also greatly reduced.

448.   **M. montezuma** ♂: 15 specimens have all the spots from a to g of fair size. One specimen has a spot in the place a_1, as shewn

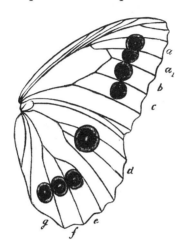

Fig. 79. *Morpho montezuma*. Abnormal specimen having an ocellus on both sides in the position a_1 (where an ocellus normally exists in *M. sulkowskii*). (From a specimen in the collection of Messrs Godman and Salvin.)

in Fig. 79. One specimen has a very faint a^1 and g_1; another has a^1 as a small ocellus, and g_1 indicated as a bulging of the spot g.

In Camb. Univ. Mus. are 4 normal males and one specimen having both a^1 and g_1 marked somewhat as shewn in the case of the abnormal *M. octavia* (Fig. 80).

449.   **M. octavia.** Mr Salvin tells me that this form has a very restricted distribution and is probably only a local form of *M. montezuma*. In addition to 12 normal males the following were seen, all being male. Specimen having g_1 as a spot of moderate size; another having g_1 very small. In another a^1 and g_1 were both present as shewn in Fig. 80. Besides these is one having g very small. All are from the Pacific slope of Guatemala. The specimen figured is from El Reposo in this district, one of the normals being from the same place.

---

[1] In each of the figures the faint lines round the ocelli should be shewn as in Fig. 81; they are omitted for simplicity.

298   MERISTIC VARIATION.   [PART I.

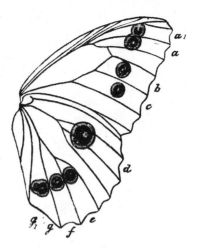

Fig. 80. *Morpho octavia*: abnormal specimen having ocelli on both sides in the positions a^1 and g_1 (where ocelli normally are in *M. sulkowskii*).
(From a specimen in the collection of Messrs Godman and Salvin.)

*450.   **M. menelaus** ♂ : ten normals, and two having no a; one having left a absent and right a very faint, c and g both absent. In addition to these, the specimen shewn in Fig. 81, having no c

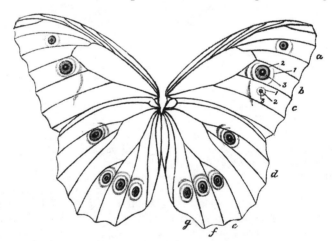

Fig. 81. *Morpho menelaus*: abnormal specimen having no ocellus c in rt. f. w. In left f. w. there is a small ocellus c, *but it wants the two innermost colours of a normal ocellus*. Compared with a normal ocellus, as that at b of the same wing, the abnormal has only the zones 1, 2 and 3, the latter colour forming the centre.
Fig. 78, I. may be taken as approximately shewing the normal for this species also.
(From a specimen in the collection of Messrs Godman and Salvin.)

on right side, while on the left the same spot is reduced as shewn in the figure, the centre being of the colour normally constituting the third band.

In connexion with the above cases it should be mentioned that in another species, *Morpho sulkowskii*, one of the more transparent species, the spots a^1, a_1, and g_1 are all *normally* present. The spot c is however sometimes absent in this species. In *M. psyche* the spot c is normally absent, though present in one specimen examined.

Complex ocelli.

Besides the simpler ocelli there are other forms of ocelli of more complex structure, having two or more centres around which the coloured zones are disposed without an accurate symmetry. Such ocelli may be seen in *Vanessa io* or in *Junonia*, and it is noticeable that they are no less variable than the simpler forms. The following examples may be given.

Vanessa io. Looking at the eye-spot on the fore wing of the Peacock-butterfly one can readily see that it is not a structure of the same nature as the other ocelli that have been already considered. The eye-spot of the hind wing does not materially differ from other eye-spots, being essentially a black spot surrounded by a pale band and containing an irregular and incomplete centre of blue. The eye of the fore wing on the contrary is not actually made up of concentric markings but is quite exceptional, being formed of a combination of patches of different colours. But whether the eye of the fore wing is a true ocellus or not it is nevertheless certain that its formation may vary with that of the eye of the hind wing, as the following examples testify.

451. Specimen, British; reared from a larva in captivity, having all the eye-spots deficient (Fig. 82). On the fore wings the series of white spots along the margin (on the creases) are present. The three which lie within the field of the normal eye-spot are *longer* than usual. The costal black mark is extended so as to cover the greater part of the

FIG. 82. *Vanessa io*, the Peacock butterfly, having all the four eye-spots deficient (No. 451). (From Newman.)

situation of the eye-spot. On the hind wings the eye-spots are entirely obliterated and their place is taken by an ill-defined patch of pale colour. NEWMAN, *Ent.*, 1872, p. 105, Fig.

452. Similar specimen described by GOOSSENS, *Bull. Ent. Soc. France*, S. 5, v. p. cxlix.

453. Similar specimen in Lord WALSINGHAM's collection in Brit. Mus. Here the blue and black of the eye-spots of the hind wing are altogether absent. The black internal border of the spot is broader than usual, and the place of the spot is lightish in colour. In the spot of the fore wing the blue is deficient, the yellow is largely absent, but the white spots are emphasized.

*454. Specimen in which the eye-spots on the hind wings are obliterated, as in the foregoing : those of the fore wings are also similarly modified, but the white spots of the marginal series are enlarged to a much greater extent. Also another specimen in which the eye-spots were partially deficient. These two specimens were from one brood reared in Germany: of this brood none were typical, and several resembled the specimens described. SOUTH, R., *Ent.*, 1889, XXII. p. 218, Pl.

455. Specimen figured in which the eye-spots are symmetrically absent from both posterior wings. In this case both the greyish yellow bordering of the eye-spots and the blue marks generally contained within them are entirely absent. The ground-colour of the hind wings is greyish brown, and upon this two black marks are placed in the situation of the normal eye-spot and a series of small black lines occurs round the margins of the hind wings. The eye-spots of the anterior wings are modified in a peculiar manner which is not easily described. MOSLEY, S. L., *Varieties of Brit. Lepid.*, Pt. III. Pl. 2, Fig. 3.

456. **Junonia clelia**, Cram. In this species there are normally two ocelli in each fore wing and a similar pair in each hind wing (TRIMEN, *S. Afr. Butterflies*, I. p. 214). In a series of nine specimens in the Cambridge University Museum very great variations in the size of the ocelli appear. The *posterior* ocellus of each wing is more constant in size than the *anterior*. One specimen wants altogether the anterior ocellus of the hind wings, which in most specimens has a diameter of about 2·5 mm. In several the anterior ocellus of the fore wings is hardly visible.

457. **Junonia cœnia**: the degree to which the two eye-spots of each wing are developed varies greatly. In a Californian specimen in Godman and Salvin's collection the spots are all very large, while in a Granada specimen they are almost entirely obliterated. Of four specimens in the same collection from the United States of Colombia (but not from the same locality), one has scarcely a trace of the anterior eye-spot of the fore wing, the second eye being very faint. In the hind wing the anterior eye-spot is very faint and the posterior is absent.

The two following cases are important from the fact that in each of them there is said to have been abnormality in neuration.

*458. **Pararge megæra** ♂ (the Wall Butterfly): specimen in which the second nervure of the median vein is wanting *in each of the four wings*. In the anterior wings the place which should be crossed by this nervure is occupied by an extra ocellus (Fig. 83), which is nearly as large as the normal large ocellus of the wing. The normal ocellus itself is incompletely doubled. In the hind wings, the two ocelli (2nd and 3rd), which in the normal insect are separated by the missing nervure, are elongated towards each other, so that their black borders touch and the usual central white dots join into a line, one-twelfth of an inch long. On the under

FIG. 83. *Pararge megæra*, the Wall; case described in No. 458. [This copy is rather too light, and the banding on the hind wing is too distinct.]
(From WEBB.)

side, the anterior wings have respectively six and five ocelli and the hind wings five and six. The arrangement of the dark colour on the upper surface of the anterior wing differs somewhat in the direction of the pattern of the female. WEBB, S., *Entomologist*, 1889, XXII. p. 289, Fig.

*459. **Saturnia carpini** ♂; variety without eye-spots. (Fig. 84.) This specimen was bred from a larva found with many others

FIG. 84. *Saturnia carpini* lacking the ocellar marks in each wing (No. 459).
(From BOND.)

feeding upon sallow in Sawston Fen, Cambridgeshire. " In the colour and markings of the specimen there was perhaps nothing worth notice excepting the absence of the ocellus in each wing and also of one of the veins in each of the anterior wings."

About 50 larvæ were collected at the same time on one large sallow. One of them, a female, was destitute of scales[1], but the remainder of the specimens reared were remarkably fine. BOND, F., *Entomologist*, X., 1877, p. 1, *fig*. [This is the specimen mentioned by HUMPHREYS, *Brit. Moths*, p. 20. It is unfortunate that no further description is given, and the figure is not sufficiently clear to enable one to see which nervure was absent. On the fore wings a narrow, elongated patch of light colour was in the place of each ocellus, and on the hind wings there was a somewhat wider

---

[1] Partial deficiency of scales, occurring evenly over all the four wings, is not very rare in *S. carpini*. I have myself reared two such specimens.

and irregularly shaped patch of pale colour. If this specimen, which was in the collection of the late Mr F. Bond, is still in existence it is greatly to be wished that a proper description of it should be published.]

460. **Saturnia carpini** ♂: wings yellowish-grey throughout, with the usual markings, *save that on the fore wings there is no ocellus*, and on the hind wings is only a small black eye, without a border, having a yellowish-grey central spot. OCHSENHEIMER, F., *Schmet. von Europa*, 1816, IV. p. 191.

From this evidence it is clear that the range of Variation of ocellar markings in Lepidoptera is very great. It is especially to be noticed that this variability affects no one family, or the species of one geographical region, or one kind of ocellus exclusively, though doubtless it is more marked in some than in others; but it seems rather to be a property belonging to ocelli in general. From the fact that they can bodily come and go, it seems clear that, as was suggested above, each ocellus is as regards its origin *one* structure made up of parts in correlation with each other.

RAIIDÆ.

The great variability of ocellar markings is probably not peculiar to Lepidoptera, but I have no evidence sufficient to produce regarding the variability of ocellar markings in other forms. I may however instance the case of the Raiidæ, many of which have been found marked with a large ocellar mark on the dorsal surface of each pectoral fin. At different times such a mark has been thought to characterize a certain species, but I believe it is now generally admitted that it may appear as a variation in several species. The best figure of this ocellar mark is that given by DONOVAN (*Brit. Fishes*, 1808, V. Pl. CIII.) in a Ray described under
461. the Linnean name *Raia miraletus*. On each "wing" was a large spot, having a dark purple centre, surrounded by a zone of silvery green enclosed by a broad dark boundary composed of five equidistant, contiguous spots of blackish purple. Donovan suspected that the fish might be a variety of the Homelyn (**R. maculata**), and it has been generally believed by other authors to have been so. Donovan states that a similar eye-spot was seen by him in various degrees of definition in several young Skates.

462. **R. clavata,** the Thornback, also sometimes has a large white spot surrounded with black on the "wings." DAY, *Brit. Fishes*, II. p. 344.

Raia circularis, the Cuckoo Ray, has *normally* on each "wing" a large black blotch banded with yellow and surrounded by yellow spots. This structure may be absent as a variation. DAY, *Brit. Fishes*, II. p. 349.

Simultaneity of Colour-variation in Parts repeated in Linear Series.

Reference was made (Introduction, Section V.) to that relation subsisting between the several members of a linear series of segments or other repeated parts, by virtue of which they may resemble each other in respect of colour or pattern of colours. From the fact that the several members do in such cases often bear the same colours or patterns it is clear that they must at some time or other have undergone similar Variation. In order to measure the possible rapidity of the process of evolution by which such parts may have reached their present condition it is important to ascertain the extent to which their several variations may be simultaneous.

Variations in colour are of course Substantive variations and a full consideration of their nature cannot be taken here. For the present we are only concerned with the consequences of the fact that the parts are repeated in series. As was pointed out in the Introduction the problem of the resemblance between the colours of such segments is only a special case of the same problem of Symmetry which is again presented in bilateral or other Repetition.

Simultaneous colour-variation taking place abruptly in a large number of organs, such as hairs, feathers, &c. is a very common occurrence, and the part that repetition of structures plays in producing the total effect is apt to be overlooked. In comparing two varieties of some whole-coloured animal, a bay horse with a chestnut for example, it must be remembered that the difference is really made up of a simultaneous variation in the pigment of each particular hair. Similarly if a caterpillar normally green appears in a uniformly brown variety we may conceive the total change as brought about by variation occurring simultaneously in the skin of the several segments, or in some smaller units. But whatever unit be taken, whether segment, or hairs, or cells, that all or any particular groups of such units should vary together and in the same direction is not a matter of necessity. That such simultaneity is not universal and that segments may vary independently of each other is a matter of common observation, and indeed is sufficiently proved by the occurrence of differentiation between segments. Nevertheless the evidence goes to shew that between parts repeated in series there may be a relationship of the kind spoken of, though its causes, nature and limitations are unknown. In the case of actual segmentation this relationship may appear either in the simultaneous variation of the colour-patterns of the segments, or of some one colour or patch borne by each, or by the appearance of some unusual mark or patch on several of them at once.

In some cases it happens that certain of the segments may vary together, the rest remaining unchanged, and, as seen in

Chiton marmoreus, (*q.v.*), the segments thus undergoing the same variation are not always even adjacent to each other.

The whole question is a very large one and it is not possible here to do more than refer briefly to a few cases illustrating some of its different aspects. Fuller treatment will be attempted in connexion with the evidence of Substantive Variation.

463. As examples of a form whose segments in their colour-variations manifest a very close agreement with each other, the Hirudinea may be taken. Figures of numerous varieties of medicinal Leeches are given by ÉBRARD, *Nouvelle monogr. des Sangsues*, 1857, and other cases are represented by MOQUIN-TANDON, *Monogr. de la famille des Hirudinées*, 1827 (see especially Pl. v. fig. 1). As these figures testify, there is a wide diversity both in the ground-colour and in the size, colour and manner of distribution of the lines and spots with which it is decorated, but so far as may be judged from the figures and descriptions the same decorations are repeated on the various segments. It cannot be doubted that a close scrutiny of the specimens would shew points of difference even between adjacent segments but substantially the patterns are the same for the segments of an individual. The patterns of the varieties may thus, like patterns of ribbon, be each represented by a drawing of a short piece of the body in the way adopted by the writers named.

As regards the larvae of Lepidoptera a good deal of information bearing on this subject exists, and some of these results, especially those relating to Sphingidae, are of interest[1].

*464. In the larvae of many species of Sphingidae there is a more or less regular dimorphism in colour. Notable examples of this are *Acherontia atropos*, *Chœrocampa elpenor*, *C. porcellus* and *Macroglossa stellatarum*, in each of which the larva is known both in a light green and in a dark form[2]. The dark form is the commonest in *C. porcellus* but in *A. atropos* it is much rarer than the green form. Judging from the figures, the ground-colour of the segments generally varies as a

---

[1] The facts which follow are chiefly taken from WILSON, *Larvae of Lepidoptera*, 1880; WEISMANN, *Studies in Theory of Descent*, Eng. Trans., 1882; POULTON, *Trans. Ent. Soc.*, 1884, 1885, 1886, 1887; BUCKLER, *Larvae of Brit. Butterf. and Moths*, Vol. III. Ray Soc., 1887.

[2] That this dimorphism is 'phytophagic' is not very likely, but the possibility should be remembered. It seems to be established that in many of the species the colour-varieties are definite and largely discontinuous. Of *M. stellatarum* WEISMANN (p. 250) bred 140 from one batch of eggs, and of these 49 were of the green form and 63 of the brown form, only 28 being transitional. The discontinuous character of the variation was illustrated by one most remarkable specimen. In it the body was *particoloured*, being partly of the green and partly of the brown form. The head, prothorax, all the abdominal segments behind the 2nd, and the *right* side of the remainder were brown, but the left side of the meso- and meta-thorax, of the 1st abdominal, and part of the left side of the 2nd abdominal were green [according to the figure 9 on Pl. III., with which the description in the text, p. 249, differs slightly]. In *A. atropos* I know no account of any intermediate form. In most of the species the dimorphic or polymorphic character appears in the later periods of larval life and especially after the last moult; but in *C. porcellus*, according to both WEISMANN (p. 188) and BUCKLER (p. 117) though the larvae are of both kinds in the penultimate state all or nearly all after the last moult turn to the dark form.

whole, shewing only slight differences in tint in different parts of the body. To this there are certain exceptions, of which *A. atropos* is especially remarkable. In the brown variety of this species the abdominal segments have a dark ground-colour composed of shades of brown, while the three thoracic segments in it are *white* "like linen" (see WILSON, Pl. VI.; BUCKLER, Pl. XXI.; POULTON, 1886, p. 149; HAMMOND, *Zool.*, 6282; BALDING, *Ent. Mo. Mag.* XXII. p. 279; GIRARD, *Bull. Soc. ent. Fr.*, 1865, S. 4, v. p. xlix. &c.).

In *M. stellatarum* though the ground-colour of the head and of all the segments varies greatly it appears that the head and prothorax vary in colour simultaneously with each other and are of one colour, while the other two thoracic segments and the abdominal segments also vary together but usually differ from the head and pro-thorax (see WEISMANN, Pl. III.).

In illustration of the degree to which simultaneity of Variation is possible over considerable areas of the body the varieties in markings are perhaps more important than those in ground-colour. Of such changes simultaneously occurring in several segments there are many examples.

465. In all the varieties of ground-colour in *M. stellatarum* the pattern of the markings remains the same though of differing intensities (WEISMANN, p. 248), but in the brown variety of *A. atropos* the pattern is quite peculiar and cannot even be recognized as a representation of the markings seen in the green form. Even the oblique stripes are absent (POULTON, 1886, p. 149; see also authors quoted above). But as in the ground-colour so in the markings, the abdominal segments have one new pattern while the thoracic segments have another.

*466. The figures of larvæ of *Deiphila euphorbiæ* given by BUCKLER and by WEISMANN are especially interesting in this connexion, shewing that in the complex variations of this polymorphic form the particular pattern of the individual is carried out with little difference in each segment behind the prothorax. Some of these changes are extensive, but to be at all appreciated the figures must be referred to. In one case all the triangles at the posterior part of each segment were red instead of green as usual, and this change was found in many specimens from one locality (see WEISMANN, p. 206, Pl. V.). This identical variation was known to and figured by HÜBNER (WEISMANN). In one specimen from the same place as the last the second row of marks which should occur just below the sub-dorsal mark of each segment was absent throughout the whole line, and the ring-spots of the upper or sub-dorsal row had, as a variation, a red centre or nucleus, well marked in the posterior spots but fading away anteriorly. The occurrence of these considerable changes is still more noteworthy if, as WEISMANN states, the members of each batch are much alike. He remarks also that the variability is great in some localities but little in others.

467. The larva of *Deilephila hippophaës* has a sub-dorsal row of red markings upon a variable number of segments from the 7th abdominal to the 3rd or even 2nd abdominal, increasing in size and distinctness from behind forwards. The size of these markings differs greatly in different specimens, varying from a mere dot to a distinct red spot with a black ring. As the figures shew, there is a considerable cor-

respondence between the segments in the extent to which the spots are developed, though in each case they fade away in the anterior segments (see WEISMANN's figs. 59 and 60).

468. Another interesting example of considerable uniformity in the colour-variation of a series of segments is to be seen in *Saturnia carpini*. In this species besides change in the tint of the green ground-colour [two chief tints being found, one dark and one light] there is immense difference in the amount of black pigment deposited, most marked in the last two stages of the larvæ. Good figures and descriptions of these are given by WEISMANN (Pl. VIII.). Though no two segments are alike and though there are differences perceptible even between the two sides of most segments, yet the general scheme of colour of each individual is carried out with fair constancy over the several segments. As I have myself seen, the lightest and darkest may both be reared from one batch of eggs and in the same breeding-cage or sleeve.

469. The colour of the tubercles of *S. carpini* also varies greatly. They may be light yellow, dark yellow, pink, violet, or white, but the yellow and pink forms are the commonest. As I have myself observed, there is generally a close agreement between the different tubercles of each larva in point of colour. In a few specimens I have seen the tubercles of the anterior and posterior segments pinkish, while the remainder were yellow, but this diversity is exceptional. The importance of this case is increased by the fact that POULTON (1887, p. 311) has found that the offspring of a pair whose tubercles had been pink shewed a high proportion of larvæ with pink tubercles. The two parents were from a lot of 80 larvæ found together, of which only 3 had pink tubercles: but of their 88 offspring 64 had pink tubercles.

470. The case of the occurrence of red spots on the larvæ of *Smerinthus ocellatus* and *S. populi*[1] may be quoted as an instance of great irregularity in the degree to which the segments agree in their colour-variations. This well-known case is also of great interest as an example of a parallel variation occurring in different species. The larvæ of both species are most commonly without any red spots, but not rarely a number of red spots are present. In extreme cases each of the spiracles is surrounded with red, and there is in addition a row of red spots in the sub-dorsal region of all segments from the 1st thoracic to 7th abdominal, and also a red spot on each clasper. The number of spots, number of rows, the size and tint and distinctness of the spots is exceedingly variable. In point of time the spots of the 3rd abdominal segment appear first and those of the 2nd thoracic next (POULTON, 1887, p. 285, &c.). Though in much spotted specimens the spots may remain till the larva is full-fed, in some cases a few spots appear at an early stage and are afterwards lost. Among the individuals of the same brood there may be great diversity, some having spots and others being without them (POULTON, 1887, p. 287). In several cases a spot present on one side of a segment has been found absent on the other side. As Poulton observes, it is especially

---

[1] I have not referred to the case of *S. tiliæ*, as it is possibly of a different nature.

remarkable that though there are no spiracular openings on the meso- and meta-thoracic segments, yet in cases of extremely spotted larvæ there are red spots at the level of and continuing the spiracular series of spots upon these segments also (*S. ocellatus*, BUCKLER, Pl. xx. fig. 1 *a*; POULTON, 1887, Pl. x. fig. 1. *S. populi*, POULTON, 1887, p. 286). As an indication of an element of definiteness in this variation may be mentioned the fact that in fully spotted larvæ of *S. populi* the sub-dorsal spot on the 7th abdominal seems to be always the smallest in that row (POULTON, 1887, p. 285; WILSON, Pl. v. fig. 2 *a*; FLEMYNG, *Ent.*, 1880, p. 243, &c.).

In our present consideration the fact that these very large variations sometimes occur simultaneously over a large range of segments and are sometimes restricted to particular segments is of considerable importance.

We may note that WEISMANN (p. 360) is prepared to believe that these spots represent a new variation arising similarly and independently in the different species of *Smerinthus*. As however is usual in cases of considerable Variation an attempt has been made to lessen the value of these indications of the magnitude of Variation by suggesting that they may be of the nature of "reversion" (POULTON, 1884, p. 28). Apart however from a general reluctance to recognize the possibility of the occurrence of large variations there seem to be no special grounds for the suggestion here. It is nevertheless true that in the case of the *Smerinthus* larvæ a complete disproof of the hypothesis of "reversion" is wanting. This is only to be obtained in cases (like that of *D. euphorbiæ*), in which a great number of complex and mutually exclusive variations exist side by side. In the absence of such complete refutation the hypothesis of reversion may still find favour.

*471. **Chitonidæ.** The following facts observed in certain Chitons are given in illustration of the existence of a similar possibility of simultaneous Variation between parts which are repeated in series but whose repetition is not of the kind commonly included in the term Metameric. Unfortunately the material at hand is very limited and I do not know what might be the result of further examination, but the facts seen suggest that the subject is worth investigating.

The dorsal plates of Chitons are eight in number. Though the colours and markings in different species are complex and various yet in many species all the plates are alike or nearly so. The question then arises do all the plates change colour together, or do they change one by one, or otherwise? From the few observations made it seems that in this respect the species differ, but variation uniformly occurring in all the plates seems to be rare. This may perhaps be due to the constitution of such specimens as separate species, but I saw little likelihood of this. On the other hand in several cases the same variation was present in more than one segment, and in particular there was strong evidence that in some species the segments 2, 4 and 7 shew a noticeable agreement with each other in colour-variation. The specimens are all in the MacAndrew Collection in the Cambridge University Museum, and I have as usual simply followed the labelling of the specimens.

C. arbustum, Australia. 10 specimens, of which the plates in 6 are nearly uniform. In one there is a white band in the centre of each plate; in 2 the plates are irregularly coloured; in one the plates 1 and 6 agree in being broadly marked with white.

Chiton hennahi, Peru. 4 specimens. 3 are uniformly dark brown; but in the other specimen there is *a strong white mark on the centre of plates* 2—7, and a faint one on plates 1 and 8.

C. elegans, Chili. 2 specimens. In one, complicated markings are repeated on each plate nearly uniformly; in the other specimen *a much simpler pattern recurs on each segment*.

On the other hand, *C. pellis-serpentis*, New Zealand, 8 specimens: great diversity of markings and no uniformity among plates in 4 specimens, but in one specimen plates 2—5 were black and the rest light-coloured. Similar want of uniformity among the plates in 2 specimens of *C. incanus*, New Zealand.

The evidence of agreement between segments 2, 4 and 7 in the following cases is very striking.

C. (Tonicia) marmoreus, "Hebrides, &c." 18 specimens, all of a light brown colour marked with dark red.

In 4 specimens the plates are uniformly marked or nearly so.

In 6 specimens plates 2, 4 and 7 are much darker than the others, being for the most part of a uniform dark red.

In 5 specimens plates 2, 4, 7 *and* 8 are darker than the rest.

In 1 specimen plates 2, 4, 5 and 7 are darker than the rest.

In 2 specimens the central parts of most of the plates have dark markings, but no segment is specially distinguished.

Of 18 specimens therefore 12 have plates 2, 4 and 7 darker than the rest.

Among 3 specimens of the same species from Gr. Manan (N. America) 2 are nearly uniform throughout, but in one plates 2, 4, 7 and 8 are much darker than the rest.

C. (Tonicia) lineatus, 2 specimens. In one the markings on all the plates are nearly similar, and the white wavy streaks characterizing the species are almost similarly distributed on the sides of all the plates. In the other specimen these lines are absent on the plates 2, 4 and 7, which are much darker than the rest; but the lines, though less extensive than in the first specimen, are present on plates 1, 3, 5, 6 and 8.

The preceding evidence may suffice to indicate the nature of this important question of the degree to which the colour-variations of parts repeated in Linear Series may be similar and simultaneous, a question which, as must be evident, is of the highest consequence in estimating the magnitude of the steps by which Evolution may proceed. To the consideration of this matter it will be necessary to return when the evidence of Substantive Variation is considered.

Meanwhile it will not be forgotten that though we have only spoken of this question in reference to colour and to Linear Series, the same question arises also with regard to other variations and in reference to all parts which are in any way repeated and resemble each other, whether such repetition is strictly serial or not. In a survey of any group of animals cases will be seen in which organs in one region are repetitions of organs in another region though not necessarily in serial homology with them in any sense in which the term is commonly used. Many such cases were spoken of by Darwin in the chapter on "Correlated Variability[1]" and are now famous. The simultaneous colour-variations of the mane and tail of horses[2], the correspondence between the large quills of the wings and those of the tail of pigeons[3] and other birds are among the most familiar of such cases.

When with such facts in mind we turn to some species which differs from an ally in the presence of some characteristic development or condition common to a number of its parts, in making any estimate of the steps by which it may have been evolved it must be remembered that it is at least possible that the common feature characterizing these several parts may have been assumed by all simultaneously. To take a single instance of this kind, the species of the genus *Hippocampus*, the Sea-horses, have the shields produced into more or less prominent tubercles or spines. The back of the head is also drawn out into a prominent knob. In an allied genus from Australia, *Phyllopteryx*, many of these spines are provided with ragged looking tags of coloured skin, like the seaweed which the fishes frequent[4], giving the animal a most fantastic appearance and no doubt contributing greatly to its concealment [probably from its prey]. If in this case it were necessary to suppose that the variations by which this form has departed from the ordinary *Hippocampi* had occurred separately, and that each spine had separately developed its tag of skin, the number of variations and selections to be postulated would be enormous; but probably no such supposition is needed. We are, as I think, entitled to expect that if we had before us the line of ancestors of *Phyllopteryx*, we should see that many and perhaps all of the spines which are thus modified in different parts of the body had simultaneously broken out, as we may say, into tags of skin, just as the feathers of the Moor-hen (*Gallinula chloropus*)[5] may collectively take on the "hairy" form, or as, to take the case

[1] *Animals and Plants under Domestication*, ed. 1885, II. chap. xxv.
[2] As Darwin mentions, simultaneity in the variations of the hair may be manifested in size and texture as well as in colour. A bay horse was lately exhibited at the Westminster Aquarium standing 16½ hands, having the hair of both mane and tail of prodigious length. The longest hairs of the mane measured 14 ft. and those of the tail 13 ft. It did not appear that the hair of the fetlocks or body was unusual in character, but these were kept closely clipped and nothing could be affirmed on this point.
[3] By the courtesy of Professor L. VAILLANT I was enabled to examine a number of specimens of the singular breeds of Gold-fish from China in the Paris Museum of Natural History. Some of these are characterized by the great length both of the appendicular fins and of the caudal fin also. Measurement shewed that there was a substantial correspondence between the lengths of these parts, those with long appendicular fins having also very long tails. The correlation between these parts is not however universal in Gold-fishes, and in many of the ordinary "Telescope" Gold-fish the tail may be longer than that of a common Gold-fish of the same size, though the length of the appendicular fins be not exceptional (v. *infra*).
[4] GÜNTHER, *Study of Fishes*, 1880, p. 682, fig. 309.
[5] See Introduction, p. 55.

of Radial Series, the petals of a flower may all together take on the laciniated condition[1].

Further study will indeed probably lead to the recognition of a principle which may be thus expressed: that *parts which in any one body are alike*, which have, that is to say, undergone similar Variation in the past, *may undergo similar variations simultaneously*; a principle which, if true at all, is true without regard to the morphological position of the parts in question.

[1] For cases see MASTERS, *Vegetable Teratology*, 1869, p. 67.

CHAPTER XIII.

LINEAR SERIES—*continued*.

MINOR SYMMETRIES: DIGITS.

ALL the cases considered in the foregoing chapters have illustrated Variation of parts whose repetition is disposed in Linear Series along the chief axis of the body, being thus arranged directly and immediately with reference to the Major Symmetry of the body. We have now to consider cases of the Meristic Variation of parts which are also repeated in Linear Series but normally possess in some degree the property of symmetry partially completed within the limits of their own series, thus forming a Minor Symmetry.

Of Linear repetitions thus occurring there is a great diversity, and evidence will here be produced regarding two of the chief examples, namely, the digits of vertebrates and the segmentation of antennæ and tarsi of Insects.

In each of these groups of organs the parts are frequently formed in such a way as to make an approach to symmetry, about one or more axes within the limits of the appendage to which they belong. This fact will be found to lead to consequences apparent in the manner in which numerical Variation takes place in limbs of the various types.

In these Minor Symmetries Linear Repetition may occur in two forms: there may be repetitions of digits or other parts in lines forming an angle with the axis of an appendage; and there may be repetitions in the form of joints &c. along the axis of the appendage itself.

The cases of Variation in number of joints in the appendages of Insects are chiefly interesting as examples of manifest Discontinuity in Variation, and from the conclusions which they suggest as to the supposed individuality of segments. This latter question arises also in considering the relation of the two phalanges of the pollex and hallux to the three phalanges of the other digits, but the evidence which can be gained from a study of Variation with reference to this question is so intimately connected with the subject of the variation of digits in general that it cannot be considered apart. Other cases referring to repetitions in the line of the axis of appendages will be taken in a subsequent chapter.

In studying numerical Variation in the digits of certain animals, especially the Horse and the Pig, we shall meet with forms of Variation which are peculiar to structures having a bilateral symmetry. In examining the evidence as to Meristic Variation of Bilateral Series further reference to these cases will have to be made, but it appears simplest to describe the facts in the first instance in connexion with the subject of digits.

From the evidence as to Meristic Variation in digits I propose to make a selection, taking certain groups of cases having a direct and obvious bearing on the general problems of Variation. It will be understood and should be explicitly stated that unless the contrary is declared the principles of form which can be perceived as operating in special cases are not of universal application in the Variation of digits, but are enunciated as applying only to the special cases in which they are perceived. In the human subject, for example, cases of polydactylism will be quoted which when arranged together form a progressive series illustrating the establishment of a novel and curious Symmetry; but though these cases are valuable as illustrations of the way in which the forces of Division and growth can dispose themselves to produce a symmetrical result, yet it must always be borne in mind that very many variations of the digits have been seen in Man, whether consisting in increase in number of digits or in decrease, of which the result is almost shapeless. The case of polydactyle Cats is thus especially interesting from the fact that in this animal the polydactyle condition, though differing in degree of expression in various specimens, yet, in the greater number of cases, occurs in ways which may be interpreted as modifications of one plan, or rather of one plan for the hind foot and of another for the fore foot.

I arrange the evidence primarily according to the animal concerned, Cat, Man and Apes, Equidæ, Artiodactyles, &c. To these are added a few facts as to digital variations in Birds, but from the scantiness of the evidence and the difficulty of determining the morphology of the parts I have not found it possible to give a profitable account of these phenomena in other vertebrates below Mammalia.

In most of the groups increase in number of digits may be seen to occur in several distinct ways; and, just as in the case of teeth, mammæ, &c., it is possible to recognize cases of division of single members of series, and cases of addition to the series either at one of its ends (often associated with remodelling of other members of the series) or in the middle of the series.

Reduction in number of digits, or ectrodactylism as it is often called, is usually so irregular in the manner of its occurrence that little could be done as yet beyond a recitation of large numbers of cases amongst which no system can be perceived. For the present therefore the interest of these observations for the student of Variation is comparatively small and they are for the most part omitted.

To the irregularity of ectrodactylism in general certain cases of syndactylism are a marked exception and of these an account will be given.

After stating the morphological evidence as to numerical Variation in digits in the several groups, reference will be made to some collateral points of interest concerning such variations.

There is a good deal of evidence respecting the recurrence of digital variations in those lines of descent wherein they have appeared. Facts of this kind have been frequently seen in the case of Man, and other examples are known in the Cat, the Pig, the Ox, Deer, Sheep, &c. References to these cases will be given.

It will be seen that the facts contained in this section of evidence are of consequence rather as indicating the limits set on Variation, and from their bearing on the question of the nature of Symmetry and of Homology, than from any more direct application to the problem of Species, but even this cannot be said with much confidence.

There are in certain groups limbs such as the pes of Macropodidæ or that of Peramelidæ whose appearance forcibly recalls what is seen in some teratological cases and the possibility that they may have had such a sudden origin may well be kept in view[1].

CAT.

The apprehension of the chief features in the evidence as to digital variation in the Cat will be made more easy if a general account of the subject be given as a preliminary. In order to understand the peculiar phenomena seen in the limbs of polydactyle cats certain points of normal structure are to be remembered. Of these the most important relate to the claws and their disposition with regard to the second phalanx; for it is by this character that the relation of digits to the symmetry of the limb may be determined.

[1] In the case named this is all the more likely from the circumstance that according to THOMAS, *Cat. Marsup. Brit. Mus.*, p. 220, there is reason for supposing that the extraordinary condition of the digits II and III was attained independently in these two groups.

Hind foot.

The phenomena seen in the case of the hind foot are in some respects simpler than those of the variations in the fore foot, and for this reason they may conveniently be described first.

If the phalanges of the index of the hind foot, for example, be examined, it will be seen that the proximal phalanx is nearly bilaterally symmetrical about a longitudinal axis, but that the second phalanx is deeply hollowed out upon the external or fibular side. Into this excavation the ungual phalanx is withdrawn when the claw is in the retracted position. The retraction is chiefly effected by a large elastic ligament running from the outside of the distal head of the second phalanx and inserted into the upper angle of the last phalanx (see OWEN, *Anat. and Phys. of Vert.*, III. p. 70, fig. 36). The same plan is found in the digits II to V both of the fore foot and of the hind foot. By this asymmetrical retraction of the claw a digit of the right side may be differentiated at a glance from one of the left side, for the claw is retracted to the *right* side of a right digit and to the *left* side of a left digit. The importance of this fact will be seen on turning to the evidence, for it is found that with variation in the number of digits there is a correlated variation in their symmetry.

With respect to the tarsus little need be said. The proximal part of the tarsus contains three bones, the calcaneum, astragalus and navicular. The distal row consists of four bones, the cuboid and three cuneiform bones. In the majority of polydactyle cats that I have seen in which the tarsus is affected, the cuboid is normal and the ecto-cuneiform is also normal and recognizable; internal to the latter there are *three* small cuneiforms articulating with the navicular instead of two, making four cuneiforms in all. In some specimens there is no actual separation between the two innermost of these cuneiforms, but the lines of division between them are clearly marked.

In the normal hind foot of the Cat there are four fully formed toes, commonly regarded as II, III, IV and V, each having three phalanges. In the place where the hallux would be there is a small cylindrical bone articulating at the side of the internal cuneiform. As usually seen, all the four digits are formed on a similar plan, each having its claw retracted to the external or fibular side of the second phalanx, the four digits of a right foot being all right digits and those of left feet being all left digits. The rudimentary hallux has of course no claw.

Starting from this normal as the least number of digits, it will be found that a large proportion of cases are such that they may be arranged in an ascending or progressive series. In this series the following Conditions have been observed.

In the schematic representations of the limbs the words 'Right' or 'Left' signify that a digit is shaped as a right or as a left. The Roman numeral indicates that the digit to which it is assigned has the tarsal or carpal relations of the digit so numbered in the normal. For brevity each is described as a *right* foot.

I. The normal, consisting of four three-phalanged digits, each retracting its claw to the external, viz. *right* side, and a rudimentary hallux with no claw. In this foot therefore the digits enumerated from the external side are

Right.	Right.	Right.	Right.	Rudiment.
V	IV	III	II	I

II. Five digits, each with three phalanges. Of these the minimus and annularis borne by a normal cuboid are normal and are formed as *right* digits. The medius is borne by a normal ecto-cuneiform and is also a true right digit. Internal to this is a full-sized digit having the relations of an index and borne by a bone placed as a middle cuneiform. But the claw of this digit cannot be retracted to the external side of the limb, for the second phalanx is not excavated on this side. There is on the contrary a slight excavation on the *internal* side of the second phalanx, but this is very incomplete and the claw cannot be fully retracted, being in fact almost upon the middle line of the digit when bent back. This digit is thus intermediate between a right and a left. Nevertheless it is truly the index of this right foot, for it has the tarsal relations of an index.

Internal to this digit is another, which by all rules of homology should be the hallux, but it has three phalanges and is fashioned as a *left* digit, retracting its claw to the left (internal) side of the digit. This digit (Fig. 85, II, d^1) is borne jointly by *two* cuneiforms, c^1 and c^2, as shewn in the figure. There is thus one cuneiform more than there is in the normal. In this foot therefore the digits enumerated from the external side are as follows:—

Right.	Right.	Right.	Indifferent.	Left.
V	IV	III	II	I

Such a specimen is No. 472, right pes.

Between this state and the normal I have as yet met no intermediate. It might perhaps have been expected that a foot having four three-phalanged digits and a hallux with *two* phalanges would be a common form of variation. Such a condition has not however been seen, so far as I know.

III. The foot shewn in Fig. 85, I exemplifies the next condition. In it the three external digits, which are structurally the minimus, annularis and medius of a normal foot are normal in form, position and manner of articulation with the tarsus. Internal to the medius are three digits, of which the innermost has *two* phalanges (Fig. 85, I, d^1) and a claw which cannot be retracted, like the pollex of the normal fore foot. The other digits, d^3 and d^2, are fashioned as *left* digits, retracting their claws to the internal or left side of the limb. It will be seen that of them d^3 has the relations to the tarsus which an index should have. The tarsus is as in the last Condition.

In the specimen seen, c^1 and c^2 were not actually separate from each other, but there was a distinct line of division between them.

Here then the digits enumerated from the external side are as follows:—

Right.	Right.	Right.	Left.	Left	Hallux-like
V	IV	III	II	digit	digit

IV. The stage next beyond the last is shewn in Fig. 87, II. [The drawing is from a *left* foot.] Here there are six digits, each with three phalanges. The three externals are normal and true rights as before. The other three are all formed as *lefts*. Tarsus as before.

This foot may be represented thus:—

Right.	Right.	Right.	Left.	Left	Left
V	IV	III	II	digit	digit

As far as I have seen the last or fourth Condition is the commonest. There are doubtless many variants on these plans. No. 477 is an especially noteworthy modification of the third Condition and the cases of the hind feet in No. 478 must also be specially studied as not conforming truly to either Condition.

Fore foot.

I. The normal right fore foot has four digits II—V each with three phalanges all differentiated as rights, and a pollex with two phalanges, the last being non-retractile but bearing a claw. It may be represented thus:—

Right.	Right.	Right.	Right.	Pollex.
V	IV	III	II	I

Departures from this normal are more irregular than they are in the case of the hind foot. Those given in this summary being only a selection. For the others the evidence must be examined.

II. One specimen, No. 474, has the four external digits normal. The pollex however has three phalanges and is formed as a digit of the other side, thus:—

Right.	Right.	Right.	Right.	Left.
V	IV	III	II	I

III. The next Condition seen was as follows:—

Right.	Right.	Right.	Right.	Meta-carpal space.	Left	Indifferent
V	IV	III	II		digit	digit

IV. In the majority of polydactyle cats the manus has the digits II—V normal in shape and symmetry. Internal to the digit II are two digits more or less united in their proximal parts; sometimes the metacarpal only, sometimes the metacarpal and first phalanx are common to both. Of these two digits the external, that is, the one next to the digit II, is in some degree shapeless and imperfect, but the external branch is as a digit of the other side in form. Internal to this double digit is a seventh digit, sometimes with two phalanges, sometimes with three, but in either case the claw is as a rule non-retractile, and the digit is in this respect not differentiated as either right or left. Such a manus may be thus represented (cp. Fig. 86 a *left* manus):—

Right.	Right.	Right.	Right.	Amorphous	Left	Indifferent
V	IV	III	II	digit	digit	digit

As regards the carpus its changes are like those of the tarsus. When there are six metacarpals there are three carpals in the distal row internal to the magnum. That next the magnum may be supposed to be trapezoid, and the other two may be spoken of as first and second trapezium. In correspondence the length of the scapho-lunar is increased.

No comment can increase the interest of these curious facts. In the pes, as has been stated, with change in the number of digits there is change in the grouping and symmetry of the series of digits, and in particular the digit having the relations of the index or digit II is formed as the optical image of its neighbour III instead of forming a successive series with it. There is thus a new axis of symmetry developed in the limb, passing between the parts which form the digits II and III of the normal.

The evidence of the above statements may now be given.

*472.   **Cat** having the digital series of each extremity abnormal, being that preserved in the Coll. Surg. Mus., *Teratological Catalogue*, 1872, Nos. 305 and 306.

Right pes (Fig. 85, II). Digits III, IV and V normal right digits. Internal to these are two digits each having three phalanges and claws. That lettered d^1 is formed as a *left* digit but d^2 is almost indifferent, the second phalanx being slightly hollowed on the *inside*. Internal to the external cuneiform there are three small bones, of which the inner two together bear the digit d^1. [This is the Condition II of the pes.]

Left pes has the same structure as the right so far as can be seen from the preparation (in which the muscles remain). The digits III, IV and V are normal left digits, but internally to them there are two digits each with three phalanges, of which the external is an indifferent digit, while the internal is formed as a right. [Condition II of the pes.]

Left manus. The digits II, III, IV and V are normal. But the carpal of the distal series (trapezoid) which bears the digit II is imperfectly separated from a similar bone placed internal to it. This second part of the trapezoid bears a metacarpal which articulates with a full-sized digit of three phalanges formed as a *right* digit. From the external side of the first phalanx of this

318   MERISTIC VARIATION.   [PART I.

digit there is given off a rudimentary digit, which has however a complete claw, but its bones do not differentiate it as right or left.

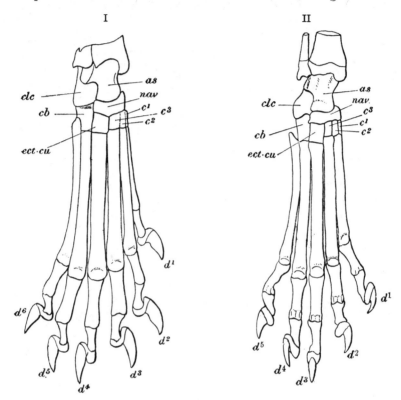

FIG. 85. I. Right pes of Cat No. 473, shewing condition III of the pes. II. Right pes of Cat No. 472 shewing Condition II of the pes.

as, astragalus. c^1, c^2, c^3, three ossifications representing the ento- and meso-cuneiforms of the normal. *cb*, cuboid. *clc*, calcaneum. d^1—d^6, the digits numbered from the inside. *ect, cu*, ecto-cuneiform. *nav*, navicular.

(From specimens in Coll. Surg. Mus.)

The "pollex", d^1, has two phalanges and is rather slender. The trapezium which bears it is not separated from the scaphoid. (Fig. 86). [Condition IV of the manus.]

Right manus. This is exactly like the left manus so far as can be seen from the dissection, except for the fact that the rudimentary digit borne by the large digit external to the "pollex" is much more reduced than in the case of the left manus. The digit which supports it is fashioned as a *left* digit. [Condition IV of the manus.]

*473.   **Cat** having digital series of all feet abnormal, being the specimen in Mus. Coll. Surg., *Teratol. Catalogue*, 306 B.

Right pes. The digits III, IV and V (Fig. 85, I) are normal and are fashioned as right digits. The cuboid and external

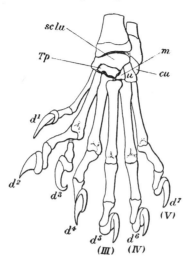

FIG. 86. Left manus of Cat No. 472, shewing Condition IV of the manus.
cu, cuneiform. d^1—d^7, digits numbered from the inside. *m*, magnum. *sclu*, scapho-lunar. *Tp*, trapezoid.
(From a specimen in Coll. Surg. Mus.)

cuneiform (*cb* and *ect. cu*) are also normal. Internally to the external cuneiform there is a long flat bone which is grooved in such a way as to divide it into three parts (c^{1-3}) and each of these bears a digit.

Of these digits, d^2 and d^3 have each three phalanges, but d^1 has only two phalanges and may therefore be called a hallux. The digits d^2 and d^3 are fashioned not as right digits but as *left* digits, and their claws are thus retracted towards the *internal* side of the second phalanges, which are hollowed out to admit of this.

The bones of the hallux are not thus differentiated as right or left, for the claw is not retractile. The navicular is enlarged in correspondence with the presence of the fourth cuneiform element and the astragalus and calcaneum are normal. (Fig. 85, I). [Condition III of the pes.]

Left pes. This foot is almost exactly like the right. As in it, the digits III, IV and V are normal and are left digits. Internal to this are three digits, viz. a hallux and two long digits with three phalanges which are both made as *right* digits. The bones of this foot have not been cleaned. [Condition III of the pes.]

Right manus. This is formed on the same plan as the manus of the last animal, differing from it in details of the carpus, chiefly in the presence of two separate trapezial elements. The four digits on the external side, II—V are shewn by their claws to be true right digits. They articulate in a normal way with the trapezoid, magnum and unciform, and are thus clearly II, III, IV and V. The metacarpals of the "pollex" and of the double digit corresponding to d^2 and d^3 of Fig. 86 articulate with two separate carpal bones of the distal row. The external of these bears a rather thick metatarsus which peripherally gives articulation to two digits. Of these the internal is well formed and bears a claw which slides up on its internal side, and thus shews it to be formed as a *left* digit. The other is misshapen in its proximal phalanx which perhaps contains two phalangeal elements compounded together and aborted; hence the relation of this digit to the symmetry of the limb is not apparent. The claw and last phalanx are well formed. The innermost carpal bone is nearly normal and bears an almost normal "pollex." [Condition IV of the manus.]

Left manus. This foot has not been dissected, but from examination it appears that the digits II, III, IV and V are normal like those of the right manus. As in it, there is a "pollex" with two proper phalanges, but the metacarpal of the "pollex" is in its proximal part united with the metacarpal of an imperfectly double digit corresponding to d^2 and d^3 of Fig. 86. The division between the two parts of this double digit is not so complete in the left manus as it is in the right and from external examination it appears that the phalanges of the two are not separate. There are two claws of which one is rudimentary and the pads of the two are separated only by a groove. There is nothing to indicate whether these digits are formed as right or left digits. [Approaches Condition IV of the manus.]

*474. **Cat** having supernumerary digits. This specimen belonged to the strain of polydactyle Cats observed by Mr POULTON (see No. 480) and I am indebted to Mr J. T. CUNNINGHAM for an opportunity of examining it.

Left manus. Five digits, the normal number. The "pollex" however is a long digit, composed of *three* phalanges, which reaches very nearly to the end of the index. The claw of this digit is not retracted to the outside of the second phalanx, like that of a normal digit, but to the *inside*, and the chief elastic ligament is on the inside of these joints instead of being on the outside as in a normal digit. This pollex therefore may be said to be fashioned as a *right* digit, bearing the same relation to the others as a right limb bears to the left. The flexors and extensors of this digit were fully developed. The carpal series was normal. [Condition II of the manus.]

Right manus. Six digits fully formed, one bearing an additional nail on the third digit from the inside. Beginning from the outer or ulnar side, there are four normal right digits, placed and formed as V, IV, III and II respectively. Internal to these are two digits, the outermost having three phalanges, being shaped as a *left* digit and bearing a minute supernumerary nail in the skin external to the normal nail. The innermost digit has two phalanges, and is formed like a normal pollex, excepting that its claw was very deep and looked as if it were formed from the germs of two claws united and curving concentrically. . The carpus as regards number of elements was normal, but the trapezium and trapezoid were both of rather large size, and the pollex articulated partly with the trapezium but chiefly with the downward process on the radial side of the scapho-lunar. [This approaches Condition IV of the manus, but in it the external of the two united digits is only represented by the minute extra nail.]

Left pes. Six digits, each having three phalanges. The three outer digits were formed as left digits, but the three inner digits were shaped like right digits. The internal cuneiform is double the normal size, but is not divided into two pieces. It bears the two internal digits, of which the innermost is ankylosed to it. [Condition IV of the pes.] Compare Fig. 87, II.

Right pes. Same as the left, except for the fact that the two internal digits are completely united in their metacarpals and first phalanges, and the cuneiform series consists of four bones, two of which correspond to the internal cuneiform of double size described in the left foot. (Compare Fig. 85, I, c^1 and c^2.) [Condition IV of the pes, save for the union of the metacarpals of the two internal digits.]

475.   **Kitten** belonging to Mr Poulton's strain (see No. 480) and kindly lent by him to me for examination. The specimen was very young and the carpus and tarsus were not dissected.

Left manus. Six digits, all with three phalanges. The two internal digits are separated by a space from the others so as to form a sort of lobe. The claw of the innermost digit is retracted on the top of the second phalanx and not to the side, so that this digit is not differentiated either as a right or a left. The next digit is a *right* and the four external digits (II, III, IV and V) are normal lefts. [Condition III of manus.]

Right manus. Same as left.

Left pes. Same as left pes of No. 474 [sc. Condition IV of the pes].

Right pes: same as the left [Condition IV of the pes].

476.   **Cat** having its extremities abnormal, the property of the Oxford University Museum and kindly lent for examination; bones only preserved.

Right pes. Like the left pes of No. 474, but c^1 not separated from c^2. [Condition IV of pes.]

Left pes. Like the right, but c^1 separate from c^2. [Condition IV of pes.]

Right manus. The four external digits II—V normal. The double digit like that of No. 472. The innermost digit with three phalanges, but the claw not retracted to one side more than to the other. [Condition IV of the manus.]

Left manus. The same as the right. [Condition IV of the manus.]

*477.   **Cat** having all extremities abnormal, also the property of the Oxford University Museum.

Left pes. Like the left pes of case No. 474 [sc. Condition IV of the pes] represented in Fig. 87, II.

Right pes a peculiar case (Fig. 87, I). The digits V, IV and III are normal right digits. The digit II marked 3 in the figure is

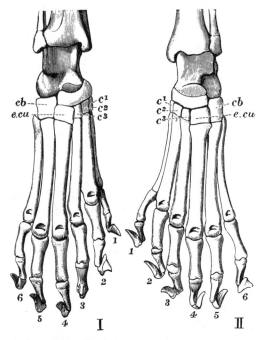

Fig. 87. Hind feet of Cat No. 477.
I. Right pes not truly conforming to any of the Conditions numbered.
II. Left pes shewing the ordinary form of Condition IV of the pes.
Lettering as in Fig. 85. (From a specimen in Oxford Univ. Mus.)

very slightly differentiated as a right digit, but the excavation on the external side is very slight, and the claw when retracted is almost on the middle of the second phalanx. The digit 2 of the figure is a *left*, and internal to it is a three-phalanged digit of which the claw is not retracted into any excavation. [Not conforming to any of the Conditions specified.]

478.   **Cat** having all feet abnormal, kindly lent to me by Mr Oldfield Thomas.

Left pes. Digits V, IV, III normal lefts. The next internally (II) is a three-phalanged digit formed as a right. The next is a thick three-phalanged digit *with a partially double nail and double pad.* This is not differentiated as either right or left. The innermost digit is a two-phalanged hallux-like digit, not differentiated as right or left. [Not conforming to any condition in my scheme.]

Right pes. The same as the left except that the digit II is only slightly differentiated as a left. The next has a double nail, and the innermost is hallux-like as described for the other foot. [Not conforming to any condition of my scheme.]

Right manus. As in No. 472. "Pollex" with two phalanges. [Condition IV of the manus.]

Left manus. Same as right, but the "pollex" is only represented by a single bone not differentiated or divided into metacarpal and phalanges and bearing no claw. [Approaches Condition IV of the manus.]

479. **Cat**. A *left pes* bearing abnormal digits. The digits II, III, IV and V are normal and are true left digits. Internal to these are two metatarsals which are united centrally and peripherally but are separate in their middle parts. These two metatarsals by their common distal end bear amorphous phalanges belonging to three digits. There are two large claws and one rudimentary one. [For details the specimen must be seen.] The navicular bone is divided into two distinct bones, of which one carries the external cuneiform and a small cuneiform for the digit II, the metatarsal of which is rather slender and compressed in its proximal part. The internal part of the navicular bone bears two cuneiforms, one for each part of the united metatarsals. The digits borne by these metatarsals are so misshapen that it is not possible to say anything as to their symmetry. Mus. Coll. Surg., *Terat. Catal.*, No. 306 A. [This specimen does not conform to any of the Conditions of my scheme.]

*480. In the case of the Cat the polydactyle condition has been observed by POULTON (*Nature*, XXIX. 1883, p. 20, *figs.*; *ibid.*, XXXV. 1887, p. 38, *figs.*) to recur frequently in the same strain. A female cat had six toes on both fore and hind feet. The mother of this cat had an abnormal number of toes not recorded. The grandmother and great-grandmother were normal. Two of the kittens of the 6-toed cat had seven toes both on the fore and hind feet [no 7-toed *pes* among specimens examined by me]. Many families produced by the 6-toed cat, and among them only two kittens with 7 toes on *all* feet, but between this and the normal numerous varieties seen. The abnormality is not in all cases symmetrical on the two sides of the body. The pads of the different toes are sometimes compounded together. In some cases an extra pad was present on the hind foot behind and interior to the central pad. The second pad was sometimes distinct from the central pad and sometimes was united with it. [From the figures it appears that the secondary central pad in the pes bore to the digits internal to the axis of symmetry a relation comparable with that which the chief central pad bears to the digits III—V, but the secondary central pad is at a higher level than the primary one.] It was especially noted that the details in the arrangement of the pads were inherited in several instances.

The history of the descendants of the 6-toed cat was followed and a genealogical tree is given shewing that the abnormality has been present in a large proportion of them. This was observed in five generations from the original 6-toed cat, so that including the mother of the 6-toed cat the family has contained polydactyle members for seven generations. It may reasonably be assumed that in most of these cases the fathers of these kittens have been normal cats and a good deal of evidence is adduced which makes this likely.

It was observed also that some normal cats belonging to this family gave birth to polydactyle kittens. In the later period of the life of the original 6-toed cat she gave birth to kittens which were all normal.

I know no case of reduction in number of digits or of syndactylism in the Cat.

Man and Apes.

Increase in number of Digits.

Increase in the number of digits occurs in Man in many forms. Among them may be distinguished a large group of cases differing among themselves but capable of being arranged in a progressive series like that described in the Cat. These cases are all examples of amplification or proliferation of parts internal to the index of the manus.

Taking the normal as the first Condition, the next in the progress is a hand having the digits II—V normal, but the thumb with three phalanges, or as the descriptions sometimes say, "like an index." (Condition II.)

In the next condition a two-phalanged digit is present internal to the three-phalanged "thumb." (Condition III.) In the next Condition the digit internal to the three-phalanged "thumb" has itself three phalanges. (Condition IV.) A variant from this occurred in the left hand of a child (No. 488) of parent having hands in Condition IV. In the child the right hand was in Condition IV, but in the left there were the usual four digits II—V, and internal to them two complete digits, each of three phalanges, but of these the external had a small rudimentary digit arising from the metacarpus. Hence the hand may be described as composed of two groups, the one containing four and the other three digits.

In one case, No. 490, the right hand was in Condition IV, but the left hand was advanced further. For in it the metacarpal of the innermost digit bore a 2-phalanged digit internally to its 3-phalanged digit. This may be considered as a Condition V.

The number of phalanges in the digits in these Conditions may be represented thus. The ‖ marks the metacarpal space. (The hand is supposed to be a right.)

Condition	I	2 ‖	3 3 3 3
,,	II	3 ‖	3 3 3 3
,,	III	2 3 ‖	3 3 3 3
,,	IV	3 3 ‖	3 3 3 3
,,	V	2 3 3 ‖	3 3 3 3

Distinct from these Conditions are the states sometimes described as "double-hand." In the full form of this there are eight digits, each of three phalanges. The eight digits are arranged in two groups, four in each group. The two groups stand as a complementary pair, the one being the optical image of the other; or in other words, the one group is right and the other is left.

Besides the double-hand with eight digits there are also forms of double-hand with six digits, arranged in two groups of three and three.

Lastly, there are cases of double-hand having seven fingers, an external group of four and an internal group of three. Thus expressed these cases seem to come very near that mentioned as a variant on Condition IV, but in one and perhaps both of these double-hands there was in the structure of the fore-arm and carpus a great difference from that found in the only recorded skeleton of Condition IV.

At first sight it would naturally be supposed that these double-hands in one or all kinds stand to the other Conditions in the some relation that Condition IV of the pes in the Cat does to the other polydactyle conditions in the Cat. But the matter is complicated by the fact that the evidence goes to shew that in the human double-hands the bones of the arm and carpus may be modified, and in DWIGHT'S example of seven digits (No. 489) at all events, and perhaps in other double-hands, an *ulna-like bone takes the place of the radius,* or in other words, *the internal side of the fore-arm is fashioned like the external side.* In the polydactyle cats the bones of the fore-arm were normal, as are they also substantially in cases of the human Conditions III and IV, which have been dissected. Further, in some of the human cases of eight digits the abnormality was confined to one hand, which is never the case in the higher condition of polydactylism in the Cat, so far as I know. These circumstances make it necessary to recognize the possibility that some at least of the human double-hands are of a different nature from the lower forms of polydactylism. This subject will be spoken of again after the evidence as to the variation of digits has been given (Chap. XIV. Section (4).)

In addition to cases more or less conforming to schemes that can be indicated are several which cannot be thus included. These will be duly noticed when the more schematic cases have been described. That any of the cases can be arranged in a formal sequence of this kind is perhaps surprising, and the relations of some of the Conditions, II and III for instance, to each other must at once recall the principle seen already in other examples of addition of a member at the end of a successive series of parts, notably in the case of Teeth (see p. 272). It was then pointed out that when a new member is added beyond a terminal member whose size is normally small relatively to that of the normal penultimate, then the member which is normally terminal is raised to a higher condition. Now this same principle is seen in Condition III of the polydactyle manus.

Attention must nevertheless be forthwith called to the fact that a two-phalanged digit[1] may be present internal to the thumb (usually arising from it) though the thumb has still but two phalanges. But generally these cases may properly be described as examples of *duplicity* of the thumb; and as was well seen in the case of Teeth, any member of a series may divide into two though the rest of the series remain unaltered. Duplicity of a member without reconstitution of the series is to be recognized as one occurrence, and change in number associated with reconstitution of other members especially, of adjacent members, is another. In Teeth and other Meristic series these two phenomena are both to be seen, though as was pointed out (p. 270) they pass insensibly into each other.

Another feature to be specially mentioned in this preliminary notice is the difference in the manner in which the higher forms of polydactylism appears in the human foot from that seen in the human hand. In the hand there is this strange group of cases forming a progress from the normal hand to Condition V, besides the distinct series of double-hands. Polydactyle feet on the contrary do not in Man, so far as they have been observed (with the doubtful exception of Nos. 499 and 500), develop a new symmetry.

CASES OF POLYDACTYLISM ASSOCIATED WITH CHANGE OF SYMMETRY.

A. *Digits in one Successive Series.*

*481. Man having a "supernumerary index" on each hand. *Left hand.* No "thumb" present. In its stead there is a digit having three phalanges which "performs its office." The middle phalanx was abnormally short. The first intermetacarpal space was not great. [Degree of opposability not stated.] *Right hand.* In addition to four normal fingers there was a three-jointed digit

---

[1] A case in which a 3-phalanged digit was placed on the radial side of the pollex is mentioned by WINDLE, *Jour. Anat. Phys.*, XXVI. p. 440, but has not yet been described. No other such case is known to me. This perhaps should be classed with double-hands. Cp. No. 502.

CHAP. XIII.] DIGITS : MAN. 327

which could be opposed to them and could perform all the movements of flexion, &c. Internal to this three-jointed digit was a rudimentary thumb having only one phalanx and no nail. [Relations of metacarpals to each other not particularly described.] GUERMONPREZ, F., *Rev. des mal. de l'enfance*, IV. 1886, p. 122, *figs*. [Left hand Condition II; right hand almost Condition III.]

482.   **Girl** having a three-jointed thumb, resembling a long forefinger. ANNANDALE, *Diseases of the Fingers and Toes*, p. 29, Pl. II. fig. 19. [Condition II.]

483.   Man having a thumb with three phalanges on each hand. Feet normal. In the thumbs the metacarpal is $2\frac{1}{4}$ in. long; the first phalanx $1\frac{3}{4}$ in., being longer than usual. The second phalanx is longer on the radial side than on the external side, causing the distal phalanx to curve towards the index. On the internal it measures $\frac{5}{8}$ in., in the middle $\frac{1}{2}$ in., and on the ulnar side $\frac{1}{4}$ in. The distal phalanx is 1 in. long. When the left thumb is straightened it passes $\frac{1}{6}$ in. beyond the joint between the 1st and 2nd phalanx of the index. In the right hand the thumb scarcely reaches that joint. The utility of the thumb is not impaired. A maternal aunt had a similar thumb on right hand. STRUTHERS, *Edin. New Phil. Journ.*, 1863 (2), p. 102, Pl. II. fig. 6. [Both hands Condition II.]

*484.   Father and three children, each having 3-phalanged thumbs shaped as indices and not opposable. [Full description *q.v.*] Paternal grandmother had double-thumb. FARGE, *Gaz. hebd. de méd. et chir.*, Ser. 2, II. 1866, p. 61.

*485.   Man having the following abnormalities of the digits. (Fig. 88). *Right hand.* The number of digits was normal, but the

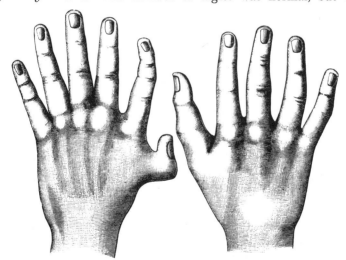

FIG. 88. Right and left hands of No. 485. Right hand in Condition II; left hand in Condition III. (After WINDLE.)

radial digit or thumb had three phalanges in addition to the metacarpal, all the articulations being moveable. Relatively to the others their digit was placed as a thumb. *Left hand.* The digit corresponding with the thumb was composed of three phalanges like that of the right side, and though finger-like in form it was functionally a thumb. On the radial side of this 3-jointed digit there was a supernumerary digit composed of two phalanges articulating with the metacarpal bone of the 3-jointed thumb. This supernumerary digit had a well-formed nail. The 3-jointed thumb of the left hand was longer than that of the right hand (measurements given), WINDLE, B.C.A., *Journ. of Anat.* XXVI. 1891, p. 100, Pl. II. [Right hand, Condition II; left hand, Condition III.]

*486. Man having 3 phalanges in the thumb of the left hand together with a supernumerary digit. (Fig. 89.) This case in several respects resembles the left hand of the subject described by WINDLE. The four fingers were normal. The thumb stood in its normal relations to them, but was finger-like in form, having three phalanges in addition to the metacarpal. On the radial side of

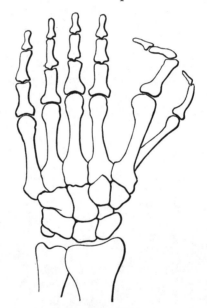

FIG. 89. Bones of left hand of No. 486, shewing Condition III.
(After RIJKEBÜSCH.)

this 3-phalanged digit there was a supernumerary digit, having two phalanges and a separate metacarpal, which articulated with the head of the metacarpal of the thumb and the trapezium. In the carpus of this hand there was a supernumerary bone which is described as an *os centrale*. The bones and muscles of this limb are described in detail. The thumb and the supernumerary digit were closely webbed together and were very slightly moveable. Specimen first described by RIJKEBÜSCH, *Bijdr. tot de Kennis der Polydactylie*, Utrecht, 1887, *Plates*, and subsequently by SPRONCK, *Arch. néerl.*, XXII. 1888, p. 235, Pl. VI.—IX. [Condition III.]

*487. Woman having 6 digits on each hand and foot as follows. In *each hand* the thumb has three phalanges, and internal to it articulating with the same metacarpal is an extra digit having two phalanges [measurements given] webbed to the three-phalanged thumb. [Condition III of the manus.] *Right foot* has six complete metatarsals and digits very regularly set, one of them being internal to but *longer* than the hallux which has two phalanges as usual. The digit internal to it has also two phalanges. *Left foot* has also an extra digit with two phalanges longer than the hallux, placed internal to and articulating with the metatarsal of the hallux which has two phalanges as usual. Many members of family polydactyle [particulars given]. STRUTHERS, *Edin. New Phil. Jour.*, 1863 (2), p. 93. [Note in this case that in the feet the digits added internally to hallux are *greater* than it, and they thus stand as the largest terms in the series, the other members being Successive to them. The series thus does not decline from the hallux both internally and externally in the way seen in most other cases of extra digits on the internal side of the limb.]

488. Man having six digits, each with three phalanges, on each hand.

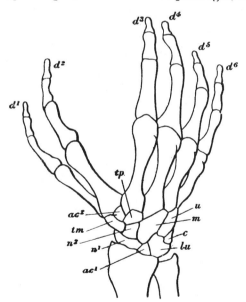

FIG. 90. Bones of right hand of No. 488 shewing Condition IV.
n^1 and n^2 represent the scaphoid. *lu*, lunar. *c*, cuneiform. *tm*, trapezium. *td*, trapezoid. *m*, magnum. *u*, unciform. ac^1, ac^2 are supernumerary bones.
(After RÜDINGER.)

The digits were arranged in two groups, which were to some extent opposable to each other. The digits II, III, IV and V stood in their normal positions and were properly formed. In the place where the thumb should stand there were two digits, each with three phalanges. Of these the external (d^2) was of about the length and form of the index finger while the internal, d^1, was a good deal shorter and more slender. The bones of the carpus are shewn in Fig. 88. The scaphoid was represented in the right hand by two bones n^1 and n^2, and there were two accessory bones, ac^1 and ac^2 placed in the positions shewn. The two hands were almost exactly alike, save for slight differences in the carpal bones [see original figures], and for the fact that in the left hand the internal of the two digits of the radial group was rather more rudimentary. RÜDINGER, *Beitr. zur Anat. des Gehörorgans, d. venosen Blutbahnen d. Schädelhöhle, sowie der überzähligen Finger*, München, 1876, Plate. [Both hands in Condition IV.]

489. A female child born to the last case, No. 488, had the right hand in the same condition as that of the father, while the left hand differed from it in the presence of an additional rudimentary finger arising from the ulnar side of the digit d^2. This additional finger bore a nail but it appeared to consist of two joints only and to be attached to the metacarpus by ligamentary connexions. RÜDINGER, *ibid*. [Right hand in Condition IV; left hand departing from the Conditions enumerated. Compare with manus of Cat, Fig. 84.]

*490. Man. *Right hand* bore six digits and metacarpals. The most external digit was a normal minimus, succeeded by digits IV and III webbed together. Next to III there was an index. Internal to this and separated from it by a small metacarpal space was a 3-phalanged long digit much as in Windle's case, No. 481, and internal to it is a 2-phalanged thumb of nearly normal form like that of No. 485. *Left hand* bore seven digits but six metacarpals. Minimus normal. IV, III and II webbed together. Internal to II was a 3-phalanged digit much as in the right hand; but internal to this there was a metacarpal bearing *two* digits, an external having 3 phalanges and an internal having 2 phalanges. Each *foot* had six digits and six metatarsals (*q. v.*). Redescribed from the account and figures given by GRUBER, *Bull. Ac. Sci. Pét.*, XVI. 1871, p. 359, *figs*. [Right hand Condition IV, left hand Condition V.]

491. Child having six fingers on each hand. The fingers were united together. In the thumb [? both] there were three phalanges and the length of the thumb was as great as that of the "other fingers." DUBOIS, *Arch. génér. de Méd.*, 1826, Ann. IV. T. XI. p. 148; this case is quoted by GEOFFROY ST HILAIRE, *Hist. des Anom.*, I. p. 227, *Note*. [? Condition IV.]

491, a. New-born male child having on the right hand two "thumbs" each with three phalanges. OBERTEUFER, J. G., *Stark's Arch. f. Geburtsh.*, 1801, XV. p. 642. [Condition IV.]

(*No more cases known to me.*)

B[1]. *Digits in two homologous groups, forming "Double-hands."*

*492. DOUBLE-HAND I. *Seven digits in two groups of four and three.* Male: left arm abnormal, having seven digits arranged in two groups, the one an external group of four normal digits, and the other an internal group of three digits[2]. (Fig. 91.) Described from a dried specimen in Mus. of Harvard Med. School. The man was a machinist and found the hand not merely very useful to him in his business, but he also thought that it gave him advantages in playing the piano.

"The fore-arm consists of the normal left ulna and of a right one in the place of a radius. The left one shews little that calls for comment, excepting that there is a projection outward at the place of the lesser sigmoid cavity to join a corresponding projection from the other ulna. The upper surface of this projection articulates with the humerus. At the lower end the styloid process is less prominent than usual, and the head rather broad. The right or extra ulna is put on hind side before, that is, the back of the olecranon projects forward over the front and outer aspect of the humerus. If the reader will place his right fore-arm on the outer side of the left one he will see that it is necessary for the

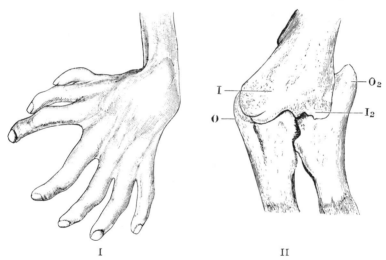

FIG. 91. I. The left hand of No. 492 from the dorsal surface.
II. The humerus and two bones of the fore-arm at the elbow of the same case. O, olecranon. O², the secondary "olecranon". I, the inner condyle of the humerus. I², the second or external "inner condyle."
(After DWIGHT.)

ulna to be thus inverted if the thumbs are to touch and the palms to be continuous. This olecranon is thinner, flatter, and longer than normal. The coronoid process is rudimentary. From the side of this process and from the shaft just behind it arises the projection already

[1] Every case known to me is given.
[2] This is the case reported by Jackson, to *Bost. Soc. of Med. Imp.*, 1852.

referred to which meets a similar one from the normal ulna [Fig. 89, II].
On the front of this there is a small articular surface looking forward
which suggests a part of the convexity of the head of the radius. The
upper articular surface shews a fissure separating it from the side of
the olecranon which is not found in the normal ulna. These projections
which touch each other are held together by a strong interosseous ligament. The lower end of this ulna is very like the other, only somewhat
broader. The mode of union of the lower ends could not be seen without
unwarrantable injury to the specimen. There can hardly have been
any definite movement between these bones. Perhaps the ligaments may
have permitted some irregular sliding, but it is impossible to know.
These bones have been described first because their nature is very clear
and, once understood, is a key to the more difficult interpretation of the
lower end of the humerus."

The upper end of the humerus presented nothing noteworthy. A
detailed description and figures are given, from which it appears that
the lower end of the humerus had such a form as might be produced
by sawing off the greater part of the external condyle and applying in
place of it the internal condyle of a *right* humerus.

The carpus seen from the dorsal side had the structure shewn in the
diagram (Fig. 92). The proximal row consisted of three bones besides
the two pisiforms (p^1 and p^2). There was a cuneiform at either side of
the wrist, and between them a bone evidently composed of a pair of
semilunars, having a slight notch in its upper border. At each end of

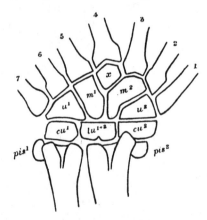

Fig. 92. Diagram of the carpal bones in the left hand of No. 492 from the dorsal surface.

pis^1, cu^1, u^1, m^1, pisiform, cuneiform, unciform and magnum of the external or normal half of the hand consisting of four fingers; pis^2, cu^2, u^2, m^2, the similar bones for the internal group of three fingers. lu^{1+2}, the compounded lunar elements corresponding to the two groups. x, bone placed as trapezoid.
(After Dwight.)

the second row is an unciform bearing the middle and ring fingers.
Next came two ossa magna very symmetrically placed, each bearing the
metacarpal of a medius. Between these is a bone which Dwight states

to have clearly represented the trapezoid of the left hand, bearing an index finger. The metacarpals and phalanges needed no description.

The muscles are described in detail [*q.v.*]. Some of the features in the distribution of the arteries and nerves are of interest, and I transcribe Dwight's account in full. It appears that, like the bones, the vessels and nerves proper to the radial side of a normal left arm have in a measure been transformed into parts proper to the ulnar side of a right arm.

"THE ARTERIES. The brachial divides at about the junction of the middle and lower thirds of the humerus. The main continuation, which is the ulnar proper, runs deeply under the band thought to represent the *pronator radii teres*, to the deep part of the fore-arm where it gives off the interosseous. Above the elbow there is a branch running backward between the internal condyle and the olecranon. The interosseous branches are not easy to trace. There seems to be an anterior interosseous and three branches on the back of the forearm, one running on the membrane and one along each bone. At least two of them share in a network on the back of the carpus. Having reached the hand the ulnar artery runs obliquely across the palm to the cleft between the two sets of fingers, supplying the four normal fingers and the nearer side of the extra middle finger. The other branch of the brachial crosses the median nerve and runs, apparently superficially, to the outer side of the fore-arm. It supplies the little and ring fingers and the corresponding side of the middle finger of the supernumerary set. There is no anastomosis in the palm between the superficial branches of the two arteries. Each gives off a deep branch at the usual place, which forms a deep palmar arch from which some interosseous arteries spring. There is also an arterial network over the front of the carpal bones. The arteries of the deep parts of the hand cannot all be seen.

THE NERVES. The ulnar nerve proper pursues a normal course and supplies the palmar aspect of the little finger and half the ring finger of the normal hand. Near the wrist it gives off a very small posterior branch, which is not well preserved, but which seems to have had less than the usual distribution. The median nerve is normal as far as the elbow, running to the inner side of the extra condyle. It is then lost in the dried fibers of the *flexor sublimis*, from which it emerges in two main divisions near the middle of the fore-arm. The inner of these soon divides into two, of which one supplies the adjacent sides of the ring and middle fingers and the other those of the middle and index fingers of the normal hand. The outer division of the median supplies the outer side of the index and both sides of the extra middle finger and one side of the extra ring finger. One of the branches to the index gives off a dorsal branch, and there is a doubtful one for the extra middle finger. The musculo-spiral nerve passes behind the humerus as usual. A nerve which is undoubtedly continuous with it emerges from the hardened muscles over the fused outer condyles. It seems to be the radial branch changed into an ulnar. It runs with the extra ulnar artery to the hand and sending a deep branch into the palm, goes to the ring finger. There is a detached branch on the other side of the little finger which in all probability came from it. The deep branch sends a twig along the metacarpal bone of the ring finger. It probably supplied the side of the ring finger left unprovided for, but this is uncertain. Assuming this to have been the case, each ulnar nerve supplies the palmar surface of one finger and a half, the median supplying the remaining fingers of both hands. Unfortunately no dorsal branches except those mentioned have been preserved."

DWIGHT, T., *Mem. Boston Soc. of N. H.*, 1892, Vol. IV. No. X. p. 473, Pls. XLIII and XLIV.

[This is a case of high significance. We shall come back to it hereafter. Meanwhile it will be noted that in it we meet again the old difficulty so often presented by cases of Meristic Variation. In this fore-arm there is already one true ulna. Internal to it is another bone also formed as an ulna. We may therefore, indeed we must, call it an ulna. But *is* it an "ulna"? To answer this we must first answer the question *what* is an ulna? Similarly, is the second pisiform a "pisiform," or is the second ulnar nerve an "ulnar" nerve? These questions force themselves on the mind of anyone who tries to apply the language of orthodox morphology to this case, but to them there is still no answer. Or, rather, the answer is given that an "ulna," a "pisiform" and the like are terms that have no fixed, ideal meaning, symbols of an order that we have set up but which the body does not obey. An "ulna" is a bone that has the form of an ulna, and a "pisiform" is that which has the form of a pisiform. If we try to pass behind this, to seek an inner and faster meaning for these conceptions of the mind, we are attempting that for which Nature gives no warrant: we are casting off from the phenomenal, from the things which appear, and we set forth into the waste of metaphysic.]

493. Boy having abnormalities in the left hand as follows. The four outer fingers II—V are normal in form and proportions. Internal to these is firstly an opposable digit with a single metacarpus and single proximal phalanx but having two distal phalanges side by side webbed together. Internal to this partially double thumb are two digits in series, each with a metacarpal and three phalanges, respectively resembling the annularis and minimus of a *right* hand. STRUTHERS, *Edin. New Phil. Jour.*, 1863 (2), p. 90, Pl. II. fig. 5. [Not representing any of the Conditions.]

494. Male infant, one year and five months, examined alive, having the right hand abnormal, possessing seven digits, arranged in two groups, an ulnar group of four and a radial group of three. Each digit had three phalanges, but the ring and middle fingers of the ulnar group are webbed in the region of the proximal phalanges. The ulnar group seemed to articulate with the carpus in the usual way. The radial group probably formed joints with more than one facet on the trapezium, and possibly also with a surface on the lower end of the radius. It did not seem that the carpal bones were increased in number, for the right wrist had the same circumferential measurement as the left, which was normal. The lower end of the ulna did not seem to articulate normally with the carpus. The elbow was also abnormal, and it seemed "as if the ulna were dislocated inwards." BALLANTYNE, J. W., *Edin. Med. Jour.*, 1893, CDLI. p. 623, *fig*. [Possibly this condition approached to that found in the last cases.]

495. DOUBLE-HAND II. *Eight digits in two groups of four and four.* Woman (examined alive) having eight fingers in the left hand arranged as follows (Fig. 93). With the exception of the left arm the body was normal. The limb was very muscular. The shoulder-joint was natural. The external condyloid ridge of the humerus was strongly defined. The muscles and tendons of the fore-arm were so prominent that it was not easy to decide whether there was a second radius or ulna, but Murray eventually came to the

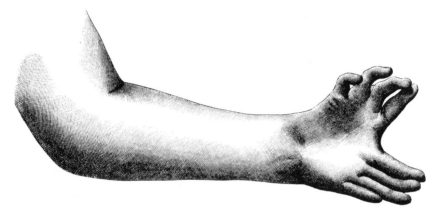

FIG. 93. Left hand of No. 495. (After MURRAY.)

conclusion that there was no such extra bone. The fore-arm could be only partially flexed. The eight fingers were arranged in two groups of four in each, one of the groups standing as the four normal fingers do, and the other four being articulated where the thumb should be. There was no thumb distinguishable as such, but it is stated that there was a protuberance on the dorsal side of the hand, between the two groups of fingers, and this is considered by Murray to represent the thumbs, for according to his view the limb was composed of a pair of hands compounded by their radial sides. In the figure of the dorsal aspect which is given by Murray taken from a photograph, this protuberance cannot be clearly made out. The four radial fingers in size and shape appeared to be four fingers of a *right* hand. In the radial group of fingers, the "middle" and "ring" fingers (6 and 7) were webbed as far as the proximal joints, and the movements of the fingers of this group were somewhat stiff and imperfect. Between the two groups of fingers there was a wide space as between the thumb and index of a normal hand, and the two parts of the hand could be opposed to each other and folded upon each other. The power of independent action of the fingers was very limited. No single finger could be retained fully extended while the other seven fingers were flexed, but if both "index" fingers (4 and 5) were extended,

336   MERISTIC VARIATION.   [PART I.

the other six fingers could be flexed, or the four fingers of either group together with the "index" of the other group may be extended, while the other three are flexed. The "index" fingers could not be flexed while the other fingers were extended, nor can the "little fingers" be extended while the others were flexed. MURRAY, J. JARDINE, *Med. Chir. Trans.*, 1863, XLVI. p. 29, Pl. II.

496.   Female child, five weeks, having a hand of eight digits on the right side (Fig. 94). The digits were disposed in two groups of four in each. [No further

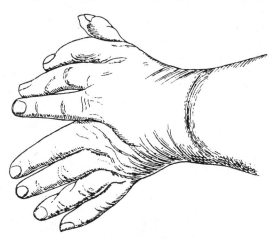

FIG. 94. Right hand of No. 496. (After GIRALDÈS.)

description.] GIRALDÈS, *Bull. soc. de Chirurg.*, *Paris*, 1866, Ser. 2, VI. p. 505, fig. The same case referred to again, GIRALDÈS, *Mal. Chir. des Enfants*, 1869, p. 42, *fig.*

497.   Female child having right hand almost exactly like MURRAY's case, but without syndactylism. The two halves could be folded on each other. The four extra digits articulated with an imperfect metacarpal which was annexed to the normal metacarpal [of the index]. FUMAGALLI, C., *Annal. Univers. di Med. Milano*, 1871, vol. CCXVI. p. 305, *fig.*

Girl's right hand having eight fingers, represented in a wax model. LANGALLI, *La scienza e la pratica*, Pavia, 1875 [Not seen: abstract from DWIGHT, *l. c.*].

498.   DOUBLE-HAND III.   *Six digits in two groups of three and three.* Man having abnormalities of left arm as follows (Fig. 95). The left hand was composed of six digits with three phalanges, which were disposed in two groups of three digits in each. The two middle digits were the longest (d^3 and d^4), and the length of the digits on either side of them diminished regularly. The appearance was as of a hand composed of the middle, ring and little fingers of a *pair* of hands united together. The two groups of fingers were to some extent opposed to each other and all the digits could be flexed and extended. The digit d^3 though single in its peripheral parts articulated with *two* metacarpals, its proximal phalanx having two heads. Upon the radial side of the

carpus of this hand there was a soft tumour about 2·5 cm. in height, resembling a cyst with a firm wall.

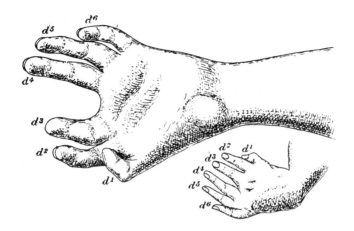

Fig. 95. Dorsal and palmar aspects of the left hand of No. 498. The digits are numbered from the inside.
(After Jolly.)

The structure of the bones of the arm and fore-arm could not be made out with certainty in the living subject, but it appeared that the humerus was formed by two bones partially united together.

As regards the skeleton of the fore-arm an ulna could be felt extending from the upper arm to the *processus styloideus*. The existence of a radius could not be made out with certainty, but a second bone could be felt which was in very close connexion [with the ulna]. JOLLY, *Internat. Beitr. z. wiss. Med.*, 1891.

499.   Male child, three years old, twin with a normal female child, having all extremities abnormal. *Right hand.* Six metacarpals arranged in two groups of three in each group. Each bore a three-phalanged digit, none resembling a thumb. The first and sixth were alike, resembling a minimus, while the two median fingers resembled middle fingers. On the radial side the three digits were completely united together. The next was free, and the two external to this were also united. *Left hand.* Like the right, but all the fingers united together in two groups of three in each group. *Feet.* Each foot had nine metatarsals and nine digits, the central being like a hallux and having two phalanges perhaps, but thicker than a hallux. The externals were like minimi. The four toes on each side of the "hallux" were united two and two. The tarsus was of about double size. The right leg was shorter than the left. GHERINI, A., *Gaz. med. ital.-lombard.*, 1874, No. 51, p. 401, *figs.*

338   MERISTIC VARIATION.   [PART I.

Complex and irregular cases of Polydactylism associated with Change of Symmetry.

*500.   Man (examined alive) having abnormalities in the digits of hands and feet (Fig. 96). The case is very briefly and inadequately described, but the condition was apparently as follows.

Right hand. Beginning from the ulnar side, there were three normal digits (6, 5, 4). Beyond the third of these, which must be

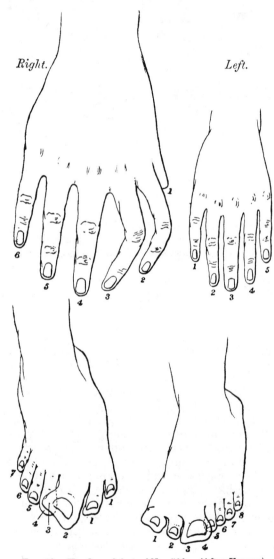

Fig. 96. Hands and feet of No. 500. (After Kuhnt.)

regarded as the medius, there were two complete digits (3, 2) each having three phalanges: and on the radial side of the innermost of these digits there was a stump-like rudiment (1), apparently representing another digit. [This case therefore differed from those of Windle and Rijkebüsch in the fact that both the digits internal to the *medius* (*m*) were disposed as though they belonged to a *left* hand, and KUHNT, in fact, states that each hand was, as it were, composed of parts of a *pair* of hands, thus agreeing with JOLLY's case, No. 499.]

Left hand. In this hand there were only five digits, each of which had three phalanges. None of them was fully opposable, but that on the radial side (1) could to some extent be moved as a thumb. Of these five digits the middle one was the longest, and on each side of it there were two similar digits, those next to the middle finger being the longest and those remote from it being a good deal shorter and having the form of little fingers, which KUHNT considers them to have been. [This hand is perhaps in Condition II.]

Right foot. The hallux (2, 3) was of abnormal width and its bones were to some extent double, the ungual phalanx being completely so. [The nail however is drawn as a single structure and the double character of the toe was not apparent in its external appearance.] On the internal (tibial) side of the hallux there were two supernumerary toes (1, 1) having, so far as could be ascertained, a single metatarsus. The number of phalanges in these toes is not distinctly stated.

Left foot. The hallux (3, 4) was to some extent double, like that of the right foot. Internally to it were two supernumerary toes (1, 2) having apparently a common metatarsal. [Of these the most internal is represented as being very wide and resembling a hallux, but this feature is not mentioned in the description and the number of phalanges is not given].

[It is greatly to be regretted that no fuller account of this important case is accessible. According to KUHNT's view each hand and each foot were structurally composed of parts of a complementary pair of hands and feet. As regards the hands the facts agree with this description and with what has been seen in other cases, but the condition of the feet is more doubtful, and without more knowledge of the details no opinion can be given. It should be remembered that the original description is very brief and Dr Kuhnt offers an apology for the imperfection of the figures.] KUHNT, *Virch. Arch. f. path. Anat. u. Phys.*, LVI. 1872, p. 268, Taf. VI.

501. Case of a foot with eight toes, stated to have resembled KUHNT's case (No. 501). EKSTEIN, *Prager Wochens.*, No. 51, 1891.

502. Man whose right arm beside the normal hand bore an extra thumb and finger. The two thumbs were united and had a common metacarpal joint. They were of equal size. They were flexed and extended together and had the power of spreading apart. The extra finger was beyond the extra thumb and was shaped like an index. Besides the radius and ulna of the normal arm there was an extra radius on the outer [? internal] side of the normal radius. This bone had a joint of its own at its elbow. The wrist was broad, suggesting the presence of additional bones. Nothing is said of a metacarpal bone for the new index. CARRÉ, *Séance publ. de la soc. roy. de Méd., Chir. et Pharm. de*

Toulouse, 1838, p. 28. [Not seen by me. Abstract taken from DWIGHT, *l. c.*, *vide* No. 492. Cp. p. 326, *Note*.]

503.   Girl, new-born, having the left foot "double," bearing eleven toes.

The left labium majus was twice as large as the right, and the left leg and thigh were much thinner than the corresponding parts on the right side [measurements given]. The extra parts were all on the *plantar* side of a foot which had toes of nearly normal shapes and sizes. This foot was bent into a position of extreme talipes equino-varus, and the great toe was bent so that it pointed inwards at right angles to the metatarsal.

Upon the plantar side of this foot there was a series of six well-formed, small toes, arranged in a series parallel to that of the 'normal' five, and having their plantar surfaces in opposition to those of the latter. Of the series of six toes that facing the normal little toe exactly resembled it. The second was the longest of the six, but did not resemble a great toe. The third and fourth were equal in length, the fifth and sixth being shorter, as are the external toes of a normal foot. None of the toes were webbed. BULL, G. J., *Boston Med. and Surg. Jour.* 1875, XCIII. p. 293, *fig.* [This figure copied by AHLFELD, *Missb. d. Menschen*, Pl. xx. fig. 2.]

[The case described by GRANDIN, *Amer. Jour. of Obstetrics*, 1887, xx. p. 425, *fig.*, is probably a case of a pair of limbs composing a Secondary Symmetry attached to and deforming the limb belonging to the Primary Symmetry and corresponding with that of the other side. The nature of this case will be better understood when evidence as to the manner of constitution of Secondary Symmetries has been given.]

*504.   **Macacus** sp. A monkey, full-grown, having nine toes on the left foot; right foot normal, upper extremities not preserved. The specimen is described as No. 307 in the *Catalogue of the Teratological Series* (1872) in the Mus. Coll. Surg. (Hunterian specimen). Though I am disposed to agree in the main with the view of the nature of the specimen given in the *Catalogue* it is not in my judgment possible to decide confidently in favour of this view to the exclusion of all others. For this reason the specimen is here described afresh. This is the more necessary as the account of the *Catalogue* is incorrect in some particulars.

Extra parts are present in the limb and in the pelvic girdle. (Figs. 97 and 98.) The names to be given to the parts depend on the hypothesis of their nature which may be preferred. In general terms it may be stated that the ventral or pubic border of the girdle and the internal (tibial) border of the limb are nearly normal. The external (fibular) border of the limb is also normal, but between these there are in addition to the normal parts other structures, whose true nature is somewhat uncertain.

The appearances may be realized best in the following way. Suppose that two similar left feet lie in succession to each other, the "posterior" having its hallux next to the minimus of the "anterior," so that the digits read I, II, III, IV, V, I, II, III, IV, V. Now if the two feet could interpenetrate so far that the minimus of the "anterior" foot took the place of the hallux of the "posterior," this

POLYDACTYLE FOOT: *Macacus.*

second hallux not being represented, the condition of this specimen would be nearly produced. In the same way the left pelvic girdle is just what it would be if two left innominate bones were placed in succession, the ischium of the "anterior" superseding the pubis of the posterior. As in the foot, so in the innominate, of the portions which coincide the parts belonging to the anterior are alone represented. Something very like this was seen in the case, for instance, of the imperfect division of vertebræ in *Python,* No. 7.

The chief difficulty attending this view of the nature of the case is the fact that as regards the tarsus the "anterior" foot

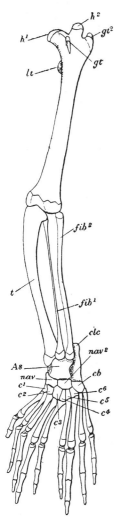

Fig. 97. *Macacus,* No. 504, left leg.
C. S. M. 307.

h^1, head by which femur articulates. h^2, supernumerary head (?). *gt*, great trochanter. gt^2, "posterior" great trochanter. *lt*, lesser trochanter. *t*, tibia. fib^1, "anterior fibula." fib^2, "posterior" fibula (?). *clc*, calcaneum. *As*, astragalus. *nav*, navicular. nav^2, supposed second navicular. c^1-c^6, six cuneiform bones. c^3, the ecto-cuneiform of "anterior" foot. *cb*, cuboid.

lacks the external (fibular) parts of a tarsus, viz. the cuboid and calcaneum. There is a cuboid, cb, and a calcaneum, c, for the "posterior" foot, but none for the "anterior." The bone c^3 might of course be called a cuboid; but if this is a cuboid there is no ecto-cuneiform for the anterior foot. The account given in the *Catalogue* avoids these difficulties by the statement that each foot has three cuneiforms and a cuboid, declaring that there is a second cuboid between the two sets of cuneiforms. This is nevertheless incorrect, for the whole distal series in the tarsus contains only seven bones and not eight. The mistake has no doubt arisen by counting c^3 twice over. The *Catalogue* is also in error in neglecting the fact that the tarsal articulation of the digit 2 is quite abnormal.

Similarly in the crus, there is no good reason to affirm that the bone fib^1 is a fibula rather than a tibia. The *Catalogue* regards it as a second tibia, but I incline to speak of it as the fibula of the 'anterior' foot following the view already indicated. As I have said, the leg is almost normal in the structure of its external border and almost normal in its posterior border, but between these the nature of the parts is problematical. All that can be done is to describe the parts as they are seen.

Beginning at the external (fibular) border of the foot there is a nearly normal series of three digits, 9, 8, 7, fashioned as V, IV and III

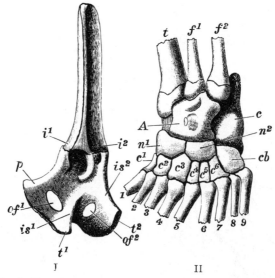

Fig. 98. I. Innominate bone of *Macacus*, No. 504. i^1, p, is^1, t^1, of^1, ilium, pubis, ischium, ischial tuberosity and obturator foramen of the supposed anterior part of the girdle; the parts marked 2 being the corresponding structures of the supposed posterior part.
II. Details of tarsus of the same. Digits numbered 1—9 from the inside. A, astragalus. c, calcaneum. n^1, navicular of "anterior" foot. n^2, navicular of "posterior" foot. cb, cuboid. c^1—c^6, six bones placed as cuneiforms.

respectively, the V and the IV articulating with the cuboid (cb) and the III with an external cuneiform, c^6, as usual. There is a middle cuneiform, c^5, bearing a digit, 6, which is almost exactly formed as a II. Internal to this point the parts can only be named with hesitation. The tarsal bone, c^4, of the distal series internal to c^5 is shaped like another c^5, but the digit which it bears rather resembles a minimus. This is succeeded by a tarsal bone, c^3, shaped like the external cuneiform, c^6, but it bears a digit of the length suited to an annularis. Internal to this are two tarsal bones of the distal row, c^2 and c^1, which bear three digits, 1, 2, and 3. Of these the most internal is undoubtedly an internal cuneiform; it bears firstly a slender but otherwise normal hallux with two phalanges, and secondly, it contributes (abnormally for an internal cuneiform) to the articulation of a digit, 2, which is thinner than all the others and resembles rather a minimus than an index. The digit, 2, also articulates with c^2 which chiefly supports the third digit.

Between the metatarsals of the digits 5 and 6 there is a considerable space, owing to the fact that the head of the metatarsal of 6 is prolonged upwards like that of a normal metatarsal V.

In addition to those described are four other tarsal bones: firstly, a calcaneum c, which is rather *smaller* than that of the normal right leg. It articulates with the cuboid, cb, with the astragalus, A, and with the bone, n^2. The astragalus is very large in its transverse dimension but its length is less than that of the normal astragalus. Peripherally it bears two bones, firstly, a navicular, n^1, and secondly, a bone of uncertain homology, marked n^2 in Fig. 96. The navicular articulates with c^1, c^2 and c^3, together with the bone n^2. The latter, n^2, articulates with c^3, c^4, c^5, c^6, and also with the cuboid, cb, the astragalus and calcaneum and navicular. From its form and relations it is probably a second navicular.

The bones of the crus are three. Firstly, a tibia, *tib.*, which is rather thinner than the normal bone and is somewhat bowed inwards. Passing as a chord to the curve of the tibia there is a thin bone, fib^1, which is tendinous in its upper part. External to this, articulating with the external condyle of the femur there is a third bone, fib^2, which has nearly the form and proportions of a normal fibula. All three bones articulate with the large astragalus.

There is a small patella.

The femur is about half as thick again as that of the right leg. Its head is nearly normal in form, articulating with the rather shallow acetabulum. The lesser trochanter and the internal border of the femur are nearly normal. Anteriorly and externally there are the following parts. Upon the external border there is a projecting callosity, clearly being a great trochanter in its nature. Internal to this there is a knob-shaped, rounded protuberance, which in texture so closely resembles the head of a femur that it is almost certainly of this nature. It is rounded and smooth as though for articulation with an acetabulum, though it stands freely. Between this tuberosity and the real head of the femur there is a third tuberosity, apparently representing the end of the great trochanter of that limb which has been spoken of as "anterior." The peripheral end of the femur is nearly normal on its inner side, while on the outside it is considerably enlarged. The external condyle is thus much larger than that of the normal femur, but there is in it only a very slight suggestion of a division into two parts.

The innominate bone has an ilium which anteriorly is normal, but which posteriorly enlarges and to some extent divides into two parts, i^1 and i^2. Of these the ventral part, i^1, unites with a nearly normal pubis, p, and bounds the shallow acetabulum with which the femur articulates. The rest of this acetabulum is made up by the ischium, is^1, of the "anterior" limb, which together with the pubis bounds an obturator foramen, of^1. Dorsal to these parts the ilium has a partly separated portion, i^2, which forms part of the wall of a cavity apparently representing the acetabulum of the "posterior" limb. Dorsal to this a complete ischium arises which bears a normal ischial tuberosity and curves round a second smaller obturator foramen, of^2.

In so far as the foregoing description involves conceptions of homology it is merely suggestive, but the structure of the innominate bone leaves little doubt that the nature of the parts is much as here described. Nevertheless the appearance of the digits 5 and 6 and of the tarsal bones c^3 to c^6 somewhat suggests that there is a symmetry about an axis passing between the digits 5 and 6; but if 5 were a minimus and if 6 were fashioned as an index, which it is, the appearance of a relation of images would to some extent exist in any case. This appearance is however confined to the dorsal aspect of the foot and is not present on the plantar aspect.

This case, if the view of it proposed be true, differs from other examples of double-hand (*e.g.* Nos. 491 to 499) in that the Repetition is Successive and is not a Repetition of images; for the digits stand I, II, III, IV, V, II, III, IV, V, and not V, IV, III, II, [I], II, III, IV, V as in those other cases. In this respect it is so far as I know unique.

Those who have treated the subject of double-hand generally make reference to the following records. RUEFF, *De conceptu*, Frankfurt, 1587, Pl. 41; ALDROVANDI, *Monstr. Hist.*, 1642, p. 495; KERCKRING, *Obs. anat.*, Amst. 1670, Obs. xx. Pl., but the descriptions are scarcely such as to be useful for our purpose. A case quoted by DWIGHT, *Mem. Bost. Soc. of N. H.*, IV. No. x. p. 474, from DU CAUROI, *Jour. des Scavans*, 1696, pub. 1697, p. 81 [originally quoted by MORAND and misquoted by many subsequent authors], is probably *not* an example of double-hand (see No. 522).

Cases of Polydactylism in Man and Apes not associated with definite change of Symmetry.

From the evidence as to polydactylism in general the foregoing cases have been taken out and placed in association as exhibiting the development of a new system of Symmetry in the limb. It will have been noticed that in all of them the external (ulnar or fibular) parts of the limb remain unchanged, and the parts not represented in the normal are on the internal (radial or tibial) sides. In the remaining cases of polydactylism, which constitute the great majority, there is no manifest change in the general symmetry of the limb.

These general phenomena of polydactylism have been observed from the earliest times and the literature relating to the subject is of great extent. Most cases known up to 1869 [not including STRUTHERS' cases] were collected by FORT, *Difformités des Doigts*, Paris, 1869, and independently by GRUBER, *Bull. Ac. Sci. Pét.*, XV. 1871, p. 352 and p. 460, and good collections of references have subsequently been published, especially by FACKENHEIM. *Jen. Zeits.*, XXII. p. 343. Of the whole number of cases the majority fall into a few types, and a great part of the evidence may thus be easily summarized and illustrated by specimen-cases. The forms of polydactylism thus constantly recurring may be dealt with conveniently under the following heads.

(1) Addition of a single digit, complete or incomplete.
    A. external to minimus, in series with the other digits.
    B. in other positions.

(2) Duplication of single digits, especially of the pollex and hallux.

(3) Combinations of the foregoing.

Besides these are a certain number of cases not included in the above descriptions, and of them an account will be given under the heading

(4) Irregular examples.

As bearing upon the frequency of the several forms of polydactylism it may be stated that in this irregular group are included all cases which I have met with that exhibit any feature of importance in departure from the cases otherwise cited. For the purpose of this list I have examined every record of polydactylism to which access could be obtained.

(1) A. SINGLE EXTRA DIGIT EXTERNAL TO MINIMUS IN HAND OR FOOT.

(*a*) *Incomplete form.*

This is one of the commonest forms of extra digit. In the great majority of such cases the extra digit is not complete from the carpus or tarsus but arises from the metacarpal or metatarsal, less often from one of the phalanges, of the minimus. The attachment may be either by a direct articulation upon the side of one of these bones, or they may give off a branch bearing the extra digit. In a not uncommon form of the variation the extra digit has no bony attachment to the hand, but is a rudimentary structure hanging from some part of the minimus by a peduncle. Of these several forms the following are illustrative cases.

505. *Extra digit hanging from minimus by a peduncle.*
    Manus. ANNANDALE, *Diseases of Fingers and Toes*, 1865, p. 30, Pl. II. fig. 20; TARNIER, *Bull. Soc. de Chir.*, Paris, VI., 1866, p. 487; and numerous other examples.
    Pes. BUSCH, quoted by GRUBER, *l.c.*, p. 470: this form in the pes is rare.

346   MERISTIC VARIATION.   [PART I.

506.   *Extra digit arising from one of the phalanges of minimus.*
ANNANDALE, *l. c.*; OTTO, *Monstr. sexc. Descr.*, Taf. xxv. fig. 7; CRAMER, *Wochens. f. d. ges. Heilkunde*, 1834, No. 51, p. 809; GAILLARD, *Gaz. méd.*, 1862. This form seems to be comparatively scarce.

507.   *Extra digit arising from metacarpus or metatarsus of minimus.*
The great majority of cases are of this nature but exhibit many differences of degree. The articulation may be on the *side* of the metacarpus V (see MORAND, *Mém. Ac. Sci. Paris*, 1770, p. 142, fig. 4; *Coll. Surg. Mus., Catal. Teratol. Ser.*, 1872, No. 308, and numerous other cases), or of the metatarsus V (see GRUBER, *l. c.*, p. 476, *Note* 28) but in the pes this is less common. Frequently also the articulation of the extra digit is on the *head* of the metacarpus V (GAILLARD, *l. c.*) or metatarsus V (*Mus. Coll. Surg., Terat. Ser.*, No. 310).

In the foregoing cases the extra digit articulates immediately with the side or head of metacarpal or metatarsal, but sometimes in the manus and often in the pes the digit articulates at the end of a branch given off by the metacarpus (MORAND, *ibid.*, fig. 3, and numerous other records), or by the metatarsus (MORAND, *l. c.*; STRUTHERS, *Edin. New Phil. Jour.*, 1863 (2), p. 89; MECKEL, J. F., *Handb. d. path. Anat.*, II. Abth. 1, p. 36, and many more.

*508.   **Hylobates leuciscus** (Fig. 99) having an extra digit in the left manus articulating externally with the metacarpus V and in the right manus articulating with a branch from it. *Mus. Coll. Surg., Teratol. Ser.*, No. 307, A.

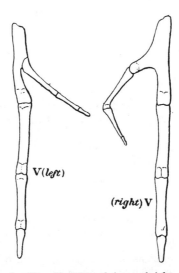

FIG. 99. *Hylobates leuciscus*, No. 508, minimus of right and left manus bearing a supernumerary digit articulating with the metacarpals.
(From specimen in Coll. Surg. Mus.)

(b) *Complete digit having metacarpus or metatarsus external to minimus.*

Extra digits external to the minimus are occasionally complete, having a metacarpal or metatarsal and three phalanges, standing truly in series with the other digits, but to judge from the records this complete form is decidedly rare. In the first of the following examples given it should be noted that the digit standing fifth,

CHAP. XIII.] DIGITS: MAN. 347

that is to say, as minimus, was itself rather longer than it should be in the normal, thus illustrating the principle with regard to the Variation of a small terminal member of a Meristic Series on becoming penultimate which was predicated especially in regard to Teeth (see p. 272). In MORAND'S case the interesting fact of the partial assumption by the sixth digit of anatomical characters proper to the minimus is commended to the attention of the reader.

*509. Girl: one extra digit on the external side of each hand. The normal little fingers are rather longer than usual and the extra fingers have nearly the same length. Each has three phalanges. Neither of the extra fingers can be moved separately from the finger adjacent to it. In the *left* hand the extra finger is borne on a supernumerary metacarpal which lies parallel with the normal metacarpal V. Each extra digit can be opposed to the pollex. In the *right* hand the extra finger is borne on the enlarged head of the fifth metacarpal. BÉRANGER, *Bull. Soc. d'Anthrop.*, Paris, 1887, Ser. 3, x. p. 600.

*510. Man (parents normal, one brother had six digits on each extremity, six other members of family normal) having an extra digit external to minimus on both hands (Fig. 100) and both feet, in series with the normal digits.

Left hand: unciform abnormally large, having two articular facets, one for the metacarpal of the fifth and the other for that of the sixth digit. The sixth metacarpal bears a digit of three phalanges of which the second and third were very short. [It does not appear that V was of increased length.] *Right hand:* metacarpals normal in number, but the fifth is very thick, having in its peripheral third on the external

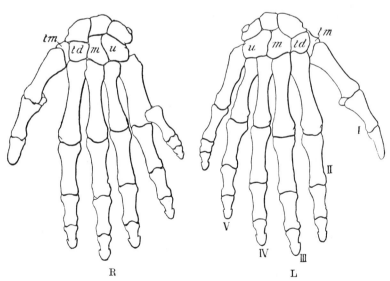

FIG. 100. Palmar views of the bones of the hands of No. 510. (After OTTO and MORAND.)

surface an articulation for a short digit of three phalanges, the second and third being very small. *Feet :* well formed ; cuboid of size greater than the normal, bearing the proximal end of two united fifth and sixth metatarsals. Each of these is separate peripherally and bears a digit [of 3 phalanges to judge from the figure (fig. 6)] in series with the normal toes, but shorter than the minimus.

Muscles. In the *left hand* the sixth digit was fully supplied with muscles. There were two extra interossei and the extensor communis sent tendons to the sixth digit. The abductor, the flexor brevis and the flexor ossis metacarpi which in the normal are proper to the minimus were all inserted into the sixth digit instead.

In the *right hand* the extensor communis gave a tendon to the sixth, which also possessed a proper abductor, but the fifth had no special extensor. Of the flexors the sublimis gave a tendon to each of the digits index, medius and annularis, none to the fifth, but a small slip to the sixth. The flexor profundus gave four tendons as usual, but from that going to the fifth a small tendon passes off laterally and piercing the sublimis is inserted as usual.

In both feet the muscles were similar. The extensor longus gave a tendon to the sixth digit, and the extensor brevis does not. The flexor longus has four tendons as usual, none going to the sixth digit ; the flexor brevis has four normal tendons and an extra one for the sixth. The two tendons proper to the fifth (minimus) go to the sixth. The interossei are normal and there are only two lumbricales, one for the second digit and one for the fourth. MORAND[1], *Mém. de l'Acad. Roy. des Sci.*, Paris, 1770, p. 142, Figs. 1, 2, 4, 5 and 6. [The condition of the muscles in regard to the fifth and sixth digits in this case is worthy of special attention. If the morphologist will here propose to himself the question *which* is the extra digit, he will find it unanswerable. In the right hand, judging from the bones, it may seem evident that the fifth with its complete metacarpal is the minimus and that the sixth is a new structure ; but the condition of the feet and the right hand taken with that of the left, make a series or progression from which the similarity of the variation in each of the three states is evident; hence, if it is thought that the most external digit in the right hand is the extra part, it must also be held that the external or sixth digit in the left hand is the extra digit. But this digit in respect of its muscles has some of the points of structure peculiar to a minimus, while the fifth digit or supposed minimus on the contrary is without these characters. Hence neither digit is *the* minimus. Just as in the Condition III (see p. 326) of the hand, we saw that on the presence of a digit internal to the pollex, the pollex itself may be promoted to be a finger-like digit with three phalanges, so may the fifth digit be partially fashioned as a more

---

[1] The similar descriptions and figures given by OTTO, *l. c.*, Pl. xxv. figs. 9—11, SEERIG, *Üb. angeb. Verwachs. d. Finger u. Zehen,* AMMON, *Die angeb. Kr. d. Mensch.,* all refer, to this one original case of Morand's, though the fact is not stated and though several authors (GRUBER, &c.) quote them as separate cases. Seerig states that his figures are from preparations in the Breslau Museum. These figures agree exactly with those of Otto, which again agree closely with those of Morand but give more detail as to the carpi, taken no doubt from the actual specimens which had been acquired by the Breslau collection. I have therefore copied Otto's figures, though taking the important descriptions from Morand.

central digit on the presence of a digit external to it. If therefore it be still called the "minimus" this term can only be applied to it by virtue of its ordinal position.]

For other cases of complete digits in this position see AUVARD, *Arch. de Tocologie*, xv. 1888, p. 633; MARSH, *Lancet*, 1889 (2), p. 739.

(1) B. SINGLE EXTRA DIGIT IN OTHER POSITIONS.

Apart from cases of extra digit external to the minimus, cases of duplication of the pollex or hallux (to be considered below), and cases of extra digits internal to the pollex or hallux associated with change of symmetry of the digital series, the remaining cases of single extra digit are very few. In other words, it is with digits as with Meristic series in general, when a new member is added, the addition taking place in such a way that homologies may be recognized, it is most often at one of the ends of the series that the addition is made. Cases of extra digits in other positions are in Man and Apes very rare, and even in some of the few recorded cases of a new digit arising on the *inner* side of the minimus (No. 511) it should be remembered that this inner digit is judged to be the extra one rather than the outer mainly by reason of its smaller size. I can only give particulars of few such cases, and of the remainder no details are available.

*511. **Simia satyrus** (Orang-utan), having a rudimentary extra digit arising from the *internal* side of the minimus of each hand: feet normal. In the *left* manus the minimus has all joints moveable as usual; the first phalanx is normal, but the second is bent outwards nearly at right angles, thus making room for an extra digit arising from the first phalanx and directed inwards. This digit is fixed and has no articulation and no nail, but it is in its outer part bent back again towards the minimus with which it is webbed. The structure in the *right* manus is almost the same but the extra digit is larger and in its outer part free from the minimus, bearing a nail. BOLAU, *Verh. naturw. Ver. Hamburg*, 1879, N. F. III. p. 119.

512. Woman: left pes bearing an extra digit articulating by an imperfect metatarsal with outside of metatarsal of IV. The extra digit stands obliquely to the others, sloping outwards and being attached by ligaments to the normal V. [The *Catalogue* states that the extra digit resembles a *right* digit, but I see no sufficient evidence of this.] C. S. M., *Ter. Cat.* 312.

[A case perhaps similar to foregoing is briefly quoted by GRUBER, *l. c.*, p. 471, note 83, as being in the Vienna Museum of Anatomy.]

512*a*. Child: left metacarpal IV bore a supernumerary digit on external side. This digit was shorter than the digit IV and was completely webbed to it. BROCA, quoted by FORT, *l.c.*, p. 66.

*512*b*. Fœtus (otherwise abnormal): left hand bore extra digit attached by peduncle to first phalanx of digit IV. The minimus was separated from IV by a metacarpal space, standing almost at right angles to it. HENNIG, *Sitzb. naturf. Ges. Leipzig*, 1888. Oct. 9.

[AMMON (*Die angeb. Krankh. d. Mensch.* p. 101, Pl. XXII. fig. 7) describes a case of rudimentary finger appended to the "ring-finger" and is so quoted by GRUBER; but the figure apparently represents the appendage as attached to the minimus.]

(2) *Duplication*[1] *of single Digits, especially of the Pollex and Hallux.*

*513. Duplication of the pollex or of the hallux is one of the commonest forms of polydactylism and numerous cases have been described by all who have dealt with the subject. It consists in the development of two digits, complete or incomplete, in the position of the usually single series of bones composing the pollex (or hallux). In the section dealing with polydactylism associated with change of Symmetry (p. 326) we saw how upon the appearance of an extra digit in this position the thumb itself may have three phalanges. In these cases the extra digit may properly be considered as arising in Successive Series with the

[1] A few cases are thought by some to shew *triplication* of digits, but it seems doubtful whether there is a case of division of one digit into three really equivalent digits, perhaps excepting the thumb of No. 521.

pollex. But in a large majority of cases of the presence of an extra digit on the radial side, the thumb has two phalanges as usual. Upon a review of the evidence it is I think clear that we shall be right in considering that in most of these cases the extra digit is not really in *Succession* to the thumb, but that the two radial digits together represent the thumb, the increase in number being achieved by duplication and not by successive addition.

Most authors (GRUBER, &c.) thus speak of these formations as "double-thumbs" and recognize them as examples of duplicity, but it should be remembered that this view of their nature is not consistent with any statement that either of the two digits is the extra one. If these thumbs are instances of duplicity then both together represent the normally single thumb.

In clear cases of double-thumb the two thumbs are equal or nearly equal in size and development, as commonly happens in cases of true duplicity. Double-thumbs are known in every degree of completeness. The division between the two may occur at any point in their length. Thus the duplicity may be confined to the nail and first phalanx (OTTO, *Monstr. sexc. Descrip.*, Taf. XXV. fig. 1; BIRNBAUM, *Monatsschr. f. Geburtsk.*, 1860, XVI. p. 467); or it may include both first and second phalanges (GRUBER, *Arch. f. path. Anat. Phys.*, XXXII. 1865, p. 223); or both phalanges and the greater part of the metacarpal (GAILLARD, *Mém. Soc. de biol.*, 1861, p. 325); or even the whole digit and metacarpus, the two thumbs separately articulating with the trapezium (JOSEPH, quoted by GRUBER, *l. c.*, p. 463, *Note* 37). It would be interesting to know which of these conditions is the most frequent, for it is likely that between the degrees of this variation there is Discontinuity, but the point is not easy to determine. As regards records the conditions first and last named are much the rarest, and the double-thumbs with two sets of phalanges articulating with one metacarpal constitute the majority of cases.

Sometimes the two thumbs are webbed together (GRUBER, *Bull. Ac. Sci. Pét.* XV. p. 480, fig.) sometimes they are separate and *may be*

FIG. 101. Right hand having a thumb double from the metacarpus, shewing the relationship of images between the two thumbs. (After ANNANDALE.)

opposed to each other (FACKENHEIM, *Jen. Zts.*, XXII. p. 358, fig. IV.; ANNANDALE, *Diseases of Fingers and Toes*, Pl. III. fig. 25). This condition is important as an indication that between these double-thumbs there may be a relation of images (Fig. 101).

The duplicity may be and often is very different in degree in the two hands, though it is very commonly present in both.

514. The description given of duplicity in the pollex applies equally to the hallux, though of duplicity in the latter perhaps fewer cases are recorded. Here too the duplicity may be in all degrees of completeness. An example from ANNANDALE (*l. c.*, Pl. III. fig. 32) is shewn in Fig. 102.

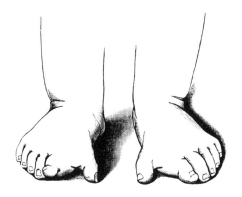

Fig. 102. Feet of infant, No. 514, having thumb-like supernumerary digits arising from the metatarsi of the great toes.
(After ANNANDALE.)

Here a thumb-like extra hallux is borne on the inner side of the metatarsal I. Several such cases are known (cp. No. 517).

515. Among the cases called by authors "double-thumb" are a certain number in which the two thumbs are not equally developed, that on the radial side being more rudimentary. In such a case we are entitled to consider the radial thumb as an extra digit formed in Succession to the normal thumb, and not as a double of it. In speaking of other Meristic Series (especially mammæ and teeth) we have seen that it is not possible accurately to distinguish between cases of duplicity and cases of change in number of the series by formation of another member in the Succession. This is extremely well seen in digits. For firstly several conditions intermediate between the two are recorded by many authors (*e.g.* a case in which the radial thumb had two phalanges "ankylosed" together [or rather not completely segmented from each other]. GRUBER, *l.c.*, p. 480; cases in which the radial thumb had only one phalanx, *ibid.*, p. 482; STRUTHERS, *Edin. New Phil. Jour.*, 1863 (2), p. 87; BOULIAN, *Rec. de Mém. de Méd. milit.*, 1865, Ser. 3, XIII. p. 67, *figs.*); and besides this there are several examples in which one hand bore a clear pair of double-thumbs, while in the other hand there is an extra radial digit in succession to the normal thumb (*e.g.*

FACKENHEIM, *l.c.*, p. 359, fig. IV.). Thus do the two conditions pass into each other, though some cases are clearly cases of duplicity and some are clearly cases of extra digit in Succession[1].

I know no case of unmistakeable duplicity in any digit but pollex or hallux; but no doubt a good many cases of extra digit arising from the minimus may be of this nature (*e.g.* ANNANDALE, Pl. III., fig. 28), though it is more likely that the extra digit is in Succession.

In digits other than I or V the only case of possible duplicity known to me as occurring in a limb not exhibiting one of the complex conditions of polydactylism, are those of STRENG (*Vierteljahrsschrift f. prakt. Heilk.*, XLIX. 1856, p. 178; original not seen by me; quoted by GRUBER, p. 476), being a case apparently of double medius on one metacarpal; and of DUSSEAU, *Cat. Mus. Vrolik*, No. 518, two terminal phalanges on right medius (together with double thumb; six fingers on left hand and peripheral duplicity of hallux in each foot). Accompanied by numerical Variation in other parts of the digital series such cases of duplicity are known in a few other cases.

(3) *Combinations of the foregoing.*

Limbs not rarely present the forms of polydactylism already named in combination. Such combination may be found in the same limb, or one or more limbs may present one form, while another form may be found in the other limb or limbs. Of these combinations the following three cases will be sufficient illustration.

Case of double hallux on each foot, and rudimentary digit attached by peduncle to the minimus of each hand.

516. A female member of a polydactyle family [particulars given] had an abortive supernumerary finger attached by a peduncle to the little finger of each hand. In the feet the two *great toes* were each partially double. In the left great toe the individual phalanges could be felt and there were two nails. In the great toe of the right foot the adjacent sets of phalanges were inseparably united by their lateral borders, forming one bone, which was correspondingly broadened. There was only one nail which was notched in the middle of its free border. MUIR, J. S., *Glasgow Med. Jour.*, 1884, N. S. XXI. p. 420, *Plate*.

Case of each extremity with double pollex or hallux and rudimentary digit attached to minimus.

517. Female infant having thumb of each hand double, the two sets of bones lying in the same skin and connective tissue. In the right hand the nails and phalanges of each were quite distinct, but it was not certain whether the metacarpals were separate or not. In the left hand the nails were not completely separate and the phalanges of the two thumbs were less distinctly separate. To the first phalanx of the little finger of each hand was appended a rudimentary bud-like finger, hanging by a peduncle.

The feet resembled the hands. From the inner border of the metatarsal of each great toe there proceeded a well-formed thumb-like toe with two phalanges. This toe was set at right angles to the great toe and could be flexed and to some extent opposed. On the external border of the right foot there was a small extra little toe hanging by a peduncle from the metacarpal V. In the left foot the supernumerary little toe was bound up with the normal little toe for its whole length. HAGENBACH, E., *Jahrb. f. Kinderheilk.*, XIV. 1879, p. 234, *figs.* [Cp. No. 514.]

[1] Compare with the largely similar series of phenomena seen in the foot of the Dorking fowl (*v. infra*). But in it if the two hallucal digits are not a true pair it is most commonly the *inner* that is the largest, conversely, to the general rule in the extra digits arising from the pollex in Man.

Case of double hallux in combination with extra digits on external side.

518. Man in Middlesex Hospital, 1834, having on the right foot two toes articulating with the first metatarsal, and on the left foot two toes articulating with the first metatarsal, and also two toes articulating with the fifth metatarsal. From the ulnar side of one of his hands two fingers had been removed. In each hand the middle and ring fingers were adherent throughout their length, as also were all the toes, except the minimi. Five brothers and three out of four sisters of this man had six toes on each foot and six fingers on each hand. The other sister had seven toes on one foot and six on the other, and had two extra fingers on each hand. *London Med. Gaz.*, 1834, April, p. 65, *figs.*

(4) *Irregular examples.*

Thus far we have considered cases of polydactylism that can be in some degree brought into order and included in general descriptions. There remain a small number of irregular cases each presenting special features which make general treatment inapplicable. These cases are instances of extremities, mostly feet, having seven, eight or nine digits. The descriptions of these cases are for the most part fragmentary, and as the bones have been examined in only one of them (MORAND) so far as I am aware, the relations of the digits to each other and to the limb are obscure. Speaking generally in these irregular examples there is an appearance of division, possibly of duplication, of several digits. It should be noticed also that in some of them (*e.g.* BLASIUS, No. 520) the digits did not lie evenly in one plane but were in a manner bunched up so as to overlie each other. In such a case it would be interesting to know whether the digits originally grew in one plane and were afterwards shifted during growth, or whether the original Repetition was thus irregular.

As all these cases differ from each other an adequate account of them could only be given at great length, and by reproducing the original descriptions in full, together with such figures as are attainable. For these reasons it would not be profitable to introduce them here, though in a study of the nature of Meristic Repetition it is important to remember that these irregular cases exist. As illustrative of several cases I have appended an account of two complex cases in the foot and of one in the hand, giving references to such others as I am acquainted with.

519*. Girl, æt. 6, having abnormal toes on the left foot as follows (Fig. 103). The total number of toes on the left foot was nine. From the position and form it appeared that the digits (6—9) representing II, III, IV and V were normal, but upon the radial side of these instead of a single hallux there were five toes. Of these 1 and 2 were imperfectly separated, articulating with the first metatarsal by their first phalanges, which were united to form a common proximal head. Each had a distinct second phalanx and in general form resembled a great toe having a separate nail. The second metacarpal bore firstly a pair of toes, 3 and 4, which were still less separate from each other than 1 and 2, the bifidity being confined to the soft parts. These two toes had one proximal and one distal phalanx in common. The second metatarsal also bore an external digit, 5, which in form rather resembled a normal *third* digit, being considerably shorter than 6 [and presumably containing three phalanges]. The toes 1, 2, 3 and 4 were found after amputation to be devoid of muscles and presented only the terminations of the flexor and extensor tendons having their normal insertions. The toes 1 and 2 were supplied by the same flexor tendon which bifurcates and passes to be inserted into the ultimate phalanx of each

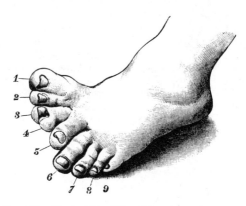

FIG. 103. Foot of No. 519. (After ATHOL JOHNSON.)

by a separate slip. The vinculum by which it is attached is common to the two bones. JOHNSON, ATHOL A., *Trans. Path. Soc.*, IX. 1858, p. 427, *fig.*

520. Male infant having supernumerary toes on the left foot. The tarsus and metatarsus were abnormally wide. The hallux appeared externally to be divided into two. This duplicity was most marked in the second phalanx and appeared in a slight infolding of the skin. The nail also shewed traces of duplicity. Next to the hallux were two toes which were bent upwards and inwards. Of these the one overlay the other. The uppermost was found after excision to have two sets of phalangeal bones enclosed in the same skin; these two articulated with a single metatarsal bone. The lower toe was thought by BLASIUS to represent the digit II. Next to this there was a rudimentary digit with a slightly developed nail. After excision it was found that this toe contained a cartilaginous basis which was partly segmented into two phalanges and articulated with a metatarsal. External to this rudimentary toe were three normal toes, representing as BLASIUS supposes, the digits III, IV and V. External to the putative V was another digit of the same size and shape. BLASIUS, *v. Siebold's Jour. f. Geburtsh.*, XIII. 1834, p. 131, figs. 1 and 2; figures copied in AHLFELD, *Missb. d. Mensch.*, Taf. xx. fig. 11. [This foot appears to contain parts of ten digits.]

521. Child having polydactyle hands as follows. In each hand the fingers were webbed to the tips, each minimus having an extra nail. In the *right* hand the pollex was triplicate, having three sets of phalanges and three nails, the whole being in a common integument. In the *left* hand the pollex was duplicate, having two sets of phalanges webbed together and two nails. Each member thus formed a prehensile paw. In right foot little toe webbed to next toe. Some (not all) of brothers and sisters had similar hands: father and grandfather had similar hands: mother and grandmother normal. HARKER, J., *Lancet*, 1865 (2), p. 389, *fig.*

522. The following are other examples of irregular polydactylism: MORAND, *Mém. Ac. Sci. Paris*, 1770, p. 139, figs. 8 and 9. (The same redescribed from Morand's figure by DELPLANQUE, *Études Tératol.*, II. Douai, 1869, p. 67, Pl. v.; and again by LAVOCAT, *Mém. Ac. Sci. Toulouse*, v. 1873, p. 281, Pl. I., who takes a different view.)
GRUBER, *Mém. Ac. Sci. Pét.*, Ser. VII. Tom. II. No. 2 (fig. copied in *Bull. Ac. Sci. Pét.*, xv. 1871, fig. 6, and by AHLFELD, *Missb. d. Mensch.*, Pl. xx. fig. 20).
GRUBER, *Bull. Ac. Sci. Pét.*, xv. 1871, p. 367, figs. 4 and 5.
OTTO, *l.c.*, Pl. XXVI. figs. 8—11.
FRORIEP, *Neue Notizen, &c.*, Weimar, No. 67, 1838, IV. p. 8, figs. 4—8 (very brief account of important case, copied by AHLFELD and others).
DU CAUROI, *Jour. des Sçavans*, 1696 (pub. 1697), p. 81 (quoted first by Morand, afterwards wrongly quoted by many writers. DWIGHT, *Mem. Bost. N. H. S.*, IV.

No. x. p. 474, supposes that this is a case of double-hand, palm to palm (as No. 503), but the original probably means that two adjacent thumbs and two adjacent annulares were united, the digits being all in one plane).

POPHAM, *Dubl. Quart. J. of Med. Sci.*, XLIV. 1867, p. 481.
DUSSEAU, *Cat. Mus. Vrolik*, 1865, p. 457 (very brief, see p. 352).
GRANDÉLÉMENT, *Gaz. des hôp.*, 1861, p. 553.
LISFRANC (see *Schm. Jahrb.*, XII. 1836, p. 263).
RÖRBERG, *Jour. f. Kinderkr.*, XXXV. 1860, p. 426.
MARJOLIN, *Bull. Soc. de Chir.*, 1866, Ser. 2, VI. p. 505, *fig.* (probably case of double-hand).
ANNANDALE, *Dis. of Fingers and Toes*, 1865, p. 39 (eight metatarsals on a foot possibly associated with change of Symmetry).
Ibid., p. 35, figs. 41 and 49 (pollex with two sets of phalanges but *three* nails, together with extra digit external to V). Cp. No. 521.
HEYNOLD, *Virch. Arch.*, 1878, LXXII. p. 502, Pl. VII.
MASON, F., *Trans. Path. Soc.*, 1879, XXX. p. 583 (foot having eight metatarsals and nine digits).
MELDE, R., *Anat. Unters. eines Kindes mit beiders. Defekt d. Tibia u. Polydactylie an Händen u. Füssen*, Inaug. Diss., Marburg, 1892 (important).

REDUCTION IN NUMBER OF DIGITS.

Though in reduction of digits the course of Variation is generally irregular and the result often largely amorphous there are still features in the evidence which may be of use to us, and a few selected cases are of some interest. These features will be spoken of under the three following heads, though for a general view of the subject reference must be made to teratological works.

(1) Reduction in number of phalanges.
(2) Syndactylism.
(3) Ectrodactylism.

(1) REDUCTION IN NUMBER OF PHALANGES.

As in certain cases of polydactylism it appeared that increase in the number of phalanges in the thumb could be regarded as a step in the direction of increase in the number of digits, so a reduction may be thought to be a step towards diminution in the number of digits. But though many cases of reduction in number of phalanges are recorded, there is in them nothing which suggests that they may be fitted into a series of gradual reduction comparable with the series of gradual increase already described. It is indeed chiefly as illustrating the possible completeness and perfection of Variation that these phenomena have a direct bearing on the subject of Meristic Variation. The following case is chosen as being especially regular and symmetrical.

*523. Man having only one phalanx in each hallux, and two in each of the other fingers and toes. The hands were almost exactly alike. The thumb had a short metacarpal ⅝ in. long, and one phalanx (1¼ in.), the joint between them being loose as if composed of soft tissue. By the length of the metacarpal (3 in.) the index is longer than the other digits. The next two metacarpals have only half that length. The metacarpal of V is 1½ in. long, but from its obliquity does not project so far as that of IV. The proximal phalanx of the index measures 1⅜, medius 1⅜, annularis 1, minimus 1¼. The distal phalanx in index and middle ⅞, ring and little ⅝ in. In left hand the distal phalanx of index is proportionally shorter. Except the index all the digits present their usual proportions. The feet are well formed as far as distal ends of metatarsals. The toes are short, pulpy and loosely articulated. Each has two phalanges except the hallux, which has only one. This case was a twin with a normal male. An elder brother and younger sister have the digits similarly formed, but in the last the feet are also turned in. STRUTHERS, *Edin. New Phil. Jour.*, 1863 (2), p. 100.

As an example of similar and simultaneous Variation in both extremities this is an instructive case.

(2) SYNDACTYLISM.

Under this name have been described those cases in which two or more digits are to a greater or less extent united together. In their bearing on the morphology of Repeated Parts some of these variations are very instructive. It will be found that the important considerations in this evidence may be divided into two parts. Of these the first concerns the *manner* of the variation and the second to the *position* in which it is most commonly found.

The manner of union between digits.

In many cases of union of digits the limb is amorphous; with these we have now no special concern. In simpler examples the digits may be of normal form but some or all of them may be united by a web of integument for a part or the whole of their length. (For records of such cases see FORT, ANNANDALE, &c.).

*524. But besides these cases of webbing are many in which the union may be of a much more intimate character. Taking the cases together a progressive series may be arranged shewing every condition, beginning from an imperfect webbing together of the proximal phalanges to the state in which two digits are intimately united even in their bones, and perhaps even to the condition in which two digits are represented by a single digit (see No. 529). That the latter condition represents a phase in this series of variations does not seem to be generally recognized by those who have dealt with the subject but it is impossible to exclude it.

The lower conditions of this variation are sufficiently illustrated by Fig. 104, I and II (from ANNANDALE, *Diseases of Fingers and Toes, figs.* 39 and 33), shewing cases of medius and annularis partially combined for the whole of their length. A higher condition is shewn in Fig. 104, III, in which the same digits are united so closely that their external appearance suggests that only four digits are present in the hand. In this specimen (ANNANDALE, *l. c.*, p. 14) there were nevertheless five metacarpals, but the first phalanges of III and IV were united peripherally and bore a second and third phalanx and one nail common to them both. The same author (*l. c. fig.* 44) gives an illustration of such a set of bones from OTTO [1].

The following cases are interesting as occurring in Apes.

*525. **Pithecia satanas** (Monkey): young male having the third and fourth digits of the hand on each side completely connected by a fold of nude skin. The remaining digits of the hands and feet were normal. FORBES, W. A., *P. Z. S.*, 1882, p. 442.

526. **Macacus cynomologus**: specimen having the fifth finger of the right hand represented by a rudiment only. On dissection the first phalanx of the fifth finger was found to be enclosed with that of the fourth. All the fingers of the abnormal (right) hand were somewhat misshapen and bore several exostoses. [? congenital variation] FRIEDLOWSKY, A., *Verh. zool. bot. Ges. Wien*, 1870, xx. p. 1017, *Plate*.

---

[1] I have failed to find the original of this figure in OTTO's works.

Before going further certain points are to be noted. First, the union as shewn in the figures is a union or compounding *as of optical*

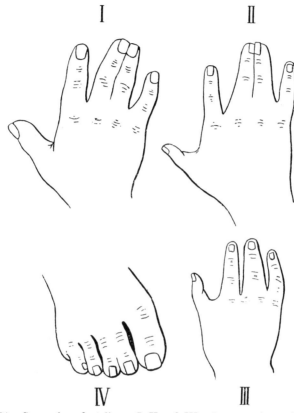

Fig. 104. Cases of syndactylism. I, II and III. A progressive series illustrating degrees in the union of medius and annularis in the hand. IV. Case of union of index and medius of the foot. The union is incomplete peripherally. (After ANNANDALE.)

images in Bilateral Series, and is not like that of parts in Successive Series. Next, the union of the bones is *more complete peripherally and less complete centrally*. The latter is a rule very commonly observed in cases of the union of the bones of digits both in Man and other mammals. This statement is made without prejudice to the other fact that in the least state of syndactylism as manifested by union of the soft parts, it is the most central phalanges which are united. Such a case of partial union between II and III in the foot[1] is shewn in Fig. 104, IV (ANNANDALE, *l. c.*, fig. 34). The rule that in the lowest condition of syndactylism of the bones it is commonly at the periphery that the union is most complete is also difficult to understand in connexion with

[1] Compare several remarkable cases of this variation in one family, LE CLERC, *Mém. soc. Linn. Normandie*, IX. p. xxvi.

the fact that the division of digits in the lowest forms of polydactylism appears also first in the *peripheral* phalanges. These phenomena appear to be in contradiction to each other, and I am not aware that the fact of the appearance of the digits early in the development of the limb throws any light on the difficulty.

The number of digits which may be thus united is not limited to two, and examples of intimate union between three and even four digits are common.

The position of union.

*527.   Those who have treated of this subject do not, so far as I am aware, notice the fact that the phenomenon of Syndactylism most frequently affects particular digits. From an examination of the recorded cases it appears that in the hand there is a considerable preponderance of cases of union between the digits III and IV. I regret that I have not material for a good analysis of the evidence on this point, but I may mention meanwhile that in a collection taken at random of some thirty-five cases of hands having only *two* digits united (chiefly those given by FORT and ANNANDALE) over 25 are cases of union of the digits III and IV [1]; in only one were the digits I and II united; the digits II and III in ?4 cases; the digits IV and V in ?3 cases.

*528.   On the other hand if two digits in the *foot* are united they are nearly always II and III.

If in the hand *three* digits are joined they may be either III, IV and V, or (perhaps less commonly) II, III and IV. In cases of union of all the digits II to V, the digits III and IV are often much more intimately united than the others, and are often recorded as having a common nail, while II and V have separate nails.

This question of the comparative frequency of the different forms of syndactylism would probably repay full investigation, and to the study of the mechanics of Division it would clearly be important. In the meantime may be noted the fact that the evidence suggests the possibility that we have here to do with a case of union of parts which are related to each other as optical images, and that the digits II to V of the hand constitute an imperfect Minor Symmetry within themselves. The fact that the subjects of most frequent union in the foot are the digits II and III, not the digits III and IV as in the hand, may be connected with the fact that the hallux stands to the foot in a different geometrical relation from that which the pollex bears to the hand and that consequently the axes of Symmetry are different in it.

(3) ABSENCE OF DIGITS (ECTRODACTYLISM).

In the conditions already described though the digits are not all clearly divided from each other yet no one whole digit can be supposed to be absent. Even in the specimen shewn in Fig. 104, II, from the presence of separate metacarpals III and IV the identity of the several digits is still easily recognized. These simplest cases however by no means exhibit all the phenomena. From a large group of cases the three following are chosen as each illustrating a distinct possibility.

---

[1] Owing to the ambiguity of some records as to the similarity of the condition in the right and left hands I cannot give exact numbers.

CHAP. XIII.] ONE DIGIT STANDING FOR TWO. 359

Upon the morphological questions arising out of these facts comment will be made when the whole subject of numerical Variation of digits is discussed.

Representation of digits II and III of the pes by one digit.

*529. Man having four digits in the right foot as shewn in Fig. 105. The calcaneum, astragalus, navicular, first (internal) cuneiform and cuboid were normal. The navicular had on its peripheral surface three facets as usual. The second and third cuneiforms were completely united to form one bone which bore no traces of its double nature as shewn in the figure (c^2+c^3). The peripheral surfaces of both form one plane. Taking the four digits in order, the minimus has its normal form and tarsal relations. The digit next to it has the normal form and relations of a digit IV.

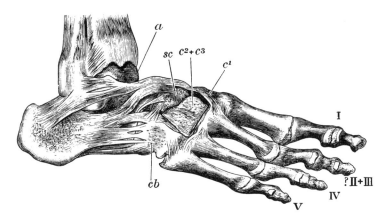

Fig. 105. Bones of the right foot of No. 529. I, hallux. II + III, digit apparently representing index and medius. IV, annularis. V, minimus. a, astragalus. sc, navicular. cb, cuboid. c^1, internal cuneiform. c^2+c^3, bone apparently representing the middle and external cuneiforms. (After GRUBER.)

Internal to this is a metatarsal of abnormal thickness articulating with the single bone presumably representing the external and middle cuneiforms. This metatarsus presented no trace of duplicity. It bore a digit of three phalanges of more than normal thickness but otherwise normal. The hallux was normal, having two phalanges. Each of the other digits had three phalanges, but the 2nd and 3rd phalanges of the minimus were ankylosed.

Of the muscles, the transversalis pedis, one of the lumbricales, one of the interossei dorsales and one of the interossei plantares were absent. The extensor and flexor longus each had three tendons. [Detailed description of bones and soft parts given.] GRUBER, W., *Virch. Arch. f. path. Anat. u. Phys.*, 1869, XLVII. p. 304, Pl. VIII.

Single digit articulating with the cuboid [probably a case of representation of digits IV and V by one digit].

*530. Man having four digits on the left foot as follows. The foot is well formed. The digits I, II and III are normal and have normal tarsal relations. The fourth digit has a well-formed metatarsal and three phalanges. The bones are perhaps rather more robust than those of a normal fifth digit, but the metatarsal has the normal tuberosity at the base strongly developed. This metatarsal articulates with a cuboid of somewhat reduced size having only one articular facet on its peripheral surface. The other parts were all normal, and even in the muscular system only a trifling abnormality was found. Parents normal. STEINTHAL, C. F., *Virch. Arch. f. path. Anat. u. Phys.*, 1887, CIX. p. 347.

Reduction of digit IV of pes.

*531. [This case is introduced here for comparison with the last.] A left foot having abnormalities as follows. Calcaneum, astragalus, internal cuneiform normal in size and shape. The second cuneiform is rather broader than usual, but the surface which it presents to the internal cuneiform has all the characters of a middle cuneiform. External to this middle cuneiform is only *one* large tarsal bone in the distal row. This bone presents no clear sign of duplicity, but from its form and relations it appeared that it represented both the cuboid and the ecto-cuneiform. The hallux and digit II have approximately normal relations. The large cuboid-like bone bears externally a metatarsal agreeing in shape with a metatarsal V; and internal to this the same tarsal bone bears another metatarsal which upon its external side gives off yet another metatarsal of reduced size. Each of the five metatarsals bore a digit, but the digits of the minimus and of the slender IV were webbed together. [Full details given.] BRENNER, A., *Virch. Arch. f. path. Anat. u. Phys.*, 1883, XCIV. p. 23, Pl. II.

532. Besides these simpler cases there are very many recorded instances of reduction in number of digits in which the identification of the parts is quite uncertain. From the point of view of the naturalist it is worthy of remark that even in some of the cases departing most widely from the normal form the limb though having only three or perhaps two digits still presents an approach to a symmetry. Examples of this kind are given by GUYOT-DAUBÈS (*Rev. d'Anthropol.*, 1888, XVII. p. 541, *figs.*) and by FOTHERBY (*Brit. Med. Jour.*, 1886 (1), p. 975 *figs.*) and many more. Fotherby's record is interesting as relating to a family among whose members feet bearing only two opposable claw-like digits of irregular form recurred for five generations. Evidence relating to limbs of this kind is so obscure that it is not possible as yet to make deductions from it, but there seems to be a general agreement among anatomists that when two digits only remain one of them has the characters of a minimus.

Reference must be made also to the fact that in cases of absence of radius the pollex is almost always absent. This seems to be established in very many cases. The only examples of a pollex present in the absence of a radius known to me are that of GRUBER, *Virch. Arch. f. path. Anat. u. Phys.* 1865, XXXII. p. 211, and that of GEISSENDÖRFER, *Zur Casuistik d. congen. Radiusdefectes*, Münch. 1890.

HORSE.

Variation in the number of digits in the Horse[1] has been repeatedly observed from the earliest times. The mode of occurrence of the change is by no means always the same, but on the contrary several distinct forms of Variation may be recognized. On inspection the cases may be divided into two groups.

A. Cases in which the extra digit (or digits) possesses a distinct metacarpal or metatarsal.

B. Cases in which the large metacarpal or metatarsal (III) gives articulation to more than one digit.

Besides these I have placed together in a third group (C) two very remarkable cases which cannot be clearly assigned to either of the other groups. These instances are of exceptional interest from the fact that in them is exhibited a condition intermediate between those of the other two groups. We have seen repeatedly that

[1] In the Mule two cases have been recorded, but in the Ass I know no instance of polydactylism. Describing a polydactyle horse seen on a journey in Rio Grande VON JHERING (*Kosmos*, 1884, XIV. p. 99) states that he believes polydactyle horses to be much more common in S. America than in Europe, and that most persons who have travelled much in that country have met with cases. Mules between the jackass and mare are bred in great numbers, but he had heard of no case in a mule.

Meristic Variation may take place by division of single members of Series, a phenomenon well seen in the B group; and we have also seen many cases of numerical Variation by addition to the Series associated with a reconstitution, or more strictly a redistribution of differentiation amongst the members of the series thus newly constituted; but here in these rare examples of the C group the nature of the parts is such that it cannot be predicated that the change is accomplished by either of these methods exclusively. From such cases it follows that the two processes are not really separable, but that they merge into each other. (Compare the similar facts seen in regard to teeth p. 269, and mammæ p. 193.)

A. EXTRA DIGITS BORNE BY DISTINCT METACARPAL OR METATARSAL.

The cases in this group may be subdivided as follows:

(1) *Two digits, one being formed by the development of the digit II.*

  a. Only *three* metacarpals or metatarsals (II—IV) as usual. Common form: fore and hind limb.

  b. *Four* metacarpals (? I—IV). Common form: anatomically described in fore limb only.

  c. *Five* metacarpals (? I—V). Single case in fore limb.

(2) *Two digits, one being formed by development of the digit IV.* Rare.

(3) *Three digits; the digits II and IV both developed.* Rare.

(4) *Two digits; the digits II and IV both developed, III aborted.* Rare.

It will appear from the evidence that though the same variation is often present in the limbs of both sides this is not always so. The fore and hind limbs also sometimes vary similarly and simultaneously, but in other cases they do not. Different forms of numerical Variation are also sometimes found on the two sides, and not rarely the variation in the fore limb is different from that in the hind limb.

(1) *Two digits, one formed by development of the digit II.*

  a. Three metacarpals or metatarsals only.

To this division and to the next, (1) *b*, belong the great majority of cases of polydactylism in the Horse. Unfortunately most of the records have been made from living animals and contain no anatomical description: in the absence of such particulars it is not possible to know whether a given case belongs to this division or to the next, and it thus is impossible to determine the relative frequency with which the two forms occur.

The following are given as specimen cases.

Fore foot.

*533. Horse of common breed, having a supernumerary digit on the inner side of the right fore foot (Fig. 106).

Humerus and radius: no noticeable variation. *Ulna* a little more developed than usual; lower end slightly broken, having probably reached to lower fourth of radius. The part of the inferior and external tuberosity of the radius which is usually supposed to represent the ulna is larger than in the normal form.

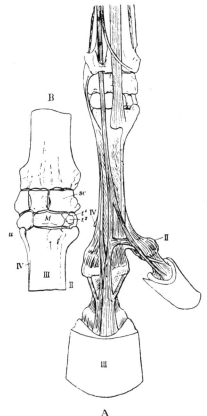

FIG. 106. Right fore foot of Horse, No. 533.
A. The leg seen from in front.
B. The carpal bones enlarged.
M, magnum. sc, scaphoid. u, unciform. t^2, trapezoid. t^1, supernumerary bone not found in normal, representing trapezium. IV, the metacarpal representing digit IV. III and II, metacarpals bearing those digits respectively. (After ARLOING.)

Carpus consisted of eight bones, instead of seven as usual. *Scaphoid* much larger than normal; lunar, cuneiform and pisiform normal. In the lower row the magnum and unciform have normal relations, but in the place of the normally single *trapezoid* are *two* bones, one anterior (t^2), the other posterior (t^1). These together bear the enlarged inner metacarpal (II). The posterior of these bones had a short pyramidal process lying beside the inner metacarpal. This process was partially constricted off and is regarded by

ARLOING as a representative of the metacarpal I, the carpal portion of the bone being the trapezium.

The outer metacarpal (IV) was perhaps slightly larger than usual.

The inner metacarpal (II) was greatly enlarged at its central end, articulating with the two bones t^1 and t^2, and partly with the magnum. In its central part this metacarpal was fused with the large metacarpal (III) and above is united to it by ligamentous fibres. Below it again separates from the large metacarpal and is enlarged, bearing an additional digit of three phalanges, the lowest bearing a hoof. [This hoof is not curved towards the large hoof as in many specimens described, but is convex on both sides, resembling the hoof of an ass.] The large central metacarpal was flattened on the side adjacent to the enlarged metacarpal II. The muscles, nerves and vessels are fully described (*q.v.*). ARLOING, M. S., *Ann. Sci. Nat., Zool.*, Ser. V. T. VIII. pp. 61—67, *Pl.*

534. Foal having two toes on each fore foot. The father and mother of this foal were both of the "*variété chevaline comtoise.*" The foal in question was the only one which this mare dropped and she died two months afterwards. The foal was in nowise abnormal excepting for the peculiarity of the fore feet. The carpus was normal and the *external* metacarpal was rudimentary as usual and ends in a small knob. The internal metacarpal is thicker than the external one and bears a digit of three phalanges, the terminal phalanx bearing a small hoof. This hoof is curved outwards towards the normal hoof. The ligaments and tendons of the foot did not suffice to keep it stiff, and as the animal walked, it not only touched the ground with the hoof but also with the posterior surface of the phalanges. This led to inflammation of the foot, in consequence of which the foal was killed. CORNEVIN, *Nouveaux cas de didactylie chez le cheval*, Lyons (1882?). [Note that this case differs from the last in the fact that the carpus was normal.]

A similar case in the right fore foot is given by KITT, *Deut. Ztsch. f. Thiermed.*, 1886, XII. *Jahresb.*, 1884—5, p. 57, *fig.*

Hind foot.

Among the many accounts of polydactyle horses I know none which gives an anatomical description of a case of a fully developed digit II in the *hind* foot. The following case, indeed, is the only one known to me in which any facts respecting the condition of the tarsus of a polydactyle horse have been ascertained. In it, as will be seen, the digit II was not fully developed.

535. Horse having the metatarsal II enlarged and bearing a rudimentary digit (Fig. 107 B and C). In the left hind foot the arrangement was as shewn in Figs. B and C. The metatarsal II was enlarged and articulated with "two united cuneiform bones" [presumably one bone with indications of duplicity]. Internal to this digit was a "first cuneiform bone," but the digit I was not developed. The metatarsal II bore peripherally a rudiment of a digit as shewn in the figure. The right hind foot was similar to the left but it is stated that the "three small cuneiform

bones" were separate[1], as shewn in Fig. 107 C. The fore feet of the same animal were in the condition described in (1) *b*. [See No. 537.] MARSH, O. C., *Am. Jour. Sci.*, XLIII. 1892, pp. 340 and 345.

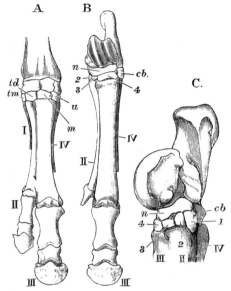

FIG. 107. Limb bones of a polydactyle horse.
A. Left fore foot. No. 537.
B. Left hind foot. No. 535.
C. Tarsus of right hind foot from the inside. No. 535.
n, navicular. *cb*, cuboid. 4, ecto-cuneiform. 1, 2, 3, *three* bones placed as cuneiforms. *td*, trapezoid. *tm*, trapezium. *u*, unciform. *m*, magnum.
I, II, III, IV, numerals affixed to the metacarpals on the hypothesis that these are their homologies. Cp. Fig. 108, which is lettered on a different hypothesis.
(After MARSH.)

b. Four metacarpals.

This condition is a higher manifestation of the variation seen in the cases just given. In No. 533 the digit II was developed and in addition the trapezium had appeared; in the cases now to be

[1] MARSH introduces this case in support of a contention that these variations are of the nature of Reversion. Upon the same page appears the statement that "in every specimen examined, where the carpal or tarsal series of bones were preserved and open to inspection, the extra digits were supported in the usual manner," *l. c.*, p. 345: this assertion is hardly in agreement with the previously stated fact that the metatarsal II is supported by *two* cuneiform bones. On p. 349 Marsh comments on the presence of five bones in the distal row of the tarsus, and from the expressions used it is implied that five such bones had been met with in other polydactyle hind feet. A number of alternative explanations are proposed; (1) that the five tarsals correspond "to those of the reptilian foot"; (2) that the first may be a "sesamoid"; (3) that the first may be a remnant of the first metatarsal, for such a rudiment "apparently exists in some fossil horses." With conjectures of this class morphologists are familiar. Into their several merits it is impossible to inquire, but it may be mentioned that the real difficulty is not the presence of the cuneiform marked 1, but the fact that the tarsal element of the digit II seems to have been double, and that the digits in reality are not supported in the usual manner.

given the digit II is extensively developed and the trapezium bears a splint bone representing the metacarpal I, like that which in the normal represents the digit II. This is a phenomenon illustrating the principle seen in the case of teeth and other parts in series (see p. 272), namely, increase in the degree of development of the normally last member of a series correlated with the appearance of a new member beyond it.

Nevertheless the same cases have sometimes been described (e.g. *Catal. Mus. Coll. Surg.*) on a different hypothesis. This is illustrated by the lettering of Fig. 108. On this other view the innermost carpal is considered to be the *trapezoid* and its splint-bone is regarded as the original metacarpal II. The second digit, *ac*, and its tarsal bone are supposed to be "accessory" or "intercalated." To these terms it is difficult to attach any definite meaning. The proposal that some digits are to be reckoned in estimating homologies and that others are to be omitted is arbitrary, and, if allowed, would make nomenclature dependent on personal choice. It is, as has been often pointed out in foregoing chapters, simpler to number the parts in order as they occur and to accept the visible phenomena as the safest index of the methods and possibilities of Variation. Nevertheless, to illustrate the point at issue I have introduced two cases of the same Variation, the one, No. 536, lettered on the view advocated by the *Catalogue* of the College of Surgeons, &c., the other, No. 537,

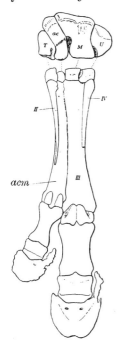

Fig. 108. Right fore foot of Horse No. 536 from behind. The upper surfaces of the carpal bones of the distal row are separately shewn above. Specimen in Coll. Surg. Mus., *Ter. Cat.*, 304.

T, trapezoid. *M*, magnum. *U*, unciform. *ac*, accessory carpal bone. II, III, IV, metacarpals. *acm*, accessory metacarpal.

This figure is lettered to illustrate the hypothesis adopted in the *Catalogue*, which is alternative to that adopted in Fig. 107, A.

lettered on the other and in the case of polydactyle horses, more usual method.

*536. Horse: right manus with extra digit (Fig. 108). The distal row of the carpus is present. It consists of four bones, the unciform, magnum and two other bones. Of these that lettered *T* on the view of the *Catalogue* must be supposed to be the normal trapezoid, while *ac* is considered to be an intercalated bone, perhaps an additional os magnum. The unciform bears a splint-bone, namely mcp. IV. The magnum bears a fully-formed mcp. and digit III. With the bone *ac* articulates a large and substantial metacarpal with a digit of three phalanges and a hoof, while the bone *T* bears another splint-bone, marked II in the figure on the hypothesis that the digit *ac* is not to be reckoned. *Cat. Mus. Coll. Surg., Terat. Series*, 1872, No. 304. As mentioned above, it would be more consistent with fact to count the bone *ac* as trapezoid with mcp. II and the bone *T* as trapezium with mcp. I.

537. Horse having both fore feet (Fig. 107, A) as in the last case, the hind feet being in the condition described in the last Section, No. 535. MARSH, *Am. Jour. Sci.*, XLIII. 1892, p. 340, *figs*. 3, 6, and 8.

538. Foal having right manus closely resembling the above, the other limbs being unknown. The mcp. I was longer than the normal mcp. II. In this case the metacarpal II was partially united to mcp. III at the central end but was free from it peripherally. WEHENKEL, J. M., *La Polydactylie chez les Solipèdes*, from the *Journal de la soc. r. des sci. méd. et nat. de Bruxelles*, 1872, *fig*. 2.

Probably the feet of a large number of polydactyle horses would be found to be in this condition if examined. MARSH, *l.c.*, mentions three other cases known to him in Yale Museum.

c. *Five metacarpals.*

*539. Horse having *five* metacarpals and one supernumerary digit in the left manus, and *four* metacarpals with a similar supernumerary digit in the right manus.

In the *left* manus with the trapezoid there articulated a well-developed metacarpal II bearing the extra digit. Internal to this was a trapezium bearing a splint-bone, 6 cm. long, 1·5 wide at proximal end, representing metacarpal I [as in Section (1) *b*] coalescing peripherally with III. On the external side of III the splint-bone IV was present as usual. The case is remarkable from the fact that external to the metacarpal IV there was another rudimentary metacarpal, presumably representing V. This bone was distinctly separated from IV at the central end, but was for the most part united with it. PÜTZ, *Deut. Ztschr. f. Thierm.*, 1889, XV. p. 224, *figs*. [The figures illustrating this paper are carefully drawn. The representation of mcp. I is quite clear, but the condition of the mcp. V cannot be well seen, as the whole foot is represented with its ligaments, &c., which partly conceal the structure. The whole account is very minute and gives confidence in the statements.]

The *right* manus of the same animal came into the possession of the University of Graz and was described independently. In it also the metacarpal II was developed and bore a well-formed digit. There was also a rudimentary metacarpal I beside it, having a length of 5·7 cm., and a breadth of 1·5 cm. at the central end. [The description is brief and makes no mention of a mcp. V: further account promised.] MOJSISOVICS, *Anat. Anz.*, 1889, IV. p. 255.

(2) *Two digits, one being formed by development of the digit IV.*

Cases of this variation are exceedingly rare. No. 540 is the only instance known to me in which a proper account exists. Most writers on the subject make a general statement that such cases exist, but give no references.

*540. Horse, having a supernumerary digit on the *outside* of each fore foot. (Fig. 109.) The animal was from Bagdad. The outer rudimentary metacarpal (IV) was well formed and of nearly even thickness throughout its length. It bore a digit of three phalanges and a well-formed hoof. The hoof was elongated and is described as being shaped like the hoof of a cloven-footed animal. [The description is very imperfect, but two good figures are given, from which it may be gathered that the inner metacarpal (II) was somewhat more developed than in an ordinary horse; and it appears that both the inner and outer metacarpals were separate throughout their course, but whether they could be detached from the large metacarpal or were ankylosed with it is not stated. The carpal bones are not described, but the figure suggests that the unciform was larger than it normally is. It is not stated that the two feet were alike in details. The large hoof (III) is represented as of the normal shape.] WOOD - MASON, J., *Proc. Asiat. Soc. Bengal*, 1871, p. 18, *Plate.*

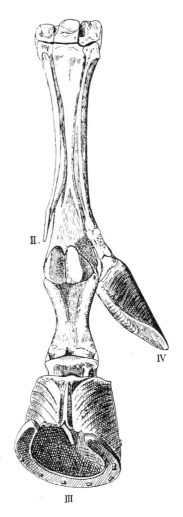

FIG. 109. Right fore foot of Horse, No. 540, the *external* metacarpal (IV being developed, bearing a digit.
(After WOOD-MASON.)

WEHENKEL, *l. c.*, p. 15, mentions a similar specimen in the Museum of the Veterinary School at Berlin described by GURLT, *Mag. f. gesam. Thierh.*, 1870, p. 297 [not seen, W. B.].

(3) *Three digits [? the digits II and IV being both developed].*

Examples of this variation are alluded to by many authors but I know of no anatomical description. The following are all very imperfectly described.

541. Foal (fœtus) : left manus having three sub-equal digits; right manus two digits and rudiments of a third more developed than usual. Hind feet normal. GEOFFROY ST HILAIRE, *Ann. Sci. Nat.*, XI. 1827, p. 224.

Similar case, BREDIN, *Froriep's Notizen*, XVIII. p. 202.

542. Horse from Texas, having extra digit on inside of each manus, and an extra digit both on the outside and on the inside of each *pes* [external view only]. MARSH, *Am. Jour. Sci.*, XLIII. 1892, p. 344, *fig.* 7.

543. Horse with both splint-bones bearing digits in each foot. FRANCK, *Handb. d. Anat.*, Stuttg., 1883, p. 228.

(4) *Two digits; the digits II and IV both developed, III aborted wholly or in part.*

Mention of these cases must be made in illustration of the possibilities of Meristic Variation, but the parts were in all three instances so misshapen that the animals could not have walked.

544. Foal having two toes on each foot, the developed toes belonging to the metacarpals and metatarsals II and IV, while the normally large III was not developed at all in the fore feet and was in the hind feet represented by a wedge of bone only.

Hind feet. Left. Bones of leg and tarsus said to have been normal. Metatarsal III represented by a wedge of bone fixed between the greatly developed metatarsals II and IV. The wedge-like bone 5 cm. wide at upper end, having usual tarsal relations. Its length about the same as its width. Laterally it is united to the metatarsals II and IV which curved round it till they met, and then curve away from each other again. Each was about 20 cm. long and bears a misshapen digit consisting of a proximal phalanx and a hoof-bearing distal phalanx. A small nodule of bone attached to the proximal phalanx may or may not represent part of a middle phalanx. *Right.* Very similar to left, but the wedge-like III was rather broader—[for details see original].

Fore feet. More misshapen and less symmetrical than hind feet: metacarpal III not developed at all. The metacarpals II and IV curved towards each other and crossed, giving an unnatural appearance to the feet. *Right foot.* Cuneiform and lunar united, and upon the surface of the bone formed by their union there was a groove occupied by two parts of the tendon of the anterior extensor metacarpi passing to mcp. II and IV respectively. Pisiform and scaphoid normal [this is not clear from the figure]. Magnum absent. Unciform and trapezoid abnormal only in respect of their relations, for whereas they should articulate with the magnum they do not do so, for both magnum and mcp. III are not represented. Metacarpal II was 11 cm. long, mcp. IV being 19 cm. long. Each bore a digit with a hoof; the digit IV having a proximal and a distal phalanx connected by a fibrous cord instead of a middle phalanx. The digit II had a rudimentary distal phalanx only. *Left foot* like the right, but with the mcp. and digit II more fully developed. [Muscles fully described. It may perhaps be thought that there is not sufficient proof that the developed digits are actually those normally represented by the splint-bones II and IV, but the condition of the hind feet is practically conclusive that this is the right interpretation.] WEHENKEL, *La Polydactylie chez les Solipèdes*, from *J. de la soc. r. des sci. méd. de Bruxelles*, 1872, Plate.

545. Foal, in which the right anterior leg possessed two metacarpals and digits.

The radius, ulna and proximal series of carpal bones were normal. In the distal series only *two* bones were present, viz., an inner bone corresponding to the trapezoid, and a magnum. There was no separate bone corresponding to the unciform, but in its stead, the head of the outer metacarpal was continued upwards to articulate with the cuneiform. Between the heads of the two metacarpals was an irregularly quadrate bone which articulated with the magnum in the place where the large metacarpal (III) should be. This bone however only extended a little way, articulating at its outer end with a notch in the external metacarpal. [This is the author's view, but the figure strongly suggests that this quadrate bone may have been originally in connexion with the external metacarpal and that it may have been separated from it by fracture. If this were so, the large metacarpal would then not be represented by a separate bone at all.] The outer metacarpal distally bore three phalanges of irregular shape, flexed backwards and outwards. The inner metacarpal articulated solely with the trapezoid. Peripherally it bore a callosity which was due to the healing of a fracture. The phalanges of the inner metacarpal were three, but the first was reduced in length, while the second was elongated and bent in a sinuous manner. The ungual phalanx of this toe was cleft. [The author regards this case as analogous to the foregoing one, No. 544, that is to say, as an instance of development of the normally rudimentary lateral metacarpals to the exclusion of the large one (III), and he considers therefore that the large metacarpal (III) is only represented by the quadrate ossification which lay between the two developed metacarpals.] ERCOLANI, G. B., *Mem. della Acc. Sci. d. Istituto di Bologna*, S. 4, T. III. 1881, p. 760, *Tav.* I. *fig.* 11.

546. Foal in which the feet were all very abnormal. In the two *fore feet* the metacarpal of the normal toe (III) was very little developed, being however somewhat larger on the left side than it was on the right. It bore no digit. The external metacarpal bone (IV) of each fore foot attained a considerable length and bore a small hoof-bone. In the *left fore foot* the inner metacarpal was present but reduced; in the right foot it was absent. *Right hind foot* also had the *external* metacarpal developed and bearing three small phalanges, but the central metacarpal (III) was fairly developed, bearing however only two phalanges. *Left hind foot* was amorphous. BOAS, J. E. V., *Deut. Ztschr. f. Thiermedecin*, VII. pp. 271—275. [For full description, measurements and figures see original.]

B. CASES IN WHICH METACARPAL III GIVES ARTICULATION TO MORE THAN ONE DIGIT.

These cases are clear examples of the representation of a single digit by two. It will be seen besides that the two resulting digits may stand to each other in the relation of optical images (see Fig. 110) and do not form a Successive Series, thus following the common method of division of structures possessing the property of Bilateral Symmetry in some degree (cp. p. 77). All cases of this variation known to me occurred in the fore limb.

*547. Foal: a right fore foot figured from a specimen in the collection of the Veterinary School of Copenhagen (Fig. 110) has two complete digits articulating with a single normal metacarpal bone. The two digits are symmetrically developed; each consists of three phalanges and bears a hoof. These two hoofs are well formed and curve towards each other like those of Artiodactyles. BOAS, J. E. V., *Deut. Ztschr. f. Thiermedecin*, VII., p. 277, *Taf.* XI., *fig.* 9.

548. Two fore feet of a foal, each being irregularly and unequally bifid. BOAS, *ibid.*, *figs.* 7 and 8.

FIG. 110. Right fore foot of Horse No. 547.
Mcp, peripheral end of metacarpal III.
ext, external side. *int*, internal side.
(After BOAS.)

549. Filly, two-year old, which had been born with left fore foot cleft like that of the Ox. Each of the two toes had three phalanges, which were completely separate as far up as the metacarpo-phalangeal joint. The division externally was carried to the same extent as in the Ox. The lower end of the great metacarpal III felt as if bifurcated like that of the Ox, so as to give separate articular support to the two toes. Upper parts normal. The lesser metacarpals, II and IV, felt through skin, seemed to terminate rather lower down in left foot than in right, but this was uncertain. Animal examined alive. No attempt at shoeing had been made, and hoofs having become elongated forwards had had their points sawn off. The whole foot was much larger or more spread than the other. STRUTHERS, J., *Edin. New Phil. Jour.*, 1863, pp. 279 and 280.

*550. Horse: right fore foot having phalanges bifid (Fig. 111). The limb was normal as far as the distal end of the metatarsal, except for some exostoses. The proximal phalanx was short and of great width; in its lower third it divided into two divergent parts, the divergence being more marked on the posterior face than on the anterior. Each of these diverging processes bears a complete second and third phalanx. The third phalanges each bear hoofs, which are convex on the outer sides but fit together on the opposed surfaces, the external hoof being slightly concave on its inner face, while the internal is slightly convex. On the plantar surface, each toe bore a half-frog. The two large sesamoids, normally present in the Horse, are in this specimen united along their inner borders to form a single bone, which was placed behind the upper part of the proximal phalanx. Two small sesamoids lay behind the third phalanx. A good deal of exostosis had taken place in all the phalangeal bones. ARLOING, M. S., *Ann. Sci. Nat.*, Ser. V., Tome VIII. pp. 67—69, *Pl.*

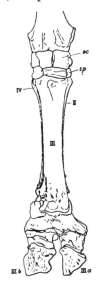

FIG. 111. Right fore foot of Horse No. 550, from in front. *sc*, scaphoid. *tp*, trapezoid. II, III, IV, metacarpals. III*a*, III *b*, internal and external sets of phalanges representing the digit III of the normal. (After ARLOING.)

551. Foal: in right fore foot the large metacarpal divided into two parts, each bearing a separate digit. The proximal row of the carpus consisted of four normal bones, but the distal row was composed of two bones only. The external splint-bone (IV) was of normal proportions, but the internal splint-bone (II) had almost completely disappeared. The large metacarpal (III) divided in its peripheral third into two equal cylindrical branches, each of which bore a digit composed of three phalanges and bearing a crescentic hoof. These two digits were bent across each other in a shapeless way. DELPLANQUE, *Mém. Soc. centr. d'Agric. du Dép. du Nord*, s. 2, IX. Douai, 1866—1867, p. 295, Pl. III. *fig.* 5.

552. Mule, having two distinct toes on each fore foot. The hoofs were shaped like those of the Ox. They were of unequal length. JOLY, *Comptes Rendus*, 1860, p. 1137. [Perhaps a case belonging to this section.]

CHAP. XIII.] DIGITS OF HORSE: SPECIAL CASES. 371

C. Intermediate cases.

We have now seen cases of increase in number of digits occurring by addition to the series, and cases occurring by division of III. It may at first sight seem impossible that there can be any process intermediate between these two. Nevertheless the word sufficiently nearly describes the condition of at least the first of the following cases, and is to some extent applicable to the second also. If the condition shewn in Fig. 112 be compared with those in Figs. 106 and 110 it will be seen that it is really intermediate between them.

*553. Horse (young): right manus with internal supernumerary digit. The bones are not in place, but have been attached with wires. The condition is as follows. The distal series of carpus remains and is normal or nearly so. Of the splint-bones, the inner (mcp. II) is thicker than the outer mcp. IV, but it is very little longer. The large metacarpal (III) is almost, but not quite, bilaterally symmetrical about its middle line. In the distal epiphysis the asymmetry is distinct, the internal side of the epiphysis being less developed than the external side. This epiphysis bears a large digit of three phalanges, but instead of being bilaterally symmetrical, like the normal toe of the Horse, each of the joints is flattened on the internal side, the flattening increasing from the first to the third phalanx. The hoof is greatly flattened on its inner face.

Internally to the epiphysis of the digit III there is a separate small bone, representing the distal end of an inner metacarpal. This bone bears a digit with *two* phalanges, and a hoof which is flat on the side turned towards the other hoof, like that of a calf, though it only reaches to the top of the larger hoof. The first phalanx of this digit is imperfectly divided by a suture into two parts. This division is *not* that of the epiphysis from the shaft. This extra digit may be thought to be that of mcp. II, but it is clear that it was in part applied to mcp. III. Note also that mcp. III is modified in correlation with its presence. *Coll. Surg. Mus.*, in *Terat. Cat.*, No. 301.

The foregoing case well illustrates the inadequacy of the view on which an individuality is attributed to members of the digital series. The smaller digit in it is as regards the Sym-

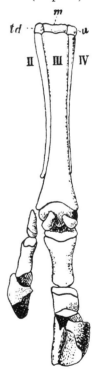

Fig. 112. Right manus of a horse, No. 553, from behind, *m*, magnum. *td*, trapezoid. *u*, unciform. (From a specimen in Coll. Surg. Mus.).

24—2

metry of the limb complementary to the larger digit. It is a partial substitute for the inner half of the digit III. If the visible Symmetry of the limb is an index of mechanical relations in which the parts stood to each other in the original division of the manus into digits it is possible that there may have been a mechanical equivalence between the two digits.

554. Mule (between jackass and mare): fœtus of about nine months having supernumerary digits. Hind limbs normal. Fore limbs normal as far as peripheral ends of metacarpals. Each manus consisted of three digits. *Right*. Metacarpals II and IV normal splint-bones. Metacarpal III normal as far as line of union with its distal epiphysis. The inner part of the sheath of the epiphysis is continued into a rod of fibro-cartilage which supports an extra toe. This rod of cartilage contains a small ossification which represents, as it were, the proximal phalanx of this internal supernumerary toe. Its outer end bears a small second phalanx, and this bears a small distal phalanx which was covered by a hoof. This extra toe, therefore, is internal to the main continuation of the leg, commences from the line of union between the large metacarpal and its epiphysis, and has three phalangeal joints.

Fig. 113. Left fore foot of Mule No. 554.
IV, the external splint-bone. III, the chief metacarpal. III *a*, III *b*, internal and external rudimentary digits borne by III. II, a supernumerary digit attached to the inner side of III.
(After JOLY and LAVOCAT.)

The epiphysis of the large metacarpal supports a normal first phalanx with which the second phalanx articulates. This second phalanx is enlarged internally [details obscure] to bear a small extra nodule of cartilage which appears to be of the nature of an extra toe. The second phalanx also bears a large third (ungual) phalanx. This ungual phalanx together with the minute supernumerary toe borne by the second phalanx are together encased in a common hoof, but the hoof is divided by a groove into two distinct lobes, corresponding with the division between the two digits which it contains. The whole foot, therefore, has one free internal toe and one large toe bearing a small internal one, which are enclosed in a common hoof.

Left fore foot. Fig. 113. The small, lateral metatarsals II and IV, and the large central metatarsal III are normally constructed; but from the inner side of the sheath of the large metatarsal, upon the line of union between the bone and its epiphysis, arises a fibro-cartilaginous rod, which contains an ossification representing the proximal phalanx of a supernumerary toe (lettered II in fig.). This rod of tissue in its proximal portion is represented in the figure as abutting on, but distinct from the end of the inner small, lateral metatarsal. It bears a cartilaginous second phalanx, containing a small ossification, which articulates with a terminal (ungual) phalanx covered by a hoof.

The distal end of the large metatarsal articulates with a large first phalanx, which at its proximal end is of normal width. At about its middle point this phalanx bifurcates into two parts, of which the inner, III *a*, is short and ends a little beyond the point of bifurcation: it bears an *ungual* phalanx only, which is encased in a hoof. The outer limb (III *b*) of the bifurcated first phalanx bears an elongated second phalanx of somewhat irregular shape which carries a larger ungual phalanx covered by a separate hoof. In this foot, therefore, there is an inner toe consisting of three phalanges attached to the inside of the large metatarsal: next, the proximal phalanx of the large toe is divided longitudinally into two parts, bearing (1) an internal toe having only the ungual phalanx and hoof; (2) an outer toe which has a second and third (ungual) phalanx.

In the case of both feet, the hoof and ungual phalanx of the outer toe are turned *inwards*, having an external curved edge and an internal straight edge; but the two inner toes in each case are turned *outwards*, having their outer edges straight and their inner edges curved. JOLY, A. et LAVOCAT, N., *Mém. de l'Ac. des Sci. de Toulouse*, S. 4, Tome III., 1853, p. 364, *Plates*. [Authors regard this case as proof of truth of certain views of the phylogeny of the Horse and employ a system of nomenclature based on these views. This is not retained in the abstract here given.]

ARTIODACTYLA.

In the domesticated animals of this order digital Variation is not rare, being in the case of the Pig especially common. Such variation has been seen in the Roebuck and Fallow Deer, but not in any more truly wild form so far as I am aware. These variations may take the form either of polydactylism or of syndactylism. Of the former a few cases are known in Ox, Sheep, Roebuck Fallow Deer, and many cases in the Pig; syndactylism has been seen only in the Ox and in the Pig. The absence of cases of syndactylism in the Sheep is a curious instance of the caprice with which Variation occurs.

The phenomena of polydactylism in Pecora may conveniently be taken separately from the similar phenomena in Pigs.

POLYDACTYLISM IN PECORA.

At the outset one negative feature in the evidence calls for notice. It is known that in the embryo Sheep rudiments of metacarpals II and V exist[1] which afterwards unite with III and IV. In view of this fact it might be expected by some that there would be found cases of Sheep and perhaps Oxen polydactyle by development of the digits II or V. In the Sheep only one case (No. 555) is known that can be possibly so interpreted; and in the Ox there is no such case unless Nos. 557, 558, and 559 should be held by any to be examples of the development of II, a view attended by many difficulties.

The two following examples are the only ones known to me in which there can be any question of reappearance of a lost digit, but in neither is the evidence at all clear.

*555. **Sheep.** Some specimens of a small Chilian breed had an extra digit on the hind foot. It was not present in all individuals and was not seen to be inherited; but normal parents were observed to have offspring thus varying. [From the description given I cannot tell whether the extra digit was internal or external. Also, though said to have been on the hind foot, in describing the bones the cannon-bone is twice called *metacarpus*; probably this is a slip for *metatarsus*.] The digit was only attached by skin. It contained a bent bone, of which the upper segment was 20 mm. long, the lower 13 mm. Proximally the

---

[1] ROSENBERG, *Z. f. w. Z.*, 1873, XXIII. pp. 126—132, *figs*. 14, &c. Sometimes these rudiments remain fairly distinct at the proximal end of the cannon-bone, especially of the fore foot. See NATHUSIUS, *Die Schafzucht*, 1880, pp. 137 and 142, *figs*.

374   MERISTIC VARIATION.   [PART I.

cartilaginous head of this bone rested in a pit on the tendon of the flexor brevis digitorum at the level of the end of first third of the cannon-bone, and peripherally it bore an end-phalanx and claw-like hoof, properly articulating. No splint-bones present. [Other details given: it was suggested that the bent bone represented an extra 'metacarpal' and first and second phalanx.] VON NATHUSIUS, H., *Die Schafzucht*, 1880, p. 143.

556.   **Capreolus caprea** (Roebuck), 2 yr. old, killed in district of Betzenstein, having a slender fifth digit on the inside of each fore foot. In the left there was a small, conical metacarpal element, bearing a digit with three phalanges. The right extra digit had a longer metacarpal piece with epiphysis, but in it there were only two phalanges. Each bore a hoof of about the size of those of II or V. The hoofs curved outwards. BAUMÜLLER, C., *Abh. naturh. Ges. Nürnb.*, IX. 1892, p. 53, *Pl.*

Other cases of polydactyle Pecora mostly fall into two groups:

(1) Examples of limbs having three digits borne by a large cannon-bone made up of three metatarsal or metacarpal elements, grouped in *one* system of Symmetry. The axis of Symmetry is then deflected from the normal position, and instead of falling between two digits it approaches more or less to the central line of the middle of the three digits. The degree to which this change of Symmetry takes place corresponds irregularly with the extent to which the innermost digit is developed. This form is known in the Ox only [? Goat].

(2) Limbs in which the series of digits has *two* more or less definite axes of Minor Symmetry. Both of the systems of Symmetry thus formed are in addition arranged about one common axis of Symmetry. The nature of this condition will be discussed later. It occurs in Ox, Sheep, Roebuck and Deer.

(1)   *Three digits in one system of Symmetry.*

*557.   **Calf.** Right manus (Fig. 114) having three digits borne by a single cannon-bone. This is an old specimen of unknown history which was kindly sent to me by Mr W. L. Sclater for examination.

Of the carpal bones only the distal row remains, containing a trapezoido-magnum and unciform not differing visibly from the normal. The cannon-bone spreads at about its middle into three sub-equal parts, each ending in a separate articular head bearing a trochlear ridge. Between these articular surfaces the only point of difference was that in that of the middle digit (*b*), the trochlear ridge was rather nearer to the outer surface of the joint, not dividing it into two halves as usual (see figure). The foramen for entrance of the nutrient vessel was in the channel between the external and middle digits. This channel was very slightly deeper than the corresponding channel between the middle and inner digits. Each articular head bore a digit, well formed, of approximately similar lengths, having a hoof. The hoofs of the outer and inner digits curved to the middle line of the limb, like the

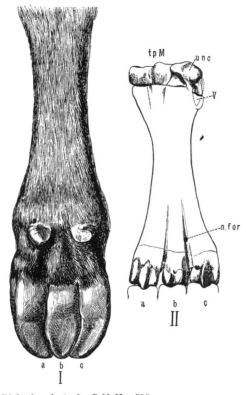

Fig. 114. Right fore foot of a Calf, No. 537.
I. The whole foot seen from behind.
II. The bones from behind.
tpM. trapezoido-magnum. *unc,* unciform. *n. for.,* nutrient foramen. *V,* dotted outline shewing position of supposed rudiment of digit V. Sesamoids not shewn.

normal hoofs of a cloven-footed animal, but the hoof of the central toe was convex on both sides. The two accessory hoofs were in place, one on each side as shewn in the figure. The whole manus was very nearly symmetrical about the middle line of this digit. It was noticeable that the outer and inner hoofs were both rather narrow in proportion to the length of the limb, but the whole width of the foot was rather greater than it should be. The small bone considered to represent the digit V articulates with the unciform as usual, being of normal size. Each of the three digits was supplied with flexor and extensor tendons.

558. **Heifer** having three fully developed toes on each hind limb. The right hind foot described (Fig. 115). The calcaneum, astragalus and cuboido-navicular presented no special abnormality. The cuneiform

series usually consisting of two pieces in the Ox, were here represented by one piece (c and c^3), though externally the bone seemed to be in two pieces. The internal portion (c) approximately corresponding in position with the normal ento-cuneiform was imperfectly and irregularly divided by a groove into two parts. The metatarsus or cannon-bone at its proximal end was almost normal, but from about its middle it spread out into three parts as shewn in the figure, each part ending in an articular surface and bearing a digit, but the trochlear ridge for the innermost digit (ac) was not quite so large as those for the others. From the skeleton it seems clear that this innermost digit could not have reached the ground.

Of the three hoofs the middle one was the largest, the other two being nearly equal to each other in size. The outermost hoof curved inwards, and the innermost hoof curved outwards. The middle hoof also curved outwards, *but less so than that of a normal digit III, being rather flatter underneath, and having its two edges more nearly symmetrical*. The accessory hoofs ('*ergots*' of French writers) were "in their usual place, on either side of and behind the foot." This specimen was originally described by Goodman, Neville, *Jour. Anat. Phys.*, 1868, Ser. 2, I. p. 109. The skeleton of the foot is in the Cambridge University Museum of Pathology.

In answer to my inquiries Mr G. Daintree of Chatteris, the owner of this animal, kindly gave me the following information. This cow was bought in 1861 and from her a three-toed strain arose, of which about ten generations were produced. The three-toed condition appeared in both males and females, but no three-toed bull was kept, so that the descent *was wholly through females*. About two in three calves born of this strain had three toes. In one case only were there three toes on the *fore feet*. The third toe was never walked on. The breed was got rid of because it was at last represented only by males, the last being sold in 1887. The beasts were as good as any other cattle of the same class.

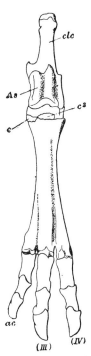

Fig. 115. Right hind foot of heifer, No. 558.

As, astragalus. *clc*, calcaneum. *c*, *c*³, parts of a large united bone representing cuneiforms.

III, IV, *ac*, letters affixed to illustrate the hypothesis that III and IV represent these digits of the normal, and that *ac* is an accessory digit.

559. **Calf.** Left manus having three digits, generally resembling the last case. The external digit is nearly normal. The middle digit is very thick, and is somewhat twisted and flexed. Its ungual phalanx is not specially curved in either direction but it is not truly symmetrical. The innermost digit is thin and short and its ungual phalanx is not much curved. In this specimen there is a decided appearance of division in the distal epiphysis of the metacarpal of the middle digit (? III). *Coll. Surg. Mus., Terat. Series*, No. 300.

The following two cases are perhaps of the same nature as the foregoing.

560.   **Goat** having three digits in each manus, described by GEOFFROY ST HILAIRE, *Hist. des Anom.* I. p. 689. The description states that a supernumerary toe was placed between the two normal toes. The middle toe was one-third of the size of a normal toe, but the lower part of the foot was larger than usual. This case was probably like No. 557; for from the shapes of the lateral hoofs that case also might seem to an observer at first sight to be an example of a toe "intercalated" between two normal toes. But in No. 560 the middle digit was reduced.

561.   **Calf** having a small supernumerary toe 'placed between the digits of the right manus.' This toe had a hoof and seemed externally to be perfect, but on dissection it was found to contain no ossification, but was entirely composed of fibrous tissue and fat. ERCOLANI, *Mem. Ac. Bologna*, S. 4, III. p. 772. [Probably case like last, the middle digit being still less developed.]

This case is probably distinct from the others given.

562.   **Calf**: right fore foot having three complete metacarpals, each bearing a digit of three phalanges. The two outer were disposed as in the normal, but the innermost metacarpal was quite free from the others and its digit stood off from the others [not grouping into their symmetry as in preceding cases] and having an ungual phalanx [of ? pyramidal shape]. DELPLANQUE, *Études Tératol.*, Douai, 1869, II. p. 33, Pl. II. *figs.* 2 and 3. [It is difficult to determine the relation of this case to the others and I am not sure that I have rightly understood the form of the inner digit; but since this digit seems to be outside the Minor Symmetry of the limb it is almost impossible to suppose that it can really be the digit II reappearing. I incline to think that it is more likely that this digit belongs to a separate Minor Symmetry. Compare the similar phenomena in Pigs, No. 570.]

On the foregoing cases some comment may be made. It may be noted that the two first (Nos. 557 and 558) present two stages or conditions of one variation. In No. 557 all three digits reach the ground and the change of Symmetry is completed; in No. 558 the internal digit is not so large in proportion and the plane of Symmetry is not deflected so far.

As to the morphology of the three digits in these cases three views are open on the accepted hypotheses. First, the internal digit (if it be admitted to be *the* supernumerary) may be simply a developed II. The existence of the normal accessory hoofs practically negatives this suggestion, for there can be little doubt that one of them represents II (*v. infra*, No. 579). The condition of the cuneiforms in No. 558 suggests further that an element is introduced into the cuneiform series between the almost normally formed ento-cuneiform and the ecto-cuneiform. But if this new element is the middle cuneiform, then the internal digit (Fig. 115, *ac*) may still be II. But the innermost *ergot* is II in the normal. Or is the inner *ergot* in this case I, and is this once more a case of the development of a normally terminal member, II, and of the addition of I beyond it in correlation, as we saw in the Horse (see p. 364)? That such a correlation may exist is unquestionable, and it is not clear that these cases are not examples of it. But even if this principle be adopted here as a means of bringing these cases into harmony with received conceptions it will presently be seen that it still will not reconcile some other cases, notably those of the presence of supernumerary digits in a Minor Symmetry apart from that of the normal series. Yet if the conception of the digits as endowed with individuality be not of universal application, we shall not save it even if by ingenuity we may represent the facts of the present case as in conformity with its conditions.

On the other hand it may be suggested that there is a division of some one digit, and undoubtedly in No. 559 there is a suggestion that the innermost digit and the central digit are both formed by division of III. But in the first place this view cannot so easily be extended to Nos. 557 and 558, for in them there is practically no indication that the digits are not all independent and equivalent. The circumstance that the nutrient vessel enters between the external and middle digits may perhaps be taken to shew that they are III and IV; but this vessel, if single, must necessarily enter in one or other of the interspaces and there is no reason for supposing that, were there an actual repetition of a digit, the vessel must also be doubled, though doubtless repetition of vessels commonly enough occurs with repetition of the organs supplied.

Next, the Symmetry of the foot, the development of the middle digit to take a median place, the position of the accessory hoofs, one on either side equidistant from the middle line of the manus, all these are surely indications that this limb was from the first developed and planned as a series of *three* digits, and not as a series of *two* digits of which one afterwards divided. The series has a new number of members, and each member is in correlation with the existence of the new number remodelled.

It is no part of the view here urged to deny that a single digit, like any other single member of a series, may divide into two (or even into three) for this phenomenon is not rare. Probably enough No. 559 is actually a case of such a division of the digit III. But here in digits as in mammæ, teeth, &c., the evidence goes to shew that there is no real distinction between the division of one member to form two, and that more fundamental reconstitution of the series seen in No. 557, for the state of No. 558 is almost halfway between them. In it we almost see the digit III in the act of losing its identity.

(2) *Limbs with digits in two systems of Minor Symmetry* (Double-foot).

In dealing with these there are difficulties. The cases are examples of limbs of Calves or Sheep bearing four or five digits arranged in two groups either of two and two, or of two and three. The members of each group curve towards each other in such a way that each group has a separate axis of Symmetry (Figs. 117 and 118). In several such cases the two groups are related to each other as right and left. Of these facts two different views are possible. For first, a limb of this kind may be a structure like the double-hands seen in Man (pp. 331 to 337), for it is certain that an almost completely symmetrical series of parts is in those cases formed by proliferation of a series normally hemi-symmetrical, however unexpected this phenomenon may be.

On the other hand it might be argued that one of the groups of digits represents the normal, and that the other group is supernumerary.

For, as will be hereafter shewn at length in the case of Insects, supernumerary appendages may grow out from a normal appendage and are then a pair, being formed as a right and a left, composing a separate Secondary Symmetry.

On the first view the digits of each group are in symmetry with each other like those of the normal limb, the two groups also balancing each other like the halves of a double-hand: on the other view one of the groups would be supposed to be made up of a right and a left digit III, or of a right and a left digit IV. The possibility of the second view being true arises of course in the Artiodactyles from the fact that in them the normal digits compose a bilateral Minor Symmetry.

There is nevertheless little doubt that the former account is the right one and that neither group is a Secondary Symmetry; for were either of the groups really in Secondary Symmetry the supposed supernumerary group should contain at least parts of *four* digits. Lastly, some of the cases, as No. 566, are clearly of the nature of double limbs, both groups having a common axis of Symmetry.

A further difficulty arises from the fact that most of these double limbs are old specimens cut off from the trunk. There is therefore no proof that such a limb is not that of a polymelian in Geoffroy St Hilaire's sense. In other words, though it is practically certain that neither of the groups of digits is itself a system of Secondary Symmetry it is quite possible, and in some cases likely that the whole limb is of this nature. In cases of duplicity, especially of posterior duplicity, the two limbs of one or both of the united bodies frequently form a compound structure somewhat resembling one of the double limbs here under consideration. Hence it is not possible to include with confidence great numbers of cases of double limbs described by various writers or preserved in museums, for it is rarely that particulars regarding the rest of the animal are to be had. This difficulty applies to almost all cases known to me and they are therefore given with this caution. This objection of course does not apply to such a case as No. 564.

The following few cases will sufficiently illustrate the different forms of limbs included in this section. They consist of two chief kinds ; first, limbs like Nos. 563 and 566, in which both groups contain two digits, and secondly, cases like No. 567, in which one of the groups contains three digits, recalling the state described in the last section (cp. Nos. 558 and 559). Besides these there are some cases of amorphous extra digits not here related.

563. **Cow**, full-grown, right fore foot with four digits arranged in two groups of two, as shewn in fig. 116. The carpus not preserved. No particulars as to the rest of the animal. This specimen is in the Museum of Douai and is described in detail by DELPLANQUE, *Études Tératologiques*, II. Douai, 1869, p. 30, Pl. I. [The possibility that this may be a limb of a pygomelian is not excluded.]

*564. **Cervus dama** (Fallow Deer). A female having each hind foot double. The division occurs in the upper part of the tarsus, which gradually diverges into two separate tarsi [? metatarsi] and two separate feet. This doe had for several successive years dropped a fawn with the same malformation, though she had been served by several bucks. WARD, EDWIN, *Proc. Zool. Soc.*, 1874, p. 90.

565. Two cases, a **Roebuck** and a **Deer**, mentioned by GEOFFROY ST HILAIRE (*Hist. des Anom.*, I. p. 697) are probably of this nature.

380   MERISTIC VARIATION.   [PART I.

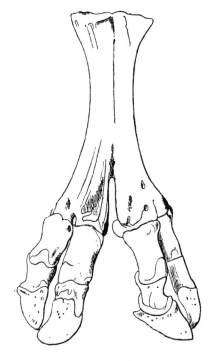

Fig. 116. Specimen stated by Delplanque to have been the right foot of a Cow (see No. 563). (After Delplanque.)

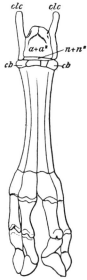

*566.   **Sheep**, having four toes, each having three phalanges, on each posterior limb (Fig. 117). In each case the toes were arranged as two pairs, the hoofs of each pair being turned towards each other. Each foot had four united metatarsals, marked off from each other by grooves on the surface of the bone, the division between the metatarsals of each pair of toes being clearly marked at the peripheral ends of the bones. In the case of each foot there were parts of a pair of tarsi arranged in a symmetrical and complementary manner about the middle line of the limb. In each tarsus there was a large bone having the structure of two calcanea, a right and a left, united posteriorly; the upward prolongation, proper to the calcaneum, was present on each side of this bone and projected upwards on each side of the tibia. The astragalus of each foot was similarly a bone double in form, uniting in itself the parts of a right and left astragalus. The left foot had a single flat bone below the astragalus, representing as it were two naviculars fused together; and four bones in a distal row, representing presumably two cuboids, and two cuneiform elements. In the right foot also there was a single bone below the astragalus, and four other bones arranged in a way slightly different from that of the other foot. ERCOLANI, *ibid.*, p. 773, *Tav.* II. figs. 7 and 8.

Fig. 117. Bones of left hind foot of a Sheep, No. 566 [*q.v.*] copied from ERCOLANI. clc, clc, the two calcanea. $a+a^2$, bone representing the two astragali. $n+n^2$, the two naviculars. cb, cb, the two cuboids.

[A case given by ERCOLANI (*l. c.*, p. 783, Tav. II., figs. 9 and 10) of similar duplicity in a lamb seems to be very possibly a case of double monstrosity. In this animal the hind limbs were altogether absent.]

567.   **Calf**, having five digits on one manus. There is nothing to shew positively whether this specimen is a right or a left, and it is even possible that it is part of a polymelian[1]. Carpal bones gone. Metacarpals four, disposed in two pairs. One pair bear the digits d^4 and d^5 (Fig. 116), which have a common proximal joint. Their ungual phalanges curve towards each other, forming a Minor Symmetry like those of a normal Calf. The other two metacarpals bear three digits; two (d^3 and d^2) articulate with one metacarpal having a divided epiphysis. The other metacarpal bears a digit (d^1) of full size curving towards d^2. The ungual phalanges of d^2 and d^3 are nearly straight [cp. Nos. 558 and 559.] C. S. M., Terat. Ser., No. 299.

568.   **Calf**: left hind foot similar case: inner group of *two* toes curving towards each other and an outer group of three toes of which the middle one was almost bilaterally symmetrical while the hoofs of the other two were each turned towards it. Five metatarsals united but marked out clearly by grooves. Tarsus much as in No. 566. ERCOLANI, *l. c.*, p. 774, Tav. I. fig. 8.

569.   **Calf**: left hind foot a somewhat different case, DREW, *Commercium Litterarium*, Nuremberg, 1736, p. 225, Taf. III. fig. 2. [Description meagre, but figure good. Beginning from the inside the five toes turned (1) outwards, (2) outwards, (3) inwards, (4) outwards, (5) inwards, respectively. There were only *four* metatarsals, (3) and (4) being both borne on one metatarsal.]

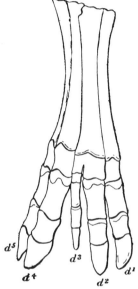

FIG. 118. Manus of a Calf, No. 567. d^1, d^2, d^3, group of three digits [? internal]; d^4, d^5, group of two digits [? external].

POLYDACTYLISM IN THE PIG.

Of the great numbers of feet of polydactyle pigs recorded or preserved in museums all I believe are *fore feet*. No case of a polydactyle hind foot is known to me in the pig. All the cases are examples of proliferation upon the internal side of the digital series. With very few exceptions the variation takes one of two forms, consisting either in the presence of *a single digit* internal to the digit II, or in the presence of *two digits*, either separate or partially compounded, in this position. A very few cases depart from these conditions[2]. The condition is very usually the same or nearly the same in both fore feet.

One extra digit, internal to digit II.

570.   Such a digit may either have a separate bone for its articulation in one or both rows of the carpus (as ERCOLANI, *l. c.*, Pl. I. fig. 3), or it may articulate with a half-separated extension of the trapezoid (as *Coll. Surg. Mus., Ter. Ser.*, 297 *A*), or with the metacarpal or other part of digit II (very common), sometimes simply branching from this digit without an articulation. In no case of which good accounts are to be

---

[1] The *Catalogue* gives no indication on these points.

[2] For example a l. fore foot in which the metacarpal of II. bears a rudimentary digit on each side of the digit II, three in all. ERCOLANI, *Mem. Ac. Bol.*, 1881, Pl. I, fig. 1.

had does such a digit group itself into the Symmetry of the normal manus; but it stands apart, or is bent or adducted behind the other digits, having a hoof which is irregularly pyramidal, curving in neither direction especially. Such a digit has generally three phalanges, and is of about the size of digit II, though not rarely it is large in size approaching more nearly to III than to II (as *Coll. Surg. Mus., Ter. Ser.*, 297).

Two extra digits internal to digit II.

571. This condition is not less common than the last. The two extra digits are borne either by two separate extra carpal bones (Fig. 119, c^1, c^2), or by one carpal imperfectly divided (ERCOLANI, *l. c.*, Pl. I., *fig.* 6); or the metacarpals of the extra digits simply articulate against the carpo-metacarpal joint of II (as in a specimen in my own possession). The extra digits may be double throughout, or the two may be compounded in their proximal parts (ERCOLANI, *l. c.*, Pl. I., fig. 5; also case in Oxford Mus.[1], 1506, *a*, in which the two extra digits were ill-formed and of unequal size, having a common metacarpal). Fig. 119 shews such a pair of extra digits in their most complete form. The central part of the metacarpal of II has either never ossified or has been absorbed. As bearing on the question of the relations of parts in Meristic Repetition the fact of most importance is the circumstance that the digits III and IV retain their normal Symmetry, but the two

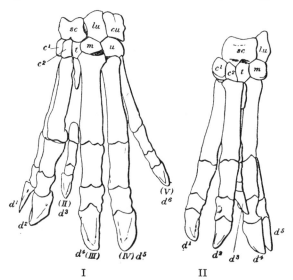

FIG. 119. Left manus of a Pig, No. 571.
I. View from in front. II. View from inside to shew the convergence of d^1 and d^2 towards each other.

d^1, d^2, two extra digits placed internally. c^1, c^2, two extra carpal bones with which they articulate. *sc*, scaphoid. *lu*, lunar. *cu*, cuneiform. *t*, trapezoid. *m*, magnum. *u*, unciform. d^3—d^6, the digits II, III, IV, V.

[1] For note of this specimen I am indebted to Mr W. H. Benham.

extra digits form another Minor Symmetry of their own. It is perhaps worth noting that the metacarpal of the digit lettered d^2 in the form of its head is nearly the optical image of that of III (d^4), but this resemblance may be misleading and must not be insisted on. *Coll. Surg. Mus., Ter. Ser.*, 298.

572   **Wild Boar.** Two cases, apparently resembling the foregoing are described, from external examination only, in the wild boar by GEOFFROY ST HILAIRE, *Hist. des Anom.*, I. p. 696.

SYNDACTYLISM IN ARTIODACTYLA.

This phenomenon is known in the Ox and is common in the Pig. In all cases the variation consists in a more or less complete union or absence of division between the digits III and IV. Among the many records of digital variation in the Pig no case relates to union between a *lateral* and a chief digit, but it is always the two chief digits III and IV that are united. (Compare the case of Man, p. 358.) In this case there is therefore an absence of a division in the middle plane of a bilateral Minor Symmetry, *and the parts that remain united are related to each other as optical images*. The phenomenon is thus the exact converse of the variation consisting in a division along a plane of bilateral symmetry which was seen in the Horses Nos. 547 and 550. As was remarked in speaking of similar variations in Man, it is to be noticed that if the union is incomplete, as it commonly is, the *peripheral* parts are the least divided, the division becoming more marked as the proximal parts are approached.

In the normal Sheep according to ROSENBERG[1] the metacarpals II and V are distinct in the embryonic state, afterwards completely uniting with III and IV. The same is presumably true of the Ox; but whether this be so or not, the digits II and V are in the normal adult not represented by separate bones in the hind foot, and in the fore foot V only is represented by the rudimentary bone articulating with the unciform. Unusual interest therefore attaches to the observations made by BOAS and by KITT of the development of lateral metacarpals and metatarsals (II and V) in Calves having III and IV united. Note also that in two of Kitt's cases there was not only a development of lateral digits but also indications of a *division* occurring in them. Besides this, in the right fore foot of one solid-hoofed Pig (No. 585) there is a slight appearance of duplicity in the ungual phalanx of the lateral digit V.

On the other hand the reduction of accessory hoofs (*ergots*) in LANDOIS' case, No. 582, seems to be an example of a contrary phenomenon; for the connexion between the developed lateral metacarpals and metatarsals in Kitt's case (No. 579) must be taken as evidence that the accessory hoofs do really represent II and V.

[1] ROSENBERG, A., *Z. f. w. Z.*, 1873, XXIII. pp. 126—132, *figs.* 14, &c.

*573. **Ox.** Young ox having the two digits of the right fore foot completely united together. At the lower extremity of the large double metacarpal (III and IV) of the normal limb a deep cleft is present, which separates the two articular extremities of the bone. In this specimen this cleft was represented only by a sort of antero-posterior channel, at the bottom of which there was a slight groove, which was all that remained as an indication of the original double nature of the bone. At the back of this metacarpal there were only three sesamoids instead of four, and in the central one there was not the slightest trace of duplicity. This sesamoid was placed opposite to the channel above mentioned. The two first phalanges were entirely united, but the vestiges of this fusion could be seen both before and behind and also in the two articular surfaces by which the bone was in contact with the metacarpal. The same was true of the second phalanges. The third phalanges however were so completely fused and so reduced in size that they had the appearance of a single bone. . The two small sesamoids were similarly united. The general appearance of this limb was remarkably like that of the Horse. BARRIER, *Rec. méd. vétér.*, 1884, Ser. 6, Tome 13, p. 490. [No particulars given as to the condition of the other feet of the same animal.]

574. **Ox** having right fore foot with a single large metacarpal and one splint-bone [? V]. The peripheral end of the metacarpal had two articular surfaces closely compressed together, and these two surfaces bore but one digit of three phalanges and one hoof like that of a Foal. The preparation was an old one, and with regard to the accessory hoofs there was no indication that could be relied on. KITT, *Deut. Ztschr. f. Thierm.*, XII. 1886, *Jahresb.*, 1884—5, p. 62, Case No. III.

575. **Calf:** each foot having only one hoof. The phalanges, sesamoids, metacarpals and metatarsals, were all normal and the hoofs alone were united. The cavity of the hoof was divided internally into two chambers, which were more distinct in front than behind. Externally each hoof was slightly bifid in front, but the soles of the feet were without trace of division. MOROT, C., *Bull. de la Soc. de méd. vét.*, 1889, Ser. VII. T. VI. p. 39. Case I.

576. **Calf:** killed at 10 weeks old. The left fore foot alone was abnormal, having only one hoof. Viewed from without, this hoof was like that of a young ass, but it bore a slight median depression, which was about 3 cm. wide and only 1 to 2 mm. deep, which was all that remained to shew its double structure. Internally the cavity of the hoof was single, but a horny ridge was present on the inside in the region of the depression. The two unequal phalanges were peripherally united into a single bone, but were separate centrally, and the two parts were not quite symmetrical [details given]. The other parts were nearly normal. MOROT, C., *l. c.*, Case 2.

*577. **Ox.** In a newly-born calf the following abnormalities were seen. In the right fore foot there was a small well formed metacarpal bone on the outside of the normal paired metacarpals, and a similar but more rudimentary structure was also present on the inside of the limb. The additional outer metacarpal bore two small phalangeal cartilages, and with them had a length of about 10 cm., but the supernumerary metacarpal on the inner side was more rudimentary and bore no trace of phalangeal structures. The toes borne by the normal metacarpal of the right fore foot were abnormal, inasmuch as the second and third phalanges were united together. The first pair of phalanges were separate, but their outer ends were modified so as to articulate with the single second phalanx. The distal (third) phalanx bore a groove indicating its double origin, but the second phalanx was without any such groove, and was to all appearance a single structure.

The left fore foot also bore an outer and an inner supernumerary metacarpal, but in this case it was the inner supernumerary metacarpal which attained the greatest size. This inner metacarpal bore two small phalangeal bones, while the outer extra metacarpal was more rudimentary and had no phalanges. The phalanges of the two normal toes were separate in the left foot, but though the bones were of the ordinary formation the two toes were enclosed in a common hoof. BOAS, J. E. V., *Morph. Jahrb.*, 1890, p. 530, *figs*.

BOAS also states that in the museum of the Agricultural School of Copenhagen are several instances of united toes in the fore foot of the Ox, and that in all these specimens the outer metacarpals (II and V) are larger than they are in normal specimens, but are not so much developed as in the case just described. BOAS, *l.c.*

578.   A case [sc. Ox (?)] is also mentioned in which the two normal toes of the *hind foot* were united, and the median and distal parts of the metatarsals II and V were developed, though they are absent in the normal form. BOAS, *l.c.*

*579.   **Calf** having the digits of each foot united and bearing a single hoof. The carpus and tarsus were not seen. *Fore foot*. The chief digits, III and IV, were completely united in the fore limbs and bore a single hoof, but, in addition to this variation, the metacarpals of the lateral digits, II and V, were developed and ossified. The length of metacarpal II was 9 cm. and its thickness at the proximal end was 1·5 cm. Metacarpal V had a length of 8 cm. and a maximum thickness of 1·3 cm. at the proximal end. The metacarpal of the united digits, III and IV, measured 13 cm. in length. The metacarpal V was slightly bifid at its distal extremity, and here presented two articular surfaces. With the internal of these there articulated a bone measuring 2 cm. by 0·5 cm., and attached by fibrous tissue to the end of this bone there was a cartilaginous nodule. The external end of metacarpal V bore a rod-like piece of cartilage, 1 cm. in length. This and the cartilaginous nodule of the other part of the digit together formed the basis of one of the accessory hoofs (*ergots*), but the horny covering itself was divided by a deep cleft into two imperfectly separate parts. To the metacarpal of II was loosely articulated a bone 2·5 cm. in length, to which a nodule of cartilage was attached. The end of this digit was covered by an accessory hoof, which was imperfectly double like that of V and contained a second cartilaginous nodule, which was distinct from the first and was not supported by any proximal bone. The union between the digits III and IV was complete, and the resulting structure with its hoof was like that of the Horse. The articulations were perfectly mobile. At the metacarpo-phalangeal joint there were two sesamoids only. [With this division in the lateral digits on fusion of III and IV compare Pig, No. 585.]

Hind foot. The digits III and IV were united as in the fore feet, but the single hoof was more pointed. The metatarsals II and V were developed. The latter was 12·7 cm. long, and was united to the large metatarsal above, but was free below, and was joined by a ligament to its accessory hoof. That of II began in the middle of the metatarsus, being cartilaginous and of about the thickness of a goose-quill; it was connected with the accessory hoof by a ligament only. KITT, *Deut. Z. f. Thierm.*, XII., 1886, *Jahresb.* 1884–85, p. 59, Case No. I, *fig.*

*580.   **Calf.**   Three of the feet had each *one* large digit (III and IV) formed much as in the last case. But in the dried preparation it could be seen that in each of these feet there were *four* accessory hoofs, and connected with them several ossicles irregularly placed, representing phalanges 1 and 2 connected by ligaments with lateral metacarpals. The fourth foot [which?] had only *three* accessory hoofs, but the phalanges 1 and 2 of the digits III and IV were partially separated from each other, and there were two distal phalanges, one for each digit; but instead of being side by side, they were placed one behind the other, both being encased in a single hoof. KITT, *l.c.*, p. 61, Case No. II.

581.   **Calf.**   A right fore foot having the two chief digits (III and IV) represented by one digit with one hoof. The distal end of the common metacarpal had two articular surfaces in close contact which bore a digit in which there were only slight traces of duplicity. The metacarpal of the digit V was represented by two small bones, one beside the upper and one beside the lower end of the large metacarpal. These two ossicles were connected together by a ligament which is prolonged downwards as far as the accessory hoof, and contains two nodules of cartilage. On the median side of the foot there is no rudiment of the metacarpal II, but the accessory hoof contains a nucleus of partly ossified cartilage. KITT, *l.c.*, p. 63, Case IV.

582.   **Calf** having a single hoof on each fore foot. In external appearance, the hoof was a single structure, but its anterior portion shewed two projections which suggested that it was really a double structure. The outer accessory hoof was present on the right foot in a very much reduced form, but the corresponding structure of the inner side of the foot was entirely absent, and a marked 'turning-point' in the hairs (*Haarwirbel*) indicated the place where it should normally have been developed. In the left foot the accessory hoofs were in the same condition as in the right foot, but the 'turning-point' was not formed at all. There were no skeletal structures corresponding to the accessory hoofs.

The skeleton of left fore foot was prepared. In it the metacarpal was 125 mm. long, having a deep cleft on its anterior face, indicating the line of union of the two metacarpals. The two articular heads, which in a normal animal of the same age are separated from each other by about 5 mm., are in this specimen united by the inner edges of their anterior borders. The proximal phalanges formed a single bone, 32 mm. long. The division between the two bones was visible as a cleft on the anterior surface, in which place the two ossifications were distinctly separated from each other; on the posterior surface the union between the two is continued for half the length. The second phalanges formed a typically single bone, as did also the distal phalanges which bore the hoof. The foramina for the two nutrient arteries of the two toes remained double and entered the single bone, one on each side. LANDOIS, H., *Verh. d. naturh. Ver. d. preuss. Rheinl.*, Bonn, 1881, S. 4, VIII. p. 127.

Pig. "Solid-hoofed" pigs have been mentioned by many writers from the time of ARISTOTLE. The fact that they have been reported as occurring in many parts of the world makes it likely that the variation has often arisen afresh. The first case (No. 583) is the only instance of *complete* union of III and IV in the pig that is known to me. The variation is most commonly simultaneous in fore and hind feet. As seen, it occurs in many degrees. Several specimens not separately mentioned below are in the Coll. Surg. Mus. and other collections.

583. A fore foot and a hind foot of the same individual, in which the two chief digits were completely united, viz. represented by a single series of bones.

In each case the two chief metacarpals and metatarsals (III and IV) were respectively represented by a single large bone, and with each a single digit of three phalanges articulated. The bones of these digits were straight, and not curved as they are in an ordinary foot in which two toes are present. There was not the slightest trace of duplicity, and the lateral digits were placed symmetrically on either side. The sesamoids were *two* in number

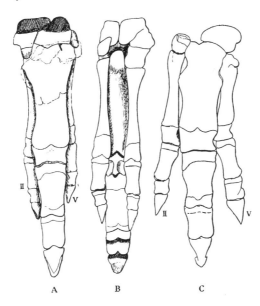

FIG. 120. Bones of feet of solid-hoofed Pig, No. 583, from specimens in the Museum at Alfort, described by BARRIER.
    A. Left manus from in front.
    B. Left manus from behind.
    C. Left pes from in front.
The numbers II and V indicate the digits so numbered in the normal.

instead of four. The carpus and tarsus appear to have also been changed in connexion with this unification of the digits, for in the distal series at least the normal number of bones was not present. [The feet had been cut off across the tarsus and carpus before being received. By kind permission of the authorities at Alfort I examined these specimens and made the sketches in Fig. 118. I could not satisfactorily identify the bones of carpus and tarsus. The proximal parts were covered by a large exostosis.] The extensor of the phalanges ended in three tendons only, and the same was true of the deep and superficial flexors. The central tendon in each case however shewed signs of its double nature. BARRIER, *Rec. méd. vétér.*, 1884, Ser. 6, Tom. XIII. p. 491.

584   A skeleton of a solid-hoofed pig exists in the Museum of the Royal College of Surgeons of Edinburgh which was presented by Sir Neil Menzies of Rannoch, Perthshire. Inquiries instituted by Struthers (1863) elicited the following facts.

"The solid-hoofed pig has been well known and abundant on the estates of Sir Neil Menzies at Rannoch for the last forty years. Most, if not all of them, were black. They were smaller than the ordinary swine, and seem to have had shorter ears. They liked the same food and pasture as the common swine, and showed no antipathy to herd with them. They were more easily fattened, though they did not attain so large a size as the ordinary swine; their flesh was more sweet and tender, but some of the Highlanders had a prejudice against eating the flesh of pigs which did not "divide the hoof," unaware apparently that the Mosaic prohibition applied to all pigs. A male and female of the solid-hoofed kind was brought to Rannoch forty years ago, by the late Sir Neil Menzies, which was the commencement of the breed there; but I have not been able to learn where they were brought from. Although they did not breed faster than the common kind, they multiplied rapidly, in consequence of being preserved, so that the flock increased to several hundred.

"At first, care was taken to keep them separate, on purpose to make them breed with each other, but after they became numerous they herded promiscuously with the common swine. As might be expected in a promiscuous flock, some of the young pigs had solid and some cloven feet, but I am unable as yet to say whether any definite result was ascertained as to the effect of crossing; whether any experiments were tried as to crossing; or whether after the promiscuous herding, some of the pigs of the same brood presented cloven and some solid hoofs.

"No pig was ever known there with some of its feet solid and some cloven; nor, so far as is known, was there any instance of young born with cloven feet, when both parents were known to be solid-hoofed. The numbers diminished—for what cause is not apparent; so that last year there was only one or two—one of them a boar, which died; and now the solid-hoofed breed appears to be extinct in Rannoch."

585.   "*Fore foot.*—The distal phalanges of the two greater toes are represented by one great ungual phalanx, resembling that of the Horse,

but longer in proportion to its breadth. The middle phalanges are also represented by one bone in the lower two-thirds of their length, presenting separate upper ends for articulation with the proximal phalanges. The proximal phalanges are separate through their entire length. The whole foot above the middle phalanges presents the usual arrangement and proportions in the hog." *Middle Phalanges.* "There is no symphysis or mark indicating a line of coalescence of the two phalanges. The surface across the middle is somewhat irregularly filled up to nearly the level of each lateral part. Each half of the phalanx, as indicated by the notch between the separate upper ends, has the full breadth of the proximal phalanx above it." *Distal Phalanx.* The middle part of this is raised above the lateral parts, and is partially separated from them by a fissure on each side, giving it an appearance as of the union of three bones. The end of the phalanx is notched like that of the horse; it bears no trace of symphysis. "The ungual phalanx of one of the lesser internal toes of the fore foot presents a bifurcation reaching half the length of the phalanx." See Fig. 121.

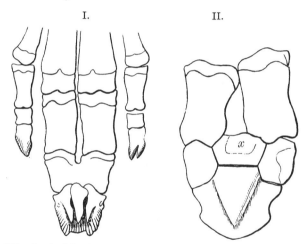

Fig. 121. I. A right fore foot of a solid-hoofed Pig, No. 585, from in front. The ungual phalanx of the digit V is bifid [cp. Nos. 579 and 580].
(After STRUTHERS.)
II. Middle digits of foot of solid-hoofed Pig, No. 587. *x*, an extra ossification wedged in between the phalanges of III and IV. (After ELLIOTT COUES.)

"*Hind foot.* In the hind foot only the *distal* phalanx is single.... There is no trace of double origin to the bone." STRUTHERS, J., *Edinburgh New Phil. Journ.*, 1863, pp. 273–279, *figs.*

586. A pair of solid-hoofed pigs received by Zoological Society of London from Cuba in 1876. The sow gave birth to a litter of six [the solid-hoofed boar being presumably the father]. The six young were three males and three females. The hoofs were solid like those of the parents in two males and one female: in the others the hoofs were cloven as in the normal pig. The feet of one of the solid-hoofed males of this litter were dissected, and it was found that "the proximal and second phalanges are separated as usual, whilst at the extreme distal ends of the ungual phalanges

these bones are completely fused together; and, further, a third ossicle was developed at their proximal ends, where they are not completely united, between and above them" [cp. No. 587]. "It might have been imagined that the deformity was simply the result of an agglutination along the middle line of the two completely-formed digits; but such is not the case, the nail-structure being absent in the interval, where it is replaced by bone with a transverse cartilage below it. The nail is continued straight across the middle line of the hoof, as in the horse." GARROD, A. H., *Proc. Zool. Soc.*, 1877, p. 33.

587. Domestic pigs having the two central hoofs compounded into a single solid hoof have been known to occur several times in America. The two other toes remain distinct in these cases. A breed of pigs having this character is said to have been established in Texas, which transmits this peculiarity in a definite way. In this breed the peculiarity is said to have been so firmly established that "no tendency to revert to the original and normal form is observable in these pigs." A cross between a solid-hoofed boar and an ordinary sow is said to produce a litter of which the majority shew the peculiarity of the male parent. "On the sole of the hoof, there is a broad, angular elevation of horny substance, apex forward, and sides running backward and outward to the lateral borders of the hoof, the whole structure being curiously like the frog of the horse's hoof. In fact it is a frog, though broad, flattened, and somewhat horseshoe-shaped, instead of being narrow, deep and acute as in the actual frog of the horse. This arcuate thickening of the corneous substance occupies about the middle third of the whole plantar surface of the foot." The terminal phalanges are united together, and above this single bone is another independent ossification lying between the second phalanges of the two digits, which remain distinct. [Cp. No. 586.] COUES, ELLIOTT, *Bull. U. S. Geol. Geogr. Surv.*, IV. p. 295, *fig.*

588. Case resembling the above reported from Sioux City, Iowa, in which these pigs were bred for some time and were advertised for sale, with the statement that they were also of superior quality. Other cases given from different parts of the United States. In one of these it is stated that one hind foot was thus formed [the others being presumably normal]. AULD, R. C., *Amer. Nat.*, 1889, XXIII. p. 447, *fig.*

589. **Pig.** In all four feet the digits III and IV partially united and covered by one hoof. The metacarpals and first phalanges were separate in each case but the second and third phalanges of the two digits were united together. The common hoofs were not compressed laterally, as in some of the cases seen in the Calf, and the small digits II and V were unmodified. KITT, *Deut. Zt. f. Thierm.*, XII. 1886, *Jahresb.*, 1884–85, p. 64, *Case* IV, *figs.*

POLYDACTYLISM IN BIRDS[1].

The whole number of cases of Polydactylism recorded in birds generally is small. The phenomena however seen in the Dorking fowl are well worthy of attention and have scarcely been adequately treated. I propose here to give an account of this case, mentioning instances seen in other birds and indicating so far as may be their relation to the facts of the Dorking.

Five-toed fowls have been known from very early times. The character is now most definitely associated with the Dorking, though it is also considered necessary in Houdans for show purposes. It is likely that the latter breed derived the fifth toe from the Dorking. Fifth toes may often be seen to occur in other breeds, but I cannot quote a satisfactory record of their appearance in pure strains.

In the foot of an ordinary four-toed fowl the hallux articulates with the tarso-metatarsus by a separate metatarsal. The hallux in such a foot most often has two phalanges. In its commonest form the five-toed foot departs from this normal in the fact that the hallucal meta-

---

[1] See also the case of *Rissa*, p. 396.

tarsus bears two digits instead of one. The morphological nature of these digits is obscure. Some have judged that one of them is a "præ-hallux;" Cowper[1] sees in the internal toe the true hallux, and argues that the digit commonly called the hallux is really the index; Howes and Hill[2] consider that the normal hallux has split into these two digits. The diversity of these views comes partly from an insufficiency of the area of fact over which the inquiry has been extended, for it will be found that the conditions are very various and shade off imperceptibly in several directions. As in all cases of Meristic Series, the first question relates to the position of these digits in the system of Symmetry of the limb. Are they in a Successive Series with the other digits, or do they balance them? Are they in Succession to each other or do they balance each other as images?

Turning to the facts with these considerations in view it will be seen that no general answer can be given, but that the condition is sometimes of the one kind and sometimes of the other. For there are not merely two conditions, a four-toed and a five-toed, but there is a whole series of conditions and according to the cases chosen so may the question be answered. By examining a few score of fowls' feet many sorts may be seen.

590. (1) The most usual five-toed foot is that figured by Cowper (*l.c.*, p. 249), in which the metatarsal of the hallux bears two digits, an outer one of two phalanges and an inner of three phalanges. For purposes of description let us call the outer the hallux. In this foot then the hallux is the least digit, and the members of the digital series *increase in size on either side of it.*

591. (2) But not rarely is found a state like the last save that the inner digit is borne by the proximal phalanx of the hallux. This is very common. The two digits may then be about equal in size, or more often the hallux is the smaller.

592. (3) Hallux more or less perfectly divided into two digits with a common base, having (*a*) two, or (*b*) three phalanges (as in Howes' case Fig. 5). This state is practically that of the human "double-thumb" (see p. 350), and, just as in that phenomenon, the duplicity may be of various extent, often affecting only the nail and distal phalanx. Between the two parts of such a double digit there is often that relation as of optical images found in human double-thumb, the curvatures of the two parts being equal and opposite. But if both digits are of good size and are separate up to the metatarsal this equality is rarely if ever found, and one of the digits, generally the innermost, is the larger. In this condition therefore there is a Succession from the hallux to the inner digit just as in (1). So the condition of double-hallux, that is to say the representation of one member of a series by two members in bilateral symmetry, shades off imperceptibly into the condition in which a new member is formed in Succession to the terminal member.

It should be noted that this case presents a remarkable difference from that seen in the like cases of variation on the radial side of the hand of Man. In Man the states of true double-thumb are just as in the Fowl; but if there is a difference or Succession between the two parts

---

[1] *Jour. Anat. Phys.*, xx. p. 593; and xxiii. p. 242.
[2] *Ibid.*, xxvi. p. 395, *figs.*

it is the *external*[1] which is the greater, being in several cases a three-phalanged digit shaped like an index (see No. 486). Nevertheless in the Fowl it is the *internal* which is the greater.

The conditions in the following cases are not far removed from those named above.

593. **Archibuteo lagopus** (Rough-legged Buzzard): specimen in good condition shot near Mainz, being otherwise normal. The toes of the *left* foot were placed as usual in a bird of prey, but on the outside[2] of the hind toe was a much smaller accessory toe. This accessory toe was attached to the hind toe almost as far as the base of the claw of the latter. The claw of the accessory toe was half the size of that of the hind toe. In the left leg the muscles of the thigh and shank were less developed than usual. Toes of *right* foot abnormally arranged, being all directed forwards. The three normally anterior toes were on the inside of the series, and the toe which should properly be single and directed posteriorly was double and was directed anteriorly. These abnormally disposed toes were not functional. The right leg was much more developed than the left, and it seemed as if the bird had habitually stood on the right leg. von Reichenau, W., *Kosmos*, 1880, vii. p. 318.

594. **Gallinula chloropus** (Moorhen): specimen killed in Norfolk in 1846. "Each of the hind toes possessed a second claw, which in the right foot merely springs from about the middle of the true toe, but in the left is attached to a second toe, which proceeds from the original one, about half-way from its junction with the tarsus." Extra toe and claw in each case attached outside[2] of the true hind toe. Gurney, J. H., and Fisher, W. R., *Zoologist*, 1601.

Guinea-hen having double hallux; of the two digits the external[2] was the longer. Geoffroy St Hilaire, *Hist. de Anom.*, i. p. 695.

Division of digits II and III.

595. **Anas querquedula**, L. (Garganey Teal): wild specimen having the left foot abnormally formed. In it there was no toe occupying the place of the hallux, but the digits II and III [using the common nomenclature] were partially bifurcated. In the digit III, the extremity only was divided, but each part bore a separate nail and there was no web between these secondary digits, which were somewhat irregular in form. The digit II divided in about its middle into two nearly similar digits, which were united by a web. The nails of these digits were hypertrophied. Ercolani, *Mem. Acc. Bologna*, S. iv. T. iii. p. 804, *Tav.* iii. *fig.* 1.

596. (4) From the condition seen in (3) it might be supposed that duplicity of the hallux is the least possible step in the progress of the four-toed form towards the five-toed. It is only one of the least possible steps. For in a few cases upon the base of the digit recognizable as the hallux, and standing in the normal place of the hallux, may be found a minute rudiment of a digit, sometimes with a nail, sometimes without. Between this and the well-formed fifth toe all conditions exist.

There are thus, as usual in the numerical variations of Meristic Series, *two* least conditions, one being found in duplicity of a single member, the other taking the form of addition of a rudimentary member beyond the last member.

597. Passing now from the simpler conditions of the variation to the more complex, several distinct states may be mentioned. The divergence from the normal may be greater either by the presence of *two* extra digits, or by change in the position of the extra digit or digits.

---

[1] The only case to the contrary is that mentioned by Windle, *Jour. Anat. Phys*, xxvi. p. 440, in which a three-phalanged digit stood on the radial side of a pollex. This case has not been described. See pp. 326 and 352.

[2] In reading these records it should be remembered that owing to the backward direction of the hallux the apparent outside is morphologically inside, and probably this is meant in each case.

Two extra digits are said to be not uncommon in the Dorking but I have myself seen only one case. A foot of this kind is figured by Cowper[1], and in it the appearance is as of an extra digit of three joints (? all phalanges) arising internally and proximally to the hallucal metatarsal, which already bears two small and sub-equal digits. In the case seen by myself there was one large internal digit with three phalanges separately articulating with the tarso-metatarsus, and the hallucal metatarsus bore a digit divided peripherally, bearing two nails related as images. Here therefore there was a *double hallux*, and internal to it a separate digit.

598. The evidence regarding extra digits in other positions, though small in amount, is of importance as a light on the morphology of these repetitions of digits. We have seen that the ordinary extra digit is, with the hallux, borne on the hallucal metatarsal. In one of Howes' cases (*l. c.* figs. 2 and 3) this metatarsal instead of simply articulating with the shank of the tarso-metatarsal *was continued up to articulate also with the tibio-tarsus*. From this state the condition in which a separate digit (or digits) articulates with the tibio-tarsus only is not far removed. Of this condition I know no detailed account in the Dorking, though it is referred to by Lewis Wright[2], but I have met with the following cases in other birds.

599. **Aquila chrysaetos** (Golden Eagle): having two extra toes borne by right metatarsus [left foot is not described]. The two extra toes attached to upper part of the back of the metatarsus. Each bears a full-sized claw which was curved backwards and upwards. One of the toes bore six scutella on the morphologically upper surface and four on the plantar surface. The other toe, which was more completely united to the metatarsus along its whole length, bore only a single scutellum on the plantar surface. The rest of the foot was normal. Jackel, A. J., *Zool. Gart.*, xv. 1874, p. 441, *fig.*

600. **Pheasant:** right foot bearing a thin and deformed digit articulating internally with the distal end of tibio-tarsus. Hallux normal. Left not seen. Specimen received from Mr W. B. Tegetmeier.

601. **Pheasant:** each leg bears a large extra digit of irregular form attached to the *middle* of anterior surface of tibio-tarsus. The two legs almost exactly alike, but in one the digit is firmly and in the other loosely attached to tibio-tarsus. Specimen kindly sent by Mr Tegetmeier.

602. **Buteo latissimus** ♂, having extra digit on right leg, the toe was well formed, with two phalanges, bearing perfectly formed claw, loosely attached internally to tibio-tarsus just above articulation with tarso-metatarsus. Coale, H. K., *Auk.* 1887, iv. p. 331, *fig.* [Cp. No. 593.]

603. **Turkey** having two imperfectly separate digits [? images] attached to process of tibio-tarsus. Two cases differing in degree: hallux normal. Ercolani, *Mem. Ac. Bologna*, Ser. iv. iii. Pl. iii. *figs.* 2 and 3.

604. **Pheasant:** somewhat similar case, in which two such digits were similarly placed, but one was large and the other small. *Ibid., fig.* 4.

605. **Larus leucopterus.** For the following case I am indebted to Professor R. Ridgway, Curator of the Department of Birds, in the United States National

---

[1] Cowper, J., *Journ. Anat. Phys.*, xxiii. p. 249.
[2] "Perhaps the most difficult point in judging Dorkings, however, is to watch against malformations of the feet which have been fraudulently removed; for ...... the abnormal structure of the Dorking foot is very apt to run into still more abnormal forms, which disqualify otherwise fine birds for the show-pen. Birds are not unfrequently produced which possess *three* back toes, or have an extra toe high up the leg; or, in the case of the cock, with supernumerary spurs, which have been known to grow in every possible direction....... We have on two occasions seen prizes awarded to birds which shewed unquestionable traces of such amputation...." *The Illustrated Book of Poultry*, 1886, p. 331.

394   MERISTIC VARIATION.   [PART I.

Museum. The specimen is No. 76,221 in that collection, marked "Greenland, Sept. 1877; Loc. Kumlien." The accompanying figures were kindly made for me

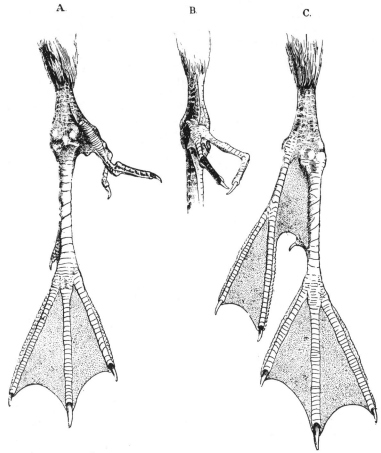

FIG. 122. *Larus leucopterus*, No. 605.
    A. Right foot seen from in front.
    B. The same from the internal side.
    C. Left foot from in front.
From a drawing of specimen in U. S. Nat. Mus., kindly made for me by Prof. RIDGWAY's direction.

under Professor Ridgway's supervision and sufficiently shew its structure (Fig. 122). [It will be seen that the hallux in A, the right foot, appears on the outside; this I conceive is due to partial rotation to shew the abnormal toes.]

Besides these there are a few amorphous cases of extensive repetition of digits in birds.

These facts shew how fruitless a work it is to try to find a general statement which shall include all the cases. There is an almost unbroken series of conditions starting from *either* duplicity of the hallux, *or* from the presence of an internal rudimentary digit, up to a condition somewhat resembling that of "double-hand" in Man. If the first digit behind the hallux is the præ-hallux, what are the digits on the tibio-tarsus? If on the other hand the appearance of an extra digit internal to the supposed hallux is to be evidence that this "hallux" is the index, it may equally be argued that if *two* digits come up internal to the "hallux" then the supposed hallux is the medius, and so on indefinitely. Again, though with HOWES and HILL we may accept the cases of double-hallux as evidence that an extra digit may appear by division of the hallux, which is indisputable, we must equally accept the cases Nos. 597 and 598 as evidence that extra digits may grow directly from the tarsus or even from the tibia, though the hallux remain single and unchanged. And between these two there is no line of distinction; they pass into each other. Do not these things suggest that we are looking for an order that does not exist? Is it not as if we should try to name the branches of a tree in their sequences?

Possibly Continuous numerical Variation in Digits: miscellaneous examples.

Under this heading are placed in connexion a few cases of great interest. Whatever may be held as to the relation to the problem of Species of the phenomena hitherto described, it can scarcely be doubted that the following are instances of Variation which at least *may* be of the kind by which new forms are evolved.

Great interest would attach to a determination whether the reduction of the digits in these cases is a continuous or a discontinuous process, but unfortunately these phenomena have been statistically studied by no one, and it is not possible to do more than make bare mention of the fact that such Variation is known to occur. There is no statistical evidence as to whether the individuals in any one locality may not fall into groups, dimorphic or polymorphic in respect of the degree to which the digits are developed (compare the case of the Earwig, Introduction, p. 40). As an inquiry into the Continuity of Variation such an investigation would be exceptionally valuable. In the case, for instance, of *Cistudo* mentioned below, such a statistical inquiry should surely not be hard to make.

*606. **Chalcides.** This is a genus of Lizards belonging to the family Scincidæ. In several genera of this family the limbs are reduced or absent, differences in this respect being frequent among species of the same genera. (See BOULENGER, *Catalogue of Lizards in Brit. Mus.*, 1887, III. pp. 398, &c.)

Mr BOULENGER kindly shewed me a number of Lizards of the genus *Chalcides* from the shores of the Mediterranean basin which strongly resemble each other in colour and general appearance, but which contained almost a complete series of conditions in respect of the development of the limbs and digits, ranging from *C. ocellatus* and *C. bedriagœ* with pentadactyle limbs fairly developed, through *C. lineatus* (tridactyle) and *C. tridactylus* to *C. guentheri* in which the limbs are minute conical rudiments. Amongst the species of this series great individual variations occur.

607. **Chalcides mionecton**: normally four digits on each foot. A specimen in Brit. Mus. kindly shewn to me by Mr BOULENGER has on each hind foot *five* digits.

608. **C. sepoides**: Mr BOULENGER tells me that the normal number of digits on each foot is five, but that specimens occur having *four* digits on each foot.

609. **Cistudo.** This genus includes the North American Box-turtles as defined by AGASSIZ (*N. Amer. Testudinata, Contrib. to N. H. of U. S.*, I. p. 444). These animals are widely distributed to the E. of Rocky Mountains. On the hind feet of some of them there are three digits, while others have four. GRAY (*P. Z. S.*, 1849, p. 16) described two Mexican specimens which agreed in having *three* large claws on the hind foot with no appearance of a fourth claw, and even scarcely any rudiment of the fourth toe, which was then believed to be present in the other members of the genus. To this three-toed form he gave the generic name *Onychotria*, but in *Brit. Mus. Cat.*, 1855, he gave up this name as a generic distinction, describing the Mexican form as *Cistudo mexicana*, giving three toes on the hind foot as a definite character.

AGASSIZ in 1857 (*l.c.*) divided *Cistudo* into four species, giving to the Mexican form the name *C. triunguis*, and he states that the western and south-western type is remarkable for having almost universally only three toes on the hind feet. The toe which is missing is the *outer* toe and "it fades away so gradually that the genus *Onychotria* cannot stand." The form found from New England to the Carolinas is called by Agassiz *C. virginea* = *C. carolina*, and he states that he received a three-toed specimen from N. Carolina which agreed in all other respects with those from New England.

PUTNAM (*Proc. Boston, N. H. S.*, X. p. 65) stated that the three-toed form found in the South is only a variety of *C. virginea*, and that he had seen two specimens which had three toes on one hind foot and *four* on the other.

610. **Rissa**[1]. The common Kittiwake (**R. tridactyla**) as found in

[1] In illustration of the possible bearing of these facts on the problem of Species reference may be made to the fact that among birds there are several examples of species differing from their near allies by reason of the absence of the hallux. Speaking of this feature in *Jacamaralcyon tridactyla*, SCLATER observes: "In the present bird we meet with another example of the same character [viz. a monotypic form], and with one, perhaps, more isolated in its structure than any of those above mentioned, *Jacamaralcyon* being notably different from all other members of the Galbulidæ in the absence of the hallux. At the same time we must be careful not to put too high a value upon this at first sight seemingly important

this country and in N. Atlantic has no hallux, but only a small knob without a nail in its place. No variation in respect of this digit is recorded[1]. Birds not distinguishable from the Atlantic Kittiwake occur in the North Pacific, but amongst these Pacific specimens birds are found occasionally as rarities having a hallux "as large as it is in any species of *Larus*" (COUES, p. 646). This feature also exhibits gradations. Specimens are described by COUES and also by SAUNDERS having the hallux including the nail ·2 in. long, with a perfect claw. These are given as extreme examples. SAUNDERS remarks that this hallux is small for the size of the bird, stating that another species of similar size, *L. canus*, had a hallux ·5 in. long. Of these specimens of *R. tridactyla* from Alaska one had the nail of the hallux developed, though less so than in the extreme case. Saunders states further that the variation is not always equal in extent on both feet of the same individual: he considers that the extreme form is probably rare and local. COUES, E., *Birds of North-West (U. S. Geol. Surv. Terr.)*, 1874, p. 646; and SAUNDERS, HOWARD, *P. Z. S.*, 1878, pp. 162—64.

611. **Rissa brevirostris:** a species from the N. Pacific distinct from *R. tridactyla* shews a similar variation in the development of the hallux, though in a smaller degree. A specimen has no claw on right hind toe and only minute speck on left; another has no hind nail whatever; another has small black nails of unequal size on the two hind toes. SAUNDERS, H., *l. c.*, p. 165.

612. **Erinaceus.** *E. europæus* has a large hallux, while in *E. diadematus* it is only 4 mm. in length, and in *E. albiventris* it is normally absent in adults. An adult female **E. albiventris** had a minute hallux in the left hind foot, represented by a claw and ligamentous structures, the phalanges being absent[2]. In a female a few months old a minute hallux with usual number of phalanges was present on both sides. The presence or absence of a hallux has often been considered a sufficient ground for the formation of a new genus. DOBSON, G. E., *P. Z. S.*, 1884, p. 402.

613. **Elephas.** In both the Indian and African elephant the number of digits represented by bones is five, both in the fore and the hind foot. The number of *hoofs* differs in the two species. The African elephant has normally four on the fore foot and three

---

character, as the same feature occurs as is well known, not only in certain genera of other allied families (such as Alcedinidæ and Picidæ), but even in a genus of Oscines (*Cholornis*), in which group the foot-structure is generally of a very uniform character." SCLATER, P. L., *Monograph of the Jacamars and Puff-Birds*, 1879—82, p. 50.

[1] Mr A. H. EVANS has called my attention to a recent paper by CLARKE (*Ibis*, 1892, p. 442) giving an account of a minute rudiment of the hallux in embryos of *R. tridactyla* from Scotland.

[2] Compare facts as to the loss of the hallux in Mungooses (Herpestidæ), THOMAS, O., *P. Z. S.*, 1882, p. 61.

on the hind foot, and I am not aware that variations from this number have been seen.

In the Indian elephant there is variation, and though I cannot give any complete account of the matter the following particulars may be of interest.

According to BUFFON the 'Elephant' has generally five hoofs on both fore and hind feet, but sometimes there are four, or even three[1]. He gives a particular case of an Indian elephant with four hoofs on each foot, both fore and hind feet.

TACHARD[2], to whom Buffon refers, was desired by the French Academy to notice on his journey in Siam, whether elephants had hoofs, and he states that all that he saw had five on each foot. Possibly the four-toed variety does not occur in Siam.

I am indebted to Mr W. T. BLANFORD for the information that the natives of India attach importance to the number of hoofs, and also for the following references. HODGSON[3] gives a sketch of elephants with four hoofs on each foot, marked "*Elephas Indicus*, var. *isodactylus* nob., Hab. the Saul forest," together with the following note: "The natives of Nepal distinguish between the breeds with four toes [sic] on all the feet and those with five to four toes." SANDERSON[4] speaking of this says that some elephants have but sixteen hoofs, the usual number being five on each fore foot and four on each hind foot; and that in the native opinion 'a less number than eighteen hoofs in all disqualifies the best animals.' FORSYTH[5] also alludes to the same fact.

Taken together these accounts seem to shew that five on the fore foot and four on the hind foot is the most usual number, but that both the number on the fore foot may diminish to four and that on the hind foot may increase to five. Several text-books mention the subject but I know no statistics regarding it. In view of the different number characteristic of the African elephant this variation has some interest. In particular it would be of use to know whether the variation exhibits Discontinuity, and also to what extent it is symmetrical.

INHERITANCE OF DIGITAL VARIATION.

614. Recurrence of digital Variation in strains or families is frequent, but though many observations on the subject have been made no guiding principle has been recognized. To the general statement that digital Variation, whether taking the form of polydactylism or other-

---

[1] BUFFON, *Hist. Nat.*, xxviii. p. 201. The mention of *three* hoofs must I think refer to the African species, which Buffon does not distinguish from the Indian. In the Cambridge Museum (*Catal.* 699) is an old preparation of the skin of an elephant's foot having three hoofs. This is declared by the Catalogue to be the fore foot of an Indian elephant. Perhaps this is a mistake.
[2] TACHARD, *Voy. de Siam*, 1687, p. 233.
[3] HODGSON, B. H., *Mammals of India*, MS. in Zool. Soc. Library.
[4] SANDERSON, G. P., *Wild Beasts of India*, p. 83.
[5] FORSYTH, J., *Highlands of India*, 1872, p. 286.

wise, does very commonly appear in the offspring or kindred of the varying individuals I can add nothing. It should be mentioned that though in families exhibiting digital Variation the forms that the change takes may differ (in some cases widely even among individuals nearly related) yet on the whole the variation, if recurring at all, more often recurs in a like form. This holds good apart from the rarity of the particular form of variation. The facts described by FARGE (*l. c., infra*) are exceptionally interesting in this connexion. In the family described by him *duplicity* of the thumbs occurred in the paternal grandmother, while the father and three children had their thumbs of the *three-phalanged* form as in No. 483. This case strikingly illustrates the well-known principle that Meristic variability may appear in the same strain or family under forms morphologically very dissimilar.

Attention is also called to the circumstance that in the case of the three toes in the ox (No. 558) the descent was wholly through *females*, and the same was almost certainly true in the polydactyle cats (No. 480). In the case of the syndactyle pigs the evidence of maintenance of the variation in the strain is very clear (No. 584). See also No. 564.

As regards digital Variation in Man the following are the best genealogical accounts:

ANDERSON, *Brit. Med. Jour.*, 1886 (1), p. 1107. BILLOT, *Mém. méd. milit.*, 1882, p. 371. BOYD-CAMPBELL, *Brit. Med. Jour.*, 1887, p. 154. FACKENHEIM, *Jen. Zts.*, 1888. FOTHERBY, *Brit. Med. Jour.*, 1886 (1), p. 975. FÜRST (see *Canst. Jahresb.*, 1881, p. 283). HARKER, *Lancet*, 1855 (2), p. 389. LUCAS, *Guy's Hosp. Rep.*, xxv., p. 417. MORAND, *Mém. Ac. Sci.*, 1770, p. 140. MUIR, *Glasg. Med. Jour.*, 1884. POTT, *Jahresb. d. Kinderh.*, xxi., p. 392. POTTON[1] quoted by GRUBER from DE RANSE, *Bull. Soc. d'Anthrop.*, 1863, iv. p. 616. STRUTHERS, *Edin. New Phil. Jour.*, 1863 (2), pp. 87 et seqq. WOLF, *Berl. klin. Wochens.*, 1887, No. 32. FARGE, *Gaz. hebd. de méd. et chir.*, Ser. 2, II. 1866, p. 61. Case given *Lond. Med. Gaz.*, 1834, p. 65.

Association of digital Variation with other forms of Abnormality.

615. In the great majority of cases of polydactylism the rest of the body is normal, the limb or limbs varying alone. There are however a certain number of examples of polydactylism in association with other abnormalities; as for instance with phocomely, cyclopia, double uterus, hare-lip, defective dentition, defect of tibia, &c., but there is nothing as yet to indicate any special connexion between these several variations. Diminution in number of digits and syndactylism is on the contrary very often associated with general deformity and with many forms of arrested development. To this no doubt is largely due the fact that cases of ectrodactylism are commonly irregular, whereas polydactylism is generally fairly regular in its manifestations, for numerous cases of diminution in number of digits occur in bodies or in limbs otherwise amorphous.

[1] The notorious case of a village in Isère where the majority of the inhabitants are said to have been polydactyle. Most modern writers on the subject quote this statement but I have never found original authority for the fact. By some it is referred to DEVAY, *Du danger des mariages consanguins*, 1862, p. 95, but I can find no mention of the facts in that work.

CHAPTER XIV.

Digits : Recapitulation.

In the remarks preliminary to the evidence of digital Variation it was stated that this group of facts is interesting rather as bearing on morphological conceptions than from any more direct relation to the problem of Species. The indications to be gained from the evidence will be treated under the following heads:

(1) Comparative frequency of digital Variation in different animals.
(2) Particular forms of digital Variation proper to particular animals.
(3) Symmetry in digital Variation.
(4) The manus and pes as systems of Minor Symmetry.
(5) Duplicity of limbs.
(6) Homœotic Variation in terminal digits when a new member is added beyond them.
(7) The absence of a strict distinction between duplicity of a given digit and other forms of addition to the Series.
(8) Discontinuity in digital Variation.
(9) Relation of the facts of digital Variation to the problem of Species.

(1) *Comparative frequency of digital Variation in different animals.*

In reviewing much of the evidence of Variation and especially in the evidence concerning the variations of teeth it has been seen that the frequency of these variations is immensely greater in some classes or species than in others. This is remarkably clear in the case of the variations of digits. Compare for instance the great frequency of polydactylism in the Horse with the complete absence of recorded cases in the Ass. It is true that the latter is the rarer animal, but it might still be expected that some record would have been found if the variation were as frequent in the Ass as in the Horse. Again polydactyle Cats are certainly not very rare and specimens are in several collections having been acquired at many dates. On the other hand digital Variation in the Dog seems to be confined to the formation of a hallux in the hind foot, and to duplicity of hallux and pollex [1]. Similarly though digital Variation is so common in the Pig it is very rare in the Sheep, only one or two clear cases being so far known to me. Note again that polydactylism is common in the Fowl and has been often seen in the Pheasant, while in other birds it is very rare.

Some one will of course remark that the Fowl is a domesticated bird and the Pheasant is partially so; but pigeons, ducks [2] and geese [3] are as much domesticated and in them digital Variation does not seem to be known. The cases in Apes deserve mention in connexion with this matter. One case of syndactylism was quoted in *Pithecia* No. 525, a case of polydactylism in *Macacus* No. 504, in Orang No. 511, and in *Hylobates* No. 508, and a case of ectrodactylism in *Macacus* No. 526. These five cases surely suggest that Meristic Variation is something more than a mere result of high feeding or of "unnatural" conditions. It is not a little strange that among Apes Meristic Variation should be frequently met with in so many systems of organs.

(2) *Particular forms of digital Variation proper to particular animals.*

Of more significance than the frequency with which digital Variation recurs in certain animals is the frequency with which in particular animals it approaches to particular forms, or to particular conditions in a series or progression of forms. This has been seen in the Cat, Man, Horse, Pig, Ox, &c. In each of these the mode of occurrence of Variation has in it something distinctive, something that marks the phenomenon as in some way different from the similar phenomena in other forms. Taking for instance the curious series of cases found in the human manus, ranging

---

[1] Both these variations are of course very common and may be seen in any walk in the streets. The hallux is very frequently present in the Dachshund and is common in Collies, Mastiffs and other large breeds. In the Mastiff dew-claws (hallux) are not a disqualification (SHAW, *Book of the Dog*). In the St Bernard the hallux is very often *double*, perhaps more often than not. This is largely due to the fact that the monks of the Hospice considered the presence of the dew-claw of the utmost importance and preferred it double if possible (SHAW, *l.c.*). The same writer states that 'the more fully the dew-claws are developed the more the feet are out-turned.' This fact suggests that there may be a change of Symmetry like that in the Cat, but I have no observations on the point. I have several times seen simultaneous duplicity of hallux and of pollex in the same individual (Dachshund, &c.). Other digital variations must be rare in dogs as there are hardly any recorded cases. A problematical case of ectrodactylism is given by BAUM, *Deut. Ztschr. f. Thierm.*, xv. 1889, p. 709, *fig.* [*q. v.*]. I once saw a mongrel Fox-terrier with no pollex on either manus, but I was not satisfied that they had not been cut off, though there was no suggestion of this.

[2] For an interesting account of a Duck with the webs of the toes almost wholly absent see MÖBIUS, *Zool. Gart.*, XVIII. 1877, p. 223. Another case of the same kind MORRIS, F. O., *Zool.*, IV. p. 1214.

[3] Pygomelian geese often recorded; *e.g.* CLELAND, *Proc. Phil. Soc. Glasg.*, XVIII. 1886, p. 193, *fig.*; WYMAN, *Proc. Bost. N. H. S.*, VIII. 1861, p. 256.

from the addition of a phalanx to the pollex up to the condition of Nos. 488 or 490, and comparing them with the essentially similar series of cases in the hind foot of the Cat, there is this remarkable difference: that though both progressions lead up to a similar kind of Symmetry in the series of digits, in the human manus an approach is made to a system of Symmetry whose axis lies internal to the index, while in the Cat's feet the axis lies external to the index (see Section (4)). The series of forms in the manus of the Cat is still more peculiar and is not like any case of polydactylism in other animals.

(3) *Symmetry in digital Variation.*

From the evidence it will have been seen that digital Variation in most of its manifestations may be similar and simultaneous in the limbs of the two sides of the body, though not rarely it affects the limb of one side only; and still more frequently the form which it assumes on one side differs in *degree* from that found on the other side. Considerable difference in kind between Variation on the right side and on the left is much rarer.

Almost the same statement may be made respecting simultaneity of Variation between the manus and the pes, though in the pes the manifestation of Variation is rarely *identical* with that in the manus of the same individual. Some variations, as for instance duplicity of pollex and hallux, or extra digit external to minimus, are not rarely found simultaneously in both pes and manus, but there are many cases in which no such agreement is found. The frequency of this simultaneous variation in the case of syndactylism in the Pig may be specially noticed.

Certain variations in certain animals seem to be almost or quite restricted either to hind limb or to fore limb. The form taken on by the pes of the Cat upon increase in number of digits is distinct from that assumed by the manus. The development of the digit II in the Horse is much more common in the manus. The extra digit (or pair of digits) in the Pig is so far as I know seen only in the manus. On the contrary the three-toed state in the Ox is found in the manus and also in the pes. Generally speaking, Meristic Variation is much commoner in fore limbs than in hind limbs.

One fact here calls for special notice. Though general statements are hazardous, we are perhaps justified in affirming the principle that *large* Meristic Variation, involving great departure from the normal, very rarely affects exclusively one side of a bilaterally symmetrical body. In cases of variation in vertebræ, in spinal nerves, in teeth, in the oviducts of *Astacus*, and many more, it is seen that on the occurrence of *great* variation the change is seldom restricted wholly to one side of the body, though the condition reached by the two sides is frequently of differing degree. Now in the extreme forms of double-hand as seen in Man there is a curious exception to this principle. For in nearly all the extreme cases the abnormality was on one side only, the other being normal. This was seen in Nos. 492—500 and 501—503, and also in *Macacus* No. 504. The case No. 500 is probably an exception to this general statement. As to the significance of this absence of correspondence between the right and left sides in extreme cases of digital Variation I can make no conjecture. It has seemed that perhaps in such cases the absence of symmetry between the two sides of the body may be connected with the fact that in these extreme forms of double-hand an approach is made to a bilateral symmetry completed within the series of digits. But against this suggestion must be noticed first the fact that a similar bilateral symmetry is established in the six-toed pes of the Cat (Condition IV of the pes, p. 316), but the variation is nevertheless found on both sides of the body; and secondly the case of double-foot in the lamb (No. 566), though for reasons stated this latter case may perhaps be open to question.

(4) *The manus and pes as systems of Minor Symmetry.*

This is a subject to which it is most difficult to give adequate treatment. Several of the phenomena have as yet been studied in far too small a range of cases to justify sound generalization, and with further knowledge the suggestions arising from the facts now before us may not improbably be found to have been misleading wholly or in part. Besides this there is a serious difficulty in finding modes of expressing with clearness even those principles of form which seem to underlie the phenomena. This difficulty proceeds first from the vague and contradictory character of the indications, and next from the total absence of a terminology by which diversities of symmetry and the form-relations of parts may be expressed. Nevertheless it has seemed best to abstain from the introduction of new terms until the ideas to be expressed shall have been more clearly apprehended. It need scarcely be said that the remarks which follow merely represent an attempt to state some of the lines of inquiry along which the facts point.

On p. 88 mention was made of the fact that in a Bilateral Symmetry the organs which occur as a pair, one on the right and the other on the left, in so far as they are symmetrical are optical images of each other, this relation of images being what is implied by the statement that these organs are bilaterally symmetrical. The hands and feet of vertebrates are organs of this kind, the right hand and the right foot being approximately images of the left hand and foot respectively. But beyond their symmetrical relations to each other in the Major Symmetry of the whole body each manus and each pes may exhibit the condition of a Minor Symmetry within the limits of its own series of digits. Not only may each limb geometrically balance the limb of the other side but its own external parts may more or less balance its own internal parts. This relation differs greatly in different animals, the Minor Symmetry being nearly complete in the Artiodactyles and in the Horse, but much less so in the human manus and pes, &c. The matter now for consideration is the influence or consequences of the existence of this symmetry in the Meristic Variation of digits; and conversely the light which the observed phenomena of Variation throw on the nature of that relation of symmetry. It will be seen that in some points the two halves of a bilaterally symmetrical limb behave just as do the two halves of the bilaterally symmetrical trunk, while in other points their manner of Variation is different.

Thus, the digit III of the Horse may divide into two halves related to each other as images, bearing hoofs *flattened on their adjacent edges*; that is to say, the two resulting parts are formed not as *copies* of the undivided digit, but as *halves* of it, a condition never seen in division occurring anywhere but in the middle line of a bilateral Symmetry.

In the syndactyle feet of the Pig or the Ox the converse phenomenon exists; for the digits III and IV, which normally stand as images of each other, are here wholly or in part compounded to form a digit to which the uncompounded digits are related as halves.

Thus far the connexion between the geometrical relations of the digits and the modes of their Variation is clear and simple, and does not differ from that maintained in the Major Symmetry. But in proceeding further there is difficulty.

If, for instance, the manus or pes of a Horse possesses within itself the properties of a bilateral Symmetry, then the splint-bone II may be supposed to be in symmetry with the similar bone IV. It would therefore be expected that on the occasion of the development of II to be a full digit, the splint-bone IV would at least not unfrequently develop, thus exhibiting that similarity and simultaneity of Variation which we have learnt to expect from parts in symmetry with each other. Nevertheless such an occurrence seems to be extremely rare. Then arises a further question: if the digit II develop simultaneously, say in the two fore feet, would the mechanical conditions of which Symmetry is the outward expression be satisfied without a corresponding change in the digit IV of the fore feet? Is the frequent absence of symmetry in the variation of the halves of the Minor Symmetry in any way connected with the possibility that the two Minor Symmetries together may be maintaining their relations to each other as parts of a Major Symmetry? Of course as to this we know nothing, but the existence of this double relation should be remembered.

In several other phenomena of digital Variation the influence of Symmetry is to be suspected. Reference may first be made to the series of changes seen in the Cat's hind foot in correlation with numerical change. The bones of this pes do not normally exhibit any very clear bilateral symmetry[1]. Yet on the appearance of new digits the foot is reconstituted and its parts are, to use a metaphor, 'deposited' in a system of bilateral symmetry[2] whose completeness is proportional to the degree of development of the new digits. What may be the meaning of this extraordinary fact one cannot yet guess. The fancy is constantly presented to the mind that there is in the normal foot a condition of strain, that the balance between the right foot and the left is a condition of imperfect stability, and that upon the introduction of some unknown disturbance this balance is upset and each foot settles down as a separate system. But I see no way of testing this fancy and no way of following it further.

Still more complex are the facts seen in the human hand. There is here first the fairly complete series of conditions ranging from the normal, through the three-phalanged thumb up to the several Conditions in which extra digits upon the internal side of the limb seem to have sprung up to balance the four normal digits; but on the contrary there is the exceptional case of the Macacque's foot (No. 504) where the extra parts are, as I believe, *external*. (Besides these there are the wholly distinct series of "double-hands," which will be spoken of below.) The former cases taken alone would certainly suggest that there is an imperfect balance or system of symmetry subsisting between the thumb and the four fingers of the normal manus, but to this suggestion there are numerous difficulties which need hardly be detailed in this preliminary glance at the phenomena.

With more confidence it can be maintained that the pollex and perhaps the hallux of Man is in itself a Minor bilateral Symmetry, apart from the four fingers, for it may divide into equal parts related as images. The same is true of the hallux of the Dorking (p. 390), and probably of the extra digit or digits sometimes arising from the tibio-tarsus of the Turkey for example (see No. 603).

Besides this the facts of the frequent syndactylism between the digits III and IV of the human manus, taken in connexion with the phenomena of the Pig and Ox, suggest that the four fingers may have among themselves again a relation of the nature of Symmetry.

---

[1] In the normal pes, though all the claws are retracted to the outside of the second phalanges, yet the claws of digits III and IV rest close together, that of III being external to its pad, while that of IV is *internal to its pad*, forming, so far, a relation of images between these two digits. In the polydactyle foot it is a remarkable feature that, though the bones are in symmetry about an axis passing between II and III, the relation of the claws of III and IV to their pads remains almost normal, still giving a superficial appearance of symmetry between these two digits. (In the polydactyle pes the pads are mostly rather narrower.)

[2] It will be remembered that this symmetry appears not merely in the lengths of the several digits but in the manner of retraction of the claws and in the corresponding form of the second phalanges, three digits being fashioned (in the case of six perfect digits) as right digits and three as lefts.

It has been mentioned that there is some evidence to shew that in the human pes it is the digits II and III which are most frequently syndactyle, even up to the point of being (in No. 529) apparently represented by a single digit, and in this connexion it will be remembered that in the polydactyle pes of the Cat it is also between these digits that the new axis of Symmetry falls.

These scanty allusions to the possible influences which Symmetry may exercise over Meristic Variation of digits will suffice to indicate the nature of the problem to those who may care to examine it. It is with hesitation that so indefinite a matter is spoken of at all. Nevertheless it is likely that if any one can find a way of interpreting these indications the result will be considerable.

(5) *Duplicity of limbs.*

In the evidence as to the digits of Man facts were given respecting the state known as Double-hand, and some similar cases were referred to in Artiodactyles. In these instances the digital series, and to some extent the limb, is in its new shape made up of the *external* parts of a pair of limbs compounded together in such a way that there is a partial *duplicity of the limb*, the two halves being more or less exactly complementary to each other and related as images[1].

This phenomenon in its perfect form must be essentially distinct from the other cases of increase in number of digits; for in the double-hands the limb developes an altogether new bilateral symmetry (see especially No. 492). Between cases of duplicity in limbs and the other forms of polydactylism confusion can only arise when the nature of the parts is ambiguous.

As has been stated, in all certain cases of double-limbs the two are compounded by their internal or præaxial borders, but the case of Macacque No. 504 was peculiar in the fact that there was in it a presumption that the two limbs were not a pair but in Succession.

In Arthropoda there are a very few cases of true duplicity in appendages comparable with the double-hands. These cases will be dealt with hereafter.

---

[1] The fact that a structure naturally hemi-symmetrical, needing the limb of the other side to balance it, may on occasion develop as a complete symmetry is most paradoxical, but no other interpretation of the facts seems possible. The phenomenon is of course comparable with that observed by DRIESCH in the eggs of *Echinus*, where each half-ovum developed into a whole larva on being separated from the other half-ovum (see p. 35, *Note*). It will be shewn that in almost every case in which such an appearance is found in the extra appendages of Insects this appearance is misleading, and that the extra parts have a Secondary Symmetry of their own; but no such way through the difficulty is here open.

(6) *Homœotic Variation in terminal digits when a new member is added beyond them.*

This is a principle that has been several times seen in Meristic Variation, and in Chapter X. Section 7, it was treated of at length in the case of teeth. Some few illustrations of the same principle occur among the evidence as to digits. It has been seen for instance how that, upon the appearance of an extra digit on the radial side, the digit which stands in the position of pollex may have three phalanges and resemble an index (No. 485, &c.). Similarly it was found that upon the formation of a large digit externally to the minimus the digit standing in the ordinal position of the minimus may have an increased proportional length (No. 509). Still more important is MORAND'S case (No. 510), in which the most external digit had muscles proper to a minimus, while the digit standing in the ordinal position of the minimus was without them.

The cases of extra digit in the Horse (No. 536, &c.) still more clearly illustrate the principle, if the view of the nature of those cases taken in the text be received.

It should be expressly stated that in digits, as in teeth, it is not always that the terminal member is promoted on becoming penultimate. Such promotion is indeed rather exceptional in digits, but the fact that it may occur is none the less a phenomenon of great significance.

(7) *The absence of a strict distinction between duplicity of a given digit and other forms of addition to the Series.*

This subject has been so often spoken of in connexion with special cases that it is unnecessary here to make more than brief allusion to it. The same principle was shewn to be true of teeth (p. 270) and of mammæ (p. 193), and there is little doubt that it is true of Meristic Series generally. Facts illustrating the matter in relation to digits will be found in the evidence as to duplication of pollex and hallux in Man (p. 351), as to duplication of the hallux in the Fowl (p. 391), in the evidence of cases in the Horse of variation intermediate between division of III and development of II (p. 371), and in the cases of three-toed Cows (p. 377).

In almost all the animals in which any considerable range of digital Variation is to be seen it is possible to find a series of cases making an insensible transition from true duplicity, or division into two equivalent parts whose positions and forms are such that they may be reasonably looked upon as both representing a normally single member, up to the condition in which while the series contains a greater number of members, each member nevertheless stands in a regular Succession to its neighbour.

Upon the proper understanding of this proposition and upon the recognition of its truth hang those corollaries before enuntiated touching the false attribution of the character of individuality to members of Meristic Series.

(8) *Discontinuity in digital Variation.*

The evidence that the Meristic Variation of digits may be discontinuous is often rather circumstantial than direct. If for example in the case of the Horse any one chooses to suppose that every polydactyle horse had in its pedigree an indefinitely long series of ancestors in which the size of the extra digit progressively increased, it would not be easy to produce direct evidence that this was not the fact. But as regards the human examples such evidence is abundant, many of the most marked cases being the offspring of normal parents and there can be no reasonable doubt that the same would be found true of other animals.

But it may fairly be replied that until it shall have been shewn that formations like those described as variations may be established in a natural race or species the contention that the Variation of digits may be discontinuous is so far weakened. To this I would reply by referring to the case of *Cistudo*, *Chalcides*, and the other similar examples; for though in respect of these forms the evidence is sadly imperfect yet it plainly indicates that very distinct and palpable variation may be found between different individuals. And since it is actually known that there may in these points be considerable differences between the two sides of the body it may safely be assumed that at least the same differences may occur between parent and offspring.

We may therefore take it that there is in these cases some Discontinuity of Variation, though until some one shall have examined statistically such cases as that of the Box-turtles or of the Kittiwakes, as to the magnitude of the Discontinuity it is not possible to speak. If hereafter Discontinuity shall be shewn to occur in many such cases it will be difficult to resist the suggestion that similar numerical diversity elsewhere characterizing the digital series of various forms may have come about by similarly discontinuous Variation.

(9) *Relation of the facts of digital Variation to the problem of Species.*

This relation is both direct and indirect: direct, inasmuch as some of the conditions seen to occur as variations are not far removed from those known as normals in other forms; and indirect, since those strange and paradoxically regular dispositions of digits which are found among the variations bear witness to the influence of the principles of Symmetry, and prove that there are modes in which Variation may be controlled and may produce a result which has the quality of regularity and order of form independently of the guidance of Natural Selection.

Of actual variations from the arrangement of digits characteristic of one form to that characteristic of another there are as yet scarcely any examples. The cases given on pp. 395 to 398 being the most evident.

For the rest, that is to say examples of arrangements happening as variations matching no normal, some may say in haste that with their like Zoology has no concern. It would be convenient if those who make this careless answer (as many do) would mark the point at which it is proposed to begin this rejection of the evidence of Variation. Few perhaps realize how impossible it is to give a real meaning to these distinctions. As regards digits, for instance, I suppose that no one who holds the doctrine of Common Descent would refuse to admit the evidence of Variation as to the hallux of Hedgehogs (No. 612) as exemplifying the way in which species *may* be built up—if indeed species are built up of variations at all. And if this case is admitted, by what criterion shall we exclude cases of the formation of a hallux in the Dog? But if these are not excluded it is difficult to shew good reason for not admitting the case of the three-phalanged digit placed as a hallux in the Cat (No. 472) with all the curious series of which that is only the first term. Are we quite sure that because there is no Carnivore with a three-phalanged hallux therefore such a creature could not exist in nature? Still more difficult is it to shew cause why duplicity of the hallux should be set apart as a variation not capable of being perpetuated or of becoming part of the specific characters of an animal, seeing that there is actual evidence both in the case of the Dorking fowl and in the St Bernard dog that it may become at least an imperfectly constant character.

In connexion with the subject of this section many suggestions with special bearing on particular cases, both positive and negative, will strike every reader. In the present imperfect state of the evidence it would be premature to pursue these. It may however be well to mention that several writers, especially JOLY and LAVOCAT (No. 554), have seen in the cases of divided digit III in the Horse an indication that the digit III of the Horse corresponds with the digits III and IV of the Artiodactyles. The evidence as to syndactylism between these two digits in Ox and Pig would probably be considered to give support to the same view. But while we may note that the relations of the digits with the carpus and tarsus of these forms, were comparative evidence absent, should absolutely prevent any one from seriously maintaining such an opinion, nevertheless the fact that such closely similar systems of Symmetry may thus arise independently of each other is of interest.

CHAPTER XV.

LINEAR SERIES—*continued*.

MINOR SYMMETRIES: SEGMENTS IN APPENDAGES.

MERISTIC Repetition along the axes of appendages is very like that along the axis of the body. Just as particular numbers of segments or repetitions along the axis of Major Symmetry characterize particular forms, so particular numbers of joints characterize particular appendages. Such numbers frequently differentiate species, genera, or other classificatory divisions from each other. In the evolution of these forms therefore there must have been change in these numbers.

Those who are inclined to the view that Variation is always continuous do not perhaps fully realize the difficulty that besets the application of this belief to the observed facts of normal structure. For in those many groups whose genera or species may be distinguished from each other by reason (amongst other things) of difference in the number of joints in some particular appendage or appendages, will any one really maintain that in all these the process by which each new number has been introduced was a gradual one? To take a case: even were evidence as to the manner of such Variation wanting, would it be expected that the Longicorn Prionidæ, most of which have the unusual number of 12 antennary joints, did, as they separated from the other Longicorns which have 11 joints, gradually first acquire a new joint as a rudiment which in successive generations increased? Or, conversely, did the other Longicorns separate from a 12-jointed form by the gradual "suppression" of a division or of a joint? If any one will try to apply such a view to hundreds of like examples in Arthropods, of difference in number of joints in appendages of near allies—forms that by the postulate of Common Descent we must believe to have sprung from a common ancestor—he will find that by this supposition of Continuity in Variation he is led into endless absurdity. Surely it must be clear that in many such cases to suppose that the limb came through a phase in which one of its divisions was half-made or one of its joints half-grown, is to suppose that in the comparatively near past it was an instrument of totally different character from that which it has in either of the two perfect forms. But no such supposition is called for. With evidence that transitions of this nature *may* be discontinuously effected the difficulty is removed.

The frequency of Meristic Variation in appendages is much as it is in the case of body-segments. On the one hand there are series containing high total numbers of repetitions little differentiated from each other (*e.g.* the antennæ of the Lobster), and in these Meristic Variation is common; on the other hand in series containing few segments much differentiated from each other, such Variation, though not unknown, is rare. Of the latter a few instances are here offered. That they are so few may perhaps be in part attributed to the little heed that is paid to observations of this class. Records of this kind might indeed be hoped for in the works of those naturalists to whom the title "systematic" has been given; but unfortunately the attention of these persons has from the nature of the case been drawn rather to features whereby species may be kept apart than to facts by which they might be brought together.

From the lack therefore of records of such variations their absence in Nature must not lightly be assumed. To quote but one case: in the common Earwig the numbers and forms of the antennary joints are exceedingly variable, but in many special treatises on Orthoptera, I cannot find that this variability is spoken of, and if alluded to at all the only notice is given in the form "antennæ 13- or 14-jointed."

ANTENNÆ OF INSECTS.

PRIONIDÆ.

I am indebted to Dr D. Sharp for the information that the number of antennary joints in certain Prionidæ varies. In Longicorns generally the number of joints is constantly 11. Dr G. H. HORN of Philadelphia who is specially acquainted with this group, has kindly written to me that of six species of N. American *Prioni* four species have 12 antennary joints constantly in both sexes. Besides these he gives the following cases of Variation. It will be seen that in both of these the normal number is much greater than it is in the other species[1].

*616. **Prionus imbricornis:** females have very constantly 18 joints; males have 18 to 20. A male in Dr Sharp's collection has only 17 joints in each antenna.

---

[1] In *Prionus imbricornis* and *P. fissicornis* doubt may be felt whether the trifid apex should be reckoned as one joint or as two, but this applies equally to each individual. I have counted it as one.

*617.   **Prionus fissicornis:** the female has 25, and the male 27—30, the note on the preceding species applying here.

618.   **Polyarthron.** A Prionid beetle, in which the male has curious many-jointed feather-like antennæ, according to SERVILLE has always 47 joints, but THOMSON (*Syst. Ceramb.*, 1866, p. 284) says the number varies with the species and individually. A male in Dr Sharp's collection has 45 joints in each antenna and a female has 31 in each.

619.   **Lysiphlebus** is a Braconid (Hymenoptera) parasite on Aphides. From a colony of Aphides on a bush of *Baccharis viminalis* 121 specimens of *Lysiphlebus* were reared: of these 57 were males and 64 were females.

The number of joints in the antennæ varied as follows:

Males.
    14 joints .... 18 specimens.
    15 .... 37 ....
    16 .... 1 ....
    15 ...... on one side and 16 on the other 1 ....

Females.
    12 .... 7 ....
    13 .... 54 ....
    14 .... 1 ....
    12 ...... on one side and 13 on the other 2 ....

In those having a different number of joints in the right and left antennæ, the last joint of the antenna which contained the fewest joints was longer than the last joint of the antenna with the larger number of joints. Nevertheless this relation did not hold throughout; for example in the case of the male with 16 joints, the last joint was of the same length proportionally as that of the males with only 14 joints. As a rule the specimens with fewer antennary joints are smaller than the others.

Variations were also seen in coloration, in the proportional length of the tarsi, and in the presence or absence of the transverse cubital nervure, but none of the characters divided the sample consistently, it was therefore inferred that the individuals belonged to one species of *Lysiphlebus*, (*L. citraphis*, Ashm.)

From another colony of Aphides living on a rose-bush 58 specimens of *Lysiphlebus* were bred, and no characters were found by which these could be separated from those bred from the Aphis of *Baccharis*. In the case of the second sample the joints of antennæ were as follows:

Males.
    14 joints................................. 10 specimens.
    15 .... 19 ....
    14 ...... on one side and 15 on the other 2 ....

Females.
    12 .... 2 ....
    13 .... 25 ....

The number of antennary joints is employed as a specific character in the classification of *Lysiphlebus* by ASHMEAD, *Proc. U. S. Nat. Mus.*, 1888, p. 664). COQUILLET, D. W., *Insect Life*, 1891, Vol. III. p. 313.

*620. **Donacia bidens.** (Phyt.) A female found by Dr D. Sharp at Quy Fen in company with many normal specimens had in each antenna *eight* joints instead of eleven as in the normal. As shewn in the figure (Fig. 123) the antennæ of the two sides were exactly

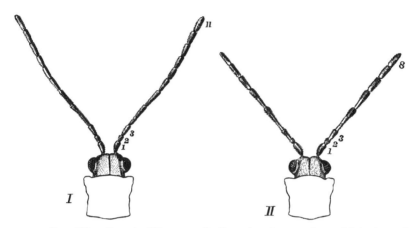

FIG. 123. *Donacia bidens* ♀. I. Normal antennæ, eleven joints in each. II. Abnormal specimen, having eight joints in each antenna. No. 620.

alike, and the insect was normal in all other respects. I am much obliged to Dr Sharp for shewing me this specimen.

Forficula auricularia, the common Earwig. In the various species of *Forficula* the number of joints in the antennæ differs, the numbers 11, 12, 13 and 14 being all found as normals in different species[1]. As regards *F. auricularia* most authors give 14 as the number of antennary joints. SERVILLE[2] gives 13 or 14. A number of adult earwigs examined by myself with a view to this question shewed that there is great diversity in regard to the number of antennary joints. The whole matter needs much fuller investigation but the preliminary results are interesting.

The commonest number is 14, which occurs in perhaps 70—80 per cent. The next commonest is 13, which was seen in a considerable number, while 12, and even 11 occur in exceptional cases. Different numbers were frequently found on the two sides.

---

[1] BRUNNER VON WATTENWYL, *Prodr. eur. Orth.*, 1882. The number in *F. auricularia* is given by Brunner as 15, but I have never seen this number. It is no doubt an accidental error. The same mistake is repeated by SHAW, E., *Ent. Mo. Mag.*, 1888—89, xxv. p. 358.
[2] *Suites à Buffon: Orthop.*, 1839.

414   MERISTIC VARIATION.   [PART I.

As is usual with appendages the whole length of the antennæ differed a good deal independently of the number of joints.

*621. On comparing antennæ with different numbers it seemed that the proportional length of the first two joints was nearly the same in all, but in the third joint there was great difference, as shewn in Fig. 124. The *left* antenna in Fig. 124, I may be taken to be the normal form with 14 joints. In it both 3rd and 4th joints are small. The *right* antenna of the same specimen has 13 joints and in most of the 13-jointed antennæ the arrange-

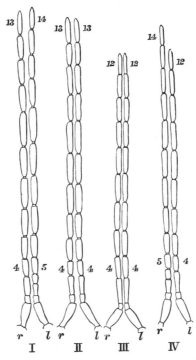

Fig. 124. Various forms of antennæ of adult Earwigs (*Forficula auricularia*), all from one garden and taken at one time.

I. Specimen having the left antenna normally 14-jointed, and the right 13-jointed. No. 621.
II. Both antennæ 13-jointed. No. 622.
III. Both antennæ 12-jointed. No. 623.
IV. Right antenna normally 14-jointed; left antenna 12-jointed. No. 624.

Note that the rights and lefts are arranged as marked by letters *r* and *l*. The antennæ were so fixed for drawing in order to bring them side by side after the bend from the first joint. This figure was drawn with the camera lucida by Mr Edwin Wilson.

ment was much as shewn in this figure. As shewn, the 3rd joint especially is here rather longer than in the 14-jointed form, but several of the peripheral joints are also a little longer, so that though the 13-jointed antenna is not as a whole so long as the 14-jointed antenna of the same individual it is longer than its first 13 joints.

622. But besides the common 13-jointed form occasional specimens are as shewn in Fig. 124, II. Here both antennæ are 13-jointed, the 3rd joint being much longer, and the 4th a little longer than the corresponding joints of the normal with 14 joints. Two specimens were seen having this structure in both antennæ, thus presenting a difference which, did it occur in a form known from but few specimens, would assuredly be held to be of classificatory importance.

*623. In another case (Fig. 124, III) each antenna contained only 12 joints, the 3rd, 4th and 5th being all of greater length than in the normal.

624. Fig. 124, IV shews a case in which there was on the right side a normally 14-jointed antenna but that of the left side was 12-jointed, agreeing nearly with those in Fig. 124, III.

In considering these facts the possibility that some or all the abnormal states may result from or be connected with regeneration must be remembered; but from the frequency of the variations, from their diversity, and from the fact that symmetrically varying individuals are not rare, it is on the whole unlikely that all can owe their origin to regeneration. It will besides be noticed that it is in the *proximal* joints that the greatest changes are seen, and it must surely be rarely that these are lost by mutilation.

The difficulty—indeed the futility—of attempting to adjust a scheme of individuality among such series of segments must here be apparent to all. We can see the change in number and the change in proportions, and we are doubtless entitled to affirm that the differences between these several kinds of antennæ are reached by changes occurring chiefly in the neighbourhood of the 3rd and 4th joints; but not only is there no proof that the changes are restricted to these joints, but the appearances suggest that there are correlated changes in many, and perhaps in all of the joints.

Tarsus of Blatta[1].

*625. Among the families of the class Orthoptera the number of tarsal joints differ. In Forficularia the number of tarsal joints

---

[1] In connexion with variation in the number of joints in legs I may mention the case of *Stenopterus rufus* ♀ (Longicorn) described by Gadeau de Kerville as having each tibia divided into two parts by an articulation (*Le Naturaliste*, 1889, s. 2, xi. p. 9, *fig.*); but upon examination it proved that each tibia had been sharply bent at each of these points, and there was no real articulation. I have to thank M. Gadeau de Kerville for lending me this insect together with many interesting specimens of which mention will be made hereafter.

is 3, in Blattodea, Mantodea and Phasmodea 5, in Acridiodea 3, in Locustodea 4, in Gryllodea 2 or 3[1].

The fact, originally observed by BRISOUT DE BARNEVILLE[2], that in various species of Blattidæ the number of tarsal joints may vary from *five* to *four* is therefore of considerable importance in a consideration of the manner in which these several forms have been evolved from each other. The species in which BRISOUT observed this variation were ten in number and belonged to four genera of Blattidæ.

At my suggestion Mr H. H. BRINDLEY has made an extended investigation of the matter and a preliminary account of the results arrived at was given in the Introduction (p. 63). It was found that of **Blatta americana** 25% of adults have one or

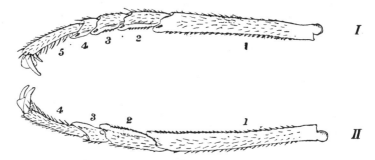

FIG. 125. I. Normal five-jointed left tarsus of *Blatta americana*. II. Right tarsus of the same having four joints.

more tarsi 4-jointed. In **Blatta orientalis** these cases amounted to 15%, and of 102 **B. germanica** examined, 16 had one or more 4-jointed tarsi.

The abnormality occurred sometimes in one leg and sometimes in another, being more frequent in the legs of the second pair than in those of the first, and much more frequent in the third pair than in either. In some specimens legs of the two sides were symmetrically affected, but this was exceptional. Only one specimen has hitherto been met with having *all* the tarsi 4-jointed. There was a slightly greater frequency in females than in males.

When the examination of these abnormal tarsi was begun it was supposed that the variation was congenital, but as explained in a note to the Introduction (p. 65) doubt subsequently arose as to this. It is well known that Blattidæ like many other Orthoptera have the power of renewing the appendages after loss, and Mr Brindley found by experiment that when the tarsus of *Blatta orientalis* is renewed after mutilation the resulting tarsus is 4-jointed. It was also found that 4-jointed tarsi were much more frequent in adults than in the young. The question therefore arises, is the 4-jointed tarsus *ever* congenital?

[1] From BRUNNER VON WATTENWYL, *Prodr. europ. Orthop.*, 1882.
[2] *Ann. ent. soc. France*, s. 2, VI. 1848, *Bull.*, p. XIX.

To this question a positive answer cannot yet be given; but as about 200 young *B. orientalis* have since been hatched from the egg and no 4-jointed tarsus was found among them, while in every instance of regeneration the new tarsus had four perfect joints, there is now a presumption that the variation does not occur congenitally. On the other hand it should be mentioned that the 4-jointed tarsus was seen in 3 specimens, found by Mr Brindley, which by their size would be judged to have been newly hatched. But even if the variation shall hereafter be found to be sometimes congenital it is certain that this occurrence must be very rare, and there can be no doubt that in the majority of cases the 4-jointed tarsus has arisen on regeneration[1].

As mentioned in the Introduction, the existence of the 4-jointed tarsus, whatever be the manner of its origin, raises two questions. Of these the first is morphological, relating to the degree to which the joints exhibit the property of individuality, and the second is of a more general nature, relating to the application of the theory of Natural Selection to such a case of discontinuous change. The interest of the case in its bearing on both of these questions arises from the Discontinuity, which was complete. All the tarsi seen were either 5-jointed or 4-jointed, and in none of the latter was any joint ever rudimentary, or any line of articulation imperfectly formed (except in a single specimen having a deformed tarsus). There were 5-jointed tarsi and 4-jointed tarsi: between them nothing.

Following the usual methods of Comparative Anatomy it must be asked *which* of the 5 joints is missing in the 4-jointed tarsus? With reference to this question careful measurements of the separate joints were made by Mr Brindley in 115 cases of 4-jointed tarsi occurring in legs of the *third* pair in *B. americana;* and for comparison the separate joints of 115 normal 5-jointed third tarsi of the same species were also measured. (It is clear that the legs compared must belong to the same pair, 1st, 2nd or 3rd, for there is considerable differentiation between them. From this circumstance it was comparatively difficult to obtain a large number of cases, and hence the smallness of the whole number measured. But though of course statistics respecting a larger number would be more satisfactory there is no reason to think that by examination of a greater number of cases the result would be materially affected.)

In the two sets of tarsi the total length of each tarsus was reduced to 1·000, the lengths of the joints being correspondingly reduced.

The arithmetic means of the ratios of the several joints to the whole lengths of the tarsi to which they belonged was as follows:

Five-jointed form.

1st joint	2nd joint	3rd joint	4th joint	5th joint
·532	·156	·095	·049	·168

Four-jointed form.

1st joint	2nd joint	3rd joint	4th joint
·574	·183	·064	·179

[1] The circumstance that in Mr Brindley's observations the variation was in all species more frequent in females than in males, and that the frequency differed in

The evidence derived from these numbers lends no support to the expectation that any one particular joint of the 5-jointed form is missing from the 4-jointed, or that any one joint of the 4-jointed form corresponds with any two joints of the 5-jointed; for if the numbers are treated with a view to either of these hypotheses it will be found impossible to make them agree with either. It appears rather that the four joints of the 4-jointed form collectively represent the five joints of the normal.

The other question upon which the statistics bear has already been stated in the Introduction. In any appendage the ratio of the length of each joint to the whole length of the appendage varies; but if it varies about one normal form it will be possible to find a normal or mean value for this ratio, and the frequency with which other values of the same ratio occur will be inversely proportional to the degree in which they depart from the normal value. The curve representing the frequency of occurrence of these values will then be a normal Curve of Error. The form of this curve will indicate the constancy with which the normal proportions of the tarsal joints are approached. If the proportional lengths of the tarsal joints vary little then the curve representing the frequency of their departure from their normal value will be a steep curve, but if these proportions are very variable and have little constancy, then the curve will be flatter. The probable error will thus in the case of each value be a measure of the constancy with which it conforms to its normal proportions. As explained in the Introduction, upon the hypothesis that all constancy of form is due to the control of Natural Selection, it would be anticipated that the 4-jointed tarsus, if a variation, would be very much less constant in the proportions of its joints than the 5-jointed tarsus. It was however found that as a matter of fact the proportions of the joints of the 4-jointed form were very nearly as constantly conformed to as those of the joints of the normal tarsus.

The evidence of this is as follows. The total length of the 5-jointed tarsus being L, and t^1, t^2, &c. being the lengths of its several joints, l, T^1, T^2, &c. representing the same measurements in the 4-jointed form, the ratios $\frac{t^1}{L}$ &c., $\frac{T^1}{l}$ &c., represent the proportional length of the several joints in each case. The values of these ratios were then arranged in ascending order in their own series and the measures occupying the positions of the first, second, and third quarterly divisions noted[1] (indicated hereafter by Q^1, M and Q^3 respectively). The probable error or variation of each ratio $\frac{t^1}{L}$, $\frac{T^1}{l}$, &c. will then be represented by the expression $\frac{Q^3 - Q^1}{2}$. Inasmuch as the joints are of different lengths, to compare the results each must be converted into percentages of the mean length of the joint concerned. These results are set forth in the accompanying tables.

the different pairs of legs may seem to point to the existence of some control other than the simple chances of fortuitous injury. As regards the latter point it is not unlikely that the legs of the third pair, being longer and less protected, may be more often mutilated than the others.

[1] As described by GALTON, F., *Proc. Roy. Soc.*, 1888–9, XLV. p. 137.

Five-jointed tarsus.

	$\dfrac{t^1}{L}$	$\dfrac{t^2}{L}$	$\dfrac{t^3}{L}$	$\dfrac{t^4}{L}$	$\dfrac{t^5}{L}$
Q^1	·521	·152	·095	·046	·162
M	·529	·156	·099	·049	·168
Q^3	·535	·160	·101	·051	·174
Mean error as percentage of M	1·3	2·6	3·0	5·0	3·6

Four-jointed tarsus.

	$\dfrac{T^1}{l}$	$\dfrac{T^2}{l}$	$\dfrac{T^3}{l}$	$\dfrac{T^4}{l}$
Q^1	·565	·178	·060	·172
M	·575	·183	·064	·177
Q^3	·584	·189	·068	·183
Mean error as percentage of M	1·6	3·0	6·2	3·1

It is thus seen that the percentage variation of the ratios of the several joints to the total length is very little greater in the case of the abnormal than it is in the normal tarsus.

As regards the longer joints these results are probably a trustworthy indication of the amount of Variation, but in the case of the shorter joints the errors of observation must no doubt be so great in proportion to the smallness of the lengths to be measured that no reliance should be placed on results obtained from them.

As evidence that in spite of the small number of instances examined the general result is satisfactory it may be mentioned that the mean obtained as the value of $\dfrac{Q^3 + Q^1}{2}$ agrees fairly well in each case both with the value of M, the middlemost value, and also with the arithmetic mean given above. It may therefore be taken that the curve is regular and the series nearly uniform.

The correlations between the lengths of the joints and that of the whole tarsus have also been examined by Mr Brindley using the method proposed by GALTON *l.c.*, the results closely agreeing with those obtained by the ordinary method here described[1].

If the 4-jointed tarsus be a congenital variation the significance of the fact that the abnormality is in its constancy to its normal hardly less true than the type-form must be apparent

---

[1] It is hoped that a fuller account of this subject will be given separately. I am indebted to Mr F. Galton for advice kindly given when this investigation was begun, and Mr Alfred Harker has most obligingly given much help in connexion with it.

to all. Yet even if, as now seems likely, the 4-jointed tarsus be not a congenital variation but is rather a result of regeneration, there is still difficulty in reconciling the now established fact that the form of the regenerated part, though different from the normal, is scarcely less constant, with any hypothesis that the constancy of the normal is dependent upon Selection.

If it were true that the smallness of the mean variation of the ratio $\dfrac{t^1}{L}$, which is ultimately the measure of the constancy and truth to type of the 5-jointed tarsus, is really due to Selection and to the comparative prosperity of specimens whose tarsal proportions departed little from the normal, to what may we ascribe the smallness of the mean variation of the ratio $\dfrac{T^1}{l}$? Are we to suppose that the accuracy of the proportions of the regenerated tarsus is due to the Natural Selection of individuals which in renewing their tarsi conformed to this one pattern?

We are told that the struggle for existence determines every detail of sculpture or proportions with such precision that individuals which fall short in the least respect are at a disadvantage so great as to be capable of being felt in the struggle, and so decided as to lead to definite and sensible effects in Evolution. If this is so, should we not expect that individuals which had suffered such a comparatively serious disadvantage as the loss of a leg or of a tarsus, would be in a plight so hopeless that even though some of them may survive, renew the limb and even breed, yet, as a class, by reason of their mutilation they must rank with the unfit? Nevertheless we find not only that there is a mechanism for renewing the limb, but that the renewal is performed in a highly peculiar way; that in fact the structure newly produced differs from the normal just as species differs from species, and is scarcely less true and constant in its proportions than the normal itself.

Now if this exactness in the proportions of the renewed limb is due to Selection, it must be due to Selection working among the mutilated alone; and of them only among such as renewed the limb; and of them only among such as bred. Moreover if the accuracy of the form of the renewed tarsus is due to Selection working on fortuitous variations in the method of renewal, and not to any natural definiteness of the variations, the number of selections postulated is already enormous. But this vast number of selections must by hypothesis have all been made from amongst the mutilated—a group of individuals that would be supposed to be at a hopeless disadvantage[1].

---

[1] The same dilemma is presented in all cases where a special mechanism or device exists (and must be supposed to have been evolved) only in connexion with regeneration. An instance is to be seen in the Lobster's antenna. As is well known the antennary filament of the Lobster when lost is renewed not as a *straight* out-

One or more of the hypotheses are thus clearly at fault. A natural, and I believe a true comment will occur to every one: that probably the injured insects are *not* at any serious disadvantage, and that these mutilations perhaps make very little difference to their chances. But can we admit that the loss of a leg matters little, and still suppose that the definiteness and accuracy of the exact proportions of the tarsal joints makes any serious difference?

The hypothesis, therefore, that the smallness of the mean variation in the proportional lengths of the tarsal joints of the 4-jointed tarsus has been gradually achieved by Selection is untenable, whether that 4-jointed tarsus be a product of regeneration or a congenital variation. But if the accuracy with which the abnormal conforms to its type be not due to a gradual Selection, with what propriety can we refer the similar accuracy of the normal to this directing cause?

RADIAL JOINTS IN ARMS OF COMATULÆ.

The number of radial joints above the basals up to the division of the rays in Crinoids is usually constant in the genera. In *Antedon* and *Actinometra* there are normally three such joints, the third radial being the axillary, and none of these bear pinnules. Both increase and decrease in the number of radials has been observed, but variations from this number are rare, more so than variations in the number of rays. CARPENTER, P. H., *Chall. Rep.*, xxvi. Pt. LX. p. 27.

626.　**Antedon alternata**: specimen having in one ray *four* radials, none bearing pinnules or united by syzygy. *ibid.*, Pl. XXXII. *fig.* 6.

627.　**Encrinus gracilis** (fossil): in one ray *four* radials. WAGNER, *Jen. Ztschr.*, 1887, xx. p. 20, Pl. II. fig. 13.

628.　**Antedon remota, A. incerta, Actinometra parvicirra** (Fig. 126); one specimen of each of these species had one ray with only *two* radials. CARPENTER, *l.c.*, Pl. XXIX. fig. 6; Pl. XVIII. fig. 4; Pl. LXI. fig. 1.

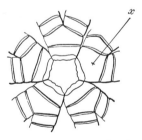

FIG. 126. *Actinometra parvicirra*, No. 628. Specimen having only two radials in the ray marked *x*. (From P. H. CARPENTER.)

growth, as the other appendages are, but when formed again it is coiled up in a tight conical spiral which cannot be extended at all, but is kept firmly in place by the shortness of the skin on the inner curvature. (For figure see HOWES, *Jour. Anat. Phys.*, XVI. p. 47.) During the process of regeneration the antenna is very soft, and were it extended it would from its great length be much exposed to injury. At the next moult after renewal the new antenna is drawn out as a straight filament like the normal, and its skin then hardens with that of the rest of the body. This strange manner of growth occurs only on regeneration. It is hard to believe both

Metacrinus. Some species have normally 5, others normally 8 radials. If there are 5, the 2nd and 3rd are united by syzygy and bear pinnules; but if there are 8, both 2nd and 3rd, and the 5th and 6th are thus united and bear pinnules. In *Plicatocrinus* the number of radials is *two*, and this is also the case in one or two fossil Comatulæ. *Pentacrinus* has normally three radials like *Antedon.*

629.   **Pentacrinus mülleri**: specimen having in one ray *four* radials, the 2nd and 3rd united by syzygy, though bearing no pinnules. CARPENTER, *l. c.*; and *Chall. Rep.* XI. Pt. XXXII. p. 311, Pl. xv. *fig.* 2.

(1) that the number of individuals that have lost antennæ—a serious injury one may judge—and have renewed them, and have bred, can have been enough to lead to the establishment by Selection of a distinct and highly special device to be invoked solely on the occasion of mutilation of an antenna; and also (2) that the least detail of normal form is of such consequence as to be rigorously maintained by Selection.

CHAPTER XVI.

Radial Series.

Little need be said in preface to the facts of Meristic Variation in Radial Series. In them phenomena analogous to those of the Variation in Linear Series are seen in their simplest form. Just as in Linear Series the number of members may be changed by a reconstitution of the whole series so that it is impossible to point to any one member as the one lost or added, so may it be in the Meristic Variation of Radial Series: and again as in Linear Series, single members of the series may divide. Between these there is no clear line of distinction.

Next, as in Linear Series, Variation, whether Meristic or Substantive, may take place either in single segments (quadrants, sixths, &c.), or simultaneously in all the segments of the body. For instance, a single eye may be divided into two, or there may be duplicity simultaneously occurring in all the eyes of the disc (see No. 634) and so on.

These phenomena are here illustrated by facts as to the Meristic Variation of Hydromedusæ and of *Aurelia*. The latter is exceptionally variable and in its changes exhibits important features.

Together with these facts as to Variation in Major Symmetries is given an instance of similar Variation in the pedicellariæ of an Echinid, and it will be seen that in this case of a Minor Symmetry the change is perfect and altogether comparable with those found in Major Symmetries of similar geometrical configuration.

The best field for the study of the variations of Radial Series is of course to be found in plants; and in the Meristic Variations of radially symmetrical flowers precisely similar phenomena may be easily seen.

I. CŒLENTERATA.

*630.   **Sarsia mirabilis**[1]: normally *four* radial canals, &c. (Fig. 127, I and III). Out of many hundreds of N. American specimens *two* were found with *six* radial canals, six ocelli, and six tentacles,

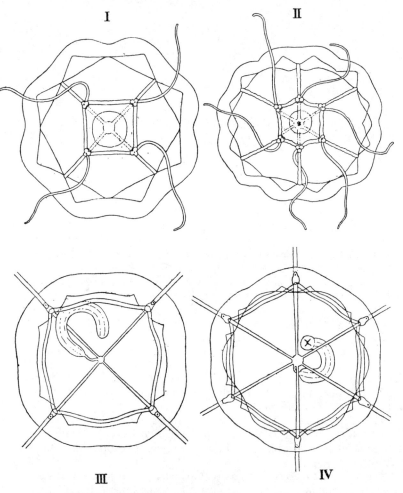

Fig. 127. *Sarsia mirabilis.* I and III, the normal form, with four radii, from below and from above. II and IV, an abnormal form with six radii, from below and from above respectively. (From AGASSIZ.)

symmetrically arranged (Fig. 127, II and IV). These specimens were of larger absolute size than the normals. AGASSIZ, L., *Mem. Amer. Ac. Sci.*, IV. p. 248, Pl. v. fig. 5.

[1] *Sarsia* is the medusa of the Gymnoblastic Hydroid *Syncoryne*.

*631. **Sarsia** sp. Among many thousands examined on the east coast of Scotland *one* was found having six radial canals, six ocelli and six tentacles. ROMANES, G. J., *Jour. Linn. Soc., Zool.*, XII. p. 527.

*632. **Sarsia** sp. A single specimen having *five* complete segments: the only abnormality met with among thousands of naked-eyed medusæ observed, *ibid.*, XIII. p. 190.

There is perhaps in the whole range of natural history no more striking case of the Discontinuity and perfection of Meristic Variation.

Is it besides a mere coincidence, that the specimens presenting this variation, so rare in the free-swimming Hydromedusæ, should have been members of the same genus?

633. **Clavatella (Eleutheria) prolifera.** This form has a medusa which creeps about on short suctorial processes borne by the tentacles. The number of these tentacles varies from 5 to 8. In the specimens examined by KROHN[1] the number was 6. Most of CLAPARÈDE'S[2] specimens had 8. FILIPPI[3] found that the majority had 6 arms, but 15 per cent. had 7. Those examined by HINCKS[4] never had more than 6. Filippi considered that the difference in number was evidence that his specimens were of a species different from Claparède's. I examined many of this form at Concarneau and found six the commonest number in the free medusæ, but those still undetached frequently had 5, possibly therefore the number increases with development. [See also *Cladonema radiatum*, &c. HINCKS, *l.c.*, p. 65, &c.]

Claparède states that the 6-armed specimens had 6 radial canals, but the 8-armed usually had *four* though occasionally six, but never eight canals.

In this case note not only the frequent occurrence of Meristic Variation, but also the suggestion that particular numbers of tentacles are proper to particular localities.

*634. Normally there is a single eye at the base of each arm. CLAPARÈDE figures (*l.c.* p. 6, Pl. I. fig. 7 *a*) a case of duplicity of an eye, and says that specimens occur in which each eye is doubled, so that there are two eyes at the base of each arm instead of one.

635. **Stomobrachium octocostatum** (Æquoridæ): variety found in Cromarty Firth, ⅔rds of size represented by FORBES (*Monogr. Br. Naked-eyed Medusæ*); ovaries bluish instead of orange, and without denticulated margins. Tentacles arranged *in double series*, long and short alternating, while in the type the series is single. The number of large tentacles same as in type. Each smaller tentacle bears vesicular body at base, without pigment or visible contents. The same variety figured by EHRENBERG, *Abh. Ak. Berl.*, 1835, Taf. VIII. fig. 7. ROMANES, G. J., *Jour. Linn. Soc.* XII. p. 526. [Simultaneous Variation of the several segments.]

With Nos. 634 and 635 compare the fact that in *Tiarops poly-*

---
[1] *Arch. f. Naturg.*, 1861, p. 157.
[2] *Beob. üb. Anat. u. Entw. Wirbelloser Thiere*, 1863, p. 5.
[3] *Mem. Ac. Torino*, S. 2, XXIII. p. 377.
[4] *Brit. Hyd. Zoophytes*, 1868, p. 71.

diademata there are *normally as a specific character* four diadems between each pair of radial tubes, making in all *sixteen* instead of *eight*, which is the usual number in the genus. ROMANES, G. J., *Jour. Linn. Soc. Zool.*, XII. p. 525.

*636.   **Aurelia aurita.** This form exhibits an exceptional frequency of Meristic Variation. In the normal there are 16 radial canals, 4 oral lobes, 4 generative organs and 8 lithocysts. The departures from this normal form have been described in detail by EHRENBERG[1] and by ROMANES[2].

Meristic Variation in *Aurelia* may occur in two distinct ways, first in the degree to which there is complete separation between the generative sacs, and second in actual numerical change.

Imperfect division of generative sacs.

In the commonest form of *Aurelia* there are four generative organs each distinct from its neighbours, but in some specimens the generative epithelium is continuous all round the mouth, and there is then one continuous generative chamber, though opening by 4 openings as usual. (Such absence of complete separation between some of the generative organs is not rarely seen in cases of numerical Variation, *v. infra.*) Though the epithelium is then continuous it does not form a true circle, but is sacculated to form 4 (as normally) 3, 6, or some other number of incompletely separated parts. EHRENBERG (*l. c.*, p. 22) saw a case in which there were 6 such sacculations, three on each side being united and having one generative pouch, but each of these pouches opened by 3 openings. There was thus a bilateral symmetry, each half containing three lobes of ovarian epithelium incompletely separated from each other. Complete union of all the generative organs was very rare.

The specimens differ greatly with regard to the degree to which the generative epithelium is folded off, and in the shapes of the generative organs. Commonly the generative epithelium is of a horse-shoe form, the two limbs of the horse-shoe not meeting each other; but in some specimens the two limbs may be to various degrees approximated, so that each generative organ is kidney-shaped or even roughly circular. (Cases figured by EHRENBERG, *l. c.*, Pl. II.) [Here note the Simultaneous Variation of the single quadrants.]

Numerical Variations.

Of these the most striking and also the most frequent are variations consisting in a perfect and symmetrical change in the fundamental number of segments composing the disc. Normally there are four quadrants (Fig. 128, I). Varieties are found having only half the usual number of organs, the disc being made up of two halves, each containing one generative organ (Fig. 128, IV). Other symmetrical varieties having three, and six, as their fundamental numbers are shewn in Fig. 128, V, and II. These figures are from ROMANES. Symmetrical forms having five segments and eight segments are described and figured by EHRENBERG. As to the comparative frequency of these forms facts are given below. In each of them all the parts normally proper to one quadrant are repeated in each segment of the disc, the number of parts being greater or less than the normal in correspondence with the fundamental number of the specimen.

---

[1] EHRENBERG, C. G., *Abh. k. Ak. Wiss.*, Berlin, 1835, pp. 199—202, *Plates.*
[2] ROMANES, G. J., *Jour. Linn. Soc., Zool.*, XII. p. 528, and XIII. p. 190, Pls. XV. and XVI.

RADIAL SERIES: *Aurelia*.

Next, the number of certain organs may vary independently of other organs. For example as seen in Fig. 128, VI the radial canal

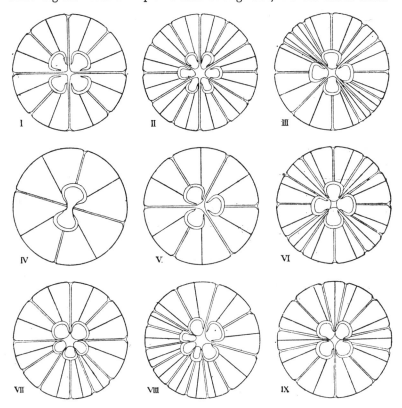

Fig. 128. Diagrams of various forms of *Aurelia aurita*, slightly simplified from ROMANES. I. The normal. II. Symmetrical form with 6 radii. III. Two additional chief radial canals in opposite interradii (where manubrial lobes also were bifid) and substitution of two canals for one in another interradius. IV. Form with two generative organs. V. Form with three generative organs. VI. Symmetrical form in which the intergenital canals are all doubled, the others remaining single. VII. Apparently upper half-disc arranged as for a symmetry of four, lower half for a symmetry of six. VIII. One of the quadrants tripled (?). IX. Form resembling VI. except that in one quadrant the intergenital canal is not doubled. The descriptions are not altogether those of ROMANES.

normally lying in a plane between each pair of generative organs may in each quadrant be represented by two canals, and in correspondence with this change the number of marginal organs is proportionately changed in the quadrants affected.

But besides these changes symmetrically carried out in each quadrant or in the whole disc, one or more quadrants or a half-disc may vary *independently*. For example Fig. 128, VII, shews a specimen in which the two upper quadrants are normal but the lower half-disc is primarily divided into three. (In the case figured the parts of the lower half-disc are not quite accurately distributed). Similarly a particular quadrant may be represented by two sets of parts or by three sets (Fig. 128, VIII), the other three quadrants being normal or nearly so. I have seen a case also in which the chief symmetry was arranged as for *three* segments (having 3 oral lobes), but one of the three segments was imperfectly divided into two.

In a case of 6 segments, 3 on one side may be large and the other 3 small, somewhat as in Fig. 128, VIII, but the whole disc was not circular, the radius on the side of the large segments being the greater.

In the figures (after ROMANES) all the discs are represented as circles, but my own experience was that when there was not a truly symmetrical distribution of the generative organs the half quadrant or other segment in which the number of parts was greatest bulged outwards, thus exemplifying the general rule that when an organ divides the two resulting parts are together larger than the undivided organ.

Besides those specified, there are also irregular cases, *e.g.*, specimens with 3 generative organs but 4 oral lobes and other parts in multiples of 4, but as EHRENBERG says in such cases it is generally possible to detect that one of the generative organs is larger than the others or even partially double. He also saw cases otherwise arranged in a symmetry of 6, but having 22 chief radial canals instead of 24, &c. Also 14 radial canals (instead of 12) were found in some cases of 3 generative organs.

As everyone will admit, it is impossible in regular threes, sixes, &c. to say that any particular segment is missing or is added rather than another.

Comparative frequency of the several forms.

*637. Among thousands of individuals seen by EHRENBERG only two were 8-rayed, 15—20 were 6-rayed, some 20—30 were 5- and 3-rayed, the remainder being 4-rayed. In percentages, 90 are 4-rayed, 3 are 3-rayed 3 are 5-rayed, 2 are 6-rayed and 2 have other numbers.

The result of an attempt to ascertain these percentages in a great shoal of *Aurelia* washed ashore on the Northumberland coast on 4 Sept. 1892 is given below. The radial canals were not counted, and the numbers apply strictly to the generative sacs only. It will be seen that the proportion of abnormals is lower than that given by Ehrenberg.

2 generative sacs	0	
3 ,,	symml.: 3 oral lobes in 4 unbroken cases ...10	$(0.57°/_\circ)$
3 ,,	2 large, 1 small: 3 or. lobes	1
4 ,,	normal...1735	
4 ,,	3 large, 1 small: 5 or. lobes	1
4 ,,	2 large, 2 small: 3 lobes	1
5 ,,	symmetrical 5 lobes in one	2
5 ,,	not quite symmetrical	1
6 ,,	sym.: 6 lobes in 2 unbroken cases	7	$(0.39°/_\circ)$
6 ,,	not symmetrical.....................................	1
6 ,,	4 large, 2 small ,, ...	1
6 ,,	4 large, 2 united: 6 lobes in 1 unbroken ...	2
6 ,,	3 large, 3 small: 6 lobes	1
		1763

CHAP. XVI.]   RADIAL SERIES: PEDICELLARIÆ.   429

There were therefore 1735 normals, 19 symmetrical varieties and 9 irregulars. It will be noted not only that the symmetrical varieties are comparatively frequent, but also that the several forms of irregularity were seen for the most part in single specimens only.

II. PEDICELLARIÆ OF ECHINODERMS.

The number of jaws in the pedicellariæ differs in different forms of Echinoderms, and I am indebted to Professor C. STEWART for information concerning them.

In Asteroidea the number of jaws is usually *two*, but in *Luidia savignii* the normal number of jaws is *three*.

In the Echinoidea the number of jaws is usually *three*, but in *Asthenosoma* the normal number is *four*.

638.   **Dorocidaris papillata:** number of jaws in pedicellariæ

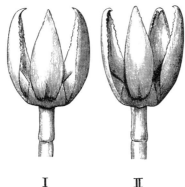

FIG. 129. Pedicellariæ of *Dorocidaris papillata*.
I. Normal form with three jaws.
II. A pedicellaria with four jaws from the abactinal region.
(From Prof. STEWART's specimens.)

normally *three* as in Fig. 129, I, but occasionally *four* in pedicellariæ of the abactinal region, as in Fig. 129, II. [Note that the variety is perfect and symmetrical.] For this fact I am obliged to Professor Stewart, who kindly allowed this figure to be made from his preparations.

639.   **Luidia ciliaris:** pedicellariæ nearly all with *three* jaws; but on Roscoff specimens a few having *two* jaws occur on the borders of the ambulacral groove. In Banyul's specimens none such were found in this position, but there is one in almost all the marginal intervals. CUÉNOT, *Arch. zool. exp.*, S. 2, V. *bis*, p. 18.

640.   **Asterias glacialis:** occasionally *three*-jawed pedicellariæ like those of *Luidia* are found among the normal two-jawed pedicellariæ. CUÉNOT, *l. c.*, p. 23.

III. Cell-Division.

*641. It was purposed at this point to have introduced an account of Meristic variations observed in the manner of division of nuclei and cells; but I have found that, to give adequate representation of these facts even in outline, it would be necessary not only to treat of a very complex subject with which I have no proper acquaintance, but also greatly to enlarge the scope of this work. But were no word said on these matters, indications most useful as comment on the nature of Meristic Variation at large would have to be foregone; and unwilling that these should be wholly lost I shall venture briefly to allude to so much of the matter as is needful to shew some ways in which the facts of abnormal cell-division can be used in reference to the wider question of Meristic Variation.

We have been dealing with cases of Radial repetition, and we have seen that with Variation in the number of parts the result may still be radially symmetrical. It therefore becomes of interest to note that in the case of abnormal cell-division the result of numerical change may in like manner be radially symmetrical. Cells which should normally contain two centrosomes and which should divide into two parts have been seen to contain three centro-

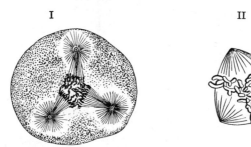

Fig. 130. Triasters. I. Tripolar division of nucleus in embryonic tissue of Trout (after Henneguy[1]). II. Triaster from mammary carcinoma. Centrosomes not shewn (from Flemming[2]).

somes (Fig. 130) prior to division into three parts, and the triangle formed by the three centrosomes may be equiangular just as may be the triangle of the segments of the abnormal *Aurelia* (Fig. 128, V), or of the jaws of the normal pedicellaria of *Dorocidaris* (Fig. 129). It is, I imagine, difficult to suppose that the radial symmetry of each of these series of organs is

---

[1] Henneguy, *Jour. de l'Anat. et Phys.*, 1891, p. 397, Pl. xix. fig. 9.
[2] Flemming, *Zellsubstanz, Kern u. Zelltheilung*, 1882, Pl. viii. fig. v. after Martin, *Virch. Arch.*, 1881, lxxxvi. Pl. iv.

different in its nature, or indeed that it is anything but a visible expression of the equality of the strains tending to part each segment from its neighbours. (The case of the triaster is taken as the simplest and most plainly symmetrical, but examples of cells with greater numbers of centrosomes, sometimes dividing symmetrically, have also been seen.)

For our purpose this fact is first of use as a demonstration of the absurdity of an appeal to "Reversion" as a mode of escape from the admission that variations in Radial Symmetry may be total and perfect though the new number of segments is one which presumably never occurred in the phylogeny; for we need scarcely expect that even conspicuous defenders of the doctrine that all perfection must have been continuously evolved, will plead that the cells of every tissue in which a triaster is found did once normally divide with three poles. Yet if it be once granted that the symmetry of these abnormal forms is a sudden and new departure from the normal, it will not be easy to put the other cases on a different footing.

Though we have repudiated all concern with the causes of abnormality, mention may be made of the fact that multipolar figures, both regular and irregular, have been observed to result from the action of reagents (*e.g.* quinine, HERTWIG[1]). Such figures are of course well known especially in the case of carcinomatous growths, and as Hertwig observes, from the resemblance of these figures to those artificially induced by chemical means it seems possible that these pathological appearances may also be the result of some chemical stimulus. But whatever be the immediate or directing causes of abnormalities in cell-division, or of those other abnormalities in the segmentation of Radial Series of larger parts, and whether any of the causes in the several cases be similar or different, we can scarcely avoid recognition that the resulting phenomena are closely alike[2].

[1] O. HERTWIG, *Die Zelle u. d. Gewebe*, 1893, pp. 192—198.
[2] See also a case of the presence of triasters in two bilaterally symmetrical tracts of the blastoderm of *Loligo* (*v. infra*).

CHAPTER XVII.

RADIAL SERIES: ECHINODERMATA.

As seen in the majority of adult Echinoderms the repeated parts are arranged with a near approach to a Radial Symmetry and it is thus convenient to consider their Meristic Variations in that connexion. But it must of course always be remembered that in their development these repetitions are in origin really a Successive Series and not a Radial Series. The segments are not all identical (as, in appearance at least, they are in many Cœlenterates &c.), but are morphologically in Succession to each other, though there may be little differentiation between them.

In the case therefore of Variation in the number of segments, resulting in the production of a body not less symmetrical than the normal body, there must be in development a correlated Variation among the several members like that seen in so many cases of additions to the ends of Linear Series.

This circumstance should be kept in view by those who seek in cases of numerical Variation, in Echinoderms to homologize separate segments of the variety with those of the type, hoping to be able to say that such a radius is added, or such other missing. As in other animals, this has been attempted in Echinoderms, and though I know well that in the complex subject of Echinoderm morphology I can form no judgment, yet it is difficult to suppose that the same principles elsewhere perceived would not be found to hold good for Echinoderms also.

All that is here proposed is to give abstracts of facts as to Variation in the numbers composing the Major Symmetries. It will of course be remembered that though the fundamental number in Echinoderms is most commonly five, other numbers also occur as normals, (*e.g.* four in the fossil *Tetracrinus*, six in some Ophiurids, &c. Examples will be given of total change from five to four and to six, and so on. It is besides not a little interesting that of the normally 4-rayed *Tetracrinus* both 5-rayed and 3-rayed varieties should be known.

Besides the examples of total Variation there are a few cases of incomplete Variation in which there is a fair suggestion that a particular ray is reduced in size (Nos. 680 and 681, &c.). There are also two cases of imperfect division of a ray in an Echinid (Nos. 688, &c.), while in Asteroids &c. this condition is common. It is of importance to observe that just as in Linear Series abnormal divisions of members of the series are commonly *transverse* to the lines of Repetition, so in radial forms the divisions of rays are commonly *radial*.

The evidence is complicated by the fact that in many Echinoderms extensive regeneration can occur, and in some genera reproduction by division of the disc and subsequent regeneration is almost certainly a normal occurrence[1]. Nevertheless it cannot be doubted that the variation seen in Echini, in *Asterina*, in the discs and stems of Crinoids, &c., are truly congenital. Similarly, though in *Asterias* &c. reduction in the number of arms might otherwise be thought to be due to mutilation, it cannot be so in Echini &c.

HOLOTHURIOIDEA.

*642. **Cucumaria planci**: among 150 half-grown specimens found at Naples five were 6-rayed. LUDWIG, H., *Zool. Anz.*, 1886, IX., p. 472. [These specimens are described in detail.] To determine which is "the intercalated ray" the following ingenious reasoning is offered, and as a good practical illustration of the conception of the individuality of segments as applied to an Echinoderm we may well consider it.

[1] It is likely that several of the Ophiurids and Asteroids which normally have more than 5 arms undergo such fission. LÜTKEN (*Œfvers. Dansk. vid. Selsk. Förh.*, 1872, pp. 108—158 : tr. *Ann. and Mag. N. H.*, 1873, S. 4, XII. pp. 323 and 391) gave an account of this phenomenon. *Ophiothela isidicola* (Formosa) generally has 6 arms, rarely equal, usually 3 large opposite to 3 small; specimens common with only 3 arms, with appearance as if corresponding half-disc cut off. There can be no doubt that the animal divides and that the other 3 arms are renewed. The same phenomenon has been seen in other small 6-armed Ophiurids, especially of genus *Ophiactis*, but Lütken never saw any trace of it in any normally 5-rayed species of the genus. There are indications that the division occurs once when the animal is very small and again when it is adult or nearly so. In *Ophiocoma pumila* the *small* specimens have 6 arms, while the *adults* have 5. Probably therefore division only occurs in the young, the last division being followed by the production of 1 or 2 arms instead of 2 or 3.

Division is probably not a usual occurrence even in Ophiurids having more than 5 arms. *Ophiacantha anomala* has normally 6 arms, and *O. vivipara* has 7—8, but no such appearances are known in them.

Similarly there is evidence [figs. given] that certain Asteroids having normally more than 5 arms viz. *Asterias problema* Stp. [= *Stichaster albulus*], *A. tenuispina* &c. undergo fission; but there is no reason for believing that other many-armed Asteroids divide. The Solasters have many rays, *Asterias polaris* has 6, but no signs of division are seen in them.

An account is also given of the comet-like specimens of *Ophidiaster cribrarius*, occasionally found, having one long arm, at the adoral end of which are present 4 or 5 arms as mere tubercles or as half-grown structures. This phenomenon is well known in *Linckia multiflora*, in which doubtless the separate arms may break off, each reproducing complete disc and arms. [See also as to *Stichaster albulus*, *Asterina wega*, &c., CUÉNOT, L., *Arch. zool. exp.*, V. bis, 1879—90, p. 128 ; and as to *Linckia*, SARASIN, *Ergeb. naturw. Forsch. auf Ceylon*, 1888, I. Hft. 2.]

In the normal there are 5 radii and interradii, and 10 tentacles: in the abnormals there are 6 and 12 respectively. In half-grown normals the 3 ambulacra of the ventral trivium have more tube-feet than the 2 ambulacra of the bivium; also the pair of tentacles corresponding to the central radius of the trivium are smaller than the rest. In the abnormals 3 ambulacra have more tube-feet and are separated by narrower interradii than the rest, and of them the central has the least pair of tentacles: therefore these are the 3 radii of the ventral trivium, and of them the central is the central of the normal. The structure of the calcareous ring bears out this correspondence. The central radius of the ventral trivium is therefore *not* the intercalated radius.

In the 6-rayed specimens there is thus a ventral trivium and a 'dorsal trivium.' (There were 2 Polian vesicles in 3 specimens, 3 in one and one in the other, but in the normal also these vary in number.) The stone-canal was single in all; but in one of them it could be seen that the canal arose in the interradius *to the left* of that which bore the madreporic plate, suggesting that the radius thus crossed was supernumerary; for in a normal the interradius of the dorsal mesentery is in the centre of the bivium. In a normal there are in the calcareous ring two radials on either side between the dorsal mesentery and the ventral median radius. In 4 of the abnormals (to which alone what follows refers) there were 3 such radii on the left and 2 on the right, while in the 5th specimen there were 3 on the right and 2 on the left.

The respiratory trees of the normal are in the right interradius of the bivium and in the left interradius of the trivium. In the 6-rayed they are in the left interradius of the ventral trivium and in the lower right interradius of the dorsal trivium, agreeing with the normal and shewing that the right radius of the ventral trivium is *not* an intercalated one. Next, the mesentery in its course traverses in the 6-rayed form 4 radii and 3 interradii, the lower right interradius of the dorsal trivium with its 2 adjacent radii alone being free. In the normal, 3 radii and 2 interradii are thus traversed, the right bivial interradius and its 2 adjacent radii being free. Therefore the right radius of the dorsal trivium and of the ventral trivium are *not* intercalated. The central radius of the ventral trivium has already been excluded; therefore the intercalated segment is either the middle or the left of the dorsal, or the left of the ventral trivium.

In a normal, the mesentery which is attached to the alimentary canal at that place where its upward portion again turns downwards comes from that interradius which bounds the ventral trivium on the left. This is the case also in the abnormals, and therefore the left radius of the ventral trivium is *not* intercalated. Of the two remaining radii the left of the dorsal trivium is in nowise abnormal, but the central dorsal radius is abnormal in that it is crossed by the sand-canal, therefore *the central dorsal* is the intercalated radius.

And since in four cases there were three radii in the calcareous ring on the left, between the interradius of the stone-canal and the central of the ventral trivium, and two on the *right*, therefore the new segment is in them intercalated on the *left* of the median interradius of the bivium; while in the fifth specimen the intercalation has been made on the *right* of the same interradius.

Now all this argument rests on the premiss that the several members of a series of differentiated parts *cannot* undergo a Substantive Variation in correlation with Meristic change in the total number of members constituting the series. It is assumed that there can be no redistribution of differentiation.

This assumption has now in many cases of Linear Series been shewn to be false. To refer to one of the simplest cases, there is, in the case of the Frog, evidence that the peculiarities of the 9th vertebra may be wholly or in part transferred to the 10th vertebra, when by Meristic Variation there are 10 vertebræ (Nos. 56, 57 and 60), and the like has been shewn in many other examples (cp. No. 35). The functions (as indicated by the structures) of the vertebræ may be redistributed on the occasion of Meristic Variation.

Will anyone affirm that similar redistribution of differentiation may not happen in the Meristic Variations of Echinoderms?

643. *Variations in organs of Holothurioidea.* LAMPERT calls attention to the great variability found in this group and the consequent difficulty in distinguishing specific characters from individual abnormalities. These variations often take the form of alterations in the number of organs. For example, the distribution of the tube-feet is liable to great alterations during the lifetime of individuals. In some forms (as *Thyone* and *Thyonidium*) the feet are confined to the ambulacral areas in the young animal, but are distributed over the whole body in more mature individuals; and in species of the genus *Stichopus*, though the tube-feet are arranged in rows, yet in old individuals this arrangement may become obliterated. On the contrary, in others, as for example, *Holothuria graeffei*, the arrangement of the feet in thoroughly mature specimens is still most sharply defined.

The number of the tentacles is generally a multiple of five, and such cases as *Amphicyclus* and *Phyllophorus* in which other numbers are found, are rare. In these forms the tentacles are said to vary both in number, position and size, but the number is always about 20. The case of *Thyonidium molle* is cited as an extreme case. Of this species 4 specimens had 20 tentacles arranged in a paired manner as in typical *Thyonidia;* other specimens had 20 tentacles of similar length; others had from 16 to 19 tentacles of nearly equal lengths, and others again had from 19 to 21, which instead of being disposed in pairs were placed irregularly, some being larger and some smaller.

Of all the organs, the Cuvierian organs are the most variable and they are of little value for purposes of classification. Their number is very inconstant and they may even be absent altogether. It is impossible to distinguish any circumstances whether of locality or of structure in which the individuals without Cuvierian organs differ from the others which possess them. The two chief appendages of the water vascular ring, namely Polian vesicles [cp. No. 642] and the stone-canal are usually constant when they are single, but in rare cases there are exceptions even to this rule. If however more than one of these organs is normally present, it may generally be assumed that there is no constancy in their numbers, and in such cases the number of the Polian vesicles is especially variable. A few species have been recorded in which, from a single Polian vesicle, secondary ones are formed by lateral outgrowths.

The calcareous plates are of all the organs the least liable to variations, but in certain cases they are stated to change with age.

LAMPERT, K., *Die Seewalzen*, in *Semper's Reisen im Archipel der Philippinen*, 1885, IV. III. pp. 6, 13, and 174; also in *Biol. Centrabl.* v. p. 102.

CRINOIDEA.

Variation from the pentamerous condition has been many times observed, though considering the vast number of specimens collected it must be a rare occurrence. In *Tetracrinus* the four-rayed condition is normal, and it is an especially interesting circumstance that in this form Variation to both a five-rayed and to a three-rayed condition has been observed. For nearly all the references to the following facts I am indebted to the useful collection of evidence on the subject given by BATHER, F. A., *Quart. Jour. Geol. Soc.*, 1889, p. 149.

Four-rayed varieties of five-rayed forms [1].

644.   **Holopus rangi.** This genus was originally described from a 4-rayed specimen by D'ORBIGNY, *Mag. de Zool.*, 1837, Cl. x., Pl. III. Subsequently, 5-rayed examples were obtained and this condition was found to be normal (see CARPENTER, *Chall. Rep.*, XI., Pt. XXXII., p. 197).

645.   **Eugeniacrinus:** departure from 5-rayed condition very rare. Among many hundreds of calyces in Brit. Mus. one only is 4 rayed, BATHER, *l. c.*, p. 155.

646.   **E. nutans:** 4-rayed specimen at Tübingen figured in QUENSTEDT'S Atlas to *Petrefactenk. Deutschl.* Taf. CV., figs. 179—181. Another case GOLDFUSS, *Petrefacta Germaniæ*, I., p. 163, Pl. I., fig. 4, now in Poppelsdorf Mus., Bour. (Bather).

647.   **E. caryophyllatus:** 4-rayed specimen seen at Stuttgart. Such a specimen [? the same] ROSINUS, *Tentaminis de Lithozois...Prodr.* &c., tab. III. (Hamb. 1719). Another case GOLDFUSS, *l.c.*, fig. 4: now in Poppelsdorf Mus. (Bather).

648.   **Pentacrinus:** a 4-rayed stem-joint from the Chalk, MANTELL, G. A., *Geol. of Sussex*, 1822, p. 183 : now in Brit. Mus., E. 5501 (Bather).

649.   **Pentacrinus jurensis:** 4-rayed specimen. The stalk had only 4 sides, one being quite flat. This flat side had an articulation for a cirrus. DE LORIOL, P., *Paléont. Franç. Terr. jur.*, Ser. 1, Paris, 1886, p. 112, Pl. CXLIV., *fig.* 6.

650.   **P. subsulcatus:** 6 joints of a 4-rayed stem, *ibid.*, p. 117, Pl. CXLV., *fig.* 2.

651.   **P. dumortieri:** 8 joints of a 4-rayed stem, *ibid.*, 1887, p. 186, Pl. CLXII., figs. 6 and 6 *a*.

652.   **P. dubius:** 4-sided stem quite regular. Basle Mus., BATHER, *l.c.*, p. 168.

653.   **Balanocrinus subteres:** 4-sided stem quite regular. *ibid.*

654.   **B. bronni:** "the articular surface shows 4 sectors quite regularly disposed; this peculiar character is continued over the whole series of joints, 26 in number." *ibid.*

655.   **Encrinus fossilis:** a 4-rayed calyx, &c., v. STROMBECK, A., *Ztschr. d. deut. geol. Ges.*, I., 1849, p. 158 *et seqq.* See also *Palæontographica*, 1855, IV., p. 169, Pl. XXXI. figs. 1 and 2.

656.   **E. fossilis:** two 4-rayed calyces with mutilated arms, v. KOENEN, *Abh. k. Ges. d. Wiss.*, Göttingen, 1887, XXXIV. *Phys. Kl.*, p. 23.

657.   **Antedon rosacea:** 4-rayed specimen, CARPENTER, P. H., *Chall. Rep.*, XXVI. Pt. LX. p. 27. Four-rayed Japanese specimen, *ibid.* Another in Brit. Mus. *ibid.*

[1] 4-sided stem joints undetermined. PUSCH, *Polens Paläont.*, 1837, p. 8, Pl. II. fig. 8, *a, b, c, d.* See also AUSTIN, *Ann. and Mag. N. H.*, 1843, XI. p. 203.

658. **Actinometra paucicirra :** 4-rayed specimen, *ibid.* "In all these [Nos. 657 and 658] the anterior ray (A) is missing, so that the mouth, instead of being radial in position is placed interradially between the rays E and B." CARPENTER, *l. c.*

Compare the following case of *imperfect* change towards the 4-rayed state :

659. **Cupressocrinus crassus :** abnormal calyx (now referred to this species, see BATHER, *l. c.*, p. 169) has one segment of the calyx reduced in size and bearing no radial plate or arm. This reduced segment is covered in by the adjacent segments so that the calyx as a whole is regularly 4-sided. GOLDFUSS, *Nova Acta Ac. C. L. C.*, 1839, XIX. p. 332, Pl. XXX., figs. 3 *a* and *b* [cp. No. 665].

Six-rayed varieties of five-rayed forms.

660. **Actinometra pulchella :** doubtful case of six rays, CARPENTER, *l. c.*
661. **Antedon** sp. Six-rayed specimen. "The additional ray is inserted between the two of the right side (D and E)." CARPENTER, *l. c.*
662. **Rhizocrinus lofotensis :** 6-rayed specimen. Four and six rays stated to be more common in *Rhizocrinus* than in other recent Crinoids ; seven rays are also found, but very rarely. In *Pentacrinus* no 6-rayed specimen seen. CARPENTER, P. H., *Chall. Rep.*, XI. Pt. XXXII. p. 38, Pl. VIII. *a*, figs. 6 and 7.
*663. **Pentacrinus jurensis** (probably) : stalk with 6 sides. [Fig. represents two adjacent lobes of the stalk as smaller and closer together than the rest, suggesting that perhaps these two may correspond with one lobe of the normal.] DE LORIOL, *l. c.*, Pl. CXLIV. fig. 7.
664. **P. jurensis :** 6-sided stalk having two adjacent lobes *larger* than the others. *ibid.*, fig. 10.

The following is a case of *imperfect* change towards the six-rayed state :

665. **Sphærocrinus geometricus :** abnormal specimen having the basal plate irregularly six-sided by reason of the flattening of the external angle of an infra-basal piece. Three of the sides are normal and each of these bears a normal parabasal ; but of the other three sides two are rather shorter than the normal sides and each of them bears a somewhat smaller parabasal. Upon the sixth side between these two, is a still smaller parabasal. The radials are five as usual, but one of them articulates with the smallest parabasal and in connexion with this its form is changed [for details see original figure]. Sculpture, &c. normal. ECK, H., *Verh. naturh. Ver. preuss. Rheinl.*, 1888, Ser. 5, v. p. 110, *fig.*

Three-rayed and five-rayed varieties of a four-rayed form.

*666. **Tetracrinus moniliformis :** normally 4-rayed (as shewn in Fig. 131, I.). A 3-rayed basal from the same locality, Birmensdorf (Fig. 131, II.). A 5-rayed basal from Oberbuchsitten (Fig. 131,

III.). DE LORIOL, P., *Mém. Soc. paléont. suisse*, 1877—1879, p. 245, Pl. XIX. figs. 39 b, 40 a, 41 a.

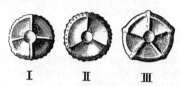

FIG. 131. I. Normal four-rayed basal of *Tetracrinus moniliformis* (from Birmensdorf). II. A three-rayed basal of the same species from the same locality as I. III. A five-rayed basal of the same species from Oberbuchsitten.
(After P. DE LORIOL.)

*667.   **Cupressocrinus gracilis.** This form has normally a 5-rayed calyx, and a 5-sided basal plate containing only 4 canals round the central canal (Fig. 132, I.). Varieties have been seen in which

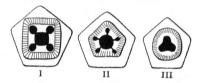

FIG. 132. *Cupressocrinus gracilis.* The normal form of the basal is shewn in I. A form with five canals round the central is represented in II, and in the specimen shewn in III there are three peripheral canals. See No. 667 a.
(After L. SCHULTZE.)

there are 5 (Fig. 132, II.), or 3 (Fig. 132, III.) such peripheral canals. The stalk is normally 4-sided, but in the varieties it is either 3- or 5-sided in correspondence with the number of canals.

667 a.   **C. elongatus:** stalk may be either 4- or 5-sided. The species *C. inflatus* has *normally* 3 canals in the (circular) stalk. SCHULTZE, L., *Denkschr. Ak. Wiss., Math.-nat. Cl.*, 1867, XXVI. pp. 130 and 136, Pl. I. fig. 2 b, and Pl. III. figs. 2 c and 2 i. [Cp. No. 667.]

668.   Abnormalities in the manner and frequency of branching in the arms of Crinoids leading to great numerical variation have been often recorded. See CARPENTER, *Chall. Rep.*, XXVI. Pt. LX. p. 28; *id. Phil. Trans.*, 1866, Pt. 2, p. 725 *Pl.*, also a case of *twelve* arms in **Antedon rosacea**, the abnormality not being symmetrical, [DENDY, *Proc. R. Phys. Soc. Edin.*, IX. p. 180, Pl.; also case of **A. rosacea** having abnormal branches in two arms symmetrically placed with regard to the axis. BATESON, W., *P. Z. S.*, 1890, p. 584, fig. 4 (now in Coll. Surg. Mus.). The abnormal arms were b_2 and e_1 of the usual nomenclature, as shewn in Fig. 133. For details see original description.

*669.   **Encrinus liliiformis:** amongst other abnormalities case given in which one of the radii bore *only one arm.* v. STROMBECK, *Palæont.*, IV. p. 169, Pl. XXXI. fig. 3.

CHAP. XVII.] ASTEROIDEA. 439

ASTEROIDEA.

670. Symmetrical change in number of rays is common in some of the forms. **Asterias rubens** and **A. glacialis** are frequently seen with 6 or with 7 arms symmetrically arranged, and I have

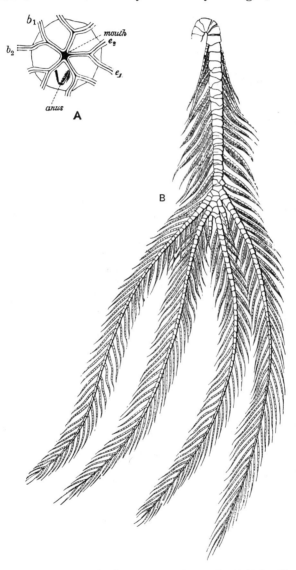

FIG. 133. *Antedon rosacea* having two arms abnormally divided. The figure A shews the relations of the two abnormal arms, b_2 and e_1, to the mouth and anus. B shews the arm b_2. (From *Proc. Zool. Soc.*)

seen one with 8. Individuals with 4 arms occur, but are much less common than those with 6. I have seen **Asterina gibbosa** with 4 rays, and a specimen (Scilly) given me by Mr S. F. Harmer has 6 rays, of which 2 are a little nearer together than the others (suggesting division of a ray). Mr E. W. MacBride tells me that he has seen several 6-rayed specimens of this species. Mr E. J. Bles kindly tells me that he dredged a 4-rayed **Porania pulvillus** in the Clyde estuary. There appeared to be no trace of a fifth ray and the specimen was as nearly as possible symmetrical.

The following cases exhibit special points.

671. **Asterias glacialis:** specimen with 8 rays possessed 3 madreporites. COUCH, J., *Charlesworth's Mag. of N. H.*, 1840, IV. p. 34.

672. **Asterias rubens:** 6-rayed specimens frequent at Wimereux. In several of these there are *two sand-canals* terminating at a common madreporite. GIARD, A., *Comptes rendus*, 1877, p. 973; cp. id. *C. R. soc. biol.*, 1888, p. 275.

673. *Partial division* of an arm is fairly common in Asteroids, but less common I believe than the total variation in number, though I know no statistics on this point. For a figure of **Asterias (Hippasterias) equestris** L. with a bifid arm, presenting no appearance as of regeneration see TIEDEMANN, *Zeitschr. f. Phys.*, 1831, IV. p. 123, Plate 1.

The two following are peculiar cases.

674. **Cribrella oculata:** one of the arms bearing a branch, not as a radius, but about (in dried specimen) at right angles to the normal arm, the property of Prof. C. STEWART, who kindly shewed it to me.

675. **Porania pulvillus,** Gray (a Starfish): Specimen 5 cm. in diameter, having five short rays. The ray opposite the madreporite *when viewed from the aboral surface* is seen to be distinctly bifurcated at

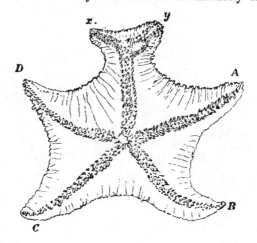

FIG. 134. *Porania pulvillus*, No. 675, having the arm opposite the madreporite abnormally divided as shewn at *x* and *y*. (From a sketch kindly sent by Prof. HERDMAN.)

about 1 cm. from its termination. The ambulacral groove of (Fig. 134) this abnormal ray divides into two branches at a distance of 2 cm. from the edge of the mouth. One of these branches runs along one of the forks of the ray to its extremity without further complication; but the other branch, belonging to the second fork, divides again 2 mm. from the first bifurcation, so as to form two tracts which unite with one another 3 mm. further on, thus inclosing a small piece of the ordinary integument in an ambulacral area. Finally, this ambulacral area divides once more close to the tip of the ray. There are no signs of injury or disease in the specimen. HERDMAN, W. A., *Nature*, 1886, XXXI. p. 596. [I am indebted to Professor Herdman for the accompanying diagram of this specimen.]

ECHINOIDEA.

In the Echinoids there are (1) cases of total Variation to a 4-rayed form with 4 ambulacra and 4 interambulacra[1]: (2) cases of partial or total disappearance of a definite ambulacrum or interambulacrum, which can be named either because part of it is present, or because two sets of similar plates thus become adjacent: (3) a case of total variation to a 6-rayed form: (4) cases of imperfect reduplication of a radius, thus forming an imperfectly 6-rayed form.

(1) *Total Variation to a 4-rayed form.*

676. **Cidarites coronatus?** : 4-rayed regular specimen (Fig. 135). MEYER, A. B., *Nova Acta C. L. C.*, XVIII. 1836, p. 289, Pl. XIII.

FIG. 135. *Cidarites coronatus?* No. 676, a regularly 4-rayed specimen from oral surface. (From A. B. Meyer.)

*677. **Echinoconus (Galerites) subrotundus:** 4-rayed specimen in Woodwardian Mus. (Fig. 136). The ambulacral and interambulacral areas are relatively wider than in a normal of the same size, the space of the areas that are wanting being as it were shared among those that are present. Apical disc roughly rectangular, and seems to be composed of 4 perforated basals (genitals) and 4 perforated radials (oculars). The basal plate corresponding to the posterior unpaired interambulacral area is perforated, though normally imperforate. Statement made that

---

[1] CUÉNOT, *Arch. de Biol.*, 1891, XI. p. 632, says that **Echinoconus vulgaris** has been seen with only *three* radii, but no authority is given.

the parts missing are those which lie on the left side of a line drawn through the middle of the anterior single ambulacrum and the posterior

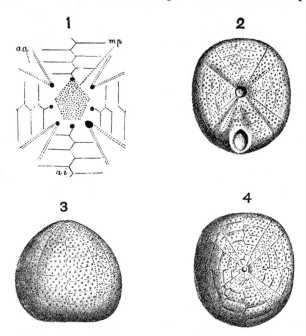

FIG. 136. *Echinoconus subrotundus* having 4 rays, No. 677. (From ROBERTS, *Geol. Mag.*, 1891.)
1. View of apical system. 2. Seen from side. 3. From apex. 4. From below. *aa*, anterior ambulacrum [?]. *mp*, madreporite. *ai*, anal interradius.
The parts are lettered after Roberts.

unpaired interambulacrum, but it is not possible to say which of the paired areas of this side are wanting, as the pores in the ambulacral plates round the peristome are indistinctly shewn. ROBERTS, T., *Geol. Mag.*, 1891, Dec. III., VIII. p. 116, figs.

678.   **Discoidea cylindrica :** a 4-rayed specimen, absolutely symmetrical. There are only 4 oculars corresponding with the 4 ambulacra. COTTEAU, G., *Pal. franç.*, 1862—67, VII. p. 31, Pl. 1011, figs. 6 and 7. [This is exactly like ROBERTS' case No. 677 and is illustrated by beautiful figures (*q.v.*). Cotteau in describing it says that the *anterior* ambulacrum is wanting. It is difficult to see any sufficient reason for the determination that this ambulacrum in particular is wanting. For in this case there are only 4 sets of interambulacral plates as well as 4 ambulacral areas in perfect symmetry. The anus of course lies between two ambulacra; and as the whole number is even and the radii are symmetrically arranged, there is thus no ambulacrum in the plane of the anus. Hence the suggestion that it is the anterior ambulacrum which is wanting. But if by Variation an Echinid has 4 symmetrical radii it would *always seem* that the

anterior ambulacrum was missing, whether it be the anterior ambulacrum, or the left anterior, or the left posterior that is wanting, or even if all 4 new ambulacra correspond with all 5 of the normal.]

679.   **Amblypneustes** sp. (S. Australia): four specimens, each with four ambulacra [no description or statement as to symmetry]. HAACKE, *Zool. Anz.*, 1885, p. 505. (See No. 687.)

(2) *Partial or total disappearance of a definite ambulacrum or interambulacrum.*

*680.   **Echinus melo**, having only four complete ambulacral areas (Fig. 137). The specimen is not spherical, for the apical system is warped over in one direction and the oral pole is pulled in an opposite direction, while the shell is much higher in the region of the apical system than it is at the opposite side. There are only four ocular plates, which are subequal, the madreporic plate and the plate opposite to it being somewhat larger than the other two. The genital plates are also four. Only four ambulacral areas leave the apical system, and at that point they are almost symmetrically disposed. Lower down however a triangular series of plates bearing ambulacral pores is intercalated between the plates of one of the interambulacral systems which it divides into two. This intercalated series is of course the representative of the ambulacral area which is wanting at the apex of the shell. The *five* ambulacra are nearly symmetrically disposed round the oral surface just as the *four* ambulacra are round the apical system. This transition from a tetramerous to a pentamerous symmetry is effected by complementary changes in the amount of divergence of the rays as they pass down the shell. Examination shews that the ambulacrum which is thus partially absent is the right posterior. PHILIPPI, *Arch. f. Naturg.*, III. p. 241, Plate.

*681.   **Amblypneustes formosus:** a 4-rayed specimen having a somewhat asymmetrical test. One of the interambulacral regions is abnormally wide, and at about 9 plates down the side of the test in this region a wedge-shaped piece composed of several partially distinct plates bearing 7 pairs of ambulacral pores. This fragment doubtless represents the deficient ambulacral area. The apical system consists of 10 plates. The two genital plates of the abnormal area are reduced in size, and the ocular plate between them is abnormally large. Considering the madreporic plate as indicating the right anterior interambulacrum, it appears that it is the left anterior ambulacrum which is thus deficient. The height of the shell at the abnormal side is less than at the other. BELL, F. JEFFREY, *Jour. Linn. Soc.*, xv. p. 126, Plate.

In each of the foregoing the missing ambulacrum is actually at some place represented by plates of ambulacral character, and the shape of the test is greatly changed in correlation with the partial disappearance of the radius. The following cases differ, in that in them one ambulacrum is *wholly* wanting in the affected radius, and the interambulacra are contiguous with each other. Curiously enough in two of these specimens the symmetry is changed little or not at all. The cases in *Hemiaster* were all Algerian fossils [1].

[1] Besides those here given in the text, GAUTHIER in the same place describes an interesting case of symmetrical reduction in the two posterior ambulacra of **Hemiaster africanus.**

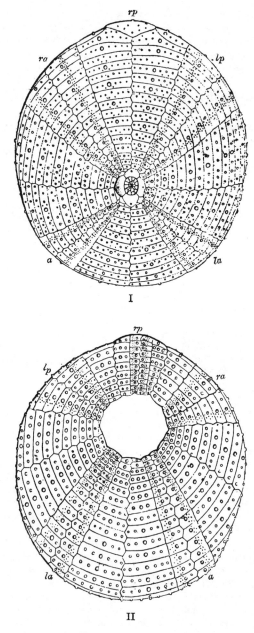

Fig. 137. *Echinus melo*, No. 680, having the right posterior ambulacrum partially absent. *a*, anterior ambulacrum. *ra, la*, right and left anterior ambulacra. *rp, lp*, right and left posterior ambulacra. I. View from apex. II. View from oral surface. (From Philippi.)

682. **Hemiaster batnensis:** specimen in which the *left posterior* ambulacrum is not present, and the ambulacral groove is only indicated by a shallow depression, beyond which there are some rounded pores which continue the ambulacral area beyond the fasciole. The corresponding ocular seems to be absent. The test is of normal form, but the median suture of the right posterior interambulacrum is not quite straight. GAUTHIER, M. V., *Comptes rendus de l'Ass. pour l'av. des sci.*, 1885, XIII. p. 258, Pl. VII. fig. 1.

683. **H. batnensis:** very similar case of absence of right anterior ambulacrum and corresponding genital and ocular plate. *ibid.*, fig. 3.

684. **Hemiaster** sp.: left anterior ambulacrum wanting and is gone without trace. There are only 4 oculars and 3 genitals. In correspondence with this variation there is considerable change in symmetry of the test, which is irregular, the anterior and right anterior ambulacra being deflected from their normal courses. [See details.] *Ibid.*, figs. 4 and 4 *bis*. [Here, where there is a clear differentiation between the several ambulacra, it is doubtless possible to affirm that such a definite ambulacrum is missing, for the two interambulacra are left adjacent to each other.]

685. **Echinus sphæra** (O. F. Müller): specimen described in which the *left posterior interambulacral* series of plates is almost entirely absent. The details of the structure are as follows: the genital plate which stands at the head of the left posterior interambulacrum is reduced in size in all directions; but the two ocular plates which should be separated by it are somewhat enlarged, bearing several extra tubercles, and meet together peripherally to the genital plate. The series of interambulacral plates which should begin from this genital plate are represented by a rudimentary row of small tubercles: the ambulacral systems which are normally separated by these plates are consequently almost contiguous. The rudimentary interambulacral series widens somewhat at a short distance from the apical series and forms a small island of interambulacral structure bearing 4 large tubercles. Beyond this, viz. at a point placed about $\frac{1}{3}$ the distance from the apex to the oral surface, the two ambulacra again unite and are continued as a single ambulacrum of double width. DÖNITZ, W., *Müller's Arch. f. Anat. u. Phys.*, 1866, p. 406, Pl. XI.

(3) *Case of total Variation to a 6-rayed form.*

*686. **Galerites albogalerus**(?): a regularly 6-rayed specimen having six symmetrical ambulacra and interambulacra (Fig. 138). MEYER, A. B., *Nova Acta Ac. Cæs. Leop. Car.*, XVIII. 1836, p. 224, Pl. XIII.

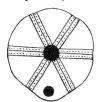

FIG. 138. *Galerites albogalerus*, No. 686. A six-rayed specimen. (After MEYER.)

687. **Amblypneustes** (S. Australia): 6-rayed specimen [no description or statement as to symmetry]. HAACKE, W., *Zool. Anz.*, 1885, p. 505. (See No. 679.)

(4) *Cases of imperfect reduplication of a radius.*

*688. **Amblypneustes griseus**: having one of the ambulacra doubled (Fig. 139); the apical system was normal. The width of the anterior ambulacral region was almost double that of the others: it contained two ambulacra lying side by side, each, as usual, composed of a double row of plates with an ambulacral area and two poriferous zones. The

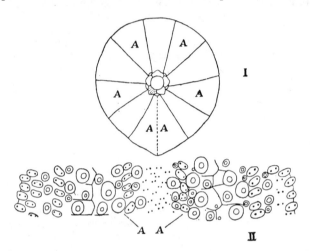

FIG. 139. *Amblypneustes griseus*, No. 688. Specimen having the anterior ambulacrum doubled. I. The test seen from the apex. II. Details of anterior ambulacrum shewing combined poriferous zones between *A* and *A*. The dotted line bisects the ambulacrum of double width. (After STEWART.)

areas and external poriferous zones are like those of a normal ambulacrum; but the poriferous zones which touch one another are fused together, with the pores irregularly arranged. The combined poriferous zones are not quite equal to the sum of two normal ones. The whole of this area, formed of the union of two ambulacra, projects as a ridge which is continued down the whole of the side of the shell. STEWART, C., *Jour. Linn. Soc.*, xv. p. 130, Pl.

689. **Hemiaster latigrunda**: right posterior ambulacrum double, the two resulting ambulacra are closely adjacent peripherally and a small interambulacral area is formed between them in their more central parts. There are 6 oculars but no extra genital. GAUTHIER, *l. c.*, figs. 5 and 5 *bis*.

690. **Hemiaster batnensis**: right anterior ambulacrum double, the two ambulacra are in contact through all their length. COTTEAU, *Pal. franç.*, 1869, p. 150, Pl. xx., and GAUTHIER, *l. c.*

[For interesting evidence as to variation in the number of *genital pores* on the costals in several genera of Echini, see LAMBERT, *Bull. Soc. Yonne*, 1890, XLIV. Sci.

nat., p. 34; also GAUTHIER, *Comptes rendus Ass. fr. pour l'av. Sci.*, Toulouse, 1887, and other references given by these authors.]

OPHIUROIDEA.

Individuals with various numbers of arms are often seen, especially in the genera *Ophiothela, Ophiocoma, Ophiacantha* and *Ophiactis,* and in many of the species there are most usually six arms. In these forms the evidence as to Meristic Variation is complicated by the circumstance that in several of them change in the number of arms may take place in the ontogeny, by division and subsequent regeneration (see note on p. 433).

CHAPTER XVIII.

BILATERAL SERIES.

OF the organs repeated in Linear Series whose variations have been illustrated, many are bilaterally repeated also; but thus far we have considered them only in their relations as members of Linear Series. It now remains to examine the variations which they exhibit in virtue of their relation to each other as members of a Bilateral Series.

Meristic Variation in this respect is manifested in two ways. A normally unpaired organ standing in the middle line of a bilateral symmetry may divide into two so as to form a pair of organs; and conversely, a pair of organs normally placed apart from each other on either side of a middle line may be compounded together so as to form a single organ in the middle line.

In animals and plants nothing is more common than for different forms to be distinguished from each other by the fact that an organ standing in the middle line of one is in another represented by two organs, one on either side. The facility therefore with which each of these two conditions may arise from the other by discontinuous Variation is of considerable importance.

Admirable instances of the bearing of this class of evidence upon the question of the origin of Species are to be seen in zygomorphic flowers. *Veronica* for example differs from the other Scrophulariaceæ especially in the fact that it has only one posterior petal, instead of two posterior petals one on each side of a middle line. But there is evidence not only that forms having normally two posterior petals may as a discontinuous variation have only one such petal, placed in the middle line, but also that the single posterior petal of *Veronica* may as a variation be completely divided into two. Similarly the single *anterior* petal of *Veronica* may also as a variation be divided into two, thus giving three posterior and two anterior petals as in for example *Salpiglossis*[1]. In these cases, which might be indefinitely multiplied,

[1] An account of several discontinuous variations in the structure of zygomorphic corollas was given by Miss A. BATESON and myself. *Jour. Linn. Soc.*, 1892, xxviii., *Botany*, p. 386.

there is thus a clear proof that so far as the variations in number and symmetry are concerned, the transition from the one form to the other *may* be discontinuous.

Analogous phenomena in animals are so familiar that general description of them is for the most part not needed, and an account will only be given of a few less known examples both of union and of division of such parts. Besides these strictly Meristic Variations in the amount of separation between the two halves a few examples are introduced in further illustration of the relationship that subsists between the two halves of a bilateral animal.

In considering the evidence both of median union and of division it must be remembered that the germs of most of the organs in question are at some time of their developmental history visibly double, and that when organs that should normally unite to form single median structures are found double in older stages, this duplicity is strictly speaking only a persistence of the earlier condition. But to appreciate this comment it should be extended. For, in every animal in which at some period of the segmentation of the ovum, the plane of one of the cleavages corresponds with the future middle line, all median organs must in a sense be paired in origin, and the distinction between paired and median organs is thus seen to be only one of the degree or amount of separation between the symmetrical halves. Nevertheless the evidence of Variation bears out the expectation that would be formed on examination of normal diversities between species or larger groups both in animals and plants, namely that whenever structures are geometrically related to each other as optical images, instability may shew itself as Variation in the degree to which such parts unite with or separate from each other. It is remarkable that this instability appears as much in the case of organs bilaterally symmetrical about an axis of Minor Symmetry as it does in the parts paired about the chief axis of Symmetry of the whole body.

Examples of such Variation in bilaterally symmetrical parts of a Minor Symmetry have been already given in the case of the feet of the Horse and of the converse phenomenon in the feet of Artiodactyles (*q.v.*).

A good illustration of the way in which duplicity about an axis of Minor Symmetry may pass into the unpaired condition is seen in the case of ocellar markings on bilaterally symmetrical feathers. By comparing different feathers on several species of *Polyplectron*, DARWIN found that it was possible to find most of the gradations between the complete duplicity shewn in Fig. 140, I, where each half of the feather bears an almost symmetrical ocellus, and the partially confluent condition shewn in Fig. 140, II, which is not far removed from the state of the ocellus in the Peacock's tail-coverts, where the whole ocellus has no peripheral

450   MERISTIC VARIATION.   [PART I.

indentation and is very nearly symmetrical about the rachis of
the feather, though each of its *halves* has no axis of symmetry.

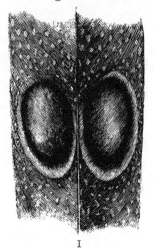

I   II

Fig. 140. I. Part of tail-covert of *Polyplectron chinquis*, with the two ocelli of
nat. size. II. Part of tail-covert of *Polyplectron malaccense*, with the two ocelli
partially confluent, of nat. size.
(From C. Darwin, *Descent of Man*, 1871, II. p. 139, *figs.* 54 and 55.)

Attention should be called to the fact that abnormal division along
a middle line may in many cases represent one of two different phenomena which are not readily distinguishable. For when a normally
single organ is represented by two, standing on either side of a middle
line it is often possible that there may be not only a division of the
organ but a partial duplicity of the axis. These two conditions are of
course morphologically distinct; for in the case of division of the organ
only, the two parts are still in symmetry about the original axis of
Major Symmetry of the body, but in the case of duplicity of the axis
there are two equivalent axes of symmetry, about which each half is
separately symmetrical. But though this distinction is in a sense a
real one it cannot be applied to cases of duplicity occurring in any
organ whose halves assume a bilaterally symmetrical form when separate. For example in the case of the foot of the Horse, or of the
hæmal spines &c. of Gold-fishes (*v. infra*), when division occurs, each of
the two halves is only hemi-symmetrical, and this duplicity is no more
evidence that the axis is double than is the ordinary double condition
of the vertebrate kidney; but in the case of duplicity of the central
neural canal in Man for instance, or in the case of the tail-spine of
Limulus described below, it is not clear that there is not a partial
duplicity of the axis.

Division or absence of union in Middle Line.

Most of the organs which in a vertebrate stand in a median
position have been seen more or less often in a divided condition.

Examples of such division in the middle line were, I believe, first put together by Geoffroy St Hilaire, and a very full collection of the evidence seen in Man is given by Ahlfeld[1]. The organs most often divided are the sternum, neural arches, uterus, penis, &c., and of these, specimens may be seen in any pathological collection. Organs more rarely divided are the tongue[2], epiglottis[3], uvula[4], and central neural canal[5]. The following are special cases of variation consisting in a median division.

Division of caudal and anal fins in Gold-fishes.

*691. **Cyprinus auratus** (Gold-fish). The following account of the multiple fins of Gold-fishes in China and Japan is taken chiefly from Pouchet[6] and Watase[7]. There is evidence to shew that these animals were first imported to Japan from China.

Three distinct breeds of Gold-fishes are kept in Japan. The first, called "*Wakin*" has a slender body closely resembling that of the common carp. The second "*Maruko* or *Ranchiu*" has a very short body, being in some cases almost globular in shape and in it the dorsal fin is generally entirely absent. The head is usually disfigured by rough-looking protuberances of the skin which often attain a considerable size.

The third or "*Riukin*" has a short body with a rounded abdomen. Of all the breeds, this has the most beautiful tail which is very large and often longer than the rest of the body.

Gold-fish breeders of the present day can freely produce the "*Riukin*" or "*Maruko*" from the "*Wakin*." Various intermediate forms between the above-mentioned breeds exist.

In all gold-fishes, irrespective of the breed to which they belong, the tail-fin is, above all other parts, subject to the greatest variation. It is to be found in one of the following three states;

(1) It is vertical and normal.

(2) It may consist of two separate halves; each of these halves is to all appearance a complete tail and the two tails pass backwards side by side, *but are united dorsally at the point where they join the body.*

(3) The two tails thus formed are united by their dorsal edges to a variable degree and their lower edges may be bent outwards, so that the two combined tails come to be spread out into a three-lobed, nearly horizontal fin.

---

[1] Ahlfeld, F., *Missb. d. Menschen*, 1880.

[2] Partsch, *Bresl. Ärztl. Ztsch.*, 1885, No. 17; Pooley, *Amer. Jour.*, 1872, N.S., cxxvi. p. 385 [from Ahlfeld, p. 119].

[3] Manifold, W. H., *Lancet*, 1851 (1), p. 10; French, *Ann. Anat. Surg. Soc. Brooklyn, N. Y.*, 1880, ii. p. 271 [not seen], from *Cat. Libr. Surg.-gen. U. S. Army.*

[4] Trélat, *Gaz. des Hôp.*, 1869, No. 125 [for others v. Ahlfeld, *Abschn.* ii. p. 175].

[5] Wagner, J., *Müll. Arch. Anat. Phys.*, 1861, p. 735, Pl. xvii. A.

[6] Pouchet, G., *Jour. de l'anat. et phys.*, vii. p. 561, Pl. xvii.

[7] Watase, S., *Jour. Imp. Coll. Sci. Tokio*, i. p. 247, *Plates*.

Besides the caudal fin, the anal fin undergoes remarkable

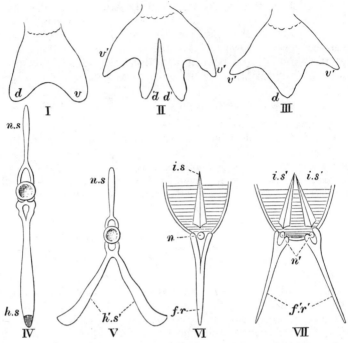

FIG. 141. Caudal and anal fins of Gold-fish (*Cyprinus auratus*).
I. Normal tail, seen from side. v, dorsal lobe. d, ventral lobe. II. Abnormal form divided as far as the notochord. $v'\ v'$, two ventral lobes. $d'\ d'$, two dorsal lobes. III. Abnormal form, the two ventral lobes, $v'\ v'$, separate. IV. Penultimate vertebra of normal Carp (*C. carpio*). $n.s$, neural spine. $h.s$, hæmal spine. V. Penultimate vertebra of a Gold-fish with trilobed caudal fin. $h'.s'$, double hæmal spine. VI. Diagram of transverse section through region of anal fin of normal Gold-fish. VII. Similar section through a specimen having the anal fin doubled. $i.s$, interhæmal spine. $f.r$, fin ray. n, bony nodule. $i.s'$, $f'.r'$, n', corresponding parts doubled. (After WATASE.)

variation. It is either median and normal; or it may be distinctly paired (Fig. 141, VII).

There are all stages of caudal and anal fins, intermediate between the normal and the completely paired states. Thus the tail-fin with its lower portion alone in a double state, or the anal fin with either its anterior or posterior portion double and the remainder single, is of quite common occurrence. These different conditions of the two fins combine in various ways in different individuals thus giving rise to manifold varieties of form.

This doubling of the tail-fin consists essentially in a longitudinal splitting of the morphologically lower lobe of the tail. The first step in the process of doubling is seen in the case of gold-fishes in which there is a slight longitudinal groove in the ventral margin of the tail-fin. This groove may be extended up through all the rays of the lower lobe of the tail, which then consists of two tails side by side. *The small dorsal lobe, which lies above the notochord, is never involved in the process, but always remains single.* There is therefore in this case no doubling of the axis of the body. Examination of the skeleton shews that in those fishes which have two tails the hæmal spines of the last three vertebræ are longitudinally split[1] and diverge to carry the two tail-fins (Fig. 141, V).

POUCHET lays stress on the fact that the size of each of the paired tails is greater than that of the normal tail of a Gold-fish; but as Watase states that in the variety "*Riukin*" the tail may be as long as the body, it is clear that this hypertrophy may exist without any repetition.

In cases where the anal fin is doubled the process is exactly the same, resulting from a longitudinal splitting of the rays of which it is composed. This may only affect the outermost parts of the fin or may be carried up further so as to divide the interhæmal spines, in which case the two anal fins arise from the body wall at separate points and diverge from each other.

POUCHET, who has extensively studied the history of Gold-fishes in Europe, believes that it is almost certain that those which were brought to Europe in the eighteenth century were all more or less of the double-tailed order. He refers especially to the figure given by LINNÆUS[2] representing the double-tailed form as a normal.

POUCHET states that the evidence goes to shew that this anomalous race is not maintained in China by any rigid selection. He quotes a Chinese encyclopædia to the effect that the double-tailed Gold-fish is found in running streams, and gives the evidence of KLEYN[3], a missionary in China during the eighteenth century, who states that "*In fluvio Sleyn Cyprini sunt qui caudam habent trifurcam et a piscatoribus* Leid-brassen *vocantur, quasi diceres aliorum Cyprinorum conductores.*"

Though the duplicity of the hæmal spines may be unaccompanied by other variations it should be noticed that the extraordinary "Telescope" Gold-fish not unfrequently has also the double tail-fin. In the Telescope Gold-fish the eyes project from the orbit to a greater or less extent, in the extreme form being entirely outside the head and attached by a small peduncle only. The various forms of abnormal Gold-fishes are generally to be seen in large quantities in the shops of the dealers in aquariums &c. which abound near the Pont Neuf in Paris. One of these dealers told me that he bred considerable numbers of them every year, and that in fish from the same parents there was little uniformity, many normals being produced for one that shewed any of the extreme variations. It is recorded that of the Gold-fish hatched in Sir Robert Heron's menagerie about two in five were deficient in the dorsal fin and two in a hundred or rather more had a "triple". [sc. three-lobed as described above] tail-fin, and as many have the anal

[1] It should be observed that there is no want of original union between the hæmal spines, for these close in the hæmal canal as usual. The phenomenon is thus altogether different from that of spina bifida in the neural spines.

[2] *Fauna suecica*, 1745, p. 331, Pl. II.

[3] KLEYN, *Miss.*, v. p. 62, Tab. XIII. fig. 1 [not seen], quoted by BASTER, *Opusc. subsec.*, Harl., 1762, p. 91, *note.*

fin doubled. The deformed fishes were separated from the others but did not produce a greater proportion of varying offspring than the normals (*Ann. Mag. N. H.*, 1842, p. 533).

For a magnificent series of plates illustrating the various forms of Goldfishes see BILLARDON DU SAUVIGNY, *Hist. nat. des Dorades de la Chine*, Paris, 1780. [In Brit. Mus. copy text wanting; I do not know if it ever appeared.]

Division of median structures in Coleoptera.

The following list includes every case known to me.

I. EPISTOME.

692. **Anisoplia floricola** (Lam.): Algerian specimen having the epistome (*chaperon*) completely divided into two parts in the middle line. Attention is called to the fact that this is a *normal* character in certain genera of Lamellicorns, for example, *Diphucephala* and *Inca*. FAIRMAIRE, L., *Ann. Soc. ent. France*, 1849, Ser. 2, VII. *Bull.*, p. LX.

II. PRONOTUM[1].

In Coleoptera the pro-thoracic shield or pronotum is normally a single plate continuous from side to side. The following is a list of cases in which this structure was composed of two lateral parts. In Nos. 695 and 706 the division was not completed through the whole length of the shield. The two halves were in most cases symmetrical, but in Nos. 700 and 703 they were unequal.

As is shewn by No. 704 &c., there is in these variations more than a mere fault of union between two chitinous plates, for in this case the adjacent or inner edges of the plates were beset with yellow hairs such as occur on the anterior and posterior margins of the normal pronotum. In No. 703 again the adjacent edges of the two plates are everted and form definite margins.

693. **Melolontha vulgaris** (Lam.), prothoracic shield consists of two symmetrical pieces which do not meet in the dorsal middle line. The prothorax is greatly reduced in length and the head consequently is almost in contact with the scutellum (Fig. 142, I). KRAATZ, G., *Deut. ent. Ztschr.*, 1880, p. 341, Pl. II. *fig.* 8.

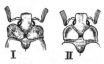

FIG. 142. *Melolontha vulgaris*, the Cockchafer, two cases of division of pronotum. (After KRAATZ.)

[1] With these cases compare the following: **Hydrobius fuscipes**, specimen having pronotum formed into *three lobes*, one being central, and two lateral. The lateral lobes projected from each side as considerable expansions. KRAATZ, G., *Deut. ent. Ztschr.*, 1889, p. 222, fig. 21.

694. A male, closely similar case (Fig. 140, II., *ibid.*, 1877, XXI. p. 57, *Taf.* I. *fig.* 2.

695. A male in which the pronotum was similarly divided, but the division was not quite complete. DE LA CHAVIGNERIE, *Ann. Soc. ent. France*, 1846, Ser. 2, IV., *Bull.*, p. XVIII., Pl. II., fig. II.

696. An almost identical specimen (male). MOCQUERYS, *Coléop. anorm.*, 1880, p. 140, fig. [Now in the Rouen Museum, where I have examined it.]

697. Another case; extent of division not specified. STANNIUS, *Müll. Arch. Anat. Phys.*, 1835, p. 304.

698. **Oryctes nasicornis** ♂ (Lam.): anterior part of pronotum divided into two parts by a longitudinal suture: posterior part of pronotum undivided. Head normal. *ibid.*, Pl. V. fig. 7.

699. **Onitis bison** (Lam.): pronotum divided in the middle by a longitudinal suture, the lateral pieces being raised up. *ibid.*

700. **Heterorhina nigritarsis** (Lam.): specimen in the Hope Collection at Oxford having the pronotum completely divided into two somewhat unequal halves, of which the left is the largest. The posterior angle of each of the pieces does not occupy its normal position, but lies internal to the outer border of the elytron. Owing to this disposition the mesothorax is exposed for a short distance on each side and for a considerable extent in the centre.

701. **Attelabus curculionides** (Rhyn.): specimen of moderate size; head, elytra and legs normal. Structure of prothorax peculiar in that the two lateral halves do not meet in the middle line, leaving betwixt them a membranous space. The prothorax is shortened and the head is pushed back into the thorax as far as the level of the eyes. The edges of the plates of the prothorax are well formed and properly finished. Scutellum present, but is not at all concealed by the prothorax. DRECHSEL, C., *Stettiner ent. Ztg.*, 1871, XXXII. p. 205.

702. **Chrysomela fucata** (Phyt.): Pronotum divided centrally into two parts, each of which is triangular. The parts of the head and scutellum which should be covered by the thoracic shield are thus exposed. KRAUSE, *Stettiner ent. Ztg.*, 1871, XXXII. p. 137.

703. **Telephorus nigricans** (Mal.): the pronotum is divided into two unequal halves. The left half is nearly twice as large as the right, and projects beyond the middle line, covering a part of the right side of the prothorax. The right portion is small and very concave. Both of these two parts of the pronotum are everted at their edges to form a definite margin. The margins are continued all round each piece, and thus two margins are adjacent in the contiguous parts of the plates. This specimen was kindly lent to me by M. H. GADEAU DE KERVILLE.

704. **Carabus scheidleri**: thorax dorsally covered by two completely separate and symmetrical plates, whose inner edges are beset with yellow hairs [as the anterior and posterior margins

normally are]. The rest of the animal was normal. KRAATZ, G., *Berl. ent. Ztschr.*, 1873, XVII. p. 430, fig.

705. **Carabus lotharingus:** thoracic shield divided in centre to form two triangular pieces which only unite at a single point. The head is drawn back into the thorax. DUPONCHEL, *Ann. Soc. ent. France*, 1841, S. 1, x., *Bull.*, p. xx., *Pl.*

706. **Lixus angustatus** (Rhyn.): thoracic shield partially divided, present a deep emargination both before and behind [description not quite clear]. DOUÉ, *Ann. Soc. ent. France*, 1851, IX. *Bull.*, p. LXXXII.

III. METASTERNAL PLATES.

707. **Rhizotrogus marginipes** ♀ (Lam.) having the abdomen deformed in a symmetrical manner. Looked at from the ventral surface the metasternal plates are seen to be divided in the middle line by a deep depression so that the abdomen consists superficially of two lobes; these two lobes are united together in the last segment in which the metasternal plate is undivided. The two lobes are of equal size and the longitudinal depression which divides them is shewn in the figure to be regularly and symmetrically formed. The animal is otherwise normal. [No dissection was made.] BAUDI, L. V., *Bull. Soc. Ent. Ital.*, 1877, IX., p. 220, fig.

IV. PYGIDIUM.

708. **Melolontha vulgaris** (Lam.): pygidium bifid, two cases. KRAATZ, G., *Deut. ent. Ztschr.*, 1880, p. 342, Pl. II., figs. 4 and 4 a; and *ibid.*, 1889, p. 222, Pl. I., fig. 19.

709. A case of "double proboscis" is recorded in **Sphinx ligustri**. The specimen was a pupa, and through the pupal skin it could be seen that the two mandibles had not united to form the single proboscis, but were divaricated. KRAATZ, *Deut. ent. Ztschr.*, 1880, xxIV., p. 345, fig.

Miscellaneous cases of doubtful nature.

710. **Ascidians.** Prof. W. A. Herdman tells me that he has several times met with Ascidians having a supplementary lateral atriopore. He regards this as a retention of a larval character, since in the young there are two atriopores which in normal individuals afterwards unite dorsally.

711. **Limulus polyphemus:** large specimen found at Fort Macon, N. Carolina, having a forked caudal spine (Fig. 143). This variation is

FIG. 143. *Limulus polyphemus* No. 711, having forked caudal spine. (After PACKARD.)

probably very rare. PACKARD, A. S., *Mem. Bost. N. H. S.*, 1872, II. p. 201, *fig.*

712. **Palamnæus borneensis** (Scorpion): specimen in which the terminal poison-spine was double, as shewn in Fig. 144. The two halves were not quite equal and there was no opening of a poison-gland on the shorter spine. This specimen, which is in the Brit. Mus. was kindly shewn to me by Mr R. I. POCOCK.

713. **Chirocephalus** ♀ : specimen having the generative sac with two horns instead of one. [Normally there is only one such horn which forms a median downward prolongation of the ovisac. No further description.] PRÉVOST, B., *Mém. sur les Chirocephales*, p. 232 ; in Jurine's *Hist. des Monocles*, Geneva, 1820.

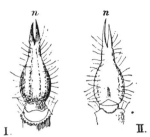

FIG. 144. Double poison-spine of a Scorpion (*Palamnæus borneensis*). I. From dorsal side. II. From ventral side. *n*, the spine which bore the openings of the poison-glands.

714. **Buccinum undatum.** A number of specimens were formerly obtained from Sandgate in Kent[1], having the operculum double. Sometimes the two opercula were separate, sometimes united. Many specimens of this variation are in the collection of Dr A. M. Norman, who kindly shewed them to me. The shells and opercula alone remain and consequently it is not now possible to determine the position of the line of division relatively to the morphological planes of the animal; but, from the fact that in several instances the two opercula were related to each other as images, it seems likely that the division was in the longitudinal median plane, though this must be uncertain. Moreover in one of Dr Norman's specimens, from the fragment of dried flesh adhering, it appeared that the apex of the foot might have been bifid. Four cases are shewn in Fig. 145. In two of them (I and II) there is a fairly close relation of images, while in III this relation is less clear and in IV it is practically destroyed, though it is of course quite possible that this may be the result of unequal growth. Several of these opercula are much contorted and without any very definite shape.

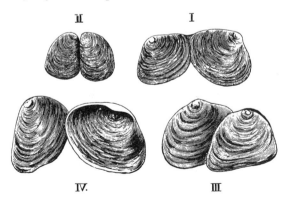

FIG. 145. Cases of duplicity in operculum of *Buccinum undatum*, from specimens in the collection of Dr A. M. Norman. I and II nat. size. III and IV enlarged. III and IV were kindly drawn for me by Mr J. J. Lister.

[1] See JEFFREYS, J. G., *Ann. Mag. N. H.*, 1860 (2), p. 152.

It was intended to have introduced here some account of the curious and very rare cases in which, for a greater or less region of the spine, corresponding half-vertebræ, on either side of the middle line, are not united together in their proper order, but I fear this would be too great a digression. For references on the subject see LEVELING, *Obs. anat. rarior.*, Norimb., 1787, Fsc. 1, cap. III. p. 145, Tab. V.; SANDIFORT, *Mus. anat.*, Leyden, 1835, IV. p. 74, Pl. CLXXVIII.; REID[1], *Jour. of Anat.*, 1887, XXI. p. 76, *fig.*; *Guy's Hosp. Rep.*, 1883, p. 132.

UNION OR ABSENCE OF DIVISION IN THE MIDDLE LINE.

This phenomenon is the converse of that described above. Examples of median union are found in many organs of different kinds. In vertebrates such union is especially well known in the case of the eyes, the ears, and the posterior limbs, producing the cyclopic, synotic and symmelian conditions respectively.

Each of these is of some interest to the student of Variation by reason of the symmetry and perfection with which the union takes place. In the cyclopian the degree to which the two eyes are compounded presents all shades intermediate between the perfect duplicity of the normal and the state in which the eye-balls are united in the middle line of the forehead and have one circular cornea[2]. These variations are closely comparable with those of the eye-spots on feathers referred to on p. 449; for there also all stages are seen between a pair of eye-spots placed one on either side of a middle line and complete union to form one eye-spot bisected by the middle line. There is of course no normal vertebrate having the eyes thus united in the middle line, but as MECKEL has remarked, the case of the cyclopian is not essentially different from that of the Cladocera in which the compound eyes, paired in other Crustacea, are united to form a single median eye. The cases No. 718 and 719 of median union of the compound eyes of Bees may also be considered in this connexion.

A very similar series of variations occurs in regard to the ears of vertebrates, which in the synotic or cephalotic condition are compounded in the middle line to a varying degree[3]. Such union of the ears is especially common in the Sheep, cyclopia being most frequent in the Pig. DARESTE[4] states that the first beginning of the cyclopian condition appears in the Chick as a precocious union of the medullary folds in the region of the fore-brain, occurring before the optic vesicles are fully formed from it. The degree to which the union of the eyes is complete then depends on the earliness with which the folds begin to meet relatively to the time of budding off of the optic vesicles. DARESTE[5] also declares that the cephalotic state is similarly first indicated by a premature union of the folds in the region of the medulla, taking place

[1] A case in Man, resembling No. 7.
[2] For an extensive collection of cases illustrating the various degrees of cyclopia see especially AHLFELD, *Missb. d. Mensch.*, Abschn. II. 1882.
[3] For figures see *e.g.*, OTTO, *Mus. anat. path. Vratisl.*, Pl. I. fig. 5, Pl. III. fig. 2 (Lambs); GUERDAN, *Monats. f. Geburtsk.*, x. p. 176, Pl. I. (Man) and many more.
[4] *Comptes rendus*, 1877, LXXXIV. p. 1038.
[5] *l. c.*, 1880, XC. p. 191.

before this part of the brain has widened out. In this way the auditory involutions are approximated. This account however cannot apply to all cases of union of ears; for the compounded ears are sometimes on the *ventral* side of the neck, as in Guerdan's case[1].

The body of the symmelian ends posteriorly in an elongated lobe made up of parts of the posterior limbs compounded together by homologous parts. The two femora are usually united to form a single bone, the tibiæ are separate and the two limbs are again compounded in the tarsal region. The axial parts posterior to the hind limbs are always greatly aborted[2].

Union of the kidneys in the middle line (Fig. 146), forming the "horse-shoe kidney" of human anatomists, is a similar phenomenon. As to the mode of development of this variation I know no evidence. Usually the kidneys together form a single horse-shoe shaped mass of glandular tissue, the union being posterior[3]; very

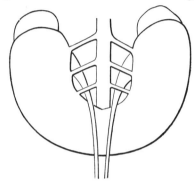

FIG. 146. Kidneys united in the condition known as "horse-shoe" kidney (Man). In this specimen there were three renal arteries on each side.
(From *Guy's Hosp. Rep.*, 1883.)

[1] See note 2, p. 458.
[2] See especially, MECKEL. *Arch. Anat. Phys.*, 1826, p. 273; GEOFFROY ST HILAIRE, *Hist. des Anom.*, ed. 1837, II. p. 23; GEBHARD, *Arch. Anat. Phys.*, 1888, *Anat. Abth.*, p. 164 (good fig.). To the determination of the morphology of the hind limb the structure of the symmelian monster is of unique importance, but I do not know that it has had the notice it deserves from comparative zoologists. From the manner of union of the parts of the two limbs may be obtained a positive proof of the morphological relations of the surfaces of the two limbs to each other. In a symmelian the feet are united by their fibular borders, *the minimi being adjacent*, the halluces exterior, and *the combined plantar surfaces ventral*. The great trochanters are *dorsal*, being often united into one in the dorsal middle line, and the patellæ are also *dorsal*, being also not rarely partly compounded. From these facts, even were other indications wanting, we have a proof that if the hind limbs were laid out in their original morphological relations to each other (as the tail-fins of a Crayfish may be supposed to be) the halluces would be external and anterior, the minimi internal and posterior, the flexor surfaces of the thigh and crus and the plantar surface of the (human) foot would be *ventral* and the extensor surfaces of the thigh and crus and the dorsum of the (human) foot would be *dorsal*. This is of course affirmed without prejudice to any question of phylogeny; but that these must be the ontogenetic relations of the parts is clearly proved by the symmelian.
[3] Sometimes anterior, *e.g.* ODIN, *Lyon méd.*, 1874, No. 12 [from *Canstatt's Jahresb.*, 1874, I. p. 19]; and FREUND, *Beitr. z. Geburtsh. u. Gyn.*, IV. 1875 [from *Canstatt's Jahresb.*, 1875, p. 340].

460   MERISTIC VARIATION.   [PART I.

rarely the posterior ends of the kidneys are joined by a bridge of ligamentous tissue[1].

A remarkable case, in which the union of the two kidneys was very complete and only indications of duplicity remained, is given by PICHANCOURT, *Gaz. hebd.*, 1879, p. 514.

Illustrative Cases.

To these familiar instances are added a few less generally known.

*715.   **Capreolus caprea** (Roebuck): specimen having the two horns compounded in the middle line, forming a common beam for almost the lower half of the horn (Fig. 147). This specimen was exhibited among a large series of abnormal horns in the German Exhibition held in London 1891. Casts of it are in the Brit. Mus. and Camb. Univ. Mus.[2].

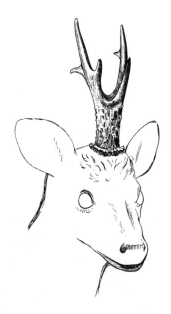

FIG. 147. A Roebuck (*Capreolus caprea*) No. 705, having the horns compounded to form one.

716.   **Limax agrestis** : specimen having the upper tentacles united into one in the middle line. The eyes were paired as usual. FORBES and HANLEY, *Hist. Brit. Moll.*, IV. p. 288 and I. Pl. JJJ, fig. 4.

[1] See GRUBER, *Virch. Arch.*, 1865, XXXII. p. 111.
[2] The original is at Darmstadt.

717. **Helix hispida:** specimen in which the tentacles were united together. They were adherent throughout, excepting for a slight cleft at the end, about one line in length. A shallow longitudinal suture was visible between the two. The animal and shell were otherwise normally formed. ROBERTS, G., *Science Gossip*, 1886, XXII. p. 259.

*718. **Apis mellifica** (Honey-bee): a worker having the two compound eyes continued up so as to unite on the top of the head (Fig. 148). The union between the eyes of the two sides was complete. There was no trace of any groove or division between them and the resulting structure was perfectly symmetrical. In a normal the three simple eyes are arranged in a triangle between

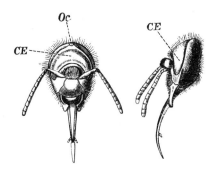

FIG. 148. A worker Bee (*Apis mellifica*) No. 708, having the two compound eyes united across the middle line, seen from in front, and from the side. *CE*, the united compound eyes. *Oc*, a single structure representing the three simple eyes of the normal. (After STANNIUS.)

the upper edges of the compound eyes, but in this specimen they were united into a single structure which was symmetrically placed in the middle line in front of the united compound eyes (Fig. 148, *Oc*). The body thus formed by the union of the simple eyes was a round projection beset with long yellowish hairs.

In a normal male the compound eyes are much larger and are in contact with each other at the top of the head, but the division between them is sharply defined. In a normal worker, however, the compound eyes are widely separated.

The facetting and the hairs on these eyes were normal and the animal was in all other respects properly formed. STANNIUS, *Müller's Arch. Anat. Phys.*, 1835, p. 297, *Pl.*

*719. **Apis mellifica** having the compound eyes completely and symmetrically fused. This individual was either a young and abnormally developed queen, or else a worker. Its structure was in several respects abnormal. The third pair of legs are like those of the workers, as is shewn by the structure of the first joint of the tarsus, the brush of hairs on the outside of the leg is not so much developed as in the workers, and this feature suggested that perhaps the specimen may be a young and abnormal queen. The abdomen is small and seems to have been arrested in its development, but its shape is that of the abdomen of the workers. The last segment of the abdomen is elongated, triangular, and slightly grooved in the middle of the posterior border, so as to permit the passage of the sting. The wings are more like those of a queen or worker than those of a male; for in the latter they generally greatly exceed the abdomen in length. The thorax is small, narrow, and contracted more than in the normal form, being also less convex. The space between the wings is less than in a fully developed bee. The antennæ are mutilated, but seem to have been normal; but their last joints are slightly reddish brown as they are in females, whether workers or queens, and not black as they are in drones. The two compound eyes were completely fused together in the middle line, across the place in which the simple eyes ought to be found. The simple eyes are not present at all. LUCAS, H., *Ann. Soc. Entom. France*, S. 4, VIII. 1868, p. 737, *Pl.*

CHAPTER XIX.

BILATERAL SERIES—*continued*.

FURTHER ILLUSTRATIONS OF THE RELATIONSHIP BETWEEN RIGHT AND LEFT SIDES.

I. *Variations in Segmentation of the Ovum of* Loligo.

The following facts, taken from WATASE[1], are introduced in further illustration of the mode of occurrence of bilaterally symmetrical Meristic Variation.

*720. **Loligo pealei.** In the blastoderm the nucleus is placed eccentrically, being rather nearer to the posterior pole, as shewn in Fig.

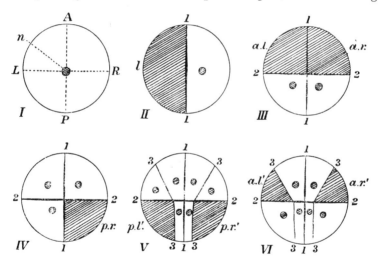

FIG. 149. Diagrams illustrating variations in segmentation of a Squid. (*Loligo pealei*).
I. Normal unsegmented ovum. *n*, the nucleus eccentrically placed. *A*, anterior. *P*, posterior. *L*, left. *R*, right. II, III, and IV. The shaded portions shew areas in which in some specimens nuclear division was precocious. V. In the two shaded areas triasters occurred in one specimen. VI. The blastomeres of the shaded areas in one specimen were not divided from each other. 1, 2, 3, successive planes of division. *ar*, anterior right quadrant. *pr*, posterior right quadrant. *ar'*, *pr'*, &c. areas separated off by the third segmentation-furrow. (After WATASE.)

[1] WATASE, S., *Jour. of Morph.*, IV. 1891, p. 247, Plates.

149, I. The first furrow, 1, 1, divides the blastoderm into two halves and corresponds with the future longitudinal middle line. The second furrow, 2, 2, is at right angles to this, dividing the blastoderm into anterior and posterior halves, and the third furrow, 3, 3, passes as shewn in Fig. 149, V.

In the subsequent segmentations various irregularities were seen in single eggs, some of the variations being bilaterally symmetrical while others were confined to a particular half or to a particular quadrant. For example, in some ova the nuclei of the cells formed from the left half of the blastoderm, excepting those next the median axis posteriorly (Figs. 149, II and 150, I), began to divide before those of the right side and reached an advanced stage of karyokinesis while the nuclei of the right half were still resting. The nuclei of each half kept time very nearly (for details see original figures). This curious variation was seen in three (perhaps four) ova all taken from one mother.

In another the nuclei of the two anterior quadrants al, ar, in their divisions kept ahead of those of the posterior quadrants. Fig. 149, IV. represents an ovum in which the nuclei of the right posterior quadrant on the contrary divided before those of the 3 other quadrants.

Another variation is shewn in Figs. 149, VI and 150, II. There the four blastomeres shaded had either been never fully divided from each other or had subsequently fused together symmetrically on each side.

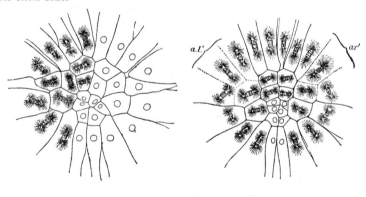

I   *II*

Fig. 150. Variations in segmentation of ovum of *Loligo pealei*. I. Case in which the nuclei of cells of the left half of the blastoderm began to divide precociously. II. Case in which the blastomeres of the areas ar' and al' were not distinct from each other. (After Watase.)

Fig. 149, V, illustrates another remarkable Meristic variation which symmetrically affected the portions shaded. In both of these shaded segments the nuclei divided into *three* by triple karyokinesis, forming "triasters."

II. *Homœosis in cases of normal Bilateral Asymmetry.*

In proportion as an animal is bilaterally symmetrical the right side is an image of the left. Nevertheless in many substantially symmetrical forms there is asymmetry in the condition of some one or more organs present on both sides. (This asymmetry, in the cases to be considered, is of course distinct from that due to asymmetrical disposition of unpaired viscera, such as the heart and liver of vertebrates, &c.) In several of these cases there is evidence that both sides may on occasion assume the form normally proper to one only.

Some one will no doubt be prepared with the suggestion that these variations are reversions: with this suggestion I shall deal after the facts have been recited.

Spiracle of Tadpole.

721. **Pelobates fuscus:** a tadpole, 7 cm. long, having two spiracles symmetrically placed (Fig. 151), one on the right side and the other on the left[1]. [No details given.] HÉRON-ROYER, *Bull. soc. zool. Fr.*, IX.

FIG. 151. A tadpole of *Pelobates fuscus*, having, as a variation, two spiracular openings, No. 721. (After HÉRON-ROYER.)

1884, p. 162, *fig.* [In the normal there is only one spiracle, that of the left side. In *Pipa* and *Dactylethra* two spiracles are normally present. See WYMAN, *Proc. Bost. N. H. S.*, IX. p. 155; WILDER, *Am. Nat.*, 1877, XI. p. 491; BOULENGER, *Bull. soc. zool. Fr.*, 1881, VI. p. 27.

Tusk of Narwhal.

722. **Monodon monoceros** (Narwhal). In normal males the left tusk alone is developed while the right remains abortive in its alveolus. In the female both tusks are in this rudimentary condition. No reliable record (1871) of a specimen having the *right tusk only* developed, but in eleven cases from various sources the two tusks were *both* developed, and in several of these the two were of about equal length. The normal asymmetry of the skull is not affected by the presence or absence of the teeth. CLARK, J. W., *P. Z. S.*, 1871, p. 42, *figs.* (full literature); see also TURNER, W., *Jour. Anat. Phys.*, 1871, p. 133 and 1874, p. 516.

Ovary and oviduct of Fowl.

It might be anticipated that development of the right ovary and oviduct in birds would be a frequent form of Variation, but as a matter of fact very few such cases are recorded. In consideration of

---

[1] In the same place is recorded a case of a tadpole of this species having the spiracle on the right side instead of the left, perhaps a case of *situs inversus*.

the large numbers of birds, wild and domesticated, annually dissected in laboratories it may perhaps be concluded that these variations are exceedingly rare[1].

723. **Hen** having a small right ovary in addition to the left ovary. The left oviduct was normal, but the left ovary was partially transformed into sacculated tissue. [Full histological details of the structure of both ovaries given.] The hen had partly assumed the plumage of the cock, having four sickle-feathers and other characters proper to the male. BRANDT, *Z. f. w. Z.*, XLVIII. 1889, p. 134, *Pls.*

724. **Hen** having a normal left oviduct and in addition a partially developed *right* oviduct which formed a large thin-walled cyst distended with gas. C. S. M., *Ter. Cat.*, 1872, 455.

Proboscis-pore of Balanoglossus *and water-pores of larvæ of* Asterias.

*725. **Balanoglossus kowalevskii.** The anterior or proboscis-body-cavity is continued backwards into the proboscis-stalk as two hollow horns. In this and most other species the *left* of these alone acquires an opening to the exterior at the proboscis-pore. In *B. kupfferi* alone there are *two* such pores, one opening into each of the two horns[2]. A specimen of *B. kowalevskii* in which *both* horns thus opened to the exterior was seen by MORGAN, T. H., *Jour. of Morph.*, 1891, v. p. 442.

726. **Asterias vulgaris.** The *Bipinnaria* larva as commonly seen resembles the usual *Tornaria* in having a left water-pore only. In several larvæ 3½ to 4 days old the presence of *two* such water-pores, a right and a left, symmetrically placed, has been observed by FIELD and BROOKS. The right pore subsequently closes. This condition is believed by Field to represent not a variation but *a normal phase of development* [though further confirmation is needed]. FIELD, G. W., *Q. J. M. S.*, 1893, XXXIV. p. 110, Pl. XIV. figs. 22 and 23.

Variations in Flat-fishes.

A curious series of variations bearing on the relations of the right side to the left occur in Pleuronectidæ. The evidence on this subject was collected by STEENSTRUP[3] in 1863.

Flat-fishes are normally coloured on the upper side and are without chromatophores in the skin of the lower side[4]. Variations in colour occur in two ways; the upper side may be white like the lower, or on the contrary the lower side may be coloured like the upper. The former change cannot well be distinguished from other cases of albinism[5] and does not call for special notice here.

---

[1] In view of the cases of the Crayfish and the Cockroach mentioned in the Preface, much stress cannot be laid on this consideration.

[2] SPENGEL, J. W., *Mitth. zool. Stat. Neap.*, 1884, v. p. 494, Pl. xxx. fig. 2.

[3] STEENSTRUP, *Overs. k. Dansk. vid. Selsk.*, 1863, p. 145, abstr. by WYVILLE THOMSON, *Ann. and Mag. N. H.*, 1865 (1), p. 361.

[4] In some species the coloured side is normally the right, in others the left, reversed specimens being common in some species (*P. flesus*), rare in others. The reversed condition concerns only the head, skin, muscles, &c., and there is no transposition of the internal viscera.

[5] Evidence collected by STEENSTRUP. GOTTSCHE (*Arch. f. Naturg.*, 1835, II. p. 139) states that *P. platessa* is not rarely wholly white on both sides. I have never

The converse variation, by which the lower side assumes the colour of the upper side is important in several aspects.

Interest has of late been drawn to this subject especially through an experiment recently made by Cunningham[1], who found that of a number of young flat-fishes reared in a vessel illuminated by mirrors from below, some became partially marked with pigmented patches on the lower side. The suggestion was made that this pigmentation was induced by the direct action of light. It is of course impossible here to enter into the theoretical questions raised in connexion with this subject and this account will be confined to description of the colour-variation as seen in nature and of the singular variation in structure commonly associated with it. Mr Cunningham has obligingly advised me in connexion with this subject.

Pigmentation of the lower side has been seen in *Rhombus maximus*, *R. lævis*, *Pleuronectes flesus*, *P. platessa*, *P. oblongus*, *Solea vulgaris* [?] and probably other forms. Attention is drawn to one feature in these changes which from our standpoint has an important bearing. When the underside of a flat-fish is pigmented, it is often not merely pigmented in an indefinite way but it is coloured and marked just as the upper side is. There are, I know, many specimens upon whose undersides a brownish yellow tint is either generally diffused or restricted to patches, but when there is pigment of a deeper shade, as in all the well marked cases of the variation, the colour and markings are closely like those of the upper side. For example, a Plaice (*P. platessa*) sent to me by Mr Dunn of Mevagissey is fully coloured over the posterior half of the lower side; but there is not merely a general pigmentation, for the coloured part of the lower side is *marked with orange markings* exactly like those of the upper side.

More than this: it was found by passing pins vertically through the body that there was in the case of most of the spots a close *correspondence in position between those of the upper and those of the lower side*. There were 13 spots on the coloured part of the lower side, which extended slightly beyond the line of greatest width. Of these, 13 spots on body and fins coincided exactly with those of the upper side; 2 coincided nearly; 2 were not represented on the upper side; and 2 spots of the upper side were not represented on the lower. From these facts it is clear that in "double" flat-fishes we have an instance of *symmetrical* variation of one half of the body into more or less complete likeness of the other half, resembling other cases of Homœosis in Bilateral Series already noticed.

This is made the more evident by the fact that in the two best described specimens of "double" Turbot (No. 727) not merely did the lower side resemble the upper side in point of colour, but upon it were also present the bony tubercles normally proper to the dark side, being only slightly less well developed on the lower side than on the upper

succeeded in seeing an entirely white specimen, though individuals partially white on the upper side are not rare. See also *Zool.*, pp. 4596, 4914. *Zeugopterus punctatus* white on both sides, Day, *Brit. Fishes*, II. p. 19.

[1] Cunningham, J. T., *Proc. Roy. Soc.*, 1893.

(Such a development of tubercles[1] on the lower side may however occur without any correlated change of colour.) It is also stated that in the "double" turbots the muscles of the lower side are thicker than they normally are, thus approximating to the upper side, a feature that may be taken as an indication that the manner of swimming is different from that of normals.

A flat-fish having pigmentation on the lower side does not necessarily present any other abnormality[2]. The Plaice, for instance, just mentioned, was, colour apart, quite normal. But some specimens of flat-fishes darkly coloured below present in addition a very singular structural variation. This consists essentially in the presence of a notch of greater or less depth occurring below the anterior end of the dorsal fin above the eye (Fig. 152). By this cleft the anterior end of the dorsal fin is separated from the back of the head and is borne on a process or horn projecting anteriorly so as to continue the contour of the body above the

FIG. 152. Head of a Brill (*Rhombus lævis*) having the dorsal fin separated from the head as described in the text. (From YARRELL.)

---

[1] The literature relating to discontinuous variations consisting in the presence of bony tubercles upon the blind side of *Rhombi* is extensive. See especially DÉMIDOFF, *Voy. dans la Russie Mérid.*, 1840, III. p. 534, Pls. 28, 29 and 30. STEINDACHNER, *Sitzb. Ak. Wiss. Wien*, 1868, LVII. (1), p. 714. RATHKE, *Mém. Ac. Sci. Pét.*, 1837, III. p. 349. GÜNTHER, *Cat. Brit. Mus. Fishes*, IV. p. 409. These cases will not be confounded with those of supposed hybrids between *R. maximus* and *R. lævis*, which bear upon both sides scales of various sizes.

[2] I know no detailed description of a flat-fish *wholly* pigmented on the underside, having the dorsal fin normal, but numerous authors (Gottsche, Duhamel, &c.) make mention of such cases. Since this chapter was written I have seen two recent papers on the subject by GIARD (*Comptes rend. Soc. Biol.*, 1892, S. 9, IV. p. 31 and *Nat. Sci.*, 1893, p. 356) contributing further evidence on the subject and giving new cases in the Turbot. According to Giard, of flounders (*P. flesus*) at Wimereux 3 $°/_o$ are fully coloured on the blind side, in addition to many that are piebald. This must be a very much higher proportion of abnormal specimens than is found in English fisheries.

head. STEENSTRUP states that the variation has, he believes, been observed in all flat-fishes[1] except the Halibut (*Hippoglossus*).

In several but *not all* cases of this abnormality the eye belonging to the lower side was not placed in its normal position on the upper surface, but stood in an intermediate position on the top of the head, so that it could be partially seen in profile looked at from the "blind" side. It seems possible that the pigmentation of the "blind" side is in some way correlated with some abnormal delay in the shifting of the eye and a consequent continuation of the power of receiving visual sensations from this side.

The abnormality of the dorsal fin is in accordance with this supposition. To understand the nature of this condition it must be remembered that the form of the flat-fish is derived from the usual "round" form by two principal changes. (1) By a twisting of the head the eye is brought over from the blind side to the upper side. (2) The dorsal fin is extended forwards *above* the eye thus shifted; for as STEENSTRUP and TRAQUAIR[2] have shewn, this anterior extension of the dorsal fin is not in the morphological middle line. It is in fact an anterior repetition of the series of dorsal fin-rays along the new contour-line of the body, and occurs irrespective of the fact that the tissues with which it is there associated are not median at all.

STEENSTRUP and TRAQUAIR shewed plainly that it is insufficient to suppose that there is a twisting of the head, for this does not explain the presence of the dorsal fin in the position in which it is found, *curving along that which was once the side of the head*. Traquair suggested that these relations could be attained by two processes; (1) a twisting of the head so as to bring over the eye from the future "blind" side, and (2) a forward growth of the dorsal fin along that which is then the upper contour-line of the head. These processes have now been actually seen by AGASSIZ[3] in several Pleuronectidæ. The first observation of a specimen at the stage when the eye is on the top of the head and the dorsal fin is not yet extended, seems to be that of MALM[4] and there can be little doubt that the normal development proceeds in this way[5]. It has been pointed out by many writers that if the upper eye were to remain in an intermediate position on the top of the head, and the dorsal fin were then to grow forwards, arching over it, the condition of these abnormal forms would be reached. That this is what has actually occurred in them seems likely.

A number of difficult questions are thus raised as to the histological

---

[1] The evidence as to the Sole seems to be doubtful (v. infra).
[2] TRAQUAIR, *Trans. Linn. Soc.*, 1865, xxv. p. 263.
[3] AGASSIZ, *Proc. Amer. Ac. Sci.* 1878, xiv. p. 1, *Pls.*
[4] MALM, *Œfvers. k. Sven. Vet. Ac.*, 1854, p. 173, see *Ann. and Mag. N. H.* 1865 (1), p. 366.
[5] Allusion should be made to the fact that in the genus "*Plagusia*" the dorsal fin acquires its forward extension at a time before the shifting of the eye occurs. When the time for this change comes the eye of the future blind side passes under the dorsal fin and above the skull, *through the tissues from one side of the head to the other*. This was first observed by STEENSTRUP, and afterwards by AGASSIZ in great detail and the fact can hardly now be questioned. This mode of development is peculiar to "*Plagusia*," though when Steenstrup wrote he expected that the same would be found to occur in other Pleuronectidæ.

processes by which the dorsal fin comes to stand where it does. We are accustomed to think of the repetition of the fin-rays as being an expression of the fundamental segmentation of the trunk, accessory to it no doubt, but still of the same nature and histologically dependent upon it. The extension of this repetition along the morphological side of the face is thus an anomaly.

Further comment on the nature of the variation will be made after the chief cases have been given.

*727. **Rhombus maximus** (Turbot). Two specimens respectively 9 in. 9 lines and 7 in. 6 lines in length, 7 in. 6 lines, and 5 in. 6 lines broad. Both sides of a similar coffee-brown colour. The smaller had a yellowish white spot, about 1 in. square, on the operculum of the lower side. The colour was more uniform than usual and the dark spots normally found on the fins of the Turbot were absent. Both sides irregularly beset with horny tubercles, only slightly more developed on the upper than on the lower sides. Fine *scales* were also found deep in the skin of both sides. All fins except the dorsal were normal in form and position. The dorsal fin was anteriorly detached from the head, being borne on a horn-like projection. The separation between the head and dorsal fin was continued backwards as a semi-circular notch to a level behind the eyes. Upon many of the fin-rays of the dorsal, anal and caudal fins there were 1—7 small knotty elevations of the size of poppy-seed. In the smaller specimen these elevations were smaller, and on the caudal fin absent. The left eye had its normal position, but the right eye [of "blind" side] was placed on the top of the head, but in such a position that it could scarcely have seen any thing not directly over it. [See further details given.] SCHLEEP, *Isis*, 1829, p. 1049, Pl. III.

Similar specimen COUCH, *Fishes Brit. Isl.*, III. p. 157. Dried specimens in Brit. Mus., Newcastle Mus., &c.

*728. Very good figures of such a Turbot are given by DUHAMEL DU MONCEAU (*Traité général des Pesches*, 1777, III. Sect. IX. p. 262, Pl. III. figs. 3 and 4). The under side was of nearly the same colour as the upper and the tubercles generally found on the upper side only were present on the lower side also, though of smaller size. A slight notch separated the dorsal fin from the head; *but the upper eye is figured as in its normal place*, not being on the top of the head, and it would of course be invisible from the "blind" side. [This important case is referred to by STEENSTRUP, but seems to be unknown to others, who attribute the separation of the dorsal fin to the persistence of the eye on the top of the head.]

729. A young turbot, similarly coloured on both sides, having the eyes still symmetrical, swimming on edge, is figured by MCINTOSH, *Fishes of St. Andrews*, 1875, Pl. VI. figs. 5 and 6. Prof. McIntosh kindly informed me that these "double" individuals swim on edge much longer than usual.

730. **Rhombus lævis** (Brill). Specimen presenting similar characters. The lower (rt.) side of a uniform dark colour with exception of a white patch on operculum. The right pectoral fin was whitish. The under side was rather darker than the upper and the mottling present on the upper side was entirely absent from the under side, which was without marking or spot. This is very probably a post-

mortem change. Right pelvic fin dark, but the left was whitish, speckled with black. Nostrils normal. The eye of the right (blind) side was placed almost entirely on the left side, but not completely so, for it could be seen to some extent in profile from the right side. The notch separating the dorsal fin from the head was rounded, and extended to about the level of the posterior margin of the *left* eye. There were about 6 chief fin-rays borne by the prominence above the eye. The fish seemed to be in all respects healthy and well grown. *Paris Mus.*, numbered ✳90✳310. [This specimen was kindly shewn to me by Prof. Vaillant.]

Similar specimen, also having white patch on operculum DUHAMEL DU MONCEAU, *l. c.* See also Fig. 152, from YARRELL, *Brit. Fishes*, 3rd ed., I. p. 643.

The specimen described by DONOVAN (*Brit. Fish.*, 1806, IV. Pl. XC.) under the name "*Pleuronectes cyclops*" was in Steenstrup's opinion a young Brill having this variation. In this specimen the right eye is seated on the top of the head and is seen in profile from the right side. The right side was coloured like the left, but was not so dark. The dorsal fin began behind the right eye. This specimen was found in a rock-pool "inveloped in a froth" said to have resembled cuckoo-spit.

731. **Zeugopterus punctatus** (Müller's Topknot). This fish is very liable to malformations of the anterior end of the dorsal fin, causing it to form an arch over the eyes. YARRELL (quoting COUCH), *Brit. Fish.*, 3rd ed., I. p. 648.

732. "**Platessa oblonga**" De Kay (American Turbot); specimen having both sides darkly coloured; upper eye placed on the top of the head; dorsal fin separated by a notch. STORER, *Mem. Amer. Ac. Sci.*, VIII. p. 396, Pl. XXXI. fig. 2 *b*.

733. **Pleuronectes platessa** (Plaice): specimen completely and similarly pigmented on both sides far from rare. In a specimen thus coloured the '*tubercula capitis*' were as strongly marked on the one side as on the other. In several examples the anterior end of the dorsal fin was separated from the head, GOTTSCHE, *Arch. f. Naturg.*, 1835, II. 1, p. 139.

734. **Pleuronectes flesus** (Flounder): several specimens found at Birkenhead, having a deep notch of this kind above the eyes. These fishes were 'very dark brown (almost black) on both sides.' In the length of the fins these examples differed somewhat from the Flounder, HIGGINS, *Zoologist*, 1855, p. 4596, *fig.* Specimen of this kind figured by TRAQUAIR, *Trans. Linn. Soc.*, 1865, XXV. p. 288, Pl. XXXI. figs. 8 and 9. See also NILSSON, *Skandin. Fauna: Fiskarna*, Lund, 1855, p. 621; COUCH, *Brit. Fishes*, 1864, III. p. 198.

735. **Solea vulgaris**. Many authors mention Soles coloured on both sides, but I know no good description of one. YARRELL (*l. c.*, p. 669) says "we have not seen the *Solea Trevelyani* of Ireland (Sander's News-letter, 16th April, 1850). It is dark-bellied and is described as bearing a projection on the head like the monstrosity figured on p. 643." DUHAMEL DU MONCEAU (*l. c.*, Pl. I. figs. 3 and 4) represents a sole darkly coloured on both sides. The dorsal fin is shewn in its normal state, not separated from the head. No special description is given, and as the author does not state that he had himself seen such a sole the figure was perhaps not drawn from an actual specimen. A sole with the under side piebald is described in *Zool.* x. p. 3660.

In connexion with this evidence STEENSTRUP refers to a small flat-fish, *Hippoglossus pinguis*, found in a few localities in Scandinavian waters, having a form almost intermediate between a "flat" and a "round" fish. The eye of the "blind" side is exactly on the top of the head and can be seen in profile from the blind side. The blind side is nearly as muscular as the upper side, and its skin is yellowish-brown in colour and is only slightly paler than that of the upper side. The dorsal fin begins behind the eye, not arching over it. Steenstrup looked on this creature as representing in a normal form the "double" condition presented as a variation in the cases we have been speaking of. See description and figures in SMIT's edition of FRIES, EKSTRÖM and SUNDEVALL's *Hist. of Scand. Fishes*, 1893, pp. 416 and 417. SMIT makes a new genus, *Platysomatichthys*, for this animal.

Comment on the foregoing cases.

In the cases preceding many will no doubt see manifest examples of Reversion. There is a sense in which this view must be true, for it can scarcely be questioned that if we had before us the phylogenetic series through which the Flat-fishes, the Narwhal, &c. are descended, it would be seen that each did at some time have a bilaterally symmetrical ancestor. But, for all that, in an unqualified description of the change as a reversion the significance of the facts is missed. By the statement that a given variation is a reversion it is meant that in the varying individual a form, once the normal, reappears. The statement moreover is especially intended to imply that the *definiteness and magnitude* of the step from normal to variety is due to the circumstance that this variety was once a normal. It is meant, in fact, that the greatness of the modern change can be explained away by the suggestion that in the past, the form now presented as a variation, was once built up by a gradual evolution, and that though in its modern appearance there is Discontinuity, yet it was once evolved gradually.

Now the attempt to apply this reasoning, especially to the case of the "double" Flat-fishes, leads to difficulty. We may admit that in so far as the varieties are bilaterally symmetrical they represent a normal. Their bilateral symmetry, as a quality apart, may be an ancestral character, if any one is pleased so to call it. But that in the contemporary resumption of a bilateral symmetry we have in any further sense a reappearance of an ancestral form is very unlikely.

First it might be fairly argued that it is improbable that there was ever a typical flat-fish having on *both* sides the peculiar pigmentation of the present upper sides of the Pleuronectidæ of our day. Such a creature would be highly anomalous. But even if in strictness we forego the assumption that since the evolution of Flat-fishes there has never been an ancestor fully pigmented on both sides, there still remains the difficulty that each species may in the "double" state have upon its lower side the *specific* colour proper to its own upper side. A notable instance of this has been mentioned in the Plaice (p. 467); and here not only was the pigmentation of the lower side, as far as it went, like that of the upper, but the spots were even almost bilaterally symmetrical. It is true that the lower side does not in every case copy the upper in colour, but it *may* do so; and, in proportion as it does so in different species, so far at least are the changes not simply reversions; for the several patterns of Turbot, Plaice &c. are mutually exclusive and it can hardly be supposed that each species had separately a "double" ancestor having the present specific pattern on both sides.

The outcome of this reasoning is to shew that the hypothesis of Reversion in the strict sense is an insufficient account of the actual variation in these Flat-fishes, and in the production of these varying forms there is thus a Discontinuity over and above that which can be ascribed to Reversion. The facts stated in connexion with the Plaice (p. 467), especially the symmetry of the spots, probably indicate the real nature of this Discontinuity, and raise a presumption that in the new resemblance of the lower side to the upper we have a phenomenon of Symmetry resembling that Homœosis shewn to occur between parts in

Linear Series. In the Flat-fish the right side and the left have been differentiated on different lines, as the several appendages of an Arthropod have been, but on occasion the one may suddenly take up all or some of the characters, whether colour, tubercles or otherwise, in the state to which they have been separately evolved in the other.

What may be the cause leading to this discontinuous change we do not know. That it is often associated with a delay in the change of position of the eye of the "blind" side seems clear from the frequent detachment of the dorsal fin in these cases. But it should be borne in mind that even in such examples the eye may still eventually get to its normal place, though probably it was delayed in the process and so led to detachment of the fin. Taken with the fact that the young "double" turbots swim on edge longer than the normals it must be concluded that the bilateral symmetry of colour is associated with reluctance or delay in the assumption of the asymmetrical state, but more than this cannot be affirmed.

I do not urge that the same reasoning should be applied in other cases, but the possibility must be remembered. In the Narwhal, for instance, it is perhaps unlikely that there was ever an ancestor which had two tusks developed to the extent now reached by the left tusk of the male; but if there ever were any such form, it is hard indeed to suppose that it could have been connected with the present species by a series of successive normals in which the right tusk gradually diminished while the left was of its present size. On the whole it seems more likely that when the right tusk now develops to be as long as the left, it is taking up at one step the state to which the left has been separately evolved.

However this may be, the fact that such Homœosis is possible should be kept in view in considering the meaning of such cases as that of a *Tornaria* with two water-pores. For while on the one hand we may suppose that *Balanoglossus kupfferi* with its normal pair of water-pores is the primitive state and that the varying *Tornaria* is a reversion, on the other hand *B. kupfferi* may be a form that has arisen by a Homœotic variation from the one-pored form, and of this variation *Balanoglossus* No. 725 may be a contemporary illustration[1].

[1] The following interesting example of a similar Variation has appeared since these pages were set up. **Eledone cirrhosa**: specimen having not only the third left arm developed as a hectocotylus, as usual, but the *third right arm also*. The right had 57, the left 66 suckers, but otherwise they were alike. APPELLÖF, A., *Bergens Museums Aarbog*, 1893, p. 14.

CHAPTER XX.

SUPERNUMERARY APPENDAGES IN SECONDARY SYMMETRY.

INTRODUCTORY.—THE EVIDENCE AS TO INSECTS.

OF all classes of Meristic variations those consisting in repetition or division of appendages are by far the most complex and the most difficult to bring into system. There is besides no animal which normally presents the condition seen in the variations about to be described, though there may be a true analogy between them and phenomena found in colonial forms. It has nevertheless seemed well to introduce some part of this evidence here for two reasons. First the subject is a necessary continuation of the evidence as to digits, which would otherwise be left incomplete; secondly it will be shewn that though many of the cases are irregular and follow no system that can be detected, there remain a large number of cases (being, indeed, the great majority of those that have been well studied) whose form-relations can be put in terms of a simple system of Symmetry. Thus not only are we introduced to a very remarkable property of living bodies, but also the way of future students of Variation may be cleared of a mass of tangled facts that have long been an obstacle; for on apprehension of the system referred to it will be seen that cases of repetition in Secondary Symmetry are distinct from those of true Variation within the Primary Symmetry and may thus be set apart.

Arrangement of evidence as to Repetition of Appendages.

In the first instance I shall give the evidence as to Secondary Symmetries in Insects and Crustacea, prefacing it with a preliminary account of the system of Symmetry obeyed by those cases which I shall call *regular,* and explaining the scheme of nomenclature adopted. Besides the regular cases of extra parts in Secondary Symmetry there are many *irregular* examples which cannot be shewn to conform to the system set forth. Of all but a few of these, details are not accessible, and of the rest many are mutilated or so amorphous that the morphological relations of the surfaces cannot be determined.

Over and above these there remain a very few cases of Repetition of parts of appendages where the arrangement is certainly not in Secondary Symmetry, but is of a wholly different nature, exemplifying in Arthropods that *duplicity of limbs* already seen in the human double-hands (p. 331) and in the double-feet of Artiodactyles (p. 378). Genuine cases of this kind are excessively rare; but owing to hasty examination great numbers of cases have been described as instances of duplicity, though in reality the supernumerary parts in them can be shewn to be of paired structure. To emphasize the distinctness of these cases they will be made the subject of a separate consideration. Logically they should of course be treated before the Secondary Symmetries; but their essential features may be understood so much more readily if the latter are taken first that I have decided to change the natural order.

In continuation of the evidence as to Secondary Symmetry in Arthropods will be given a brief notice of similar phenomena in vertebrates. This evidence is comparatively well known and accessible and I shall attempt no detailed account of it, referring to the facts chiefly with the object of shewing how the principles found in Arthropods bear on the vertebrate cases.

It will then be necessary to consider how repetitions in Secondary Symmetry are related to other phenomena of Repetition. Lastly something must be said with regard to the bearing of these facts on the general problems of Natural History.

Preliminary account of paired Extra Appendages in Secondary Symmetry (Insects).

Supernumerary appendages in Insects are not very uncommon, perhaps 120 cases of this kind being recorded[1]. Nearly all known examples are in beetles, but this may be due to the greater attention paid to the appendages in that order. They do not seem to appear more often in one family than in another, but perhaps the rarity of instances in Curculionidæ is worth noting. They are found in both sexes, in all parts of the world, and in species of most diverse habits.

Supernumerary parts may be antennæ, palpi or legs. (Extra wings are probably in some respects distinct. They have already been considered. See p. 281.) Extra appendages may be either outgrowths from the body in the neighbourhood of the part repeated, or, as in the great majority of cases, they occur as outgrowths from an appendage, extra legs growing from normal legs, extra antennæ from antennæ, &c. In every case there are two essentials to be determined: first the *constitution*

[1] Not including some 110 cases of alleged *duplicity* of appendages given later.

of the extra parts, and secondly the *symmetry* or relation of form subsisting on the one hand between the extra parts themselves, and on the other between the extra parts and the normal parts.

In few cases of extra appendages arising from the body itself have these essentials been adequately ascertained.

For brevity I shall describe the phenomena as seen in extra legs. The same description will apply generally to the antennæ. Recorded cases of extra palpi are very few, but probably are not materially different.

Structure of Paired Extra Legs.

The parts composing extra legs do not as a rule greatly differ from those of the normal legs which bear them. Though in many instances extra legs are partially deformed, they are more often fairly good copies of the true leg. Not rarely the extra parts are more slender or a little shorter than the normal appendage, but in form and texture they are real appendages, presenting as a rule the hairs, spurs, &c. characteristic of the species to which they belong.

The next point is especially important. *The parts found in extra legs are those parts which are in the normal leg peripheral to the point from which the extra legs arise, and, as a rule no more.* Though in extra legs parts may be deficient or malformed, structures which in the normal leg are central to the point of origin of the extra legs are not repeated in them[1]. For instance, if the extra legs spring from the trochanter they do not contain parts of the coxa, if from the second tarsal joint, the first tarsal joint is not represented in them, and so on.

Extra legs may arise from any joint of the normal leg, and are not much commoner in the peripheral parts than in the central ones, but there is a slight preponderance of cases beginning from the apex of the tibia. It is rather remarkable that cases of extensive repetition are not much less rare than others, the contrary being for the most part true of the limbs of vertebrates.

It does not appear that extra legs arise more commonly from either of the three normal pairs in particular.

Supernumerary legs of double structure are sometimes found as two limbs separate from each other nearly or quite from the point of origin, but in the majority of cases their central parts may be so compounded together that they seem to form but one limb, and the essentially double character of the limb is not then conspicuous except in the periphery. For example it frequently happens that the femora of two extra legs are so compounded together that they seem to have only a single femur in common,

---

[1] Particular attention is therefore called to one case of extra antennæ, which did actually contain parts normally central to the point of origin. (See No. 804.)

and careless observers have often thus declared them to be two legs with one femur. Similarly the two tibiæ or the two tarsi may be more or less compounded. In the case of *Silpha nigrita* (No. 769), the two extra legs which arose from a femur were compounded throughout their length, having a compound tibia and tarsus (see Fig. 167). Even in cases when the two extra legs appear to arise separately it will generally be found that they articulate with a double compound piece of tissue which is supernumerary and is fitted into the joint from which they appear to arise. This is especially common in cases of two extra tarsi, which seem to spring directly from a normal tibia. As a matter of fact in all such cases these extra tarsi articulate with a supernumerary piece of tissue, as it were let into, and compounded with, the apex of the normal tibia. These bodies are themselves double structures, composed of parts of two tibiæ. In determining the morphology of the limbs they are of great importance, but unfortunately they are not generally mentioned by those who describe such formations. But though extra parts are generally present in the leg centrally to the point from which the extra legs actually diverge, it should be expressly stated that if this point is in the periphery of the leg, the central joints are normal: if for example, there are two extra tarsi, there may be parts of two extra tibial apices, but the base of the tibia, the femur, &c. are single and normal.

Symmetry of Paired Extra Legs.

To appreciate what follows it is necessary to have a distinct conception of the normal structure of an insect's leg, and to understand the use of the terms applied to the morphological surfaces.

If the leg of a beetle, say a *Carabus*, is extended and set at right angles to the body, the four surfaces which it presents are respectively dorsal, ventral, anterior and posterior. In the femur, tibia and tarsus the dorsal is the extensor, and the ventral is the flexor surface. The anterior surface is seen from in front and the posterior from behind. (The terms 'internal' and 'external' are to be avoided as they denote different surfaces in the different pairs.) Difficulty as to the use of terms arises from the fact that as the beetle walks or is set in collections, the legs are not at right angles to the body but are rotated on the coxæ, so that the plantar surface of the first pair of legs is turned forwards, but the plantar surfaces of the second and third pairs are turned backwards[1].

[1] Attention is directed to the fact that in a beetle there is a complementary relation not only between the legs of the right and left sides but also imperfectly between the legs of the first pair and those of the second and third pairs, which are in some respects images of the first leg of their own side. For instance, in *Cerambyx* (see Fig. 160) the trochanter of the fore leg is kept in place by a process of the coxa which goes down *behind* it, but the corresponding process in the second and third legs is *in front* of each trochanter. Again in *Melolontha* &c. the tibial serrations of

Extra legs may arise from any one of the morphological surfaces, but more often their origin is in a position intermediate between them, *e.g.*, antero-ventral, or postero-dorsal.

The next question is that of the determination of parts which are extra from the parts which are normal. Two extra legs spring from a normal leg. The appearance is often that of a leg single proximally, but triple peripherally. All three limbs are often equally developed and at first sight it might well be supposed that the three collectively represent the single leg of the normal.

In many cases of Meristic Variation I have contended that the facts are only intelligible on the view that there has been such collective representation. But in these Secondary Symmetries this supposition is [? always] inadmissible. On closer examination it is generally more or less easy to see that the three legs do not arise in the same way, but that one arises as usual while the other two are, as it were, ingrafted upon it. It is thus possible in all but a very few cases to determine the normal leg from the others by tracing the surfaces from apex to base, when it will be found that some surface of the normal is continuous throughout the appendage while those of the extra legs end abruptly at some part of the normal leg.

Nearly always besides, as has been mentioned, the extra legs are more or less compounded together at their point of origin even if separate peripherally. In a few very exceptional cases it happens that one of the extra appendages is compounded with the normal and not with the other extra appendage. A remarkable case of this in an antenna may be seen in *Melolontha*, No. 800, and in a leg in *Platycerus caraboides*, (*q.v.*)

We have now to consider the positions of the paired extra legs in regard to the normal leg and in regard to each other. At first sight their dispositions seem entirely erratic; but though it is true that scarcely two are quite identical in structure, yet their divers structures may for the most part be reduced to a system. This system, though far from including every case, still includes a large proportion and even the remainder do not much depart from it except in very few instances. The comprehension of the general system will also greatly help to make the aberrant cases appreciated with comparatively few words. For simplicity therefore, the consideration of exceptional cases will be deferred and the principles stated in a general form. It will be remembered that we are as yet concerned only with *double* extra legs.

When extra appendages, arising from a normal appendage, are thoroughly relaxed and extended, the following rules will be

the first legs curve backwards, but those of the other legs curve forwards. This circumstance is mentioned lest it might be thought to have been neglected in what follows, but this complementary relation has nothing to do with that which will be shewn to exist between the extra legs.

found to hold good with certain exceptions to be hereafter specified.

I. *The long axes of the normal appendage and of the two extra appendages are in one plane:* of the two extra appendages one is therefore nearer to the axis of the normal appendage and the other is remoter from it.

II. *The nearer of the two extra appendages is in structure and position formed as the image of the normal appendage in a plane mirror placed between the normal appendage and the nearer one, at right angles to the plane of the three axes; and the remoter appendage is the image of the nearer in a plane mirror similarly placed between the two extra appendages.*

Transverse sections of the three appendages taken at homologous points are thus images of each other in parallel mirrors.

As the full significance of these principles may not be at once seen it may be well to add a few words of general description. The relation of images between the *extra* legs is easy to understand. They are a complementary *pair*, a right and a left. This might indeed be predicted by any one who had considered the matter.

The other principles, which concern the relations of the extra legs to the normal leg, are more novel. For first it appears not that either of the extra legs indifferently may be adjacent to the normal, but that of the extra pair the adjacent leg is that which is formed as a leg of the other side of the body. If therefore the normal leg bearing the extra legs be a right leg, the nearer of the extra legs is a left and the remoter a right. This principle holds in every case of double extra appendages of which I have any accurate knowledge, where the structure of the parts is such that right limbs can be distinguished from left.

But perhaps of greatest interest is the fact that the inclination of the surfaces of each extra leg to those of its fellow and to those of the normal are determined with an approach to uniformity in the manner described.

These principles of arrangement may be made clear by a simple mechanical device (Fig. 153). A horizontal circular disc of wood has an upright rod fixed in its centre. This rod passes through one end of a vertical plate of wood which can be turned freely upon it as an axle, so as to stand upon any radius of the horizontal circle. The head of the axle bears a fixed cog-wheel. In the vertical wooden plate are bored two holes into which two rods each bearing a similar cog-wheel are dropped, so that each can rotate freely on its own axis. The three cog-wheels are geared into each other. They must have the same diameter and the same number of teeth. Three wax models of legs are fixed on the head of each wheel as shewn in Fig. 153. In that figure, R represents the apex of the tibia and tarsus of a normal right leg. The anterior surface is dark, and the posterior is white. The anterior and posterior spurs of the tibia are shewn at A and P. SL and SR represent the two supernumerary legs, SL being a left, SR a right. (They are supposed to arise from the leg R at some proximal point towards which they converge.) When the wooden plate is put so that the arrow points to the word "Posterior" on the disc, the models will then take the positions they would have if they arose from the posterior surface, all the ventral surfaces coming into one plane. If the arrow be

480   MERISTIC VARIATION.   [PART I.

set to "Ventral" the two supernumeraries will turn their dorsal surfaces to each other, and so on. The model SL thus rotates twice on its own axis for each

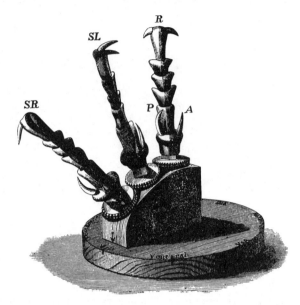

Fig. 153. A mechanical device for shewing the relations that extra legs in Secondary Symmetry bear to each other and to the normal leg from which they arise. The model R represents a normal right leg. SL and SR represent respectively the extra right and extra left legs of the supernumerary pair. A and P, the anterior and posterior spurs of the tibia. In each leg the *morphologically anterior* surface is shaded, the posterior being white. R is seen from the ventral aspect and SL and SR are in Position VP.

revolution round R, but the surfaces of the model SR always remain parallel to those of the model R. In every possible position therefore each model is the image of its neighbour in a mirror tangential to the circle of revolution. In the figure the models have the position they should have if arising postero-ventrally. Here the plantar surface of SL is at right angles to the plantar surfaces of the other two legs.

Since at each radius the relative position of the legs differs, it is possible to define these positions by naming the radius. This will be done as shewn in Fig. 154. In this diagram imaginary sections of the legs are shewn in the various positions they would assume at various radii. The central thick outline shews a section of the normal leg, a longer process distinguishing the anterior surface from the posterior. The radii are drawn to various points D, A, V, P, representing the dorsal, anterior, ventral and posterior positions respectively. Intermediate positions may be marked by combinations, DA, VVP, &c., using the system employed in boxing the Compass.

On several of the radii ideal sections of the extra legs are shewn in thin lines, the shaded one being the nearer and the plain one the remoter. M^1 and M^2 shew the planes of the imaginary mirrors.

The manner in which the pair of extra limbs are compounded with each other in their proximal parts, and with the normal limb at their

CHAP. XX.]   SECONDARY SYMMETRY: SCHEME.   481

point of origin is most extraordinary. It does not appear that the surfaces compound together along any very definite line or that the

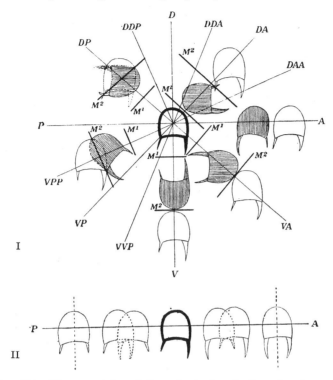

Fig. 154. Diagrams of the relations of extra legs in Secondary Symmetry at various positions relatively to the normal leg from which they arise.

The legs are represented in transverse section, the morphologically anterior side of each being indicated by the longer spur. The section of the normal leg, in which the radii converge, is shewn with a thick black line. The section of the *nearer* extra leg in Diagram I is shaded, while the *remoter* is blank. The radii shew them in various positions, anterior, posterior, dorsal, ventral, &c. relatively to the normal leg.

M^1, the plane of reflexion between the nearer extra limb and the normal.
M^2, plane of reflexion between the nearer and the remoter extra limbs.

Diagram II is constructed in the same way to illustrate special cases of extra legs arising anteriorly or posteriorly. If the two extra legs diverge from each other centrally to the tibial apex each tibial apex is then complete, as on radius A of Diagram I. In Diagram II are shewn two degrees of composition of the two tibial apices, illustrating how, in cases of complete composition, the extra parts may consist wholly of two morphologically posterior or anterior surfaces according as they arise posteriorly or anteriorly to the normal leg. (See for instance Nos. 750 and 764.)

line of division between the several limbs is determined by the normal structure of the limbs. The homologous parts seem to be compounded at any point, almost as an object partly immersed in mercury compounds with its image along the line to which it is immersed, wherever that line may be.

B.   31

From this some curious results follow. For instance, if two extra limbs arise anteriorly and are separate at their tibial apices, they bear *four* spurs as shewn at radius A in the upper diagram of Fig. 154. But if the two are fully compounded at the tibial apices in the anterior position the compounded limb will only have two spurs, both being shaped as anterior spurs (as shewn in the lower diagram) and conversely for the posterior position (see No. 764). The parts, in fact, where the pair may be supposed to interpenetrate (dotted in the diagrams) are not represented.

Those who have described these phenomena have in consequence often made the following error. Observing a limb giving off a morphologically double limb with a common proximal part subsequently separating into its two components, they speak of this as a "primary and secondary dichotomy." When the facts are understood it is clear that there is no dichotomy between the extra legs and the normal, for the parts are not equivalent and the normal is undivided.

Such are the principles followed. *It would not be true to assert that these rules are followed with mathematical precision, but in the main they hold good.* Special attention will be given to cases departing from them, but the number of such cases is small. The cases of slight deviation from the schematic positions are besides mostly those of extra limbs in the Positions A and P, and generally the deviation in them takes the same form, causing the ventral surfaces of the extra parts to be inclined to each other downwards at an obtuse angle instead of forming one plane.

In all cases possible I have examined the specimens myself, and I am under obligation to numerous persons who have very generously given me facilities for doing so. Amongst others I am thus greatly indebted to M. H. Gadeau de Kerville, Dr G. Kraatz and Dr L. von Heyden for the loan of many valuable insects, and also to Messrs Pennetier, Giard, Dale, Mason, Westwood, Waterhouse, Janson, Harrington, Bleuse, &c. In this part of the work I am under especial obligation to Dr D. Sharp, for without his cooperation it would not have been possible for me to have undertaken the manipulations needed. He has most kindly given up his time to the subject, and in the case of almost every one of the specimens examined at Cambridge I have had the benefit of his help and advice.

Of cases not seen by me few are described in detail sufficient to warrant a statement as to the planes in which the parts stood, but sometimes the figures give indications of this. Some of the accounts are quite worthless, merely recording that such an insect had two extra legs: in such cases I have thought it enough to give the reference and the name of the insect for statistical purposes. But every case known to me is here recorded: there has been no rejection of cases.

The cases will be taken in order of the Positions, beginning with the Position V and taking the other radii in order, going round against the hands of a watch.

CASES OF EXTRA LEGS IN SECONDARY SYMMETRY.

(1) *Position V.*

*736. **Carabus scheidleri** ♀ : pair of extra legs having a common femoral portion arising from the trochanter of the right fore leg (Fig. 155). This case is of diagrammatic simplicity. The troch-

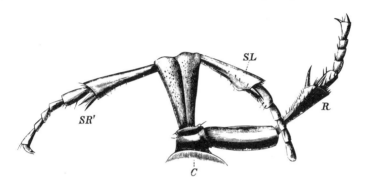

FIG. 155. *Carabus scheidleri*, No. 736. The normal right fore leg, R, bearing an extra pair of legs, SL and SR', arising from the ventral surface of the coxa, C. Seen from in front. (The property of Dr Kraatz.)

anter bears a normal leg (R) articulating as usual. Immediately ventral to this articulation there is a second articulation upon a small elevation. This bears a double femur made up of parts of a pair of femora compounded by their dorsal borders. The double femur has thus two structurally ventral surfaces opposite to each other.

The apex presents two articular surfaces in the same plane as that of the normal leg, each bearing a tibia, both tibiæ flexing in the same vertical plane.

Since the double femur of the extra legs stands vertically downwards at right angles to the normal femur, it will be seen that both the extra tibiæ flex *upwards*, but one of them is a left leg (SL), bending to meet the normal leg, while the other is a right (SR'), bending towards the ventral surface of the body. The tibia of the left extra leg is a little shorter than that of the normal, and the tibia of the right extra leg is a little longer than it. All three tarsi are thinner than a normal tarsus; and the claws are a good deal reduced in the case both of the normal and the right extra leg, while in the left extra leg they are absent altogether. This is an example of a pair of extra legs arising

31—2

in the position marked V in the Scheme and having precisely the relations there shewn. Specimen first described by KRAATZ, G., *Berl. ent. Zt.*, 1873, p. 432, fig. 9. I am greatly indebted to Dr Kraatz for an opportunity of examining it.

737. **Carabus marginalis**: penultimate joint of left hind tarsus is enlarged and presents two articular surfaces, a proximal one on the ventral surface, and another at the apex. The latter bears the normal last joint with its claws. From the proximal articular surface arises a thick joint shorter than the normal last joint, bearing at its apex *two pairs* of claws set back to back, as in the Position V. Specimen redescribed from KRAATZ, G., *Deut. ent. Zt.*, 1880, XXIV. p. 344, Pl. II. fig. 29.

738. **Carabus granulatus** ♂, left posterior tibia bearing an amorphous rudiment of two extra tarsi arising from the ventral surface of its apex. The apex of the tibia is produced at the dorsal border to form an irregular process which bears a tarsus of normal form but reduced size and immediately ventral to this tarsus is a pair of tibial spurs. Ventral to these spurs is another deformed pair of spurs and below them again is a deformed 3-jointed rudiment which probably represents two tarsi. Ventral to the rudiment of the extra tarsi is a third deformed pair of spurs. It was not possible to recognize the surfaces of the tarsal rudiment, but the presence of *two* extra pairs of spurs indicates plainly that the extra parts are morphologically of double structure; and as the spurs indicate the morphologically ventral surfaces, it follows that the surfaces adjacent in the extra tarsi are *dorsal*. This specimen was originally described by Dr L. VON HEYDEN, who was so good as to lend it to me for examination, see *Deut. ent. Zt.*, 1881, XXV. p. 110, *fig.* 26.

739. **Prionus coriarius** (Longicorn): three legs in region of right posterior leg. The proximal relations not quite clear and hence it is not easy to distinguish the normal. Presumably it was the most dorsal. This leg was of normal form but of reduced size and it wanted the claw-joint. Internal to it, arising by a double coxa, trochanter and femur, were the other two legs. The remoter was a normal right, but the nearer was a left leg of reduced size, slightly crooked and lacking three apical tarsal joints. The compound femur was just as in No. 736. The normal leg must either have been the most dorsal or the most ventral. If the former, the extra parts are in the Position V; if the latter, they are in the Position D, but in this event the normal would be compounded with one of the extra legs. [Redescribed from description and figure given by KRAUSE, *Sitzb. nat. Fr. Berl.*, 1888, p. 145, *fig.*]

740. **Melolontha vulgaris** ♀ (Lamellicorn): right posterior femur bears a supernumerary pair of limbs having a double tibia in common. The supernumerary parts are rather smaller than the normal ones. [The position of origin and symmetry, according to the figure, must have been approximately V.] KOLBE, H. J., *Naturw. Wochens.*, 1889, IV. p. 169, fig.

741. **Carabus perforatus** ♂: from the ventral or plantar surface of the 5th tarsal joint of left hind leg project an extra pair of claw-joints compounded in Position V, each bearing a pair of claws, set back to back. This is a diagrammatic case, well and clearly described by ASMUSS, *Monstr. Coleop.*, 1835, p. 54, Tab. IX.

(2) *Position VAA.*

742. **Feronia (Pterostichus) mühlfeldii** ♀ (Carabidæ): left middle tibia bearing two supernumerary tarsi arising by a common proximal joint (Fig. 156). As in other cases of supernumerary tarsi arising from the tibia, the apex of the tibia itself is really a triple structure, containing parts of the apices of a pair of tibiæ in addition to the normal apex. This is shewn by the presence of three pairs of spurs, &c. The additional parts are in this case anterior and ventral to the normal apex and a complementary pair. All three are completely blended together, forming in appearance a single apex. The relations of the three component parts are almost exactly those indicated in the Scheme for the Position VAA.

Fig. 156. *Pterostichus mühlfeldii*, No. 742. Semidiagrammatic representation of the left middle tibia bearing the extra tarsi upon the antero-ventral border of the apex. *L*, the normal tarsus; *R*, the extra right; *L'* the extra left tarsus. (The property of Dr Kraatz.)

The two extra tarsi (*R, L'*) arise by a common proximal joint of double structure having two complete ventral surfaces inclined to each other as in the Position VAA. Peripherally to this the two tarsi are separate. The tarsus which is nearer to the normal tarsus is perfect, and stands in the schematic position. The second joint of the remoter arises in the position shewn for VAA, but its apex is slightly shrivelled and in consequence the remainder of this tarsus, though perfect in size and form is thrown a little out of position. This specimen was kindly lent to me by Dr KRAATZ, and was originally described and figured by him in *Deut. ent. Zt.*, 1877, XXI. p. 56, fig. 21.

743. **Aromia moschata** ♀ (Greece) (Longicorn): right anterior tibia enlarged at apex bearing anteriorly a supernumerary pair

of tarsi. The widened apex bears three supernumerary spurs of which the middle one is thicker than a normal posterior spur. This is no doubt a double spur representing the two *posterior* spurs of the extra tibiæ. The other two extra spurs are ordinary *anterior* spurs. The relative positions of these spurs are exactly those marked VAA in the Scheme. Of the extra tarsi 3 joints only remain and the two tarsal series are so closely compounded that superficially they seem to form one tarsus only. In their first joints the inclination of the ventral surfaces to each other is at an *acute* angle, thus departing from the Scheme, but in the second and third joints, where they are more separate from each other, the inclination is at approximately the same angle as that of the lines joining their respective spurs. Specimen in General Collection of the British Museum.

744. **Carabus græcus** ♀ : trochanter of right middle leg bears a supernumerary pair of legs having trochanter, femur, tibia and 1st tarsal joint common. The coxa of the normal leg is enlarged and the trochanter has two heads, of which the anterior belong to the extra pair of legs. The femur of the extra pair is a single piece but is morphologically double, presenting *two structurally anterior surfaces and two structurally ventral surfaces*, the latter being inclined to each other at an angle of about 120°. From the apex of this femur there arises a double tibia, also composed of two anterior and two ventral surfaces. This fact is especially clear in the case of the tibia and is proved by the arrangement of the spines and spurs. In a normal tibia there are two spurs, one posterior and one anterior, and the posterior spur is longer than the anterior. Now in this tibia there are *three* spurs, two shorter ones at either margin of the apex, and *one longer one with a bifid point* between them, which is clearly therefore *a pair of posterior* spurs not completely separated from each other. This view of the structure of the double tibia is equally evident from the arrangement of the remainder of the spines on its surfaces. In it the inclination of the ventral surfaces is about the same as in the femur, but is perhaps rather more acute. The 1st tarsal joint is similarly a double structure. Its apex presents two articulations, but while the posterior bears a complete 4-jointed continuation, the anterior bears only a single aborted joint, from which possibly some portion has been detached, but this is not certain.

The relations of the parts are a little obscured by the fact that the normal tibia is slightly bent. The double part of the trochanter lies very nearly anterior to the single part but it is also somewhat *dorsal* to it. This gives to the base of the double femur a trend dorsalwards: but from the base the femur curves ventralwards so that the nett result is that its apex is actually ventral to the apex of the single femur when both limbs are extended. This curve of course gives the femur an abnormal form which is increased by the fact that it is perceptibly shorter than the single femur. Now the relative position of the pair of extra limbs is that marked VAA, and as it stands when extended the apex of the double femur and the peripheral parts of the double limb stand in the Position VAA with regard to the single limb;

but as has been mentioned, by the curvature of the double femur its base is somewhat dorsal to the single limb. This specimen was very kindly lent to me by Dr L. VON HEYDEN and was first described and figured by him in *Deut. ent. Zt.*, 1881, xxv. p. 110, fig. 25.

(3) *Position A.*

*745. **Eurycephalus maxillosus** (Longic.): right anterior femur divides at base into two parts, of which the posterior bears a normal leg. The other part of the femur is *bilaterally symmetrical*, being made up of the anterior surfaces of two femora, for both sides present the same convexity (Fig. 157), neither being flattened as the posterior surface of a normal first femur is. With the apex of this joint articulates a bilaterally symmetrical tibia of extra width, bearing a 1st and 2nd tarsal joint, each of nearly double width.

The 2nd tarsal joint bears two 3rd tarsal joints, which are both much wider than the normal 3rd joint of the tarsus. (This is exaggerated in the diagram.) One of these in 1891 bore a perfect terminal joint with a pair of claws; but the terminal joint and claws of the other side were gone, though Mocquerys' figure shews that they were originally present. Mocquerys' statement that "*la cuisse antérieure du côté droit se bifurque dès son origine en deux branches ayant chacune le volume d'une cuisse normale*" is misleading, as suggesting that the two femora are similar, while upon closer examination they are seen to be dissimilar. Here a pair of extra legs arising from the anterior surface of the normal limb, are compounded together as in the position marked A in the Scheme. Specimen originally described by MOCQUERYS, *Col. anorm.*, 1880, p. 54, *fig.*

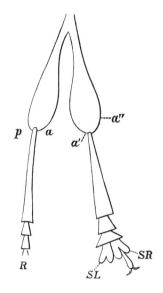

FIG. 157. *Eurycephalus maxillosus.* Right anterior leg bearing an extra pair arising from the femur. R, the normal right. SL, supernumerary left. SR, supernumerary right. p, posterior surface. a, anterior surface of normal femur. a', a'', the two *structurally* anterior surfaces of the extra legs. (In Rouen Mus.)

*746. **Eros minutus** (Malacoderm): right anterior tibia slightly divided at apex, forming two apices (Fig. 158). The posterior apex bears a normal tarsus. The anterior apex bears a double tarsus having the first three joints simple (3rd being enlarged). The 4th joint is of nearly double width and bears peripherally two claw-joints each with a pair of claws. From the structure of these it was clear that they are a *pair*. When extended the three plantar surfaces are not truly in a horizontal plane, as they

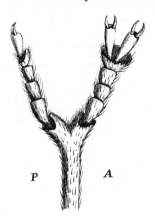

FIG. 158. *Eros minutus*, No. 746. The right fore leg seen from dorso-posterior aspect. *P*, posterior face. *A*, anterior face. This figure was drawn from the microscope and has been reversed. (From a specimen the property of Dr Mason.)

should be in Position A, but they are nearly so. This deviation is exaggerated in the figure. Specimen very kindly lent by Dr MASON.

747. **Aleochara mæsta** (Staph.): middle left tibia has two articulations at apex. The posterior bears a tarsus normal in form but without claws. The anterior bears an extra tarsal series with a *pair* of rudimentary terminal joints, each having a pair of claws. Of this double tarsus the 3rd and 4th joints are not distinctly separated. The parts are in Position A. Specimen kindly lent by Dr MASON.

748. **Meloe proscarabæus** ♂ (Heteromera). The apex of the femur of right hind leg is extended on the anterior side so as to form a second apex in the same horizontal plane. With this second apex articulates the common head of a pair of extra tibiæ each bearing a complete tarsus. As usual they are a right and a left respectively. The two extra legs are twisted out of their natural position so that they turn their ventral surfaces upwards. The tibia which in origin is remoter from the normal tibia is moreover bent over the nearer tibia so that it stands actually nearer to the normal tibia. In this way the morphological relations are obscured, but nevertheless on tracing the ventral surfaces up to the point of articulation with the femur it is clear that they arise in the normal position and that they have the relations marked in the Scheme for the Position A, which is their position of origin. As this case is a somewhat obscure one, I may add that Dr Sharp, who has kindly examined this specimen, gives me leave to state that he concurs in the above description. This is the specimen described by von HEYDEN, *Isis*, 1836, IX. p. 761 and by MOCQUERYS, *Col. Anorm.* p. 52, *fig.*, and was kindly lent to me by Dr L. von HEYDEN in whose possession it remains.

749. **Cetonia opaca** (Lamell.): [right fore leg bears a pair of extra terminal tarsal joints very nearly in Position A, arising from 4th tarsal joint. All the claws are turned ventralwards, but those of the extra joints are turned away from each other as well as downwards]. MOCQUERYS, *l.c.*, p. 61, *fig.*

*750. **Prionus coriarius** ♀ (Longicorn), having parts of a supernumerary pair of tarsi arising from the middle right tarsus, and also a similar double structure arising from the posterior right tarsus (Fig. 159). This is a very important case as a clear

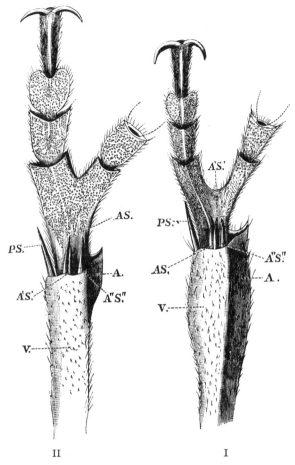

Fig. 159. *Prionus coriarius*, No. 750. I. Apex of tibia of right hind leg with its tarsus. II. Similar parts of right middle leg. (The property of Dr von Heyden.)
PS, *AS*, posterior and anterior tibial spurs belonging to the normal leg. $A'S'$, $A''S''$, the spurs of the extra legs, all structurally *anterior* spurs. *A*, anterior surface. *V*, ventral surface. (The property of Dr von Heyden.)

illustration of the mode in which double supernumerary limbs may be compounded together so as to closely simulate a single limb. Each of the extra parts in this case in the original account was described as a *single* extra limb, but as will be shewn, each is really composed of parts of a complementary pair. Cases of this kind suggest very strongly that other cases of supposed single extra limbs are really instances of double extra limbs in which the duplicity is disguised.

Right hind leg (Fig. 159, I), the tibia is dilated towards the apex which presents dorsally two emarginations instead of one as usual. On the ventral aspect of the apex there are two whole spurs PS, $A''S''$ and a double one AS, $A'S'$, between them.

These spurs give the key to the nature of the structure. The proximal tarsal joint gives off a process on its anterior side and is then continued to bear a normal termination as shewn in the figure. The process from the first tarsal bears a second tarsal from which the termination has been broken off. The extra parts are as in the figure, being covered ventrally from edge to edge with papillæ, and having *no longitudinal cleft* in the middle line like the normal tarsus.

Looking at these tarsal joints alone, the real nature of the extra parts does not appear, for the anterior and posterior surfaces of the normal tarsi are not differentiated from each other, and hence it is not possible to say of what parts the supernumerary limb is made up. Fortunately, however, the tibial spurs are normally distinguishable from each other, for the anterior spur is a short spur while the posterior is a long thin spur. Now the spurs present in this case are firstly *one* long posterior spur PS, and then *three short anterior* spurs, of which two are united for part of their length AS, $A'S'$. *The extra spurs are thus both anterior spurs*, that of the extra tarsus which is nearer to the normal being united to the normal anterior spur. Hence this case is a case of a supernumerary pair of appendages compounded together in the Schematic Position A, having the posterior surfaces adjacent and suppressed.

Right middle leg. (Fig. 159, II.) In this case there would have been more difficulty in making out the real nature of the parts; for in the normal middle leg the anterior spur is not so much differentiated from the posterior one as it is in the hind leg: but having this case for comparison it is easy to see that this also is a case of a pair of appendages similarly compounded in Position A. This case differs from that of the hind leg in the fact that the parts are not so fully formed, and especially the anterior spur of the nearer extra tarsus is scarcely separated from the anterior spur of the normal. By turning the specimen over in the light however, its form can be made out to be that shewn in the figure. When the specimen was received by me the parts present were as shewn in the figure, but when originally described by von Heyden there was a third joint in the extra appendage which was small and elongated, and to all appearance it was the original termination and nothing had been broken off. For the loan of this specimen I am indebted to Dr L. von Heyden, who originally described and figured it in *Deut. ent. Zt.*, 1881, xxv. p. 110, *figs.* 27 and 28.

In the two following cases there was nothing to differentiate the normal limb from the two supernumeraries, and the Position may either have been P or A.

751.   **Fœnius tarsatorius** (Ichneumon): tibia of left posterior leg bears a pair of supernumerary limbs. This is rather a remarkable case by reason of the great similarity in the modes of origin of the three limbs, whence it is difficult to determine positively which is the normal one. The tibia divides into three parts which lie in a horizontal plane and are separate from each other for about $\frac{1}{3}$ of the length of the tibia. Of these the anterior is a good deal more slender than the other two which are similar and about of normal size. The middle of the three is shewn by its spurs to be a *right* limb. Each bears a complete tarsus The ventral surface of the most anterior tibia is horizontal while those of the other two are not quite so, but converge downwards at a very obtuse angle. From this fact, and from the equality in size between them, it seems probable that the two posterior limbs are the supernumerary pair. The Position is therefore very nearly P or perhaps A. This specimen was described by Mr HARRINGTON in *Can. Ent.*, 1890, p. 124, who was so kind as to lend it to me.

752.   **Agestrata dehaanii** (Lamellicorn): the coxa of the right anterior leg has two articulations, one anterior and the other posterior. With the anterior there articulates a single trochanter, bearing a normal right leg. The posterior articulation bears a large structure which is composed of two trochanters united together. This double trochanter bears two legs and is placed in such a way that the two do not lie in the same horizontal plane; but the posterior extra leg is in the same horizontal plane as the normal leg while the anterior extra leg is wedged out towards the ventral surface, between the normal leg and the posterior extra leg. The posterior extra leg is a normally shaped right leg having its structurally anterior surface forwards as normally. The anterior extra leg is fashioned as a *left* leg and the surface of it which is structually anterior faces backwards towards the other extra leg. These two are therefore a complementary pair, having their structurally anterior surfaces adjacent: all three legs are normal and similar in form, size and colour. [Specimen kindly lent by Mr E. W. Janson.]

(4)   *Positions DAA and DA.*

*753.   **Cerambyx scopolii** (Longicorn.): pair of extra legs arising from the coxa of the right anterior leg. As this is a remarkably simple and perfect case it will be well to describe it in some detail, as it will serve to illustrate the arrangement of such cases in general.

A normal leg of such a beetle as *Cerambyx* consists of coxa, trochanter, femur, tibia and four tarsal joints. To a proper understanding of the mode of occurrence of the extra legs in this case it is essential that the forms of these parts and their mode of movement with regard to the body and to each other should be accurately known.

Of the large, irregularly pear-shaped *coxa* only the hemispherical face is seen from the exterior. It is chiefly enclosed by embracing outgrowths from the sternum, forming a socket in which it can be rotated like a ball. Upon its broad, exposed surface it is itself hollowed out to form a socket for the ball of the *trochanter*. For our purposes it is necessary to find some means of distinguishing the anterior face of the coxa from the posterior face. The structure which at once enables us to do this is the process (Fig. 160, *p*), which goes down from the coxa to embrace the neck of the ball of trochanter and lock it into its socket. Now in the case of an anterior leg, this process is *posterior* to the trochanter (but in a middle or hind leg it is *anterior* to the trochanter). The next point to be considered is the position of the *femur*. The

492   MERISTIC VARIATION.   [PART I.

femur itself is flattened antero-posteriorly, having two broad surfaces, morphologically anterior and posterior, and two narrow surfaces which are extensor and flexor surfaces, or morphologically dorsal and ventral.

By rotation of the coxa the whole leg may assume a great variety of positions, and it is thus of the utmost consequence that the nature of the surfaces be truly recognized. If the front leg be placed with the

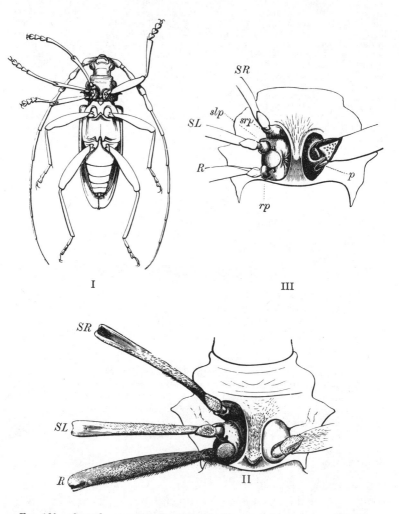

Fig. 160. *Cerambyx scopolii*, No. 753. I. The whole beetle seen from ventral surface. II. Details of right anterior coxa bearing extra trochanters and legs. In this figure the legs are rotated so as to shew that *SR* is an image of *SL*. III. The same, placed so as to shew that *SL* is an image of *R*.

p, process of coxa locking in the trochanter. *srp*, *slp*, corresponding processes for the extra trochanters. (From a specimen belonging to M. H. Gadeau de Kerville.)

femur at right angles to the body it may either be placed so that the ventral surface is downwards, or by rotation of the coxa through 90° the broad posterior surface may be downwards. The rotations of the middle and hind legs are complementary to this.

In the abnormal specimen the extra pair of legs arise from the anterior side of the normal coxa, forming with it a solid mass and preventing its free rotation in its socket, so that the normal leg can scarcely be moved from the first position with the ventral surface downwards. The common coxal piece is about half as large again as the normal. Posteriorly it bears the trochanter of the normal leg, which is of full size and of proper proportions. The process of the coxa locking in the ball of the trochanter is *posterior*, as in the normal front leg.

Anteriorly the legs SL and SR articulate with the coxa by separate trochanters. Each is separately closed in by a process of the coxa, slp and srp, respectively. Of these processes that of the leg SR is *posterior*, but that of SL is *anterior*. Hence the two legs are complementary to each other, and SR is a *right* leg while SL is a *left*. This complementary relation is maintained in all the other parts of these legs. In size the two extra legs are rather more slender than the normal leg.

It was explained in the introduction to the subject of supernumerary legs that the relations of form between them depend upon the surface of the normal leg from which they arise. Here the point of origin is chiefly anterior to the normal leg, but is also slightly nearer to the extensor or dorsal surface of the coxa. This is not at first sight evident owing to the rotation of the normal leg due to the great outgrowth from its anterior surface; but nevertheless if the plane of the ventral surface of the normal femur were produced, it would pass ventrally to the ventral surface of the remoter extra leg SR, and therefore this leg is morphologically dorsal to the leg R. The positions of the extra legs are approximately those of the Scheme for the radius marked DAA, and while the surfaces of SR are parallel to those of R when both are extended, the surfaces of SL are inclined slightly to them as in position DAA. In the enlarged Figure III the coxa is rolled forwards so as to exhibit the relation of images between R and SL, and the figure II shews the coxa rolled back to shew the similar relation between SR and SL.

For the loan of this beautiful specimen I am greatly obliged to M. Henri Gadeau de Kerville.

754.   **Harpalus rubripes** (Carabidæ): left posterior tibia bears a supernumerary pair of tarsi. The apex of the tibia is widened and presents two articulations, of which the posterior bears a normal tarsus. The anterior articulation bears a pair of complete tarsi having proximal joints compounded. The two extra tarsi are a complementary pair, the posterior being fashioned as a *right*. The surfaces adjacent in these two tarsi are structurally posterior surfaces, but they are a little supinated, so that the ventral surfaces are also partly turned towards each other. The position of origin and the relations of the surfaces to each other are almost exactly those which are

494   MERISTIC VARIATION.   [PART I.

indicated in the Scheme for the position DAA. This specimen was described by M. A. FAUVEL (*Rev. d'Ent.*, 1889, p. 331) and was kindly lent by him for further examination.

755.   **Chrysomela banksii** (Phytophagi): right hind tibia bearing an extra pair of tarsi. The border of the tibia which corresponds in position to the ventral or flexor border of the normal tibia is covered with the hairs which characterize it in the normal limb; but the opposite border of this abnormal tibia is similarly covered with hairs, shewing that the anterior parts of at least two tibiæ are included in it. A rigid process projects from the wide apex of the tibia. Upon the inner side of this process is the articulation for the tarsus, which from its direction and position appears to be the normal tarsus of the limb. Outside the process articulates a slightly smaller tarsus, which from its form and from the plane in which it moves is a *left* tarsus, flexing away from the normal one. At a point slightly external to this is the third tarsus, which is again a *right* tarsus and moves in a plane complementary to the middle one. The two are therefore a pair. The position of origin is anterior and dorsal, being nearly that marked DA, but the relative positions of the extra tarsi are approximately DDA. As to the nature of the tibial process I can make no conjecture. (Fig. 161.)

FIG. 161. *Chrysomela banksii*, No. 755. View of right hind tibia from posterior surface. A normal right hind tibia is shewn for comparison. (From *Proc. Zool. Soc.*, specimen the property of Dr D. Sharp.)

This specimen is the property of Dr Sharp, who was good enough to lend it to me. It was briefly described and figured by me *P. Z. S.*, 1890, p. 583, but I was not at that time aware of the complementary relation existing in these cases and failed to notice the somewhat inconspicuous differences which are evidence of it in this case.

756.   **Hylotrupes bajulus** (Longic.): right middle tibia bears a supernumerary pair of limbs having proximal parts in common. From the antero-dorsal surface of the base of the normal tibia, there arises a slender tibial piece which is not so long as the normal tibia and bears no spurs. At the apex of this supernumerary tibia, which is doubtless a double structure, articulate a pair of tarsi having their first and second joints compounded together. After the second joint the two tarsi separate from each other and each bears a pair of claws. The relative position of the two tarsi when they separate from each other is almost exactly that marked DA. It should be mentioned that the supernumerary parts central to the 3rd tarsal joints are not fully formed, being deficient in thickness, and the transverse separation between the 1st and 2nd tarsal joints is incomplete. Specimen first described by MOCQUERYS, *Col. anorm.*, 1880, p. 53, fig. I am indebted to Dr L. VON HEYDEN for an opportunity of examining it.

(5)   *Position D.*

*757.   **Aphodius contaminatus** ♂ (Lamellicorn.): left middle tibia bearing two supernumerary tarsi which stand very nearly in the position DDP, being rather nearer to D. The relative positions are shewn in Fig. 162. The articular surface at the apex of the tibia is extended along an elongated process which projects on the dorsal side of the tibia. Upon this extension of the apex articulate two extra tarsi. They stand with their ventral or plantar surfaces facing each other, and the tarsus RT is placed so that its dorsal surface is very nearly opposed to the dorsal surface of the normal tarsus LT, and the three tarsi thus flex almost in the same vertical plane. It is to be observed, however, that the tarsus LT is not actually in the same plane as the other two, but is a little deflected from it so as to flex rather more towards the posterior surface of the line than it would do if it stood actually as $L'T'$ stands. This may be made clear by reference to the Scheme (p. 481): for while the two extra tarsi are placed relatively to each other as if they were in the position D, the position of RT to LT is that which it would have if it stood in DDP.

In this species the middle tibia in the male bears one large spur, namely, the posterior one, while the anterior spur is rudimentary. PS in the figure represents the large posterior spur of the normal tarsus LT, while a large double spur $RP'S'$, $LP'S$, standing posteriorly and between the two extra tarsi represents their two posterior spurs. The double nature of this spur is seen

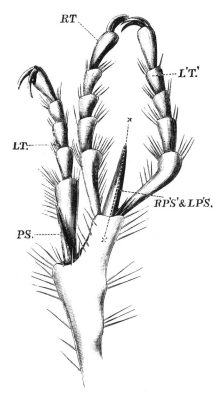

FIG. 162. *Aphodius contaminatus*, No. 757, left middle tibia bearing extra parts. LT, normal left tarsus. RT, LT, right and left extra tarsi. PS, normal posterior spur. $RP'S'$ and $LP'S$, spur representing compounded spurs of RT and LT corresponding with the single spur PS of normal. x, x, line of suture between these two spurs. The limb is seen from the posterior surface. (Specimen the property of Dr Kraatz.)

when it is examined from the anterior side, for upon that surface it is marked by a longitudinal ridge-like suture. This specimen was first described by KRAATZ, *Deut. ent. Zt.*, 1876, xx. p. 378, fig. 13, and I am indebted to Dr Kraatz for an opportunity of examining it.

758.   **Galerita africana** (Carab.): (Fig. 163) right middle leg normal as far as the last tarsal joint, which bears three additional claws arising dorsally to the normal pair. The extra claws are three in number, two of them being small and standing at the anterior border of the limb, while at the posterior border there is one claw of larger size. This larger claw is really a double structure, which is clearly shewn by the

presence of *two* channels on its concave surface. Position of origin is therefore D, while the inclination of the extra pairs of claws to each other is about that required

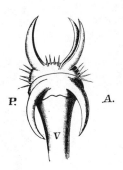

FIG. 163. *Galerita africana*, No. 758. Apex of right middle tarsus. *A*, anterior. *P*, posterior. *V*, ventral. (Specimen in Rouen Mus.)

for the position DDA; for the planes of the two pairs are not parallel but incline to each other at an acute angle. Specimen originally described by MOCQUERYS, *Col. anorm.*, p. 64, *fig.*

(6) *Position DP.*

*759. **Pyrodes speciosus** (Longic.); having two supernumerary legs articulating with the thorax by a common coxal joint, which is distinct from the coxa of the left middle leg, but is enclosed in the same socket with it. In this remarkable case the normal leg is complete, though slightly pushed towards the middle line. The socket in the mesothorax is enlarged posteriorly and dorsally so as to form an elongated, elliptical articulation, which lies obliquely, so that its ventral end is anterior to its dorsal end. The anterior and ventral end is occupied by the coxa of the normal leg, while the coxal joint of the two extra legs fills the space dorsal and external to it. Both are capable of being moved independently in the relaxed insect. The extra legs articulate with their coxa by a common double trochanter which has two apices, from which point the legs are distinct. Their position is dorsal and posterior to the normal leg, being practically that marked DP in the Scheme, and the relative positions of the extra legs are very nearly those indicated for the Position DP. The leg nearest the normal leg is of course a *right* leg in structure, and its plantar and a little of its structurally anterior surfaces are turned posteriorly. On the other hand, the remoter leg is a true left leg and the ventral surface of its femur is placed almost exactly horizontally. All three legs are complete, but they are a little shorter and more slender than the middle leg of the other side.

This specimen is in the Hope Collection at Oxford.

*760. **Carabus irregularis** ♀ ; left middle leg and right hind leg bearing supernumerary tarsal portions. In the *left middle leg*, Fig. 164, I, the 2nd tarsal joint is short and thick; the 3rd joint is partially double, as shewn in the figure. One of its apices bears a tarsus of reduced size, and the other apex, which is *postero-dorsal*, bears a double tarsus having common 4th and 5th joints. The 5th joint of the latter bears two pairs of claws which curve ventrally and partly *towards* each other. The figure I shews the appearance from the ventral or concave side of the claws, while the figure II is drawn from the convex or dorsal side. The disposition and small number of the spines on the ventral side of the extra 5th joint shew that the ventral surfaces are partly suppressed, and in fact that the surfaces which are adjacent in the extra tarsi are in part *ventral* surfaces. This view is also borne out by the direction and curvature of the claws. Relatively to each other and to the normal the extra parts have nearly the Position DP.

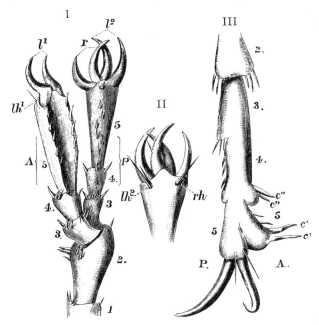

Fig. 164. *Carabus irregularis*. I. Semi-diagrammatic view of left middle leg from antero-ventral surface. l^1, the claws of normal left tarsus. r, l^2, claws of extra tarsi compounded together. lh^1, hair marking the dorsal surface between the claws. A, anterior. P, posterior. II. Dorsal view of apex of extra tarsus rh, lh^2, two hairs marking dorsal surfaces. III. Dorsal view of right hind leg. $c'c'$, $c''c''$, claw-like spines, perhaps representing extra claws. (Specimen the property of Dr Kraatz.)

The *right hind tarsus* has the form shewn in Fig. 164, III. The 3rd, 4th and 5th joints are not fully separated from each other. Both the 4th and 5th joints bear extra parts, but their nature is obscure. The 5th joint is partly double, and the anterior part bears two shapeless claw-like spines ($c'c'$). The 4th joint bears a similar pair of claw-like structures of smaller size ($c''c''$). Probably these should be considered as rudiments of extra tarsi; but if this view is correct, it appears that two extra tarsi are present, arising from different joints. For the loan of this specimen I am indebted to Dr G. KRAATZ, who first described and figured it in *Deut. ent. Zt.*, 1877, XXI. pp. 57 and 63, fig. 27.

761.   **Chrysomela graminis** (Phytophagi): the femur of the right middle leg bears a supernumerary pair of legs attached to the posterior and dorsal side of its apex. At this point there is an articulation with which the single proximal part of the extra pair of tibiæ articulates. This piece, which is common to the two supernumerary tibiæ, is a sub-globular, amorphous mass from which the two tibiæ diverge. Each of the two tibiæ bears a complete tarsus, except that the most posterior has only one claw. In colour the two supernumerary tarsi differ from the normal, being brown instead of metallic green, but the tibiæ are normal in colour. From the shape of the articulations and the arrangement of the pubescence, it is clear that the surfaces of the legs which are naturally adjacent are constructed as *posterior* surfaces, and the forms of the two are complementary to each other, the hindmost of the extra legs being formed as a left leg, while the foremost is a right leg. As they stand, however, the two tibiæ are not in the same position relatively to the body, for the foremost is placed normally, having its plantar surface turned downwards, but the hindmost is rotated so that its plantar surface is partially turned *forwards*. The relative positions are nearly those marked DP in the Scheme, but the most posterior tarsus is more rotated than it should be according to that diagram. This condition may be to some degree connected with the presence of the amorphous growth at the base of the extra tibiæ. This specimen was kindly lent for description by Dr Mason.

762.   **Pimelia interstitialis** (Tenebrion.): left posterior femur bears two supernumerary tibiæ arising from the postero-dorsal surface of its apex. These two are a pair, for the tibia nearest to the normal tibia is a *right* tibia, the remoter being a left. The adjacent surfaces are chiefly anterior surfaces in structure, but the ventral surfaces are inclined to each other at an obtuse angle. The position of the extra legs is almost that marked DP in the Scheme, but the inclination of the ventral surfaces of the extra legs is rather more acute than it would be in the Position DP. The tarsi are all broken off. Specimen originally described by MOCQUERYS, *Col. anorm.*, p. 44, *fig*.

763.   **Acinopus lepelletieri** (Carab.): two extra legs arising from posterior surface of base of femur of *l.* middle leg. From position it seems that the most anterior is the normal, but this is doubtful. The arrangement is nearly that of Position DP, but as one of the femora is constricted and bent, the relations are rather irregular. Specimen first described by MOCQUERYS, *Col. anorm.*, p. 41, *fig*.

(7)   *Position P.*

*764.   **Silis ruficollis** ♂ (Malacoderm): right anterior femur bearing a supernumerary limb (Fig. 165). The coxa and trochanter normal. The femur is of about twice the antero-posterior thickness of a normal femur and at its apex presents two articulations in the same horizontal plane. Of these the anterior bears a normal tibia and tarsus, but the posterior bears an extra tibia which appears at first sight to be a single structure. This tibia is more slender than the normal one and is provided with four tarsal joints, the terminal one being withered and without claws. Upon closer examination it is found that this extra tibia is in reality made of the *posterior surfaces of a pair of tibiæ* not separated from each other. In this case the morphological duplicity of the extra tibia is capable of proof. For, as shewn in Fig. 165, II, the normal tibia is not bilaterally symmetrical about its middle line. On the contrary the anterior surface is differentiated from the posterior by several points. This may be seen in the spurs at the apex of the tibia, for the anterior spur (*a*) is long, but the

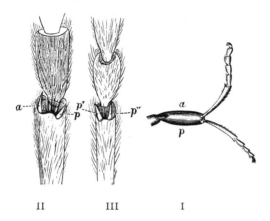

Fig. 165. *Silis ruficollis*, No. 764. I. The right anterior leg seen from ventral surface. *a*, anterior. *p*, posterior. This figure was drawn with the microscope and is *reversed*. II. Detail of apex of tibia of the anterior or normal tibia, shewing *a*, the anterior, and *p*, the posterior spurs. III. Similar detail of apex of the tibia of the extra limb, shewing *p'* and *p''*, *two* structurally posterior spurs.

posterior (*p*) is short (as is usual in the *front* leg of many beetles). The hairs on the surface of the tibia are also directed asymmetrically and the parting or division between them is not median, but is nearer to the anterior border (see figure).

But in the extra part there is no such differentiation, and *both surfaces are structurally posterior surfaces*. The hairs part in the middle, and both spurs (*p'*, *p''*) are formed as posterior spurs. This extra structure is therefore made up of the two posterior borders of a right and a left tibia compounded together in Position P. (See diagram, Fig. 165, II.)

This specimen was found by Dr Sharp amongst a number of insects collected by myself in his company at Wicken Fen on Sunday[1], July 26, 1891.

Such a case taken in connexion with others (*e.g.* No. 801) makes it certain that many cases of supposed "single" extra appendages are really examples of *double* extra parts.

[1] A day or two before, the manuscript of this part of the subject had been put by with the remark that no good opportunity of thoroughly investigating a case of "single" extra leg had occurred, but that it could scarcely be doubted that traces of duplicity would be found in them. Considering the great rarity of extra appendages in Insects, and remembering that even of the whole number very few are of the supposed "single" order, I have thought the occurrence of this capture a coincidence of sufficient interest to be worthy of mention. Dr Sharp tells me that amongst all the beetles that have gone through his hands only one case of extra appendage (No. 755) was seen.

765.   **Scarites pyracmon** (Carab.). At base of posterior face of the trochanter of left normal front leg, immediately above the cotyloid articulation was implanted an elongated lanciform joint. This joint was directed backwards and represented a pair of trochanters compounded by their anterior surfaces. With each of the two apices of this double trochanter was articulated a complete leg, in all respects formed as an anterior leg. The two moved as a complementary pair. [Details given. This is one of the earliest and best described cases. Asmuss[1] in quoting it points out that the description and figure plainly shew that the two extra legs were a pair, a right and a left, respectively. They were in fact a pair, arising from the posterior surfaces of the normal leg, and presenting their anterior surfaces to each other.] LEFEBVRE, A., *Guérin's Mag. de Zool.*, 1831, Tab. 40.

766.   **Geotrupes mutator** (Lamellicorn): two supernumerary limbs arising from femur of right anterior leg. Femur greatly widened, upon posterior border giving off a large prominence which divides into two processes at right angles to each other. Each of these processes bears a normal tibia and tarsus, but the foremost of these tibiæ is shaped as a *left* tibia, having its serrated border placed anteriorly, while the other extra tibia is formed as a *right* tibia, having its serrated border placed *posteriorly*. [The pair of limbs arise from the posterior surface of the normal limb and have their anterior surfaces adjacent, as in Position P.] FRIVALDSKY, J., *Term. Füzetek.*, 1886, x. p. 79, Pl.

767.   **Pterostichus lucublandus** ♀ (Carabidæ): third tarsal joint of left middle leg at apex presents wide articular surface. On this stands a triple 4th joint, made up of a single anterior portion, bearing the rest of the normal tarsus and a posterior portion, double in structure, the two parts being completely united. The single anterior part of this 4th joint bears a normal 5th joint with claws. The double posterior part of the 4th joint bears a pair of separate 5th joints, each having a pair of claws. Of these the anterior is perfect, but the peripheral part of the posterior 5th joint is crumpled, so that its claws are twisted out of position, but at its base it stands exactly as the normal 5th joint, and as the 5th joint of the anterior extra tarsus, all three being in the same horizontal plane. These extra parts, therefore, are in the Position marked P in the Scheme and have the relations there indicated for that position. This specimen was kindly lent to me by Mr HARRINGTON, who first described it *Can. Ent.*, 1890, xxii. p. 124.

(8)   *Positions VPP to VVP.*

*768.   **Ceroglossus valdiviæ**, Chili (Carabidæ): left anterior tibia bearing a pair of supernumerary legs. The tibia widens, and in its middle part gives off posteriorly and ventrally a wide branch having the form of a pair of tibial apices compounded together. The double tibia bears two tarsi (Fig. 166, R', L') having a common proximal joint, but these have unfortunately been broken, two joints being missing from the one and three from the other. The legs are a right and left as usual, and they stand in the relative positions marked VPP in the Scheme. This is a very simple and striking case, for the animal is of good size and the parts are well formed. The two tibial spurs which are adjacent in the two extra tibiæ are compounded so as to form a double spur with two points as shewn in the figure. As shewn for the Position

[1] *Monstrositates Coleopterorum*, 1835, p. 44, Pl.

VPP in the Scheme, the *compounded* parts of the extra appendages, viz., the double tibia and the double first tarsal joint have two complete ventral surfaces inclined to each other at an obtuse angle, while there are only two *halves* of dorsal surfaces.

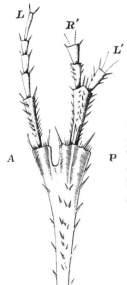

FIG. 166. *Ceroglossus valdiviæ*, No. 768. Left anterior tibia with extra parts seen from the ventral surface. L, the normal left tarsus. R' and L', the extra tarsi, compounded in their proximal joint. A, anterior. P, posterior.

Note that the anterior spur of the normal is curved and that the double spur representing the two anterior spurs of the extra tibial apices has thus a bifid point. (Specimen the property of Mr E. W. Janson.)

Similarly there are two structurally posterior surfaces, but no structurally anterior surfaces, for these are adjacent and undeveloped. This specimen was kindly lent by Mr E. W. JANSON.

769.   **Silpha nigrita** (Heteromera): from right middle femur arises a pair of legs which are completely united as far as the apex of the last tarsal joint. The point of origin of the supernumerary limbs is on the anterior and ventral border of the femur. The form of the extra limbs is shewn in Fig. 167. The surfaces V and V' are structurally ventral surfaces. They are turned chiefly forwards, but are inclined to each other at an acute angle. The surfaces, therefore, which are adjacent in this pair of legs, and which are consequently obliterated, are chiefly the morphologically anterior surfaces and to some extent the dorsal surfaces. The plantar or ventral surfaces of the last tarsal joints are inclined to each other rather more obtusely than those of the tibiæ, so that the curvatures of the two pairs of claws are very nearly turned forwards as well as away from each other. This is not fully brought out in the figure. The position of origin is about VP, but the claws are in Position VPP. Specimen first described and figured by MOCQUERYS, *Col. anorm.*, p. 43, *fig.*

502   MERISTIC VARIATION.   [PART I.

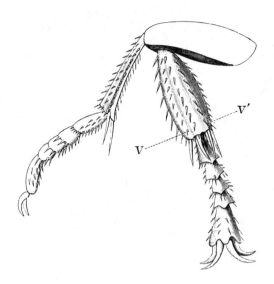

FIG. 167. *Silpha nigrita*, No. 769. Right middle femur bearing a compounded pair of extra legs. *V*, ventral surface of nearer extra leg. *V'*, ventral surface of the remoter extra leg. (In Rouen Mus.)

770.   **Tenthredo solitaria** (Sawfly): tibia of right middle leg divides in peripheral third to form two branches, of which the anterior bears the normal tarsus. The posterior branch arises from the postero-ventral surface of the normal and bears a double tarsus consisting of the posterior parts of a pair compounded in Position VP, almost exactly. Tibial spurs as in Fig. 166. The compound tarsus has only 4 joints, the 5th being apparently broken off. In Cambridge Univ. Mus., history unknown.

771.   **Telephorus rusticus** (Malacoderm): tibia of left middle leg dilated and somewhat deformed in its peripheral portion. It presents two apical processes, the one anterior and the other posterior. The anterior of these bears a normal, backwardly directed tarsus, but the posterior process bears two tarsi by separate articulations. The anterior of these two tarsi is directed forwards to face the tarsus of the other apex, but the posterior tarsus is backwardly directed. [From its attitude it is clear that the middle of these tarsi is a structure complementary to one of the others, but there is no evidence to shew whether it is a pair to the anterior or to the posterior. Position either VPP, or DAA, probably the former.] KRAATZ, *Deut. ent. Zt.*, 1880, p. 344, fig. 33.

772.   **Anthia** sp. (Carabidæ): left posterior tibia bearing two supernumerary tarsi. The postero-ventral side of the apex of the tibia is dilated so as to form a triangular projection, causing the point of articulation of the normal tarsus to be raised upwards. The projection bears two tarsi of which the posterior curves downwards and backwards, being fashioned as a left tarsus while the anterior curves forwards and slightly upwards being a *right tarsus*. These two tarsi have unfortunately been broken but were presumably complete. The whole apex of the tibia bears five spurs instead of two, but the relation of the spurs to the separate tarsi was not clear. The

tarsi are very nearly in the Position VPP. Specimen very kindly lent by Mr E. W. Janson.

773. **Julodis æquinoctialis** (Buprestidæ): the extra legs arise from the posterior and ventral side of the base of the tibia of left middle leg. They are a pair, and are compounded together by their lateral and dorsal surfaces in such a way that the morphologically ventral surfaces of the two are almost in contact along the anterior border of the compound limb. The ventral surfaces here converge at an acute angle. The two extra legs are compounded together throughout the tibiæ and first 4 tarsal joints. The 5th tarsal joints are free, but only one of them remains. The former presence of the other is only shewn by a socket. The normal tibia is constricted and bent at one point so that it does not stand in its normal position. The femoro-tibial articulation is rigid.

This is a case of a pair of legs compounded as in the position marked VVP in the Scheme but the point of origin is more nearly that of VPP. Specimen originally described by MOCQUERYS, *Col. anorm.*, p. 47 *fig.*

774. **Metrius contractus** (Carab.) Esch.: specimen in which the middle left femur bears an incomplete pair of legs in addition to the normal one. The femur is of normal length. The tibia of the normal leg is articulated with the end of the femur as usual, but is somewhat shorter, stouter and more curved than the tibia of the corresponding leg of the other side. A supernumerary tibia arises from the posterior [and ventral ?] side of the femur a short distance within the apex, and is articulated with it by a separate cotyloid cavity; the two articular cavities for the two tibiæ are confluent, being connected by a groove. The end of this tibia is dilated at its outer end, and bears two articular surfaces, one on each side; with each of these surfaces, a complete tarsus is articulated, nearly normal in form but somewhat stouter than a normal tarsus. There are four terminal spurs to this tibia, two being below the outer tarsus and two being below the inner tarsus. [It therefore seems that this tibia is made up of parts corresponding with the ventral side of a right tibia and the ventral side of a left tibia, and it is hence probable that if the disposition of the claws of the tarsi had been examined, it would have been found that they too were a pair, one being a right foot and the other a left. Position probably VVP.] JAYNE, H. F., *Trans. Amer. Ent. Soc.*, 1880, VIII. p. 156, Pl. IV. figs. 3 and 3 *a*.

775. **Aromia moschata** ♀ (Longicorn): right anterior coxa bearing a pair of supernumerary legs having trochanter and the proximal half of the femur in common. The normal leg and the extra ones were all somewhat reduced in size but were complete. The extra leg adjacent to the normal is a left leg. [From the figure it appears that the legs arose in the Position P, or VPP, and their relative positions seem to have been those indicated in the Scheme. Of course it is not possible to state this definitely without examination, but it is clear that there was at least no great departure from the position shewn in the Scheme.] It is remarked that in this specimen the right mandible was abnormally small. KOLBE, H. J., *Naturw. Wochens.*, 1889, IV. p. 169, *figs.*

---

(9) *Two cases not conforming to the Scheme.*

Two cases of double supernumerary tarsi require separate consideration. The arrangement in both of these cases departs from that which is usually followed, but it will be seen that there is considerable though imperfect agreement between the two exceptions. Both of these occur in the anterior legs of males of the genus *Calathus*, and it happens that in the normal form the apex of the tibia presents a considerable modification from the simple structure of other beetles. This modification affects the anterior legs only, and is found in several genera of Carabidæ, being especially pronounced in *Calathus*.

In order to appreciate the nature of these cases it is necessary that the anatomy of the parts should be understood.

The apex of the tibia in the simple form, *e.g.* the second or third leg in *Carabus*, bears two large articulated spurs. The two spurs are ventral to the articulation of the tarsus, and one of them is placed at the anterior border of the tibial apex while the other is posterior. In these unmodified legs both spurs are placed at the same level in the limb, so that the bases of both are in the same transverse section (cp. Fig. 166). In the forms presenting the sexual modification, the anterior spur is of somewhat small size but occupies the same position relatively to the other parts that it does in a simple leg.

The posterior spur however, which is large, does not stand at the same level on the tibial apex, but has, as it were, travelled up the tibia so that it stands at a considerable distance central to the apex, and instead of marking the posterior border of the limb it is placed nearly in the middle of the actual ventral surface. A long channel runs from the posterior spur to the anterior one, and the appearances suggest that the modified form is reached by a deformation of the original apical surface, which is twisted so that the posterior spur is thus drawn up into the secondary position. In the fore leg of a male *Carabus* the beginning of such a change can be seen, but in *Pterostichus* and especially in *Calathus* it reaches a maximum. The change may be briefly described by saying that a section to include the two spurs must be taken in a plane which is oblique to the long axis of the limb instead of transverse to it.

As a result of this modification the morphological surfaces of the anterior tarsus of *Calathus* &c. have a peculiar disposition relatively to the same surfaces of the tibia when compared with other forms. Commonly the ventral surface of the tarsus is parallel to a line taken through the bases of the spurs, but owing to the rotation of the posterior spur into its secondary position this plane is here oblique to the ventral surface of the tarsus. These points will be at once evident if the front leg of a male *Calathus* is examined.

It was laid down as a principle generally followed in cases of double extra appendages, that the three terminations, when extended, stand in the same plane, and the chief feature which distinguishes the two following cases is that the three terminations are not in one plane.

Moreover, though the two supernumerary tarsi are a complementary pair, and together with the normal tarsus are arranged as a series of images, yet in order to produce the arrangement of the present cases the planes of reflexion would not be parallel to each other (as in Fig. 154) but inclined in the manner to be described.

*776.   **Calathus græcus** ♂ (Carabidæ): left anterior tibia bearing a pair of supernumerary tarsi compounded together. The diagram, Fig. 169, I, shews, in projection, the relations of the parts round the tibial apex. As has been explained, the posterior spurs P^1, P^2 and P^3 are really much central to the apex, but they are here represented as if they were projected upon the apex. The head of the tibia is produced posteriorly into a long and narrow process which is formed of the united parts of the two extra limbs and bears the articulation common to the two extra tarsi. The two tarsi stand with their ventral surfaces almost at right angles to each other, but the united dorsal surfaces are almost in a continuous plane. The fifth joints alone are separate, that of RT being small (Fig. 168).

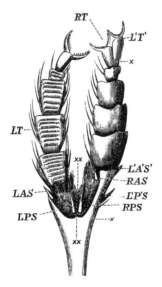

Fig. 168. *Calathus græcus*, No. 776. Left anterior tibia bearing a double extra tarsus. *LT*, normal tarsus. *RT*, *L'T*, extra pair of tarsi. *LAS*, *LPS*, normal anterior and posterior spurs. *L'A'S'*, *L'P'S'*, anterior and posterior spurs belonging to *L'T*. *RAS*, *RPS*, anterior and posterior spurs belonging to *RT*. *x, x*, dotted line indicating plane of morphological division between extra tarsi. *xx, xx*, plane of division between the normal and *RT*. (Specimen the property of Dr Kraatz.)

In studying this case one source of confusion should be specially referred to. It is seen that though the origin of the extra tarsi is posterior to the normal tarsus, the extra tarsi are as a fact united along their morphologically posterior borders. Nevertheless the position of the spurs shews that it is the anterior surfaces which are morphologically adjacent to each other, for the spurs are arranged in the series A^1P^1, P^2A^2, A^3P^3, and the union of the posterior borders of the tarsi is a result of the modification in the form of the tibia consequent on the rotation of the posterior spur.

To produce the arrangement here seen, the planes of reflexion would be M^1 and M^2 respectively, and these are almost at right angles to each other. The present case therefore is very different from those hitherto described, for in them the planes of reflexion were nearly or quite parallel. Whether this difference in the Symmetry of the extra parts may be connected with the departure of the normal tibia from its own customary symmetry cannot be affirmed, but such a possibility should be borne in mind.

This specimen was kindly lent to me by Dr G. KRAATZ, who first described it in *Deut. ent. Zt.*, 1877, XXI. p. 62, fig. 23.

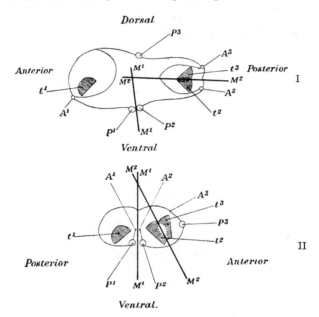

FIG. 169. I. Ground-plan of tibial apex of *Calathus græcus*, No. 776. II. Similar ground-plan of the tibial apex of *Calathus cisteloides*, No. 777.

In each case the spurs are conceived as projected upon one plane. t^1, the normal tarsus. A^1, P^1, its anterior and posterior spurs. t^2, A^2, P^2, similar parts of nearer extra tarsus. t^3, A^3, P^3, similar parts of remoter extra tarsus. M^1, plane of reflexion between t^1 and t^2. M^2, plane of reflexion between t^2 and t^3.

*777.   **Calathus cisteloides** ♂ (Carabidæ): right anterior tibia bearing a pair of supernumerary tarsi compounded together. In this case the extra parts were anterior to the normal tarsus. The parts were arranged as in the diagram, Fig. 169, II, which is a projection of the tibial apex. The apex is produced anteriorly so as to form a wide expansion which bears the common articulation for the double tarsus. This produced portion is of course formed by the composition of parts of a pair of tibiæ. It is noticeable that the three tibial apices which enter into the formation of the general apex are in one respect not actually images of each other. For the angular distances between A^1 and P^1, and between A^2 and P^2, are exceedingly small, being far less than in a normal tibia of the species, and in fact the grooves running from each anterior spur to the corresponding posterior one are almost parallel to each other and to the long axis of the tibia. The tarsi t^2 and t^3 separate in the first joint.

The relative positions are shewn in the diagram, and it is thus seen that the planes of reflexion M^1 and M^2 are inclined to each other at an acute angle.

This specimen was kindly lent to me by Dr L. VON HEYDEN and was first described and figured by MOCQUERYS, *Col. anorm.*, 1880, p. 65, *fig.*

It is difficult to observe the two foregoing cases without suspecting that the fact that they deviate from the normal symmetry of extra parts may be connected with the normal modification of the anterior tibia in these *Carabidæ*. It should be remembered that the tibia and tarsus of the unmodified leg of a beetle are very nearly bilaterally symmetrical about the longitudinal median plane of the limb, but in this leg of these forms the symmetry is lost. Possibly then the upsetting of the ordinary rules for the Symmetry of extra parts may follow on this modification. The difference between the two cases moreover is possibly due to the fact that in one the extra parts are on the posterior surface of the leg, while in the other they are on the anterior. Since the normal limb is not bilaterally symmetrical it is reasonable to expect that the results would differ in the two cases. One other case of a pair of extra tarsi in the fore leg of a male *Calathus* is recorded (No. 777 a), but insufficiently described. It is to be hoped that a few cases of extra tarsi in the fore leg of male *Calathus* or *Pterostichus* may be found, and it is very possible that such a case even in *Carabus* would help to clear up these points.

777 *a*. **Calathus fulvipes** ♂ (Carabidæ): tibia of right fore leg bears pair of extra tarsi. [Fig. and description inadequate.] PERTY, *Mitth. nat. Ges. Bern*, 1866, p. 307, *fig.* 5.

(10) *Nine other cases departing from the Schematic Positions.*

Each of these needs separate consideration.

*778. **Platycerus caraboides** (Lucan.): left hind tarsus has form shewn in Fig. 170. The terminal joint had only one claw. R and L' are presumably the extra pair, but it will be seen that they arise at separate places from the 3rd tarsal joint. Otherwise, they stand approximately in Position V. Described originally by MOCQUERYS, *Col. anorm.*, p. 67, *fig.*

779. **Philonthus ventralis** (Staphylinidæ): third joint of right posterior tarsus bearing supernumerary termination of double structure. The apex of the third joint is enlarged, and at a point anterior and slightly dorsal to the articulation of the normal fourth joint the super-

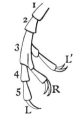

FIG. 170. *Platycerus caraboides*, No. 778. Left hind tarsus from posterior surface. L, the presumably normal apex, has only one median claw. R and L', arise separately from the 3rd joint. (In Rouen Mus.)

numerary parts arise. The fourth and fifth joints of the supernumerary tarsi are of double structure, but are not separated from each other. The double fifth joint bears two pairs of claws, of which the two adjacent members are compounded together at their bases. The plane in which one pair of claws stands is about at right angles to the plane in which the other pair is placed, the opposed surfaces being ventral surfaces. Stated in terms of the Scheme on p. 481, the supernumerary tarsi are placed as in the position DDA, whereas their position of origin is DAA. It is noticeable that the normal fifth joint does not stand quite in its usual position, but is a little twisted so that it partially turns its ventral surface in an anterior direction. This specimen was described and figured by FAUVEL, *Rev. d'Ent.*, 1883, II. p. 93, Pl. II. No. 2. It was kindly lent to me by M. Bleuse, to whom it belongs.

780. **Alaus sordidus** (Elateridæ): Ceylon, femur of right middle leg bears two supernumerary legs arising from its postero-dorsal surface. All three legs are somewhat abnormal in form and the principal femur is partly shrivelled at its base. At a point on the postero-dorsal surface about halfway from the apex there is a large, irregular boss from which the two extra femora diverge. Of these that which is nearest to the normal leg may be distinguished as a *left* leg by the planes of movement of its tibia and tarsus, while the remoter leg is a right leg. The tarsus of the latter is broken but was probably complete. The surfaces which the extra legs present to each other are structurally anterior surfaces, but the relative positions of the three legs do not correspond with any of the positions shewn in the Scheme. It should however be noticed that this fact may be connected with the presence of the amorphous thickening at the point of origin of the extra femora. Specimen in Hope Collection first described and figured by WESTWOOD, *Oriental Entomology*, Pl. xxv. fig. 9, and mentioned *Proc. Linn. Soc.* 1847, p. 346.

781. **Clythra quadripunctata** (Phyt.): left anterior trochanter bears two supernumerary legs. Both the normal leg and the two extra ones are complete. The position of the latter is very peculiar; for, arising from the anterior surface of the trochanter, they turn their structurally dorsal surfaces towards the anterior surface of the normal leg, which thus stands *between* them, one of them being above it and the other ventral to it. Of these that which is placed dorsally is structurally a *right* leg, while the lower one is a *left*, like the normal one. Both the extra legs are also partly rotated so that their ventral surfaces are partially directed *upwards*. From these facts it appears that the position of these extra legs relatively to the normal one does not correspond with any of the positions indicated in the Scheme, and it did not seem to be possible to refer this deviation from the usual arrangement to any special malformation of any of the parts. Specimen originally described by MOCQUERYS, *Col. anorm.*, p. 42, *fig.*

782. **Clytus liciatus** (Long.) : right tibia reduced and thickened, being shapeless and bent. Its apex presents two articulations, the one anterior and the other posterior, the latter bearing a normal, 4-jointed tarsus. The anterior articulation bears a slender double tarsus, the two parts of which are compounded in the 1st, 2nd, and 3rd joints but separate in the 4th or terminal joints. The supernumerary tarsi are very

CHAP. XX.] EXTRA LEGS : UNCONFORMABLE CASES. 509

slender and the whole thickness of their common proximal joint is even less than that of the proximal joint of the normal tarsus. The

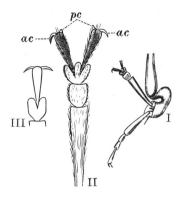

FIG. 171. *Clytus liciatus*, No. 782. I. View of right tibia. II. Detail of the extra parts, from plantar surface. *ac, ac*, claws supposed to be morphologically anterior. *pc*, rudiments supposed to represent posterior claws. III. Enlarged view of the end of the normal tarsus. (The property of Dr Kraatz.)

terminal joints of the extra tarsi are well formed, but they each bear only *one* fully developed claw, the claw of the adjacent side of each being only represented by a rudimentary knob. It appears at first sight that these extra tarsi are at their origin from the tibia only a single appendage and that their double nature only begins from the third joint. This however is not the case, for there are five spurs on the tibia, together with a small brown knob which perhaps represents the sixth spur. The tibia is greatly misshapen and the arrangement of the spurs is so amorphous that I did not succeed in determining their morphological relations. This specimen was kindly lent by Dr KRAATZ, having being first described by him in *Berl. ent. Zt.*, 1873, XVII. p. 433, figs. 17 and 17 *a*.

783. **Cryptohypnus riparius** (Elater.). The tibia of the right anterior leg is enlarged at its apex and bears one very large tarsal joint: this joint has two apical articulations, of which the posterior bears the remaining 4 joints of what is presumably the normal tarsus. The other articulation bears a large tarsal joint, common to a pair of complete extra tarsi. This pair of tarsi stand with their lateral parts closely adjacent and their plantar surfaces downwards, but the other tarsus which is posterior to them, and is presumably the normal, stands with its plantar surface turned *backwards*. This disposition differs considerably from that indicated in the Scheme. For the place of origin of the extra tarsi and their position relatively to each other is A; but the normal tarsus is twisted so that it turns its dorsal surface forwards, towards the posterior surface of the nearer extra tarsus. For this specimen I am obliged to Dr Mason.

784. **Taurhina nireus** (Lamell.): right middle tibia bearing two extra tarsi. [In the normal leg of this beetle the tibia is like that of many other Lamellicorns, presenting at its apex two sharp processes, the one anterior and the other dorsal: and ventrally two articulated spines, one anterior and the other posterior to the tarsus. The abnormal tibia of this specimen is considerably widened at its apex, and bears in addition to the normal two processes two other processes of a similar kind separated from each other by a pair of articulated spines. Instead of a single

pair of articulated spines, this tibia bears *five* such spines, of which a pair stand between the two extra processes. The disposition of these spines could not be made clear without several figures. There are two complete tarsi and both have their ventral surfaces turned downwards. The anterior tarsus is somewhat the smaller. I did not succeed in definitely determining the homologies of the parts in this specimen. It should be specially observed that while the tarsi are only *two* in number, suggesting that the supernumerary part is *single*, the spines indicate that there are here at least some elements of further repetition.] Specimen figured by KRAATZ, *Deut. ent. Zt.*, 1889, xxxiii, p. 221, fig. 18, and kindly lent by him.

785.   **Ranzania bertolonii** (Lamellicorn): in the right posterior foot the last joint of the tarsus is curved outwards and bears six claws instead of two, and three onychia instead of one. The arrangement of the parts is somewhat complex and could not well be made clear without elaborate figures. Speaking generally, the last (fifth) tarsal joint presents at its apex a large articular surface of irregular shape. This surface bears four large claws disposed in the same direction as the normal pair of claws. Of the four claws the two adjacent ones are in solid continuity for a part of their length, being joined together by chitin much as the extra dactylopodites are in Fig. 184, III. It is clearly shewn that the conjoined claws are respectively the fellows of the two free claws, for the two extra onychia stand one upon either side of and opposite to the curvature of the conjoined claws. Terminally the fifth tarsal joint bears also a small pair of somewhat deformed claws with which an enlarged and misshapen onychium corresponds. This specimen was kindly lent to me by M. Henri GADEAU DE KERVILLE and was mentioned by him in *Bull. Soc. Ent. France*, Ser. 6, vi. 1886, p. CLXXX.

*786.   **Rhizotrogus æstivalis** ♀ (Lamellicorn), bearing supernumerary parts of double structure upon the right posterior 5th tarsal joint (Fig. 172). The structure found in this case is very remarkable and is, I believe, in some respects unique. The tarsus is normal as far as the extremity of the terminal joint, and the abnormality consists entirely in repetition of claws and pulvillus. The normal formation is shewn from the ventral surface in Fig. 172, A. There is an anterior claw, a posterior claw and a small pulvillus, placed ventrally to the claws, bearing two hairs. Fig. 172, B, shews the abnormal foot from the ventral side. Each claw gives off from its base a ventrally-directed supernumerary claw, and each supernumerary claw is bifid at its point. Examined from below each of these extra claws is seen to bear *two grooves separated by a ridge*, and is therefore morphologically a double structure. The next structure of importance is the pulvillus. The normal pulvillus (pul) is in place and of the usual form, but *dorsally* to it there is a supernumerary pulvillus (pul^2) of cylindrical form and rather longer than the normal pulvillus. At its apex this extra pulvillus bears a median bifid hair with another hair on each side of it; these hairs thus prove that the extra pulvillus is morphologically double.

In this foot, therefore, a supernumerary pair of claws and a supernumerary double pulvillus are intercalated between the normal claws and the normal pulvillus. Hence though the repetition affects both claws and pulvilli, and the structures found are sufficient for an incomplete pair of extra feet, yet the extra parts are disposed in the system of symmetry of the normal foot, forming, all taken together, *one* foot only. Specimen very kindly lent by Dr G. KRAATZ.

CHAP. XX.] EXTRA LEGS: MISCELLANEOUS CASES. 511

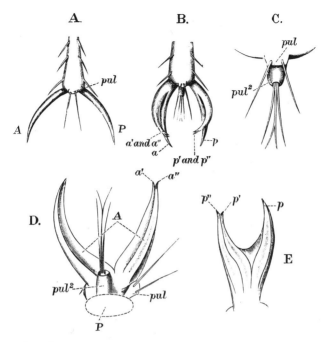

Fig. 172. *Rhizotrogus æstivalis*, No. 786. A, normal hind foot from ventral surface. B, right hind foot of No. 786 from ventral surface. C, enlarged view of pulvilli. D, inside view of the claws at the anterior border of the apex. E, inside view of the claws of the posterior border of the apex. *A*, anterior. *P*, posterior. *a*, normal anterior claw of abnormal foot. *p*, normal posterior of the same. *a'*, *a''*, the two points of extra claw of anterior side. *p'*, *p''*, the two corresponding posterior points. In D the posterior group of claws is supposed to be cut off at *P*. *pul*, normal pulvillus. *pul*², extra double pulvillus.

(11) *Cases in which the legs were either mutilated, or in part amorphous, or insufficiently described.*

787. Want of space prevents me from giving more than a list of references to these cases. Most of them besides are imperfectly known. Of those seen by myself the case of *Hister* would, I think, be interesting, but I regret that my notes of this case are imperfect. In the following list the letters R and L shew the leg affected; the † means that the case probably *did not* agree with the Scheme, the ‖ that it probably *did* agree; the ‡ means that the parts were either mutilated, or imperfect, or deformed. Of those unmarked, the accounts are inadequate.

  Ichneumon luctatorius   R₃ Tischbein, *Stet. ent. Ztg.*, 1861, xxii. p. 428.
‡ **Carabus auratus**   Kraatz, *Deut. ent. Zt.*, 1889, xxxiii. p. 222, fig. 17.
  C. auronitens   L₁ Gredler, *Korresp. zool.-min. Ver. Regensb.*, 1877, xxxi. p. 139.
  C. cancellatus   R₂ Landois, *Zool. Gart.*, 1884, xxv. p. 288 [q. v.]
‡   ditto ♂   L₂ Kraatz, *Berl. ent. Zt.*, 1873, xvii. p. 432.

* ‖ † **C. catenulatus**	L₃	*Brit. Mus.*
C. italicus	R₃	BAUDI, *Nat. Sicil.*, VIII. No. 9, p. 199.
Dyschirius globulosus (Car.)	R₁	JAYNE, *Trans. Amer. Ent. Soc.*, 1880, VIII. p. 157, Pl. IV. *figs.* 6, 6a.
Calopus cisteloides (Het.)		VON HEYDEN, *Isis*, 1836, IX. p. 761.
Pterostichus prevostii (Car.)	L₃	MÜLLER, A., *Proc. Ent. Soc.*, 1869, p. XXVIII.
* ‡ **Chlænius nigricornis** (Car.)	L₃	MOCQUERYS, *Col. anorm.*, 1880, p. 62, *fig.*
‡ **Agra catenulata** (Car.)	L₃	STANNIUS, *Müll. Arch. f. Anat. Phys.*, 1835, p. 306, *fig.* 13.
‡ **Prionus coriarius** (Long.)	R₃	PERTY, *Mitth. nat. Ges. Bern*, 1866, p. 308, *fig.* 11.
Prionus sp.	?₃	*Ann. and Mag. N. H.*, 1841, p. 483.
‡ **Aromia moschata**[1] (Lam.)	L₁	KRAATZ, *Deut. ent. Zt.*, 1877, XXI. p. 56, Pl. I, 2, *fig.* 11.
‡ **Dorcadion rufipes** (Long.)	L₃	PERTY, *l.c.*
Blaps sp. (Het.)	R₃	LABOULBÈNE, *Bull. Soc. Ent. Fr.*, S. 4, v. 1865, p. XLIX.
‡ **Ptinus latro** (Plin.)	L₁	VON FRICKEN, *Ent. Nachr.*, 1883, IX. p. 44.
‖ ‡ **Dytiscus marginalis** (Dyt.)	R₁	RITZEMA BOS, J., *Tijds. v. Ent.*, 1879, XXII. p. 206, Pl.
‖ **Colymbetes sturmii** (Dyt.)	L₁	STANNIUS, *l.c.*, p. 307, *fig.* 9.
Strategus antæus (Lam.)	L₂	JAYNE, *l.c.*, p. 159, *fig.* 10.
‖ **Rutela fasciata** (Lam.)	R₃	SPINOLA, *Ann. Soc. ent. Fr.* 1835, IV. p. 587, *Pl.*
* ‖ **Hister cadaverinus** (Clav.)	R₁	MOCQUERYS, *l.c.*, p. 59, *fig.*
Cetonia morio[2] (Lam.)	L₁	SARTORIUS, *Wien. ent. Monats.*, 1858, II. p. 50.
Melolontha vulgaris (Lam.)	L₃	TREUGE, *Ent. Nachr.*, VIII. 1882, p. 177.
‡ ditto	R₁	DOUMERC, *Ann. Soc. ent. Fr.*, 1834, III. p. 171, Pl. I A, *fig.* 1.
ditto	L₃	BOULARD, *Bull. Soc. ent. Fr.*, 1846, S. 2, IV. p. XLVIII. *fig.*
‖ ditto	R₃	TIEDEMANN[3], *Meckel's Arch. f. Phys.*, 1819, v. p. 125, Pl. II. fig. 1.
* † ‡ ditto	L₃	MOCQUERYS, *l.c.*, p. 68, *fig.*
‖ **Rhizotrogus castaneus** (Lam.)	R₁	BASSI, *Ann. Soc. ent. Fr.*, 1834, III. p. 373, Pl. VII A.
† **R. æstivalis**	L₁ and R₃	PERROUD, *Ann. Soc. Linn. Lyon*, 1854, II. p. 325.
Oryctes nasicornis (Lam.)	R₁	AUDOUIN, *Bull. Soc. ent. Fr.*, 1834, III. p. IV.
Enema pan. (Lam.)	L₃	TASCHENBERG, *Zts. f. ges. Naturw.*, 1861, XVIII. p. 321.

---

[1] As Kraatz suggests, this is presumably the case given by SARTORIUS, *l. c.*

[2] Probably same specimen as that of GREDLER, *Korresp. zool.-min. Ver. Regensb.*, 1869, XXIII. p. 35.

[3] Tiedemann's grave comment is of interest as recalling past phases of thought. He says: "*Was die Entstehung der oben beschriebenen Missbildung betrifft, so lässt sich wohl annehmen, dass die Phantasie der Mutter des Maikäfers durch ein vorausgegangenes Versehen aufgeregt, hier nicht als Ursache beschuldigt werden kann, theils weil wir überhaupt keine Beweise für eine lebhafte Phantasie der Maikäfer haben, und theils weil die Bildung des Embryo ausserhalb des Leibes der Mutter nur sehr langsam geschieht, und die Mutter ohnehin gleich nach Legung der Eier stirbt*" *l. c.*, p. 126.

Paired Supernumerary Antennæ.

In dealing with extra antennæ there is more difficulty in determining the true nature of the parts than there is in the case of extra legs. We have seen that the real duplicity of compounded extra parts often appears only in the fact that they have a bilateral symmetry, while in the normal appendage one side is differentiated from the other. Now in very many species of Insects the antenna seems to be a bilaterally symmetrical filament, having joints cylindrical or elliptical in section. When from such an antenna there proceeds an extra filament, itself bilaterally symmetrical, it is almost impossible to determine whether the extra filament is really a *single* repetition of the normal or whether it is made up of two homologous borders of a *pair*. (Cp. Nos. 801 and 764.) In speaking of actual cases of duplicity in Arthropodan appendages we shall have to return to this subject.

Meanwhile evidence will be given as to examples of obvious duplicity in extra antennæ. It will be seen that in species having normally a marked differentiation between the anterior[1] and posterior borders of the antennæ (Lamellicorns, Lucanidæ, &c.), and the case has been really studied, there is often clear proof not only of the duplicity of the extra parts but also that they are arranged as images, almost as described for legs.

We shall moreover meet cases where of the paired extra parts one springs free from the normal at a point proximal to the point of origin of its fellow. Among extra legs there is scarcely any certain example of this phenomenon, *Platycerus caraboides* No. 778 being perhaps the clearest case. But among antennæ there are several where no other interpretation seems possible. These cases I have set in a separate section.

Of the remainder, little can be said with confidence. Probably if they were carefully examined microscopically it would be found that differentiation between the two sides exists in respect of the distribution of sense-organs or hairs, and that thus the duplicity and symmetry might be traced.

After giving the clear cases I have thought it enough to give a list of those of this doubtful order. As has been said, there is little doubt that with careful study of the specimens many of the cases now included in the list of supposed single extra appendages might be shewn to be cases of extra parts in Secondary Symmetry.

---

[1] This term is used, as in the case of legs, to denote the border which is anterior when the appendage is extended horizontally at right angles to the body. The upper surface will then be dorsal, the lower ventral. These terms are thus applied without any intention of affirming that they are morphologically correct.

(1) *Clear cases of Supernumerary Antennæ in Secondary Symmetry.*

(a) *The extra parts arising together.*

*788. **Phyllopertha horticola** (Lamellicorn): specimen in which the right antenna bears a supernumerary pair of clubs. This specimen may conveniently be described in detail as it furnishes a good example of the mode in which repetition of the antennæ occurs in the Lamellicorns. The left antenna is normal and possesses nine joints (Fig. 173, L). The first is a large pear-shaped joint, articulating with the head by its narrow end. The

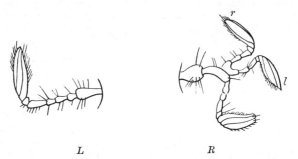

FIG. 173. *Phyllopertha horticola*, No. 788. *L*, the normal left antenna. *R*, the normal right antenna. *l*, *r*, extra left and right clubs.

second joint is also a pear-shaped joint, of about half the size of the first. The third, fourth and fifth joints are elongated and cylindrical. The sixth is short and wide. The seventh, eighth and ninth are each expanded into a lamella. These three lamellæ are generally kept firmly closed together and form the sensory organ, or "club." In *Melolontha* (*v. infra*) and several other genera of Lamellicorns, there are ten joints, of which seven are developed as lamellæ, forming the club.

In the *right* antenna (Fig. 173, R), which bears the extra pair of clubs, the basal joint is rather thick. The second joint is longer than it normally is, and curves slightly backwards and downwards. At its apex it bears the rest of the normal antenna, which is in all respects well formed. In addition to the normal antenna, the second joint upon its anterior surface gives attachment to a large joint which is imperfectly constricted into two parts in a vertical plane at right angles to the general direction of the normal antenna. Each of these half-joints bears a structure containing in itself all the parts proper to an antenna peripherally to the third joint, the clubs being well-formed and normal. In absolute size they are equal, but are a little smaller than the normal antenna.

These two antennæ curve in opposite directions and are in all respects complementary to each other, forming a true *pair*. The most anterior of them, *r*, is disposed as a *right* antenna, while the posterior, *l*, is disposed as a *left*. This specimen was taken by M. Albert Mocquerys, and was kindly lent to me by M. Henri Gadeau de Kerville.

789. **Melolontha vulgaris** ♀ (Lamellicorn): left antenna bearing a pair of supernumerary clubs. The extra pair arises from the second joint of the normal antenna, and they have their third joints united at the base. The relative positions of the extra clubs and the normal one are those marked VP in the Scheme. All these three clubs are perfect and of the same size, but each is a little smaller than a normal club. At the thoughtful suggestion of Prof. Howes this specimen was very kindly lent to me by Mr E. E. Green, and has been placed in the Museum of the Royal College of Surgeons.

790. **Melolontha vulgaris**: [right antenna bearing a supernumerary pair of clubs in Position P. For details see original, where a different and I think untenable view is taken] LEREBOULLET, *Rev. et Mag. de Zool.*, S. 2, III., 1851, *fig.*

791. **Melolontha vulgaris** ♀, with a pair of supernumerary antennæ arising from the left antenna. [The figure shews that the proximal joint or scape was of abnormal thickness and had two peripheral articulations in the same horizontal plane. The anterior articulation bore a normal antenna. The posterior articulation bore a single large first funicular (2nd) joint which in its turn bore a pair of clubs in the same horizontal plane, the anterior being a *right* club and the posterior a *left*, having their anterior surfaces adjacent: they are therefore a complementary pair in Position P.] KRAATZ, G., *Deut. ent. Zt.*, 1880, XXIV. p. 341, figs. 7 and 7 *a*.

792. **Amphimallus solstitialis** (Lamellicorn): left antenna bearing a supernumerary pair of imperfect antennæ articulating by a common stalk on the *anterior* surface of the second joint. The two extra clubs are an imperfect pair, complementary to each other, being set on back to back, in Position A. The most anterior of the clubs has only two lamellar joints, one small and one large. The posterior has three lamellæ. The normal club has three lamellæ as usual. Originally described by MOCQUERYS, *l. c.*, p. 15, *fig.*

793. **Anomala junii** (Lamellicorn): left antenna bears 3 clubs, each having 3-jointed stem articulating with elongated 2nd joint of antenna. [Symmetry not clear: possibly Position DPP.] KRAATZ, *Deut. ent. Zt.*, 1881, XXV. p. 111, Pl. III. fig. 4.

*794. **Geotrupes typhæus** ♂ (Lamellicorn): left antenna bearing a pair of supernumerary clubs compounded together. The antenna is normal up to the 7th joint which is dilated. The 8th is still more dilated and bears posteriorly the normal club composed of three lamellæ; and anteriorly by a separate articulation a supernumerary structure (Fig. 174, *mr*, *ml*) consisting of three joints, each of which has the form of a complementary pair of lamellæ joined by their morphologically posterior (sc. external) edges. The whole supernumerary structure is thus morphologic-

33—2

ally a *pair* of clubs, a right and a left, compounded together. The histology of the supernumerary lamellæ is just the same as

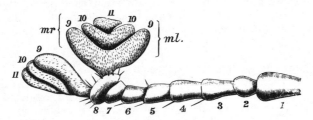

Fig. 174. *Geotrupes typhæus*, No. 794. Left antenna bearing a compounded pair of clubs. *ml*, *mr*, morphological left and right of the extra parts. (The property of Dr Kraatz.)

that of the normal lamellæ, all being covered with pubescence. The form of the compound eleventh joint is somewhat irregular. The extra parts are in the Position A of the Scheme. Specimen kindly lent by Dr KRAATZ, and first described and figured by him in *Deut. ent. Zt.*, 1889, XXXIII. p. 221, fig. 13.

5. **Melolontha hippocastani** ♂ having supernumerary parts of double structure upon both the right and the left antenna.

Right Antenna. Third joint elongated, thickened and presenting two articular surfaces; of these one is terminal and bears a normal antennary club, while the other is dorsal and bears a supernumerary double club. This structure has the form shewn in the drawings. Fig. 175, A, shews its appearance when looked at from above, B shews the structure when seen from below and externally. It consists of seven pieces shaped like half-funnels, fitted into each other.

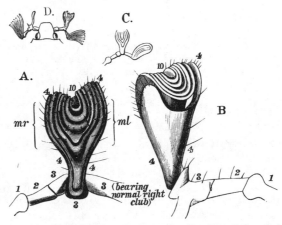

Fig. 175. *Melolontha hippocastani*, No. 795. D, view of the whole head and antennæ after von HEYDEN. C, view of right antenna. B, detail of right antenna from below. A, detail of the same from above.

The morphological nature of this supernumerary organ may be determined thus. The upturned edges of the folds bear hairs as shewn in the figure A; since in the normal antenna the dorsal edges of the lamellæ alone bear hairs, these edges are in this case dorsal morphologically as well as by position.

Since the outermost lamella (marked 4) is articulated into the third joint of the funiculus, it is therefore the 4th joint, or proximal lamella, and the remaining lamellæ are therefore 5th, 6th, 7th, 8th, 9th and 10th respectively. Next, the surface *mr* is structurally like that of the internal (sc. anterior) surface of the proximal lamella of a normal club, and the surface *ml* is a similar surface: but *ml*, being an *internal* surface, faces towards the right and is therefore morphologically a *left*; while *mr*, being an internal surface and facing towards the left, is a *right*; hence this club consists of two clubs compounded together by their external or posterior borders, and the two are a right and a left, the left being next the normal right club.

Lastly, since the upper free edges of the lamellæ are structurally dorsal, it follows that their lower edges are structurally ventral: but these lower edges do not exist as free edges, for the lamellæ are continuous upon their ventral aspect; therefore the surfaces which are adjacent in the extra right and left clubs, and by which they are compounded together, are partly *ventral* surfaces. This is approximately Position DP of the Scheme.

Left Antenna. Second joint thickened and presenting three articulations as follows. 1. a peripheral articulation bearing the normal club; 2. a ventral articulation bearing a 4th joint and club composed of 3 formless lamellæ; 3. a dorsal articulation bearing a small cylindrical joint only. The shape and formation of these extra parts is so indefinite that their morphology could not be determined.

For the loan of this specimen I am indebted to Dr L. von Heyden, who first described it in *Deut. ent. Zt.*, 1881, xxv. p. 105, fig. 1.

796.   **Rhizotrogus æquinoctialis** (Lam.) : 4th joint of right antenna bears a supernumerary structure projecting forwards and lying in the same horizontal plane as the normal club. This structure is lanceolate in form and its outer surface is in texture similar to the external surfaces of a normal club. On the ventral aspect it presents a simple ridge, but on the dorsal side its outer coating is divided by a spindle-shaped slit through which part of the internal structure protruded. The edges of this opening and the protruding portion of the interior bear a few hairs. There can be little doubt that this supernumerary body represents an imperfectly formed pair of clubs, and that it is in fact a more rudimentary condition of the parts found in No. 795. Specimen originally described and figured by Mocquerys, *Col. anorm.*, p. 16, *fig.*

797.   **Lichnanthe vulpina** (Lam.) : right antenna bears in addition to normal club a small spherical club made up of three joints, arising from posterior border of a long

joint apparently representing the normal 4th, 5th, and 6th joints not segmented from each other. [As this supernumerary part is in itself symmetrical it probably contains within itself parts of a pair of clubs compounded in Position P. Cp. No. 795.] JAYNE, H. F., *Trans. Amer. Ent. Soc.*, 1880, VIII. p. 158, Pl. IV. fig. 8.

798.   **Polyphylla decemlineata** (Lamellicorn). A specimen in which the right antenna bears a partially double supernumerary branch in addition to the normal antenna. This additional structure articulates with the second joint of the antenna by means of a single large joint. This joint carries a double club consisting of two sets of lamellæ, seven being in each set. The two sets of lamellæ are united at their bases at an angle of forty-five degrees. The plane of the normal club is perpendicular to that of the abnormal ones. The normal club itself is ⅕th shorter than that of the other side. [The details of the structure of this specimen are difficult to follow and the reader is referred for further particulars to the description and figures given in the original.] JAYNE, H. F., *Trans. Amer. Ent. Soc.*, 1880, VIII. p. 158, *figs.*

(*b*)   *The extra parts arising from the normal at separate points.*

*799.   **Odontolabis stevensii** ♀ (Lucanidæ). As the repetition in this specimen is almost complete and the relations of the parts fairly clear though in some respects peculiar, a detailed account will be useful.

The body, legs, &c. are normal, save that the back of the head and thorax have been crushed by some accident. The antennæ are both abnormal in the way shewn in Fig. 176. The condition will be better understood if the normal antenna is first described.

FIG. 176. *Odontolabis stevensii*, No. 799. The head seen from below, and enlarged views of the two antennæ. *R*, right. *L*, left. There is some doubt as to which of the branches is the normal and which the supernumeraries. See description in text.

The normal antenna of *Odontolabis* is much like that of its ally *Lucanus cervus*, the Stag-beetle. It is made up of 10 joints composing three parts differentiated from each other.

The first, or "scape," is a single joint as long as the rest of the antenna. It widens a little from its central end or base towards the apex, and is slightly flattened from above downwards. The second part, or "funiculus," has six simple joints. The last three joints form the club. They are flattened from above downwards and lie in a horizontal plane. The anterior ("inner") border of each of these three joints is produced into flat expansions, covered with sensory pores, which together form a series of serrations along the anterior border. When in its natural position the serrated border of the right antenna faces towards the left side, and that of the left is turned towards the right. The structure of the abnormal specimen is as follows.

Left Antenna. Scape normal. Its plane however is not quite horizontal as usual, but is a little oblique, the anterior border being slightly higher than the posterior. In the funiculus the 1st and 2nd joints (2nd and 3rd of the whole antenna) are a little thicker than usual but otherwise normal. The 3rd joint of the scape is enlarged and presents at its apex two sockets, each bearing a continuation as shewn in the figure. The two sockets are not in a horizontal plane, but their plane is oblique and nearly at right angles to the plane of the scape, the socket bearing the branch l^1 being the higher. It is important that the precise relations of these parts should be clearly understood.

This *outer* socket of the 3rd funicular joint bears the branch l^1, made up of three more funicular and three club-joints, turning their serrated border *in the direction of the right antenna:* l^1 is therefore structurally a *left* antenna. Its surface is of the same nature as that of a normal antenna, but its size is a little smaller. It is in an oblique plane inclined to the horizontal at about 45°, the posterior (outer) border being the higher.

The *inner* socket of the 3rd funicular joint bears a cylindrical joint not quite fully segmented off from the next joint peripheral to it. These two are 4th and 5th funiculars. The 5th again presents two sockets, bearing respectively the branches l^2 and l^3. The branch l^2 has one small joint (6th funicular) and three club-joints, turning their serrated border towards l^1. This branch is therefore structurally a *right* antenna. It stands in the same oblique plane as l^1, the serrated border being the higher. In size it also agrees with l^1, being rather smaller than the normal. The branch l^3 is a normal *left* in size and shape, and it lies in a horizontal plane.

Here therefore there is a *left* antenna and a pair, one a *right* and the other a *left*. Which then is the normal, l^1 or l^3? Inasmuch as l^3 and l^2 arise by a common stalk it may seem that they are the extra pair and that l^1 is the normal. We have now seen in many cases that extra parts in Secondary Symmetry are compounded together as l^3 and l^2 are here. But considering the fact that l^3 is of normal size and in the normal horizontal plane, whereas l^1 and l^2 are both smaller and are in an oblique plane complementary to each other, I incline to the view that if one branch is the normal, it is l^3, and that l^1 and l^2 are the extra pair in Secondary Symmetry, *though they do not arise together.* They are then nearly in Position DPP, but depart from that position in the fact that l^1 is not horizontal (cp. No. 757).

If l^1 and l^2 are really the extra parts, in the fact that they do not arise together, but spring separately from different points on the normal, we meet with a condition rarely seen, but that this is a possible condition is proved beyond doubt by the succeeding case.

Right Antenna. Scape precisely as in left antenna. The 1st funicular (2nd antennary) has two sockets at its apex, placed like those on the 3rd funicular of the left side, the anterior socket being the lower and the posterior socket being the higher. The anterior socket bears a normal right antenna, r^3. The posterior bears the structure shewn in the figure. This appendage has unfortunately been broken, but enough remains to suggest the original structure. It consists of five funicular and a 1st club-joint. The 5th joint of the whole funiculus bears a large socket looking downwards and forwards, its other socket looking backwards and upwards. From the former the original continuation has been lost. The latter bears the 6th funicular and its 1st club-joint, this again having an empty socket.

The plane of the two sockets of the 5th funicular is oblique to the horizon, like that of l^1 and l^2. Though it is clearly impossible to shew how this antenna was in its unbroken state, we may note that if it were continued in the way suggested by the dotted lines it would have borne a complementary pair of clubs, r^1 and r^2, like l^1 and l^2 of the other side, placed like them in an oblique plane nearly corresponding with DPP of the Scheme.

This specimen was kindly entrusted to me by M. Henri Gadeau de Kerville. He tells me that he believes a description of it has already appeared, but this I have failed to find. I have therefore ventured to describe it again, with apologies to the original describer. The specimen bears a label in the handwriting of the late Major Parry and was no doubt in his celebrated collection of Lucanidæ.

800.   **Melolontha vulgaris:** right antenna bearing a pair of incomplete supernumerary antennæ (Fig. 177). The first joint is normal; it bears a second joint of singular form, consisting of a long anterior branch, and a short posterior branch $\frac{1}{3}$ the length of the anterior. The anterior bears two clubs in the manner shewn in the figure (Fig. 177). Of these

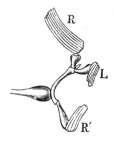

Fig. 177. *Melolontha vulgaris*, No. 800. Lettered according to the view that R is the normal right club. L, the supernumerary left, and R' the supernumerary right. (From Wesmael.)

one (R) is inwardly directed and is as wide as, but only $\frac{3}{4}$ the length of a normal club. The posterior of the two clubs (L) is directed backwards and has only *four* lamellæ which are apparently united together. The other small club (R') is also composed of only four lamellæ which are similarly united together. In both L and R' the middle lamellæ shew traces of further subdivision. The figure represents the three clubs as being all in one plane, but the club R' is really below L, which stands up from the normal antenna. It is mentioned that some of the tarsi were mutilated or defective. [Here L and R' are clearly a complementary pair, though separately arising from the normal. It will be observed that as in Lereboullet's case (No. 790) the second joint, which is common to two clubs, is greatly elongated.] WESMAEL, *Bull. Ac. Belg.*, 1850, XVI. 2, p. 382, *fig.*

*801.   **Navosoma** sp. (Longic.) Left antenna abnormal. The joints of the normal are a little flattened from above downwards and are nearly elliptical in section. But the anterior border is differentiated from the posterior by the presence of two elongated patches of tissue covered with sensory pores. The two patches are both on the anterior border, one being on the dorsal surface and one on the ventral, separated from each other by a chitinous ridge. Upon the general surface of the peripheral joints of the antennæ are several other such patches, but none are so distinct as those of the anterior border. The abnormal left antenna has the form shewn in Fig. 178. So far as the 8th joint it does

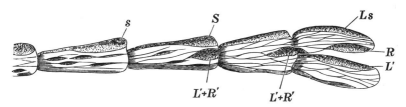

Fig. 178. *Navosoma*, No. 801. Left antenna seen from below. Lettered on the view that R and L' are the extra parts. S, sensory patch. (In Hope Collection.)

not differ from the normal. The 9th and 10th joints have besides their chief patches of sensory pores (S) on the anterior border, an additional patch ($L' + R'$) posterior to the chief patch. But up to the 10th joint there is no vertical division. The 10th joint however has two articular surfaces, anterior and posterior, in the same horizontal plane. The posterior bears an apical (11th) joint of normal form, having anteriorly a sensory patch. But the apical joint borne by the anterior articular surface has *two* such sensory patches, an anterior and a posterior. This joint therefore contains in itself parts of a *pair* of joints. It is not quite fully segmented off from the 10th joint.

Nevertheless it is difficult to suppose that the anterior joint is the extra pair in Secondary Symmetry, for its anterior patch, Ls, seems to continue the normal series of patches, S, S, &c. Therefore the patches R and L' seem to be the patches of the extra pair, though one of them is on a separate joint and the other is applied to the normal. Taken with the case of *Odontolabis* No. 799 and *Melolontha* No. 800, this must, I think, be judged to be a possible account, and in this case R and L' are, as regards symmetry, in Position P. It is of course possible that Ls and R are really the extra pair in Position A, but the presumption is rather the other way [1]. Specimen in Hope Collection at Oxford.

(2) Cases of double extra antennæ, Symmetry unknown.

802.   In none of the following can any confident statement be made as to the symmetrical relations of the parts. Several of the cases I have myself seen, but I noticed no clear indications as to their symmetry. A good many of them however were examined before I was fully alive to the importance of these matters in the case of filamentous antennæ, and perhaps if they were studied with proper regard to the question of symmetry more might be made of them. Many cases that follow are mutilated or partly amorphous, and of almost all the descriptions are very imperfect. For our purpose some value attaches to these records as evidence of the distribution of such abnormalities, and to any person who may hereafter pursue the subject a fairly complete list of the references may be of use. To this therefore I shall confine myself; for on reviewing the abstracts that I have made of these examples it is clear that they only give the results of superficial examination.

Speaking generally, in these cases, from some one joint of an antenna there arises either a pair of extra antennæ compounded for a greater or less extent of their proximal parts, or two extra antennæ distinct from their point of origin.

The letters R and L indicate the side affected, and the number following is approximately that of the joint from which the extra parts spring. In the greater number of sound cases the three branches lie in or nearly in a horizontal plane and are, I anticipate, in Positions A or P.

Cases which seem from the indications to conform to the Scheme are marked ‖. Mutilated or partially amorphous cases are marked ‡.

	Blaps attenuata (Het.)	R 3	MOCQUERYS, *Coléoptères anormaux*, 1880, p. 5, *fig.*
* ‖	**Malachius marginellus** (Mal.)	L 2	*ibid.*, p. 7, *fig.*
	Timarcha tenebricosa (Phyt.)	R 9	*ibid.*, p. 13, *fig.*
* ‖	**Clytus tricolor** (Long.)	L 7	*ibid.*[2], p. 19, *fig.*
* ‡	**C. arcuatus**	L 1	*ibid.*, p. 20, *fig.*
‖	**Calopteron reticulatum** (Mal.)	L 1	*ibid.*, p. 25, *fig.*
*	**Carabus monilis** (Car.)	L 3	*ibid.*, p. 3, *fig.*
*	**C. auronitens**	L 7	*ibid.*, p. 9, *fig.*
*	**Ptinus latro** (Ptin.)	L 5	*ibid.*, p. 8, *fig.*
	Elater murinus (Elat.)	L 2	*Ann. and Mag. of N. H.*, 1831, IV. p. 476.
	Zonites præusta (Het.)	R 3	STANNIUS, *Müll. Arch. f. Anat. Phys.*, 1835, p. 303.

[1] This is perhaps too strongly put.
[2] Description and figure incorrect. Apical joint of extra branch is *bifid*.

Helops cæruleus (Het.)	R 5	Séringe, *Ann. Soc. Linn. de Lyon*, 1836, Pl. I. *fig.*
‖ **Dendarus hybridus** (Het.)	L 4	Romano, *Atti Ac. sci. Palermo*, 1845, N. S., I. *fig.*
‡ **Scraptia fusca** (Het.)	L 5	Rouget, *Ann. soc. ent. France*, 1849, S. 2, VII. p. 437.
‡ **Carabus sacheri** (Car.)	R 7	Letzner, *Jahresb. schles. Ges. f. vaterl. Kultur*, 1854, p. 86.
Pimelia scabrosa (Het.)	R 2	Blackmore, *Proc. Ent. Soc.*, 1870, p. xxix.
Anchomenus sex-punctatus (Car.)	L 6	Kraatz, *Deut. ent. Zt.*, 1877, XXI. p. 56, *fig.* 19.
Calosoma investigator (Car.)	R 5	*ibid.*, 1889, XXXIII. p. 221, *fig.*
‖ **Dromæolus barnabita** (Eucn.)	L 5	von Heyden, *ibid.*, 1881, XXV. p. 108, *fig.* 16.
* ‡ **Carabus arvensis**	L 4	Specimen kindly lent by M. A. Fauvel.

803. **Meloe violaceus** ♀: between right eye and the base of the right antenna arise two supernumerary antennæ *from the head*. Of these one has 3 joints and the other has one. Kraatz, *Deut. ent. Zt.*, 1877, XXI. p. 57, Pl. I. *fig.* 22.

The following example is mentioned here, though its nature is quite obscure. In it there is a suggestion that parts of two extra antennæ are present, but the extra parts seem to be peripheral to the parts which they repeat.

As my stay in Rouen was short I was not able to give as much time to this specimen as I should have wished[1].

804. **Melolontha vulgaris** ♂: left antenna abnormal. This case differs wholly from any other that I know of. I can only describe it in a most tentative way. The appearance when the lamellæ were cleaned and separated was as shewn in Fig. 179. Joints 1—8 are fairly normal, but peripheral to this place there were

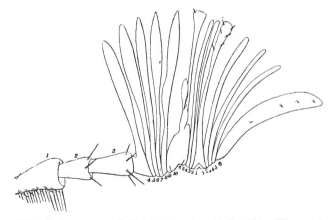

Fig. 179. Left antenna of *Melolontha vulgaris*, No. 804. The numbers are set in tentative suggestion of the possible nature of the parts. (In Rouen Mus.)

[1] This antenna was when I saw it covered with mould and dirt. In washing it I accidentally detached it from the head, but I mounted it again carefully with the specimen.

a number of lamellæ, some like normal lamellæ, others quite irregular. As far as I could make out, the divisions were as shewn in the figure, and I have affixed numbers to the several parts in illustration of their possible nature. The appearance suggests that there is an irregular repetition of a pair of clubs *peripheral* to the normal antenna, but I can form no opinion as to the morphology of the parts. Originally described by MOCQUERYS, *Col. anorm.*, 1880, p. 12, *fig.* [Description and figure altogether misleading.]

PAIRED EXTRA PALPI.

805.   **Bembidium striatum** (Carabidæ): left maxillary palp arises by a first joint enlarged towards its apex, bearing *three* separate terminal joints instead of one. Of these joints one stands apart on a small process of the first joint, but the other two are placed close together, on either side of the apex of the first joint, and diverge from each other at about a right angle. JACQUELIN-DUVAL, *Ann. Soc. Ent. France*, 1850, Sér. 2, VIII. p. 533, *Plate* XVI.

806.   **Helops sulcipennis** (Het.): supernumerary, partially double apical joint arises from the 2nd joint of right maxillary palp. It is set on *at right angles* to the plane of the normal palpus. JAYNE, H. F., *Trans. Amer. Ent. Soc.*, 1880, VIII. p. 161, *fig.* 14.

807.   **Euprepia purpurea** (Arctiidæ): a specimen in which the right wings and antenna were male and the left wings and antenna female, is declared to have possessed *an extra pair of palpi*. [No sufficient description of this extraordinary occurrence is given; and as the repetition of the palpi is only incidentally mentioned, it may be doubted whether a full examination was made.] FREYER, C. F., *Beitr. zur Schmetterlingskunde*, 1845, *Vol.* v, p. 127, *Tab.* 458, *fig.* 4.

CHAPTER XXI.

Appendages in Secondary Symmetry—*continued*.

The Evidence as to Crustacea[1].

The facts as to Secondary Symmetries in Crustacea are so similar to those already detailed in Insects that, were it not for their value as confirmation of the principles indicated, it would be scarcely necessary to describe them at large. Some few of the cases have besides a special interest, as in them may be seen rudimentary or bud-like structures apparently presenting the lowest condition of paired parts in Secondary Symmetry.

Precisely as in Insects there are a number of cases (including those last mentioned) where it would at first sight be supposed that the extra parts are single, but on inspection most of them prove double. Nevertheless there remain some few where this cannot be shewn, and strange as it may seem, these must be admitted to be genuine examples of duplicity of limbs. Of them a special account will be given in another chapter.

There are besides, as in Insects, a considerable number of cases in which the nature of the parts is not clear, though the majority of such cases are not examples of extra parts, but are normal appendages mutilated or deformed.

One specimen (No. 821) is the only case known to me in which *two* pairs of supernumerary parts arise from one appendage.

Another (No. 827) is unique in the fact that according to the description *three* separate appendages are repeated upon a single appendage. It is not clear that this is in any strict sense an instance of Secondary Symmetry, but for convenience it is taken in this chapter.

[1] Useful bibliography given by Faxon, *Harv. Bull.*, 1880—1, viii. p. 271.

Of the whole number, two affect antennæ, four are in non-chelate ambulatory legs, one is in a chelate ambulatory leg and the rest, being the great majority, are all in chelæ.

With reference to these extra parts several false views have from time to time been held. For example, in some of the commonest cases there is an extra pair of dactylopodites, or of indices, curving towards each other. The extra parts may then greatly resemble the dactylopodite (or "pollex") and index of a normal chela, and many authors have not unnaturally supposed that the extra parts were actually an extra pair of forceps repeating those of the normal chela. This may easily be shewn to be an error, from the fact that it is often possible by some slight structural difference between the pollex and the index to detect that both extra parts are either both pollices or both indices.

But the fullest disproof of this supposition is found in the fact that the great majority of the phenomena will be readily seen to conform to the principles enuntiated for Secondary Symmetries in Insects (p. 479).

A good many authors from the time of RÖSEL VON ROSENHOF[1] onwards have said that these cases are a result of injury, or of regeneration after injury. For this belief I know no ground. It should be remembered as an additional difficulty in the way of this belief, that when the limb of a Crab or Lobster is injured it is usually thrown off bodily, while the extra parts most often spring from the periphery of the chela. But since, according to HEINEKEN[2], such mutilated parts are sometimes retained, this must not be insisted on.

In the case of an ambulatory leg the surfaces may be named as in an insect (without any suggestion that these names denote true homologies between the surfaces so named). In describing chelæ I propose to use the following arbitrary terms. The border upon which the dactylopodite articulates is the *pollex-border*, the opposite border being the *index-border*. It should be noted that in the Crab the pollex-border is *superior*, but in a Lobster[3] it is *internal*.

(1) *Clear cases of Extra Parts in Secondary Symmetry.*

A. Legs.

*808. **Palinurus vulgaris:** left penultimate ambulatory leg bore two supernumerary legs (Fig. 180). Coxopodite of great width. The basipodite had three articular surfaces as shewn in Figure 180,

---

[1] RÖSEL VON ROSENHOF, *Insekten-Belustigung*, 1755, III. p. 344.
[2] HEINEKEN, *Zool. Jour.* 1828—29, IV. p. 284.
[3] It is worth noticing that in the chela of a Scorpion though a close copy of that of a Decapod, the arrangement is *reversed*, the articulated pincer being external.

each bearing a complete leg. When seen by me the leg marked L' was lost.

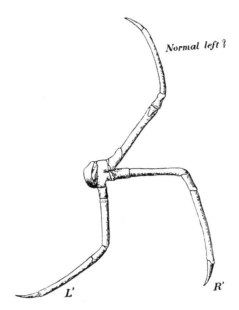

FIG. 180. *Palinurus vulgaris*, No. 808. Left penultimate walking leg. (After Léger.)

I could not quite satisfy myself as to which of the three was the normal, but it was clear that R' was in form a *right* leg and that the other two were lefts. If the leg L' is the normal, it has been pushed out of place by a pair of extra legs in Position DAA, but if R' and L' be the extra legs, then the most anterior leg is the normal and has been pushed out of place by a pair in Position VPP. For an opportunity of examining this specimen, I am obliged to the courtesy of Prof. A. MILNE EDWARDS. Originally described and figured by LÉGER, M., *Ann. Sci. Nat.*, *Zool.*, 1886, S. VII. I, p. 111, Pl. 6.

*809. **Lithodes arctica:** 2nd leg on right side has terminal joint as shewn in Fig. 181, II. If R be the normal then R' and L' are a pair in Position V, but if R' be the normal then R and L' are a pair in Position D. Attention called to the great diminution in size of all three terminations as compared with the normal (Fig. 181, I). Original description, HERKLOTS, J. A., *Bijdr. tot d. h. Genootsch. Nat. Artis Mag.*, 1852, IV. p. 37, *Pl.*; repeated *Arch. néerl.*, 1870, v. p. 410, Pl. XI.

810. **Cancer pagurus:** last left leg closely like last case [in Position D]. RICHARD, *Arch. Zool. exp.*, 1893, p. 102, *fig.*

811. **Carcinus mænas:** 2nd amb. leg as in Fig. 181, III. A pair of

compounded extra points in Position D. DUNS, *Proc. R. Phys. Soc. Edin.*, IX. p. 75, *Pl.*

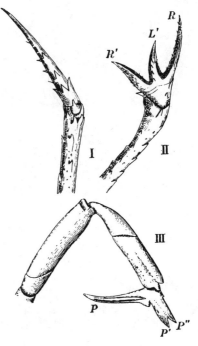

FIG. 181. I. *Lithodes arctica*, normal terminal joint of ambulatory leg. II. Second right leg of No. 809. (Both after Herklots.) III. *Carcinus mænas*, No. 811, second ambulatory leg. (After Duns.) P, normal terminal point. P′, P″, extra terminal points in Position D.

B. *Chelate Appendages.*

(a) *Two extra dactylopodites and double extra index.*

812.   **Eriphia spinifrons** ♀ : specimen of unusually large size, normal but for left chela shewn in Fig. 182, I and II[1]. The chela bore normal left dactylopodite, LD, and index, LI; also, upon pollex-border the structures shewn. These consisted of two dactylopodites, $R'D$, $L'D$, working opposite each other on a compounded double index, $R'I$, $L'I$, which had *two* toothed borders, one for each of them. This is therefore a pair of chelæ repeated in Position D [if indeed the dactylopodite mark the dorsal surface]. Taken from HERKLOTS, *Arch. néerl.*, 1870., V. p. 412, Pl. XI.

[1] In connexion with this case HERKLOTS states that the rt. chela in the normal is the larger and otherwise differs from the left (1 in 8 being reversed in this respect). It does not seem from the figure that there was such differentiation between the extra pair, but in future cases this point should be looked for.

CHAP. XXI.] SECONDARY SYMMETRY: CRUSTACEA. 529

813. **Astacus fluviatilis:** about 3 years old according to Sou-
BEIRAN'S (*Comp. Rend.* 1865, LX. p. 1249) account. Right chela apparently deformed by injury or disease. Left chela had all normal

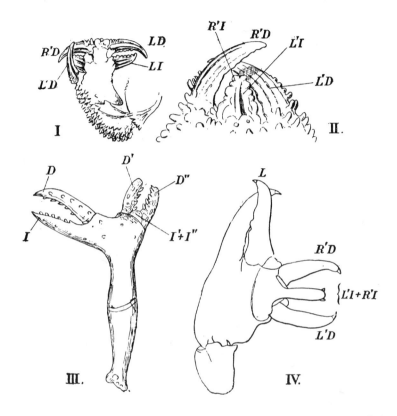

Fig. 182. I and II. *Eriphia spinifrons*, No. 812. I. A view of the left chela. II. An enlarged view of the extra parts from the other side. *LD, LI*, normal left dactylopodite and index. *R'D, L'D*, right and left extra dactylopodites. *R'I, L'I*, right and left extra indices not separated from each other. (After Herklots.)

III. Cheliped of *Homarus americanus*, No. 814. (After Faxon.) *D, I.* normal dactylopodite and index. *D', D''*, extra dactylopodites. *I', I''*, perhaps an indication of double extra indices.

IV. *Astacus fluviatilis*, No. 813, left chela. *L*, normal left dactylopodite. *R'D, L'D*, right and left extra dactylopodites. *L'I+R'I*, left and right extra indices not separated from each other. (After Maggi.)

parts and in addition the structure shewn in Fig. 182, IV upon the pollex-border of propodite. Here was a boss, separated by a groove. It was observed that the structure was that of a rt. and l. dactylopodite working upon a double index [as in last case]. Structure of muscles, fully described, was also in agreement with the view that the extra parts were a complementary

B. 34

pair [similarly in Position D]. MAGGI, L., *Rend. R. Ist. Lomb.*, 1881, XIV. p. 333, *figs.*

814. **Homarus americanus**: small cheliped as shewn in Fig. 182, III. It bears normal dactylopodite (D) and index (I), but this part is bent almost at rt. angles. From the outer angle arise the parts shewn. Apparently D' and D'' are a complementary pair of extra dactylopodites in Position D. The piece $I' + I''$ is not described; from the figure it seems possible that it may represent parts of the indices proper to D' and D''. Case given by FAXON, *Harv. Bull.*, 1880—1, VIII. p. 261, Pl. II. fig. 2.

815. **Cancer pagurus**: right chela as shewn in Fig. 183. This is a case of some complexity. The figure will best make it clear. The dactylopodite D' is single and so also is the index P. D is a double dactylopodite, and P' having teeth on two sides may be judged to be a double index. But if D' and P are the normal chela they each stand opposite

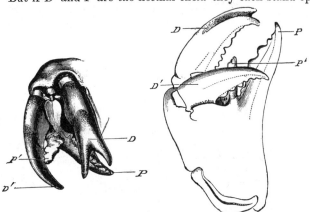

FIG. 183. *Cancer pagurus*, No. 815. Right chela seen from the apex, and from the outside. The lettering is arranged on the hypothesis that D' is the normal dactylopodite, P the normal index. D, the double extra dactylopodite, P', small double extra index. (From *Proc. Zool. Soc.*)

the pincers to which they do not belong. Nevertheless I see no other interpretation possible. (This case is curiously like that of the tarsal claws in *Rhizotrogus* No. 786.) Specimen incorrectly described by myself, *P. Z. S.*, 1890, p. 581, fig. 2. C.

816. **Cancer pagurus**: right chela in a condition not far removed from that of the last case, LE SÉNÉCHAL, *Bull. Soc. Zool. France*, 1888, p. 123, *figs.*

817. **Uca una**: a chela having complex repetition of parts somewhat as in No. 815. JAEGER, G., *Jahresh. d. Ver. vaterl. Naturk.*, 1851, XVII. p. 35, Pl. I. *figs.* 12 and 13.

Perhaps of this nature is the case in *Astacus fluviatilis*, ROESEL v. ROSENHOF, *Ins.-Belust.*, III. Tab. LX. *fig.* 28.

(b) *Two extra dactylopodites arising from normal dactylopodite.*

*818. To this and the next division belong the great majority of cases of repetition of parts in Crustacea. Including examples recorded by various authors and specimens in different Museums there are nearly fifty cases of this class known to me.

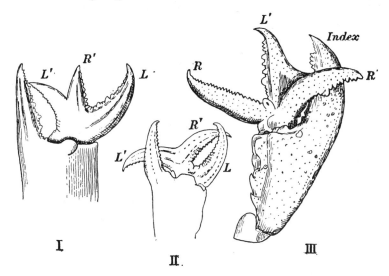

Fig. 184. Three cases of two extra dactylopodites arising from a normal dactylopodite. I. Left chela of *Carcinus mænas* in Brit. Mus. II. Left chela of *C. mænas* after Lucas, *Ann. Soc. ent. France*, S. 2, II. p. 42, Pl. I. *fig.* 2. III. Right chela of *Homarus*, after van Beneden, *Bull. Ac. Belg.*, S. 2, XVII. p. 371.

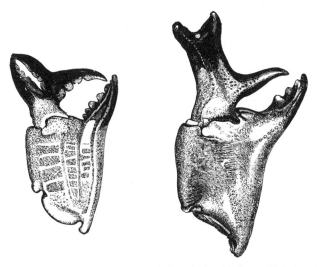

Fig. 185. *Cancer pagurus*. Two chelæ of the kind specified in No. 818, described by myself in *Proc. Zool. Soc.*, 1890, p. 581, whence figs. are taken.

34—2

The various simple forms taken are illustrated by the eight cases shewn in Figs. 184, 185 and 186. It will be seen that when such extra processes arise on the toothed border of the dactylopodite they turn their *smooth* borders to each other, but when

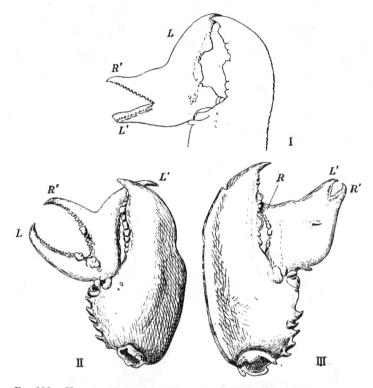

Fig. 186. *Homarus americanus.* Three chelæ whose dactylopodites bear double extra dactylopodites. I. A left. II. A left. III. A right. *R*, normal right. *L*, normal left. *R'*, extra right. *L'*, extra left. (From Faxon.)

they arise on the smooth border they turn their toothed borders to each other, thus fulfilling the conditions of the Scheme given at p. 481. Though from the close agreement between the three prongs in some of the specimens it is not always possible to tell the normal dactylopodite with certainty, it will be seen that in these the rules hold whichever of the two possible prongs be supposed to be the normal.

819.   **Astacus leptodactylus:** left chela has dactylopodite as shewn in Fig 187, II. Presumably D is the normal pushed out of place, and D' and D'' are the two extra dactylopodites. They are so placed that none meets the index. Károli, J., *Term. Füzetek*, 1877, I. p. 53, Pl. II.

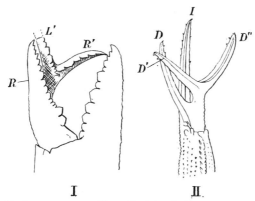

Fig. 187. I. *Cancer pagurus*, No. 820, right chela. Specimen in Coll. Surg. Mus. II. *Astacus leptodactylus*, left chela, after Károli.

820. **Cancer pagurus**: somewhat similar case in rt. chela (Fig. 187, I); but here the normal, R, stands in its normal place. In Coll. Surg. Mus.

*821. **Homarus americanus**: dactylopodite only of right chela preserved. It is bent sharply downwards, out of the plane of the "hand," and bears upon its upper surface *two pairs* of blunt, toothed processes [probably being rudiments of two pairs of extra dactylopodites]. FAXON, *l.c.*, p. 261, Pl. II. fig. 1.

822. **Homarus americanus**: dactylopodite (*a*) bent upwards and outwards, crossing index without meeting it (Fig. 188). From the smooth border of dactylopodite arise two toothed processes

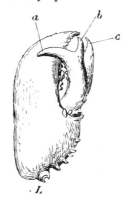

Fig. 188. *Homarus americanus*, No. 822, left chela. *a*, normal point of dactylopodite. *b, c*, extra points. (After Faxon.)

(*b* and *c*) curving towards index. [I take it that this is something like the cases of Position A in Insects (p. 481) but from the original figure the relations cannot be quite decided.] FAXON, *l.c.*, p. 260, Pl. I. fig. 15.

534   MERISTIC VARIATION.   [PART I.

(c) *Two extra indices arising from a normal index.*

*823.   This again is a fairly common form, though much less frequent

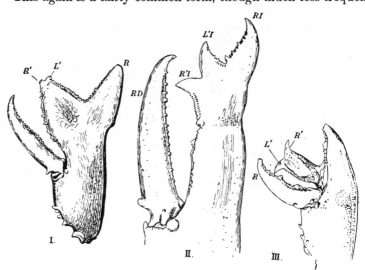

Fig. 189. I. Right chela of *Homarus americanus*. R', L', right and left extra indices not separated from each other. (After Faxon.) II. *Homarus vulgaris*, right chela in Brit. Mus. III. *H. vulgaris*, right chela bearing extra double index. R' and L', not separated. (After Lucas, *l.c.*)

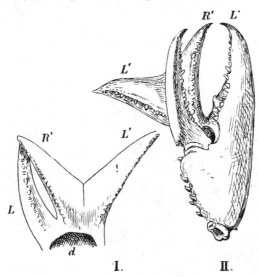

Fig. 190. I. Left chela of *Carcinus mænas*, indices only shewn. d, place of articulation of dactylopodite. In Coll. Surg. Mus. II. A similar case in *Homarus americanus*, after Faxon. L, normal left index. R', L', extra right and left indices.

CHAP. XXI.] SECONDARY SYMMETRY : CRUSTACEA. 535

than the last. The cases known to me amount to about ten
or fifteen. Seven cases are illustrated in Figs. 189, 190, and 191.

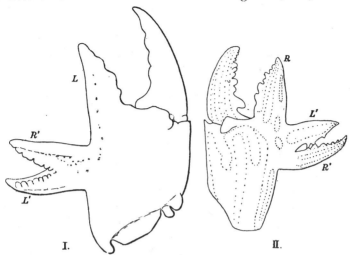

FIG. 191. Two cases of extra indices in *Cancer pagurus*. I. In Coll. Surg.
Mus. II. After le Sénéchal. R, normal right index. L, normal left. R', L',
extra rights and lefts.

(d) *Simple processes, probably being rudimentary extra pairs
of indices or of dactylopodites.*

*824. Many such are described, but of
few can anything be said with confidence. A comparatively simple case
is shewn in Fig. 192, where there is
a decided suggestion that the process L'
+ R' is morphologically a pair of indices
that have not separated from each other
but stand compounded by their toothed
borders. On comparing this case with
for instance, Fig. 191, II, it will be seen
that the two conditions might readily
pass into each other in the way so often
seen in Insects.

Other cases of a more doubtful character are shewn in Fig. 193. Though
in each the nature of the extra part is
obscure, it is probable that they are all
rudimentary states of the repetitions
described. The alternative view that
they are *single* repetitions certainly cannot be applied to all, for in many the
extra process, though in the plane of the
index and dactylopodite, is similar on

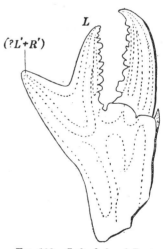

FIG. 192. Left chela of *Portunus puber* from LE SÉNÉCHAL, *Bull.
Soc. Zool. France*, 1888, XIII. p. 125.
L, normal index. $L' + R'$, ? pair
of extra indices in Position V.

536   MERISTIC VARIATION.   [PART I.

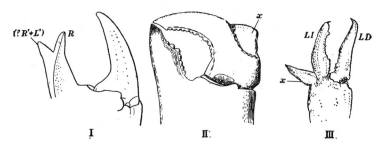

FIG. 193. I. Right chela of *C. pagurus* in Coll. Surg. Mus. *R*, right index.
II. Similar specimen whose dactylopodite bears *x*, a supernumerary process. In
Coll. Surg. Mus. III. *Astacus fluviatilis*, left chela bearing *x*, a supernumerary
process. *RI, RD*, right index and dactylopodite. (After LUCAS.)

both its faces in this plane. There is however no doubt that the
distinction between these cases and true duplicity is hard to trace
and possibly enough it is not really absolute.

825. As each case differs from the others I give a list of those not in private collections[1]. The ? indicates that the case perhaps approaches the condition of true duplicity.

R, *right*.   L, *left*.   D, *dactylopodite*.   I, *index*.

Astacus fluviatilis	RI	TIEDEMANN, *Meckel's Arch.*, 1819, v. p. 127, Pl. II. *fig*. 2.
? **A. fluviatilis**	RI	JAEGER, G., *Jahresh. Ver. vaterl. Naturk.*, 1851, XVII. p. 35, Pl. I. fig. 7.
A. fluviatilis	LD	*id., Meckel's Arch.*, 1826, p. 95, Pl. II. *fig*. 3.
A. fluviatilis	RI	RÖSEL V. ROSENHOF, *Ins.-Belust.*, III. p. 344, *fig*. 31.
? **A. fluviatilis**	LI	*ibid.*, fig. 30.
A. fluviatilis (Fig. 193, III.)	LI	LUCAS, *Ann. soc. ent. Fr.*, 1844, Sér. 2, II. p. 45, Pl. I. *fig*. 6.
Homarus americanus	LI	FAXON, *Harv. Bull.*, VIII. p. 259, Pl. I. *fig*. 11.
H. americanus	RD	*ibid.*, Pl. I. fig. 6.
? **Cancer pagurus**	LD	RICHARD, *Ann. sci. nat.*, 1893, p. 106.
C. pagurus (Fig. 193, I.)	LI	*Coll. Surg. Mus.*
C. pagurus (Fig. 193, II.)	LD	*Coll. Surg. Mus.*

(*e*)   *Exceptional Cases.*

*826.   **Homarus americanus**: Right chela. Meropodite subcylindrical instead of flattened; peripherally divides into two parts each bearing an articulated appendage as shewn in Fig. 194. [The appendage *R* is a normal chela. What is $R' + L'$? FAXON, carefully describing the case, thinks that $R' + L'$ is a rudimentary and reversed copy of *R*, and that the case is one of duplicity. But from the particulars given, and especially from the circumstance that the carpopodite was "much more spiny" than the normal, I think it likely that $R' + L'$ is morphologically a double structure formed of a *pair* of carpopodites compounded together.

[1] With these may perhaps be mentioned the following: **Apus cancriformis**, having upon the 40th foot a second small flabellum shaped like the normal flabellum. The bract was greatly reduced in size. LANKESTER, E. R., *Q.J.M.S.*, 1881, XXI. p. 350, Pl. xx. fig. 12. [In explanation of Plate the abnormal foot is called the 30th.]

Without having seen the specimen it is impossible to say much, but the parts should be examined with a view to this possibility. I conceive that the large spine marked by Faxon sp' stands on

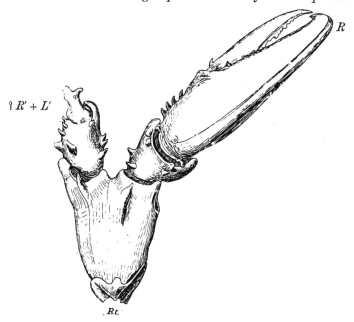

FIG. 194. *Homarus americanus*, No. 826. A right chela. (After Faxon.)

the morphologically middle line between the two extra half-meropodites.] FAXON, *Harv. Bull.* VIII. p. 262, Pl. II. fig. 6.

*827.   **Astacus fluviatilis** ♀ : large adult. Abdomen wide in comparison with slender chelæ: otherwise normal except left chela. This was formed as in Fig. 195. All normal except carpopodite, from which arose a fixed piece seeming to be an extra misshapen carpopodite, bearing *three extra chelæ, L', R' and x*. [R' and L' are a clear pair of images being right and left respectively. But between R' and the normal L there is the third extra chela x. As to the nature of this nothing can be said. Whether it is a left or a right cannot be told from fig. So far as I know, this case is unique. Full description and measurements given in original, $q. v.$] CANTONI, *Rend. R. Ist. Lomb.*, 1883, XVI. p. 771, *fig.*

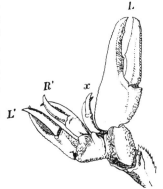

FIG. 195. *Astacus fluviatilis*, No. 827, left chela. L, the normal. R', L', presumably extra right and left chelæ. x, extra chela of uncertain nature. (After Cantoni.)

C. *Antennæ.*

*828. **Palinurus vulgaris:** right antenna bore three complete filaments. So far as last spiny joint (merocerite) normal. Of this joint the peripheral portion much enlarged, presenting two articulations. The most posterior bore a normal carpocerite and filament (Fig. 196, I). The anterior articulation bore a double carpocerite with two filaments (II and III). As author points out, II is structurally a *left* antenna. [By the kindness of M. Alphonse Milne Edwards I have been allowed to examine this specimen. I am not sure that I succeeded in correctly determining the surfaces of the extra antennæ, for the basal parts were not very fully formed; but according to my determination their relations differed markedly from those of any of the Schematic positions, for while the position of origin is VVA the two extra antennæ stand very nearly in the Position DA.] LÉGER, *Ann. sci. nat.*, *Zool.*, 1886, S. 7, I. p. 109, Pl. 6.

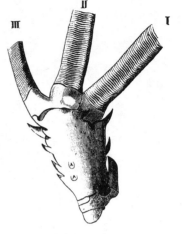

Fig. 196. Proximal parts of the right antenna of *Palinurus vulgaris*, No. 828. I, the normal. II, extra left. III, extra right. (After Léger.)

*829. **Astacus fluviatilis:** exopodite of left antenna (Fig. 197) bears two supernumerary points, R' and L', which seem to have been inserted upon the internal border of the normal exopodite. STAMATI, G., *Bull. Soc. Zool. France*, 1888, XIII. p. 199, *fig.*

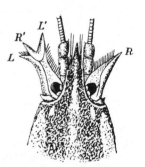

Fig. 197. *Astacus fluviatilis* having extra points to exopodite of left antenna. R, normal right. L, normal left. R', L', extra right and left. (After Stamati.)

AMORPHOUS CASES.

As has been stated, there are many cases, recorded or preserved, in which the nature of the parts cannot be made out. The majority of these are, I believe, injured or deformed limbs, and not cases of repetition of parts. Nevertheless of the latter class there are undoubtedly some amorphous cases, though they are far less common than regular ones, even as normal structures are more common in their regular shapes than in a deformed state. I mention the following as being, I think, the earliest record of abnormalities of this class.

830. **Homarus:** left chela having irregular process on inner border of dactylopodite, and two irregular processes on inner border of index. [No description.] BERNHARDUS À BERNIZ, *Miscell. Curios.*, Jena, II. 1671, p. 175, Obs. CI. *Pl.*

CHAPTER XXII.

DUPLICITY OF APPENDAGES IN ARTHROPODA.

THAT there should be such a thing as a limb double in the sense in which the following are double, has always seemed to me most strange. We know that a segment of an Annelid, or a vertebra, may be on one side of the body divided to form two segments or two vertebræ (as in No. 88 or No. 7) while on the other side of the middle line the segment is single. This is in keeping with all that we know of Division of parts in Linear Series. So might we suppose that a parapodium, or a rib, or perhaps a limb-bud might divide into two; but the two half-segments or half-vertebræ are in Succession to each other, and are not complementary images of each other as these double-limbs are.

That a parapodium may divide into two Successive parapodia is possible enough, though, apart from division of the segment bearing it, I know no clear case. But it may be stated at once that in Arthropods and Vertebrates such a phenomenon as the representation of one of the appendages by two identical appendages standing in Succession is unknown. No right arm is ever succeeded on the same side of the body by another arm properly formed as a right, and no Crustacean has two right legs in Succession, where one should be. The only cases at all approaching this state are those of *Macacus* No. 504 (*q. v.*), a case that must be interpreted with great hesitation; and of the Frogs described by CAVANNA and by KINGSLEY, also doubtful cases (see Chapter XXIII).

But though such repetition is probably unknown and is perhaps against Nature, there are still these strange double-limbs: two limbs, always I believe imperfect, placed not in Succession, but as complementary images of each other, more or less exact. These we have seen in the hand of Man and in the feet of Artiodactyles; we have now to study them in Insects and in Crustacea[1].

[1] With mistrust I name cases in Amphibia and Fishes, perhaps of this nature. **Lissotriton punctatus** (Newt): left pes having 10 digits in two groups, 6 and 4. *Coll. Surg. Mus., Ter. Ser.,* 293, A [not dissected]. **Protopterus annectens**: rt.

On the morphology or significance of duplicity in limbs I can make no comment beyond the few remarks given on p. 406. It is just possible that in Nos. 832 to 834 the duplicity of the chela or of the index is a division in the middle line of a Bilateral Minor Symmetry; for some chelæ are peripherally very nearly symmetrical about the plane of the dactylopodite and index.

In Arthropods double-limbs are no less rare than in Vertebrates, for though in various works there are some scores of cases to be found, the great majority may be safely rejected as being almost certainly cases of double extra parts in Secondary Symmetry having their duplicity disguised as we saw it in Nos. 750, 764, or 801. By most of those who have dealt with these things the possibility of disguised duplicity in the extra part has been unheeded; and ignorant of the special difficulties of these cases they have thus set down specimens as examples of duplicity of appendages at a casual glance. For this reason therefore I shall only give particulars of those few cases which are better established or otherwise of special interest, letting the rest follow as a list of references.

It will not be forgotten that whenever an extra part is in itself symmetrical it always *may* be a double structure, and the special application of this fact to cases of extra filamentous antennæ must in particular be borne in mind.

CRUSTACEA.

*831. **Hyas araneus**: a left chela having the form shewn in Fig. 198, II and III. Fig. 198, I shews a normal left chela of this species from the outside in the same position as II. In the abnormal specimen the dactylopodite D is normal save that

pectoral fin double, the division being in a *horizontal* plane, so that the two filaments were dorsal and ventral to each other [cp. No. 503]. ALBRECHT, *Sitzb. Ak. Wiss. Berl.*, 1886, p. 545, Pl. VI. **Silurus glanis**: extra fin attached to pelvic girdle and partly to rt. pelvic fin. WARPACHOWSKI, *Anat. Anz.*, 1888, III. p. 379, *fig.* **Rana esculenta**: left hind foot double; rt. not seen [a very clear case]. ERCOLANI, *Mem. Acc. Bologna*, 1881, S. 4, III. p. 812, Pl. IV. *fig.* 11.

In Raiidæ a group of cases of extra fin are known. They are upward projections from the dorsal surface near the middle line. They are often spoken of as "dorsal" fins, but in the only case I have seen (*Paris Mus. N. H.*, $\frac{7902}{A}$, kindly shewn me by Prof. L. Vaillant) the attachment is not really median but is slightly oblique, and seems, from external examination, to spring from some part of the pectoral girdle (? left scapula). See LACÉPÈDE (who named such a fish "*Raja cuvieri*"), *Hist. nat. des Poiss.*, 1798, I. p. 141, Pl. VII.; NEILL, *Mem. Wern. Soc.*, 1808, I. p. 554; MOREAU, *Poiss. de la France*, 1881, I. p. 206. In these fishes the real dorsal fins were in the proper place (though in some species they may be far forward, FORSKÅL, *Descr. Anim. in itin. Orient.*, 1775, I. p. 18). This repetition is of course quite distinct from that other curious and also Discontinuous variation in which the pectorals are partly divided into two lobes (**R. clavata**, YARRELL, *Brit. Fish.*, ed. RICHARDSON, 1859, II. p. 585); or are separated from the head so as to project like horns on either side, as in last case; and also in **R. clavata**, YARRELL, *ibid.*; p. 384; DAY, *Brit. Fish.*, II. p. 345, Pl. CLXXI. *fig.* 2; in **R. batis**, DAY, *l. c.*, p. 337; in **R. asterias**, BUREAU, *Bull. soc. zool. France*, 1889, XIV. p. 313, *fig.*

its point is rather worn. Where the index should be, there is a great eminence, bearing apically a second articulated dactylopodite D', complementary to D. Between the two dactylopodites

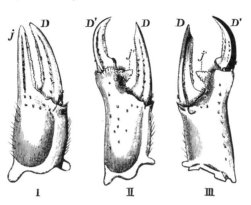

Fig. 198. *Hyas araneus*. I. A normal left chela. II. The left chela of No. 831 from the outside. III. The same from the inside. D, normal dactylopodite. D', extra dactylopodite. j, normal index. j', a small index toothed on both sides. (In Brit. Mus.)

at the inner side of the eminence there is a fixed short process, j', which is toothed upon both the edges which it presents to the two dactylopodites. Round the articulation of D' are setæ like those round the place of articulation of D. Specimen in *Brit. Mus.*, kindly shewn to me by Mr R. I. Pocock.

832.   **Cancer pagurus:** right chela. Dactylopodite and index each double in the way shewn in Fig. 199. Each is toothed on the side presented to the other half-pincer. Note that there is no *proof* that one or other of these points is not a pair compounded in Position A or P, but since both seemed equally to diverge from the normal plane of the propodite this is most unlikely. Specimen in Museum of Newcastle-upon-Tyne.

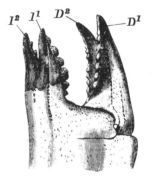

Fig. 199. Right chela of *Cancer pagurus*, No. 832. D^1, D^2, two partially separate dactylopodites. I^1, I^2, two partially separate indices. (In Newcastle Mus.)

833.   **Homarus americanus:** right chela shewn in Fig. 200, I. Two dactylopodites separately articulating. Index bifid at apex and bearing two rows of teeth, one on each edge. Dactylopodites did not meet index. FAXON, *Harv. Bull.*, VIII. p. 260, Pl. I. fig. 13.

834.   **? Hyas** sp. Right chela. Dactylopodite single and in normal plane. Two separate and similar indices, each toothed as usual,

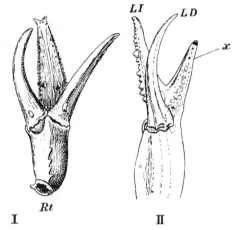

Fig. 200, I. *Homarus americanus*, right chela, No. 833. (After Faxon.)
II. *Lupa dicantha*, left chela, No. 836. *LD, LI*, left dactylopodite and index. *x*, supernumerary index. (After Lucas.)

making angle of about 45° with each other. This angle almost exactly bisected by the plane in which dactylopodite moves. *Bell Collection*, Oxford.

835.   **Maia squinado**: from inner side of base of index of right chela arises a second index as shewn in Fig. 201. It is about half as large as the supposed normal index. The latter is displaced outwards. Dactylopodite moves in approximately normal plane, missing both indices and falling between them. Specimen kindly lent by Prof. C. Stewart.

Fig. 201. Right chela of *Maia squinado*, No. 835.

The following are cases very similar to Nos. 834 and 835.

836.   **Lupa dicantha,** left chela (Fig. 200, II). Lucas, *Ann. Soc. ent. France*, 1844, S. 2, II. p. 43, Pl. I. fig. 1.
837.   **C. pagurus,** right chela, 2 cases, LE SÉNÉCHAL, *Bull. Soc. Zool. France*, 1888, XIII. p. 125, *fig.* 2.
838.   **Xantho punctulatus,** left chela (Fig. 202) in which the index divided at about its middle to form two similar and equally diverging blunt processes. HERKLOTS, *Arch. néerl.*, 1870, v. p. 410, Pl. X.
839.   **Homarus americanus**: right chela bearing an extra index. Dactylopodite does not meet the normal index. [Very doubtful if of same nature as foregoing cases.] FAXON, *l.c.*, Pl. I. fig. 14.

The following cases are exceptional.

840.   **Homarus vulgaris**: right chela has coxopodite single; but basi-

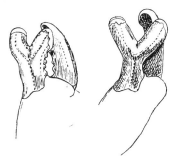

Fig. 202. *Xantho punctulatus*. Two views of left chela of No. 838, shewing the division of the index. (After Herklots.)

podite is wrinkled and has two apical articulations, each bearing a small chela; both are soft and not calcified, having articulations indicated by furrows only. [No information as to planes.] RICHARD, *Ann. Sci. Nat.*, 1893, p. 106.

841. **Homarus americanus**: right chela having a short articulated process below the dactylopodite moving in plane at right angles to it. [?a double structure]. FAXON, *Harv. Bull.*, VIII. Pl. I. fig. 12.

842. **H. americanus**: toothless process *articulating below* dactylopodite, moving in plane at right angles to its plane of motion. It articulates upon a separate process given by the propodite. [It is difficult to suppose that this extra process can be double.] FAXON, *l. c.*, Pl. I. fig. 16.

Mr G. DIMMOCK of Canobie Lake, N. H. has kindly sent me word of a **Gelasimus** having a chela of very anomalous form. Both index and dactylopodite are said to have been bifid, but the plane of division was *at right angles* to the plane of the dactylopodite and index, so that all four points were in one plane. This specimen has unfortunately been destroyed; but Mr Dimmock tells me that the arrangement was certainly thus, and that the unusual difficulty of bringing this case into agreement with others was recognized in examining it.

INSECTS.

Among the following 110 cases which all either have been or might be called cases of "duplicity" of legs, antennæ, or palpi, there is, I think, not one clear case of unmistakeable duplicity, such as for instance those of the chelæ in Nos. 831 or 832. They should thus be considered as cases in which the extra parts have not been or cannot be shewn to be double, rather than as examples of proved duplicity of normal appendages. In every case that I have myself properly examined, it is either possible to prove the duplicity of the extra parts; or else essential features (*e.g.* spurs &c.) by which a right appendage may be told from a left are wanting. Nevertheless the few straightforward cases of double-limbs in Crustacea keep one alive to the possibility that some of these also may be the same. The most probable cases of true duplicity of limbs are Nos. 844, 846 and 851.

544   MERISTIC VARIATION.   [PART I.

*843. 1. *Legs.*

Prionus californicus (Longic.): *each* femur bore two tibiæ and tarsi; both maxillary palps and also the left labial palp were partially double (Fig. 203). [No statement as to *right* labial palp. This shewn in fig. much thicker than left, but on com-

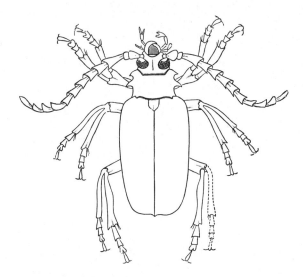

Fig. 203. *Prionus californicus*, No. 843, having extra legs and palpi. (After Jayne.)

paring with a specimen it seems to be of normal thickness.] In some of the legs the two tibiæ are compounded at their bases, in others they articulate separately. [Several details given; and in particular, enlarged views of the palpi and of the bases of the tibiæ. But as no details are given regarding the *apices* and apical spurs of the tibiæ nothing can be said as to symmetry.

It will be remembered that we have already had a case of a *Prionus*, No. 750, which similarly was supposed to have two of its legs double; but there by means of the tibial spurs it was shewn that the extra part was in Secondary Symmetry. Possibly enough the same could here be shewn. It is much to be hoped that this specimen can be traced.] JAYNE, H. F., *Trans. Amer. Ent. Soc.*, 1880, VIII. p. 159, fig. 12.

844.   **Allantus** sp. (Tenthred., Sawfly): extra leg borne by coxa of right middle leg. This coxa is imperfectly double, bearing two separate trochanters. Of these the anterior bears a small leg which, though ill formed, is complete in all its parts, but has the tarsal joints of abnormally small size. The posterior trochanter bears a leg of full size. Its femur curves forwards and then backwards. The femur of the smaller leg curves for-

wards, but its tibia curves backwards. The femora are so twisted that I failed to determine the symmetry of these legs; and while it was clear that neither was a normal left it was equally doubtful whether either was shaped as a right. Of all cases in Insects this is one of the nearest to the condition of true duplicity. *Hope Collection*, Oxford.

845. **Carabus intricatus:** middle right femur is partially bifid, presenting two apices in the same horizontal plane. The anterior apex bears a tibia and tarsus of nearly normal form. The other apex bears a tibia and tarsus of full length but much more slender than a normal one. This leg was ill-formed. The tibia bore no spurs, and there was no indication as to its symmetry, and nothing shewed that it was a right or a left leg. It is stated in the original description that the two legs could be separately moved and that both assisted in locomotion. Originally described by MOCQUERYS, *Col. anorm.*, p. 45, *fig.*

*846. **Melolontha vulgaris:** right anterior leg divided to form two legs. The femur dilates in peripheral third to form two apices, each bearing a tibia. These two tibiæ are at right angles to the femur and are together in the same straight line, the one pointing forwards and the other backwards, each tibia turning its ventral or flexor surface towards the femur. The anterior tibia carries a tarsus of 4 joints with claws, while the posterior tibia has a normal tarsus of five joints. For a figure of this specimen and particulars concerning it I am indebted to Professor ALFRED GIARD.

847. **Leptura testacea** (Longic.): in tarsus of left middle leg the 2nd joint presents two apices (Fig. 204). The posterior bears normal 3rd and 4th (terminal) joints with a proper pair of claws. The anterior apex bears a narrow 3rd and 4th joint, the latter having only a single median claw [cf. No. 848]. KRAATZ, *Deut. ent. Zt.*, 1876, xx. p. 378, *fig.* 14.

848. **Tetrops præusta** (Longic.): right anterior femur widened towards apex, which presents two articulations in same horizontal plane. Each of these bears a tibia. The posterior tibia and tarsus are complete in all respects, but they flex downwards and backwards. The anterior tibia has a normally 4-jointed tarsus, but the apical joint bears only *one* claw, and there is no sign of mutilation [cp. No. 847]. Were it not for the closely similar case of *Silis* No. 764 there would be no reason to doubt that this is a true case of duplicity, but that example

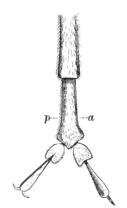

FIG. 204. *Leptura testacea*, No. 847. Tarsus of left middle leg from the plantar surface. (The property of Dr Kraatz.)

shews how masked may be the doubleness of extra parts; and though I could not prove either of these legs to be double I feel no certainty that one of them is not double. Specimen very kindly lent for description by Mr F. H. WATERHOUSE.

B.

546   MERISTIC VARIATION.   [PART I.

849.   **Chlænius holosericus** (Carab.): left anterior tibia enlarged and dividing close to base into two branches of similar form and length [curving towards each other], both equally furnished with hairs and bearing spines characteristic of the species. Anterior branch bears a complete tarsus like that of a leg *of the other side*, but posterior branch bears only one tarsal joint. CAMERANO, *Atti Ac. Sci. Torino*, 1878, XIV. *fig.*

*850.   **Brachinus crepitans** (Carab.): 3rd joint of right posterior tarsus enlarged; 4th joint divides to form two apices (Fig. 205), each bearing separate 5th joint in same horizontal plane. Each of these has a pair

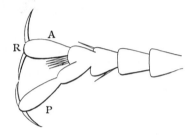

FIG. 205. Right hind foot of *Brachinus crepitans*, No. 850. A, anterior. P, posterior. R, the supposed normal right apex. (In Rouen Mus.)

of claws curving ventralwards. The two apical joints are not identical, the anterior being the shorter and continuing the general direction of the tarsus. I could not determine the symmetry. When examined by me the specimen was intact, but in cleaning it I broke this abnormal leg. First described by MOCQUERYS, *Col. anorm.*, 1880, p. 63, fig.

The two following cases differ from the rest in that the extra leg arose from the body separately from the normal leg. Among the cases of extra limbs in Secondary Symmetry were a few in which the coxa of the extra limbs was in the same socket as the coxa of a normal leg, though not united to it; but in the first, and perhaps in both of the two cases that follow, the extra leg was wholly separate. The first case, No. 851, is the only one of the kind that I have seen.

*851.   **Tenthredo ignobilis** (Tenthred., Sawfly): extra leg arising from *prothorax*, on the left side of the body, at some distance behind the proper left anterior leg. Behind the anterior legs the prothorax of a normal specimen presents ventrally an elevation on each side of the middle line; the point of origin of the extra leg is about halfway between this elevation and the socket of the coxa of the normal left anterior leg. The specimen had been a good deal injured by being pinned very nearly through the point of origin of the extra leg, and on relaxing the specimen and attempting to restore the parts to their former positions I unfortunately broke off the extra leg from the body[1]. The leg is fairly well formed, but is a little shorter and a good

[1] The specimen has been mended as nearly as possible in the position originally occupied by the leg. As it may pass hereafter into other hands, it may be well to

deal more slender than the normal anterior leg. Owing to the slight degree to which the anterior legs of this insect are structurally differentiated from the middle legs, it cannot be positively stated that the extra leg is in form an anterior or a middle leg, but in size and general conformation it approaches very nearly to that of an anterior leg. It is complete in all its joints, having normal ciliation and claws, but *the spurs are entirely absent from the apex of the tibia* and probably have never been formed. This is an unfortunate circumstance; for, inasmuch as the anterior spur of a normal anterior tibia in this species is markedly differentiated from the posterior spur, it would have been easy to determine the surfaces of this leg had the spurs been present. As it is, the matter cannot be positively decided, and it must suffice to say that the general form of the leg and the shape and curvature of its joints are such as to make it appear to be fashioned as an anterior leg and as a leg of the side upon which it occurs, namely, the left. This specimen was most kindly lent for description by Mr C. W. DALE, of Glanville's Wootton, Dorsetshire. It is the specimen mentioned in *Ann. and Mag.*, 1831, IV. p. 21.

852. **Elater variabilis** (Elat.): complete extra leg articulating by separate coxa close to right anterior leg. GERMAR, E. F., *Mag. der Ent.*, II. p. 335, Pl. I. fig. 12. [This case has been copied by many authors. The figures represent the right fore leg and the extra one as normal right legs, but they are not sufficiently detailed to give confidence that this was so. If the specimen still exists it is to be hoped that it may be properly described.]

853. This is a list of all remaining cases in which it is in any way possible that there is duplicity of a leg. The point of origin is shewn approximately.

*, seen by myself. ‡, partly amorphous or mutilated. 0, no description. R, right. L, left. tr., trochanter. f, femur. tb, tibia. ts, tarsus.

* ‡ **Osmoderma eremita**[1] (Lamell.)		L 1. c.	MOCQUERYS, *Col. anorm.*, 1880, p. 46, *fig.*
Mallodon sp. (Longic.)		R 3. c.	*ibid.*, p. 50, *fig.*
Pasimachus punctulatus (Carab.)		L 2. tr.	JAYNE, *Trans. Amer. Ent. Soc.*, 1880, VIII. p. 156, Pl. IV. *fig.* 4.
Broscus vulgaris (Carab.)		R 1. tr.	IMHOFF, *Ber. Verh. nat. Ges. Basel*, 1838, III. p. 3.
Agonum sexpunctatum (Carab.)		R 3. f.	SCHNEIDER, *Jahresb. schles. Ges. vaterl. Kultur*, 1860, p. 129.
‡ **Carabus septemcarinatus** ♂		R 3. f.	KRAATZ, *Deut. ent. Zt.*, 1877, XXI. p. 57, Pl. I. *fig.* 32.
‡ **Carabus nemoralis**		L 3. f.	OTTO, HERM., *Term. füzetek*, 1877, I. p. 52, Pl. II.
Carabus creutzeri ♀		L 1. f.	KRAATZ, *l. c.*, *fig.* 31.
Procrustes coriaceus[2] (Carab.)		R 3. f.	MOCQUERYS, *l. c.*, p. 55, *fig.*
Meloe coriaceus (Het.)		L 1. f.	STANNIUS, *Müll. Arch. Anat. Phys.*, 1835, p. 306, *fig.* 11.
0 **Carabus helluo**		R 1. f.	REY, *Ann. Soc. Linn. de Lyon*, 1882, XXX. p. 423.
0 **Trichodes syriacus** (Cleridæ)		R 1. f.	*ibid.*
‡ **Chrysomela hæmoptera** (Phyt.)		? 3. f.	CURTIS, *Brit. Ent.*, Pl. 111, *fig.* 5*.

state explicitly that there was no conceivable doubt as to the genuineness of the abnormality. When received by me it was absolutely natural and had not been in any way mended.

[1] Probably this is the specimen mentioned by BELLIER DE LA CHAVIGNERIE, *Bull. Soc. ent. France*, 1851, S. 2, IX. p. LXXXII.
[2] See also KLINGELHOFER, *Stet. ent. Zt.*, 1844, v. p. 330.

‡ **Chlænius diffinis** (Carab.)	L 2. tb.	JAYNE, *l. c.*, p. 157, Pl. IV. *fig.* 7.
Rhagium mordax (Longic.)	R 2. tb.	KRAUSE, *Stet. ent. Zt.*, 1871, XXXII. p. 136.
Agabus uliginosus (Dytisc.)	R 3. tb.	PERTY, *Mitth. nat. Ges. Bern*, 1866, p. 307, *fig.* 6.
* ‡ **Acanthoderes nigricans** (Longic.)	L 2. tb.	MOCQUERYS, *l. c.*, p. 48, *fig.*
Colymbetes adspersus (Dytisc.) ♂	L 3. tb.	KRAATZ, *Deut. ent. Zt.*, 1877, XXI. p. 56, Pl. I. *fig.* 14.
‡ **Procrustes coriaceus** (Carab.)	R 3. tb.	OTTO, HERM., *l. c.*, 1877, I. p. 52, Pl. II.
‡ **Carabus melancholicus** ♂	R 3. tb.	KRAATZ, *Deut. ent. Zt.*, 1880, XXIV. p. 344.
* ‡ **Tenebrio granarius** (Het.)	L 3. tb.	MOCQUERYS, *l. c.*, p. 49, *fig.*
* ‡ **Calosoma auropunctatum** (Carab.)	R 1. tb.	Lent by M. H. GADEAU DE KERVILLE[1].
0 **Silpha granulata** (Clav.)	R 3. tb.	RAGUSA, *Nat. Sicil.*, I. p. 281, *fig.*
* **Philonthus succicola** (Staph.)	R 3. ts.	Lent by Dr MASON.
* ‡ **Telephorus excavatus** (Mal.)	R 2. ts.	MOCQUERYS, *l. c.*, p. 60, *fig.*
Chlænius vestitus (Carab.)	L 2. ts.	*Ann. and Mag. N. H.*, 1829, II. p. 302, *fig.*
0 **Telephorus fuscus** (Mal.)	? 2 ?	BASSI, *Ann. Soc. ent. France*, 1834, S. 1, III. p. 375.
0 **Prionus coriaceus** (Longic.)	?	VON HEYDEN, *Isis*, 1836, IX. p. 761.
0 **Prionus** sp. (Longic.)	? ? f.	*ibid.*

2. *Antennæ.*

The remarks made in preface to the last section apply here also, and with additional force from the consideration pointed out (p. 513), that many antennæ are without obvious differentiation between their anterior and posterior surfaces. As Kraatz has pointed out, it is especially in such forms as Lamellicorns or Lucanidæ that extra antennæ are found *double*, and I think there is an obvious inference that this greater frequency in them is due to the fact that the two borders are so markedly differentiated that the duplicity cannot easily be disguised. I have sometimes fancied too that perhaps the existence of this great differentiation between the two borders may actually contribute to the physical separation of the two extra parts in the Positions A and P and thus prevent that masking of the duplicity which is seen for instance in *Navosoma* No. 801.

However this may be, special importance must be attached to the few cases in Lamellicorns, Lucanidæ and the like, where there seems to be a *single* extra part, making that is to say a duplicity of the antenna. Cases of this kind that I have myself seen I therefore treat more fully, and it may be stated that in none of them is there anything that can be called clear duplicity. In many on the contrary the extra part is nearly cylindrical, and thus symmetrical in itself. Hence it may possibly be morphologically double. Of the remainder I can give no confident account. For as has been said, though many, *e.g., Zonabris 4-punctata* (in No. 858), do look very like cases of true duplicity

---

[1] Originally described by FLEUTIAUX, *Rev. d'Ent.*, 1883, p. 228.

I feel no certainty that they are so. Nothing but careful microscopical examination can shew this, and it would in every case be necessary to begin by fixing upon some definite character differentiating the anterior from the posterior border in the normal antenna.

In the majority of cases one of the branches has less than the normal number of joints.

Special attention is called to No. 854, for in it is seen not only an extra branch, but an *extra joint* in the course of the chief antenna.

N.B. At the end of this list I have set three cases of extra antenna arising from the *head*.

*854. **Lucanus cervus** ♂ (Lucanidæ) : left antenna normal, practically same as that described for *Odontolabis* No. 799. Right antenna shews a rare condition. Scape and 2nd joint normal. Then follows a piece as long as the 3rd, 4th and 5th joints of a normal, together. This joint has a complex form. It has no *transverse* division and is clearly one segment from base to apex, but the posterior border is divided from the anterior by an irregular, crescentic suture, giving it the look of two joints spliced together. The *posterior* portion gives origin to a small, backwardly directed branch made up of two nearly spherical joints, the apical having a minute depression whence a fragment may have been broken.

The long third joint just described bears at its apex the rest of the antenna, which is abnormal in structure and diverges a little forward of the normal direction. In the normal there are only 7 joints peripheral to the 3rd, making 10 in all; here there are 8, making 11 in all. The four apical flattened joints are normal, but the joint preceding them (7th in this antenna) is more produced on the anterior border than in the normal, and it is thus in form *almost intermediate between a funicular and a lamellar joint*. The other three are simple funicular joints. For this singular specimen I am indebted to the kindness of M. Henri GADEAU DE KERVILLE.

855. **Nigidius** sp. (Lucanidæ) New Guinea: the second joint of the right antenna bears a small supernumerary three-jointed branch directed forwards and upwards. The terminal joint of the branch, which morphologically stands fifth from the body, bears a long hair of the kind which is borne in the normal antenna only by the seventh and subsequent joints.

There appears to be no deformation in the normal antenna in correspondence with the presence of this extra branch. The position of the antenna with reference to the second joint is a little altered, but it is not in any other way changed. This specimen was kindly lent to me by M. Henri GADEAU DE KERVILLE.

856. **Lucanus cervus** ♂ : the second (1st funicular) joint of the left antenna bears a four-jointed, pointed filament. The lower

parts of the head on the left side are also greatly deformed.
Von Heyden, *Deut. ent. Zt.*, 1881, xxv., p. 110, fig. 24.

857. **Melolontha vulgaris** (Lamell.): from ventral surface of 2nd joint of left antenna a separate joint projects vertically downwards. This joint bears a forwardly-directed process which is about as long as a normal club and is imperfectly divided into lamellæ. Nothing could be definitely determined as to the symmetry of this structure. Originally described by Mocquerys, *Col. anorm.*, p. 22, *fig.*

858. In this list * means that I have seen the specimen, ‡ that it is partly amorphous or mutilated, 0 that there is no description. The number is a rough indication of the joint from which the extra part arose.

* ‡ **Cicindela sylvatica** (Cicind.)	R 3.	*Mus.* H. Gadeau de Kerville.
Carabidæ		
Carabus sylvestris ♂	R 8.	Kraatz, *Deut. ent. Zt.*, 1877, xxi. p. 55, *fig.* 9, and Sartorius, *Wien. ent. Monats.*, 1861, v. p. 31.
C. auratus	R 2.	*ibid.*, *fig.* 8.
ditto	R 5.	Doumerc, *Ann. Soc. ent. Fr.*, 1834, S. 1, iii. p. 174, Pl. i.
ditto	L 8.	Perty, *Mitth. nat. Ges. Bern*, 1866, p. 307, *fig.* 4.
C. italicus	8.	Gredler, *Corr.-Bl. zool.-min. Ver. Regensb.*, 1877, xxxi. p. 139.
C. exaratus	L 5.	*ibid.*
C. intricatus	9.	*Ann. and Mag. N. H.*, 1841, p. 483.
C. emarginatus ♀	L 2.	von Heyden, *Deut. ent. Zt.*, 1881, xxv. p. 109, *fig.*
C. cancellatus	{R 10. L 10.	Sartorius, *Wien. ent. Monats.*, 1858, ii. p. 49.
* ‡ **C. catenulatus** ♂	L 8.	*Brit. Mus.*
Pterostichus planipennis ♀	R 9.	Kraatz, *l. c.*, p. 56, *fig.* 17.
Procrustes coriaceus ♂	L 7.	*ibid.*, *fig.* 10.
ditto ♀	5.	*ibid.*, 1881, xxv. p. 112.
Harpalus calceatus ♀	R 9.	*ibid.*, 1877, xxi. p. 57, *fig.* 24.
Calosoma sycophanta	L 9.	Gredler, *l. c.*, 1858, xii. p. 195.
C. triste	R 6.	Jayne, *Trans. Amer. Ent. Soc.*, 1880, viii. p. 155, Pl. iv. *fig.* 1.
* ‡ **Anchomenus albipes**	L 10.	Mocquerys, *Col. anorm.*, 1880, p. 17, *fig.*
* **A. angusticollis**	R 8.	*ibid.*, p. 10, *fig.*
0 **Nebria** sp.	?	Gredler, *l. c.*, 1869, xxiii. p. 35.
Agonum viduum	R 6.	von Heyden, *Deut. ent. Zt.*, 1881, xxv. p. 109, *fig.* 19.
‡ **Ditomus tricuspidatus**	R 8.	*ibid.*, *fig.* 18.
Colymbetes coriaceus (Dytisc.)	R 5.	Lucas, *Ann. Soc. ent. Fr.*, 1843, S. 2, i. p. 55, Pl.
Thylacites pilosus (Rhyn.)	L.	Kraatz, *l. c.*, 1876, xx. p. 378, *fig.*
* **Rhynchites germanicus** (Rhyn.)	{R 10. L 9.	Lent by Dr Mason.
Cryptophagus scanicus? (Clav.)	R 9.	Kraatz, *l. c.*, 1877, xxi. p. 57, *fig.* 25.
C. dentatus	L 3.	Sartorius, *Wien. ent. Monats.*, 1861, v. p. 31.
0 **Monotoma quadricollis** (Clav.)	R.	Rey, C., *Ann. Soc. Linn. de Lyon*, 1882, xxx. p. 424.
Chrysomela cacaliæ ♂ (Phyt.)	L 7.	Letzner, *Jahresb. schles. Ges. vaterl. Kultur*, 1855, p. 106.
Adimonia tanaceti (Phyt.)	L 5.	Schneider, *ibid.*, 1860, p. 129.

CHAP. XXII.] SUPPOSED CASES OF DOUBLE ANTENNA. 551

HETEROMERA

‡ Sepidium tuberculatum	L 5.	PERTY, *l. c., fig.* 10.
Zonabris quadripunctata	L 6.	KRAATZ, *l. c.*, 1889, XXXIII. p. 221, *fig.* 14.
Eleodes pilosa	R 9.	JAYNE, *l. c.*, p. 161, *fig.* 13.
* Blaps chevrolati	L 7.	MOCQUERYS, *l. c.*, p. 11, *fig.*
B. cylindrica	L 3.	*ibid.*, p. 6, *fig.*
B. similis	R 8.	VON HEYDEN, *l. c.*, p. 109, *fig.* 22.
Akis punctata	L 3.	BAUDI, *Bull. Soc. ent. ital.*, 1877, IX. p. 221, *fig.*

LONGICORNIA

0 Prionus[1] sp.		10. *Ann. and Mag. N. H.*, 1841, S. 1, p. 483.
Aromia moschata		6. KRAATZ, *l. c.*, 1889, XXXIII. p. 221, *fig.* 15.
ditto	R 2.	MOCQUERYS, *l. c.*, p. 18, *fig.*
* ditto	L 5.	Lent by Mr JANSON.
‡ Cerambyx cerdo ♀	L 6.	VON HEYDEN, *l. c.*, p. 109, *fig.* 23.
‡ C. scopolii ♂	R 3.	KRAATZ, *l. c.*, 1877, XXI. p. 56, *fig.*
‡ Lamia textor	L 1.	SMITH, F., *Zool.*, VI. p. 2245.
* ‡ Strangalia atra	L 1.	MOCQUERYS, *l. c.*, p. 14, *fig.*
S. calcarata	?	GREDLER, *l. c.*, 1858, XII. p. 195.
* ‡ Solenophorus strepens[2]	R 2.	MOCQUERYS, *l. c.*, p. 23, *fig.*
Clytus arcuatus	R 5.	VON HEYDEN, *fig.* 21.
Hammaticherus heros	L 7.	KLINGELHOFER, *Stet. ent. Zt.*, 1844, V. p. 330.
Callidium variabile	L 3.	MOCQUERYS, *l. c.*, p. 24, *fig.*
Lycus sp. (Mal.)	L 1.	VON HEYDEN, *l. c.*, p. 109, *fig.* 17.
* Telephorus lividus (Mal.)	L 2.	Lent by Mr F. H. WATERHOUSE.
T. rotundicollis	R 2.	JAYNE, *l. c.*, p. 159, *fig.* 11.
0 Elater hirtus (Elat.)		9. BASSI, *Ann. Soc. ent. Fr.*, 1834, S. 1, III. p. 375.
Ampedus ephippium (Elat.)	R 6.	KAWALL, *Stet. ent. Zt.*, 1858, XIX. p. 65.
Chiasognathus grantii (Lucan.)	L 6.	WESTWOOD, *Proc. Linn. Soc.*, 1847, I. p. 346.
Macrognathus nepalensis (Lucan.)	R 3.	KRAATZ, *l. c.*, 1880, XXV. p. 342, *fig.* 10.
Julodis clouei (Bupr.)	R 5.	BUQUET, *Ann. Soc. ent. Fr.*, 1843, S. 2, I. p. 97, Pl. IV.

Extra antenna arising from the head.

*859.
Callidium violaceum ♀ (Longic.)	R.	VON RÖDER, *Ent. Nachr.*, 1888, XIV. p. 219.
Saperda carcharias (Longic.)	L.	RITZEMA BOS, *Tijds. v. Ent.*, 1879, XXII. p. 208, Pl.
* Cerambyx cerdo (Longic.)	L.	KRAATZ. *Deut. ent. Zt.*, 1889, XXXIII. p. 222, *fig.* 23.

3. *Palpi.*

Subject to the reservations made in regard to instances of duplicity in antennæ, &c., the following examples of supposed duplicity in palpi are given.

*860.   **Nebria gyllenhalli** ♂ (Carab.): maxillary palps abnormal.

---
[1] I suspect that this is *Navosoma* No. 801.
[2] Doubtless the specimen mentioned by LUCAS, *Bull. Soc. ent. France*, 1848. S. 2, VI. p. XIX.

Fig. 206, I, shews the normal form of a right maxillary palp. Fig. 206, II, represents the *right* palp of this specimen. The 1st and 2nd joints are much thickened and the latter has 8 hairs (instead of 4) and two apical articulations, the anterior bearing

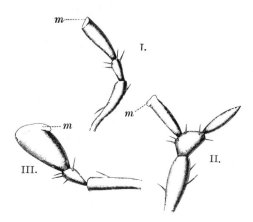

FIG. 206. *Nebria gyllenhalli*, No. 860. I. Normal right maxillary palp. II. Right palp of this specimen. III. Left palp of the same. *m*, terminal membrane. (The property of Dr Kraatz.)

an apparently normal terminal joint, the posterior bearing a symmetrical piece ending in a sharp point with no membrane like that at the apex of the normal. The *left* palp of this specimen is shewn in Fig. 206, III. In it the 2nd joint has 8 hairs instead of 4, and the terminal joint though very much enlarged is not divided at all. For the loan of this specimen I am indebted to Dr G. KRAATZ who first described it in *Berl. ent. Zt.*, 1873, XVII. p. 433, *fig.* 12.

861. **Carabus splendens**: penult. jt. of l. labial palp enlarged, and bearing two nearly similar jts. [broken before seen by me]. MOCQUERYS, *l. c.*, p. 29, *fig.*

862. **C. auratus**: 1st. jt. of l. maxillary palp bears two similar branches at rt. angles to each other, each with two jts. [Specimen not seen.] MOCQUERYS, *l. c.*, p. 30, *fig.*

863. **C. purpurascens**: extra labial palp on l. side. [Specimen not seen.] MOCQUERYS, *l. c.*, p. 32, *fig.*

4. *Mandibles.*

864. **Lucanus.** Three cases are recorded in which one of the mandibles bore an extra process of considerable size. Whether any of these are examples of duplicity, or whether the jaw, morphologically single, has in them varied towards a state of greater complexity, cannot well be said. The cases are **L. cervus** ♂, MOCQUERYS, *l. c.*, p. 106 [figure fairly true]; **L. cervus** ♂, KRAATZ, *Deut. ent. Zt.*, 1881, XXV. p. 111, *fig.*; **L. capreolus** ♂, *id.*, *l. c.*, 1876, XX. p. 378, *fig.*

CHAPTER XXIII.

Secondary Symmetry in Vertebrates.

Remarks on the Significance of Repetitions in Secondary Symmetry: Units of Repetition.

The evidence as to repetition of appendages in vertebrates is of great extent and has been studied by many, but in the morphology of these repetitions there is still much that is obscure. Speaking generally, the phenomena are similar to those seen in Arthropods, but there is no approach to the same regularity. Nevertheless when two extra limbs are present, it is usually possible to recognize that they are together a complementary pair; and if the extra part is apparently a single limb it is, I believe, never a normal limb and may very often be shewn to contain parts of a pair of limbs. The fact that the geometrical relations of the parts are less regular than they are in Arthropods may probably be ascribed in some measure to the circumstance that the surfaces of the vertebrate limbs do not maintain their original relations but are more or less rotated in the course of their development.

In Insects it appeared that repetition of the peripheral parts in Secondary Symmetry was not much more common than repetitions of whole limbs, but apparently this is not the case in vertebrates. Perhaps it would be more true to say that in vertebrates it is only in those extensive repetitions which include the greater part of the limbs beginning from the girdles, that the parts are clearly in Secondary Symmetry. From this circumstance doubt suggests itself whether some of the phenomena of polydactylism, at present regarded as repetitions of digits in Series, may not really be of the nature of Repetitions in Secondary Symmetry (see p. 378). But however this may be, there are, with the exception of some Artiodactyle cases, no examples of paired repetitions of digits or phalanges at all suggesting a comparison with the double extra tarsi &c. of Insects, or the double extra dactylopodites of Crustacea.

In the most usual forms of extra limbs in vertebrates a more or less amorphous pair of limbs, compounded together for a great part of their length, are attached to a supernumerary piece fitted into some part of the shoulder-girdle, or more often into the pelvic girdle.

It is important to notice that though, as many (especially ERCOLANI) have shewn, a complete series can be constructed, ranging for instance from the ordinary pygomelian up to complete posterior duplicity, yet repetition of limbs may be and often is wholly independent of any axial duplicity, being truly a repetition of appendicular parts only.

The question naturally arises whether there is ever an extra limb placed as a single copy of a normal limb of the same side as that on which it is attached. As to this the evidence is not wholly clear, but I incline to think that no case known to me can properly be so expressed. Perhaps the condition which comes nearest to this is exemplified by a case of a Frog fully described by KINGSLEY[1], where a single extra left hind leg is said to have been attached to the left side of the pelvis. It is difficult to question that this was actually the fact, for the figure clearly represents the extra limb as a left leg; but though the muscles are fully described, the bones are not, and it still seems possible that there was in reality some duplicity in the limb. The leg was admittedly abnormal in its anatomy and the naming of the muscles must in part have been approximate.

But though perhaps it should not be positively stated that no *single* extra limb is ever formed in a vertebrate in Succession to the normal limb of the same side of the body, it is certainly true that in the enormous majority of polymelians the extra repetition consists of parts of a complementary pair. These phenomena are thus of interest as bearing upon the morphology of repetitions in Secondary Symmetry, but in all probability are not of the nature of variations in the constitution of the Primary Symmetry.

A just view of the details of these phenomena can only be gained from the specimens or from numerous drawings. The cases of extra limbs in Batrachia may be conveniently studied as exhibiting most of the different kinds of Secondary Symmetries both in the fore and hind limbs. In all, some fifty cases are recorded. These may be found from the following references. The evidence up to 1865 was put together by DUMÉRIL, and an abstract of it is given also by LUNEL, and by KINGSLEY. A fuller bibliography is given by ERCOLANI. The best papers on the subject are marked with an asterisk. I have added a few references of less importance not included in the other bibliographies.

* DUMÉRIL, *Nouv. Arch. Mus. Paris*, 1865, I. p. 309, Pl. xx.
* LUNEL, *Mém. soc. phys. d'hist. nat. de Genève*, 1868, xix. p. 305, Pl.
[1] *Proc. Bost. N.H.S.*, 1881—2, xxi. p. 169, Pl. II.

* KINGSLEY, *Proc. Boston N.H.S.*, 1881—2, XXI. p. 169, Pl. II.
* CAVANNA, G., *Pubbl. del R. Ist. di Studi super. in Firenze*, 1879, p. 8, Tav. I. Four important cases; one, *fig.* 2, apparently resembling Kingsley's in some respects.
* MAZZA, *Atti Soc. ital. sci. nat.*, 1888, XXXI. p. 145, Pl. I.
TUCKERMAN, *Jour. Anat. Phys.*, 1886, p. 517, Pl. XVI.
Cat. Terat. Ser. Coll. Surg. Mus., 1872, No. 23.
HÉRON-ROYER, *Bull. soc. Zool. France*, 1884, IX. p. 165.
BERGENDAL, *Bihang k. svensk. vet. Ak.*, 1889, XIV. Afd. IV. Pl. I.
* ERCOLANI, *Mem. Acc. Bologna*, 1881, IV. p. 810, Pl. IV. Four important cases and very good bibliography.
SUTTON, *Trans. Path. Soc.*, 1889, XL. p. 161, *fig.*
[Three cases in Newts: *Triton cristatus*, JÄCKEL, *Zool. Gart.*, 1881, XXII. p. 156. *Triton tæniatus*, LANDOIS, H., *ibid.*, 1884, XXV. p. 94; CAMERANO, *Atti Soc. ital. sci. nat.*, 1882, XXV].

From these Batrachian cases most of the chief features of the phenomena may be learnt. To those wishing to get a general view of the subject of repetition of Vertebrate limbs in a comparatively small compass the valuable memoir of ERCOLANI quoted above is especially recommended.

Before proceeding to a consideration of the significance of the phenomenon of Repetition in Secondary Symmetry it must be expressly stated that there are in vertebrates a certain number of cases, perhaps even classes of cases, which it is likely differ widely from the rest; but as was said above, the chief difference between the Vertebrate and Arthropod cases lies in the comparative simplicity of the latter. It may be stated further that this greater simplicity of the Arthropod cases consists especially in the maintenance of the relation between the extra pair and some normal limb.

Remembering always the existence of unconformable cases we may, I think, safely gather up from the simple cases several points relating to the problems of Natural History at large. I only propose here to make allusion to those considerations which are not developed in the ordinary teratological treatises.

Of the fact that any regularity can be discerned in these strange departures from normal structure, and of the bearings of this fact on current conceptions of the causes determining the forms of animals it is now hardly necessary to speak further. Other points not before noticed remain.

In the Arthropod cases that were spoken of as 'regular' it was seen that the polarity of the Secondary Symmetries has a definite relation to that of the body which bears them. This is quite in harmony with the supposition that they are related to the normal body somewhat as buds are related to a colony, for in most colonial forms the morphological axes and planes of the buds are definitely related to those of the stock.

But in the Vertebrate cases though there is generally a relation of images between the extra pair, a definite geometrical relation between them and a normal limb is seen more rarely.

That this is so may, I think, be in part at least attributed to the normal twisting of the vertebrate limb, especially of the hind limb, from its original position (see Note on p. 459).

A question brought into prominence by facts of this kind is that of the nature of the control which determines *how much* of a body shall be repeated, or be capable of repetition, in a Secondary Symmetry.

What is a *unit* of repetition?

With repetition of a whole body we are familiar. Apart from the processes of sexual reproduction, we know this total repetition in the many forms of asexual reproduction, whether occurring by budding, or by division either of adult bodies or of embryos[1], and we thus commonly look on the whole body of any organism as in a sense a unit, capable of repetition or of differentiation—the latter especially in gregarious and colonial forms. Again, we familiarly use the conception of cells as units of repetition or of differentiation. Besides these we have come to recognize that members of series of segments are, in their degree, similar units. And generally, the same attribute of separateness may in undefined senses be properly attached to all organs that are repeated in Series, and to appendicular parts especially.

The attribution of some of the undefined properties of "unity[2]" to some at least of these various groups is very ancient, and there can be no doubt that it is in the main a right and useful induction.

The chief interest of repetitions in Secondary Symmetry lies in the fact that they give a glimpse of new light upon the nature of this unity, shewing a new form in which it may appear.

For in Secondary Symmetry there is not a simple repetition of a part in Series, taking its place as a member of that series, but an addition of paired parts, whose intrinsic relation to each other is the same as that of any pair of parts occurring in the Primary Symmetry.

The addition is thus a *unit*, is in form complete in itself, and seems to have no place in the Primary Symmetry of the whole body any more than a late side-chapel—also a unit with its own focus and polarity—had a place in the design of the original architect of the Cathedral.

From analogy, and from general knowledge of vital processes it would I think have been impossible to foresee the very curious indefiniteness of the *quantity* of the parts repeated in systems of Secondary Symmetry. It seems, especially in Arthropod cases,

---

[1] As a normal occurrence notably in the case of Cyclostomatous Polyzoa of the genus *Crisia* described by HARMER, S. F., *Q. J. M. S.*, 1891, p. 127, *Plates*.

[2] This somewhat incorrect term is used here to express some of the meanings commonly still more incorrectly rendered by the word "individuality"—a word etymologically most unhappy in this application to things endowed with divisibility as a conspicuous attribute.

that the repetition may begin from any point in an appendage and include all the parts peripheral to the point of origin. Seeing that the repeated parts are, in their degree, comparable with a whole organism, this indefiniteness is remarkable. We have thus to recognize that the property of morphological "unity" may attach not only to a pair of appendages beginning from the body, or from some definite surface of articular segmentation, but also to a pair of parts having no semblance of morphological distinctness.

Strangest of all is the repetition of the index of Crabs and Lobsters in Secondary Symmetry. The dactylopodite is of course a separate joint. Double extra dactylopodites in Secondary Symmetry present no feature different from double extra tarsi, &c. But the index we think of as merely a large spine or tubercle. It is in no sense a joint or segment. Yet a pair of indices may be added to a normal body. The interest of this fact is in its value as a comment on the principle given on p. 476 that extra parts in Secondary Symmetry contain the structures *peripheral* to their point of origin. The case of extra indices shews that the term peripheral, if it is to include the case of indices, must be interpreted as meaning not morphologically but *geometrically peripheral*[1].

We have spoken of parts in Secondary Symmetry as having no place in the Primary Symmetry of the body. This is on the whole a true statement, but there are a few cases which make it uncertain whether it is absolutely true. These cases are those few where repetitions in Secondary Symmetry were present on appendages of both sides of the body.

Cases of this class were *Odontolabis stevensii*, No. 799, and *Melolontha hippocastani*, No. 795, where such extra parts were present on both antennæ, suggesting that the similarity of the repetition of the two sides is due to the relation of Symmetry between the right side and the left. But against this view may be mentioned the cases *Prionus coriarius*, No. 750, and *Carabus irregularis*, No. 760, where two legs of the *same* side each bore extra parts, and the Lobster, No. 821, having *two* pairs of extra points on one dactylopodite. These cases suggest that bilateral simultaneity in such repetition may perhaps represent merely a general capacity for this form of repetition. The case of *Prionus californicus*, No. 843, would no doubt bear on this question, but unfortunately the facts in that case are scarcely well enough known to justify comment.

---

[1] A case is given by Faxon (*Harv. Bull.*, VIII. Pl. II. fig. 8) of **Callinectes hastatus** in which the left lateral horn of the carapace, instead of being simple as in normal specimens, had *three* spines. It is just possible that two of these may have been in Secondary Symmetry. All other cases known to me are in appendicular parts.

One further point remains to be spoken of. We have said that a system of parts in Secondary Symmetry is in a sense analogous with a bud, but in one respect the condition of these parts differs remarkably from all phenomena of budding or reproduction that are seen elsewhere. In a bud the various organs always present the same surfaces to each other, or in other words, the planes of division always pass between similar surfaces. In Secondary Symmetries this is not the case. As illustrated by the diagram on p. 481, the extra parts may present to each other, or remain compounded by *any* of their surfaces, whether anterior, posterior, or otherwise. This seems to be altogether unlike anything ever met with in animals and plants. It is as if in a bud on a plant two leaves on opposite sides of the axis could in their origin indifferently present any of their surfaces to each other.

It will be remembered that the symmetry cannot be the result of subsequent shiftings, but must represent the original manner of cleavage of the two extra limbs from each other. We must therefore conceive that in the developing rudiment of the two extra limbs either surface may indifferently be external, the polarity being ultimately determined by the relation of the bud or rudiment to the limb which bears it.

CHAPTER XXIV.

Double Monsters.

Of the evidence as to double and triple "monstrosity" and of the classification of the various forms no account can be given here. This may be found in any work on general teratology. In this chapter are put together a few notes on points respecting these formations of interest to the naturalist, and having relation to what has gone before.

It is now a matter of common knowledge that in animals [and plants] division may occur in such a way that two or more bodies may be formed from what is ostensibly one fertilized ovum (cp. multipolar cells). But by a similar division, imperfectly effected, the resulting bodies instead of being complete twins or triplets may remain united together, frequently having a greater or less extent of body in common. In other words, speaking of simple cases in bilateral animals, the whole body, resulting from the development, may contain more than one bilaterally complete group of those parts which normally constitute the Primary Symmetry of an "individual."

If well developed, the component groups are most often united by homologous parts, so that there is a geometrical relation of images between the groups together, forming the compound structure, the whole being one system of Symmetry. Concerning the relations of the several parts of such a system to each other numerous questions of interest arise, but with these it is not now proposed to deal.

To those unacquainted with facts of this class it may be of use to point out in the fewest words the direction in which this importance lies. It arises, briefly, from the fact that in the resemblance between a pair of homologous twins, whether wholly or partially divided, there is once again an illustration of the phenomenon of Symmetry, and of the simultaneous Variation of structures related to each other as symmetrical counterparts.

The frequency of close resemblance between twins is a matter of common knowledge. If it be true that such twins may result from the development of one ovum—a fact that cannot be doubted in face of the complete series of stages intermediate between total and partial duplicity—the resemblance between these twins is then of the same nature as that subsisting between the two halves of any other bilaterally symmetrical system. A wide field of inquiry is thus opened up. For, as suggested in the Introduction (p. 36) if the very close resemblance of twins to each other is a phenomenon dependent on Symmetry of Division, the less close resemblance between members of families may be a phenomenon similar in kind.

It will be remembered that the resemblance between twins is a true case of similar and simultaneous Variation of counterparts. This is clearly proved by the fact that when distinct Meristic Variations are exhibited by one twin they are not rarely present in the other also. Cases of this simultaneous Variation are familiar to all who have studied this subject. A useful list of examples in completely separate twins is given by WINDLE[1]. One of the best known cases in twins incompletely separated, is that of the Siamese Twins[2], who had each only eleven pairs of ribs (instead of twelve).

Reference must lastly be made to a particular corollary which may naturally be deduced from the fact that the bodies of incompletely separated twins are grouped as a single system of Symmetry. If the whole common body were bilaterally symmetrical, one twin must be the optical image of the other. But if the organs of one twin are normally disposed, the organs of the other must be *transposed* in completion of the Symmetry. This theoretical expectation is in part borne out by the facts. With a view to this question EICHWALD[3] examined the evidence as to thoracopagous double monsters (including xiphopagi, &c.), and found that in almost every case one of the bodies shewed some transposition of viscera, though to a varying extent[4].

There are nevertheless a few cases even of thoracopagi where neither body exhibits any transposition[5]. Moreover, contrary to natural expectation, it does not appear that in ordinary cases of completely separate twins either twin has its viscera transposed; and conversely, of 152 cases of transposition collected by Küchenmeister only one could be shewn to have been a twin[6]. It seems therefore that the frequency of transposition in double monstrosity depends in some way upon the *maintenance* of the connexion between the twins; and that if the separation be completed early, as it must be supposed to be in cases of homologous twins born separate, then both bodies as a rule develop upon the normal plan, like the bodies of multiple births of other animals. But as the evidence now stands there is no reason to suppose that individuals with transposition of viscera, born as single births, have ever had a counterpart any more than individuals whose viscera are normally placed, tempting as it is to imagine that both may have had some counterpart which in the ordinary course does not develop.

For the present we need not go beyond the fact that between complete duplicity resulting in "homologous twins," and the least forms of axial duplicity, consisting in a doubling of either extremity of the longitudinal axis almost all possible degrees have been seen[7]. By persons unfamiliar with abnormalities it

---

[1] WINDLE, B. C., *Jour. Anat. Phys.*, XXVI. p. 295.

[2] For full abstracts of all evidence relating to this case, see KÜCHENMEISTER, *Die angeb. Verlagerung d. Eingeweide d. Menschen*, Leipzig, 1883, p. 204.

[3] EICHWALD, *Pet. med. Ztsch.*, 1870, No. 2, quoted from abstr. *Virch. u. Hirsch*, *Jahresb.*, 1871, p. 167.

[4] Eichwald supports the view that in these cases it is the *right* twin which shews the transposition. As KÜCHENMEISTER (*l. c.*) points out, this cannot by the nature of the case be a universal rule; for the relative position of xiphopagous twins may result simply from the way in which they happen to be laid by the mother or the midwife. Of the Siamese Twins, besides, it was Chang, the left twin, in whose body there were indications of transposition. The twins may also remain face to face. The expression "right twin" must always need further definition, and it should be qualified as the right when the livers are adjacent, or when the hearts are adjacent, as the case may be. Whether the rule is wholly or partially true for either of these positions seems to be very doubtful.

[5] For example, BÖTTCHER, *Dorpater med. Ztschr.*, II. p. 105, quoted from *V. u. H.*, *Jahresb.*, *l. c.* In the specimen *Terat. Cat. Coll. Surg. Mus.*, 1872, No. 114, there is no transposition, but here the hearts were not separate.

[6] *l. c.*, p. 268. One, however, was a child of a mother who had before borne twins, *l. c.*, p. 313.

[7] The fact that some of the degrees are much more common than others has an obvious bearing on the question of Discontinuity, which might with profit be pursued. A statistical examination as to the angles at which the bodies are most frequently inclined to each other would also probably lead to an interesting result.

is sometimes supposed that axial duplicity is a phenomenon more or less peculiar to Man and to domesticated animals [and plants], and the occurrence is looked on as a part of that Meristic instability which is ascribed to absence of the control of a strict and Natural Selection. This view is far from sound. Such phenomena have on the contrary been found in many classes of animals, vertebrate and invertebrate, and the unquestionable frequency in domesticated animals may in great measure be fairly attributed to the comparative ease with which the births of these creatures can be observed. As considerations of this kind have weight with many it has seemed worth while to give references to examples taken from a variety of different groups, shewing not only that such compound bodies may be produced in wild animals, but also that they may sometimes be able to carry on the business of life without artificial help.

In Mammals and Birds I do not know an authentic case of a double monster that had *grown up* in the wild state.

*865.   In Reptiles many such cases are known and are referred to by most of the older writers. Of Snakes having complete or partial duplicity, nearly always of the head, some twenty cases are recorded. Several of these were animals of good size, and must have had an independent existence for some considerable time.

Some of the cases have special points of interest, but into these it is not now proposed to enter. As bearing on the question of the frequency of Meristic Variation in families and strains attention is called to the circumstance that MITCHILL's three specimens were all found in one brood of 120 which were taken with the mother. The following is a list of records of snakes having the head wholly or partially double.

Coluber constrictor. WYMAN, J., *Proc. Bost. N. H. S.*, 1862, IX. p. 193, *fig*.

Coluber constrictor. MITCHILL, S. L., *Amer. Jour. of Sci.*, x. 1826, p. 48, *Pl.* (3 specimens).

Ophibolus getulus. YARROW, *Amer. Nat.*, 1878, XII. p. 470.

Pityophis. *ibid.*, p. 264.

Pelamis bicolor. [Remarkable case[1]: the duplicity appearing only in the fact that there were 4 nasal plates instead of 2, each with a nostril] BOETTGER, O., *Ber. üb. d. Senck. nat. Ges. in Frankf. a. M.*, 1890, p. LXXIII.

In the remainder the species is not clear. REDI, *Osserv. int. agli anim. viventi*, &c., 1778, p. 2, Tav. I. [very good account]; LACÉPÈDE, *Hist. nat. des Serpens*, II. 1789, p. 482; BANCROFT, *Nat. Hist. of Guiana*, 1769, p. 214, *Pl.*; LANZONI, *Miscell. curios.*, 1690, Obs. CLXXI. p. 318, Fig. 36; *Boston Soc. Med. Imp., Catal. of Mus.*, No. 856, quoted from WYMAN, *l. c.*; EDWARDS, *Nat. Hist. of Birds*, &c., Pt. IV. 1751, p. 207, *Pl.*; DORNER, *Zool. Gart.*, 1873, XIV. p. 407; *Coll. Surg. Mus., Terat. Cat.*, 1872, Nos. 24—27.

[1] Compare with Mitchill's two last cases, and also with a case in **Alytes obstetricans.** HÉRON-ROYER, *Bull. Soc. Zool. France*, 1884, IX. p. 164.

Fig. 207. *Chrysemys picta*, 2 or 3 days old. I, II, normal. III and IV, two-headed specimen. In the latter the nuchal and two pygal plates are normal. Between them are 12 plates on each side, 11 being the most usual number. Among the costals an extra plate is wedged in on the rt. First vertebral divided by suture; fifth is made up of 4 irregular plates. In the plastron there is a doubling of the gular plate. The rt. femoral has a suture. (From Barbour.)

See also, GEOFFROY ST. HILAIRE, *Hist. des Anom.*, ed. 1838, II. p. 197; DUMÉRIL et BIBRON, *Erpét. générale*, 1884, VI. p. 209.

866. Duplicity of the head is less common in Lizards, but several examples are known. See GEOFFROY ST. HILAIRE, *l. c.*, p. 195; *Cosmos*, Paris, 1869, S. 3, v. p. 136, &c.

*867. In Chelonia also are several such instances. See EDWARDS, *Nat. Hist. of Birds, &c.*, Pt. IV. 1751, p. 206; MITCHILL, *l. c.*; BARBOUR, E. H., *Amer. Jour. of Sci.*, 1888, S. 3, XXXVI. p. 227, Pl. v. The last is a particularly interesting case from the circumstance that the behaviour during life was observed to some extent, though only a popular account is given. The two heads seemed to act independently, and it is said that there was no concerted action between the feet of the two sides. BARBOUR's figures are reproduced in Fig. 207.

In fish-hatching establishments double monstrosity is of frequent occurrence among young Salmon and Trout. A two-headed embryo of a Shark is preserved in Coll. Surg. Mus. (*Terat. Cat.* 1872, No. 22).

The following cases relate to invertebrates.

Chætopoda. Duplicity in this Class has been often seen, but that any of the cases are truly congenital cannot be stated. There is evidence that in many Annelids regeneration[1] both of head or tail may freely occur, and it is quite possible that the second head or second tail may have grown out from an injured place, though of this there is no actual proof. In cases of posterior bifurcation each tail generally contains all the parts proper to the normal, but in No. 871 one of the tails was without the terminal cirri usual in the species. So far as can be gathered from the evidence it does not appear that the two continuations of the body have always the *same* number of segments, which might perhaps be expected were both the result of a natural division of the developing body. On the other hand, they do seem generally to have a *nearly* equal development, and are almost always (in cases of double tails, at least) fairly equal in length, which would not be anticipated if one only were a new growth. Moreover, if the double tail is in some way due to regeneration one would expect to find such duplicity in its minor conditions much more commonly.

Into the details of the structure it is not now proposed to enter, and indeed of most of the cases there is little to be told. The evidence is mentioned here simply in further proof of the power of these individuals, thus greatly departing from the normal of their species, to maintain themselves with no apparent difficulty. It will be noticed that the species concerned are most various, and include not only Errantia, but two cases also in Serpulidæ.

The literature of the subject was collected by COLLIN[2], and a list of the references was independently collected and published with abstracts by ANDREWS[3]. This list, with a few additions, was republished by FRIEND[4]. Though many of the accounts are imperfect they are referred

[1] The evidence on this point does not come within the scope of this work. References to it may be obtained from ANDREWS, ZEPPELIN, &c. (*v. infra*).
[2] COLLIN, A., *Naturw. Wochens.*, 1891, No. 12, p. 113.
[3] ANDREWS, E. A., *Amer. Nat.*, 1892, XXVI. p. 729.
[4] FRIEND, H., *Nature*, 1893 (1), p. 397.

to below, in evidence that the total number of cases is considerable. There are only two certain cases of double head (see *Typosyllis*, No. 868, and *Allolobophora*, No. 873).

POLYCHÆTA.

*868.   **Typosyllis variegata**: individual having two small heads, as shewn in Fig. 208. Heads of unequal size, that on the left having 4 segments behind the eyes, while that on the right had two. The

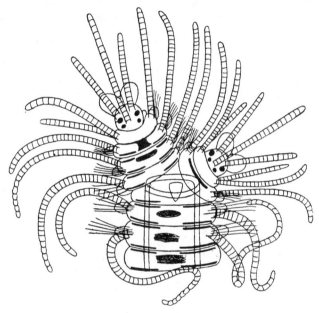

Fig. 208. *Typosyllis variegata*, No. 868, having two small heads. (After LANGERHANS.)

appearance suggested that the original head had been broken off and that two new ones had grown in its place. LANGERHANS, P., *Nova Acta Ac. C. L. C.*, XLII. p. 102, *Pl.*

869.   **Nereis pelagica**: bifid posteriorly. BELL, F. JEFFREY, *Proc. Zool. Soc.*, 1886, p. 3.
870.   **Salmacina incrustans** (Serpulidæ): posterior end double. [Two tails shewn in figure as of equal length and in the same straight line, at right angles to the body. The arrangement of the segmentation at the junction is not clearly shewn.] CLAPARÈDE, *Mém. soc. phys. et d'hist. nat. Genève*, xx. 1869—70, p. 177, Pl. xxx. fig. 5 F.
871.   **Proceræa tardigrada** (Syllidæ): tail double; two specimens. In one of these the tails were nearly equal, but one had no anal cirri. ANDREWS, E. A., *Proc. U. S. Nat. Mus.*, 1891; xiv. p. 283, and *Amer. Nat.*, 1892, xxvi. p. 729, Pl. xxi.
872.   **Branchiomma** sp. (Sabellidæ): two posterior ends, *one being rudimentary*. BRUNETTE, *Trav. Stat. Zool. de Cette*, 1888, p. 8 [quoted from ANDREWS, *l. c.*]

[With these conditions compare *Syllis ramosa*, a form found by the *Challenger* in two localities, inhabiting a Hexactinellid Sponge. The body of this creature consisted of vast numbers of branches, about as thick as thread, passing off at right angles, coiling upon each other and forming inextricable masses. In some specimens no head was found, but a single head was afterwards discovered. It seemed likely that large tracts of the body have no head, but there was no evidence to shew how many heads occur in the colony. Many female buds were found, and a single complete male. McINTOSH, *Chall. Rep.*, XII. p. 198, Pl. xxxi.]

OLIGOCHÆTA.

*873. **Allolobophora longa**: specimen represented as bearing a second head on the right side of the first segment behind the peristomium. The second head is represented with prostomium, peristomium and one more segment which rests on the peristomium of the normal body. FRIEND, H., *Science-Gossip*, 1892, July, p. 161, *fig*.

874. **Ctenodrilus monostylos:** double tail; in many hundreds examined, three cases seen, ZEPPELIN, *Z. f. w. Z.*, 1883, XXXIX. p. 621, Pl. 36, figs. 18 and 19.

875. **Lumbriculus variegatus:** similar cases. VON BÜLOW, *Arch. f. Naturg.*, 1883, XLIX. p. 94.

876. **Acanthodrilus** sp. : case of two tails arising from a much thicker anterior portion. Such worms were believed or alleged to be common in a particular district in New Zealand. KIRK, T. W., *Trans. N. Zeal. Inst.*, XIX. p. 64, *Pl*.

877. EARTHWORMS generally, belonging to genera **Lumbricus,** or **Allolobophora**: cases of double tail recorded, as follows: ROBERTSON, C., *Q. J. M. S.*, 1867, p. 157, *fig.*; HORST, *Notes Leyd. Mus.*, VII. p. 42; THOMPSON, W., *Zool.*, XI. p. 4001; BELL, F. JEFFREY (2 cases), *Ann. & Mag. N. H.*, 1885 (2), p. 475, fig.; FRIEND, H., *Sci.-Gossip*, 1892, p. 108, *figs.*; MARSH, C. D., *Amer. Nat.*, XXIV. 1890, p. 373; FITCH, A., *Eighth Rep. upon Insects of State of N. Y.*, Append., 1865, p. 204 [from ANDREWS, *l.c.*]; *Terat. Cat. Mus. Coll. Surg.*, 1872, No. 20. BREESE, *West Kent N. H. S.*, 1871; BROOME, *Trans. N. H. S. Glasgow*, 1888, p. 203; FOSTER, *Hull. Sci. Club*, 1891; [the last three quoted from FRIEND, *Nature*, 1893 (1), p. 397]; COLLIN, A., *Naturw. Wochens.*, 1891, No. 12, *figs*. I have also a specimen with two nearly symmetrical tails kindly sent by Mr W. B. BENHAM.

ARTHROPODA.

Three cases.

*878. **Chironomus** (Gnat): larva with two heads, duplicity beginning from the 5th segment behind the head [important details given, *q. v.*]. WEYENBERGH, H., *Stet. ent. Ztg.*, 1873, XXXIV. p. 452, *fig*.

879. **Euscorpius germanicus** (Scorpion): tail double from 4th præ-abdominal segment [figure represents each abdomen with *one segment too few*, presumably an error]. PAVESI, P., *Rend. R. Ist. Lomb.*, S. II., XIV. 1881, p. 329, *fig*.

880. [**Scorpio africanus:**] specimen with two tails. SEBA, *Rerum Naturalium Thesaurus*, 1734, I. p. 112, Pl. LXX. *fig*. 3. This example was kindly sent me by Mr R. I. Pocock, who tells me that the figure shews the animal to be of the species named.

CESTODA.

Conditions, perhaps akin to duplicity, have been seen to occur under three forms.

881. **Tænia cœnurus:** specimen whose head had 6 suckers instead of 4, and 32 hooks instead of 28. Proglottides were 3-sided prisms, in section triangular. Longitudinal vessels 6 instead of 4, two being in each angle. Absolute size of head greater than normal. This abnormal

form is known to occur in many kinds of Tapeworms, and especially in Cysticerci. LEUCKART, *Parasiten d. Menschen*, pp. 501—2, cp. p. 577. [Case with *five* suckers mentioned, *ibid.*, p. 578.]

In another form of abnormality the chain of segments has three longitudinal flanges, formed, as it were, by the union of two chains of proglottides having one edge in common. Head not found, but several cases known. Genital openings in one case all upon the common edge. LEUCKART, *ibid.*, p. 574. Cp. COBBOLD, *Trans. Path. Soc.*, XVII. p. 438; LEVACHER, *Comptes rendus*, 1841, XIII. p. 661.

882. *Bifurcated* chains of proglottides have also been seen, *e.g.* specimen of **Tænia (cysticerci) tenuicollis**, which bifurcated several times in terminal portion, though normal in front of this. MONIEZ, *Bull. Sci. du Nord*, X. p. 201. See also **Tænia saginata**? LEUCKART, *l. c.*, p. 573.

BRACHIOPODA.

883. **Acanthothyris spinosa** (Rhynchonellidæ): case of duplicity

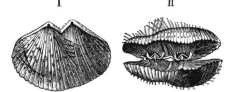

FIG. 209. *Acanthothyris spinosa*, No. 883. Case of duplicity. (From P. FISCHER.) I. Seen from ventral valve. II. Looking between the valves.

as shewn in Fig. 209. FISCHER, P., *Jour. de Conchyl.* S. 3, XIX. p. 343, Pl. XIII. *figs* 4—7.

HOLOTHURIOIDEA.

884. **Cucumaria acicula**: specimen made up of two individuals cohering laterally at posterior ends. SCHMELTZ, *Verh. d. Ver. f. naturw. Unterhaltung*, Hamb., 1877, IV. p. XV.

885. **Cucumaria planci**: case of second mouth and ring of tentacles borne on a lateral bud-like projection. LUDWIG, H., *Z. f. w. Z.*, LIII. *Supp.* p. 21, Pl. V.

---

886. CŒLENTERATA. Forms which are commonly simple, such as *Actinia* or *Sagartia*, are rarely found with two discs seemingly due to incomplete division, which in these forms may take place longitudinally [?] as well as by ordinary budding. GOSSE, P. H., *Sea-Anemones*, p. XXI., &c. See also GUYON, *Zoologist*, p. 7026, *fig.*

Similar occurrences, not distinguishable from budding, have been seen in Medusæ, *e.g.*, *Phialidium variabile*, DAVIDOFF, *Zool. Anz.*, IV. p. 620, *fig.*; *Gastroblasta raffaeli*, LANG, A., *Jen. Ztschr.*, XIX. p. 735. An interesting case of this kind
* was seen in **Cordylophora lacustris**. Several polystomatous specimens were found *on a particular mass* of *Cordylophora*, but were not found on all colonies gathered with this mass and had not been seen previously in specimens from the same locality. [Further particulars.] PRICE, H., *Q. J. M. S.*, 1876, p. 23, *figs*.

PROTOZOA. Double and triple monstrosity has been seen in several Foraminifera, see *e.g.*, DAWSON, *Canad. Nat.*, 1870, p. 177, *figs.*; BALKWILL and WRIGHT, *Trans. R. Irish Ac.*, 1885, XXVIII. p. 317, Pl. XIV., &c. [As to cases in *Stentor*, see BALBIANI, *J. de l'anat.*, 1891, No. 3, but these are doubtless examples of regeneration and duplicity following injury.]

CHAPTER XXV.

Concluding Reflexions.

To attempt at this stage any summary of conclusions would be misleading. The first object of this work is not to set forth in the present a doctrine, or to advertise a solution of the problem of Species, but rather to bring together materials that may help others hereafter to proceed with the solution of that problem. A general enumeration of particular conclusions is therefore to be avoided. Indeed, from the scantiness of the evidence, its present value is chiefly in suggestion, and the facts must therefore be themselves still studied in detail. The reader must interpret as he will.

But, as often happens, that which may not shew the right road is enough to shew that the way taken has been wrong, and so is it with this evidence. Upon the accepted view it is held that the Discontinuity of Species has been brought about by a Natural Selection of particular terms in a continuous series of variations. Of the difficulties besetting this doctrine enough was said in the introductory pages. These difficulties have oppressed all who have thought upon these matters for themselves, and they have caused some anxiety even to the faithful. And if in face of the difficulties reasonable men have still held on, it has not been that the obstacles were unseen, but rather that they have hoped a way through them would be found.

Now the evidence, of which a sample has been here presented, gives hope that though there be no way through the difficulties, there is still perhaps a way round them. For since all the difficulties grew out of the assumption that the course of Variation is continuous, with evidence that Variation may be discontinuous, for the present at least the course is clear again.

Such evidence as to certain selected forms of variations has, I submit, been given in these chapters, and so far a presumption is created that the Discontinuity of which Species is an expression has its origin not in the environment, nor in any phenomenon of Adaptation, but in the intrinsic nature of organisms themselves, manifested in the original Discontinuity of Variation.

But this evidence serves a double purpose. Though some may

doubt whether the variations here detailed are such as go to the building of Specific Differences (a doubt which, it must be granted, does fairly attach to some part of the evidence), yet the existence of sudden and discontinuous Variation, the existence, that is to say, of new forms having from their first beginning more or less of the kind of *perfection* that we associate with normality, is a fact that disposes, once and for all, of the attempt to interpret all perfection and definiteness of form as the work of Selection. The study of Variation leads us into the presence of whole classes of phenomena that are plainly incapable of such interpretation.

The existence of Discontinuity in Variation is therefore a final proof that the accepted hypothesis is inadequate. If the evidence went no further than this the result would be of use, though its use would be rather to destroy than to build up. But besides this negative result there is a positive result too, and the same Discontinuity which in the old structure had no place, may be made the framework round which a new structure may be built.

For if distinct and "perfect" varieties may come into existence discontinuously, may not the Discontinuity of Species have had a similar origin? If we accept the postulate of Common Descent this expectation is hard to resist. In accepting that postulate it was admitted that the definiteness and Discontinuity of Species depends upon the greater permanence or stability of certain terms in the series of Descent. The evidence of Variation suggests that this greater stability depends primarily not on a relation between organism and environment, not, that is to say, on Adaptation, but on the Discontinuity of Variation. It suggests in brief *that the Discontinuity of Species results from the Discontinuity of Variation.*

This suggestion is in a word the one clear and positive indication borne on the face of the facts. Though as yet it is but an indication, there is scarcely a problem in the comparison of structures where it may not be applied with profit.

The magnitude and Discontinuity of Variation depends on many elements. So far as Meristic Variation is concerned, this Discontinuity is primarily associated with and results from the fact that the bodies of living things are mostly made up of repeated parts—of organs or groups of organs, that is to say, which exhibit the property of "unity," or, as it is generally called, "individuality." Upon this phenomenon depends the fact that Meristic Variation in number of parts is often integral, and thus discontinuous.

The second factor that most contributes to the Discontinuity of Variation is Symmetry, manifesting its control in the first place directly, leading often to a result that we recognize as definite and perfect because it is symmetrical.

But besides this direct control that we associate with Symmetry, other effects greatly contributing to the magnitude of Variation can be traced to a factor not clearly to be distinguished from Symmetry itself. For, as has been explained, Symmetry, whether Bilateral or Radial, is only a particular case of that phenomenon of Repetition of Parts so universally characteristic of living bodies; and that resemblance between two counterparts, which we call Bilateral Symmetry, is akin to the resemblance between parts repeated in Series, though, as is shewn by their geometrical relations, the processes of division by which the parts were originally set off, must be in some respects distinct. Bilateral Symmetry of Variation is thus only a special case of the similar and simultaneous Variation of repeated parts.

The greatness of the observed change from the normal is often largely due to this possibility of simultaneity in Variation, the change thus manifesting itself not in one part only, but in many or all of the members of a series of repeated parts. Instances of such similar and simultaneous Variation of serial parts in animals have now been given. Examples still more marked may be seen abundantly among plants. A variation, for example, in the form or degree of fission of the leaf, slight perhaps by itself, when taken up and repeated in every leaf in its degree, constitutes a definite and conspicuous distinction. Everyone has observed this common fact. Few illustrations of it are more evident than that of the common Hawthorn. In a quickset hedge soon after the leaves begin to unfold almost each separate plant can be recognized even at a distance, and its branches can be traced by their special characters, by the shapes and tints of the leaves, by the angles that they make with the stem, by the manner of unfolding of the buds, and so forth. These variations, sometimes slight in themselves, by their similarity and simultaneity build up a conspicuous result.

The phenomenon of serial resemblance is in fact an expression of the capacity of repeated parts to vary similarly and simultaneously. In proportion as in their variations such parts retain this capacity the relationship is preserved, and in proportion as it is lost, and the parts begin to vary independently, exhibiting differentiation, the relationship is set aside. It will be noticed that to render the converse true we must extend the conception of Serial Homology in special cases to organs not commonly regarded as serially homologous with each other, but which having assumed some common character thereafter may vary together (cp. p. 309).

In the power of independent Variation, members of series once more exhibit the property of "unity" that we have already noticed as appearing in the manner in which the number of the members is changed. The fact that members of series should be capable of varying as "individuals" is paradoxical. Such members, teeth, digits, segments of Arthropods, and the like, are each made up of various tissues endowed with miscellaneous functions and dissimilar in their morphological nature. Nevertheless each group is capable of independent division and of separate Variation. Single digits for instance may thus be independently hypertrophied as a whole, single segments or single appendages or pairs of appendages may be differentiated in some special way, and so forth.

At this point reference may again be made to that extraordinary Discontinuity of Variation appearing in what I have called Homœosis, so strikingly seen in the few Arthropod cases given (p. 146), and so common in flowering plants. In these changes a limb, a floral segment, or some other member, though itself a group of miscellaneous tissues, may suddenly appear in the likeness of some other member of the series, assuming at one step the condition to which the member copied attained presumably by a long course of Evolution.

Many times in the course of this work we have had occasion to consider the modifications in the conception of Homology demanded by the facts of Variation. It is needless to speak further of this matter here, and the reader is referred to pp. 125, 191, 269, 394 and 417, where the subject is discussed in relation to Linear Series of several kinds, and to the facts given in Chapter XVI and at p. 433 bearing on the same questions in their application to Radial Series. The outcome of these considerations shews, as I think, that the attribution of strict individuality to each member of a series of repeated parts leads to absurdity, and that in Variation such individuality may be set aside even in a series of differentiated members. It appears that the number of the series may be increased in several ways not absolutely distinct, that a single member of the series may be represented by two members, that a terminal member may be added to the series, and also that the number of the members may change, no member precisely corresponding in the new total to any one member of the old series: in short, that with numerical change resulting from Meristic Variation there may be a redistribution of differentiation.

But though this is, in my judgment, a fact of great consequence, its relation to the Study of Variation is merely incidental. It is not so much that to enlarge the conception of Homology so as to include the phenomena of Meristic Variation is a direct help, as that to maintain the old view is a hindrance and keeps up an obstacle in the way of any attempt to apprehend the real nature of the phenomena of Division, and hence of Meristic Variation. So long as it is supposed that each member of a series of repeated parts is literally *individual*, it is impossible to form any conception of Division that shall include the facts of Meristic Variation, for in Variation it is found that the members are divisible.

It is an unfortunate thing that the study of Homology has been raised from its proper place. The study of Homologies was at first undertaken as a means of analyzing the structural evidences of relationship, and hence of Evolution. This is its proper work and use; but the pursuit of this search as an aim in itself has led to confusion, and has tended to conceal the fact that there are phenomena to which the strict conception of individual Homology is not applicable.

This exaggerated estimate of the fixity of the relationship of Homology has delayed recognition of the Discontinuity of Meristic Variation, and has fostered the view that numerical Variation must be a gradual process.

This view the evidence shews to be wrong, as it was also improbable.

Brief allusion may be made to three separate points of minor imimportance.

It is perhaps true that, on the whole, series containing large numbers of undifferentiated parts more often shew Meristic Variation than series made up of a few parts much differentiated, but throughout the evidence a good many of the latter class are nevertheless to be seen.

Reference may be made to a point that might with advantage be examined at length. The fact that Meristic Variation may take place suddenly leads to a deduction of some importance bearing on the expectation that the history of development is a representation of the course of Descent. In so far as Descent may occur discontinuously it will, I think, hardly be expected that an indication of the previous term will appear in the ontogeny. For example, if the four-rayed *Tetracrinus* may suddenly vary to both a five-rayed and also to a three-rayed form (see p. 437) it is scarcely likely that either of these should go through a definitely four-rayed stage; and if the origin of the four-rayed form itself from the five-rayed form came similarly as a sudden change, it would not be expected that a five-rayed stage would be found in its ontogeny. Similarly, if a flower with five regular segments arise as a sport from a flower with four, it would not, I suppose, be expected that the fifth segment would arise in the bud later than the other four. I suggest these examples from Radial Series, as in them the question is simpler, but similar reasoning may be applied to many cases of Linear Series also.

It will be noted that the attempt to apply to numerical variations the conception of Variation as an oscillation about *one* mean is not easy, difficulty arising especially in regard to the choice of a unit for the estimation of divergence. In few cases can facts be collected in quantity sufficient even to sketch the outline of such an investigation; but, to judge from the scanty indications available, it seems that in cases of numerical change variations to numbers greater than the normal number, and to numbers less than it are not generally of equal frequency. Probably no one would expect that they should be so.

As was stated in the Introduction, we are concerned here with the manner of origin of variations, not with the manner of their perpetuation. The latter forms properly a distinct subject. We may note however, in passing, how little do the few known facts bearing on this part of the problem accord with those ready-made principles with which we are all familiar. Upon the special fallacy of the belief that great Variation is much rarer in wild than in domesticated animals we have often had occasion to dwell. As was pointed out in the discussion of the evidence on Teeth (p. 266) this belief arises from the fact that domesticated animals are for the most part variable, and that we have every opportunity of observing and preserving their variations. To compare rightly their variability with that of wild animals choice should be made of animals that are also variable though wild. Taken in this way the comparison is fair, and as I have already said, if we examine the variation in the vertebræ of the Sloths, in the teeth of the Anthropoid Apes, in the colour of the Dog-whelks (*Purpura lapillus*), &c., we find a frequency and a range of Variation matched only by the most variable of domesticated animals.

It is needless to call attention to the fact that in hardly any cases even of extreme variations in wild creatures is there evidence that the animal was unhealthy, or ill nourished, or that its economy was in any visible way upset; but in almost every example, save for the variation, the body had the appearance of normal health.

After all that has been said few perhaps will still ask us to believe that the fixity of a character is a measure of its importance to the organism. To try to apply such a doctrine in the open air of Nature leads to absurdity. Let one more case be enough. I go into the fields of the North of Kent in early August and I sweep the Ladybirds off the thistles and nettles of waste places. Hundreds, sometimes thousands, may be taken in a few hours. They are mostly of two species, the small *Coccinella decempunctata* or *variabilis* and the larger *C. septempunctata*. Both are exceedingly common, feeding on Aphides on the same plants in the same places at the same time. The former (*C. decempunctata*) shews an excessive variation both in colours and in pattern of colours, red-brown, yellow-brown, orange, red, yellowish-white and black, in countless shades, mottled or dotted upon each other in various ways. The colours of pigeons or of cattle are scarcely more variable. Yet the colour of the larger *C. septempunctata* is almost absolutely constant, having the same black spots on the same red ground. The slightest difference in the size of the black spots is all the variation to be seen. (It has not even that dark form in which the black spreads over the elytra until only two red spots remain, which is to be seen in *C. bipunctata*.) To be asked to believe that the colour of *C. septempunctata* is constant because it matters to the species, and that the colour of *C. decempunctata* is variable because it does not matter, is to be asked to abrogate reason.

But the significance of the facts does not stop here. When, looking further into the variations of *C. decempunctata* it is found that most of its innumerable shades of variation are capable of being grouped round some eight or ten fairly distinct types, surely an expectation is created in the mind that the distinctness of these forms of varieties, all living [and probably breeding] together, may be of the same nature as the distinctness of Species; and since it is clear that the distinctness of the varieties is not the work of separate Selection we cannot avoid the suspicion that the same may be true of the specific differences too.

An error more far-reaching and mischievous is the doctrine that a new variation must immediately be swamped, if I may use the term that authors have thought fit to employ. This doctrine would come with more force were it the fact that as a matter of experience the offspring of two varieties, or of variety and normal, does usually present a mean between the characters of its parents. Such a simple result is, I believe, rarely found among the facts of inheritance. It is true that with regard to this part of the problem there is as yet little solid evidence to which we may appeal, but in so far as common knowledge is a guide, the balance of experience is, I believe, the other way. Though it is obvious that there are certain classes of characters that are often evenly blended in the offspring, it is equally certain that there are others that are not.

In all this we are still able only to quote case against case. No one has found general expressions differentiating the two classes of characters, nor is it easy to point to any one character that uniformly follows either rule. Perhaps we are justified in the impression that among characters which blend or may blend evenly, are especially certain quantitative characters, such as stature; while characters depending upon differences of number, or upon qualitative differences, as for example colour, are more often alternative in their inheritance. But even this is very imperfectly true, and as appeared in the case of Earwigs (p. 40) there may be a definite dimorphism in respect of a character which to our eye is simply quantitative. Nevertheless it may be remembered that it is especially by differences of number and by qualitative differences that species are commonly distinguished. Specific differences are less often quantitative only.

But however this may be, whatever may be the meaning of alternative inheritance and the physical facts from which it results, and though it may not be possible to find general expressions to distinguish characters so inherited from characters that may blend, it is quite certain that the distinctness and Discontinuity of many characters is in some unknown way a part of their nature, and is not directly dependent upon Natural Selection at all.

The belief that all distinctness is due to Natural Selection, and the expectation that apart from Natural Selection there would be a general level of confusion, agrees ill with the facts of Variation. We may doubt indeed whether the ideas associated with that flower of speech, "Panmixia," are not as false to the laws of life as the word to the laws of language.

But beyond general impression, in this, the most fascinating part of the whole problem, there is still no guide. The only way in which we may hope to get at the truth is by the organization of systematic experiments in breeding, a class of research that calls perhaps for more patience and more resources than any other form of biological inquiry. Sooner or later such investigation will be undertaken and then we shall begin to know.

Meanwhile, much may be done to further the Study of Variation even by those who have none of the paraphernalia of modern science at command. Many of the problems of Variation are pre-eminently suited for investigation by simple means. If we are to get further with these problems it will be done, I take it, chiefly by study of the common forms of life. There is no common shell or butterfly of whose variations something would not be learnt were some hundreds of the same species collected from a few places and statistically examined in respect of some varying character. Any-one can take part in this class of work, though few do.

At the present time those who are in contact with the facts and material necessary for this study care little for the problem, or at least rarely make it the first of their aims, and on the other hand those who care most for the problem have hoped to solve it in another way.

These things attract men of two classes, in tastes and temperament distinct, each having little sympathy or even acquaintance with the work of the other. Those of the one class have felt the attraction of the problem. It is the challenge of Nature that calls them to work. But disgusted with the superficiality of "naturalists" they sit down in the laboratory to the solution of the problem, hoping that the closer they look the more truly will they see. For the living things out of doors, they care little. Such work to them is all vague. With the other class it is the living thing that attracts, not the problem. To them the methods of the first school are frigid and narrow. Ignorant of the skill and of the accurate, final knowledge that the other school has bit by bit achieved, achievements that are the real glory of the method, the "naturalists" hear only those theoretical conclusions which the laboratories from time to time ask them to accept. With senses quickened by the range and fresh air of their own work they feel keenly how crude and inadequate are these poor generalities, and for what a small and conventional world they are devised. Disappointed with the results they condemn the methods of the others, knowing nothing of their real strength. So it happens that for them the study of the problems of life and of Species becomes associated with crudity and meanness of scope. Beginning as naturalists they end as collectors, despairing of the problem, turning for relief to the tangible business of classification, accounting themselves happy if they can keep their species apart, caring little how they became so, and rarely telling us how they may be brought together. Thus each class misses that which in the other is good.

But when once it is seen that, whatever be the truth as to the modes of Evolution, it is by the Study of Variation alone that the problem can be attacked, and that to this study both classes of observation must equally contribute, there is once more a place for both crafts side by side: for though many things spoken of in the course of this work are matters of doubt or of controversy, of this one thing there is no doubt, that if the problem of Species is to be solved at all it must be by the Study of Variation.

INDEX OF SUBJECTS.

Acanthoderes nigricans, double (?) leg, 548
Acanthodrilus, double tail, 565
Acanthothyris, double monster, 566
Accessory hoofs of Ox, connected with supernumerary digits, 285
Acherontia atropos, colours of larvæ, 304, 305
Acinopus lepelletieri, extra legs, 498
Actinometra, variation in number of radial joints, 421; 4-rayed specimen, 437; 6-rayed, 437
Adaptation, Study of, as a method of solving problems of species, 10; logical objection to the method, 12; speculations as to, avoided, 79; of species, approximate only, 11
Adimonia tanaceti, double (?) antenna, 550
Agabus uliginosus, double (?) leg, 548
Agestrata dehaanii, extra legs, 491
Agonum sexpunctatum, double (?) leg, 547; *viduum*, double (?) antenna, 550
Agra catenulata, extra legs, 512
Akis punctata, double (?) antenna, 551
Alaus sordidus, extra legs, 508
Aleochara mæsta, extra legs, 488
Allantus, extra appendage, 544
Allolobophora, generative organs of, 160, 162, 165; duplicity of head and tail, 565; *lissäensis*, spermathecæ, 165
Allurus, generative organs of, 164, 165; *putris*, 165; *hercynius*, *tetraedrus*, 164
Alytes, vertebræ, 127; axial duplicity, 561
Amblypneustes, 4-rayed, 443; 6-rayed, 446; partial reduction of a ray, 443; partial duplicity of a ray, 446
Ammocœtes, alleged case of eight pairs of gill-openings, 174
Ampedus ephippium, double (?) antenna, 551
Amphicyclus (Holothurian), tentacles not in multiples of five, 435
Amphimallus solstitialis, extra antennæ, 515
Amphioxus, number of gill-slits, 174
Anagallis arvensis, colour-variation, 44
Anas querquedula, division of digits, 392
Anchomenus sexpunctatus, extra antennæ, 523; double (?) antenna, *albipes*, *angusticollis*, 550
Angora breeds, 55
Anisoplia floricola division of epistome, 454
Annelids, segmentation compared with that of Chordata, 86; imperfect segmentation, 156; spiral segmentation, 157; variation in generative organs, 159; axial duplicity, 563
Anomala junii, extra antennæ, 515
Anser, spinal nerves, 130, 133
Antedon, variation in number of radial joints, 421; 4-rayed specimen, 436; 6-rayed specimen, 437; abnormal branching, 438
Antenna developed as foot, 146, 147
Antennæ, variation in number of joints, Prionidæ, 411; *Polyarthron*, 412; *Lysiphlebus*, 412; *Donacia*, 413; *Forficula*, 413
  extra, in Secondary Symmetry, 513–522; symmetry unknown, 522; arising from head, 551
  supposed double, 548
Anthia, extra legs, 502
Anthocharis cardamines, colour-variation, 45
  eupheno, 45; *ione*, 72
Anthropoid Apes, Variation in Vertebræ, 116; teeth, 199; digits, 349
Aphodius, extra legs, 494
Apis mellifica, union of compound eyes, 461
Appendages, joints of, 410
  supernumerary, arrangement of evidence, 474
  in Secondary Symmetry, 475; mechanical model illustrating relations, 480; duplicity of, 406, 539
Apteryx, brachial plexus, 130
Apus, extra flabellum, 536
Aquila chrysaetos, extra digits, 393
Archibuteo lagopus, extra digit, 392
Arctia, colour-variation, 46
Arctocephalus australis molars, 243
Arge pherusa, eye-spots, 295
Arion, sinistral, 54
Aromia moschata, extra legs, 485, 503, 512; double (?) antenna, 3 cases, 551
Artemia, *salina* and *milhausenii*, 96; *gracilis*, 100; relation to *Branchipus*, 96; segmentation of abdomen, 100
Arteries, renal, 277; in a case of double-hand, 333
Arthronomalus, number of segments, 94
Arthropoda, variation in number of segments, 87; Homœosis in appendages, 146; axial duplicity, 565
Articular processes, change from dorsal to lumbar type, 109; variations in position of change, 110, 112, 114, 117, 122

INDEX OF SUBJECTS. 577

Artiodactyla, polydactylism, 373; syndactylism, 383; teeth, 245, 246
Ascidia plebeia, specimens having every fourth vessel of branchial sac dilated, 172
Ascidians, variation in branchial structures, 171, 172; extra atriopore, 456
Ass, canines, 245; molars, 246; absence of digital variation in, 360
Astacus fluviatilis, colour-variation, 44
  variation in number of oviducal openings, 84, 152
  absence of male opening, 154
  absence of oviducal opening, 152, 153
  absence of opening from green gland, 154
  extra chelæ, 529, 537
  extra processes from chelæ, 536
  repetition of exopodite of antenna, 538
A. leptodactylus, extra dactylopodites, 532
A. pilimanus and *braziliensis*, apparent presence of female opening in males, 155
Asterias, variation of pedicellariæ, 429; arms, 439
  with 8 rays and 3 madreporites, 440; extra water-pore, 466
  polaris, normally 6-rayed, 433
  problema, *tenuispina*, undergo fission, 433
Asterina, 4-rayed and 6-rayed specimens, 440
Asteroidea, arms, 439–441
Ateles, teeth, 205, 206, 207
Atriopore, extra, in Ascidians, 456
Attelabus, division of pronotum, 455
Aulastoma gulo, asymmetrical variation in generative organs, 167
Aurelia aurita, Meristic Variation of, 426; statistics as to, 428
Auricles, cervical, in Man, 177; in Pig, 179; in Sheep and Goats, 180; are repetitions of ears, 180

B.

Baer, von, Law of, 8; its proper scope, 9; probably not applicable to cases of Discontinuous Meristic Variation, 571
Balance between mammæ, 189; between teeth, 213
Balanocrinus, 4-rayed specimens, 436
Balanoglossus, two methods of development, 9; number of gill-slits, 174; extra proboscis-pore, 466; supposed relation to Chordata, 86
Batrachia, extra limbs, 554; spinal nerves, 141; vertebræ, 124; extra

atrial opening, 465; axial duplicity, 561
Bdellostoma, individual and specific variations in number of gill-sacs, 173, 174
  cirrhatum, *heptatrema*, *heterotrema*, *hexatrema*, 173; *bischoffi*, *polytrema*, 174
Beech, fern-leaved, 25
Bees, hermaphrodite, 68; union of eyes, 461; antenna modified as foot, 147
Beetles, variation of horns, 38; antennæ, 411, 413; extra appendages in Secondary Symmetry, 475; legs, 483; antennæ, 513; palpi, 524; division of pronotum, 455
  supposed double legs, 544; supposed double antennæ, 548; supposed double palpi, 551
Bembidium striatum, extra palpi, 524
Bettongia, variation in molars, 258; *cuniculus*, *lesueri*, *penicillata*, 258
Bilateral asymmetry, Homœosis in cases of, 465
Bilateral Series, nature of, 88; Meristic Variation of, 448
Bilateral Symmetry, 19; in variation of vertebræ, 128; in variation of Annelids, 167; in variation of mammæ, 183; in variation of teeth, 267; in cervical fistulæ, 175; in variation of ocelli, 292; in variation of digits, 402; in variation in antennæ of *Forficula*, 414; in variation of Radial Series, 427; in abnormal branching of *Antedon*, 438; in distribution of triasters in segmenting egg, 464; in abnormal union of blastomeres, 464
  as found in manus and pes, 369, 403
  influence on Secondary Symmetries doubtful, 557
Bipinnaria, extra water-pore, 466
Birds, spinal nerves, 129; digital Variation, 390, 396
Blaniulus, mode of increase in number of segments, 93
Blaps, extra legs, 512; extra antennæ, *attenuata*, 522; double antenna, *chevrolati*, *cylindrica*, *similis*, 551
Blatta, variation in number of tarsal joints, discussion of, 63; facts, 415; regeneration of tarsus with 4 joints, 416
Blue, as variation of red, 44
Boar, Wild, extra digits, 383
Bombinator, vertebræ, 127

INDEX OF SUBJECTS.

Bombus variabilis, antenna developed as foot, 147
Bombyx, extra wing, *quercus*, 284, *rubi*, 282
Box-turtle, digital variation, 396
Brachial plexus, birds, 129; Man, 113, 135; Bradypodidæ, 141
Brachinus crepitans, double (?) leg, 546
Brachiopod, double monster, 566
Bradypodidæ, vertebræ, 118; brachial plexus, 141
Brachyteles, teeth, 205
Branchiæ, variations in number, 172
Branchiomma, double tail, 564
Branchipus, segmentation of abdomen, 97; relation to *Artemia*, 96—101; species distinguished by sexual characters of male, 100
ferox, *spinosus*, 97, 100; *stagnalis*, 100
Brill, pigmentation of blind side, 468, 470
Brimstone butterfly, variation in colour, 45; nature of pigment, 48
Broscus vulgaris, double (?) leg, 547
Buccinum, teeth, 262; double operculum, 457
Bucorvus, brachial plexus, 131, 132
Bulldog, teeth, 210, 221
Bulldog-headed races of Dogs, 57; of Fishes, 57
Buteo latissimus, extra digit, 393; *vulgaris* brachial plexus, 131
Buzzard, extra digit, 392, 393

Calathus fuscus, extra eye, 280; extra legs, *cisteloides*, 506, *fulvipes*, 507, *græcus*, 505
Callidium variabile, double (?) antenna, 551; *violaceum*, extra antenna arising from head, 551
Callimorpha, colour-variation of species of this genus, 46
Callinectes hastatus, extra spines on lateral horn of carapace, 557
Callithrix, teeth, 208
Callorhinus ursinus, teeth, 343
Caloptenus spretus, colour-variation, 44
Calopteron reticulatum, extra antennæ, 522
Calopus cisteloides, extra legs, 512
Calosoma investigator, extra antennæ, 523; *auropunctatum*, double (?) leg, 548; double (?) antenna, *sycophanta*, *triste*, 550
Cancer pagurus, maxillipede developed as chela, 149, 150; extra parts of limbs, 527; variations in chelæ, 530—536
CANIDÆ, digits, 401; mammæ, 189
teeth, 209—222; incisors, 210; canines, 210; premolars, 211; molars, 217

CANIDÆ, teeth,
Canis antarcticus, 215; *azaræ*, 217; *cancrivorus*, 218; *corsac*, 214; *dingo*, 212, 215; *javanicus*, 209; *lagopus*, 220; *lateralis*, 212; *lupus*, 212, 213, 217, 220; *magellanicus*, 218; *mesomelas*, 212, 217; *occidentalis*, 214, 219; *pennsylvanicus*, 210; *primævus*, 209; *procyonoides*, 215, 220; *vetulus*, 217; *viverrinus*, 212; *vulpes*, 210, 212, 213, 214, 219, 220; *zerda*, 220
vertebræ, 122; cervical rib, 122
Canines, supernumerary, Tiger, 225; Ass, 245; divided in Dog, 211
Capreolus, horns, 286; union of horns, 460; polydactylism, 374, 379
Caprimulgus, brachial plexus, 131
CARABUS, antenna, supposed cases of double, *auratus*, *cancellatus*, *catenulatus*, *emarginatus*, *exaratus*, *intricatus*, *italicus*, *sylvestris*, 550
antennæ, paired extra, *arvensis*, 523; *auronitens*, 522; *monilis*, 522; *sacheri*, 523
leg, supposed cases of double, *creutzeri*, 547; *helluo*, 547; *intricatus*, 545; *melancholicus*, 548; *nemoralis*, 547; *septemcarinatus*, 547
legs, extra in Secondary Symmetry, *auratus*, 511; *auronitens*, 511; *cancellatus*, 511; *catenulatus*, 512; *græcus*, 486; *granulatus*, 484; *irregularis*, 497; *italicus*, 512; *marginalis*, 484; *perforatus*, 484; *scheidleri*, 483
palpi, supposed cases of double, *auratus*, *purpurascens*, *splendens*, 552
pronotum, division of, *lotharingus*, 456; *scheidleri*, 455
Carcinomata, multipolar cells in, 431
Carcinus mænas, external segmentation of abdomen changed by parasites, 95; extra parts in limbs, 527, 531, 534
Carnivora, teeth, 209; vertebræ, 122
Carp, bulldog-headed, 57
Cassowary, feathers partially without barbules, 55
Castration, parasitic, of crabs, 95
Cat, variation in colours of, 48
digits, 312, 313; polydactylism inherited, 323
spinal nerves, 138
teeth, 222
vertebræ, 122
Caterpillars, segmental Repetition of pattern in, 25
Catocala nupta, colour-variation in hind wings, 44, 46

INDEX OF SUBJECTS. 579

Caudal fin, division of, in Gold-fishes, 451
Cebidæ, teeth, 205
Cebus, teeth, 205
Cell-division, variations in, 430
Centrosomes, variations in number of, 430
Cephalotia, 458
Cerambyx, extra legs, 491; double (?) antenna, *cerdo, scopolii*, 551; extra antenna arising from head, *cerdo*, 551
Cercocebus, teeth, 204
Cercopithecus, teeth, 204; abnormality in, 204
Ceroglossus valdiviæ, extra legs, 500
Cervical vertebræ, assumption of dorsal characters, Man, 107
Cervus axis, molar, 246
 rufus, premolar, 246
 dama, extra digits, 379
Cestoda, variation in segmentation of, 168; bifurcation and other conditions allied to duplicity, 565
Cestracion, teeth, 261
Cetonia, extra legs, *opaca*, 488, *morio*, 512
Chalcides, digital variation in the genus, 395
Chamois, extra horns, 286
Charadrius, brachial plexus, 130, 132
Chelæ, extra parts in Secondary Symmetry, 528; amorphous cases, 538; duplicity of, 540 developed from third maxillipede in *Cancer*, 149
Chelonia, axial duplicity, 563
Cheraps preissii, apparent presence of female openings in males, 155
Chiasognathus grantii, double (?) antenna, 551
Chilognatha and Chilopoda, variation in segmentation of, 93
Chirocephalus, supernumerary horn to generative sac, 457
Chimpanzee, vertebræ, 116
 spinal nerves, 139
 teeth, 202
Chionobas, eye-spots, 295
Chironomus, double head, 565
Chiropotes, 208
Chitons, repetition of eyes in, 26; variation in colours of scutes, 307
Chlænius nigricornis, extra legs, 512; double (?) leg, *holosericus*, 546, *diffinis*, 548, *vestitus*, 548
Chærocampa, colours of larvæ, 304
Cholæpus, vertebræ, 118, 120; brachial plexus, 141
Cholornis, hallux absent, 397
Chordata, segmentation of, 86
Chroicocephalus, brachial plexus, 130
Chrysemys, axial duplicity, 563
Chrysomela, division of pronotum, *fucata*, 455

Chrysomela, extra legs, *banksii*, 494; *graminis*, 498
 double (?) leg, *hæmoptera*, 547
 double (?) antenna, *cacaliæ*, 550
Chub, bulldog-headed, 58
Cicindela sylvatica, double (?) antenna, 550
Cidarites, 4-rayed specimen, 441
Cimbex axillaris, antenna developed as foot, 146
Cimoliasaurus, imperfect division of vertebræ, 103
Ciona intestinalis, variation in number of stigmata, 172
Cistudo, digital variation in, 396
Cladocera of salt lakes, 101
Clausilia bidens, extra eye, 280
Clavatella, variation in number of segments, 425; in number of eyes in each segment, 425
Clupea pilchardus, scales, 274
Clythra quadripunctata, extra legs, 508
Clytus liciatus, extra legs, 508; extra antennæ, *arcuatus*, 522, *tricolor*, 522; double (?) antenna, *arcuatus*, 551
Coccinella decempunctata, bipunctata and *septempunctata*, colour-variation, 49, 572
Cochin fowls, "silky" variety, 55
Cockroach, variation in number of tarsal joints, 63, 415
Cœlenterata, imperfect division, 566
Colias, colour-variation, 44; intermediates between *edusa* and *helice*, 44; varieties of *hyale*, 45
Colobus, teeth, 204
Colour and Colour-patterns, variations in, 42, 288, 572
Colour-variation, discontinuity of, perhaps chemical, 72; simultaneous, in segments, &c., 303
Coluber, double monster, 561
Columba, brachial plexus, 131, 134
Colymbetes sturmii, extra legs, 512; *adspersus*, double (?) leg, 548; *coriaceus*, double (?) antenna, 550
Colymbus, brachial plexus, 130
Conepatus chilensis and *mapurito*, teeth, 232
Continuity, use of term as applied to Variation, 15; of differences in Environment, 7
Copepoda, of salt lakes, 101
Cordylophora lacustris, polystomatous specimens, 566
Correlation, between variations of nerves and vertebræ, 145; between Meristic and Substantive Variation, 126
Corvus, brachial plexus, 131
Corymbites cupreus, colour-variation, 43
Counterparts, simultaneous variation of, 560

37—2

580　INDEX OF SUBJECTS.

Cow, variation in number of teats, 188
Crab, extra parts of appendages, 527—536; variation in segmentation of abdomen, 95
Crateronyx, extra wing, 285
Crayfish, variation in number of generative openings, 152; repetition of parts of chelæ, 529, 532, 537; extra parts in antennæ, 538
Cribrella, abnormal branching of an arm, 440
Crinoids, radial joints, 421; variation in number of rays, 435; 4-rayed varieties, 436; 6-rayed varieties, 437; 3-rayed and 5-rayed varieties of a 4-rayed form, 437; variation in number of canals in stems, 438; abnormal branching, 438
Crisia, division of embryos, 556
Crossarchus, teeth, 227—231; zebra, 230, 231
Crustacea, theory of descent of Vertebrata from, 29; of salt lakes, 100; Secondary Symmetry in, 525; Homœosis in, 149
Cryptohypnus riparius, extra legs, 509
Cryptophagus scanicus, dentatus, double (?) antenna, 550
Ctenodrilus, double tail, 565
Cucumaria planci, with six radii, 433; double monster, 566
aciculi, double monster, 566
Cuon, one lower molar absent, 209
Cupressocrinus, imperfect variation to 4-rayed state, 437; variation in number of canals in stalk, 438
Curve of Frequency of Variations, 37, 64
Cuvierian organs, variation in number of, 435
Cyclopia, 458
Cygnus olor, cervical vertebræ, 33; colour-variation of young, 44; atratus, brachial plexus, 130
Cyllo leda, variability of ocelli, 289
Cynælurus, teeth, 222, 224
Cynocephalus porcarius, extra molar, 204
Cyprinus carpio, bulldog-headed varieties, 57; hungaricus, ditto, 58; auratus, division of fins, 451
Cyprus, 4-horned sheep, 285
Cypselus, brachial plexus, 131
Cystophora cristata, premolars, 238; molars, 243

Dachshund, hallux in, 401; duplicity of hallux and pollex, 401
Dactylopodites, extra, 528
Dactylopsila, premolars, 255
Darwin's solution of problem of Species, 5; views on Reversion, 77; on sudden Variation in eye-spots, 289
Dasyuridæ, incisors, 247
Dasyurus, incisors, 247; premolars, 255; molars, 256

Dasyurus, viverrinus, variation in molars, 256
maculatus, molars, 256
Deilephila euphorbiæ, colours of larvæ, 305; hippophäes, 305
Dendarus hybridus, extra antennæ, 523
Descent, Doctrine of, assumed to be true, 4
Diaptomus, colour-variation of eggs, 44
Dicotyles torquatus, incisors, 245
Didelphyidæ, incisors, 246
Didelphys, teeth, 246, 258
Digits, Variation of
MAMMALS. Capreolus, 374; Cat, 313; Cervus, 379; Dogs, 401; Erinaceus, 397; Elephas, 397; Goat, 377; Herpestidæ, 346; Horse, 360; Hylobates, 346; Macacus, 340; Man, 324; Mule, 360, 370; Ox, 374, 383; Pig, 381, 387; Sheep, 373, 380
BIRDS. Anas, 392; Aquila, 393; Archibuteo, 392; Buteo, 393; Fowl, 390; Larus, 393; Pheasant, 393; Rissa, 396; Turkey, 393
REPTILES. Chalcides, 395; Cistudo, 396
Reduction in number, Man, 355, 358; Artiodactyla, 383
Union of, Ox, 383; Pig, 387; Man, 355
Variation in, associated with other variations, 399
Inheritance of Variation in, 398
Recapitulation of evidence, 400
Dimorphic condition, its relation to the monomorphic condition, 37
Dimorphism in Spinal nerves, 138; in position of generative openings in Pachydrilus, 165, 168; in secondary sexual characters, 38
Diopatra, abnormal repetition, 159
Discoidea (Echinid), 4-rayed specimen, 442
Discontinuity of Species, 5
in Variation, a possibility, 17; suggestion as to its nature, 68, 568
in chemical processes, 16, 48, 72
in colour-variation, 43, 48, 72; in colour-patterns, 48
in states of matter, 16
of Meristic Variation perhaps mechanical, 70
of Substantive Variation perhaps in part chemical, 71
in the Variation of spinal nerves, 145
in the Variation of the generative organs of Annelids, 168

Discontinuity in the Variation of digits, 407
  in Meristic Variation of Radial Series, 423
  partly dependent upon Symmetry, 568
Discontinuous Variation, use of the term, 15
Disease, analogy with Variation, 74
Ditomus tricuspidatus, double (?) antenna, 550
Division of organs, a process of reproduction, 193
  of teeth, 268; of mammæ, 193; of digits, 349, 369; of tentacles, 280; of radius of Echinid, 446; median, 454
Dog, cervical rib, 122; hairless, 57; bulldog, 210, 221; digits, 401; nipples, 189
  teeth, 209—222; incisors, 210; canines, 211; premolars, 213, 215; molars, 220; deficiencies in Esquimaux, 215; in Inca, 216
Dog-whelk, colour-variation, 48
Domestication, variability falsely ascribed to, 266, 401
Donacia bidens, Variation in antennæ, 413
Dorcadion rufipes, extra legs, 512
Dorking Fowl, digital variations, 390—395
Dorocidaris papillata, variation in pedicellariæ, 429
Double-foot, Artiodactyles, 378; Frog, 540; *Macacus*, 340; Man, 337, 338
Double-hand, 325, 331
Double Monsters, 559
Double-thumb, 349
Dromæolus barnabita, extra antennæ, 523
Duck, no variation in number of digits recorded, 401; cases of absence of webs between toes, 401
Duplicity of single members of series not distinct from other modes of addition, 193, 407
  of appendages, 406; in Arthropoda, 539; in Vertebrata, 539
  axial, 559
Dutch pug, 57
Dyschirius globulosus, extra legs, 512
Dytiscus marginalis, extra legs, 512

Eagle, extra digits, 393
Ears, repetitions of, known as cervical auricles, 180
Earthworms, variation of generative organs, 159; of segmentation, 157; asymmetrical arrangement of generative organs, 160, 161; table of arrangement of ovaries, 162; duplicity of head, 565, of tail, 565

Earwig, variation of forceps, 40; of antennary joints, 413
Echinoconus, 4-rayed specimen, 441; alleged case of 3 rays, 441
Echinodermata, Meristic Variation in, 432; variations of pedicellariæ, 429; duplicity,
Echinoidea, Meristic Variation of, 441; 4-rayed specimens, 441; partial disappearance of a ray, 443; partial duplicity of a ray, 446; 6-rayed specimen, 445; pedicellariæ, 429; variation in number of genital pores, 446; symmetrical reduction of two rays, 443
Echinus melo, partial reduction of an ambulacrum, 443
Echinus sphæra, partial reduction of an interambulacrum, 445
Ectrodactylism, Man, 355, 358
Elater murinus, extra antennæ, 522; *variabilis*, extra leg, 547; *hirtus*, double (?) antenna, 551
Eledone, supernumerary hectocotylus, 473
Eleodes pilosa, double (?) antenna, 551
Elephas, tusks, 244; hoofs, 397
Elytra, said to have been replaced by legs in *Pionus*, 148
Embryology, as a method of investigating problems of Descent, 7
Emperor moth, ocelli absent, 289, 301; colour-variations of larva, 306
Emu fowls, 55
Enchytræidæ, generative openings, 165
Encrinus, variation in number of radial joints, 421; 4-rayed calyces, 436; radius bearing only one arm, 438
Enema pan, extra legs, 512
Entoniscians, alter segmentation of some crabs but not of all, 95
Enhydris, incisors, 211
Epiglottis, division of, 451
Epipodites, variation of in *Hippolyte*, 151
Epistome (of Beetle), division of, 454
Erebia blandina, ocelli, 289
Erinaceus, variation in hallux, 397
Eriphia spinifrons, extra chelæ, 528
Eros minutus, extra legs, 487
Esox lucius, bulldog-headed, 58
Esquimaux Dog, absence of first premolars, 214, 215, 221
Euchloe, pigments of, 72
Eugeniacrinus, 4-rayed specimens, 436
Euprepia purpurea, extra palpi (alleged), 524
Eurycephalus maxillosus, extra legs, 487
Euscorpius, double tail, 565
Eye of *Palinurus* developed as antenna, 150
Eye-colour of Man, 43
Eye-spots, 288; Variation as a whole, 291; outer zones first to appear, 291; analogy with chemical phenomena,

292; in Linear Series, 288, 293;
simultaneous Variation of, 293; correlated with variation of neuration, 293, 301
    Arge, 295; Chionobas, 295; Hipparchia, 294; Satyrus, 295; Morpho, 296; Vanessa, 299; Junonia, 299, 300; Pararge, 300; Saturnia, 301, 302; Raiidæ, 302; Polyplectron, 450
Eyes of Clavatella, variation in number, 425
Eyes of Molluscs, 279; of Insects, 280; union of, 458, 461

Feathers, of "hairy" Moorhen, without barbules, 55
FELIDÆ, digits, 313
    teeth, 223—226
        Felis brachyurus, 224; caligata, 223; caracal, 224; catus, 224; chaus, 224; chinensis, 224; concolor, 223; domestica, 223, 224, 225, 226; eyra, 223; fontanieri, 225, 226; inconspicua, 223; jaguarondi, 224; javanensis, 224; jubata, 224; leo, 226; lynx, 226; maniculata, 223; manul, 224; minuta, 223; nebulosa, 224; onca, 224; pajeros, 224; pardalis, 226; pardus, 223, 226; tetraodon, 223; tigrina, 226; tigris, 224, 225
    vertebræ, 122
Feronia mühlfeldii, extra legs, 485
Fins, division of, in Gold-fish, 451
Fishes, undifferentiated teeth in certain, 32; bulldog-headed, 57
    division of caudal fins, 451; scales, 274; flat-fishes, 466
Fistulæ, cervical, 174; morphology of, 176; aural, 177
    in Man, 175; in Pig, 179; in Horse, 180; unknown in Sheep, Goats and Oxen, 180
Flat-fishes, reversed varieties, 54, 466; "double" varieties, 466
Fœnius tarsatorius, extra legs, 491
Foot, double, Artiodactyla, 378; Frog, 540; Macacus, 340; Man, 337, 338
Foraminifera, duplicity, 566
Forficula auricularia, variation of forceps, 40; of antennary joints, 413
Fowls, silky variety of, 55
    digital variation in, 390; ovary and oviduct, 465
Frog, vertebræ, 124; extra legs, 554; double foot, 540; Secondary Symmetry, 554
Fusus antiquus, sinistral, 54

Galerita africana, extra legs, 495

Galerites albogalerus, 6-rayed specimen, 445
    subrotundus, 4-rayed specimen, 441
Galictis, teeth, barbara, vittata, 232
    vertebræ, 123
Gallinula chloropus, hairy variety, 55; extra digits, 392
Gallus, brachial plexus, 130; digits, 390; oviduct, 465
Garganey Teal, division of digits, 392
Garrulus, brachial plexus and ribs, 135
Gasterosteus, scales, 276
Gavialis, change in number of vertebræ, 123
Gecinus, brachial plexus, 131
Generative openings, repetition of, in Astacus, 152; absence of, in Astacus, 152, 154; of Earthworms, 159; of Hirudo, 166
Generative organs of Earthworms, variations in, 159; of Leeches, 165
Genital pores, variation in number in Echini, 446
Geophilus, variation in number of segments, 94
Geotrupes mutator, extra legs, 500; typhœus, extra antennæ, 515
Gill-slits, of Ascidians, 171; of Myxine, 172; of Bdellostoma, 173; of Ammocœtes, 174; of Notidanidæ, 174; of Balanoglossus and Amphioxus, 174
Glaucium luteum, colour-variation, 47
Gmelin's test for bile-salts, 292
Goat, incisor, 245; horns, 286; digits, 377; cervical auricles, 180
Gold-fish, simultaneous variation in length of tail and fins, 309; division of anal and caudal fins, 451; "Telescope," 453
Gonepteryx rhamni, similarity of fore and hind wings, 25
    colour - variation, 45; nature of the yellow pigment, 48
    extra wing, 283
Goose, brachial plexus, 133; pygomelian, 401
Gorilla, vertebræ, 117; spinal nerves, 139; teeth, 202
Goura, brachial plexus, 130
Grus, brachial plexus, 130
Guinea-hen, double-hallux, 392
Guinea-pig, inversion of layers in, 9
Gulo, teeth, 231

Hæmal spines, division of, in Goldfishes, 453
Hair, absence of, in Mouse, Horse, Shrew, 56; silky in Mouse, 55; excessive length in mane and tail of a horse, 309

INDEX OF SUBJECTS. 583

"Hairy" Moorhen, 55
Halichœrus, vertebræ, 123; molars, 242, 243
Haliotis, extra row of perforations, 287; perforations occluded, 287; perforations confluent, 287
Halla, imperfect segmentation in, 156
Hallux, duplicity in Man, 349; Fowl, 390; variations in Kittiwake (*Rissa*), 396; *Erinaceus*, 397; Herpestidæ, 397, normally absent in certain birds, 396
Hammaticherus heros, double (?) antenna, 551
Hand, digital variations in, 324; double, 325, 331
   progressive series of Conditions, 324
Hapalidæ, teeth, 208
Harpalus, rubripes, extra legs, 493; *calceatus*, double (?) antenna, 550
Hawthorn, variation of, 569
Hectocotylus, supernumerary, in *Eledone*, 473
Helictis orientalis, teeth, 233, 234
Helix kermovani, extra eye, 280
   hispida, union of tentacles, 461
Heloderma, vertebræ, 123
Helops cæruleus, extra antennæ, 523
   sulcipennis, extra palpi, 524
Hemiaster; cases in which one ambulacrum wanting, 445; two ambulacra reduced, 443; duplicity of ambulacra, 446
Hepialus humuli, males like females in Shetland, 254
Heptanchus, seven gills, 174
Heredity, objection to use of term, 75; in digital variation, 398
Hermaphroditism, 67; in bees, 68
Hermodice carunculata, abnormal segmentation, 158
HERPESTIDÆ, hallux, 397
   teeth, 227-231
   Herpestes galera, 229; *gracilis*, 227, 228, 229; *griseus*, 229; *ichneumon*, 229, 230, 231; *microcephalus*, 229; *nipalensis*, 227; *nyula*, 228; *persicus*, 227; *pulverulentus*, 228, 229; *smithii*, 228, 229
Herring, supposed hybrid with Pilchard, 275
Heterocephalus, a naked Rodent, 56
Heterogeneity, universal presence of in living things, 18
   symmetrically distributed around centres or axes, 19
Heterorhina nigritarsis, division of pronotum, 455
Hexanchus, six gills, 174
Hipparchia tithonus, eye-spots, 293, 294
Hippocampus compared with *Phyllopteryx*, 309
Hippoglossus pinguis, 471

Hippolyte fabricii, variation in epipodites of legs, 151
Hirudinea, variation in generative organs, 165
   in colours, 304
Hirudo medicinalis, variation in number of testes, 165, 166
   officinalis, supernumerary penis, 166
Hister cadaverinus, extra legs, 512
Holopus rangi, 4-rayed specimen, 436
Holothurioidea, variation in number of radii, 433; variations in numbers of organs, 435; double monsters, 566
Homarus, repetition of parts in cheliped, 530; in chelæ, 531-538; colour variation, 44; hermaphrodite, 155
"Homodynamy," 133
Homœosis, use of the term, 85
   between vertebræ, 106-127;
   backward and forward, use of terms, 111; forward in vertebræ, 112; backward in vertebræ, 111; in spinal nerves, 144; of appendages in Arthropoda, 146; in segments of Annelids with respect to genital organs, 162, 163, 167, &c.; in teeth, 272; in bilateral asymmetry, 465; in parts of flowers, 111
Homology between members of Series of Repetitions, 30
   individual, not attributed if series is undifferentiated, 32; attempt to trace in mammæ, 191; discussed in the case of teeth, 269; in the case of digits, 351, 391, 371, 377; in the case of joints of tarsus of *Blatta*, 418; in the case of radii of Holothurioidea, 433
Horns, Sheep, 285; Goat, 286; Roebuck, 286, 460; Chamois, 286; of Roebuck united in middle line, 460
Horse, similarity of fore and hind legs, 25, 26
   naked variety, 56; teeth, 244, 245
   cervical fistulæ, 180
   simultaneous variation of mane and tail, 309
   extra digits, 360; by development of digit II, 361—367
      by development of digit IV, 367
      by development of digits II and IV, 368
      by division of digit III, 369
      by intermediate process, 371

584 INDEX OF SUBJECTS.

Hyas, double chela, 540; double index, 541
Hybrids, supposed, between Herring and Pilchard, 274; supposed, between Turbot and Brill, 468; supposed, in genus *Terias*, 52, 53
Hydrobius fuscipes, pronotum having three lobes, 454
Hylobates, vertebræ, 118; teeth, 204; *leuciscus*, extra digit, 346
Hylotrupes bajulus, extra legs, 494
Hypsiprymnus, teeth, 258

Iceland, 4-horned sheep, 285
Ichneumon luctatorius, extra legs, 511
Ichneumonidæ, extra legs, 491, 511
Icticyon venaticus, teeth, 220
Ictonyx, teeth, 233
Images, relation of, the basis of Symmetry, 19
  between upper and lower jaws, 196, 267; between right and left sides, 88; in the case of the manus and pes, 404
  division and union of parts related as, 449
  principles of, followed in the structure and position of parts in Minor Symmetry, 479
Inca Dogs, a bulldog found amongst, 57; variation of premolars and molars, 216, 222
Incisors, supernumerary, *Gorilla*, 203; *Ateles*, 207; Canidæ, 210; Felidæ, 222; Herpestidæ, 227; Pecora, 245; *Dicotyles*, 245; Horse, 244
  division of, Canidæ, 210; *Elephas*, 244
  absence of, Canidæ, 211; Felidæ, 222; Herpestidæ, 227; Phocidæ, 235; Horse, 244
Index of crabs and lobsters, peculiarity in repetition of, 557
Individuality, attributed to members of Meristic Series, 31; such individuality not respected in Variation, 32; cases illustrating the absence of supposed individuality in Members of Meristic Series, 104, 115, 124, 191, 269, 407, 433; an unfortunate term, 556

Jacamaralcyon tridactyla, distinguished by absence of hallux, 396
Jackal, vertebræ, 122; teeth, 217
Japanese pug, probable independent origin of, 57
Jaws, relation of upper to lower, 196
Julodis æquinoctialis, extra legs, 503; *clouei*, double (?) antenna, 551
Julus terrestris, mode of increase in number of segments, 93

Kallima inachys, colour-variation, 53

Karyokinesis, symmetry in, 20; variations in, 430; bilaterally symmetrical variation of, in the segmentation of an egg, 464
Kidney, supernumerary, 277; horseshoe, 278, 459
Kittiwake, variations in hallux, 396

Laciniation, simultaneous, of petals, 310
Lady-birds, colour-variation, 49
Lagorchestes, teeth, 258
Lagothrix, teeth, 208
Lamarck's solution of problem of Species, 4
Lamellibranchs, sinistral, 54
Lamia textor, double (?) antenna, 551
Larus leucopterus, digits, 393
Larvæ of Lepidoptera, variations in colours of, 304
Leaf-butterfly, colour-variation, 53
Leeches, variation in generative organs of, 165; in colours, 304
Legs, extra, in Secondary Symmetry, general account, 475, cases in Insects, 483; in Position V, 483; in Position VAA, 485; in Position A, 487; in Position DA, 491; in Position D, 494; in Position DP, 496; Position P, 498; Position VP, 500. Unconformable cases, 503; miscellaneous cases, 511; in Crustacea, 526; in vertebrates, 554
  supposed double in Insects, 544
Leopard, two cases of dental variation in a Chinese, 225, 226
Lepidoptera, colour-variation of larvæ, 304; ocellar markings, 288; nature of yellow pigments, 73
Leptura testacea, double (?) leg, 545
Leuciscus dobula, bulldog-headed variety, 58
Lichnanthe vulpina, extra antennæ, 517
Ligula, absence of segmentation in, 168
Limax, union of tentacles, 460
Limenitis populi, extra wing, 283
Limulus, division of caudal spine, 450, 456
Linaria, many symmetrical variations of, 76
Linckia multiflora, fission, 433
Linear Series, Meristic Variation in, 63; simultaneity in colour-variations of, 303
Lissotriton, supposed double limb, 539
Lithobius, number of segments, 93
Lithodes arctica, extra legs, 527
Littorina rudis, colour-variation, 49
Littorina, sp., extra eye, 280
Lixus angustatus, division of pronotum, 456
Lizards, digital variation in, 395, 396
Lobster, colour-variation, 44; hermaphrodite, 155; variations in chelæ, 530—538

INDEX OF SUBJECTS. 585

Local Races, evidence as to, not a direct contribution to Study of Variation, 17
Locusts, variation in colour of tibiæ, 44
Loligo, variations in segmentation of egg, 463
Lucanus cervus, extra antennary branch, 2 cases, 549; extra branch on mandible, 2 cases, 552; *capreolus*, ditto, 552
Luidia ciliaris, variation of pedicellariæ, 429
Lumbo-sacral plexus, 138
Lumbriconereis, imperfect segmentation, 156
Lumbriculus, double tail, 565
Lumbricus, undifferentiated segments in, 32; imperfect segmentation, 156; spiral segmentation, 157; repetition of ovaries, 160; asymmetrical arrangement of organs, 160, 161; variation of genital openings, 162; duplicity, 565
agricola, 162; *herculeus*, 160; *purpureus*, 160; *terrestris*, 156, 157; *turgidus*, 160
Lupa dicantha, extra index, 542
Lurcher, teeth, 221
Lutra, teeth, 228, 233, 234, 235
constancy of p^1 in *L. vulgaris*, 228
Lycæna icarus, extra wing, 284
Lycalopex group of Foxes, frequency of extra molars in, 217
Lychnis, repetition of fimbriation in petals, 26
Lycus, double (?) antenna, 551
Lynx, teeth, 224
Lysimachia, Meristic variation in flower of, 61
Lysiphlebus, variation in number of antennary joints, 412

Macacus cynomologus, syndactylism, 356
inuus, spinal nerves, 137, 139
teeth, 204
radiatus, doubtful extra molars, 204
rhesus, extra molar, 204
Macroglossa, colours of larvæ, 304, 305
Macrognathus nepalensis, double (?) antenna, 551
Macropodidæ, teeth, 259
Macrorhinus leoninus, teeth, 243
Madreporites, repetition of, 440
Maia squinado, extra index, 542
Major Symmetry, 21, 87
Malachius marginellus, extra antennæ, 522
Males, high and low, 39
Mallodon, double (?) leg, 547
Mammæ, numerical Variation in, 181; along mammary lines, 181; in other positions, 186; in axilla, 185; below and internal to normal mammæ, 186; above and external, 185
Mammæ, variation in Cow, 161; Dog, 189; Pig, 190; Man, 181; Apes, 188
comment on facts, 191
development of, 194
Mammary extensions to axilla, 185
lines, 181
tumours, 185, 187
Man, cervical fistulæ, 174; cervical auricles, 177
digits, increase in number, 324; reduction in number, 355; polydactylism in general, 344; double-hand, 331; symmetry of manus and pes, 403
kidneys, union of, 459; renal arteries, 277; ureters, 278
mammæ, 181
nerves, spinal, 135; brachial plexus, 135; notable variation in, 137, 113; lumbo-sacral plexus compared with that of Chimpanzee, etc., 138
teeth, 198
transposition of viscera, 559
uterus, double, Darwin's comment on, 77
vertebræ, Meristic and Homœotic variation in, 103, 106—116, 458
Mandibles, supposed duplicity of, in *Lucanus*, 552
Manus, variations in, compared with those of pes, 405; as a system of Minor Symmetry, 403
Marsupialia, teeth, 246—258
Mastiff, teeth, 210, 221; hallux permitted in, 401
Maternal impressions and extra legs in a beetle, 512
Maxillipede developed as a chela, 149, 150
Median nerve, variations in composition of, 136
Medicago, repetition of brown spot in leaflets, 26
Medusæ, Meristic Variation of, 423; duplicity in, 566
Melanoplus packardii, colour-variation, 44
Meles, teeth, 232, 233, 235
Mellivora, teeth, 233, 235
Meloe coriaceus, double (?) leg, 547; *proscarabæus*, extra legs, 488; *violaceus*, extra antennæ, 523
Melolontha vulgaris, division of pronotum (5 cases), 454
division of pygidium, 456
extra legs, 484, 512
extra antennæ, 515, 520, 533, 550

Melolontha vulgaris, double (?) leg, 545
   hippocastani, extra antennæ, 516, 557
Mephitis, teeth, 232
Merism, 20 ; importance of, to Study of Variation, 23
   indirect bearing of, on the magnitude of Variations, 25
Meristic Repetition, 20; kinship of parts so repeated, 26; similar Variation of parts in, 27, 310, 464; compared with asexual reproduction, 34
Meristic Variation, distinguished from Substantive Variation, 22 ; compared with Homœotic Variation, 84
Metacarpals, development of lateral, in Artiodactyla in correlation with syndactylism of metacarpals III and IV, 383
Metameric Segmentation, not distinguishable from other forms of Repetition, 28 ; errors derived from such distinction, 30
Metasternal plates, division of, 456
Metazoa, comparison with Protozoa, 35
Metrius contractus, extra legs, 503
Middle Line, division by images in, 404, 450 ; union of images in, 383, 458
Minnow, bulldog-headed specimen, 58
Minor Symmetry, 21, 88 ; Meristic Variation in, 311, 410 ; in manus and pes, 403
Molars, supernumerary, *Simia*, 200; *Troglodytes*, 202 ; *Gorilla*, 203; *Cynocephalus*, 204; *Macacus*, 204; *Cebus*, 205; *Ateles*, 205 ; *Mycetes*, 207 ; Canidæ, 217, 220 ; Felidæ, 226 ; Herpestidæ, 230; Mustelidæ, 234; Phocidæ, 242 ; Ungulata, 245, 246; *Dasyurus*, 256 ; *Bettongia*, 258
   special frequency in Anthropoid Apes, 200; in *Lycalopex* group of Foxes, 217
   absent, *Simia*, 200; *Ateles*, 207 ; *Pithecia*, 208; Canidæ, 219, 221; Felidæ, 226 ; Herpestidæ, 231; Mustelidæ, 235; Phocidæ, 243 ; *Bettongia*, 258
   division of, *Canis cancrivorus*, 219 ; *Crossarchus*, 230
   Variation in form, *Crossarchus*, 231; *Dasyurus*, 256
   in *Icticyon* and *Otocyon*, 220
Monkeys, Old World, teeth, 204; New World, 205
Monodon, development of *right* tusk, 465
Monomorphism, 33
Monotoma quadricollis, double (?) antenna, 550
Moorhen, hairy variety, 55; extra digits, 392

Morpho, eye-spots, 296—299
   achilles, 297 ; *menelaus*, 298 ; *montezuma*, 297 ; *psyche*, 299 ; *sulkowskii*, 299
Mouse, colour-variation, 44 ; with silky hair, 55 ; black variety, 55 ; naked, 56
Mugil capito, bulldog-headed, 58
Mule, rarity of digital variation in, 360 ; case of, 370
Mullet, bulldog-headed variety, 58
Multipolar cells, 430
Mustelidæ, teeth, 231, 235; premolars in *M. foina, martes, melanopus, zibellina,* 231
Mycetes, teeth, 207, 208
Mycomelic acid, relation to yellow colouring matters, 73
Mydaus, teeth, 232
Myriapoda, variation in number of segments, 91, 93
Myrmecobius, incisors, 247, 248
Myxine, variations in number of gill-sacs, 172

Nænia typica, extra wing, 284
Narcissus, Substantive and Meristic Variation in, 23 ; colour-variation, 46
Narwhal, development of tusks, 465
Nasalis, teeth, 204
Natural Selection, chief objection to theory of, 5 ; misrepresentations of the theory, 80; difficulty in connexion with regeneration, 420
Navosoma, extra antennæ, 521
Nebria, double (?) antenna, 550 ; *gyllenhalli*, double (?) maxillary palp, 551
Nectarine, discontinuous variation in, 59
Nereis, double tail, 564
Nerves, spinal, Birds, 129—135; Man, 135; Apes, 138 ; Cat, 138; Dog, 140; Bradypodidæ, 141; *Pipa*, 141
   attempt to homologize, 32
   variations, 129–145
   correlation with vertebræ, 145
   relation to limbs, 143
Neural canal, division of, 451
Neuration of wings varying with eye-spots, 293
Nigidius, extra branch on antenna, 549
Nipples, supernumerary, on normal breast, 184; on normal areola bifid, 184; on mammary lines, 186; in Pig, 190; in Dog, 189
Notidanidæ, number of gills, 174
Nuclei, multipolar division, 430, 464; precocious division, 464
Nyctereutes procyonoides, teeth, 215
Nyctipithecus, teeth, 208

INDEX OF SUBJECTS. 587

Ocellar markings, 288, 449
    Lepidoptera, 288; Raiidæ, 302; Birds, 449
Odontolabis stevensii, extra antennæ, 518, 557
Oligochæta, axial duplicity, 563; generative organs, 159; segmentation, 156
Ommatophoca rossii, premolars, 237
Onitis bison, division of pronotum, 455
Operculum, double in *Buccinum*, 457
Ophiacantha anomala, normally 6 arms, not known to divide, 433
Ophiactis, fission, 433
Ophibolus, double monster, 561
Ophidia, vertebræ, 103, 123
Ophidiaster cribrarius, fission, 433
Ophiocoma pumila, young with 6 arms, adults with 5 arms, 433
Ophiothela isidicola, fission, 433
Ophiuroidea, variation in number of arms, 447; fission, 433
Opisthocomus, brachial plexus, 130
Orang, vertebræ, 118; spinal nerves, 139; teeth, 200; extra digit, 349; extra mamma, 188
Organic Stability, 36
Orthosia lævis, extra wing, 284
Oryctes nasicornis, division of pronotum, 455; extra legs, 512
Osmoderma eremita, double (?) leg, 547
Otaria cinerea, molars, 240
    jubata, premolars, 240, 243
    ursina, premolars, 239, 241
Otocyon, teeth, 220, 221
Ouakaria, teeth, 208
Ovaries, variations in number and position in Earthworms, 160, 162 not always correlated with variations in oviducts, 167
Ovary, right, developing in Fowl, 466
Oviduct, right, case of, in Fowl, 466
Oviducts of *Astacus*, variation in number, 152; in Earthworms, 167
Ox, incisors, 245; molar, 246; polydactylism, 374–381; syndactylism, 384–387; syndactylism together with development of digits II or V, 385; with duplicity of II and V, 386; cervical auricles and fistulæ unknown, 180

Pachydrilus sphagnetorum, dimorphic in respect of position of generative openings, 165, 168
Painted Lady butterfly, colour-variation, 49
Palæornis torquatus, colour-variation, 43
Palamnæus borneensis, division of poison-spine, 457
Palinurus penicillatus, eye developed as antenna, 150
    vulgaris, extra legs, 527; extra antennæ, 538

Palloptera ustulata, abnormal growth from thorax, 285
Palpi, paired extra, in Insects, 524 supposed double, 551
Pangenesis, 75
"Panmixia," 573
Papaver nudicaule, colour-variation, 46; pigment of, 47, 72
Parakeet, colour-variation, 43
Pararge megæra, eye-spots, 289, 300
"Parhomology," 133
Pariah dog, teeth, 221
Parnassius, ocelli, 292
Parra, feathers, 55
Pasimachus punctulatus, double (?) leg, 547
Patella, extra tentacle and eye, 279
Pattern, universal presence of, in organisms, 19–21; difficulties arising from, 21
Peach, discontinuous variation in, 59
Peacock, ocelli, 449
Peacock butterfly, repetition of eye-spots in, 26; variation of, 299
Pecora, polydactylism, 373
Pecten, double eyes, 280
Pedicellariæ, Meristic Variation of, 429
Pelamis bicolor, imperfect division of vertebræ, 105; axial duplicity, 561
Pelecanus, brachial plexus, 130
Pelobates fuscus, extra spiracle, 465
Penis, supernumerary, in *Hirudo*, 166; in *Aulastoma*, 167
Pentacrinus mülleri, increase in number of radial joints,
    4-rayed specimens, *dubius*, *dumortieri*, *jurensis*, *subsulcatus*, 436
    6-rayed specimens, *jurensis*, 437
Penthina salicella, extra wing, 285
Peramelidæ, digits of pes, 313
Perichæta, variation in number of spermathecæ, 165; *forbesi*, *hilgendorfi*, 165
Pericrocotus flammeus, colour-variation, 46
Perionyx, generative organs, *excavatus*, 163, 167, 168
    grünewaldi, 164
Peripatus, variation in number of segments, 84, 91, 94
Petaurus, premolars, 255
PHALANGERIDÆ, incisors, 248; premolars, 248—255
Phalanger maculatus, incisors, 248; premolars, 253; females spotted in Waigiu, 254
    orientalis, incisors, 248, 250; premolars, 250
    ornatus, first premolar two-rooted as variation, 254
    ursinus, first premolar normally two-rooted, 254

Phalanges, reduction in number, 355
Phascologale dorsalis, teeth, 257
Pheasant, digits, 393
Philonthus succicola, double (?) leg, 548
    ventralis, extra legs, 507
Phoca barbata, incisors, 235, 236
    cristata, premolars, 238
    grœnlandica, premolars, 238, 240, 242; molars, 243
    vitulina, premolars, 238, 241, 242; molars, 242
Phœnicopterus, brachial plexus, 130
Phoxinus lævis, bulldog-headed, 58
Phratora vitellinæ, colour-variation, 43
Phreoryctes, generative organs, 162
Phyllopertha horticola, extra antennæ, 514
Phyllophorus, tentacles not in multiples of five, 435
Phyllopteryx, compared with *Hippocampus*, 309
Physa acuta, tentacle bifid, 280
Picus viridis, colour-variation, 43; *medius*, brachial plexus, 131
Pieridæ, colours of, 73; eye-spots in, 292
Pig, digits, 381; syndactylism, 387; syndactylism with division of digit V, 389; cervical auricles, 179; cervical fistulæ, 180
Pigeon, cervical vertebræ, 33; brachial plexus, 134
Pigments, definite variations proper to certain, 43; nature of yellow, in Pieridæ, 48.
Pike, bulldog-headed, 58
Pilchard, variation in scales, 274; supposed hybrid with Herring, 275
Pilumnus, not altered by Entoniscians, 95
Pimelia interstitialis, extra legs, 498
    scabrosa, extra antennæ, 523
Pinnipedia, Teeth, 235—243
Pipa, spinal nerves, 141
Pithecia, absent molar, 208
    satanas, syndactylism in, 356
Pityophis, axial duplicity, 561
Plaice, symmetrical spotting of blind side, 467
Plant, compared to the body of Man, 29
Platycerus caraboides, extra legs, 507
Platyonychus, not altered by Entoniscians, 95
Platysomatichthys, 471
Pleuronectes, pigmentation of blind side, 467, 471
Plume moths, repetition of pattern in wings, 26
Pluteus, double, 35
Podargus, brachial plexus, 131
Pœcilogale, 232
Pointer, teeth, 221
Polian vesicles, variation in number of, 434, 435

Pollex, duplicity in, Man, 349; Dogs, 401
Polyarthron, variation in number of antennary joints, 412
Polychæta, axial duplicity in, 564
Polydactylism, Cat, 312; Man, 324; *Macacus*, 340; *Hylobates*, 346; *Simia*, 349; irregular cases in Man, 353; Horse, 360—371; Artiodactyla, 373
Polydesmus, mode of increase in number of segments, 93
Polyodontophis, vertebræ, 123
Polyphylla decemlineata, extra antennæ, 518
Polyplectron, eye-spots, 449, 450
Polyzoa, division of embryos, 556
Poppy, Iceland, colour-variation, 46; Horned, colour-variation, 47
Porania, 4-rayed specimen, 440; irregular division of an arm, 440
Portunion, change in *Carcinus* produced by, 95
Portunus puber, extra parts on chela, 535; not altered by Entoniscians, 95
Potorous, teeth, 358
Premolars, nomenclature, 199
    supernumerary, *Brachyteles*, 205; *Ateles*, 206; *Mycetes*, 208; Canidæ, 212—214; Felidæ, 225; Herpestidæ, 229; Mustelidæ, 231—234; Phocidæ, 237—242; *Cervus*, 246; Phalangeridæ, 248; *Phascologale*, 257
    absence of, Canidæ, 214—216; Felidæ, 224; Herpestidæ, 229; Mustelidæ, 231—234; Phocidæ, 237—242
    apparent division, *Brachyteles*, 205; Canidæ, 213; *Dasyurus*, 255; Phocidæ, 237
    displacement and other variations, *Simia*, 201
PRIONUS, supposed development of elytra as legs, *coriarius*, 148
    variation in number of antennary joints, *imbricornis*, 411, *fissicornis*, 412
    extra legs, *coriarius*, 488, 512, *californicus*, 544, 557, *coriaceus*, 548
    double (?) antenna, 551
    double (?) legs, *californicus*, 544
Procerœa, double tail, 564
Protozoa, supposed relation to Metazoa, 35; duplicity, 566
Pseudochirus, premolars, 250, 255; incisors, 248
Pterostichus, extra legs, *lucublandus*, 512; *mühlfeldii*, 485; *prevostii*, 512; double (?) antenna, *planipennis*, 550
Ptinus latro, extra legs, 512; extra antennæ, 522

INDEX OF SUBJECTS. 589

Puffinus, brachial plexus, 130
Pug, breeds of, 57; teeth, 221
Purpura lapillus, colour-variation, 48
Putorius, teeth, 231, 234
Pygæra anastomosis, extra wing, 284
Pygidium, division of, in *Melolontha*, 456
Pygomelian geese, 401
Pyrameis cardui, aberrations of, 49, 52;
  var. *kershawi*, 49, 52; var. *elymi*, 50,·51
Pyrodes speciosus, extra legs, 496
Python, imperfect division of vertebræ, 103, 105

Radial joints of Crinoids, 421
Radial Series, Meristic Variation in, 60;
  evidence, 422; in Echinodermata, 432
Radii, variations in number, Holothurioidea, 433; Crinoidea, 435; Asteroidea, 439; Echinoidea, 441; Ophiuroidea, 433, 447
Radius, absence of, 360
Radulæ of *Buccinum*, 262
Raiidæ, eye-spots, 302; extra fin, 540;
  division of fin into lobes, 540; separation of fin from head, 540
Rana, vertebræ, 124, 126; double foot, 540; spinal nerves,142; extra limbs,554
Ranzania bertolonii, extra legs, 510
Raspberry, yellow variety, 47
Red, variations of, 44—48; as variation from blue, 44
Renal arteries, 277
Repetition of Parts, association of these phenomena, 21
Repetition, Linear, Bilateral or Radial, distinctions between, 88
  Units of, 556
Reptilia, vertebræ, 103, 123
Reversion, hypothesis made in order to escape recognition of Discontinuity in Variation, 76
Rhagium mordax, double (?) leg, 548
Rhea, brachial plexus, 130
Rhinoptera, teeth, 259—261, *javanica*, 261, *jussieui*, 259
Rhizocrinus, 6-rayed specimen, 437
Rhizotrogus, extra legs, *æstivalis*, 510; *castaneus*, 512
  extra antennæ, *æquinoctialis*, 517
  division of metasternal plates, 456.
Rhombus, pigmentation of blind side, 467—471; *lævis*, 467, 468, 470; *maximus*, 467, 470
  variations in scales, 468
Rhynchites germanicus, double (?) antenna, 550
Rhyttirhinus, supposed case of extra eye, 281
Ribs, division of, in Man, 105
  cervical, in Man, 108, 112, 115; in Dog, 122; Bradypodidæ, 119; on 6th vertebra in Man, 108

Ribs, variations in dorso-lumbar region, Man,109—116; Anthropoid Apes, 116—118; Bradypodidæ, 121; *Felis*, 122; *Canis*,122; *Galictis*, 123; *Halichœrus*, 123; eleven in Siamese Twins, 560
Rissa, variations of hallux in the genus, 396, 397
Roebuck, horns, 286; polydactylism, 374, 379; union of horns, 460
Rubus idæus, colour-variation, 47
Rupicapra tragus, horns, 286
Rutela fasciata, extra legs, 512

Sacculina, effect of, on segmentation of *Carcinus* and other Crabs, 95
St Bernard dog, duplicity of hallux in, 401
Salinity, doubtful relation of variations of Crustacea to changes in, 100
Salmo fario, salar, trutta, 58
Salmon, bulldog-headed variety,58; axial duplicity, 563
Salmacina, double tail, 564
Salt lakes, Crustacea of, 96, 100
Samia cecropia, extra wing, 283
Sand-canals, repetition of, in *Asterias*, 440
Saperda carcharias, extra antenna arising from head, 551
Sarcophilus, teeth, 255
Sarsia, Meristic Variation in, 424; with six segments, 424; with five segments, 425
Saturnia carpini, repetition of eye-spots in wings, 26; extra wing, 282; variation of eye-spots, 289, 301, 302; colours of larvæ, 306
Satyrus hyperanthus, eye-spots, 294
Sawfly, extra legs, 502, 546
Scales of Pilchard, 274; of *Gasterosteus*, 276; of Snakes, 276
Scarites pyracmon, extra legs, 500
Scheme, shewing the relations of parts in Secondary Symmetry, 481
Scolopendra, number of segments, 94
Scorpion, double poison-spine, 457; double tail, 565
Scraptia fusca, extra antennæ, 523
Seals, variations in dentition, 235
Segmentation, metameric, not in kind distinct from other forms of Repetition, 28; two ways by which a full segmentation may have been achieved in phylogeny, 86
  of Arthropoda, variation in, 91; imperfect in Annelids, 156; spiral in Annelids, 157; variation of in Cestoda, 168, 170
  of mammæ, 191
  of ovum, variations in, 463

Selachians, teeth in, 259
Semnopithecus, teeth, 204
Sepidium tuberculatum, double (?) antenna, 551
Sex, analogy with Discontinuous Variation, 66
Sexual characters, statistics as to, in Beetles, 38; Earwigs, 40
 of *Hepialus* in Shetland, 254
 of *Phalanger* in Waigiu, 254
Sheep, cervical auricles, 180; incisors, 245; change in form of canines, 245; molar, 246; extra horns, 285; polydactylism, 373, 380
Sheep-dog, teeth, 221
Shetland, variety of *Hepialus* in, 254
Shrew, naked variety, 56
Siamese Twins, 560
Siberia, Crustacea of salt lakes, 97, 100
Silis ruficollis, extra legs, 498
Silky fowls, 56
Silpha nigrita, extra legs, 501; *granulata*, double (?) leg, 548
Silurus, extra fin, 540
Simia, vertebræ, 118; teeth, 200; extra digit, 349; extra mamma, 188
Simultaneity of Variation, possibilities of, 25, 26, 308; in fore and hind wings of Lepidoptera, 293; in counterparts, 569; in colours of segments of Lepidopterous larvæ, 303; in Chitonidæ, 307; in limbs, 402; in homologous twins, 559; in radial segments, 423; not clearly distinguishable from Symmetry, 569
Sinistral varieties, 54
Situs transversus, 465, 560
Sledge-dog, absence of first premolar, 215; division of premolar, 214
Smerinthus, colours of larvæ, *ocellatus*, *populi, tiliæ*, 306, 307
Snakes, vertebræ, 103, 123; axial duplicity, 561
Solea, pigmentation of blind side, 471
Solenophorus strepens, double (?) antenna, 551
Sorex, naked variety, 56
Spaniel, teeth, 221
Species, the problem of, 2. Methods of attacking, 6
 Discontinuity of, a fact, 2
Specific Differences, indefinite, 2
Spermathecæ of Earthworms, variation in number, &c. of, 160, 165
Sphærocrinus, imperfect variation to 6-rayed state, 437
Sphingidæ, repetition of markings in larvæ of, 26; variation in, 304
Sphinx ligustri, division of proboscis, 456
Spinal nerves, 129; Birds, 130; Man, 135; Primates, 138; Bradypodidæ, 141; *Pipa*, 141; *Rana*, 142
 dimorphism in respect of, 138; distribution to limbs, 143; Homœosis, 144; recapitulation, 144
Spinal nerves, principles of distribution, 143
Spiracle, extra in tadpole of *Pelobates*, 465
Stability, Organic, 36
Starfishes, theory of origin of repetition in, 29; variations in number of rays, 439; multiplication by fission, 433
Stentor, duplicity, 566
Stichopus, arrangement of tube-feet changes with age, 435
Stickleback, variation in number of bony plates, 276
Stomobrachium octocostatum, variety having tentacles in double series, 425
Strangalia, double (?) antenna, *atra*, 551, *calcarata*, 551
Strategus antæus, extra legs, 512
Struthio, brachial plexus, 130
Styela, variations in branchial sac, 172
Subemarginula, extra eye, 279
Substantive Variation, distinguished from Meristic, 23; correlated with Meristic in vertebræ, 125
 in size, 38, 40; in colour, 43—48; in colour-patterns, 48—54; miscellaneous, 54—60
Swan, cervical vertebræ, 33; colour-variation of young, 44; brachial plexus, 130
Symmelian "monster," 459
Symmetry, the conception of, 19, 569
 a relation between optical images, 19
 almost universal presence of in living organisms, 21
 of mammæ, 191
 in dental Variation, 267
 in digital Variation, Man, 324, 402; Cat, 314; in manus and pes, 403
 in nuclear division, 430; in variations in segmentation of ovum, 463
 in variations of homologous twins, 559, 560
 in double monstrosity, 559
 Bilateral, characters of, 88; as appearing in variations of flat-fishes, 467
 Major and Minor, 21, 86
 Primary and Secondary, 90
 Radial, characters of, 89

INDEX OF SUBJECTS. 591

Symmetry, Secondary, preliminary account, 475; principles, 479
　Scheme of relations of parts in, 481; parts repeated in, *geometrically* peripheral to points of origin, 557; relation to Primary, 556, 557
　in Insects, 475; Crustacea, 525; Vertebrates, 553; Batrachia, 554; *Triton*, 555
Syndactylism, Man, 355, 356; *Pithecia*, 356; *Macacus*, 356; Ox, 384—387; Pig, 387—390

Tadpole, extra spiracle in, 465
Tænia cœnurus, transposition of generative organs, 170; case of six suckers and segments prismatic, 565
　elliptica, asymmetrical arrangement of genital pores, 170
　saginata, "intercalated" segments, 169; repetition of generative organs in proglottides, 169; two genital pores at the same level, 170; consecutive genital pores on same side, 170; bifurcation of chain, 566
　solium, changes in position and alternate arrangement of genital pores, 170
　tenuicollis, bifurcation of chain, 566
Tail-fin, division of, in Gold-fish, 451
Tail-spine, division of, *Limulus*, 456; Scorpion, 457
Tapeworms, variations, 168—170; duplicity in, 565
Tarsus, in some beetles with only four joints appearing, 25; variation in number of joints, *Blatta*, 63, 415; various numbers of joints in families of Orthoptera, 415
Taurhina nireus, extra legs, 509
Taxidea, teeth, 233
Taxus baccata, colour-variation, 47
Teal, Garganey, division of digits, 392
Teeth, in undifferentiated series not credited with individuality, 32
　numerical Variation, 195; division of, 268; duplicate, 268; statistics of Variation, 200, 209, 222, 235
　relation of upper to lower, 196
　of Primates, 199—208; Canidæ, 209—222; Felidæ, 223—226; Viverridæ, 227—231; Mustelidæ, 231—235; Pinnipedia, 235—243; Ungulata, 243—246; Marsupialia, 246—258; Selachians, 259—262; *Buccinum*, 262
Teeth, terminal, least size of, 270; presence and absence of, 269; Homœotic Variation in, 272
　Recapitulation, 265
Telephorus, colour-variation, *lividus*, 43
　division of pronotum, *nigricans*, 455
　double (?) antenna, *lividus*, *rotundicollis*, 551
　double (?) leg, *excavatus*, *fuscus*, 548
　extra legs, *rusticus*, 502
" Telescope " Gold-fish, 453
Tellina, sinistral variety, 54
Tenebrio granarius, double (?) leg, 548
Tentacles of Molluscs united, *Helix hispida*, 461; *Limax agrestis*, 460
　repeated, *Patella vulgata*, 279
　bifid, *Physa acuta*, 280
　of Holothurians, 435
Tenthredo solitaria, extra legs, 502
　ignobilis, extra leg, 546
Terias, colour-variation, 52, 53; *anemone, hecabe, mandarina, mariesii*, 52; *betheseba, constantia, jaegeri*, 53
Terminal members of Series, variation of, 79, 269, 271, 272, 407
　teeth, 269, 272; digits, 407
Terrier, absence of premolar, 215
Testes, variation in number in *Hirudo*, 165, 166
Tetraceros, horns not as in 4-horned Sheep, 285
Tetracrinus, normally 4-rayed, 5-rayed and 3-rayed varieties, 437
Tetrops præusta, double (?) leg, 545
Thoracopagous twins, transposition in, 560
Thumb, variation in number of phalanges, 324
　double, 349
Thylacinus, teeth, 255
Thylacites pilosus, double (?) antenna, 550
Thyonidium, variation in number of organs, 435
Tiarops polydiademata, specific character of, 426
Timarcha tenebricosa, extra antennæ, 522
Tomato, colour-variation, 47
Tonicia, variation in colour of scutes, 308
Tongue, division of, 451
Toxotus, extra eye, 280
Transposition of viscera, 560
Triasters, symmetry of, 430; found in bilaterally symmetrical areas of segmenting ovum, 464
Trichodes syriacus, double (?) leg, 547
Trichosurus vulpecula, premolars, 254

Triænophorus, segmentation of, 168
Triopa clavigera, rhinophore trifid, 280
Triton, legs repeated, 555
Troglodytes, vertebræ, 116; teeth, 202
Tropidonotus, vertebræ, 123; scales, 276
Trout, bulldog-headed, 58, 59; axial duplicity, 563
Tulip, Meristic Variation in, 60
Turbot, pigmentation of lower side, 467, 470
Turdus, brachial plexus, 131
Turkey, digits, 393
Twins, homologous, 559; Simultaneous Variation of, as a case of Bilateral Symmetry, 559
  Siamese, peculiarities of, 560; thoracopagous, 560
  in Echinoderms and in Amphioxus, 35
Typosyllis, double head, 564

Uca una, extra parts in chela, 530
Ulna, a second, 331
Ulnar nerve, variations in composition of, 136; a second, 333
Ungulata, teeth, 243; digits, 360—390, 397
Union, median, 458; of horns of Roebuck, 460; of eyes of Bee, 461; of kidneys, 459; of tentacles of *Limax*, 460; of tentacles of *Helix*, 461; of posterior limbs of Vertebrates, 459; of digits in Ox, 383, 386; of digits in Pig, 387—390
Units, of Repetition, 556
Uraetos, brachial plexus, 131
Ureters, supernumerary, 278
"Useless" parts, supposed variability of, 78
Uterus, double, Darwin's comment on, 77; is a case of median division, 451
Utility, fallacies of reasoning from, 12
Uvula, division of, 451

Vanessa atalanta, colour-variation, 46
  urticæ, extra wing, 283
  io, eye-spots, 299, 300
Variation, defined, 3
  the Study of, as a method of attacking the problem of Species, 6
  Continuous and Discontinuous, 15
  Meristic and Substantive, distinguished, 23, 24
  magnitude of integral steps affected by Merism, 25
  about a Mean form, 37
  perfection in, 60, 64
  causes of, 78
  Homœotic, 85, in vertebræ, 106; in Arthropoda, 146; in teeth, 272
  Simultaneity of, in repeated parts, 303, 402, 425, 464

Variations, minimal, questionable utility of, 16
Vertebræ, Meristic Variation in, 102
  imperfect division, 103, 458
  Homœotic Variation, 106
  reduction in numbers, Man, 111
  numerical variation, 102
  Man, 103, 106—116; Anthropoid Apes, 116; Bradypodidæ, 118; Carnivora, 122; Reptilia, 123; Batrachia, 124; features of Variation recapitulated, 127; correlation with spinal nerves, 113, 115, 139, 145
Vesperus luridus, extra eye, 280
Veronica buxbaumii, numerous symmetrical variations in, 76; illustrating variations of Bilateral Series, 448
Viverridæ, teeth, 227—231

Waigiu, female *Phalanger maculatus* coloured like male in island of, 254
Wall butterfly, variation in ocelli and neuration, 300
Water-pore, extra, in *Bipinnaria*, 466
Webs, between toes of Duck, absent, 401
Weevils, four visible joints in tarsus, 25
Wing, supposed to replace a leg in *Zygæna*, 148
Wings, supernumerary in Insects, 281
  fore and hind, varying simultaneously in Lepidoptera, 293
  quills of, varying with quills of tail in Pigeons, 309
Woodpecker, Green, colour-variation, 43

Xantho punctulatus, duplicity of index, 542
Xiphopagous twins, transposition of viscera in, 560
Xylotrupes gideon, variation of horns in, 38

Yellow, variations of, 43–48, 73
Yew, yellow-berried, 47

Zalophus californianus, molars, 243
  lobatus, premolars, 238, 242; molars, 243
Zebra, repetition of stripes in, 26
Zeugopterus, white varieties, 467; variation in dorsal fin, 471
Zonabris quadripunctata, double (?) antenna, 551
Zonites præusta, extra antennæ, 522
Zygæna filipendulæ, colour variation, 46
  supernumerary wing, 148
  minos, colours, 46; extra wing, 284

INDEX OF PERSONS.

Acton, 286
Adolphi, 124, 127, 142
Agassiz, A., 469
Agassiz, L., 396, 424
Ahlfeld, 340, 354, 451, 458
Albrecht, 105, 540
Aldrovandi, 344
Allen, J. A., 243
Alston, 286
Ammon, 348, 349
Anderson, 399
Andrews, 563, 564
Annandale, 327, 345, 346, 350–352, 355–358
Appellöf, 473
Arloing, 363, 370
Ascherson, 174
Ashmead, 413
Asmuss, 484, 500
Asper, 167
Assheton, 152
Audouin, 512
Auld, 390
Austin, 436
Auvard, 349
Auzoux, 203

Babington, 47
Bacon, 29, 146
Baird, 223, 232
Balbiani, 566
Balding, 305
Balkwill, 566
Ballantyne, 334
Bancroft, 561
Barbour, 563
Bardeleben, 183
Barr, P., 46
Barrier, 384, 388
Bartels, 187
Barth, 187
Bartlett, 216
Bassi, 512, 548, 551
Baster, 453
Bateson, Miss A., 77, 468
Bather, 436
Baudi, 456, 512, 551
Baudon, 54
Baum, 401
Baumüller, 374
Baur, 103, 105, 123, 124
Beddard, 159, 162, 163, 165

B.

Bedriaga, 127
Bell, F. J., 443, 564, 565
Bellamy, 113
Belt, 56, 57
Beneden, van, 531
Benham, 152, 159, 161, 565
Béranger, 347
Bergendal, 555
Bergh, 160
Bernhardus à Berniz, 538
Betta, de, 43
Bibron, 563
Bicknell, 45
Billardon de Sauvigny, 454
Billott, 399
Birkett, 178
Birnbaum, 350
Blackmore, 523
Blainville, de, 118, 119, 205, 224
Blanchard, 187
Blanford, W. T., 398
Blasius, 354
Bles, 440
Bleuse, 482, 508
Boas, 369, 383, 385
Boettger, 561
Boisduval, 45
Bolau, 349
Bond, 301
Bonnier, 95
Böttcher, 560
Boulard, 512
Boulenger, 123, 276, 277, 395, 396, 465
Boulian, 351
Bourne, A. G. 125, 127
Boyd-Campbell, 399
Bramson, 52
Brandt, 466
Bredin, 368
Breese, 565
Brenner, 360
Brindley, 38, 39, 63, 280, 416
Brisout de Barneville, 416
Brooks, 466
Broome, 565
Bruce, 181, 185
Brulerie, de la, 280
Bruner, 44
Brunette, 564
Brunner von Wattenwyl, 41, 413, 416
Buchanan, Miss F., 156, 157
Buckler, 304, 305, 307

38

Buffon, 286, 398
Bull, 340
Bülow, von, 565
Buquet, 551
Bureau, 540
Burmeister, 123, 232
Busch, 198, 345
Butler, A. G., 52, 53

Camerano, 127, 546, 555
Cameron, 185
Canestrini, 58
Cantoni, 537
Carlet, 59
Carpenter, P. H., 421, 422, 436–438
Carré, 339
Cassebohm, 178
Cauroi, du, 344, 354
Cavanna, 539, 555
Cazeaux, 185
Champneys, 139, 185
Chapman, J., 244
Charcot, 184
Chavignerie, de la, 455, 547
Chworostansky, 165
Claparède, 425, 564
Clark, J. A., 51
Clark, J. W., 465
Clarke, E., 397
Claus, 80, 100
Cleland, 401
Coale, 393
Cobbold, 566
Cockerell, 44
Cocks, 55
Colin, 169
Collin, 563, 565
Cooke, A. H., 262, 263
Coquillet, 413
Cori, 156, 157, 158
Cornevin, 363
Cornish, 150
Cotteau, 446
Couch, 440, 470, 471
Coues, 232, 390
Cowper, 391, 393
Cramer, 346
Cuénot, 429, 433, 441
Cunningham, 320, 467
Curtis, 547
Cusset, 176

Daintree, 376
Dale, 482, 547
Dareste, 458
Darwin, C., 1, 5, 13, 56, 57, 59, 77, 121, 288, 449
Davidoff, 566
Dawson, 566
Day, 275, 276, 302, 467, 540
Delplanque, 354, 370, 377, 379
Démidoff, 468
Dendy, 438
Desmarest, 152

Devay, 399
Dimmock, 543
Dobson, 397
Dohrn, 86
Donceel, de, 51
Dönitz, 212, 217, 220, 246, 445
Donovan, 302, 471
Dorner, 561
Doué, 456
Doumerc, 512, 550
Drechsel, 455
Drew, 381
Driesch, 35
Dubois, 330
Duhamel du Monceau, 470, 471
Duméril, 554, 563
Dunn, 374
Duns, 528
Duponchel, 456
Dusseau, 352, 355
Duval, 184
Dwight, 325, 334

Ébrard, 166, 304
Eck, 437
Edward, T., 43, 174, 563
Ehrenberg, 425, 426, 428
Eichwald, 560
Ekstein, 339
Ekström, 471
Elwes, 45
Engramelle, 46
Ercolani, 369, 377, 380, 381, 392, 393, 540, 554, 555
Eudes-Deslongchamps, 180

Fackenheim, 345, 351, 352, 399
Failla-Tedaldi, 295
Fairmaire, 454
Farge, 327, 399
Fauvel, 44, 494, 508, 523
Faxon, 152, 530, 532, 533, 536, 537, 541, 542, 557
Field, 466
Filippi, 425
Fischer, 41
Fischer, G., 174
Fischer, P., 54, 279, 566
Fischer de Waldheim, 97
Fisher, W. R., 392
Fitch, 565
Fitch, E. A., 44
Fitzinger, 200
Flemming, 430
Flemyng, 307
Fleutiaux, 548
Flower, W. H., 106, 119, 217, 220, 233
Forbes, E., 54, 425, 460
Forbes, W. A., 356
Forgue, 143
Forskål, 540
Forsyth, 398
Fort, 344, 356, 358
Foster, 565

Fotherby, 360, 399
Franck, 368
French, 451
Freund, 459
Freyer, 524
Fricken, von, 512
Friedlowsky, 244, 356
Friele, 262, 264
Friend, 563, 565
Fries, 471
Frivaldsky, 500
Froriep, 354
Fumagalli, 336
Fürbringer, 131, 133, 135, 142
Fürst, 399

Gadeau de Kerville, 415, 455, 482, 510, 548, 549
Gaillard, 346, 350
Galton, F., 36, 40, 43, 418, 419
Garrod, 390
Gaskell, 86
Gaskoin, 56
Gauthier, 443, 445—447
Gebhard, 459
Gegenbaur, 77
Geissendörfer, 360
Gené, J., 127
Geoffroy St Hilaire, I., 57, 205, 330, 368, 377, 379, 383, 392, 451, 459, 563
Gercke, 285
Gervais, 203
Gherini, 337
Giard, 95, 440, 468, 482, 545
Gibbons, Sir J., 44
Gibson, 166
Giebel, 234
Gifford, 44
Giraldès, 336
Girard, 305
Godman, 53, 297
Godwin-Austen, 286
Goldfuss, 436, 437
Goodman, 376
Goossens, 300
Gordon, 56
Gorré, 187
Gosse, 566
Gosselin, Mrs, 44
Götte, 127
Gottsche, 466, 471
Goubaux, 180, 244, 245
Grandélément, 355
Grandin, 340
Gray, J. E., 56, 242, 287, 396
Gredler, 286, 511, 512, 550, 551
Green, 515
Grobben, 169
Gruber, W., 108, 111, 119, 122, 330, 345, 346, 350, 352, 354, 359, 360
Guerdan, 458
Guermonprez, 327
Günther, 173, 174, 260, 309, 468
Gurlt, 368

Gurney, J. H., 43, 55, 392
Guyon, 566
Guyot-Daubès, 360

Haacke, 443, 446
Hagen, 148
Hagenbach, 352
Hammond, 305
Hanley, 460
Hannæus, 184
Harker, A., 419
Harker, J., 354, 399
Harmer, 440, 556
Harrington, 482, 491, 500
Harrison, 211
Hartung, 187
Harvey, 178
Haworth, 45
Heineken, 526
Helbig, 187
Heller, 170
Henneguy, 430
Hennig, 349
Hensel, 203, 209, 212—216, 220, 223, 226, 232, 244, 246, 269
Herdman, 171, 172, 439, 456
Herklots, 527, 528, 529, 542
Heron, Sir R., 453
Héron-Royer, 465, 555, 561
Herrich-Schäffer, 51
Herringham, 137, 138
Hertwig, O., 431
Heuglin, von, 234, 235
Heusinger, 174, 179
Hewett, 55
Heyden, H. von, 488, 512, 548
Heyden, L. von, 484, 487—490, 494, 517, 523, 550, 551
Heynold, 355
Higgins, 471
Hill, 391
Hincks, 425
His, 177
Hodgson, 209, 398
Hoeven, van der, 218
Hoffmeister, 162
Honrath, 284
Hopkins, 48, 73
Horn, 411
Horst, 565
Howes, 126, 153, 210, 391, 421, 515
Hübner, 305
Hudson, 56
Hügel, Baron A. von, 39
Humphreys, 221
Humphreys, H. N., 301
Humphry, 200
Huxley, 217, 218, 219

Imhoff, 547

Jäckel, 393
Jackson, 331
Jacquelin-Duval, 524

Jaeger, G., 536
Janson, 482, 491
Jayne, 503, 512, 518, 524, 544, 547, 548, 550, 551
Jeffreys, G., 54, 457
Jekyll, Miss, 46
Jentink, 248, 252, 253
Jhering, von, 140, 142, 360
Johnson, Athol, 354
Jolly, 337
Joly, 370, 372
Joseph, 350

Károli, 532
Kawall, 551
Kerckring, 344
Kiesenwetter, von, 281
Kingsley, 539, 554
Kirk, 565
Kitt, 363, 383, 384, 386, 390
Kleyn, 453
Klingelhofer, 547, 551
Klob, 187
Koenen, von, 436
Kolbe, 484, 503
Kölliker, 142
Kostanecki, von, 175
Kraatz, 146, 454, 456, 484, 485, 494, 498, 502, 506, 509—511, 515, 516, 523, 545, 547, 548, 550, 551, 552
Krause, 455, 548
Kriechbaumer, 147
Krohn, 425
Kröyer, 151
Küchenmeister, 560
Kuhnt, 339

Laboulbène, 512
Lacaze-Duthiers, 171
Lacépède, 540, 561
Lafosse, 244
Lamarck, 4
Lambert, 446
Lampert, 435
Landois, 58, 383, 387, 511, 555
Lane, 113
Lang, 566
Langalli, 336
Langerhans, 564
Lankester, 536
Lannegrace, 143
Lanzoni, 561
Lataste, 127
Laurent, 184
Lavocat, 354, 372
Le Clerc, 357
Leech, J. H., 46
Lefébvre, 500
Le Gendre, 184
Léger, 527, 538
Legge, 46
Leichtenstern, 181—185
Lereboullet, 515
Le Sénéchal, 530, 535, 542

Letzner, 280, 523, 550
Leuckart, 168—170, 566
Levacher, 566
Leveling, 458
Lidth de Jeude, van, 58
Linnæus, 453
Lisfranc, 355
Lister, 218, 457
Loriol, P. de, 436, 437, 438
Loudon, 47
Lucas, 399
Lucas, H., 462, 536, 542, 550, 551
Ludwig, H., 433, 566
Lunel, 58, 554
Lütken, 433
Lydekker, 105, 217, 233

Macalister, A., 112, 278
MacAndrew, 54
MacBride, 440
McCoy, 52
McIntosh, 470, 564
Maggi, 530
Magitot, 198, 203, 205, 210, 221, 244, 245, 270
Malm, 469
Manifold, 451
Mantell, 436
Marjolin, 355
Marsh, 349
Marsh, C. D., 565
Marsh, O. C., 364, 366, 368
Marshall, 86
Martens, von, 155
Martin, 430
Mason, 282, 488, 498, 509, 548, 550
Mason, F., 355
Masters, 60, 84, 310
Mayer, 200
Mazza, 555
Meckel, 278, 346, 458, 459
Melde, 355
Meldola, 284
Meyer, A. B., 441, 445
Michaelsen, 162, 164, 165
Mielecki, von, 175
Milne-Edwards, 151, 202, 527
Mitchill, 561, 563
Mivart, 212, 217, 219
Möbius, 401
Mocquerys, 455, 487, 488, 494, 496, 498, 501, 503, 507, 508, 512, 515, 517, 522, 545—548, 550—552
Mojsisovics, 367
Moniez, 169, 566
Moquin-Tandon, 280, 304
Morand, 346, 348, 354, 399
Moreau, 0
Morgan, T. H., 157, 466
Morot, 245, 384
Morris, F. O., 44, 401
Mortillet, de, 186
Mosley, S. L., 45, 300
Muir, 352, 399

Müller, A., 512
Müller, J., 173
Murray, 336

Nathusius, H. von, 285, 373, 374
Nehring, 57, 123, 210, 212, 216, 221, 235, 242
Neill, 540
Neugebauer, 183, 186
Neuhöfer, 176
Newman, 51, 295, 300
Newport, 94
Newton, A., 44, 55
Nicholls, 155
Nilsson, 471
Norman, 100, 457
Notta, 185

Oberteufer, 330
Oberthür, 44
Ochsenheimer, 46, 284, 302
Odin, 459
Olliff, 51, 52
Otto, 58, 278, 346, 348, 350, 354, 356, 458
Otto, H., 547, 548
Owen, 119, 188, 211, 261

Packard, 100, 457
Paget, Sir J., 175, 177
Pallas, 180
Parry, 520
Partsch, 451
Paullinus, 184, 187
Pavesi, 565
Pelseneer, 280
Pennetier, 482
Percy, 184, 187
Perroud, 512
Perty, 512, 548, 550, 551
Peters, 200, 277
Philippi, 443
Pichancourt, 460
Pocock, 93, 457, 565
Pooley, 451
Popham, 355
Porritt, 295
Pott, 399
Potton, 399
Pouchet, 451
Poulton, 304—307, 320, 321, 323
Prackel, 184
Prévost, 457
Price, 566
Puech, 181
Pusch, 436
Putnam, 174, 396
Pütz, 366

Quenstedt, 436
Quinquaud, 185

Rabl, 176
Ragusa, 548

Rambur, 50
Ramsay, R. G. W., 46
Ranse, de, 399
Rapp, 120
Rathke, 97, 468
Redi, 561
Reichenau, von, 392
Reid, 458
Reitter, 281
Rey, 547, 550
Richard, 150, 536, 543
Richardson, 148
Richmond, 279
Ridgway, 393
Rijkebüsch, 329
Ritzema Bos, 512, 551
Rivers, 47
Röber, 283
Roberts, G., 461
Roberts, T., 442
Robinson, H., 99
Röder, von, 551
Rogenhofer, 284, 285
Romanes, 425—428
Romano, 523
Rörberg, 355
Rösel von Rosenhof, 526, 530, 536
Rosenberg, 116—118, 138, 373, 383
Rosinus, 436
Rouget, 523
Rousseau, 152
Roux, 35
Rüdinger, 330
Rudolphi, 207, 244
Rueff, 344
Rütimeyer, 246

Saage, 148
Salvin, 53, 297
Sanderson, 398
Sandifort, 458
Sarasin, 433
Sartorius, 512, 550
Saunders, Howard, 397
Sauvage, 276
Schäff, 210
Schleep, 470
Schlegel, 226
Schmankewitsch, 96
Schmeltz, 566
Schmitz, 177
Schneider, 547, 550
Schneider, A., 174
Schultze, L., 438
Schultze, O., 194
Sclater, P. L., 396
Sclater, W. L., 374
Scudder, 50
Seba, 565
Sedgwick, 84, 92, 93, 173, 197
Seerig, 348
Seidel, 176
Séringe, 523
Serville, 412, 413

Shannon, 185
Sharp, D., 43, 53, 149, 411, 482, 494, 499
Shaw, E., 413
Shaw, V., 401
Sherrington, 137, 138, 144, 168
Siebold, von, 148
Sinéty, 184
Smit, 471
Smith, E. A., 287
Smith, F., 551
Smith, S. J., 151
Solger, 141
Soubeiran, 529
South, 300
Spengel, 466
Speyer, 283
Spinola, 512
Spronck, 329
Stamati, 538
Stannius, 142, 455, 461, 512, 522, 547
Staudinger, 44
Steenstrup, 466, 469
Steindachner, 58, 468
Steinthal, 359
Stevens, 41
Stewart, C., 180, 429, 440, 446
Storer, 471
Strahl, 154
Strauch, 123
Strecker, 51, 283, 295
Streng, 352
Strombeck, von, 436, 438
Struthers, 103, 105—119, 122, 140, 327, 329, 334, 346, 351, 356, 370, 389
Studer, 277
Sundevall, 471
Sutton, 105, 176, 179, 180, 188, 555

Tachard, 398
Tarnier, 185, 345
Taschenberg, 512
Tegetmeier, 57, 393
Testut, 187
Thielmann, 277
Thomas, O., 56, 120, 199, 228, 230, 246—249, 254, 257, 258, 313, 322, 397
Thompson, W., 565
Thomson, 412
Thomson, Wyville, 466
Tiedemann, 184, 512, 536, 540
Tischbein, 511
Traquair, 469
Treitschke, 284
Trélat, 451

Treuge, 512
Trimen, 51, 300
Trinchese, 118
Tuckerman, 170, 555
Turner, Sir W., 465

Urbantschitsch, 177

Vaillant, 309, 471
Viborg, 180
Virchow, 74, 177, 178, 17
Voigt, 58
Vrolik, 58

Wagner, 421
Wagner, J., 451
Walsingham, Lord, 300
Ward, E., 379
Warpachowski, 540
Watase, 451, 463
Waterhouse, F. H., 45, 545
Webb, S., 301
Wehenkel, 366, 368
Weir, J. Jenner, 45, 51, 52, 254
Weismann, 76, 304—307
Welcker, 118, 120
Weldon, 172
Werner, 170
Wesmael, 521
Westwood, 283, 284, 508, 551
Weyenbergh, 565
White, 44
Wilde, 178
Wilder, 465
Williams, 181, 185, 191
Wilson, 304, 305, 307
Wilson, E. B., 35
Windle, 221, 326, 328, 392, 560
Wiskott, 285
Wolf, 399
Woodgate, 284
Wood-Mason, 367
Woodward, M. F., 160, 162
Woodward, Smith, 259
Wright, 566
Wright, L., 55, 393
Wyman, 57, 203, 226, 401, 465, 561

Yarrell, 59, 469, 471, 540
Yarrow, 561
Youatt, 285, 286

Zeppelin, 565
Zündel, 180

EVOLUTION.

A THEORY OF DEVELOPMENT AND HEREDITY. By HENRY B. ORR, Ph.D., Professor at the Tulane University of Louisiana. Crown 8vo. 6s. net.

> SCOTSMAN.—'Professor Orr does not profess to have proved, or worked out in its detail, his new "Theory of Development and Heredity." But he has stated it with admirable perspicuity, and the solution which he offers of the problem, "How are changes brought about in the structure of organisms, and how are these changes transmitted to succeeding generations?" deserves thoughtful examination as offering a key to certain of the deeper secrets of life and growth, both bodily and mental.'

ORGANIC EVOLUTION AS THE RESULT OF THE INHERITANCE OF ACQUIRED CHARACTERS ACCORDING TO THE LAWS OF ORGANIC GROWTH. By Prof. G. H. EIMER. Translated by J. T. CUNNINGHAM, M.A., F.R.S.E. 8vo. 12s. 6d.

THE COLLECTED WORKS OF THOMAS HENRY HUXLEY, F.R.S. In Monthly Volumes. Globe 8vo. 5s. each Volume.
[*The Eversley Series.*]

    Vol. I.—METHOD AND RESULTS.

    Vol. II.—DARWINIANA.

    Vol. III.—SCIENCE AND EDUCATION.

    Vol. IV.—SCIENCE AND HEBREW TRADITION.

    Vol. V.—SCIENCE AND CHRISTIAN TRADITION.

    Vol. VI.—HUME.

    Vol. VII.—ETHICAL AND PHILOSOPHICAL ESSAYS.

    Vol. VIII.—MAN'S PLACE IN NATURE.

    Vol. IX.—ESSAYS IN SCIENCE.

SOME VOLUMES OF "NATURE" SERIES.

Crown 8vo. Cloth.

CHARLES DARWIN. MEMORIAL NOTICES reprinted from "Nature." By THOMAS H. HUXLEY, F.R.S., G. J. ROMANES, F.R.S., Sir ARCHIBALD GEIKIE, F.R.S., and W. T. DYER, F.R.S. 2s. 6d.

THE SCIENTIFIC EVIDENCES OF ORGANIC EVOLUTION. By GEORGE J. ROMANES, F.R.S. 2s. 6d.

ARE THE EFFECTS OF USE AND DISUSE INHERITED? An Examination of the View held by Spencer and Darwin. By W. PLATT BALL. 3s. 6d.

ON THE ORIGIN AND METAMORPHOSES OF INSECTS. With Numerous Illustrations. By Sir JOHN LUBBOCK, M.P., F.R.S. 3s. 6d.

MACMILLAN AND CO., LONDON.

BOOKS BY DR. A. R. WALLACE, F.R.S.

DARWINISM : An Exposition of the Theory of Natural Selection, with some of its Applications. Illustrated. Extra Crown 8vo. 9s.

SATURDAY REVIEW.—" Mr Wallace's volume may be taken as a faithful exposition of what Darwin meant. It is written with perfect clearness, with a simple beauty and attractiveness of style not common to scientific works, with a dignity and freedom from anything like personal bitterness worthy of Darwin him-. self, and with an orderliness and completeness that must render misconception impossible."

ATHENÆUM.—" Mr Wallace adds so much that is new, and he writes in so charming and simple a style, that his readers more than he are to be congratulated on the latest service he has rendered to the science he has served so well."

Prof. Ray Lankester in NATURE.—" No one has so strong a claim as Mr Wallace to be heard as the exponent of the theory of the origin of species, of which he is—with Darwin—the joint author. . . . The book is one which has interest not only for the general reader, to whom it is primarily addressed, but also for the more special students of natural history. The latter will find in its pages an abundance of new facts and arguments which, whether they prove convincing or not, are of extreme value and full of interest."

THE GEOGRAPHICAL DISTRIBUTION OF ANIMALS: with a study of the relations of living and extinct faunas as elucidating the past changes of the earth's surface. With Maps and Illustrations. In two Vols. Medium 8vo. 42s.

NATURAL SELECTION AND TROPICAL NATURE: Essays on Descriptive and Theoretical Biology. New Edition with corrections and additions. Extra Crown 8vo. 6s.

ISLAND LIFE: OR, THE PHENOMENA AND CAUSES OF INSULAR FAUNAS AND FLORAS. Including a revision and attempted solution of the problem of Geological climates. With Illustrations and Maps. Second Edition. Crown 8vo. 6s.

THE MALAY ARCHIPELAGO: THE LAND OF THE ORANG-UTAN AND THE BIRD OF PARADISE. A Narrative of Travel. With Studies of Man and Nature. With Maps and Illustrations. Fourth Edition. Extra Crown 8vo. 6s.

GLASGOW HERALD.—" There is probably no more interesting book of travel in the language. . . . For one-and-twenty years it has held its place as a monograph in a region of the East which is full of fascination, not only for the naturalist and ethnographer, but for the ordinary reader of travels."

MACMILLAN AND CO., LONDON.